Biomedical Applications of Polymeric Materials

Biomedical Applications of Polymeric Materials

Guest Editor

Andreea-Teodora Iacob

MDPI

Basel • Beijing • Wuhan • Barcelona • Belgrade • Novi Sad • Cluj • Manchester

Guest Editor
Andreea-Teodora Iacob
Department of Pharmaceutical
and Therapeutic Chemistry
Grigore T. Popa University of
Medicine and Pharmacy Iasi
Iasi
Romania

Editorial Office
MDPI AG
Grosspeteranlage 5
4052 Basel, Switzerland

This is a reprint of the Special Issue, published open access by the journal *Polymers* (ISSN 2073-4360), freely accessible at: https://www.mdpi.com/journal/polymers/special_issues/Biomedical_Applications_Polymeric_Materials.

For citation purposes, cite each article independently as indicated on the article page online and as indicated below:

Lastname, A.A.; Lastname, B.B. Article Title. *Journal Name* **Year**, *Volume Number*, Page Range.

ISBN 978-3-7258-5671-8 (Hbk)
ISBN 978-3-7258-5672-5 (PDF)
https://doi.org/10.3390/books978-3-7258-5672-5

Contents

About the Editor

Andreea-Teodora Iacob

Andreea-Teodora Iacob is a PhD Lecturer and Specialist Pharmacist in the Pharmaceutical Laboratory, Clinical Pharmacy and Pharmaceutical and Cosmetic Industry at the Faculty of Pharmacy from the Grigore T Popa University of Medicine and Pharmacy, Iasi, Romania. Her contributions to the research field of wound management, heterocycle synthesis and polymeric formulations, as well as nanotechnology (formulation and biological evaluation of niosomes, electrospun nanofibers), are highlighted by the publication of a total of 47 ISI articles, an H-index of 13 (WOS) and 16 (Google Scholar), and 1000+ citations. Dr. Iacob's work integrates pharmaceutical chemistry, nanotechnology, and regenerative medicine, with a strong focus on translational applications like wound healing, drug delivery, and tissue regeneration. This interdisciplinary strength is aligned with global healthcare priorities. Her innovations in wound dressings and polymeric systems have earned her prestigious accolades, including EuroINVENT medals, Best Oral Presentations or Best Poster Awards at International or National Conferences and the L'Oréal-UNESCO for Women in Science Award (Romania). This affirms the real-world applicability of her research. She plays a vital academic role as a lecturer, mentor, and journal editor for different Special Issues in *Polymers* and *Pharmaceutics* and peer reviewer of several Q1 journals. Her contribution to academic curriculum development and international collaborations also demonstrates leadership in her field. Her active participation in multiple national and international research projects (10+) shows consistent funding success and a strong collaborative network across Europe.

polymers

MDPI

Editorial

Biomedical Application of Polymeric Materials

Andreea-Teodora Iacob

Department of Pharmaceutical and Therapeutic Chemistry, Faculty of Pharmacy, University of Medicine and Pharmacy "Grigore T. Popa" of Iasi, 16 University Street, 700115 Iasi, Romania; andreea.panzariu@umfiasi.ro

1. Introduction

The term 'polymer' was first introduced by Swedish chemist Jöns Jacob Berzelius to describe macromolecules consisting of significant repeating structural units [1]. In polymer and materials research, a distinction is maintained between bio-macromolecules and biopolymers, such as polysaccharides or proteins, and wholly synthetic polymers, such as polyethylene or polyethylene terephthalate. The disparity mostly arises from their origins and production methods; biopolymers are derived from natural sources like plants or animals, whereas synthetic polymers are produced from petroleum-based monomers. Both bio- and synthetic polymers have significantly influenced our lives over the past century and are crucial in sectors such as textiles, healthcare, food packaging, and construction [2]. Advanced polymer materials are essential elements in the global economy and high-end industry [3]. Numerous industries, including packaging, cosmetics, sutures, dentistry (including prosthetic teeth and filling materials), material manufacture, and the food and beverage sectors, have significantly embraced polymer-based materials [4].

Currently, it is well acknowledged that polymers serve as the foundation for a diverse array of materials. These materials are extensively employed in biomedicine owing to their multifaceted characteristics. Biodegradable polymers, capable of natural decomposition via biological processes, are being evaluated for various applications. Biodegradable polymers are acknowledged as superior materials for biomedical applications due to their remarkable biocompatibility. Biomaterials are a multidisciplinary research domain that integrates material science with medical applications, emphasizing mechanical qualities and biocompatibility [1,5].

The versatility of chemistry facilitates the development of polymers with diverse mechanical and physical properties, rendering them particularly advantageous in biomedical research. The intrinsic capacity of biomaterials to break down and be assimilated or eliminated by the body without surgical intervention renders biodegradable polymers especially advantageous. Due to their low immunogenicity and biocompatibility, natural source polymers are commonly employed in biomedical applications [6]. They are selected for medicinal applications due to their minimal likelihood of inducing allergic responses. Synthetic polymers offer alternatives for persons susceptible to allergies, as they reduce immune reactions and chronic inflammation [4]. Biomedical polymer materials can efficiently encapsulate a large variety of small molecules, proteins, and functional nucleic acids by surface modification. Biopolymeric materials serve a vital function as scaffold materials in tissue engineering and regenerative medicine, facilitating cell adhesion, proliferation, and differentiation to repair or replace damaged tissues [7,8].

The methods employed in the preparation of polymer matrices significantly affect the properties of the polymer. Preparation processes affect factors like the stability, thermal conductivity, mechanical properties, barrier properties, and surface morphology of polymer matrices [9]. Selecting preparation processes according to the intended application guarantees the optimal performance of polymer matrices, facilitating the development of innovative materials with customized features [8]. Hyper-branched polymers (HPBs), a category of extensively branched three-dimensional (3D) macromolecules, exhibit attributes like ease of chemical manipulation and multi-cavity architectures. By means of suitable design, they can meet the specifications of biomedical materials in medical applications [10].

The purpose of this Special Issue is to describe the latest discoveries and research regarding polymeric materials that have applicability in the biomedical and pharmaceutical fields. This Special Issue includes 21 papers (of which 15 are articles and 6 are reviews) that will be briefly described below in order to spark readers' interest in studying them in detail.

2. An Overview of the Published Articles

The articles published in this Special Issue, which are on such a wide-ranging topic, cover a broad spectrum of biomedical applications, which are summarized in Figure 1.

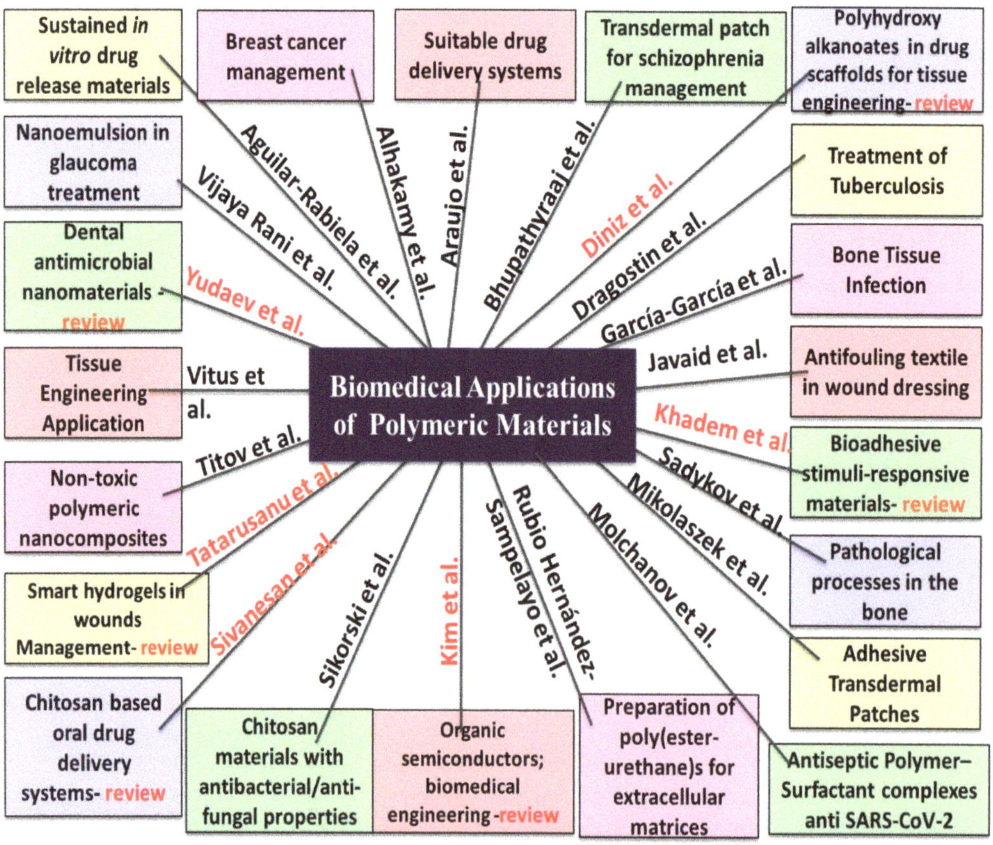

Figure 1. Schematic representation of the subject of each article published in the Special Issue "Biomedical Application of Polymeric Materials" found in the list of contributors (1–21), where the authors in red contributed with a review type of article.

In the context of their research, Aguilar-Rabiela et al. articulate a comparison between the astaxanthin (ASX) release profile from MBGNs (mesoporous bioactive glass nanoparticles) and from the composite microspheres PHBV/MBGNs (where PHBV stands for poly (3-hydroxybutyrate-Co-3-hydroxyvalerate)). Because of the regulated release of both ASX and advantageous ions from MBGNs as well as dissolving byproducts from the MBGN/PHBV composite material, it was reported that this composite approach increases cell activity. It was shown that the composite material demonstrated an extended release of ASX for up to 80 h, in contrast to the 40 h observed with the MBGN/ASX. The fluorescent cell morphology measurements and WST-8 cell viability results indicated that the variation in release behavior influenced NIH-3T3 cell viability over the initial 72 h of incubation. The results indicated that the release behavior of the composite microspheres can be adjusted by altering the MBGN content within them. Both the MBGN/ASX and composite microspheres showed optimal potential for controlled phytotherapeutic delivery required for particular applications.

The investigation carried out by Alhakamy et al. reveals the outstanding development of scorpion venom (SV)-functionalized quercetin (QRT) phytosomes for breast cancer treatment, accompanied by an anti-proliferative and anti-apoptotic assessment using the human breast cancer MCF-7 cell line. The improved QRT-SV phytosome formula demonstrated a markedly enhanced expression of p53 mRNA, Bax, Bcl-2, and caspase-9 and significantly diminished NF-κB and TNF activity in comparison to both the basic formula and QRT alone.

Through their research, Araújo et al. shed light on the optimization of chitin–glucan complex (CGC) hydrogels and their evaluation in terms of drug loading (caffeine) and release ability. The loaded hydrogels exhibited enhanced rheological and mechanical characteristics, with caffeine release profiles adhering to a Fickian diffusion mechanism in PBS solution and a non-Fickian diffusion in 0.9% NaCl solution. This study shows that CGC may be transformed into hydrogels by a straightforward technique, and the resultant structures exhibit appropriate characteristics for application as drug delivery vehicles.

The research presented by Bhupathyraaj et al. investigates the impact of polymers and permeation enhancers (1,8-cineole, linalool, and DMSO) on the release of quetiapine fumarate (QTP-F) from transdermal patches via a dialysis membrane. The best QTP-F formulation proved to be the F2 transdermal patch, utilizing a 50:250 ratio of the polymers PVP K30 (polyvinylpyrrolidone K30) and HPMC K100 (hydroxypropyl methylcellulose K100) in conjunction with the natural permeation enhancer 1,8 cineol. The stability research results indicated no substantial alteration from its initial state over a three-month period at both temperatures (room temperature and 40 °C). The F2 was reported to successfully attain cost-effectiveness, extended-release capabilities, reduced dosage, and diminished administration frequency, potentially enhancing patient adherence to the schizophrenia treatment.

Diniz et al. offer an in-depth analysis of the contemporary biotechnological utilizations of PHA (poly-hydroxy-alkanoates) in the biomedical field in a 17-page review. The review emphasizes the recent research concerning various PHA in the biomedical field, with the objective of elucidating the approaches employed and highlighting the biopolymer's versatility. Their use in drug carriers is prevalent, owing to their biocompatibility, their degradation yields products that are natural to the body, and their biodegradability facilitates the direct application of drugs in targeted tissues. This enhances the compound's applicability while mitigating the inherent toxicity of certain drugs, rendering them more appealing than alternative polymers due to their ability to be recognized by the human body as degradation products that are naturally eliminated. Consequently, these polyesters are demonstrated as a feasible alternative to petroleum-derived plastics, mitigating the environmental effect associated with them. The economic impediment posed by the elevated production costs of these polymers remains a significant challenge, as evidenced by this review. Therefore, the investigation showed an approach that integrates the bioremediation of wastewater and optimizes the utilization of microorganisms that necessitate lower cultivation expenses, such as microalgae, which can thrive in wastewater and sludge while requiring fewer nutrients due to their autotrophic nature.

In the study by Dragostin et al., the authors explore the in vitro and in vivo evaluation of chitosan (CS) microparticles loaded with isoniazid (INH) derivatives for tuberculosis (TB) treatment. Their work demonstrated the benefits of microencapsulating INH derivatives on liver damage caused by anti-TB medication through mitigating the hepatotoxicity of the compounds and facilitating the correction of cell necrosis and microvesicular steatosis in contrast to their administration in suspension. The findings indicated a substantial decrease in tissue modifications, the elimination of cellular necrosis and microvesicular steatosis, and reduced levels of alkaline phosphatase and the liver enzymes TGO and TGP when employing encapsulated medication formulations.

The article written by García-García et al. provides insight into the in vivo response of a polymeric construct composed of PLGA, poly(lactic-co-glycolic acid), within the solid phase of Palacos R® combined with antibiotics for the treatment of methicillin-resistant staphylococcal infections such as daptomycin, vancomycin, and/or linezolid. The conducted experiments confirmed that the addition of PLGA microspheres in Palacos R® bone cement and the incorporation of the antibiotic daptomycin along with vancomycin enhance the tissue response to bone infection. Nevertheless, the antibiotic linezolid combined with vancomycin yielded suboptimal outcomes in our model, irrespective of the cement type employed.

Javaid et al. provide a detailed report on a novel method for producing nanoencapsulated cotton textile fabrics by integrating antifouling functional group finishing, utilizing a layer-by-layer technique through the alternating antifouling polymeric formulations (APF) and dip coating of oppositely charged polyelectrolyte solutions. The nanoencapsulated finished antifouling cotton textile fabric may be utilized in diverse industrial applications, especially in wound dressings, for the prevention of skin infections.

In their 37-page review, Khadem et al. focus on an extensive examination of advanced smart adhesives, their constraints, and the prospective trajectories and obstacles for the forthcoming generation of smart bioadhesives. Initially, the primary requirements for bioadhesives and classifications of smart bioadhesives were examined. Secondly, the use of smart bioadhesives in diverse applications, including tissue engineering, wound healing, and medication administration, was investigated. The final section addressed the shortcomings and difficulties of the present research, as well as the potential paths for smart bioadhesives.

The review conducted by Kim et al. provides an overview of organic semiconductors in biomedical applications, encompassing their characteristics, the device architectures, fabrication techniques, and uses in documented organic material-based healthcare devices. The assessment of material properties and fabrication advantages indicates that organic semiconductors possess significant promise for biomedical applications, particularly in flexible and wearable medical devices. Due to the superior mechanical characteristics of organic semiconductors, inherently flexible electronic devices can be created. This review also highlights the limitations of these organic semiconductors, such as unsatisfactory robustness which hinders their commercialization, reduced reliability attributable to the erroneous transmission of information and sensory signals over an extended duration, and the focus on biodegradability and biocompatibility, important challenges that must be addressed.

In the findings published by Mikolaszek et al. in this Special Issue, attention is drawn to a comparative investigation of pressure-sensitive adhesive polymers (PSA), specifically silicone and acrylate matrices (DuroTak® 387-2287, DuroTak® 87-4098, DuroTak® 87-2852, Bio-PSA MD7-4502, and SoftSkinAdhesive MG7-9850) designed for transdermal patch formulations. Three active pharmaceutical ingredients (APIs) (indomethacin, cytosine, and testosterone) with varying lipophilicity and transdermal transport capabilities were examined. While optical microscopy can assess the API state in the matrix, as well as the morphology, distribution, and particle size of the undissolved fraction, surface-focused SEM offers supplementary insights into the 3D positioning of the API, which may aid in elucidating the release kinetics.

Molchanov et al. describe the antiseptic polymer–surfactant complexes incorporating the cationic disinfectant cetylpyridinium chloride (CPC) exhibiting prolonged efficacy against SARS-CoV-2. The release rate of the CPC from the oppositely charged polymer gels (copolymers of (i) vinyl pyrrolidone and sodium methacrylate; (ii) acrylamide and sodium 2-acrylamido-2-methylpropane sulfonate; and (iii) acrylamide and sodium methacrylate) can be adjusted over a considerable range by altering the disinfectant concentration, swelling degree, cross-linking extent of the gel, and the content/type of anionic repeat units in the polymer matrix. Polymer–surfactant complexes were shown to decrease SARS-CoV-2 titer by seven orders of magnitude within just 5 s. The complexes maintained significant virucidal efficacy against SARS-CoV-2 for a minimum duration of one week.

In the study reported by Rubio Hernández-Sampelayo et al., emphasis is placed on the synthesis of a novel aromatic isocyanate derived from diamino-PABA (diamine 1,3-propanediol bis(4-aminobenzoate)) and the characterization of linear polyurethanes (PUs) obtained from it. Unhydrolyzable 1,4-butanediol (BD), together with hydrolyzable N,N-ethylene-bis(6-hydroxycaproamide) (EDA2 CL), were used as chain extenders, whereas polycaprolactones (PCLs) of different molecular weights were used as soft segments. Cell survival after 28 days for all poly(ester-urethane)s containing isoPABA motifs and EDA-2CL chain extenders exceeded 80%, indicating that these polymers were non-toxic. Furthermore, Alamar Blue data demonstrated favorable cell adhesion and cytotoxicity on the surfaces of these non-toxic biodegradable polyurethanes.

In their study, Sadykov et al. address the development of cryogel composite materials derived from polyvinyl alcohol and calcium phosphates, together with the examination of their physicochemical, functional, and mechanical attributes. The investigation into monocyte viability revealed that approximately 60–80% of the cells persist in the presence of pure HA, while the survival rates in the presence of pure PVA and composites are equivalent to the control sample. The proposed materials can be deemed biocompatible, as immune response monocytes exhibit high viability in their presence. This material can be utilized in biomedical applications to fill holes created by pathological processes in bone, as well as between the bone and the mechanically loaded implant.

Sikorski et al. examine the way in which the acid salts of chitosan (CS) can influence the antibacterial/antifungal potential of CS-based materials. For this study, different acid-based CS nonwovens (acetic, butyric, formic, hydrochloric, propionic, valeric) were taken into account in the analysis. ^{1}H-NMR spectroscopy and FTIR-ATR spectroscopy were employed to verify the integration of acid groups with the amino groups of chitosan. The developed materials underwent microbiological testing. Each modified material was inoculated with bacteria. Antimicrobial activity was detected in CS salts with acetic acid and hydrochloric acid. A decrease in the quantity of bacterial cells was noted for the *S. aureus* strain treated with CS salt modified with 10% acetic acid in ethanol. The antibacterial efficacy of CS salts increased with the proportion of acid salts present on the solid material's surface, resulting in a reduction or absence of bacterial colonies. No decrease in growth was noted for the *E. coli* strain. The CS samples were either inactive or entirely eradicated the bacterial cells. The toxicity to human erythrocytes was also examined. The toxicity data for human red blood cells, quantified as a percentage of hemolysis, indicate significantly greater toxicity in the samples that were not rinsed. The maximum toxicity recorded was 21.18% for the sample subjected to acetic acid treatment. A total lack of hemolysis was noted in the samples treated with valeric acid and formic acid.

The paper written by Sivanesan et al. discusses the current state of knowledge on examining the diverse new CS compounds available for drug delivery applications. A complete evaluation has been conducted on the milestones attained by applying CS derivatives in oral medication administration. The absence of utilizing various CS derivatives for oral medication administration has been emphasized. Additionally, potential explanations for the disparity in the utilization of distinct CS derivatives for oral medication administration were examined. The potential achievements attainable by using available resources were discussed in the future perspectives section.

The review realized by Tatarusanu et al. focuses on the recent literature concerning polymer-based wound dressings as modern approaches in wound healing. This overview succinctly outlines the physiopathology and healing mechanics of chronic wounds, along with contemporary therapy strategies. In this study, the rationale for employing standard and smart hydrogels (SHs) in wound healing was also discussed, along with current research trajectories aimed at generating SHs with new characteristics while also addressing their limitations and prospects for industrial-scale production.

In the findings published by Titov et al., attention is drawn to the morpho-functional condition of the sensori-motor zone of the cerebral cortex and the hepato-renal system following the acute and sub-acute administration of silver selenide nanoparticles (AgSe NPs) encapsulated in a natural arabino galactan (AG) polymer. The investigation of acute toxicity, conducted through the oral administration of nanocomposites at a dosage of 2000 mg/kg, has demonstrated that the material in question is classified as a low-toxicity substance within the fifth hazard class. Sub-acute oral administration of nanocomposites at a dosage of 500 g/kg results in minor alterations in the brain tissue and liver of experimental subjects, suggesting the onset of compensatory adaptive responses. The Shumlyansky–Bowman chamber in the kidneys exhibits a 40.5% reduction compared to the control group. The utilization of selenium's protective capabilities, present in the composite, demonstrates a reduction in the toxicity of silver.

The study authored by X et al. outlines the influence of polymers on drug release (timolol maleate) kinetics in nanoemulsion developed by an in situ gel formulation with applicability in the treatment of glaucoma. This study sought to enhance drug accessibility in glaucomatous conditions by incorporating varying proportions of the polymer Carbopol 934p, a polyacrylic acid polymer that exhibits a sol-to-gel transition in aqueous solutions when the pH exceeds its pKa of approximately 5.5, and it is extensively utilized in ophthalmology to improve pre-corneal retention in the eye. The NEI_5 formulation maintained relatively stable effective drug levels in the ocular cavity for 24 h, and in vivo results demonstrated that the NEI_5 offered superior sustained drug release compared to the conventional marketed dosage form. The timolol maleate nanoemulsion in situ gel formulation exhibited stability under storage conditions, showing no discernible alteration in appearance.

The study conducted by Vitus et al. highlights the production of a carbon-based bioscaffold (CHAK) utilizing carbon sourced from human hair and composite materials comprising konjac glucomannan (KGM) and agar by gelation, fast freezing, and ethanol immersion techniques. The developed CHAK exhibited comparatively moderate degradability (less than 50% after 28 days of incubation), substantial swelling (higher than 1000%), and exceptional water absorption capacity (higher than 90%). The 3AHC bioscaffold exhibited the highest electrical conductivity (21.14 ± 2.29 S/m) among the produced bioscaffolds. Finally, the biocompatibility assessment with Wharton's jelly-derived mesenchymal stem cells (WJMSCs) indicated that the CHAK demonstrates compatibility, with over 90% cell survival on day 7. The results of this study demonstrated the viability of generating carbon from human hair as a material for the fabrication of a sustainable carbon-based bioscaffold for tissue engineering applications.

Yudaev et al., in their review paper, aim to delineate and rigorously evaluate research published in recent years regarding the utilization of polymeric dental nanoparticles as antibacterial agents across several domains of dentistry such as dental treatment, orthodontics, dental implantology, dental prostheses, and maxillofacial surgery regarding their antimicrobial efficacy. The breakthroughs outlined in the paper establish a new trajectory for the advancement of restorative dental materials and composites, with the objective of enhancing patients' quality of life.

3. Conclusions

This Special Issue brings together a wide range of studies based on polymers with various applications in the biomedical field, ranging from cancer treatment, glaucoma, tuberculosis management, wound healing, the treatment of neuropsychiatric diseases such as schizophrenia, a reduction in bone tissue infection, and the development of materials with enhanced antibacterial, antifungal, or improved anti-SARS-COV-2 activity. Therefore, this compilation of articles encompasses a diverse range of research, elucidative of the richness of the topic [11].

The increased interest shown by renowned researchers from all corners of the world has led to this Special Issue having two more successive issues, now reaching Biomedical Application of Polymeric Materials III. I would therefore like to conclude this editorial by inviting you to be part of this Special Issue through your innovative research contributions in this broad and promising field.

Conflicts of Interest: The author declares no conflicts of interest.

List of Contributors:

1. Aguilar-Rabiela, A.E.; Homaeigohar, S.; González-Castillo, E.I.; Sánchez, M.L.; Boccaccini, A.R. Comparison between the Astaxanthin Release Profile of Mesoporous Bioactive Glass Nanoparticles (MBGNs) and Poly(3-Hydroxybutyrate-Co-3-Hydroxyvalerate) (PHBV)/MBGN Composite Microspheres. *Polymers* **2023**, *15*, 2432. https://doi.org/10.3390/polym15112432.
2. Alhakamy, N.A.; Fahmy, U.A.; Eldin, S.M.B.; Ahmed, O.A.A.; Aldawsari, H.M.; Okbazghi, S.Z.; Alfaleh, M.A.; Abdulaal, W.H.; Alamoudi, A.J.; Mady, F.M. Scorpion Venom-Functionalized Quercetin Phytosomes for Breast Cancer Management: In Vitro Response Surface Optimization and Anticancer Activity against MCF-7 Cells. *Polymers* **2021**, *14*, 93. https://doi.org/10.3390/polym14010093.
3. Araújo, D.; Rodrigues, T.; Alves, V.D.; Freitas, F. Chitin-Glucan Complex Hydrogels: Optimization of Gel Formation and Demonstration of Drug Loading and Release Ability. *Polymers* **2022**, *14*, 785. https://doi.org/10.3390/polym14040785.
4. Bhupathyraaj, M.; Vijaya Rani, K.R.; Sridhar, S.B.; Shareef, J.; Thomas, S.; Halligudi, N.; Sockalingam, A.; Mohandoss, K.; Sundar, S. Effect of Polymers and Permeation Enhancers in the

Release of Quetiapine Fumarate Transdermal Patch through the Dialysis Membrane. *Polymers* **2022**, *14*, 1984. https://doi.org/10.3390/polym14101984.

5. Diniz, M.S.D.F.; Mourão, M.M.; Xavier, L.P.; Santos, A.V. Recent Biotechnological Applications of Polyhydroxyalkanoates (PHA) in the Biomedical Sector—A Review. *Polymers* **2023**, *15*, 4405. https://doi.org/10.3390/polym15224405.
6. Dragostin, I.; Dragostin, O.-M.; Iacob, A.T.; Dragan, M.; Chitescu, C.L.; Confederat, L.; Zamfir, A.-S.; Tatia, R.; Stan, C.D.; Zamfir, C.L. Chitosan Microparticles Loaded with New Non-Cytotoxic Isoniazid Derivatives for the Treatment of Tuberculosis: In Vitro and In Vivo Studies. *Polymers* **2022**, *14*, 2310. https://doi.org/10.3390/polym14122310.
7. García-García, J.; Azuara, G.; Fraile-Martinez, O.; García-Montero, C.; Álvarez-Mon, M.A.; Ruíz-Díez, S.; Álvarez-Mon, M.; Buján, J.; García-Honduvilla, N.; Ortega, M.A.; et al. Modification of the Polymer of a Bone Cement with Biodegradable Microspheres of PLGA and Loading with Daptomycin and Vancomycin Improve the Response to Bone Tissue Infection. *Polymers* **2022**, *14*, 888. https://doi.org/10.3390/polym14050888
8. Javaid, S.; Mahmood, A.; Nasir, H.; Iqbal, M.; Ahmed, N.; Ahmad, N.M. Layer-By-Layer Self-Assembled Dip Coating for Antifouling Functionalized Finishing of Cotton Textile. *Polymers* **2022**, *14*, 2540. https://doi.org/10.3390/polym14132540.
9. Khadem, E.; Kharaziha, M.; Bakhsheshi-Rad, H.R.; Das, O.; Berto, F. Cutting-Edge Progress in Stimuli-Responsive Bioadhesives: From Synthesis to Clinical Applications. *Polymers* **2022**, *14*, 1709. https://doi.org/10.3390/polym14091709.
10. Kim, K.; Yoo, H.; Lee, E.K. New Opportunities for Organic Semiconducting Polymers in Biomedical Applications. *Polymers* **2022**, *14*, 2960. https://doi.org/10.3390/polym14142960.
11. Mikolaszek, B.; Jamrógiewicz, M.; Mojsiewicz-Pieńkowska, K.; Sznitowska, M. Microscopic and Spectroscopic Imaging and Thermal Analysis of Acrylates, Silicones and Active Pharmaceutical Ingredients in Adhesive Transdermal Patches. *Polymers* **2022**, *14*, 2888. https://doi.org/10.3390/polym14142888.
12. Molchanov, V.S.; Shibaev, A.V.; Karamov, E.V.; Larichev, V.F.; Kornilaeva, G.V.; Fedyakina, I.T.; Turgiev, A.S.; Philippova, O.E.; Khokhlov, A.R. Antiseptic Polymer–Surfactant Complexes with Long-Lasting Activity against SARS-CoV-2. *Polymers* **2022**, *14*, 2444. https://doi.org/10.3390/polym14122444.
13. Rubio Hernández-Sampelayo, A.; Navarro, R.; González-García, D.M.; García-Fernández, L.; Ramírez-Jiménez, R.A.; Aguilar, M.R.; Marcos-Fernández, Á. Biodegradable and Biocompatible Thermoplastic Poly (Ester-Urethane)s Based on Poly(ε-Caprolactone) and Novel 1,3-Propanediol Bis(4-Isocyanatobenzoate) Diisocyanate: Synthesis and Characterization. *Polymers* **2022**, *14*, 1288. https://doi.org/10.3390/polym14071288.
14. Sadykov, R.; Lytkina, D.; Stepanova, K.; Kurzina, I. Synthesis of Biocompatible Composite Material Based on Cryogels of Polyvinyl Alcohol and Calcium Phosphates. *Polymers* **2022**, *14*, 3420. https://doi.org/10.3390/polym14163420.
15. Sikorski, D.; Bauer, M.; Frączyk, J.; Draczyński, Z. Antibacterial and Antifungal Properties of Modified Chitosan Nonwovens. *Polymers* **2022**, *14*, 1690. https://doi.org/10.3390/polym14091690.
16. Sivanesan, I.; Tasneem, S.; Hasan, N.; Shin, J.; Muthu, M.; Gopal, J.; Oh, J.-W. Surveying the Oral Drug Delivery Avenues of Novel Chitosan Derivatives. *Polymers* **2022**, *14*, 2131. https://doi.org/10.3390/polym14112131.
17. Tatarusanu, S.-M.; Lupascu, F.-G.; Profire, B.-S.; Szilagyi, A.; Gardikiotis, I.; Iacob, A.-T.; Caluian, I.; Herciu, L.; Giscă, T.-C.; Baican, M.-C.; et al. Modern Approaches in Wounds Management. *Polymers* **2023**, *15*, 3648. https://doi.org/10.3390/polym15173648.
18. Titov, E.A.; Sosedova, L.M.; Novikov, M.A.; Zvereva, M.V.; Rukavishnikov, V.S.; Lakhman, O.L. The Analysis of Acute and Subacute Toxicity of Silver Selenide Nanoparticles Encapsulated in Arabinogalactan Polymer Matrix. *Polymers* **2022**, *14*, 3200. https://doi.org/10.3390/polym14153200.
19. Vijaya Rani, K.R.; Rajan, S.; Bhupathyraaj, M.; Priya, R.K.; Halligudi, N.; Al-Ghazali, M.A.; Sridhar, S.B.; Shareef, J.; Thomas, S.; Desai, S.M.; et al. The Effect of Polymers on Drug Release Kinetics in Nanoemulsion In Situ Gel Formulation. *Polymers* **2022**, *14*, 427. https://doi.org/10.3390/polym14030427.

20. Vitus, V.; Ibrahim, F.; Shamsuddin, S.A.A.; Razali, N.; Noor Azlan, N.A.B.; Zaman, W.S.W.K. Carbonised Human Hair Incorporated in Agar/KGM Bioscaffold for Tissue Engineering Application: Fabrication and Characterisation. *Polymers* **2022**, *14*, 5489. https://doi.org/10.3390/polym14245489.
21. Yudaev, P.; Chuev, V.; Klyukin, B.; Kuskov, A.; Mezhuev, Y.; Chistyakov, E. Polymeric Dental Nanomaterials: Antimicrobial Action. *Polymers* **2022**, *14*, 864. https://doi.org/10.3390/polym14050864.

References

1. Suhas, P.; Mahesh, B.; Divakara, S.G.; Nanjundaswamy, G.S.; Mahadevaprasad, C.S.; Sionkowska, A.; Popat, K.C.; Channe Gowda, D. Synergistic Approaches in Natural and Synthetic Polymer Blends for Biomedical Applications-A Review. *Eur. Polym. J.* **2025**, *236*, 114161. [CrossRef]
2. Jäck, N.; Nagel, S.; Hartmann, L. Sequence-Defined Polymers for Biomedical Applications. *Prog. Polym. Sci.* **2025**, *167*, 101993. [CrossRef]
3. Yan, Z.-H.; Huang, H.-X.; Ding, G.-Q.; Dong, S.-S.; Jiang, K.-J.; Li, Y.-Y.; Liu, Y.; Jiang, Y.; Wang, S.-F.; Hu, G.-H.; et al. Research Progress of High-Strength Self-Healing Polymer Materials: Balance between Mechanical Strength and Self-Healing Efficiency. *Chem. Eng. J.* **2025**, *518*, 164609. [CrossRef]
4. Ornaghi, H.L.; Monticeli, F.M.; Agnol, L.D. A Review on Polymers for Biomedical Applications on Hard and Soft Tissues and Prosthetic Limbs. *Polymers* **2023**, *15*, 4034. [CrossRef] [PubMed]
5. Yadav, K.; Sahu, K.K.; Dubey, A.; Pradhan, H.K.; Sucheta; Pradhan, M. Bioprinting Functional Constructs for Women's Reproductive Health: Utilizing Tailored Biomaterials and Biopolymer Macromolecules for Drug Delivery and Tissue Regeneration. *Int. J. Biol. Macromol.* **2025**, *312*, 143990. [CrossRef] [PubMed]
6. Kalirajan, C.; Dukle, A.; Nathanael, A.J.; Oh, T.-H.; Manivasagam, G. A Critical Review on Polymeric Biomaterials for Biomedical Applications. *Polymers* **2021**, *13*, 3015. [CrossRef] [PubMed]
7. Xie, D.; Cao, Q.; Fang, H.; Li, Y.; Tian, H. Applications and Challenges of Biomedical Polymer Materials in Pulmonary Diseases. *Chin. Chem. Lett.* **2025**, 111032. [CrossRef]
8. Julius, A.; Malakondaiah, S.; Pothireddy, R.B. Polymer and Nanocomposite Fillers as Advanced Materials in Biomedical Applications. *Nano Trends* **2025**, *9*, 100087. [CrossRef]
9. Guo, J.; Tao, Y.; Du, Z.; Zhang, S.; Zheng, W.; Wang, Z.; Yi, Z.; Gou, Y.; Tang, W. Stimuli-Responsive Antimicrobial Polymer Systems: From Structural Design to Biomedical Applications. *Giant* **2025**, *24*, 100366. [CrossRef]
10. Jiang, H.; Zhang, Z.; Liu, Y.; Li, Y.; Hu, Y.; Yu, B.; Hu, H.; Wang, K. The Hyperbranched Polymers as a Medical Rising Star: From Synthesis to Biomedical Applications. *Mater. Today Chem.* **2025**, *48*, 102937. [CrossRef]
11. Guimarães, P. Cities and Retail: Sustainable Transformation of Retail in Urban Environments. *Sustainability* **2023**, *15*, 12743. [CrossRef]

polymers

MDPI

Article

Comparison between the Astaxanthin Release Profile of Mesoporous Bioactive Glass Nanoparticles (MBGNs) and Poly(3-hydroxybutyrate-*co*-3-hydroxyvalerate) (PHBV)/MBGN Composite Microspheres

Arturo E. Aguilar-Rabiela [1,2,*], Shahin Homaeigohar [3], Eduin I. González-Castillo [4,5], Mirna L. Sánchez [1,6] and Aldo R. Boccaccini [1,*]

1 Institute of Biomaterials, Department of Materials Science and Engineering, University of Erlangen-Nuremberg, Cauerstrasse 6, 91058 Erlangen, Germany; mirna.sanchez@fau.de
2 Tissue Engineering Research Group, Department of Anatomy & Regenerative Medicine, Royal College of Surgeons in Ireland (RCSI), D02 YN77 Dublin, Ireland
3 School of Science & Engineering, University of Dundee, Dundee DD1 4HN, UK; shomaeigohar001@dundee.ac.uk
4 Polymer Institute, Slovak Academy of Sciences, Dúbravská Cesta 9, 845 41 Bratislava, Slovakia; eduin.ivan@gmail.com
5 AO Research Institute Davos, Clavadelerstrasse 8, 7270 Davos, Switzerland
6 Laboratorio de Farmacología Molecular, Departamento de Ciencia y Tecnología, Universidad Nacional Quilmes, Bernal B1876, Argentina
* Correspondence: aguilar.rabiela@exatec.tec.mx (A.E.A.-R.); aldo.boccaccini@fau.de (A.R.B.)

Abstract: In recent years, composite biomaterials have attracted attention for drug delivery applications due to the possibility of combining desired properties of their components. However, some functional characteristics, such as their drug release efficiency and likely side effects, are still unexplored. In this regard, controlled tuning of the drug release kinetic via the precise design of a composite particle system is still of high importance for many biomedical applications. This objective can be properly fulfilled through the combination of different biomaterials with unequal release rates, such as mesoporous bioactive glass nanoparticles (MBGN) and poly(3-hydroxybutyrate-*co*-3-hydroxyvalerate) (PHBV) microspheres. In this work, MBGNs and PHBV-MBGN microspheres, both loaded with Astaxanthin (ASX), were synthesised and compared in terms of ASX release kinetic, ASX entrapment efficiency, and cell viability. Moreover, the correlation of the release kinetic to phytotherapeutic efficiency and side effects was established. Interestingly, there were significant differences between the ASX release kinetic of the developed systems, and cell viability differed accordingly after 72 h. Both particle carriers effectively delivered ASX, though the composite microspheres exhibited a more prolonged release profile with sustained cytocompatibility. The release behaviour could be fine-tuned by adjusting the MBGN content in the composite particles. Comparatively, the composite particles induced a different release effect, implying their potential for sustained drug delivery applications.

Keywords: Astaxanthin; composite microspheres; sustained release; controlled release

1. Introduction

Natural compounds have gained a high interest in recent years due to their beneficial properties, which are being continuously discovered year by year [1–3]. In this context, ASX is a well-studied phytotherapeutic reported for its remarkable antioxidant activity [4–6]. ASX is known as an xanthophyll carotenoid and was for the first time extracted from lobster by Kuhn and Sorensen [7] and later introduced into the market as a pigmentation material for the aquatic farm industry [8,9]. ASX, 3,3′-dihydroxy-β,β-carotene-4,4′-dione, is classified as a tetraterpene comprising 40 carbon atoms, with the molecular formula of $C_{40}H_{52}O_4$ and a molecular mass of 596.85 g/mol [10,11].

In recent years, ASX has drawn extensive attention for dermatological and cosmetic applications due to its outstanding antioxidant activity, which prevails over that of tocopherol, its supportive effect on skin health, and its protective role against UV radiation [12,13]. However, ASX is poorly soluble in water and is chemically unstable, and shows insufficient bioavailability, challenging its cosmetic and pharmaceutical applications [14]. The creation of delivery systems that can protect ASX and optimise its bioavailability is thus a highly demanded research topic. Regarding drug delivery systems, nanoparticles can be considered a proper candidate to carry drugs and release them locally for various medical applications [15,16]. However, safe assimilation and minimising the side effects of the administrated nanoparticle carriers are still existing challenges [17,18]. In this regard, mesoporous bioactive glass nanoparticles (MBGNs) have been proven to be bioactive and biocompatible drug carriers and, thus, have been proposed for various biomedical purposes [19–21]. Moreover, they can release certain ions upon dissolution that can drive complex gene transduction pathways in cells, thereby improving cell activities [22,23]. Despite such merits, there is a need to further extend the time span of drug release from MBGNs in a controlled manner [24]. In this regard, the development of particle carrier systems based on a formulation composed of biodegradable biopolymers, e.g., polyhydroxyalkanoates (PHAs), and MBGNs has been explored as a proper strategy [24,25]. PHAs are a family of biotechnology-produced polymers frequently used for biomedicine, thanks to their outstanding biocompatibility and biodegradability [23]. Particularly, poly(3-hydroxybutyrate) (P(3HB)) and poly(3-hydroxybutyrate-*co*-3-hydroxyvalerate) (PHBV) are well-studied types of PHAs that have been largely employed for the development of drug delivery microspheres [26–29]. Compared to P(3HB), PHBV has a lower melting temperature and is partially amorphous, thus is considered a more suitable candidate for drug delivery purposes [30,31]. Li et al. [24], for instance, developed vancomycin-loaded PHBV microspheres for coating 45S5 BG bone-tissue-engineering scaffolds. Different PHBV-based microsystems used for drug delivery microsystems have been reported for the release of some phytotherapeutics and natural components for different objectives [32–34]. In the current study, ASX-loaded MBGNs are incorporated in PHBV microspheres to extend the ASX release profile of the drug delivery system compared to when ASX is solely loaded into MBGNs. This composite strategy is assumed to enhance cell activities due to the controlled release of both ASX and beneficial ions from MBGN and dissolution byproducts from the MBGN/PHBV composite material.

2. Materials and Methods

2.1. Materials

PHBV was purchased from Goodfellow (Huntington, UK), dichloromethane and Astaxanthin (ASX) from Sigma-Aldrich (Steinheim, Germany), and Polyvinyl Alcohol (PVA) from Baxter Healthcare (Opfikon, Switzerland). Tetraethyl orthosilicate (TEOS; 99%), triethyl phosphate (TEP; 99%), and calcium nitrate were bought from Aldrich (Steinheim, Germany). Furthermore, ethyl acetate and cetyl-trimethylammonium bromide (CTAB) were provided by Merck (Darmstadt, Germany), ammonium hydroxide (28%) was provided by VWR (Fontenay Sous Bois, France), and distilled water (MilliQ) and absolute ethanol (99.8%) were provided by Alfa Aesar (Kandel, Germany). All the used chemicals were of analytical grade.

2.2. Synthesis of MBGN and Composite Microspheres

The mesoporous bioactive glass nanoparticles (MBGN) were produced following a modified Stöber procedure, which was previously reported in [35]. For the loading of ASX into MBGNs, a technique based on previous reports was designed [21,32]. Briefly, 2 mg of synthesised MBGN and 20 µg of ASX were suspended on 5 mL of non-ASX solvent (ultrapure water) and stirred at 800 rpm. Then, to facilitate the incorporation of ASX into the MBGN, the suspension was centrifuged at 8000 rpm. The supernatant was carefully removed to let the precipitated MBGN be loaded with ASX at 10% (*w*/*w*) and dried for 6 h in an incubator at 60 °C.

Simultaneously, composite PHBV-MBGN-ASX microspheres were produced using a solid-in-oil-on-water (S/O/W) method, which was previously reported in [21,32]. Briefly, 2 mg of the synthesised MBGN and 20 µg of ASX were added to a 10 mL of PHVB-DCM solution and stirred at 800 rpm to produce S/O phase. In parallel, an aqueous solution of 1 mg/mL of PVA was prepared and mixed at 600 rpm to form the W phase. Afterward, the S/O phase was dropwise added to the W phase while homogenising at 19,000 rpm using a T18 (IKA, Staufen, Germany) to produce composite microspheres comprised of 10% of MBGN and 1% of ASX (*w*/*w*) of loading. There was a PHBV/MBGN ratio of (10:1) for all the samples. Subsequently, the emulsion was centrifuged at 6000 rpm, and the supernatant was removed entirely to let the precipitated microspheres dry overnight in an incubator at 60 °C. All the particle synthesis was carried out in the beakers and tubes protected against light.

2.3. Particle Size and Zeta Potential

The average particle diameter and zeta potential measurement of the Astaxanthin-loaded MBGN (MBGN/ASX), the PHBV-MGBN-ASX (composite microspheres), blank MBGN, and blank PHBV particles were carried out by using a Zetasizer Nano ZS (Malvern, Worcestershire, UK) in deionized water for MBGN conditions and ethanol for PHBV conditions, similar to previous reports [21,36].

2.4. Surface Morphology of Particles

Micrographs of the surface morphology of the MBGN/ASX and the composite microspheres were obtained via Scanning Electron Microscopy (SEM) using an SEM Auriga (Zeiss, Munich, Germany). All samples were sputtered with gold in a turbomolecular-pumped coater Q150T Plus (Quorum, Laughton, UK) prior to SEM examination.

2.5. Chemical-Structural Characterization of Particles

Absorbance infrared spectra were obtained for the neat components, MBGN/ASX, and composite microspheres by using a Fourier Transform Infrared spectrometer (FTIR) Nicolet 6700 (Thermo-Scientific, Waltham, MA, USA). Measurements were carried out in the middle-infrared (MIR) region from 4000 to 400 cm^{-1} (40 scans and 4 cm^{-1} spectral resolution).

2.6. Astaxanthin Entrapment Efficiency

The ASX Entrapment Efficiency (EE) was calculated according to the supernatant method previously reported [37]. The ASX-MBGN and composite microspheres were immersed in 10 mL of ethanol, respectively, and after release depletion, the supernatant concentration was measured using a UV/Vis spectrophotometer Specord 250 (Analytik-jena, Jena, Germany) at λ = 478 nm [38]. The entrapment efficiencies were calculated by using the initially added ASX amounts during the fabrication of particles, the concentration value was measured via UV/Vis, and the following equation was obtained [21,37]:

$$EE = \frac{(M_{the} - M_{sup}) \cdot 100\%}{M_{the}} \quad (1)$$

where M_{sup} is the ASX mass measured in the supernatant in mg, and M_{the} is the initial ASX mass added during the fabrication in µg.

2.7. Astaxanthin Release Kinetics

The ASX-release kinetics were calculated by adding 5 mg of each sample to 5 mL of ethanol in triplicate [21,37]. The release kinetic curve was obtained through a technique previously reported [21,37,39]. Briefly, several measurements of the supernatant concentration were taken along the release time by using a Specord 250 device (Analytik Jena, Jena, Germany) at 478 nm. Concentration curves were normalized with respect to the calculated theoretical drug loaded during the fabrication of the samples, and all release kinetics were carried out in the tubes protected from light prior to each measurement. Calibration curves for ASX were obtained prior to each release kinetic assay.

2.8. Cell Viability Assays

NIH-3T3 mouse fibroblasts cells (Sigma-Aldrich, Germany) were grown in Dulbecco's Modified Eagle Medium (DMEM) supplemented with BCS, sodium pyruvate, and L-glutamine (Gibco, Dreieich, Germany). Afterward, 50×10^3 cells were taken from 90% confluency flasks and seeded in 24-well plates. Considering the EE of each condition, master solutions of ASX/MBGN, composite microspheres, and free ASX were so prepared to have a similar content through each condition. Afterward, 10-fold serial dilutions of 10 µL, 100 µL, and 1000 µL from each master solution were added to the 24-well plates previously seeded with cells. A similar amount of blank MBGN and blank PHBV microspheres was used as additional controls, and untreated cell samples were used as control. All experiments were carried out in triplicate. The cell viability was determined via WST-8 assay (Sigma-Aldrich, Steinheim, Germany) after 24 and 72 h post-administration according to the provider protocol. Briefly, the supernatant was carefully removed from all well plates and washed with HBSS; later, freshly prepared cell culture medium containing 1% v/v WST-8 solution was added to the 24-well plates and incubated for 3 h. After incubation, 100 µL of supernatant from each well plate was transferred, by duplicates, into a 96-well plate for the absorbance measurement at λ = 450 nm in a PHOmo Autobio (Labtec Instruments, Zhengzhou, China), similar to previous reports [21].

2.9. Cell Staining and Imaging

Cell samples were stained with Life & Dead Hoechst protocol using Calcein AM, Propidium Iodide, and Life & Dead assays (Merck, Darmstadt, Germany) for cell morphology observation after 72 h of incubation. Cell staining was carried out according to the provider's protocol. After staining, the samples were stored and protected from light prior to observations. Fluorescent micrographs were obtained and processed using a fluorescent optical microscope Primovert with AxioCam ERc 5s (Zeiss, Munich, Germany).

2.10. Bioactivity Assessment

MBGN/ASX and composite microsphere samples were immersed in SBF, produced according to the protocol of Kokubo [40], and incubated for 14 days. Afterward, SBF was removed, and the samples were dried in an incubator at 60 °C for 4 h, then stored at room temperature prior to examination.

The surface morphology of the samples was analysed via SEM according to Section 2.4. The X-ray diffraction analyses (XRD) were performed by using a Rigaku MiniflexTM 600 (Rigaku, Tokyo, Japan) X-ray diffractometer, provided with a Cu-Kα radiation (λ = 1.5406 Å) X-ray source and scintillation counter (Kβ filter) detector, in the angular range (2θ) from 10° to 40° with a resolution of 0.01° and speed scanning of 1°/min.

2.11. Statistics

Experiments were carried out in triplicate. The analysis of statistical significance was conducted by analysis of variance One-Way ANOVA followed by the Tukey test ($p < 0.05$), using the software Origin 8 (OriginLab, Northampton, MA, USA), and the data were presented as mean ± standard deviation for each measurement.

3. Results and Discussion

3.1. Particle Size, Morphology, and Zeta Potential

The synthesised MBGN/ASX nanoparticles and PHBV/MBGN/ASX microspheres exhibited a homogeneous, spherical shape (Figure 1d,e). In contrast to the composite microspheres, which possessed quasi-smooth surface morphology with low porosity (Figure 1e), MBGN/ASX particles featured a mesoporous surface, as similarly reported by Nawaz et al. [35]. The spherical shape of the composite microspheres was not affected by either the incorporation of MBGN or ASX. In both formulations, the spherical shape and surface morphology can promote a homogeneous drug release by diffusion, as observed in similar systems [37,41,42]. The particle size distributions of MBGN/ASX and the composite microspheres are shown in Figure 1a and 1b, respectively. Both systems were monodisperse in terms of particle size distribution, with a diameter between 400 to 600 nm for MBGN/ASX and 2 to 6 μm for the composite microspheres. The size difference, Figure 1f, might affect the ASX release rate due to the non-comparable surface area to volume ratio of the particle carriers [21,41]. The zeta potential of MBGN/ASX nanoparticles and composite microspheres is shown in Figure 1c. The values showed moderate stability, which may reduce the probability of undesired agglomeration for in vivo and in vitro applications [43,44]. Similar Zeta potential values have been observed in previous formulations using MBGN- and PHBV-based particles [21,45,46].

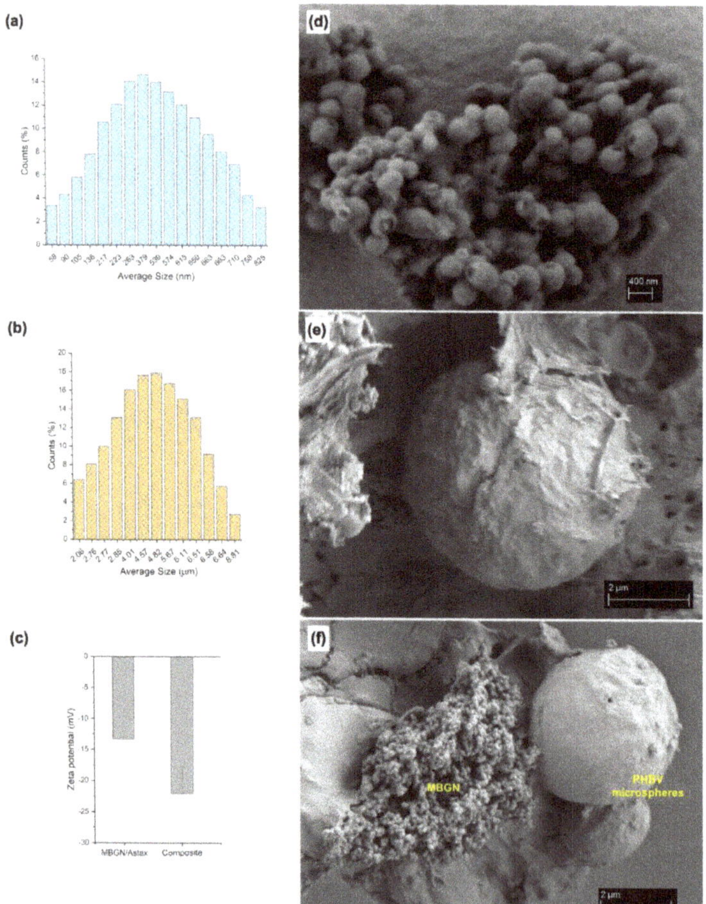

Figure 1. Particle size distribution of (**a**) MBGN/ASX and (**b**) composite microspheres; (**c**) Zeta potential of MBNG/ASX and composite particles. SEM micrographs showing the surface morphology of (**d**) MBGN/ASX and (**e**) composite microspheres and (**f**) comparison of MBGN and composite particle dimensions.

3.2. Composition Analysis

The FTIR absorbance spectra of the neat components (ASX, MBGN, and PHBV), MBGN/ASX, and composite microspheres recorded in the mid-infrared region (MIR) from 4000 to 400 cm^{-1} are shown in Figure 2. Three characteristic bands associated with ASX are observed at 1654, 1552, and 974 cm^{-1} in the MBGN/ASX spectrum, which are assigned, respectively, to the C=O stretching vibration, the stretching vibration of C=C in the hexatomic ring, and C–H in the C,C conjugate [47,48]. Except for the band at 1654 cm^{-1}, which is overlapped with that of PHBV, the other two characteristic bands of ASX are readily distinguished in the FTIR spectrum of the composite microspheres. Additionally, in the FTIR spectrum of the composite microspheres in Figure 2, the band at 1100 cm^{-1} corresponds to the Si–O–Si bending mode of the MBGN [21]. The bands at 1282 and 1179 cm^{-1} correspond to the C–O–C stretching modes of the crystalline and amorphous parts of PHBV, while the band peak at 1731 cm^{-1} is ascribed to the C=O crystalline phase of PHBV [21], which is overlapped with the carbonyl vibration in ASX. as has been previously indicated. The FTIR spectra confirm the successful loading of ASX in MBGN/ASX and the incorporation of ASX and MBGN in the composite microspheres, which may overlap with the 1654 cm^{-1} band peak of MBGN. These results support the loading of ASX in MBGN/ASX and the integration of ASX and MBGN in the composite microspheres.

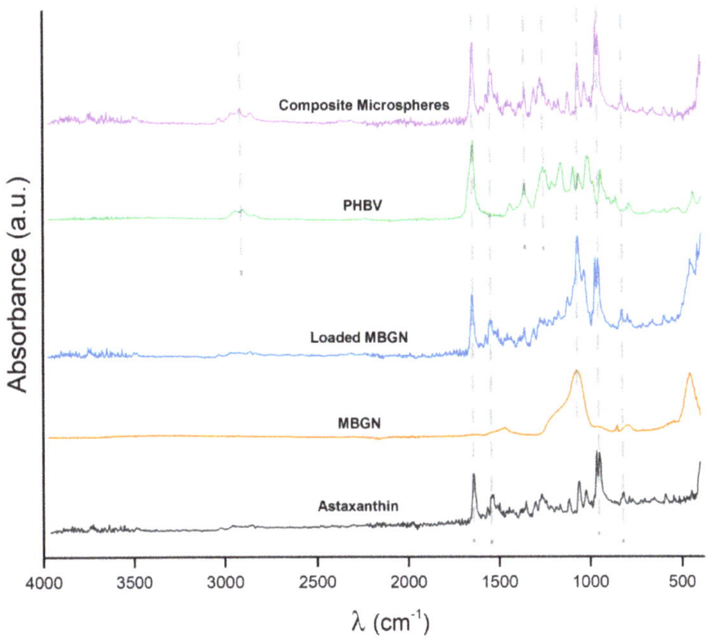

Figure 2. FTIR spectra of ASX, MBGN, MBGN/ASX, blank PHB, and composite microspheres. (λ is wavenumber).

3.3. Entrapment Efficiency and Release Kinetics

The ASX entrapment efficiencies (EE) of the MBGN/ASX and composite microspheres are shown in Figure 3a. The EE was calculated to be 60% for the MBGN/ASX and 80% for the composite microspheres. This difference could arise from the additional ASX entrapped into the PHBV polymer matrix, improving the EE of the composite in comparison with the open mesoporous structure of the MBGN/ASX, which may facilitate the diffusion of the phytotherapeutic agent through the porous structure and reduce the loading capacity. The ASX release kinetics of MBGN/ASX and the composite microspheres in an organic solvent (ethanol) are compared in Figure 3b, in which the final concentration of ASX released is comparable for both particle systems (~4 μg/mL).

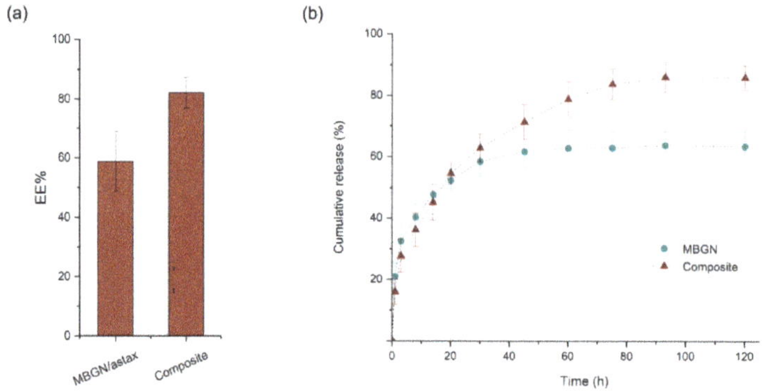

Figure 3. (**a**) ASX Entrapment Efficiency and (**b**) cumulative release kinetic of MBGN/ASX and composite microspheres (SD as error bars, n = 3).

The cumulative release with respect to the initial theoretical amount of ASX in both drug delivery systems exhibited a characteristic biphasic controlled release curve. Nonetheless, the final concentration in MBGN/ASX was slightly lower than that in the composite microspheres, a behaviour that corresponds to the EE difference between the systems. Additionally, a steeper slope was observed in the ASX-MBGN kinetic, which is related to a higher release ratio during the burst phase and resulted in an apparent complete depletion of the release during the first 40 h. In contrast, the composite release kinetic exhibited a lower slope during the burst phase and an extended controlled phase up to the first 80 h after administration, allowing a prolonged ASX release. This dissimilarity in the release behaviour may be explained based on the morphology and composition features of the systems, as the MBGN exhibited a mesoporous surface that favours the release as soon as the nanoparticles are in contact with the medium. In addition, the surface contact area/volume ratio is higher in the MBGNs because of their nanometric dimensions, promoting a fast drug release.

On the contrary, in the composite microspheres, the surface contact area/volume ratio is low. Furthermore, the homogeneous polymer surface acts as an extra phase where the medium needs to permeate through the polymeric barrier to trigger the diffusion, and once the process starts, the diffusion may be delayed in comparison to pure MBGN [21,37]. The combination of these effects results in a difference in the release kinetics between the composite microspheres and the MBGN carriers.

Nevertheless, both systems exhibited a proper potential for the ASX-controlled release applications, and this behaviour may be extended to other phytotherapeutics and other applications in which high initial concentrations must be avoided [21,49,50]. Additionally, a delayed release may be useful in applications that require a time gap before the full release of the entrapped compound, such as imagining diagnostics applications, among others [32]. Possibly, the variation in the MBGN concentration in the composite microspheres may bring a pathway to tailor the release kinetic and achievement of prolonged release.

3.4. Bioactivity Assessment

After 14 days of immersion in Simulated Body Fluid (SBF), the structure and morphology of MBGN/ASX and composite microspheres were further analysed via XRD and SEM observation (Figure 4a and 4b, respectively). In both graphs, it is possible to observe XRD patterns, namely, two peaks at 2θ values of 29° and 32°, characteristic of the formation of hydroxyapatite (HA) [21,51]. It is expected that in the MBGN/ASX and composite microspheres, calcium ions released from the systems interact with the ions present in

SBF, promoting hydroxyapatite mineralisation on the surface. This HA layer was also confirmed via SEM observation (Figure 4c,d), where the characteristic morphology of HA was present all over the surface of both particles. Similar morphology has been observed in other BG-based structures after immersion in SBF [21,52]. These results support the bioactivity of both ASX/MBGN and the composite microsphere systems.

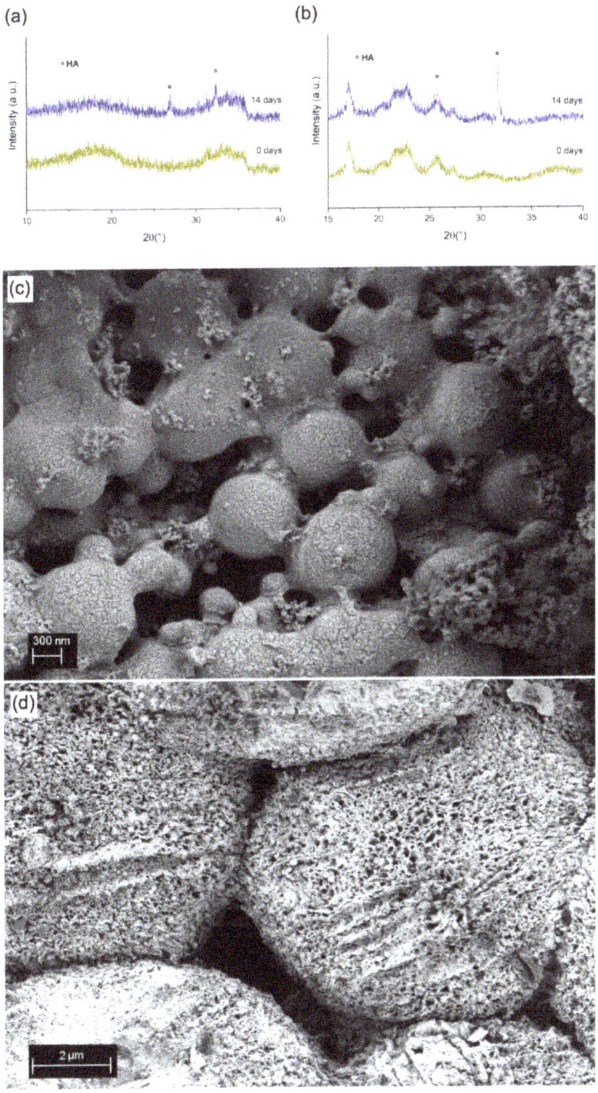

Figure 4. XRD patterns of (**a**) MBGN/ASX and (**b**) composite microspheres. SEM micrographs of (**c**) ASX/MBGN and (**d**) composite microspheres after 14 days of immersion in SBF.

3.5. Cell Viability of Particle Systems

The cell viability of both MBGN/ASX and the composite microspheres administrated to NIH-3T3 cells was assessed after 24 and 72 h; blank MBGN, blank PHBV particles, and free ASX were used as additional conditions, and cells without treatment were considered as control (shown in Figure 5). The ASX amount was calculated to have a similar final concentration in both particle systems (4 μg/mL approx.), the same as the release kinetics assay. In the WST-8 cell assay after 24 h (Figure 5a), it was possible to observe that all conditions exhibited cell viability similar to the control, and only some dilutions of free ASX (10^{-1}), MBGN/ASX (10^{-2} and 10^{-3}), and composite microspheres (10^{-1}) presented significant differences with respect to the control. On the other hand, after 72 h of incubation (Figure 5b), some dilutions of MBGN blank (10^{-1} and 10^{-2}) and MBGN/ASX (10^{-1}) exhibited a significant decrease in cell viability, while the composite microspheres (10^{-1} and 10^{-2}) sustained an increase in cell viability. At this time, free ASX and blank PHBV presented no significant difference with respect to the control.

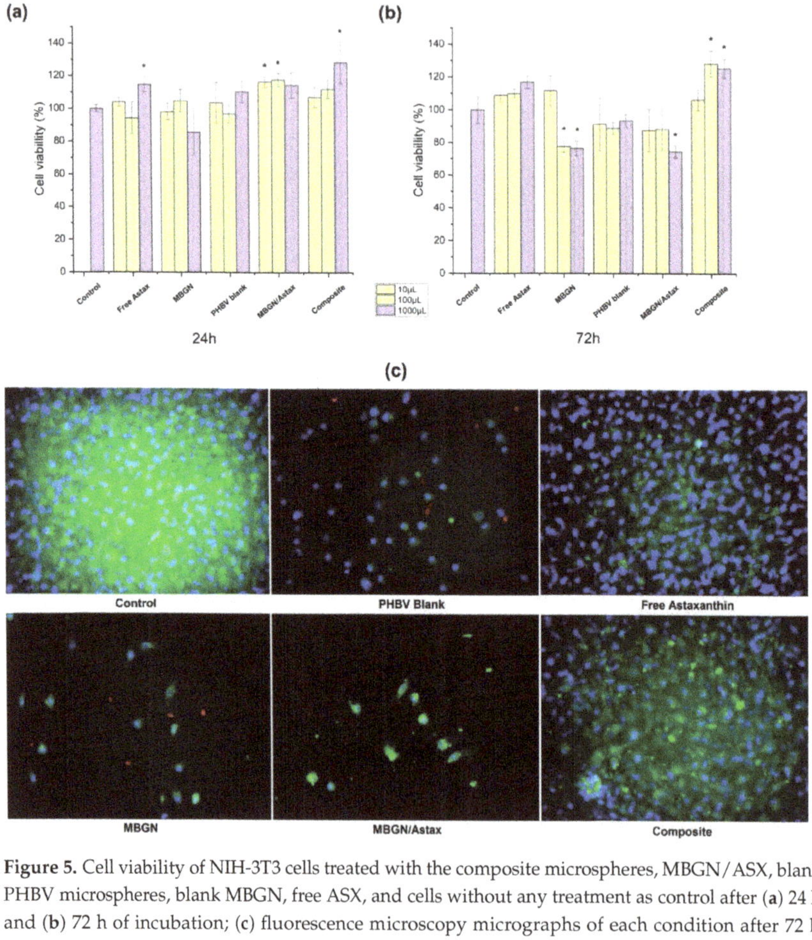

Figure 5. Cell viability of NIH-3T3 cells treated with the composite microspheres, MBGN/ASX, blank PHBV microspheres, blank MBGN, free ASX, and cells without any treatment as control after (**a**) 24 h and (**b**) 72 h of incubation; (**c**) fluorescence microscopy micrographs of each condition after 72 h. (* indicates a significant difference of $p \leq 0.5$ Tukey test, SD as error bars, n = 3).

Fluorescent microscopy observations after 72 h (Figure 5c) corresponded to WST-8 cell viability results, in which it was possible to observe a decrease in the population of living cells on the samples treated with MBGNs and PHBV blank particles. This behaviour has been reported before due to the accumulation of high concentrations of nanoparticles on cells that may increase the difficulty for the assimilation of the particles and modify the cell viability as a consequence [21,53]. Nevertheless, cell viability values for all conditions were above 70%, which may indicate good cytocompatibility of this category of devices [38,46]. On the other hand, it was noticed that the administration of both free ASX and the ASX released from the composite microspheres promoted sustained cell viability after 72 h, where the antioxidant properties of ASX might support this behaviour by reducing the oxidative stress due to the contact and assimilation of the particles [5,6]. However, this behaviour was not observed in the MBGN/ASX system, which may be explained based on the release kinetic of this system (Figure 3b), whereby a higher release ratio compared to the composite microspheres may result in a faster depletion of the loaded ASX, which during the first hours helps in damping the side effects due to the assimilation of MBGN. Nevertheless, as previously mentioned, the promoted cell viability cannot be sustained after the full release of ASX. On the other hand, the prolonged release through the composite microspheres may extend the beneficial effects of the phytotherapeutic up to 72 h, as a similar population of living cells was observed for the samples treated with free ASX and the composite microspheres.

The results mentioned above show the importance of the release kinetics for the effective administration of the model drug by the two studied delivery systems. Another important aspect was the difference in the EE of both systems, indicating that a higher amount of MBGN/ASX is required to release equivalent amounts of ASX for extending the drug's beneficial effects. Nonetheless, this also will increase the amount of MBGN in contact with cells, which, as discussed before, may impact biocompatibility. Furthermore, it could be interesting to characterise the influence of the MBGN content in the composite release kinetic as an easy alternative for tailoring the release in these kinds of carriers to determine if this behaviour can be extended to other model drugs. In any case, both the MBGN/ASX and composite microspheres exhibited an optimum potential for controlled phytotherapeutic delivery needed for specific applications.

4. Conclusions and Outlook

FTIR measurements confirmed the successful incorporation of ASX in MBGNs and composite microspheres. The MBGN/ASX and composite microspheres exhibited an optimum ASX entrapment efficiency (over 60% and 80%, respectively). The release kinetic curves showed a significant difference between the release behaviour of the MBGN/ASX and composite microspheres, where the latter exhibited prolonged release of ASX up to the first 80 h compared to half (40 h) of the MBGN/ASX.

The WST-8 cell viability results and fluorescent cell morphology observations showed that the difference in the release behaviour affected the NIH-3T3 cell viability during the first 72 h of incubation. These results suggested that the release behaviour of the composite microspheres may be tuned by modifying the MBGN content in the composite microspheres. Additionally, the exhibited release kinetics may be extended to other phytotherapeutics similar to ASX.

Although the composite microspheres are promising for sustained in vitro drug release applications, additional characterisation on these systems could be interesting, such as the effect of varying the MBGN content on the drug release kinetics and the stability and antioxidant activity of ASX during the particle synthesis process and the release.

Author Contributions: Conceptualization, A.E.A.-R.; Methodology, A.E.A.-R.; Validation, A.R.B.; Investigation, A.E.A.-R. and S.H.; Resources, A.R.B.; Data curation, S.H. and E.I.G.-C.; Writing—original draft, A.E.A.-R.; Writing—review & editing, S.H., E.I.G.-C., M.L.S. and A.R.B.; Visualization, A.E.A.-R.; Supervision, A.R.B.; Project administration, A.E.A.-R.; Funding acquisition, A.E.A.-R. and A.R.B. All authors have read and agreed to the published version of the manuscript.

Funding: This work was funded by the *Deutscher Akademischer Austauschdienst* (DAAD) under the postdoctoral program Ref—no. 57552340.

Institutional Review Board Statement: Not applicable.

Data Availability Statement: Data is available from the author.

Acknowledgments: We acknowledge the technical staff of the Institute of Biomaterials of *Friedrich-Alexander Universität* (FAU) for the technical support and the *Deutscher Akademischer Austauschdienst* (DAAD) for the funding (Ref. 57552340). MLS acknowledges CONICET and *Universidad Nacional de Quilmes*, (Argentina).

Conflicts of Interest: The authors declare no conflict of interest.

References

1. Schuhladen, K.; Roether, J.A.; Boccaccini, A.R. Bioactive glasses meet phytotherapeutics: The potential of natural herbal medicines to extend the functionality of bioactive glasses. *Biomaterials* **2019**, *217*, 119288. [CrossRef]
2. Hewlings, S.; Kalman, D. Curcumin: A review of its' effects on human health. *Foods* **2017**, *6*, 92. [CrossRef]
3. Thakur, K.; Tomar, S.K.; Singh, A.K.; Mandal, S.; Arora, S. Riboflavin and health: A review of recent human research. *Crit. Rev. Food Sci. Nutr.* **2017**, *57*, 3650–3660. [CrossRef] [PubMed]
4. Sztretye, M.; Dienes, B.; Gönczi, M.; Czirják, T.; Csernoch, L.; Dux, L.; Szentesi, P.; Keller-Pintér, A. Astaxanthin: A Potential Mitochondrial-Targeted Antioxidant Treatment in Diseases and with Aging. *Oxid. Med. Cell. Longev.* **2019**, *2019*, 3849692. [CrossRef] [PubMed]
5. Pertiwi, H.; Nur Mahendra, M.Y.; Kamaludeen, J. Astaxanthin as a Potential Antioxidant to Improve Health and Production Performance of Broiler Chicken. *Vet. Med. Int.* **2022**, *2022*, 4919442. [CrossRef] [PubMed]
6. Sj, S.; Veerabhadrappa, B.; Subramaniyan, S.; Dyavaiah, M. Astaxanthin enhances the longevity of Saccharomyces cerevisiae by decreasing oxidative stress and apoptosis. *FEMS Yeast Res.* **2019**, *19*, 113. [CrossRef]
7. Kuhn, R.; Sörensen, N.A. Über Astaxanthin und Ovoverdin. *Ber. Der Dtsch. Chem. Ges. (A B Ser.)* **1938**, *71*, 1879–1888. [CrossRef]
8. Davinelli, S.; Nielsen, M.E.; Scapagnini, G. Astaxanthin in Skin Health, Repair, and Disease: A Comprehensive Review. *Nutrients* **2018**, *10*, 522. [CrossRef]
9. Vollmer, D.L.; West, V.A.; Lephart, E.D. Enhancing Skin Health: By Oral Administration of Natural Compounds and Minerals with Implications to the Dermal Microbiome. *Int. J. Mol. Sci.* **2018**, *19*, 3059. [CrossRef]
10. Higuera-Ciapara, I.; Félix-Valenzuela, L.; Goycoolea, F.M. Astaxanthin: A review of its chemistry and applications. *Crit. Rev. Food Sci. Nutr.* **2006**, *46*, 185–196. [CrossRef]
11. Jannel, S.; Caro, Y.; Bermudes, M.; Petit, T. Novel Insights into the Biotechnological Production of Haematococcus pluvialis-Derived Astaxanthin: Advances and Key Challenges to Allow Its Industrial Use as Novel Food Ingredient. *J. Mar. Sci. Eng.* **2020**, *8*, 789. [CrossRef]
12. Sotiropoulou, G.; Zingkou, E.; Pampalakis, G. Redirecting drug repositioning to discover innovative cosmeceuticals. *Exp. Derm.* **2021**, *30*, 628–644. [CrossRef] [PubMed]
13. Couteau, C.; Coiffard, L. Microalgal Application in Cosmetics. In *Microalgae in Health and Disease Prevention*; Academic Press: Cambridge, MA, USA, 2018; pp. 317–323. [CrossRef]
14. Lima, S.G.M.; Freire, M.C.L.C.; Oliveira, V.d.S.; Solisio, C.; Converti, A.; de Lima, A.N. Astaxanthin Delivery Systems for Skin Application: A Review. *Mar. Drugs* **2021**, *19*, 511. [CrossRef] [PubMed]
15. Edlund, U.; Albertsson, A.C. Degradable polymer microspheres for controlled drug delivery. *Degrad. Aliphatic Polyest.* **2002**, *157*, 67–112.
16. Mignani, S.; el Kazzouli, S.; Bousmina, M.; Majoral, J.-P. Expand classical drug administration ways by emerging routes using dendrimer drug delivery systems: A concise overview. *Adv. Drug Deliv. Rev.* **2013**, *65*, 1316–1330. [CrossRef] [PubMed]
17. Vakili-Ghartavol, R.; Momtazi-Borojeni, A.A.; Vakili-Ghartavol, Z.; Aiyelabegan, H.T.; Jaafari, M.R.; Rezayat, S.M.; Bidgoli, S.A. Toxicity assessment of superparamagnetic iron oxide nanoparticles in different tissues. *Artif. Cells Nanomed. Biotechnol.* **2020**, *48*, 443–451. [CrossRef] [PubMed]
18. Hashemi-Beni, B.; Khoroushi, M.; Foroughi, M.R.; Karbasi, S.; Khademi, A.A. Cytotoxicity assessment of polyhydroxybutyrate/chitosan/nano-bioglass nanofiber scaffolds by stem cells from human exfoliated deciduous teeth stem cells from dental pulp of exfoliated deciduous tooth. *Dent. Res. J. (Isfahan)* **2018**, *15*, 136.
19. Wu, J.; Zheng, K.; Huang, X.; Liu, J.; Liu, H.; Boccaccini, A.R.; Wan, Y.; Guo, X.; Shao, Z. Thermally triggered injectable chitosan/silk fibroin/bioactive glass nanoparticle hydrogels for in-situ bone formation in rat calvarial bone defects. *Acta Biomater.* **2019**, *91*, 60–71. [CrossRef]
20. Reakasame, S.; Trapani, D.; Detsch, R.; Boccaccini, A.R. Cell laden alginate-keratin based composite microcapsules containing bioactive glass for tissue engineering applications. *J. Mater. Sci. Mater. Med.* **2018**, *29*, 185. [CrossRef]
21. Aguilar-Rabiela, A.E.; Leal-Egaña, A.; Nawaz, Q.; Boccaccini, A.R. Integration of Mesoporous Bioactive Glass Nanoparticles and Curcumin into PHBV Microspheres as Biocompatible Composite for Drug Delivery Applications. *Molecules* **2021**, *26*, 3177. [CrossRef]
22. Xynos, I.D.; Edgar, A.J.; Buttery, L.D.K.; Hench, L.L.; Polak, J.M. Ionic products of bioactive glass dissolution increase proliferation of human osteoblasts and induce insulin-like growth factor II mRNA expression and protein synthesis. *Biochem. Biophys. Res. Commun.* **2000**, *276*, 461–465. [CrossRef] [PubMed]
23. Xynos, I.D.; Edgar, A.J.; Buttery, L.D.; Hench, L.L.; Polak, J.M. Gene-expression profiling of human osteoblasts following treatment with the ionic products of Bioglass 45S5 dissolution. *J. Biomed. Mater. Res.* **2001**, *55*, 151–157. [CrossRef]

24. Li, W.; Ding, Y.; Rai, R.; Roether, J.A.; Schubert, D.W.; Boccaccini, A.R. Preparation and characterization of PHBV microsphere/45S5 bioactive glass composite scaffolds with vancomycin releasing function. *Mater. Sci. Eng. C* **2014**, *41*, 320–328. [CrossRef] [PubMed]
25. Xu, S.; Yi, S.; He, J.; Wang, H.; Fang, Y.; Wang, Q. Preparation and Properties of a Novel Microcrystalline Cellulose-Filled Composites Based on Polyamide 6/High-Density Polyethylene. *Materials* **2017**, *10*, 808. [CrossRef] [PubMed]
26. Xin, H.Z.; Wang, C.H.; Tong, Y.W. In vitro characterization of hepatocyte growth factor release from PHBV/PLGA microsphere scaffold. *J. Biomed. Mater. Res. A* **2009**, *89*, 411–423.
27. Huang, W.; Shi, X.; Ren, L.; Du, C.; Wang, Y. PHBV microspheres–PLGA matrix composite scaffold for bone tissue engineering. *Biomaterials* **2010**, *31*, 4278–4285. [CrossRef]
28. Riekes, M.K.; Barboza, F.M.; Vecchia, D.D.; Bohatch, M.; Farago, P.V.; Fernandes, D.; Silva, M.A.S.; Stulzer, H.K. Evaluation of oral carvedilol microparticles prepared by simple emulsion technique using poly(3-hydroxybutyrate-co-3-hydroxyvalerate) and polycaprolactone as polymers. *Mater. Sci. Eng. C* **2011**, *31*, 962–968. [CrossRef]
29. Yang, C.; Plackett, D.; Needham, D.; Burt, H.M. PLGA and PHBV microsphere formulations and solid-state characterization: Possible implications for local delivery of fusidic acid for the treatment and prevention of orthopaedic infections. *Pharm. Res.* **2009**, *26*, 1644–1656. [CrossRef]
30. Pouton, C.W.; Akhtar, S. Biosynthetic polyhydroxyalkanoates and their potential in drug delivery. *Adv. Drug Deliv. Rev.* **1996**, *18*, 133–162. [CrossRef]
31. Hazer, D.B.; Kiliçay, E.; Hazer, B. Poly(3-hydroxyalkanoate)s: Diversification and biomedical applications: A state of the art review. *Mater. Sci. Eng. C* **2012**, *32*, 637–647. [CrossRef]
32. Aguilar-Rabiela, A.E.; Unterweger, H.; Alexiou, C.; Boccaccini, A.R. Reduction of the in vitro toxicity of elevated concentrations of SPIONLA by its administration through PHBV/curcumin composite microspheres. *Front. Biomater. Sci.* **2022**, *1*, 13. [CrossRef]
33. Macías-Andrés, V.I.; Li, W.; Aguilar-Reyes, E.A.; Ding, Y.; Roether, J.A.; Harhaus, L.; León-Patiño, C.A.; Boccaccini, A.R. Preparation and characterization of 45S5 bioactive glass-based scaffolds loaded with PHBV microspheres with daidzein release function. *J. Biomed. Mater. Res. A* **2017**, *105*, 1765–1774. [CrossRef] [PubMed]
34. Mutlu, G.; Calamak, S.; Ulubayram, K.; Guven, E. Curcumin-loaded electrospun PHBV nanofibers as potential wound-dressing material. *J. Drug Deliv. Sci. Technol.* **2018**, *43*, 185–193. [CrossRef]
35. Nawaz, Q.; Rehman, M.A.U.; Burkovski, A.; Schmidt, J.; Beltrán, A.M.; Shahid, A.; Alber, N.K.; Peukert, W.; Boccaccini, A.R. Synthesis and characterization of manganese containing mesoporous bioactive glass nanoparticles for biomedical applications. *J. Mater. Sci. Mater. Med.* **2018**, *29*, 64. [CrossRef] [PubMed]
36. Zaloga, J.; Janko, C.; Nowak, J.; Matuszak, J.; Knaup, S.; Eberbeck, D.; Tietze, R.; Unterweger, H.; Friedrich, R.P.; Heimke-Brinck, R.; et al. Development of a lauric acid/albumin hybrid iron oxide nanoparticle system with improved biocompatibility. *Int. J. Nanomed.* **2014**, *9*, 4847–4866. [CrossRef] [PubMed]
37. Aguilar-Rabiela, A.E.; Hernández-Cooper, E.M.; Otero, J.A.; Vergara-Porras, B. Modeling the release of curcumin from microparticles of poly(hydroxybutyrate) [PHB]. *Int. J. Biol. Macromol.* **2020**, *144*, 47–52. [CrossRef]
38. Senthilkumar, P.; Dawn, S.; Samanvitha, K.S.; Kumar, S.S.; Kumar, G.N.; Samrot, A.V. Optimization and characterization of poly[R]hydroxyalkanoate of Pseudomonas aeruginosa SU-1 to utilize in nanoparticle synthesis for curcumin delivery. *Biocatal. Agric. Biotechnol.* **2017**, *12*, 292–298. [CrossRef]
39. Zidan, A.S.; Rahman, Z.; Khan, M.A. Product and process understanding of a novel pediatric anti-HIV tenofovir niosomes with a high-pressure homogenizer. *Eur. J. Pharm. Sci.* **2011**, *44*, 93–102. [CrossRef]
40. Kokubo, T.; Takadama, H. How useful is SBF in predicting in vivo bone bioactivity? *Biomaterials* **2006**, *27*, 2907–2915. [CrossRef]
41. Monnier, A.; Rombouts, C.; Kouider, D.; About, I.; Fessi, H.; Sheibat-Othman, N. Preparation and characterization of biodegradable polyhydroxybutyrate-co-hydroxyvalerate/polyethylene glycol-based microspheres. *Int. J. Pharm.* **2016**, *513*, 49–61. [CrossRef]
42. Nguyen, T.T.; Jeong, J. Development of a single-jet electrospray method for producing quercetin-loaded poly (lactic-co-glycolic acid) microspheres with prolonged-release patterns. *J. Drug Deliv. Sci. Technol.* **2018**, *47*, 268–274. [CrossRef]
43. Hill, E.P.; Power, G.G.; Gilbert, R.D. Rate of pH changes in blood plasma in vitro and in vivo. *J. Appl. Physiol.* **1977**, *42*, 928–934. [CrossRef] [PubMed]
44. Kaewsichan, L.; Riyapan, D.; Prommajan, P.; Kaewsrichan, J. Effects of sintering temperatures on micro-morphology, mechanical properties, and bioactivity of bone scaffolds containing calcium silicate. *ScienceAsia* **2011**, *37*, 240–246. [CrossRef]
45. Masood, F.; Chen, P.; Yasin, T.; Hasan, F.; Ahmad, B.; Hameed, A. Synthesis of poly-(3-hydroxybutyrate-co-12 mol % 3-hydroxyvalerate) by Bacillus cereus FB11: Its characterization and application as a drug carrier. *J. Mater. Sci. Mater. Med.* **2013**, *24*, 1927–1937. [CrossRef]
46. Francis, L.; Meng, D.; Knowles, J.; Keshavarz, T.; Boccaccini, A.R.; Roy, I. Controlled Delivery of Gentamicin Using Poly(3-hydroxybutyrate) Microspheres. *Int. J. Mol. Sci.* **2011**, *12*, 4294–4314. [CrossRef] [PubMed]
47. Yuan, C.; Jin, Z.; Xu, X. Inclusion complex of astaxanthin with hydroxypropyl-β-cyclodextrin: UV, FTIR, 1H NMR and molecular modeling studies. *Carbohydr. Polym.* **2012**, *89*, 492–496. [CrossRef]
48. Wang, Y.; Wu, Y.; Wang, T.; Qiua, D. Isolation and Identification of 9-cis Astaxanthin by HPLC, FT-IR, and NMR Spectra. *J. Appl. Spectrosc.* **2021**, *88*, 97–107. [CrossRef]
49. Mehrotra, N.; Gupta, M.; Kovar, A.; Meibohm, B. The role of pharmacokinetics and pharmacodynamics in phosphodiesterase-5 inhibitor therapy. *Int. J. Impot. Res.* **2006**, *19*, 253–264. [CrossRef]

50. Liechty, W.B.; Kryscio, D.R.; Slaughter, B.V.; Peppas, N.A. Polymers for Drug Delivery Systems. *Annu. Rev. Chem. Biomol. Eng.* **2010**, *1*, 149–173. [CrossRef]
51. Conoscenti, G.; Carfì Pavia, F.; Ciraldo, F.E.; Liverani, L.; Brucato, V.; La Carrubba, V.; Boccaccini, A.R. In vitro degradation and bioactivity of composite poly-l-lactic (PLLA)/bioactive glass (BG) scaffolds: Comparison of 45S5 and 1393BG compositions. *J. Mater. Sci.* **2018**, *53*, 2362–2374. [CrossRef]
52. Zheng, K.; Wu, J.; Li, W.; Dippold, D.; Wan, Y.; Boccaccini, A.R. Incorporation of Cu-Containing Bioactive Glass Nanoparticles in Gelatin-Coated Scaffolds Enhances Bioactivity and Osteogenic Activity. *ACS Biomater. Sci. Eng.* **2018**, *4*, 1546–1557. [CrossRef] [PubMed]
53. Balakrishna Pillai, A.; Jaya Kumar, A.; Kumarapillai, H. Biosynthesis of poly(3-hydroxybutyrate-co-3-hydroxyvalerate) (PHBV) in Bacillus aryabhattai and cytotoxicity evaluation of PHBV/poly(ethylene glycol) blends. *3 Biotech* **2020**, *10*, 32. [CrossRef] [PubMed]

polymers

MDPI

Article

Scorpion Venom-Functionalized Quercetin Phytosomes for Breast Cancer Management: In Vitro Response Surface Optimization and Anticancer Activity against MCF-7 Cells

Nabil A. Alhakamy [1,2,3], Usama A. Fahmy [1,*], Shaimaa M. Badr Eldin [1,4], Osama A. A. Ahmed [1], Hibah M. Aldawsari [1,2], Solomon Z. Okbazghi [5], Mohamed A. Alfaleh [1,6], Wesam H. Abdulaal [7], Abdulmohsin J. Alamoudi [8] and Fatma M. Mady [9]

1 Department of Pharmaceutics, Faculty of Pharmacy, King Abdulaziz University, Jeddah 21589, Saudi Arabia; nalhakamy@kau.edu.sa (N.A.A.); smbali@kau.edu.sa (S.M.B.E.); oaahmed@kau.edu.sa (O.A.A.A.); haldosari@kau.edu.sa (H.M.A.); maalfaleh@kau.edu.sa (M.A.A.)
2 Center of Excellence for Drug Research and Pharmaceutical Industries, King Abdulaziz University, Jeddah 21589, Saudi Arabia
3 Mohamed Saeed Tamer Chair for Pharmaceutical Industries, King Abdulaziz University, Jeddah 21589, Saudi Arabia
4 Department of Pharmaceutics and Industrial Pharmacy, Cairo University, Cairo 11562, Egypt
5 Global Analytical and Pharmaceutical Development, Alexion Pharmaceuticals, New Haven, CT 06510, USA; solomon.z.okbazghi@gmail.com
6 Vaccines and Immunotherapy Unit, King Fahd Medical Research Center, King Abdulaziz University, Jeddah 21589, Saudi Arabia
7 Department of Biochemistry, Faculty of Science, Cancer and Mutagenesis Unit, King Fahd Medical Research Center, King Abdulaziz University, Jeddah 21589, Saudi Arabia; whabdulaal@kau.edu.sa
8 Department of Pharmacology and Toxicology, Faculty of Pharmacy, King Abdulaziz University, Jeddah 21589, Saudi Arabia; ajmalamoudi@kau.edu.sa
9 Department of Pharmaceutics, Faculty of Pharmacy, Minia University, Minia 61519, Egypt; fatmamady@hotmail.com
* Correspondence: uahmedkauedu.sa@kau.edu.sa

Abstract: Breast cancer is a dangerous type of cancer in women. Quercetin (QRT), a naturally occurring flavonoid, has wide biological effects including antioxidant, anticarcinogenic, anti-inflammatory, antiallergic, and antiviral activities. The anticancer activity is considered the most valuable effect of QRT against several types of cancer, including prostate, liver, lung, colon, and breast cancer. Scorpion venom peptides (SV) has been found to induce apoptosis and aggravate cancer cells, making it a promising anticancer agent. QRT, SV, and Phospholipon® 90H (PL) were incorporated in a nano-based delivery platform to assess QRT's cellular uptake and antiproliferative efficacy against a lung cancer cell line derived from human breast cancer cells MCF-7. Several nanovesicles were prepared and optimized, using four-factor Box–Behnken, in an experimental design. The optimized phytosomes showed vesicle size and zeta potential values of 116.9 nm and 31.5 mV, respectively. The IC50 values revealed that MCF-7 cells were significantly more sensitive to the optimized QRT formula than the plain formula and raw QRT. Cell cycle analysis revealed that optimized QRT formula treatment resulted in significant cell cycle arrest at the S phase. The results also indicated that treatment with QRT formula significantly increased caspase-9, Bax, Bcl-2, and p53 mRNA expression, compared with the plain formula and QRT. In terms of the inflammatory markers, the QRT formula significantly reduced the activity of TNF-α and NF-κB, in comparison with the plain formula and QRT only. Overall, the findings from the study proved that a QRT formulation could be a promising therapeutic approach for the treatment of breast cancer.

Keywords: breast cancer; nanoparticles; apoptosis; nanocomplex; optimization

1. Introduction

Phytosomal delivery systems result mainly from the interaction of naturally occurring phospholipids and natural active ingredients [1]. Phytosomes are formed upon the attachment of the polar groups of the natural active ingredient to the polar head group of the phospholipid, phosphate group [2,3]. Phytosomes are very similar in shape to liposomes, but differ in that the active ingredients are fixed to the polar head group, becoming an intrinsic part of the membrane [4]. This is not the case for liposomes in which the active ingredients can be incorporated in the aqueous region, or are freely soluble at the membrane bilayer [5]. Phytosomes enhanced the pharmacokinetic properties of most botanicals or nutraceuticals with improved bioavailability, activity, and reduced toxicity [6,7]. A mitomycin-C-Soybean phosphatidylcholine complex (mitomycin-c loaded phytosomes) has shown enhanced activity for cancer therapy and drug delivery [8]. Rutin-loaded transdermal systemic delivery was found to greatly improve upon loading into phytosomal systems, avoiding oral administration problems [9]. An emerging approach has been evolving over the last few years, in which the lipophilicity of highly hydrophilic drugs such as insulin is increased to improve the cell uptake that is lipophilic in nature, and consequently enhance the therapeutic activity [10,11].

In 2008, the World Health Organization announced that cancer resulted in 7.6 million deaths, exceeding deaths caused by heart disease [12]. Therefore, the development of effective, potent, and safe anticancer formulation is of great importance to enhance patient life, decrease mortality, and increase survival. Quercetin (QRT), a naturally occurring flavonoid, has wide biological effects including antioxidant, anticarcinogenic, anti-inflammatory, antiallergic, and antiviral activities [13]. Anticancer activity is considered the most valuable effect of QRT against several types of cancer, including prostate, liver, lung, colon, and breast cancer [14,15]. The anticancer activity of QRT is attributed to several mechanisms, such as interactions with numerous receptors, i.e., death receptors, growth factor receptors, and androgen receptors, as well as the inhibition of the enzyme system that could initiate cancer, and the modification of the signaling system involved in cancer development [16,17]. Furthermore, quercetin augments the therapeutic activity of chemotherapeutic agents such as cisplatin, when combined [18]. Similarly, scorpion venom (SV), which is a mixture of a large number of molecules, has shown a potential effective anticancer activity in vitro, against several cell lines [19,20]. The full anticancer mechanism of SV is not yet fully elucidated, but some reports have shown that SV acts on several ion channels which are essential for the activity of the cell [16,17]. The activity of SV on the ion channels, located on the cell membrane, restrict cell proliferation and might potentiate apoptosis [21].

Our hypothesis is based on the combination of QRT and SV producing synergistic anticancer activity. We attempted to load these natural anticancer drugs into a drug delivery system; phytosomes were chosen as a perfect drug delivery system for naturally occurring drugs, such as QRT and SV. The antiproliferative activity of the developed QRT and SV phytosomal were examined through cytotoxicity tests, such as cell cycle analysis, apoptosis tests, caspase-9, and Bax and BCL-2 levels to evaluate and elucidate the superior activity of the systems when compared with free QRT or SV. The antiproliferative and anti-apoptotic activities were evaluated with the use of human breast cancer MCF-7 cell line. A four-factor Box–Behnken experimental design was used to prepare and optimize the different formulations of phytosomal systems, and the optimized formulations were chosen to evaluate the combination of QRT and SV into the phytosomal systems.

2. Materials and Methods

QRT, SV were obtained from Sigma-Aldrich Inc. (St. Louis, MO, USA), Phospholipon® 90H (hydrogenated phosphatidylcholine from soybean origin, content 90%) was obtained as a gift sample from Lipoid GmbH (Ludwigshafen, Germany). The human tumor cell line MCF-7 used in this study was obtained from the VACSERA (Giza, Egypt) cell culture unit that was originally acquired from ATCC (Manassas, VA, USA).

2.1. Experimental Design and Optimization of QRT-PHM-SV

The response surface, specifically the four-factor Box–Behnken experimental design was implemented for optimizing the proposed QRT-PHM-SV formulation. The studied independent variables included two formulation factors, namely PL amount (mg, X_1) and SV amount (mg, X_4), in addition to two process variables, namely, process temperature (°C, X_2) and reflux time (h, X_3). The upper and lower coded and actual levels of each variable are compiled in Table 1. It is worthy to note that the amounts of PL representing the lower, mid, and upper levels used are calculated based on QRT: PL molar ratios of 1:1, 1:2 and 1:3 molar ratios, respectively. Vesicle size (nm, Y) and zeta potential (mV) were considered as the response parameters in this study. As per the selected design, 27 experimental runs, including three center points were generated by Design-Expert software (Version 12; Stat-Ease Inc., Minneapolis, MN, USA) with the corresponding combination of levels listed in Table 2. Adequate precision ratio, as well as predicted and adjusted determination coefficients, were computed and utilized for choosing the optimal fitting model for each measured response. Diagnostic plots were generated to assess the goodness of the fit. The equation representing the best fitting model was generated and the coefficient of each term was used to predict the relative magnitude of the effect of the corresponding variable on each response. Analysis of variance (ANOVA) was applied for statistical analysis of the measured responses to estimate the significance of the studied variables at $p < 0.05$. Perturbation plots and two-dimensional contour plots were produced to demonstrate the effect of the investigated variables and explore the interaction between them.

Table 1. Independent variable levels and response constraints employed in the four-level Box–Behnken design for the optimization of QRT-PHM-SV formulations.

Independent Variables	Levels		
	(−1)	(0)	(+1)
X_1: PL amount (mg)	78	156	234
X_2: Process temperature (°C)	40	50	60
X_3: Reflux time (h)	1	2	3
X_4: SV amount (mg)	34	102	170
Responses	**Desirability constraint**		
Y_1: Vesicle size (nm)	Minimize		
Y_2: Zeta potential (mV)	Maximize		

Abbreviations: QRT, quercetin; PHM, phytosomes; PL, Phospholipon® 90H; SV, scorpion venom peptide.

Table 2. Combination of variable levels in QRT-PHM-SV experimental runs and their corresponding measured responses.

RUN #	Independent Variables				Dependent Variables	
	PL Amount (X_1, mg)	Process Temperature (X_2, °C)	Reflux Time (X_3, h)	SV Amount (X_4, mg)	Vesicle Size * ± SD (Y_1, nm)	Zeta Potential * ± SD (Y_2, mV)
1	156	50	3	170	241.6 ± 8.9	16.5 ± 0.3
2	156	50	3	34	227.9 ± 6.9	10.6 ± 0.2
3	156	40	1	102	233.8 ± 7.9	16.5 ± 0.4
4	234	60	2	102	269.6 ± 11.3	17.6 ± 0.6
5	156	60	2	34	224.5 ± 7.3	13.2 ± 0.2
6	234	50	3	102	278.5 ± 9.2	13.7 ± 0.4
7	78	40	2	102	146.4 ± 4.1	22.1 ± 0.5
8	78	50	3	102	152.4 ± 4.9	20.7 ± 0.6
9	78	50	1	102	139.4 ± 6.3	19.8 ± 0.3
10	156	50	1	170	241.6 ± 8.4	19.4 ± 0.4

Table 2. *Cont.*

RUN #	Independent Variables				Dependent Variables	
	PL Amount (X_1, mg)	Process Temperature (X_2, °C)	Reflux Time (X_3, h)	SV Amount (X_4, mg)	Vesicle Size * ± SD (Y_1, nm)	Zeta Potential * ± SD (Y_2, mV)
11	156	60	3	102	197.5 ± 7.7	13.5 ± 0.2
12	156	40	2	34	215.7 ± 8.3	12.1 ± 0.1
13	156	50	1	34	211.7 ± 6.9	11.3 ± 0.1
14	234	50	2	170	295.6 ± 12.6	17.4 ± 0.3
15	156	40	3	102	239.4 ± 10.5	15.9 ± 0.4
16	156	50	2	102	228.1 ± 9.2	12.7 ± 0.3
17	156	40	2	170	247.3 ± 9.3	19.3 ± 0.6
18	78	60	2	102	131.2 ± 4.2	21.9 ± 0.5
19	234	40	2	102	283.5 ± 8.5	16.3 ± 0.4
20	156	60	2	170	209.5 ± 6.3	21.4 ± 0.5
21	234	50	1	102	291.4 ± 11.9	15.7 ± 0.3
22	156	60	1	102	231.5 ± 6.9	14.8 ± 0.4
23	78	50	2	34	123.4 ± 3.6	10.6 ± 0.1
24	156	50	2	102	252.9 ± 10.1	13.5 ± 0.3
25	156	50	2	102	254.3 ± 9.2	14.6 ± 0.3
26	234	50	2	34	253.5 ± 9.9	10.7 ± 0.2
27	78	50	2	170	157.5 ± 5.8	30.5 ± 1.1

Abbreviations: QRT, quercetin; PHM, phytosomes; PL, Phospholipon® 90H; SV, scorpion venom peptide; SD, standard deviation. * Data represents mean (n = 5).

2.2. Preparation of QRT-PHM-SV Formulations

QRT-PHM-SV formulations were prepared using reflux followed by anti-solvent precipitation, as described before with some modification [22]. Briefly, weighed amounts of QRT (30 mg) and Phospholipon® 90H (78, 156, or 234 mg) in the specified molar ratios were dissolved in dichloromethane (20 mL). The solution was refluxed at the temperature and time specified according to the experimental design, and then evaporated to obtain a concentrate of about 5 mL. The concentrate was lyophilized for 72 h to obtain the QRT-PHM. SV was dissolved in distilled water utilizing the SV amounts specified in the design. The SV aqueous solution was used as the hydration medium for the dried QRT-PHM preparation to prepare the QRT-PHM-SV nanovesicles, that was then stored in airtight amber colored glass containers at 4 °C until further use. Plain PHM-SV was prepared with the same procedure mentioned in this section, but without the use of QRT.

2.3. Vesicle Size and Zeta Potential Determination

QRT-PHM-SV size and zeta potential were determined by appropriate dilution in double-distilled water using a Zetasizer Nano ZSP particle size analyzer instrument (Malvern, UK). The results were expressed as the average of five determinations. The parameters were the following: laser wavelength of 633 nm, scattering angle of 173, temperature of 25 °C, medium viscosity of 0.8872 cP, and medium refractive index of 1.33.

2.4. Predicting Optimized Variables' Levels for QRT-PHM-SV Formulation

The examined independent variables were optimized by a numerical method following the desirability function approach. The optimization aimed primarily at minimizing the size, and maximizing the absolute zeta potential of the proposed QRT-PHM-SV formulation. Nonetheless, the predicted optimized formulation was prepared for further characterization.

2.5. Transmission Electron Microscope Investigation of QRT-PHM-SV Formulation

The optimized QRT-PHM-SV formulation was investigated utilizing JEOL GEM-1010 (JEOL Ltd., Akishima, Tokyo, Japan) transmission electron microscope (TEM) at 80 kV at the Regional Center for Mycology and Biotechnology (RCMB) Al-Azhar University, Cairo, Egypt. One drop of the sample was spread on a carbon-coated grid, dried, and then 1% phosphotungistic acid was used for negative staining of the sample. The sample was then dried at ambient temperature for 15 min before visualization.

2.6. Cytotoxicity of Optimized QRT-PHM-SV

The cytotoxicity efficacy of optimized QRT-PHM-SV was performed on the MCF-7 cell line using MTT assay. For this experiment, selected cells were grown in 96-well plates at the density of 5×10^3 cells per well, and incubated overnight. After stabilization, cells were treated with plain PHM-SV, QRT-raw, and QRT-PHM-SV and incubated for 24 h. Then, previously treated cells were further treated with 5.0 mg/mL MTT solution (10 μL), and then incubated for 4 h at 37 °C. Additionally, the collected supernatant was dispersed in 100 μL of DMSO to solubilize the formazan crystal. Samples were analyzed employing a microplate reader at 570 nm. Studies were carried out in triplicate. The half-maximal inhibitory concentration (IC50) for MCF-7 cells was determined for cells treated with plain PHM-SV, QRT-raw, or QRT-PHM-SV. The IC50 values were estimated after plotting the percent of the variation in cell viability versus drug concentration.

2.7. Cell Cycle Analysis

To analyze the effects of samples on the cell cycle, the flow cytometry method was utilized. The cells were treated with various sample formulations: plain PHM-SV, QRT-raw, and QRT-PHM-SV, and incubated for 24 h. After completion of incubation, cells were separated by centrifugation and fixed with 70% cold ethanol. Prior to the washing of samples with PBS, samples were again collected by centrifugation. Separated cells were stained with propidium iodide supplemented with RNAse before starting flow cytometry analysis [23].

2.8. Analysis of Apoptosis by Annexin V Staining

In order to analyze the comparative apoptotic activity of plain PHM-SV, QRT-raw, and QRT-PHM-SV, the Annexin V method was implemented. For this purpose, selected cells were grown in 6-well plates at the density of 1×10^5 cells per well, then incubated overnight with IC50 concentration of samples for 24 h at 37 °C. All samples were then centrifuged at 200× *g* for 5 min, and collected cells were resuspended in PBS at room temperature after dual washing. Furthermore, 10 μL Annexin V and 5 μL propidium iodide solution supernatant was dispersed in the previously prepared samples and incubated at 25 °C for 5 min. Final samples were analyzed using a flow cytometer (FACS Calibur, BD Bioscience, San Diego, CA, USA) in triplicate [24].

2.9. Analysis of Caspase 9

The Caspase 9 determination was carried out through the Caspase 9 Colorimetric Assay Kit (BioVision, Milpitas, CA, USA). In this case, MCF-7 cells were grown in the density of 3×10^6 cells per well and treated with plain PHM-SV, QRT-raw, and QRT-PHM-SV. Then samples were resuspended in ice-chilled lysate buffer and incubated in an ice medium for 10 min before centrifugation (10,000× *g* for 1 min). The analysis method for the Caspase 9 assay was carried out according to the instructions of the manufacturer, and the developed color was determined by a microplate reader at 405 nm.

2.10. Determination of Bax and Bcl-2 Proteins

Bax and Bcl-2 proteins were quantified via their corresponding ELISA kits, Bax ELISA kit (DRG Instruments GmbH, Marburg, Germany) and Zymed® Bcl-2 ELISA Kit, 24 h after drug/phytosomal formulations treatment according to the instructions of their manufacturers. The untreated cells were used as a negative control.

2.11. Mitochondrial Membrane Potential Changes

MCF-7 cells were incubated with drug/phytosomal formulations after being cultured in 96-well plate at a density of 1 × 105 cells/well. After that, Mitochondrial membrane potential (MMP) changes were investigated via the use of MitoProbe™ TMRM Assay Kit. The untreated cells were used as a negative control.

Real-time polymerase chain reaction (RT-PCR) for estimation of p53 and TNF-α:

The expression of p53 and TNF-α was determined by using RT-PCR. MCF-7 cells were treated with the different formulas incubated. Total RNA was extracted from the cell fraction and then from the synthesis of cDNA. Primers for p53 and TNF-α were designed using Gene Runner software. Samples in triplicates were used to estimate the expression, and they were normalized to β actin.

Western Blot investigation of Caspase 9, Bax, Bcl-2 and p 53 proteins expression:

Western blot assay of plain PHM-SV, QRT-raw, and QRT-PHM-SV was performed utilizing the previously reported protocol and by our laboratory [25,26]. The band intensity of the target proteins was normalized against the band intensity of β-actin (ChemiDoc™ MP imager, Bio-Rad Inc., Hercules, CA, USA).

2.12. Statistical Analysis

Values were expressed as mean ± standard deviation (SD). One way ANOVA, followed by Tukey's multiple comparison test, was utilized for statistical analysis in which p-value < 0.05 was considered significant.

3. Results

3.1. Experimental Design and Fit Statistics

Fit statistical analysis results for vesicle size and zeta potential are summarized in Table 3. Amongst the polynomial models under investigation (linear, two-factor interaction, and quadratic), the quadratic model was found to be the best fitting model for both responses on the basis of its highest R^2 and lowest PRESS. The adjusted R^2 and the predicted R^2 showed good coincidence with a difference of less than 0.2. Moreover, both responses showed adequate precision values greater than the desirable value (21.3759 and 14.4582 for vesicle size and zeta potential, respectively) indicating an appropriate signal to noise ratio. Thus, the quadratic model could be considered adequate to navigate the experimental design space for both responses.

Diagnostic plots were constructed for establishing the goodness of fit of the selected model for both responses, as illustrated in Figure 1. A Box-Cox plot for power transforms, illustrated in Figure 1(AI,BI), showed favored lambda (λ) values of 1.04 and 0.05 for vesicle size and zeta potential, respectively. The current λ value of 1 was included within the 95% confidence limits (marked by the red lines); thereby, no specific transformation for either response was needed [27,28]. The absence of transformation requirement was supported by the computed maximum to minimum response ratio of 2.36 and 2.87 for vesicle size and zeta potential, respectively. It is worthy to note that a ratio greater than 10 urges for transformation. Scattered and random distribution of the points in the externally studentized residuals vs. run plots, illustrated in Figure 1(AII,BII), indicate the absence of any lurking variables that might affect the measured responses. Moreover, the normal probability plots, displayed in Figure 1(AIII,BIII), revealed a linear pattern that indicated normal distribution residuals, thus confirming the absence of transformation requirement [29].

Table 3. Fit statistics of QRT-PHM-SV responses according to quadratic model.

Responses	Sequential *p*-Value	Lack of Fit *p*-Value	R^2	Adjusted R^2	Predicted R^2	Adequate Precision	PRESS	Significant Terms
Y_1: Vesicle (nm)	0.0013	0.8854	0.9810	0.9587	0.9138	21.3759	5591.89	X_1, X_2, X_4, X_2X_4, X_{12}, X_{22}, X_{42}
Y_2: Zeta potential (mV)	0.0038	0.2201	0.9299	0.8481	0.6381	14.4582	209.79	X_1, X_4, X_1X_4, X_{12}, X_{22}

Abbreviations: QRT, quercetin; PHM, phytosomes; SV, scorpion venom peptide; PRESS, predicted residual error sum of squares.

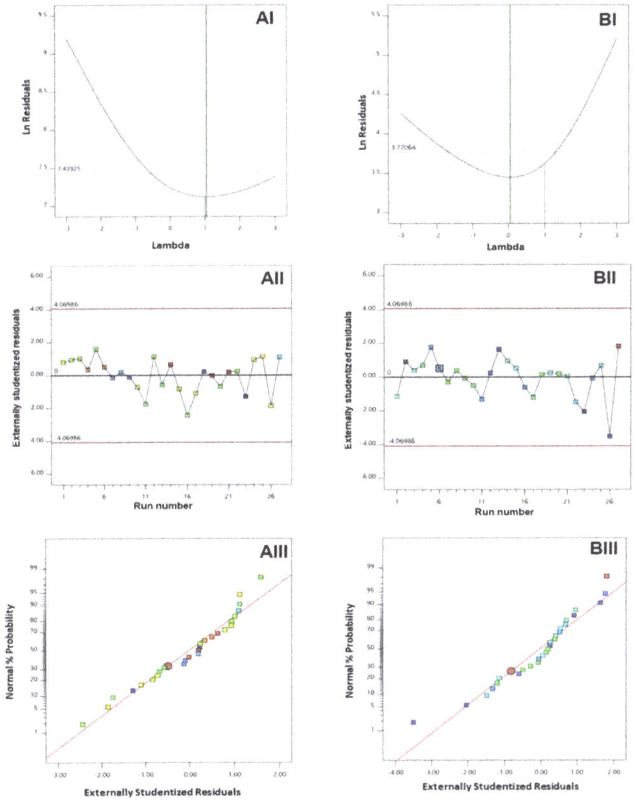

Figure 1. Diagnostic plots for responses of QRT-PHM-SV formulations. (**AI**) Box-Cox Plot for VS; (**AII**) Externally studentized residuals vs. run number plot for VS, (**AIII**) Normal probability plot for VS, (**BI**) Box-Cox Plot for ZP; (**BII**) Externally studentized residuals vs. run number plot for ZP, (**BIII**) Normal probability plot for ZP. Abbreviations: QRT, quercetin; PHM, phytosomes; SV, scorpion venom peptide, VS: vesicle size, ZP, zeta potential.

3.2. Influence of Investigated Variables on Responses

The restricted penetration of drug delivery systems into cancerous tissues represent a major obstacle in the field of cancer therapy, as this lessens their clinical efficacy. Thus, continuous research is being conducted to modify the physicochemical properties of drug delivery systems to target tumor penetration [19].

Lipid-based delivery systems have recently attracted attention in the field of cancer therapy. A particle size smaller than 400 nm was reported for preferential distribution within solid malignant tissues [30,31]. Accordingly, the prepared SV-functionalized phytosomal formulations showed acceptable vesicles size that ranged from 123.4 $\pm$ 3.6 to 295.6 $\pm$ 12.6 nm. Nevertheless, the favored accumulation of nano-sized delivery systems and their related clinical efficacy could probably be countered by inefficacious penetration owing to the pathological aspects created by the cancerous growth [32]. Improvement of tumor penetration could be achieved via minimizing vesicle size to boost the surface area available for permeation [28]. Accordingly, to ensure worthwhile tumor penetration, the vesicle size of the proposed QRT-PHM-SV formulation was optimized to minimized value.

Zeta potential is an indication of the surface charge of the nanoparticulate systems. It is reported that cationic nanoparticulate systems could exhibit a significantly increased permeation into cancerous cells and accumulate in tumor tissue and tumor vasculature, compared with the surrounding tissue [18,33,34]. Accordingly, SV was used for functionalization of the phytosomal structure via imparting positive charge. All the prepared QRT-PHM-SV formulations exhibited positive charge ranging from 10.6 $\pm$ 0.1 to 30.5 $\pm$ 1.1 mV. However, it should be considered that vesicular dispersions with an absolute zeta potential value of above 30 mV were regarded as stable systems due to the electric repulsion that prevents aggregation [35]. Therefore, the prepared systems were optimized to maximize the absolute zeta potential value.

Analysis of variance (ANOVA) for both vesicle size and zeta potential confirmed the significance of the quadratic model, as depicted by the corresponding F-values of 44.16 and 11.37 ($p < 0.0001$), respectively. The lack of fit F-value of 0.33 ($p = 0.8854$) for vesicle size and 3.92 ($p = 0.2201$) for zeta potential indicated non-significant lack of fit in relation to pure error, thereby assuring data fitting to the selected model. The software was employed to generate the equations demonstrating the quadratic model for both responses in terms of coded factor, as follows:

$$Y_1\ (\textit{vesicle size}) = 245.10 + 64.48\ X_1 - 8.52\ X_2 - 1.01\ X_3 + 11.37\ X_4 + 0.33\ X_1X_2 - 6.47\ X_1X_3 + 2.00\ X_1X_4 - 9.90\ X_2X_3 - 11.65\ X_2X_4 - 4.05\ X_3X_4 - 25.77\ X_1^2 - 12.33\ X_2^2 - 5.23\ X_3^2 - 9.84\ X_4^2$$

$$Y_2\ (\textit{zeta potential}) = 13.60 - 2.85\ X_1 + 0.017\ X_2 - 0.55\ X_3 + 4.67\ X_4 + 0.38\ X_1X_2 - 0.73\ X_1X_3 - 3.30\ X_1X_4 - 0.18\ X_2X_3 + 0.25\ X_2X_4 - 0.55\ X_3X_4 + 3.60\ X_1^2 + 2.05\ X_2^2 + 0.02\ X_3^2 + 0.59\ X_4^2$$

The statistical analysis revealed that the linear terms X_1, X_2, and X_4 corresponding to PL amount, temperature process, and SV amount exhibited a significant effect on vesicle size ($p < 0.0001$, $p = 0.0131$, and 0.0022, respectively). The interaction term X_2X_4 corresponds to the interaction between process temperature and SV amount, in addition to the quadratic terms X_1^2, X_2^2, and X_4^2, which were also found to be significant at 95% significance level. Regarding zeta potential, the linear terms X_1 ($p = 0.0001$), and X_4 ($p = 0.0001$) were significant, in addition to the interaction term X_1X_4 ($p = 0.0029$), and the quadratic terms X_1^2 ($p = 0.0005$) and X_2^2 ($p = 0.0204$) were significant. Figure 2 illustrates the perturbation graph demonstrating the effect of the independent variables on the studied responses, whereas Figures 3 and 4 illustrate the 2D contour and the 3D response plots demonstrating the interaction between the significant investigated variables on both responses.

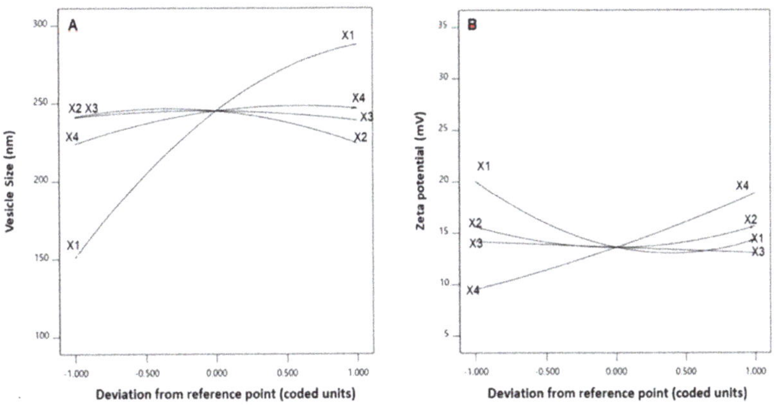

Figure 2. Perturbation graph for the effect of critical attributes; PL amount (X1), process temperature (X_2), reflux time (X_3), and SV amount (X_4) on (**A**) vesicle size, and (**B**) zeta potential of QRT-PHM-SV formulations. Abbreviations: QRT, quercetin; PHM, phytosomes; SV, scorpion venom peptide.

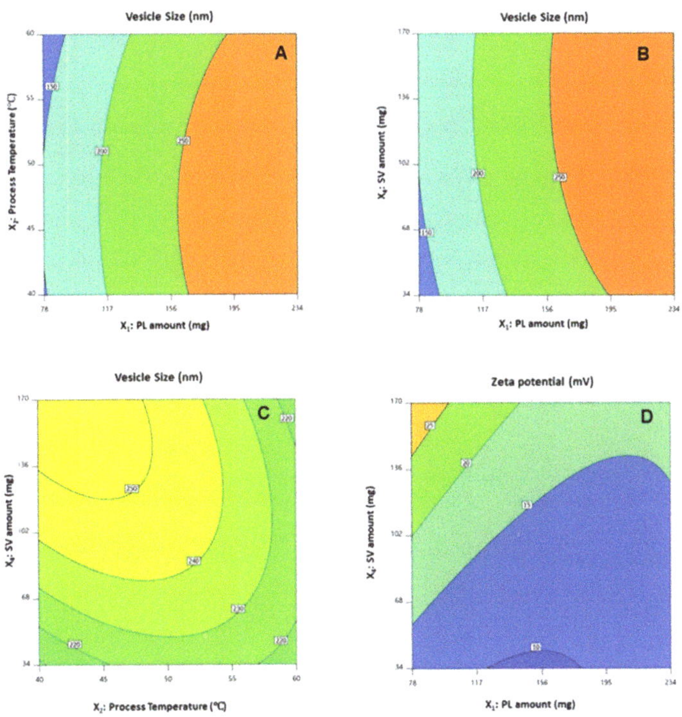

Figure 3. Contour 2D plots showing the effect and interaction between the significant factors on the vesicle size (**A–C**) and zeta potential (**D**) of QRT-PHM-SV formulations. Abbreviations: QRT, quercetin; PHM, phytosomes; SV, scorpion venom peptide.

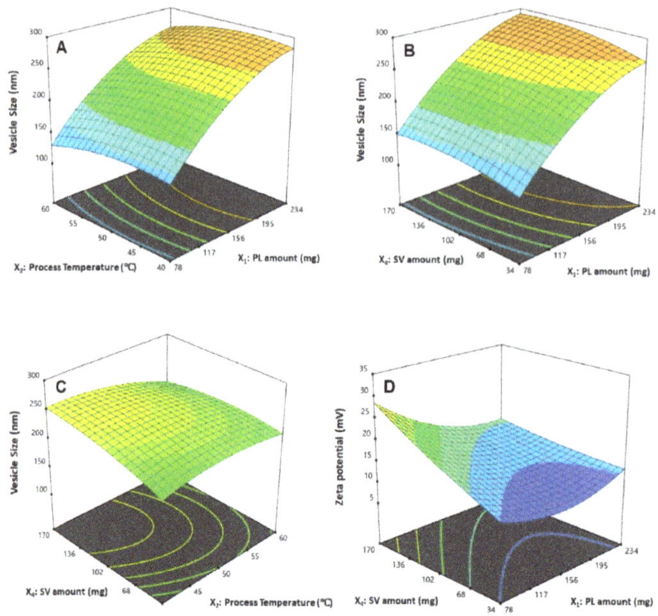

Figure 4. Response surface 3D plots showing the effect and interaction between the significant factors on the vesicle size (**A–C**) and zeta potential (**D**) of QRT-PHM-SV formulations. Abbreviations: QRT, quercetin; PHM, phytosomes; SV, scorpion venom peptide.

3.3. Optimization of QRT-PHM-SV Formulation

Design Expert software® was implemented for predicting the optimized levels of each studied independent variable by a numerical optimization technique. The optimized levels that could yield minimized vesicle size and maximized zeta potential upon combination with a desirability of 0.996 are presented in Table 4. The measured responses were in good harmony with the predicted values as depicted by the relatively low percentage error, thereby confirming the applicability of the design and the validity of the optimization process.

Table 4. Optimized independent variable levels and the c predicted and observed responses of the optimized QRT-PHM-SV formulation.

Variables	**X_1: PL Amount (mg)**	**X_2: Temperature (°C)**	**X_3: Reflux Time (h)**	**X_4: SV Amount (mg)**
Optimum values	78.00	60.00 °C	2.4 h	170.00
	Predicted value	Observed value	Error %	
Vesicle size (nm)	113.3	116.9	3.17%	
Zeta potential (mV)	30.3	31.5	3.96%	

TEM investigation of QRT-PHM-SV formulation.

The results of the TEM investigation for the optimized formula showed spherical vesicles (Figure 5). The size of the vesicles was comparable with the size data obtained from the particle size analyzer, taking into consideration the reduction of vesicles size in the case of the TEM investigation, as a result of the drying process that was subjected to the sample during preparation for the TEM imaging.

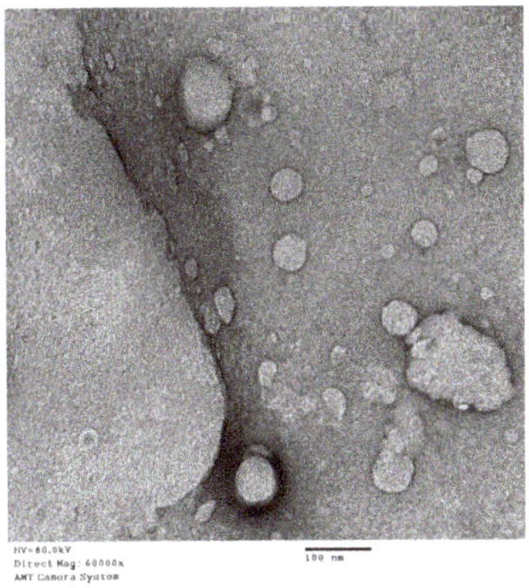

Figure 5. TEM image of the optimized QRT-PHM-SV formulation.

QRT-PHM-SV formulation showed potent cytotoxicity to MCF-7 cells:

The IC_{50} value of the QRT-PHM-SV formulation was determined based on the cytotoxic activity of the applied drug or drug formulations on MCF-7 cells. The plain formula, PHM-SV, induced very limited cytotoxicity to MCF-7 cells, as reflected by its high IC_{50} value (Figure 6). Plain PHM-SV was significantly more cytotoxic than Free QRT, indicating a good ability to inhibit cell proliferation. Yet, the combination of QRT into the phytosomal systems (QRT-PHM-SV) produced a highly cytotoxic formulation with the least IC_{50} value observed in this study (Figure 6).

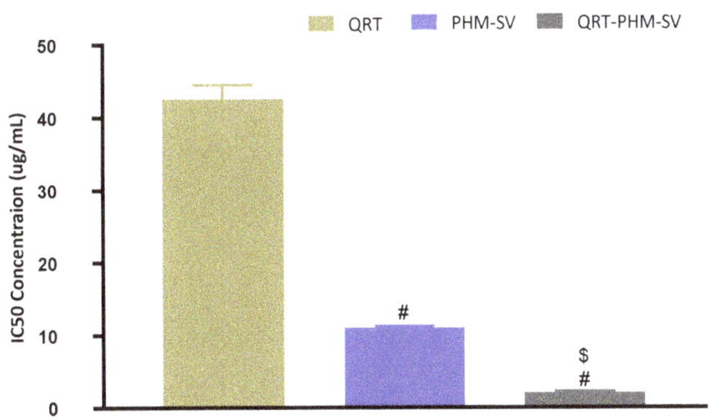

Figure 6. Representation of the IC_{50} values of QRT, PHM-SV, and QRT-PHM-SV in MCF-7 cells. # Significantly different from QRT at $p < 0.05$. $ Significantly different from PHM-SV at $p < 0.05$.

3.4. *QRT-PHM-SV Formulation Inhibited the Proliferation of MCF-7 Cells*

Results of cell cycle analysis of the different samples in this study are shown in Figure 7. It can be seen that untreated MCF-7 cells exhibited rapid proliferation, in which the largest population of cells were at the G0-G1 phase, with minimal cell accumulation at the other phases. In contrast, the free QRT increased the cell population in the G2-M and pre G1 phases, whereas plain PHM-SV increased the percentage of cells in the S and pre G1 phases with a reciprocal decrease in the G2-M population. Yet, the combination of QRT with the phytosomal system (QRT-PHM-SV) induced the most significant increase in cell population in the S and pre G1 phases, indicating significant cell-killing ability (Figure 7). These results confirmed the higher cytotoxic effects of the QRT-PHM-SV formulation compared with the plain formula and free QRT.

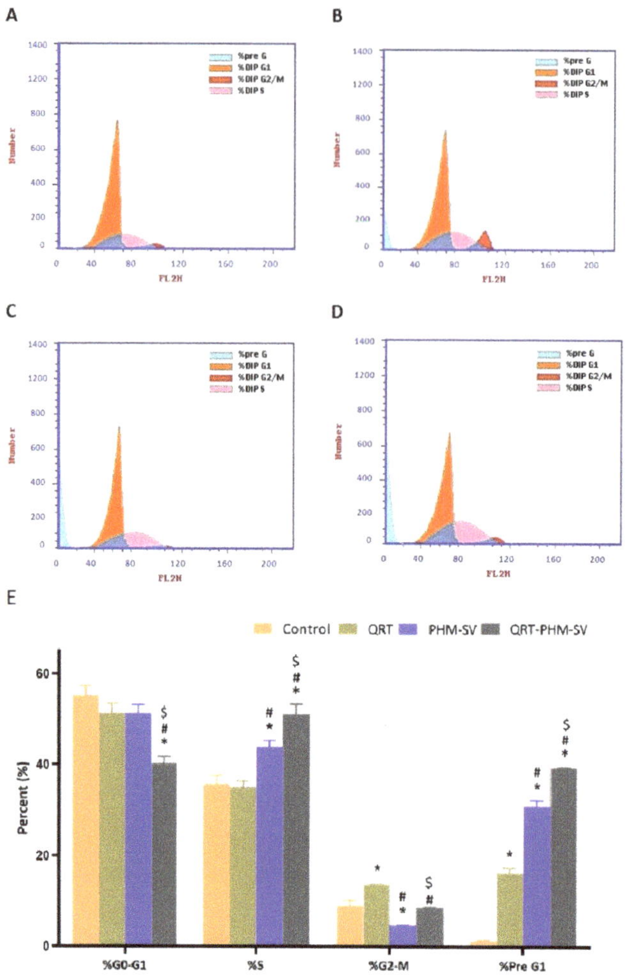

Figure 7. Flow cytometric analysis of control (**A**); QRT (**B**); PHM-SV (**C**); and QRT-PHM-SV-treated cells (**D**); and the percentages of cells in the G1, S, and G2/M phases of the cell cycle (**E**). * Significantly different from control at $p < 0.05$; # significantly different from QRT at $p < 0.05$.; $ significantly different from PHM-SVat $p < 0.05$.

3.5. QRT-PHM-SV Formulation Enhance the Apoptotic Activity of QRT toward MCF-7 Cells

As shown in Figure 8, QRT, PHM-SV and QRT-PHM-SV showed enhanced proapoptotic activities toward MCF-7 cells when compared with the control untreated MCF-7 cells. The plain formula, PHM-SV, significantly increased late and total cell death over free QRT. Yet, the most significant increase in the total population of apoptotic cells was observed following the QRT-PHM-SV treatment. At the early apoptotic phase, no significant changes in cell population were observed between QRT, PHM-SV, and QRT-PHM-SV. Similarly, significant necrotic cell death was also induced with no significant differences in the percentages of necrotic cells associated with the different treatments, compared with the untreated cells (Figure 8).

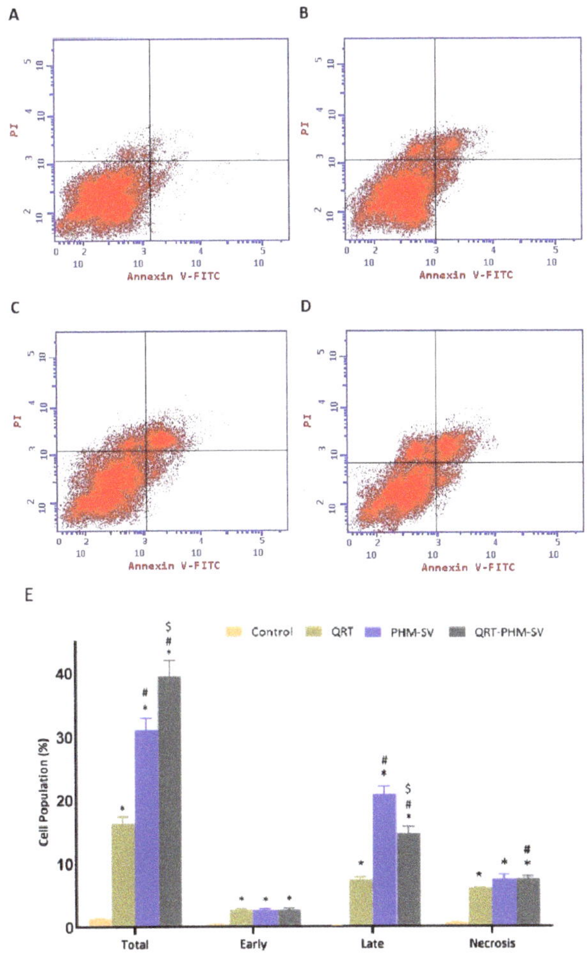

Figure 8. Assessment of MCF-7 cell death in control (**A**); QRT (**B**); PHM-SV (**C**); and QRT-PHM-SV-treated cells (**D**); and the percentages of cells early, late, and total cell death (**E**) following annexin V staining. * Significantly different from control at $p < 0.05$; # significantly different from QRT at $p < 0.05$; $ Significantly different from PHM-SV at $p < 0.05$.

3.6. Mitochondrial Membrane Potential (MMP) Changes Induced by the QRT-PHM-SV Formulation

Apoptosis is usually associated with abnormal changes in the integrity of the mitochondrial membrane; hence, the loss of MMP could be used as an indicator for apoptosis [36]. Figure 9 shows that free QRT resulted in no significant changes in MMP, with respect to control. However, PHM-SV and QRT-PHM-SV significantly reduced the value of MMP to reach about 80% and 75% of the control level, respectively. The decline of MMP is a characteristic sign of apoptosis and thus, these results highlight the proapoptotic potential of the QRT-PHM-SV formula.

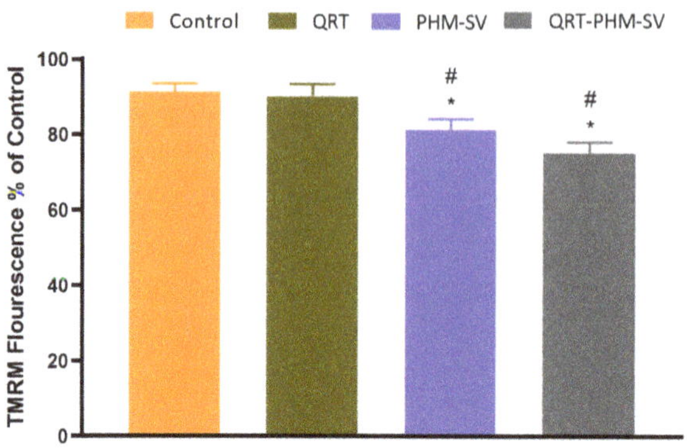

Figure 9. Effect of the QRT-PHM-SV formula on mitochondrial membrane potential (MMP) in MCF-7 cells. Data presented in bar charts are ± SD (n = 3). * or #, statistically different from control or QRT, respectively, at $p < 0.05$.

3.7. Modulation of the Expression of Caspase-9, Bax, Bcl-2 and P53

Caspase-9 is a crucial regulator of apoptosis that is frequently required for catalyzing the specific cleavage of key cellular proteins [37]. Hence, measuring caspase-9 content provides a reliable estimate of the therapeutic potential of cytotoxic agents. In this regard, a caspase-9 colorimetric assay was conducted to investigate the proapoptotic activity of the developed formulation relative to the free drug, QRT. As can be seen from Figure 10, treatment of MCF-7 cells with the different formulations resulted in differential changes in their caspases-9 content. QRT-PHM-SV treatment induced the content of caspase-9 by about 9-fold over the control. This contrasts with the 2-fold and 5-fold increase in caspase-9 content associated with free QRT and the PHM-SV formulation, respectively. These results further confirm the proapoptotic activity of the QRT-PHM-SV formula.

Bax and Bcl-2 are cytoplasmic proteins that promote and inhibit apoptosis, respectively [37]. The obtained results showed that QRT-PHM-SV induced the expression of the Bax protein by about 10-folds over the control, whereas free QRT and PHM-SV increased its levels by only 2 folds and 5 folds, respectively. Regarding Bcl-2 expression, QRT-PHM-SV reduced the expression of the Bcl-2 protein to about 0.2-fold of control, whereas QRT and PHM-SV reduced the expression of Bcl-2 to about 0.4-fold and 0.75-fold, respectively. As decreased cellular resistance to apoptotic stimuli is associated with high levels of Bax and low levels of Bcl-2, these results indicate that treatment with the QRT-PHM-SV triggered apoptotic death in MCF-7 cells.

To confirm the obtained results about the expression levels of Bax and Bcl-2, the expression of p53 was also investigated. P53 is a transcription factor that that is heavily involved in the induction of apoptosis [23]. It is known that the direct activation of Bax by p53 induces apoptosis and mitochondrial membrane permeabilization [38]. Hence, increased expression of p53 is associated with higher cytotoxic activity. As shown in Figure 10, treatments used in this study induced differential changes in the expression of p53. The expression of this transcription factor was significantly elevated upon treatment with QRT-PHM-SV by about 6-fold over the control. Yet, QRT and PHM-SV increased the level of the p53 mRNA by about 2-fold and 4-fold, respectively. Therefore, these results demonstrated the superior cytotoxic activity of QRT-PHM-SV, as it was found to induce significantly higher levels of p53 compared with free QRT and PHM-SV.

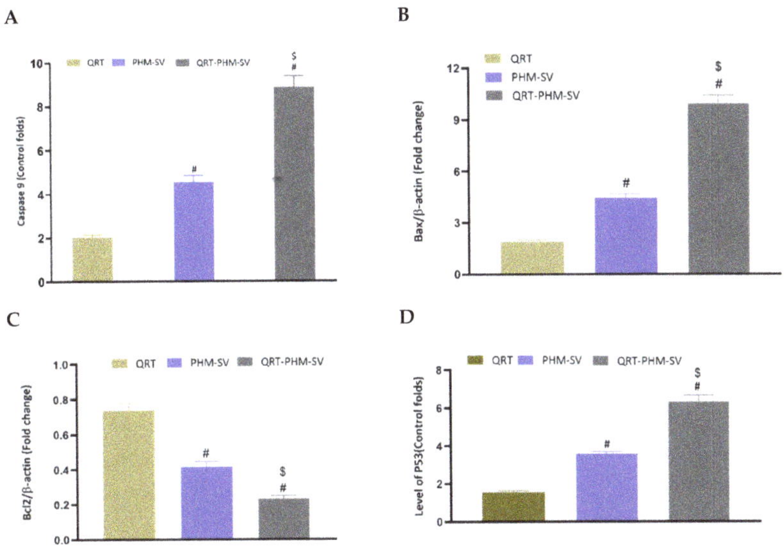

Figure 10. Effect of the QRT-PHM-SV formula on the expression of (**A**): caspase-9, (**B**): Bax, (**C**): Bcl-2, and (**D**): p53. Data are presented as Mean ± SD (n = 3); # significantly different from QRT at $p < 0.05$; $ significantly different from plain formula (PHM-SV) at $p < 0.05$.

3.8. Western Blot of Caspase 9, Bax, Bcl-2 and p 53 Proteins Expression

To investigate the effect of QRT-PHM-SV formula on Bax, Bcl2, P53 and Casp9, western blot assay was carried out (Figure 11). The results revealed a significant ($p < 0.05$) increase in Bax expression of QRT-PHM-SV formula when compared with the investigated groups: untreated control, QRT, and PHM-SV (Figure 11A,B). The results also indicated that both P53 and Casp9 showed significant ($p < 0.05$) increase in the expression for the QRT-PHM-SV formula when compared with untreated control and QRT.

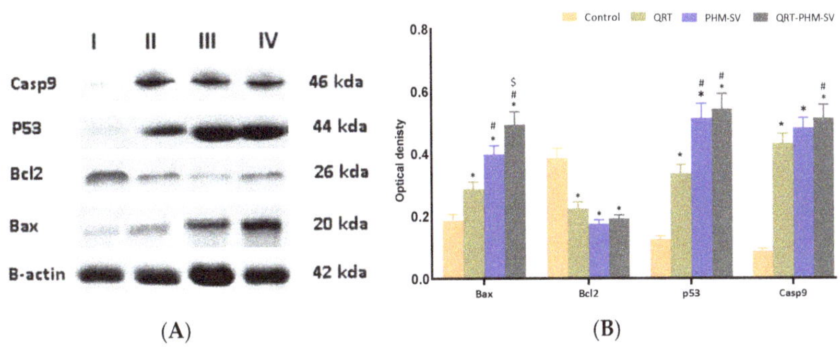

Figure 11. Western blots (**A**); histogram of proteins expression of Bax, Bcl2, P53 and Casp9 (**B**) for the four groups in MCF-7 cells; (**I**) untreated control; (**II**) QRT; (**III**) PHM-SV; (**IV**) QRT-PHM-SV. * Significantly different from control at $p < 0.05$; # significantly different from QRT at $p < 0.05$; $ Significantly different from PHM-SV at $p < 0.05$.

3.9. Changes in NF-κB and TNFα upon the Use of Different Formulations

TNF-α is a cytokine that can induce apoptosis in a variety of tissues and cell types including breast cancer cells [39]. On the other hand, NF-κB is known to inhibit apoptosis, particularly that triggered by TNF-α [39]. In this regard, the study results showed that free PHM-SV and QRT-PHM-SV were able to increase the expression of TNFα by about 1.5-fold with respect to the control (Figure 12). Yet, free QRT resulted in no significant changes in the expression of TNFα. In contrast to TNF-α levels, NF-κB activation was reduced upon treatment with PHM-SV and PHM-QRT-SV to about 0.4-fold of the control value. Treatment with free QRT reduced NF-κB activation to about 0.5-fold relative to the control. Taken together, the increased TNFα expression and reduced NF-κB activation associated with QRT-PHM-SV confirm the obtained results highlighting the superior activity of QRT-PHM-SV to induce cell apoptosis when compared with the plain formula.

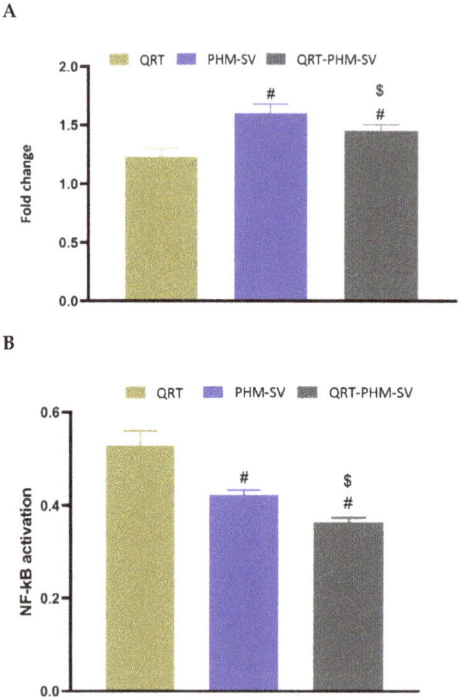

Figure 12. Effect of the QRT-PHM-SV formula on the mRNA expression of TNF-α (**A**) and the activation of NF-κB (**B**) in MCF-7 cells. Data presented in bar charts are ± SD (n = 3). # or $, statistically different from QRT or plain formula (PHM-SV), respectively, at $p < 0.05$.

4. Discussion

Phytosomes, when compared with liposomes, enhance pharmacokinetic properties and hence improve the bioavailability of most botanicals [6,7]. The illustrations show that the vesicle size significantly increases with increasing PL and SV amounts, and decreases with increasing process temperature. The positive sign of the X_1 and X_4 coefficients and the negative sign the X_2 coefficient supports this observation. In addition, the PL amount was the most significant factor affecting the vesicle size, as evidenced by the highest coefficient of its corresponding linear term. In general, the increase in size with increasing PL amount observed in our study was consistent with the results reported for previously developed vesicular systems [40,41]. For instance, an increase in ethosomal size at higher PL% was reported by Dubey et al. [42]. Moreover, Alhakamy et al. [22] reported an increase in icariin phytosomes vesicular size with increasing icariin to PL molar ratio. Regarding the process temperature, Nandhini and Ilango [20] reported similar reduced size for vasaka-loaded phytosomes at higher process temperatures. The observed increase in size with SV amount could be attributed to the possible increase in the induced repulsion between phospholipid bilayer of the vesicles by the increased positive charge. This increase in spacing is caused by pushing the polar heads of the phospholipids outwards, leading to a size increase. It is worthy to note that previous studies have reported an increase in the size of cationic vesicular systems compared with neutral ones [43,44]. Concerning the zeta potential, it was evident that the absolute value decreased with increasing PL amount, and increased with increasing SV amount. This observation was confirmed by the negative and positive signs of their corresponding linear term coefficients, respectively. The effect of SV amount was more pronounced, as evidenced by its higher linear term coefficient. This could be attributed to the positive charge on the scorpion venom peptides, and consequently its role in inducing a positive charge to the phytosomal surface [45,46].

The findings from this study have significant practical applications. It has been previously shown that QRT induces apoptosis in breast cancer cells, and it also has the potential to work synergistically with the cytotoxicity of conventional cytostatic agents [47]. Previous studies have indicated that SV peptides are K(+)-channels blockers and/ or Na(+)-channel modifiers [46,48]. The combination of QRT with SV had a significant influence on enhancing the cytotoxicity of the developed novel formulation. Our data demonstrate that the QRT-PHM-SV had the most cytotoxic activity when compared to free QRT. This underscores the importance of the phytosomal formulation in enhancing the cytotoxicity of QRT on MCF-7 cells. Such enhanced activity of drug-loaded phytosomes compared with the pure drug have previously been reported with QRT, and could be due to improved intracellular bioavailability [47]. In addition, the incorporation of SV into the novel formulation has also contributed to the improved cytotoxicity of QRT [49–51].

The developed formula, QRT-PHM-SV, exhibited a significantly larger cell population in the S, G2-M and pre-G1 phases demonstrating an enhancement in the cytotoxicity of QRT by the developed formula. A characteristic feature of apoptosis is a high cell fraction at the pre-G1 phase. Hence, the induction of cell-cycle arrest could provide an explanation for the improved cytotoxic activity of the QRT-SV-phytosomes. This is in agreement with other reports on nanoparticles [47]. Apoptosis induced by silver nanoparticles was reported to be enhanced via liposomal encapsulation [47]. In addition, the obtained results indicated that QRT-PHM-SV triggered intrinsic apoptosis. This was evidenced by annexin V staining of MCF-7 cells that demonstrated increased early, late, and total apoptotic cell death instigated by the developed formula. Moreover, the findings of this study showed that QRT-PHM-SV induces a significant loss of MMP, suggesting compromised integrity and enhanced permeability of the mitochondrial membrane, which is an initial step in apoptosis [52,53]. These reports are consistent with our findings on cell-cycle and annexin V-staining analyses. Taken together, these results suggest that the formulation of QRT and SV to phytosomes significantly enhances the disruption of the mitochondrial membrane potential, leading eventually to apoptosis induction in MCF-7 cells.

An excellent marker of apoptosis is caspase, as it is a cell death protease that regulates the breakdown of DNA and cellular proteins in apoptotic cells [54]. Thus, the enhancement of QRT cytotoxicity by the phytosomal formulation gained further support by caspase-9 findings, which are also consistent with previous reports on nanostructured formulations [55,56]. In addition, caspase-9 findings are in harmony with the observed typical apoptotic features associated with QRT-PHM-SV, including alterations in the expression of Bax, Bcl-2 and p53. Treatment with QRT-PHM-SV induced the expression of the proapoptotic protein Bax, and reduced the expression of Bcl-2, which is an inhibitor of apoptosis. Treatment with the developed optimized QRT-PHM-SV also increased the expression of p53, which is a proapoptotic transcription factor. These changes in expression, instigated by the developed formula, eventually lead to the activation of caspase-9 and the induction of apoptosis [38]. It seems that QRT-PHM-SV induces apoptosis in MCF-7 cells via different molecular pathways, including p53 activation followed by BCL-2 confirmation alterations, and hence Bax activation, leading ultimately to caspase-dependent apoptosis through cytochrome-c release, as a result of the loss of mitochondrial membrane potential (Figure 13) [38].

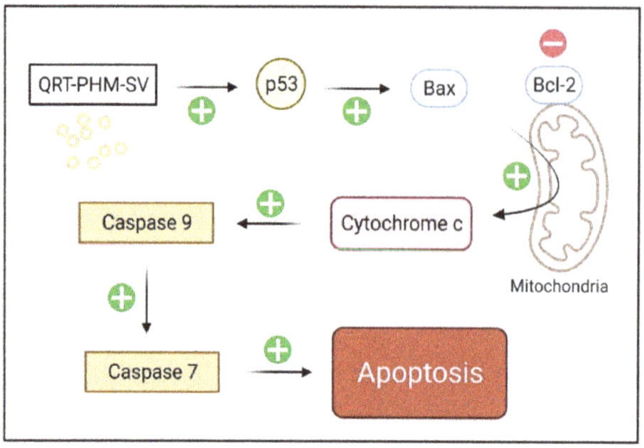

Figure 13. The proposed mechanism of apoptosis triggered by QRT-PHM-SV in MCF-7 cells. The flow chart demonstrates that QRT-PHM-SV induces apoptosis via the mitochondrial pathway and caspase-9-dependent signaling.

Furthermore, QRT-PHM-SV-treated cells demonstrated an increased TNF-α expression and a decreased NF-κB activation. This is in agreement with the enhanced cytotoxic and proapoptotic activities of the QRT-PHM-SV. TNF-α is known to promote apoptosis in breast cancer cells, whereas NF-κB can inhibit apoptosis triggered by TNF-α [38]. Hence, these findings further confirm the augmentation of QRT cytotoxicity via the developed delivery system.

5. Conclusions

The experimental design was used in this study to formulate and optimize QRT-SV phytosomes. The prepared formulations were nano-sized, and had a high zeta potential. The optimized QRT-SV phytosomes revealed that QRT formula administration significantly boosted caspase-9, Bax, Bcl-2, and p53 mRNA expression, as compared with the plain formula and QRT. In terms of inflammatory indicators, the QRT formula dramatically reduced TNF- and NF-B activity when compared with the basic formula and QRT alone. Overall, the study's findings demonstrated that the developed optimized QRT formulation could be a promising therapeutic method for the treatment of breast cancer.

Author Contributions: Conceptualization, N.A.A. and U.A.F.; methodology, O.A.A.A., U.A.F., S.M.B.E. and H.M.A.; software, S.M.B.E.; validation, H.M.A., S.Z.O. and M.A.A.; formal analysis, W.H.A.; investigation, A.J.A.; resources, F.M.M.; data curation, U.A.F., F.M.M., O.A.A.A. and W.H.A. writing—original draft preparation, A.J.A. and O.A.A.A.; writing—review and editing, N.A.A. and M.A.A.; visualization, A.J.A., S.M.B.E. and U.A.F.; supervision, M.A.A., S.M.B.E. and H.M.A.; project administration, N.A.A.; funding acquisition, N.A.A., S.Z.O. and U.A.F. All authors have read and agreed to the published version of the manuscript.

Funding: This project was funded by the Deanship of Scientific Research (DSR) at King Abdulaziz University, Jeddah, under grant no. (RG-1-166-42). The authors, therefore, acknowledge with thanks DSR for technical and financial support.

Institutional Review Board Statement: Not applicable.

Informed Consent Statement: Not applicable.

Data Availability Statement: Not applicable.

Conflicts of Interest: The authors declare no conflict of interest. The funders had no role in the design of the study; in the collection, analyses, or interpretation of data; in the writing of the manuscript, or in the decision to publish the results.

References

1. Bombardelli, E.; Curri, S.; Della, R.; Del, N.; Tubaro, A.; Gariboldi, P. Complexes between phospholipids and vegetal derivatives of biological interest. *Fitoterapia* **1989**, *60*, 1–9.
2. Hou, Z.; Li, Y.; Huang, Y.; Zhou, C.; Lin, J.; Wang, Y.; Cui, F.; Zhou, S.; Jia, M.; Ye, S.; et al. Phytosomes loaded with mitomycin C-soybean phosphatidylcholine complex developed for drug delivery. *Mol. Pharm.* **2013**, *10*, 90–101. [CrossRef]
3. Das, M.K.; Kalita, B. Design and evaluation of phyto-phospholipid complexes (phytosomes) of Rutin for transdermal application. *J. Appl. Pharm. Sci.* **2014**, *4*, 51–57. [CrossRef]
4. Peng, Q.; Zhang, Z.R.; Gong, T.; Chen, G.Q.; Sun, X. A rapid-acting, long-acting insulin formulation based on a phospholipid complex loaded PHBHHx nanoparticles. *Biomaterials* **2012**, *33*, 1583–1588. [CrossRef] [PubMed]
5. Cui, F.; Shi, K.; Zhang, L.; Tao, A.; Kawashima, Y. Biodegradable nanoparticles loaded with insulin-phospholipid complex for oral delivery: Preparation, in vitro characterization and in vivo evaluation. *J. Control. Release* **2006**, *114*, 242–250. [CrossRef]
6. Jemal, A.; Bray, F.; Center, M.M.; Ferlay, J.; Ward, E.; Forman, D. Global cancer statistics. *CA Cancer J. Clin.* **2011**, *61*, 69–90. [CrossRef] [PubMed]
7. Aguirre, L.; Arias, N.; Macarulla, M.T.; Gracia, A.; Portillo, M.P. Beneficial effects of quercetin on obesity and diabetes. *Open Nutraceuticals J.* **2011**, *4*, 189–198. [CrossRef]
8. Yi, L.; Zongyuan, Y.; Cheng, G.; Lingyun, Z.; Guilian, Y.; Wei, G. Quercetin enhances apoptotic effect of tumor necrosis factor-related apoptosis-inducing ligand (TRAIL) in ovarian cancer cells through reactive oxygen species (ROS) mediated CCAAT enhancer-binding protein homologous protein (CHOP)-death receptor 5 pathway. *Cancer Sci.* **2014**, *105*, 520–527. [CrossRef] [PubMed]
9. Yin, J.; Xie, X.; Jia, Q.; Wang, J.; Huang, G.; Zou, C.; Shen, J. Effect and mechanism of quercetin on proliferation and apoptosis of human osteosarcoma cell U-2OS/MTX300. *Zhongguo Zhongyao Zazhi* **2012**, *37*, 611–614. [CrossRef]
10. Moon, Y.J.; Wang, X.; Morris, M.E. Dietary flavonoids: Effects on xenobiotic and carcinogen metabolism. *Toxicol. In Vitro* **2006**, *20*, 187–210. [CrossRef]
11. He, M.-L.; Yuan, H.-Q.; Jiang, A.-L.; Gong, A.Y.; Chen, W.-W.; Zhang, P.-J.; Young, C.Y.F.; Zhang, J.-Y. Gum mastic inhibits the expression and function of the androgen receptor in prostate cancer cells. *Cancer* **2006**, *106*, 2547–2555. [CrossRef] [PubMed]
12. Kim, W.K.; Bang, M.H.; Kim, E.S.; Kang, N.E.; Jung, K.C.; Cho, H.J.; Park, J.H. Quercetin decreases the expression of ErbB2 and ErbB3 proteins in HT-29 human colon cancer cells. *J. Nutr. Biochem.* **2005**, *16*, 155–162. [CrossRef] [PubMed]
13. Chen, W.; Wang, X.; Zhuang, J.; Zhang, L.; Lin, Y. Induction of death receptor 5 and suppression of survivin contribute to sensitization of TRAIL-induced cytotoxicity by quercetin in non-small cell lung cancer cells. *Carcinogenesis* **2007**, *28*, 2114–2121. [CrossRef] [PubMed]
14. Gomes, A.; Bhattacharjee, P.; Mishra, R.; Biswas, A.K.; Dasgupta, S.C.; Giri, B.; Debnath, A.; Gupta, S.D.; Das, T.; Gomes, A. Anticancer potential of animal venoms and toxins. *Indian J. Exp. Biol.* **2010**, *48*, 93–103.
15. Brito, A.; Ribeiro, M.; Abrantes, A.M.; Pires, A.; Teixo, R.J.M.; Tralhão, J.; Botelho, M. Quercetin in Cancer Treatment, Alone or in Combination with Conventional Therapeutics? *Curr. Med. Chem.* **2015**, *22*, 3025–3039. [CrossRef]
16. Heinen, T.E. Gorini da Veiga AB. Arthropod venoms and cancer. *Toxicon* **2011**, *57*, 497–511. [CrossRef]
17. Goudet, C.; Chi, C.W.; Tytgat, J. An overview of toxins and genes from the venom of the Asian scorpion Buthus martensi Karsch. *Toxicon* **2002**, *40*, 1239–1258. [CrossRef]
18. Saadat, M.; Zahednezhad, F.; Zakeri-Milani, P.; Heidari, H.R.; Shahbazi-Mojarrad, J.; Valizadeh, H. Drug targeting strategies based on charge dependent uptake of nanoparticles into cancer cells. *J. Pharm. Pharm. Sci.* **2019**, *22*, 191–220. [CrossRef]
19. Han, D.; Qi, H.; Huang, K.; Li, X.; Zhan, Q.; Zhao, J.; Hou, X.; Yang, X.; Kang, C.; Yuan, X. The effects of surface charge on the intra-tumor penetration of drug delivery vehicles with tumor progression. *J. Mater. Chem. B* **2018**, *6*, 3331–3339. [CrossRef]
20. Sundaresan, N.; Kaliappan, I. Development and characterization of a nano-drug delivery system containing vasaka phospholipid complex to improve bioavailability using quality by design approach. *Res. Pharm. Sci.* **2021**, *16*, 103–117. [CrossRef]
21. Ng, W.K.; Yazan, L.S.; Ismail, M. Thymoquinone from Nigella sativa was more potent than cisplatin in eliminating of SiHa cells via apoptosis with down-regulation of Bcl-2 protein. *Toxicol. In Vitro* **2011**, *25*, 1392–1398. [CrossRef]
22. Alhakamy, N.A.; Fahmy, U.A.; Badr-Eldin, S.M.; Ahmed, O.A.A.; Asfour, H.Z.; Aldawsari, H.M.; Algandaby, M.M.; Eid, B.G.; Abdel-Naim, A.B.; Awan, Z.A.; et al. Optimized icariin phytosomes exhibit enhanced cytotoxicity and apoptosis-inducing activities in ovarian cancer cells. *Pharmaceutics* **2020**, *12*, 346. [CrossRef]
23. Alhakamy, N.A.; Ahmed, O.A.; Aldawsari, H.M.; Alfaifi, M.Y.; Eid, B.G.; Abdel-Naim, A.B.; Fahmy, U.A. Encapsulation of Lovastatin in Zein Nanoparticles Exhibits Enhanced Apoptotic Activity in HepG2 Cells. *Int. J. Mol. Sci.* **2019**, *20*, 5788. [CrossRef]
24. Fahmy, U.A.; Aldawsari, H.M.; Badr-Eldin, S.M.; Ahmed, O.A.A.; Alhakamy, N.A.; Alsulimani, H.H.; Caraci, F.; Caruso, G. The encapsulation of febuxostat into emulsomes strongly enhances the cytotoxic potential of the drug on HCT 116 colon cancer cells. *Pharmaceutics* **2020**, *12*, 956. [CrossRef] [PubMed]
25. Mahmood, T.; Yang, P.C. Western blot: Technique, theory, and trouble shooting. *N. Am. J. Med. Sci.* **2012**, *4*, 429–434. [CrossRef] [PubMed]
26. Eid, B.G.; Alhakamy, N.A.; Fahmy, U.A.; Ahmed, O.A.; Shadab, M.; Abdel-Naim, A.B.; Caruso, G.; Caraci, F. Melittin and diclofenac synergistically promote wound healing in a pathway involving TGF-β1. *Pharmacol. Res.* **2022**, *175*, 105993. [CrossRef] [PubMed]

27. Singh, B.; Bhatowa, R.; Tripathi, C.; Kapil, R. Developing micro-/nanoparticulate drug delivery systems using "design of experiments". *Int. J. Pharm. Investig.* **2011**, *1*, 75. [CrossRef]
28. Badr-Eldin, S.M.; Aldawsari, H.M.; Ahmed, O.A.; Alhakamy, N.A.; Neamatallah, T.; Okbazghi, S.Z.; Fahmy, U.A. Optimized semisolid self-nanoemulsifying system based on glyceryl behenate: A potential nanoplatform for enhancing antitumor activity of raloxifene hydrochloride in MCF-7 human breast cancer cells. *Int. J. Pharm.* **2021**, *600*, 120493. [CrossRef] [PubMed]
29. Fahmy, U.A.; Badr-Eldin, S.M.; Ahmed, O.A.A.; Aldawsari, H.M.; Tima, S.; Asfour, H.Z.; Al-Rabia, M.W.; Negm, A.A.; Sultan, M.H.; Madkhali, O.A.A.; et al. Intranasal niosomal in situ gel as a promising approach for enhancing flibanserin bioavailability and brain delivery: In vitro optimization and ex vivo/in vivo evaluation. *Pharmaceutics* **2020**, *12*, 485. [CrossRef]
30. Sharma, S.; Shukla, P.; Misra, A.; Mishra, P.R. Interfacial and colloidal properties of emulsified systems: Pharmaceutical and biological perspective. In *Colloid and Interface Science in Pharmaceutical Research and Development*; Elsevier Inc.: Amsterdam, The Netherlands, 2014; pp. 149–172. [CrossRef]
31. Yingchoncharoen, P.; Kalinowski, D.S.; Richardson, D.R. Lipid-based drug delivery systems in cancer therapy: What is available and what is yet to come. *Pharmacol. Rev.* **2016**, *68*, 701–787. [CrossRef]
32. Zhang, Y.R.; Lin, R.; Li, H.J.; He, W.L.; Du, J.Z.; Wang, J. Strategies to improve tumor penetration of nanomedicines through nanoparticle design. *Nanomed. Nanobiotechnol.* **2019**, *11*, e1519. [CrossRef]
33. Krasnici, S.; Werner, A.; Eichhorn, M.E.; Schmitt-Sody, M.; Pahernik, S.A.; Sauer, B.; Schulze, B.; Teifel, M.; Michaelis, U.; Naujoks, K.; et al. Effect of the surface charge of liposomes on their uptake by angiogenic tumor vessels. *Int. J. Cancer* **2003**, *105*, 561–567. [CrossRef] [PubMed]
34. Wang, H.; Zuo, Z.-Q.; Du, J.-Z.; Wang, Y.-C.; Sun, R.; Cao, Z.-T.; Ye, X.; Wang, J.-L.; Leong, K.W.; Wang, J. Surface charge critically affects tumor penetration and therapeutic efficacy of cancer nanomedicines. *Nano Today* **2016**, *11*, 133–144. [CrossRef]
35. Zhou, X.; Chen, Z. Preparation and performance evaluation of emulsomes as a drug delivery system for silybin. *Arch. Pharm. Res.* **2015**, *38*, 2193–2200. [CrossRef] [PubMed]
36. Sakamuru, S.; Attene-Ramos, M.S.; Xia, M. Mitochondrial membrane potential assay. *Methods Mol. Biol.* **2016**, *1473*, 17–22. [CrossRef] [PubMed]
37. Porter, A.G.; Jänicke, R.U. Emerging roles of caspase-3 in apoptosis. *Cell Death Differ.* **1999**, *6*, 99–104. [CrossRef]
38. Chipuk, J.E.; Kuwana, T.; Bouchier-Hayes, L.; Droin, N.M.; Newmeyer, D.D.; Schuler, M.; Green, D.R. Direct Activation of Bax by p53 Mediates Mitochondrial Membrane Permeabilization and Apoptosis. *Science* **2004**, *303*, 1010–1014. [CrossRef]
39. Lu, L.; Shi, W.; Deshmukh, R.R.; Long, J.; Cheng, X.; Ji, W.; Zeng, G.; Chen, X.; Zhang, Y.; Dou, Q.P. Tumor necrosis factor-α sensitizes breast cancer cells to natural products with proteasome-inhibitory activity leading to apoptosis. *PLoS ONE* **2014**, *9*, e113783. [CrossRef]
40. Saoji, S.D.; Raut, N.A.; Dhore, P.W.; Borkar, C.D.; Popielarczyk, M.; Dave, V.S. Preparation and Evaluation of Phospholipid-Based Complex of Standardized Centella Extract (SCE) for the Enhanced Delivery of Phytoconstituents. *AAPS J.* **2016**, *18*, 102–114. [CrossRef]
41. Ahmed, O.A.A.; Badr-Eldin, S.M. Development of an optimized avanafil-loaded invasomal transdermal film: Ex vivo skin permeation and in vivo evaluation. *Int. J. Pharm.* **2019**, *570*, 118657. [CrossRef]
42. Dubey, V.; Mishra, D.; Dutta, T.; Nahar, M.; Saraf, D.K.; Jain, N.K. Dermal and transdermal delivery of an anti-psoriatic agent via ethanolic liposomes. *J. Control. Release* **2007**, *123*, 148–154. [CrossRef]
43. Narayan, R.; Singh, M.; Ranjan, O.; Nayak, Y.; Garg, S.; Shavi, G.V.; Nayak, U.Y. Development of risperidone liposomes for brain targeting through intranasal route. *Life Sci.* **2016**, *163*, 38–45. [CrossRef]
44. Mehanna, M.M.; El-Kader, N.A.; Samaha, M.W. Liposomes as potential carriers for ketorolac ophthalmic delivery: Formulation and stability issues. *Braz. J. Pharm. Sci.* **2017**, *53*, 1–10. [CrossRef]
45. Liu, G.; Yang, F.; Li, F.; Li, Z.; Lang, Y.; Shen, B.; Wu, Y.; Li, W.; Harrison, P.L.; Strong, P.N.; et al. Therapeutic potential of a scorpion venom-derived antimicrobial peptide and its homologs against antibiotic-resistant Gram-positive bacteria. *Front. Microbiol.* **2018**, *9*, 1159. [CrossRef]
46. Ortiz, E.; Gurrola, G.B.; Schwartz, E.F.; Possani, L.D. Scorpion venom components as potential candidates for drug development. *Toxicon* **2015**, *93*, 125–135. [CrossRef] [PubMed]
47. Minaei, A.; Sabzichi, M.; Ramezani, F.; Hamishehkar, H.; Samadi, N. Co-delivery with nano-quercetin enhances doxorubicin-mediated cytotoxicity against MCF-7 cells. *Mol. Biol. Rep.* **2016**, *43*, 99–105. [CrossRef] [PubMed]
48. Gao, B.; Harvey, P.J.; Craik, D.J.; Ronjat, M.; De Waard, M.; Zhu, S. Functional evolution of scorpion venom peptides with an inhibitor cystine knot fold. *Biosci. Rep.* **2013**, *33*, 513–527. [CrossRef]
49. Gómez Rave, L.J.; Muñoz Bravo, A.X.; Sierra Castrillo, J.; Román Marín, L.M.; Corredor Pereira, C. Scorpion venom: New promise in the treatment of cancer. *Acta Biol. Colomb.* **2019**, *24*, 213–223. [CrossRef]
50. Akef, H.; Kotb, N.; Abo-Elmatty, D.; Salem, S. Anti-proliferative Effects of Androctonus amoreuxi Scorpion and Cerastes cerastes Snake Venoms on Human Prostate Cancer Cells. *J. Cancer Prev.* **2017**, *22*, 40–46. [CrossRef]
51. Rapôso, C. Scorpion and spider venoms in cancer treatment: State of the art, challenges, and perspectives. *J. Clin. Transl. Res.* **2017**, *3*, 233–249. [CrossRef]
52. Xue, Y.; Chen, Q.; Ding, T.; Sun, J. SiO2 nanoparticle-induced impairment of mitochondrial energy metabolism in hepatocytes directly and through a Kupffer cell-mediated pathway in vitro. *Int. J. Nanomed.* **2014**, *9*, 2891–2903. [CrossRef]

53. Costantini, P.; Jacotot, E.; Decaudin, D.; Kroemer, G. Mitochondrion as a novel target of anticancer chemotherapy. *J. Natl. Cancer Inst.* **2000**, *92*, 1042–1053. [CrossRef] [PubMed]
54. Anwar, M.M.; El-Karim, S.S.A.; Mahmoud, A.H.; Amr, A.E.G.E.; Al-Omar, M.A. A comparative study of the anticancer activity and PARP-1 inhibiting effect of benzofuran-pyrazole scaffold and its nano-sized particles in human breast cancer cells. *Molecules* **2019**, *24*, 2413. [CrossRef]
55. Kim, C.G.; Castro-Aceituno, V.; Abbai, R.; Lee, H.A.; Simu, S.Y.; Han, Y.; Hurh, J.; Kim, Y.-J.; Yang, D.C. Caspase-3/MAPK pathways as main regulators of the apoptotic effect of the phyto-mediated synthesized silver nanoparticle from dried stem of Eleutherococcus senticosus in human cancer cells. *Biomed. Pharmacother.* **2018**, *99*, 128–133. [CrossRef] [PubMed]
56. Yassemi, A.; Kashanian, S.; Zhaleh, H. Folic acid receptor-targeted solid lipid nanoparticles to enhance cytotoxicity of letrozole through induction of caspase-3 dependent-apoptosis for breast cancer treatment. *Pharm. Dev. Technol.* **2020**, *25*, 397–407. [CrossRef] [PubMed]

polymers

MDPI

Article

Chitin-Glucan Complex Hydrogels: Optimization of Gel Formation and Demonstration of Drug Loading and Release Ability

Diana Araújo [1,2], Thomas Rodrigues [1,2], Vítor D. Alves [3] and Filomena Freitas [1,2,*]

[1] Associate Laboratory i4HB, School of Science and Technology, Institute for Health and Bioeconomy, NOVA University Lisbon, 2819-516 Caparica, Portugal; df.araujo@campus.fct.unl.pt (D.A.); ta.rodrigues@campus.fct.unl.pt (T.R.)
[2] UCIBIO, Applied Molecular Biosciences Unit, Department of Chemistry, School of Science and Technology, NOVA University Lisbon, 2819-516 Caparica, Portugal
[3] LEAF, Linking Landscape, Environment, Agriculture and Food Research Center, Laboratório Associado TERRA, Instituto Superior de Agronomia, Universidade de Lisboa, Tapada da Ajuda, 1349-017 Lisboa, Portugal; vitoralves@isa.ulisboa.pt
* Correspondence: a4406@fct.unl.pt; Tel.: +351-212948300

Abstract: Chitin-glucan complex (CGC) hydrogels were fabricated through a freeze–thaw procedure for biopolymer dissolution in NaOH 5 mol/L, followed by a dialysis step to promote gelation. Compared to a previously reported methodology that included four freeze–thaw cycles, reducing the number of cycles to one had no significant impact on the hydrogels' formation, as well as reducing the total freezing time from 48 to 18 h. The optimized CGC hydrogels exhibited a high and nearly spontaneous swelling ratio (2528 ± 68%) and a water retention capacity of 55 ± 3%, after 2 h incubation in water, at 37 °C. Upon loading with caffeine as a model drug, an enhancement of the mechanical and rheological properties of the hydrogels was achieved. In particular, the compressive modulus was improved from 23.0 ± 0.89 to 120.0 ± 61.64 kPa and the storage modulus increased from 149.9 ± 9.8 to 315.0 ± 76.7 kPa. Although the release profile of caffeine was similar in PBS and NaCl 0.9% solutions, the release rate was influenced by the solutions' pH and ionic strength, being faster in the NaCl solution. These results highlight the potential of CGC based hydrogels as promising structures to be used as drug delivery devices in biomedical applications.

Keywords: hydrogels; chitin-glucan complex; freeze–thaw cycles; swelling ratio; caffeine; drug delivery

1. Introduction

Hydrogels are three-dimensional network structures fabricated from synthetic or natural polymers capable of absorbing large amounts of water [1–3]. Biopolymer hydrogels have attracted increasing interest due to their biocompatibility, biodegradability, environmentally friendly features, and tissue-mimicking consistency. These valuable characteristics make them suitable materials for utilization in a wide range of applications from food and agriculture [4] to cosmetics [5] and biomedicine [6].

Depending on the method used to crosslink the polymer chains, hydrogels can be classified as chemical or physical. Chemical hydrogels are mostly connected through a covalently cross-linked network, in which the addition of crosslinking agents promotes the reaction between the functional groups of the polymer chains [7,8]. However, those chemical agents are often toxic compounds, and their presence may have adverse effects, such as undesirable reactions with bioactive substances or affect the hydrogels' biocompatibility [9]. On the other hand, physical hydrogels are obtained by crosslinking the polymer chains through non-covalent interactions such as ionic interactions, hydrogen bonds, chain entanglements, van der Waals forces, or hydrophobic interactions [8]. Therefore, physically crosslinked hydrogels, especially biopolymer-based ones, are promising materials for use in the biomedical field due to the use of mild conditions during their fabrication, and the absence of organic solvents and toxic crosslinking agents [10,11].

The water insoluble and highly hydrophilic biopolymer chitin-glucan complex (CGC) is the main component of the inner cell wall of yeast and fungi, composed of N-acetyl-glucosamine and glucose monomers [12–14]. Owing to its biocompatibility and biodegradability characteristics—along with intrinsic antioxidant, anti-inflammatory, and antibacterial properties [15,16]—CGC has been applied as a food additive [17], an anticholesterol agent [18], and for wound healing [19]. Nevertheless, due to the numerous hydrogen bonds between its polymeric chains, similarly to chitin and others chitin-derived polymers, CGC is insoluble in the most common solvents [3,15]. Recently, alkali solvents based on NaOH or KOH have emerged as alternative solvents systems for CGC dissolution, through the freeze–thaw method [20]. In this process, the presence of a hydrated alkali component, below the freezing point, promotes the disruption of the polymer chain matrix by breaking inter and intramolecular hydrogen bonds, allowing for polymer dissolution [21]. CGC based physical hydrogels can be obtained by dialyzing the CGC dissolved in the alkali systems. During the dialysis process, gelation is induced by the interactions established between the CGC molecules promoted by the reduction of the ionic forces [22].

In this study, the impact of the number of freeze–thaw cycles and the freezing time during the procedure on the hydrogel-forming capacity of CGC was evaluated in terms of hydrogels' chemical composition, morphology, and mechanical properties. The optimized CGC hydrogel was characterized as to its rheology, swelling properties, drug loading, and drug release capacity.

2. Materials and Methods

2.1. Materials

Yeast biomass was obtained by cultivation of the yeast *Komagataella pastoris* (DSM 70877) using glycerol as the sole carbon source, as described by Farinha et al. [12]. CGC was extracted from *K. pastoris* biomass by the hot alkaline procedure described by Araújo et al. [23] and it represented 20 wt% of the cell dry mass. CGC presented a chitin content of 35.6% and a degree of acetylation (DA) of 63.4%.

2.2. Preparation of CGC Hydrogels

The CGC hydrogels were prepared as described by Araújo et al. [22], with slight modifications. Briefly, the CGC powder (0.5 g) was dispersed in a NaOH 5 mol/L solution (25 g), and the suspensions were kept at −20 °C, for either 18 or 48 h. During that period, different number of freeze–thaw cycles (0, 1, 2, or 3) were performed, being the thawed suspensions extensively stirred (500 rpm, 1 h), at room temperature in each cycle. After centrifugation (20,000× *g*, 30 min, 4 °C) to eliminate the undissolved material, the hydrogels were prepared by dialyzing the soluble fractions with a 12–14 kDa MWCO membrane (Spectra/Por®, Spectrum Laboratories Inc., Piscataway, NJ, USA), in deionized water, at room temperature, for 48 h. The obtained hydrogels were labelled according to the solvent system used (NaOH 5 mol/L) and the number of freeze–thaw cycles performed. The hydrogels prepared by freezing during 48 h using 0, 1, 2, or 3 freeze–thaw cycles were identified as $Na5_0$, $Na5_1$, $Na5_2$, and $Na5_3$ hydrogels, respectively, while the hydrogel prepared by 1 freezing cycle of 18 h was coded as $Na5_1$* hydrogel (Table 1).

Table 1. Chemical characterization of CGC hydrogels obtained by different number of freeze–thaw cycles: 1 cycle ($Na5_1$ hydrogel), 2 cycles ($Na5_2$ hydrogel), 3 cycles ($Na5_3$ hydrogel) and 1 cycle with reduced freezing time ($Na5_1$* hydrogel); n.a., data not available.

Samples	**$Na5_1$**	**$Na5_2$**	**$Na5_3$**	**$Na5_1$***	**Na5 [22]**
No. of cycles	1	2	3	1	4
Freezing time (h)	48	48	48	18	48
Polymer content (wt%)	1.68 ± 0.17	1.58 ± 0.04	1.42 ± 0.02	1.66 ± 0.11	2.28
Water content (wt%)	98.22 ± 0.20	98.45 ± 0.06	98.58 ± 0.02	97.63 ± 0.12	97.72
Chitin content (%)	25.63 ± 0.78	24.71 ± 2.98	23.85 ± 0.14	21.51 ± 1.49	n.a.

2.3. CGC Hydrogels Characterization

2.3.1. Chemical Characterization

The water content of the hydrogels was assessed gravimetrically by freeze drying the hydrogel samples, using the equation

$$\text{Water content} = ((W_{wet} - W_{dry})/W_{wet}) \times 100 \tag{1}$$

where W_{dry} (g) represents the dry mass of a pre-weighed amount of the hydrogel (W_{wet}, g).

The chitin content and the degree of acetylation were determined by elemental analysis as described by Araújo et al. [20].

2.3.2. Morphology, Density, and Porosity

The structure and morphology of the CGC hydrogels were characterized by scanning electron microscopy (SEM). The hydrogels were analyzed with a TM3030 tabletop microscope (Hitachi in High Technologies, U.S.) equipped with a sample holder with refrigeration. Hydrogels' samples were observed at a temperature of −4 °C, using a magnification of 500×.

The density (ρ, g/cm^3) of the freeze-dried CGC hydrogels was determined by equation

$$\rho = W_{dry}/V_{dry} \tag{2}$$

where W_{dry} and V_{dry} represent the weight (g) and volume (cm^3) of the hydrogel, respectively.

The porosity of the CGC hydrogels was determined using the solvent replacement method [24]. Pre-weighed freeze-dried CGC hydrogels (W_0, g) were immersed in absolute ethanol, for 30 min, in sealed tubes. After 30 min, excess ethanol on the surface was blotted and the samples were weighed. The porosity was calculated using the equation

$$\text{Porosity (\%)} = (W_{30} - W_0)/\rho V_T \tag{3}$$

where W_{30} and W_0 represent the hydrogel weight (g) at 30 min and 0 min, respectively, ρ is the density of ethanol (0.790 g/cm^3) and V_T (cm^3) is the total volume of the hydrogel sample.

2.4. Compressive Mechanical Analysis

The compressive mechanical properties of the CGC hydrogels were assessed with a texture analyzer TMS-Pro (Food Technology Corporation, England, UK) equipped with a 50 N load cell. Cylindrical hydrogels samples in the wet state (13.8 cm diameter, 0.7–1.1 cm height) were subjected to a compression of up to 80% strain of the samples original weight, at a speed rate of 60 mm/min, using an aluminum plunger with 60 mm diameter. The maximum tension of the compression corresponds to the hardness (kPa) and the toughness (kJ/m^3) was calculated by measuring the area underneath the stress–strain curve of each sample. Compressive modulus (kPa) was obtained as the slope of initial linear region. All the experiments were performed at room temperature (20 ± 0.2 °C).

2.5. Rheological Properties

The rheological properties of the $Na5_1$* and $Na5_1$* loaded hydrogels were analyzed using a controlled stress rheometer (HAAKE MARS III, Waltham, MA, USA Thermo Scientific), equipped with a plate–plate serrated geometry (diameter 20 mm) with a 1.5 mm gap. Hydrogel samples with a similar thickness (~3 mm) were equilibrated at 25 ± 0.03 °C, for 5 min. The viscoelastic properties were determined by applying frequency sweeps at a constant tension within the linear viscoelastic region, for a frequency range from 0.01 to 1 Hz.

2.6. Swelling and Water Retention Behavior

To assess the swelling properties, pre-weighed cylindrical freeze-dried samples of the $Na5_1$* hydrogel were immersed in deionized water, NaCl 0.9% or phosphate buffered saline (PBS), at 37 °C. At different time intervals, samples were carefully taken out from the solutions, blotted with a filter paper, and weighed (W_{wet}, g). The swelling ratio was determined using the equation

$$\text{Swelling ratio (g/g)} = (W_{wet} - W_{dry})/W_{dry} \quad (4)$$

where W_{dry} (g) represents the initial mass of dry hydrogel.

To evaluate the water retention behavior of the structures, the equilibrated hydrogels were taken out from the solutions and weighed (W_e), after blotted with a tissue paper. Swollen hydrogels were incubated at 37 °C and weighed (W_t) over time. Water retention was calculated by the equation

$$\text{Water retention (\%)} = (W_t/W_e) \times 100 \quad (5)$$

2.7. Drug Loading

Caffeine (Alfa Aesar, 99%) was used as model drug to assess drug loading and drug release behavior of the $Na5_1$* hydrogels. For drug loading, pre-weighed cylindrical freeze-dried hydrogel samples were immersed in a caffeine solution (1.0 wt%), for 24 h, at room temperature. After that period, the loaded hydrogels' samples were carefully taken out from the solution, blotted with a filter paper, and weighed (W_L, g). Drug loading (DL, g) was determined by the equation

$$\text{DL (g)} = (W_L - W_{dry}) \times C_{caf} \quad (6)$$

where W_{dry} (g) represents the initial mass of dry hydrogel and C_{caf} (wt%) corresponds to the concentration of caffeine solution.

The entrapment efficiency (EE, %) of caffeine in the hydrogels was calculated using the equation

$$\text{EE (\%)} = (DL/W_{caf}) \times 100 \quad (7)$$

where W_{caf} (g) represents the mass of caffeine.

2.8. Characterization of the Loaded Hydrogels

The $Na5_1$* hydrogels and the caffeine loaded $Na5_1$* hydrogels were characterized by Fourier-transform infrared spectroscopy (FT-IR). The analysis was conducted with a Spectrum II spectrometer (Perkin-Elmer, Llantrisant, UK) and the spectra were obtained between 500 and 4000 cm^{-1} after 10 scans, at room temperature.

The mechanical and rheological properties of $Na5_1$* loaded hydrogels were assessed as described in Sections 2.4 and 2.5, respectively.

2.9. In Vitro Drug Release Studies

The freeze-dried $Na5_1$* loaded hydrogels' samples were immersed in 100 mL of different physiological media: PBS (pH 7.4) and NaCl 0.9% (pH 5.5), at 37 °C, for 3 h, under constant stirring (100 rpm). Periodically, 2 mL of the release medium were withdrawn, and 2 mL of fresh medium, preheated at 37 °C, were added to keep the volume of the solution constant. The caffeine concentration in the withdraw solution was determined by UV–vis spectrophotometer (CamSpec M509T, Leeds, UK) at a wavelength of 273 nm [25], for a concentration range of 0.16–10 mg/L. Caffeine release was obtained by the equation

$$\text{Caffeine release } (\%) = ((C_W \times V)/DL) \times 100 \tag{8}$$

where C_W (g/L) is the caffeine concentration in the withdraw solution, V (L) is the volume of the release media, and DL (g) is the amount of loaded drug. The caffeine cumulative release was fitted to the Korsmeyer–Peppas model [26].

2.10. Statistical Analysis

The experimental data from all the studies were analyzed and the results were expressed as mean ± standard deviation (SD). Error bars represent the standard deviation ($n \geq 3$).

3. Results

3.1. Hydrogels Formation

The freeze–thaw procedure followed by dialysis, recently reported by Araújo et al. [22], was used to dissolve CGC in NaOH 5 mol/L (freezing at −20 °C, 4 freeze–thaw cycles, 48 h total freezing time) and prepare CGC hydrogels (labelled as Na5 hydrogels), which exhibited a dense and stiff gel structure. Following those results, the present study aimed at optimizing the procedure by assessing the impact of reducing the number of freeze–thaw cycles and of the freezing time on the hydrogels' properties.

Firstly, the effect of the number of cycles was studied by applying 1, 2, or 3 freeze–thaw cycles (samples $Na5_1$, $Na5_2$, and $Na5_3$, respectively) (Table 1). The procedures' performance was compared to that previously reported for four cycles. An experiment with no freeze–thaw cycles (sample $Na5_0$), in which CGC was simply contacted with the NaOH solution (at room temperature, for 48 h), was also performed for comparison. In this case, a viscous slurry was formed, with very low CGC dissolution, and no hydrogel formation upon dialyzing the supernatant recovered from the mixture. This outcome may be due to the non-deacetylation of chitin that the freeze–thaw procedure induces [20], and consequently, the low dissolution of CGC in the solvent system.

Except for $Na5_0$, all CGC solutions in NaOH 5 mol/L formed hydrogels upon coagulation by dialysis (Figure 1). For the same freezing time (48 h), increasing the number of freeze–thaw cycles, from 1 to 3 cycles, led to a slight decrease in the polymer's content (from 1.68 ± 0.17 to 1.42 ± 0.02 wt%, respectively). These results show that a single freeze–thaw cycle is sufficient for CGC gelling and the obtained hydrogels have a higher polymer content. Subsequently, the experiment was repeated for 1 freeze–thaw cycle, but the freezing time was reduced from 48 to 18 h (sample $Na5_1$*). The $Na5_1$ and $Na5_1$* hydrogels, both prepared with 1 freeze–thaw cycle, presented similar polymer content (1.68 ± 0.17 and 1.66 ± 0.11 wt%, respectively), indicating that freezing time had no significant impact on CGC dissolution in the NaOH solvent system.

These results show that the procedure for CGC gelling can be simplified by reducing both the number of freeze–thaw cycle and the total freezing time. All the resulting hydrogels were characterized to evaluate their physicochemical properties and select the most suitable procedure for yielding structures with superior performance.

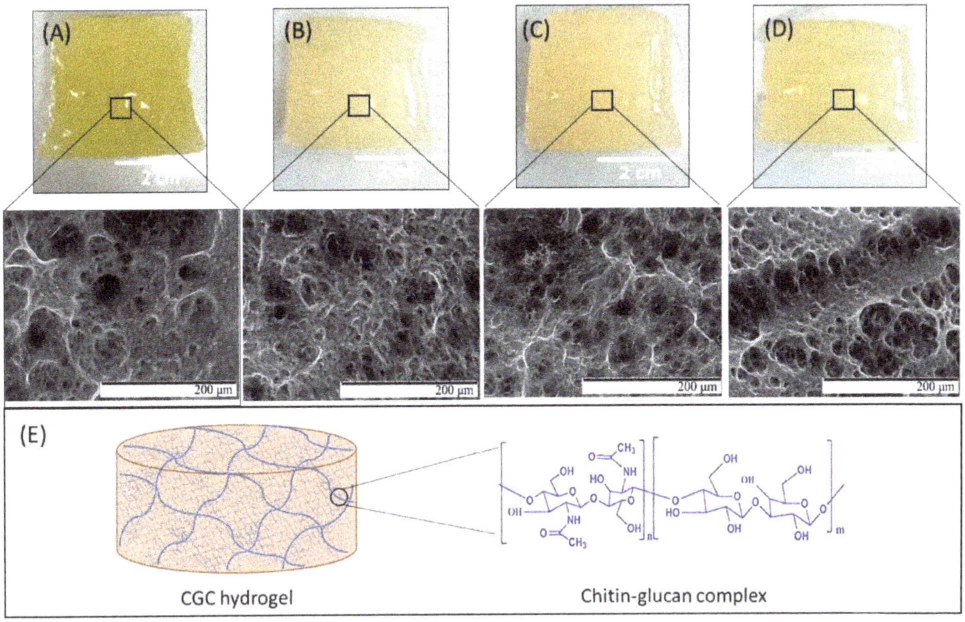

Figure 1. Macroscopic aspect and SEM images of $Na5_1$ (**A**), $Na5_2$ (**B**), $Na5_3$ (**C**), and $Na5_1$* (**D**) hydrogels under magnification 500×; chemical structure of CGC and schematic representation of the CGC hydrogel (**E**).

3.2. Chemical Characterization of the Hydrogels

Table 1 shows the chemical characterization of the CGC hydrogels. It can be observed that all hydrogels exhibited a water content above 97%, characteristic of these structures. No sodium was detected in the hydrogels, which demonstrates the efficient removal of the ion from the structure during the dialysis process.

As shown in Table 1, the chitin content was similar for all CGC hydrogels, however a slight decreasing was observed as the freezing cycles increased. In fact, the $Na5_1$ hydrogels presented a chitin content of 25.63 ± 0.78%, while the $Na5_3$ hydrogels shown 23.85 ± 0.14% of chitin. Moreover, the $Na5_1$* hydrogels exhibited the lowest chitin content (21.51 ± 1.49%), suggesting that extended freezing time might be required to dissolve the enriched chitin CGC macromolecules in the NaOH solution.

3.3. Morphological Characterization

As shown in Figure 1, all CGC hydrogels were translucid, presented a yellow coloration and their shape was molded by the dialysis tubing. Furthermore, despite the different approaches applied for their preparation, all hydrogels exhibit similar macroscopic characteristics.

The morphological features of the CGC hydrogels were evaluated by SEM analysis (Figure 1). Similarly to the Na5 hydrogels [22], all structures presented a heterogeneous, compact, and dense three-dimensional network microstructure made of polymeric chains. The porous structures of the obtained hydrogels seem to be slightly affected by the number of freeze–thaw cycles, with the pore size increasing as the number of cycles increases. As shown in Figure 1, the $Na5_3$ hydrogels exhibited a microstructure composed by larger pores when compared to $Na5_1$ hydrogels. This fact might be explained by the lower polymer concentration present in the $Na5_3$ hydrogels (Table 1). It has been reported that pore volume and pore size distribution are affected by the polymer content present during hydrogel formation [27]. Indeed, the increasing intermolecular crosslinks and physical entanglements present in hydrogels with high polymer concentrations leads to the formation of smaller pore volumes and pore sizes. Similar results were reported for chitin hydrogels [28], where the average pore size decreased from 8 to 5 μm in diameter as the chitin concentration decreased from 1 to 2 wt%.

Additionally, the SEM micrographs demonstrated that reducing the freezing time affected the hydrogels' microstructure (Figure 1). Despite the similar polymer content of the $Na5_1$ and $Na5_1$* hydrogels (1.68 ± 0.17 and 1.66 ± 0.11 wt%, respectively), the lower freezing time (18 h) applied during preparation of the $Na5_1$* hydrogels induced the formation of microstructures with larger pores. Analogous results were reported by Figueroa-Pizano et al. [29] for chitosan-poly(vinyl alcohol) hydrogels, where those produced with 4 h of freezing time presented larger pores than those formed with 12 h of freezing time.

3.4. Hydrogels' Porosity and Density

The porosity and density of the hydrogels are significantly dependent on their morphological characteristics and are important parameters for controlling the hydrogels' physicochemical properties and kinetics of drug release [24]. As shown in Figure 2, the porosity values were similar for all CGC hydrogels, ranging from 53.8 ± 10.3 to 62.6 ± 3.9%. Even so, the $Na5_3$ hydrogels presented the highest porosity value (62.6 ± 3.9%) which suggests that the porosity might have increased with increasing number of freeze–thaw cycles. Porosity depends on the size and number of pores per unit of volume. From the SEM micrographs, the pore size is easier to observe, and the structure of the $Na5_3$ hydrogels comprised larger pores than $Na5_1$ and $Na5_2$ (Figure 1), which is consistent with the higher porosity measured. On the other hand, the $Na5_1$* hydrogels exhibited the lowest porosity values (53.8 ± 10.3%), demonstrating that reducing the freezing time might have induced a decrease in the hydrogels' porosity. Overall, the prepared CGC hydrogels exhibited interesting porosity levels that render them suitable for application in areas such as drug delivery [30] and/or tissue engineering [31].

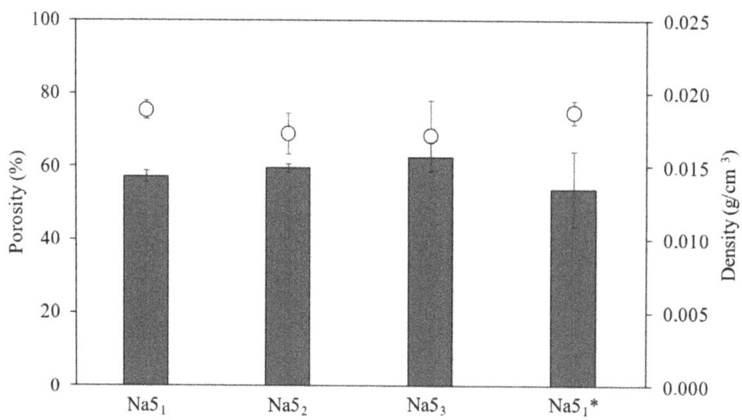

Figure 2. Porosity (■) and density (○) of the CGC hydrogels.

The density of the CGC hydrogels is shown in Figure 2. It can be observed that the density of the hydrogels slightly decreased with the number of freeze–thaw cycles. Thus, the $Na5_1$ and $Na5_1$* hydrogels exhibited similar density values (0.019 ± 0.001 g/cm^3) and the highest ones. These results are consistent with the higher polymer content present in such hydrogels (Table 1), and the resulting increase in the degree of crosslinking observed in the SEM micrographs (Figure 1). Hence, the denser CGC hydrogels ($Na5_1$ and $Na5_1$*) showed the lowest porosity, similar to the results reported for several polymer-based hydrogels, including chitosan [24] and collagen hydrogels [31].

3.5. Mechanical Properties

The mechanical properties of the CGC hydrogels (Figure 3)—namely, their hardness, compressive modulus, and toughness—were obtained by applying a single compression (80% of the initial height) to wet CGC hydrogel samples. As shown in Figure 3, the mechanical characteristics of the CGC hydrogels were not significantly influenced by the number of freeze–thaw cycles. The compressive stress–strain curves of the CGC hydrogels are represented in Figure 3A. The maximum compressive stress obtained represents the force required to produce the deformation of the hydrogels and corresponds to the hardness value (Figure 3B). It can be observed that similar stress–strain profiles were obtained for all hydrogels, with rupture strain occurring between 50% and 60% for compressive stress values of 80% (Figure 3A). Additionally, identical hardness values were achieved for the $Na5_1$, $Na5_2$ and $Na5_3$ hydrogels (3.55 ± 0.23, 3.85 ± 0.38 and 3.69 ± 0.21 kPa, respectively), while higher values were presented by the $Na5_1$* hydrogels (5.04 ± 0.14 kPa). The hardness values obtained were lower than those previously reported for the CGC Na5 hydrogels (7.23 ± 0.78 kPa) [22]. This difference is mainly related to the higher polymer content of those hydrogels (2.28 wt%), which increased the crosslinking between the polymer chains and, consequently, improved the hydrogel's mechanical properties [32].

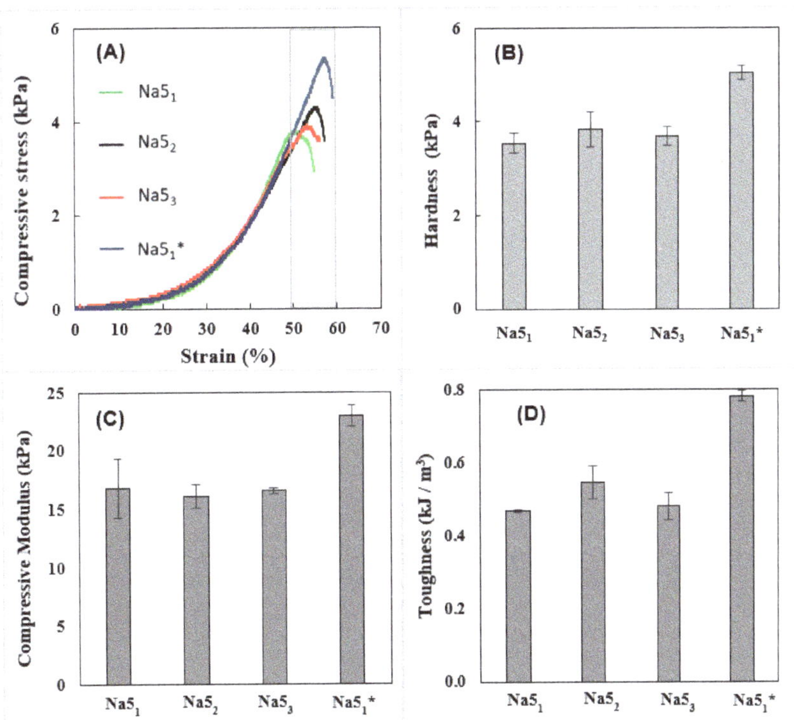

Figure 3. Compressive mechanical properties of the CGC hydrogels: compressive stress (**A**) stress–strain curves, hardness (**B**), compressive modulus (**C**), and toughness (**D**).

The compressive modulus and toughness of the CGC hydrogels, obtained from the stress–strain curves, are represented in Figure 3C,D, respectively. As shown in Figure 3C, similar compressive moduli were displayed by the $Na5_1$, $Na5_2$, and $Na5_3$ hydrogels (16.8 ± 2.6, 16.2 ± 1.0, and 16.6 ± 0.3 kPa, respectively), demonstrating that the number of freezing cycles had no significant impact on their stiffness. Nonetheless, the $Na5_1$* hydrogels were the stiffest material, characterized by a considerably higher compressive modulus (23.0 ± 0.89 kPa). As expected, analogous behavior was obtained for the hydrogels' toughness values. Figure 3D shows that the $Na5_1$* hydrogels exhibited the highest toughness value (0.78 ± 0.015 kPa), while the remaining hydrogel samples displayed values between 0.47 ± 0.003 and 0.55 ± 0.045 kPa. Despite the similar polymer content, the wispy porous structure of the $Na5_1$* hydrogels might have resulted in higher strength, thus improving their compressive modulus.

Given these results, the $Na5_1$* CGC hydrogels were selected for further characterization, including rheological properties, swelling behavior, and drug delivery capability.

3.6. Rheological Properties

The viscoelastic properties of the $Na5_1$* hydrogels (Figure 4A) show that the storage modulus (G′) displayed values one order of magnitude higher than the loss modulus (G″) over the entire range of frequencies, characteristic of their solid-like nature [22]. This behavior also indicates that the hydrogels exhibited predominately elastic characteristics [33]. Moreover, both dynamic moduli are independent of the frequency, thus revealing the formation of a stable gel [34]. Analogous behavior and similar G′ and G″ values (~100 and ~10 Pa, respectively) were obtained for hydrogels prepared with a chitin nanofiber content of 0.4 wt%, in the same range of frequency [35]. Recently, Ferreira et al. [36] also described an identical profile for gels prepared with *Aspergillus niger* CGC dissolved in ionic liquids (ILs). However, those gels presented higher values of both dynamic moduli which might be explained by the presence of ILs in the gel structure [36].

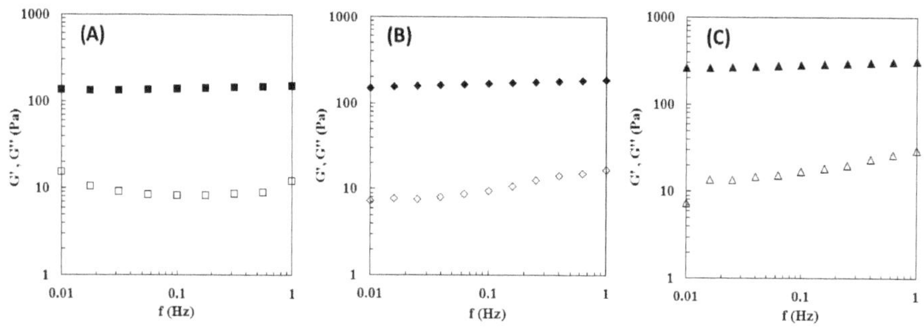

Figure 4. Rheological properties of the $Na5_1$* hydrogel (**A**), $Na5_1$* rehydrated hydrogel (**B**) and $Na5_1$* loaded hydrogel (**C**), at 25 °C. Mechanical spectrum storage (G′, solid symbols) and loss moduli (G″, open symbols).

A similar profile was reported for the CGC Na5 hydrogels [22], however this hydrogel presented significantly higher values for both G′ and G″. In particular, for the same frequency (0.1 Hz), the $Na5_1$* hydrogel presented a G′ of 136.8 ± 11.1 Pa, while a higher value (389 Pa) was found for the Na5 hydrogel. This fact is explained by the increased polymer content of the Na5 hydrogels (2.28 wt%) compared to the $Na5_1$* hydrogels (1.66 wt%), which directly improve their mechanical properties.

3.7. Swelling Behavior and Water Retention Kinetics

The hydrogels' water absorption capacity (swelling behavior) directly affects their drug loading and delivery capability [37]. The swelling ratio of the CGC hydrogels in the different tested media (PBS, NaCl 0.9%, and deionized water) at 37 °C is shown in Figure 5A. For the three tested media, the $Na5_1$* hydrogels displayed excellent water absorption capacity, reaching the swelling equilibrium immediately (less than 1 min). This can be explained by the high hydrophilicity of CGC macromolecules that possess numerous hydrophilic groups, namely, hydroxyl and amino groups, capable of establishing hydrogen bonds with the water molecules [15]. It can be noticed that the swelling capacity of the $Na5_1$* hydrogel was slightly higher in deionized water than in the other tested media

(Figure 5A). Indeed, after 30 s, the hydrogel placed in water reached a swelling ratio of 25.3 ± 0.28 g/g, while lower values were obtained in PBS and NaCl 0.9% (21.5 ± 2.42 and 18.6 ± 1.20 g/g, respectively), for the same time period. This fact is probably related to the higher ionic strength of the PBS and NaCl 0.9% solutions, which decreases the swelling ability of polyelectrolyte hydrogels, due to the decreased osmotic pressure difference between the hydrogel structure and the solution [38].

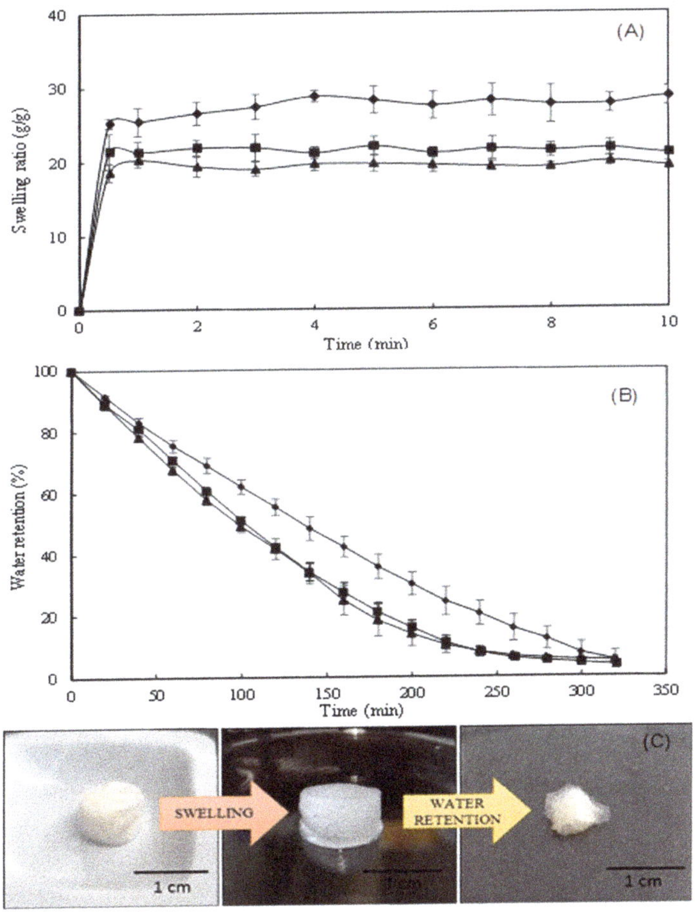

Figure 5. Swelling behavior (**A**) of Na5$_1$* hydrogels in PBS (■), NaCl 0.9% (▲) and deionized water (◆), water retention kinetics (**B**) of water-swollen hydrogels, at 37 °C and macroscopic aspect of hydrogels (**C**) after the two processes.

Significantly lower values were reported by Wu et al. [39] and Udeni Gunathilake et al. [40] for chitin-based hydrogels (~6 g/g) and chitosan-based hydrogels (4.1 g/g), respectively. In fact, the nearly spontaneous swelling of Na5$_1$* hydrogels render them a sponge-like behavior which might enhance the drug loading capacity [41]. Interestingly, as shown in Figure 5C, despite their high swelling ratio, the size of the hydrogels remained similar upon swelling in all the tested media.

The water retention kinetics of the Na5$_1$* hydrogels are shown in Figure 5B. It can be observed that, after placing at 37 °C for 320 min, over 95% of their water content was evaporated and the hydrogels considerably reduced their volume (Figure 5C).

Additionally, the water-swollen hydrogels exhibited a lower water loss rate than those swollen in PBS or NaCl 0.9% (Figure 5B). The largest difference was noticed after 160 min, where the water-swollen hydrogels still had retained 42% of their initial water content, while the PBS- and NaCl-swollen hydrogels had only kept 27% and 25%, respectively. The faster evaporation of water in these hydrogels may indicate the presence of higher levels of free water in their structures since free water has the highest mobility and is the first to evaporate [42].

The macroscopic aspect of water-swollen hydrogels is shown in Figure 5C. It was observed that the visual characteristics of the hydrogel swollen in all the three tested media were similar, with all hydrogels showing a white color.

3.8. *Hydrogels Loading and Release Ability*

3.8.1. Loading $Na5_1$* Hydrogels with Caffeine

The $Na5_1$* hydrogels were loaded with caffeine as a model drug (Figure 6). The procedure involved soaking the freeze-dried structures with a caffeine solution (1.0 wt%). The resulting loaded $Na5_1$* hydrogels were opaque with a whitish color and their dimensions remained similar to those before soaking.

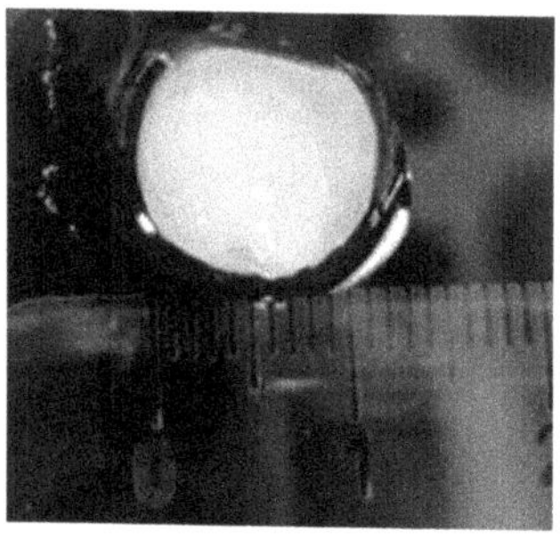

Figure 6. Macroscopic aspect of $Na5_1$* hydrogels loaded with caffeine.

The entrapment efficiency of caffeine in the $Na5_1$* hydrogels was found to be $5.82 \pm 0.89\%$, with caffeine representing $1.02 \pm 0.03\%$ of the total weight of the hydrogel. This result is lower than the values reported in the literature for caffeine entrapment in cellulose-based hydrogel membranes (100%) [25] and β-glucans microparticles ($96.52 \pm 0.63\%$) [43]. This low EE% might be explained by the chemical structure of caffeine, that in water tends to protonate, and by the lower acetylation degree of the *N*-acetylglucosamine monomers of CGC. In fact, it was reported that the use of alkali solvent systems and freeze–thaw cycles promote the deacetylation of CGC chitin molecules [20]. The $Na5_1$* hydrogels presented a degree of acetylation of $27.93 \pm 2.82\%$, which indicates that chitin was converted into chitosan. Thus, the interactions between protonated caffeine and positively charged deacetylated chitin groups lead to repulsion due to similar charges. Similar behavior was reported for nanocarriers of chitosan where a caffeine entrapment efficiency of $17.25 \pm 1.48\%$ was observed [44].

3.8.2. Characterization of the $Na5_1$* Loaded Hydrogels

The presence of caffeine in the $Na5_1$* hydrogel was detected by FTIR analysis. Figure 7 shows the FTIR spectra of caffeine, the $Na5_1$* hydrogel and the $Na5_1$* loaded hydrogel. The caffeine spectrum (Figure 7A) displayed the typical bands of heterocyclic compounds, namely, at 3115 cm^{-1} and 2952 cm^{-1} which depict the stretching of C-H bonds. The absorption peaks at 1697 cm^{-1} and 1650 cm^{-1} are characteristic of the carbonyl group (C = O) of amide I and the additional adsorption peak at 1549 cm^{-1} can be attributed to amide II [45].

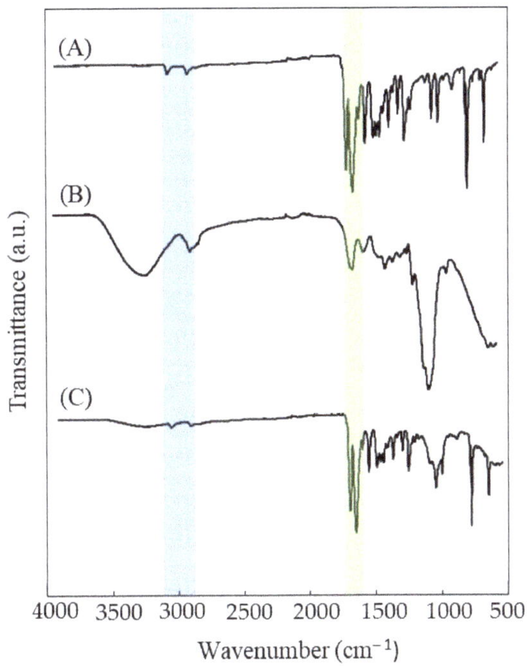

Figure 7. FTIR spectra of (**A**) caffeine, (**B**) $Na5_1$* hydrogels, and (**C**) $Na5_1$* loaded hydrogels.

As expected, the $Na5_1$* hydrogel spectrum (Figure 7B) is similar to previously reported CGC spectrum [20], presenting a characteristic a broad and intense band between 3000–3500 cm^{-1}, typical of the O-H stretching of hydroxyl groups, which overlaps the stretching peaks of N-H. The C-H stretching corresponding to CH_3 and CH_2 appeared at wavenumbers 2919 and 2852 cm^{-1} respectively. The small peaks characteristics of β-1,3-glucans are noticed at 890, 1156, and 1370 cm^{-1} while β-1,6-glucans are represented by peaks at 922, 1045, and 1730 cm^{-1}. The incorporation of caffeine in the hydrogel structure led to a general decrease in the intensity of $Na5_1$* hydrogel spectrum bands, namely the O-H band around 3400 cm^{-1} and the C-O stretching of the saccharide structure at 1020 cm^{-1} (Figure 7C). This impact on the bands intensity might be explained by the high content of caffeine in the hydrogels structure (30.58 ± 4.28%, on a dry basis). Furthermore, peaks appearing at 3113 and 2955 cm^{-1} (heterocyclic compounds), 1697 and 1650 cm^{-1} (amide I) confirm the presence of caffeine, indicating a successful loading. Similar results were reported in several studies where caffeine was encapsulated in alginate beads [45] and chitosan nanoliposomes [46].

The effect of dehydration and the presence of caffeine on the mechanical and rheological properties of the $Na5_1$* hydrogels is presented in Table 2. It can be seen that both the rehydration of the freeze-dried structure and its loading with caffeine apparently improved the mechanical parameters of the $Na5_1$* hydrogels, in particular the compressive modulus, which increased significantly from 23.0 ± 0.89 to 38.06 ± 4.46 and 120.0 ± 61.64 kPa, respectively, thus demonstrating that caffeine increased the structure's rigidity. Consequently, the rehydrated and the caffeine loaded hydrogels presented higher toughness (1.67 ± 0.09 and 1.8 ± 0.33 kJ/m^3, respectively) and hardness (11.50 ± 0.58 and 15.6 ± 2.53 kPa, respectively) values than the original $Na5_1$* hydrogel (0.78 ± 0.01 kJ/m^3, 5.04 ± 0.14 kPa, respectively). These results suggest that the freeze-drying process reinforces the hydrogel structure, thus inducing a more rigid structure upon rehydratation. It is attributed to an increase of interactions (hydrogen bonds) between macromolecules upon drying, forming pore walls made of a more tightly packed and ordered hydrogen-bonded network structure, similarly to what is referred when producing polysaccharide films by solution casting and drying [47]. Loading caffeine into the same structures further strengthened the hydrogel's network and endows enhanced mechanical properties. According to the literature [48], caffeine molecules are able to bind to saccharide molecules (e.g., glucose), which may promote the observed reinforcement of the hydrogel pore walls. Though, the nature of caffeine interactions with the polymeric CGC is likely to be more complex than simple hydrophobic biding. Further studies (e.g., NMR) would be needed to fully characterize CGC-caffeine bonds.

Table 2. Effect of caffeine on the mechanical and rheological properties of the $Na5_1$* hydrogel.

	Sample	$Na5_1$* Hydrogel	$Na5_1$* Rehydrated Hydrogel	$Na5_1$* Loaded Hydrogel
Mechanical properties	Compressive modulus (kPa)	23.0 ± 0.89	38.06 ± 4.46	120.0 ± 61.64
	Toughness (kJ/m^3)	0.78 ± 0.01	1.67 ± 0.09	1.8 ± 0.33
	Hardness (kPa)	5.04 ± 0.14	11.50 ± 0.58	15.6 ± 2.53
Rheological properties	Storage modulus$_{1\ Hz}$ (G′, kPa)	149.9 ± 9.8	186.8 ± 22.0	315.0 ± 76.7
	Loss modulus$_{1\ Hz}$ (G″, kPa)	11.9 ± 0.5	16.8 ± 2.5	29.3 ± 8.4

In the same way, the rheological properties of hydrogels were also improved by the presence of caffeine within the structure (Table 2, Figure 4B,C). The $Na5_1$* rehydrated and the loaded hydrogels presented a rheological profile similar to the original $Na5_1$* hydrogels, exhibiting a predominant elastic behavior (Figure 4B,C). This fact shows that rehydration of the freeze-dried hydrogels and the incorporation of caffeine into their structure had no significant impact on the viscoelastic degree of the structure and the crosslinked network remained homogenous [33]. As a solid-like structure, the storage moduli values were one order of magnitude higher than the loss moduli values (Figure 4). Nevertheless, both the rehydration process and caffeine loading improved the storage and the loss moduli, which is consistent with the observed enhanced mechanical properties. As demonstrated in Table 2, at a frequency of 1 Hz, G′ value was improved from 149.9 ± 9.8 kPa to 186.6 ± 22.0 and 315.0 ± 76.7 kPa and values of G″ increased from 11.9 ± 0.5 to 16.8 ± 2.5 and 29.3 kPa, for rehydrated and loaded hydrogels, respectively. The strengthening of hydrogels network of polycaprolactone-co-lactide (PCLA) by loading certain drugs into the structure was reported by Prince et al. [49]. Additional linkages promoted polymer-polymer interactions which increased G′ values and new polymer–drug interactions that raised G″ values.

3.8.3. Release of Caffeine from the $Na5_1$* Hydrogels

The caffeine release profile of the $Na5_1$* loaded hydrogels was essayed in PBS (pH 7.4) and NaCl 0.9% (pH 5.5), at 37 °C (Figure 8). It can be observed that a similar caffeine release profile was obtained for both media, comprising an initial phase where caffeine was rapidly released (burst phase) followed by a second phase where a steady slower release was achieved. The initial burst can be explained by the release of caffeine loaded close to the surface of the hydrogel [50,51]. Moreover, the maximum caffeine released achieved was similar in both PBS and NaCl 0.9% solutions (71.4 ± 4.1% and 71.5 ± 3.5%, respectively), which might be explained by the similar swelling behavior of $Na5_1$* hydrogels in each media (Figure 5A).

However, it can be noticed that the release of caffeine from the hydrogel was faster in NaCl 0.9% solution than in PBS solution. In the NaCl 0.9% solution, 50% of the loaded caffeine was release in the first 6 min, whereas in PBS solution it took 14 min to reach the same release. This behavior might be related to the difference in pH of the solutions that might have affected the solubility of caffeine [52]. Due to its alkalinity, caffeine solubility is enhanced by low pH values, which explains its rapid release rate in NaCl solution (pH 5.5). Additionally, the slightly higher ionic strength of the NaCl 0.9% solution promoted a faster release of caffeine due to decreased osmotic pressure within the hydrogel structure and weakened interactions between caffeine and polymer [53].

The mechanism of caffeine release from hydrogels is predominantly controlled by diffusion [54]. To evaluate the caffeine release kinetics from $Na5_1$* loaded hydrogels, the first 60% of the released caffeine were analyzed using the Korsmeyer–Peppas model [26], according to the equation

$$Mt/M\infty = k\ tn \tag{9}$$

where Mt and M∞ represent the amount of caffeine (g) released at time t and infinite time, respectively; k is the kinetic constant characteristic of the drug-polymer interaction and n is an empirical parameter for the release mechanism. According to this model, the diffusion mechanism can be classified as controlled diffusion (Fickian diffusion), anomalous transport (non-Fickian diffusion) and controlled swelling, as a function of the relationship between the diffusion rate and the polymer relaxation process [26,55]. For cylinder samples, a value of $n \leq 0.45$ indicates a Fickian diffusion, $0.45 \leq n \leq 0.85$ is a non-Fickian diffusion and $n \geq 0.85$ represents a relaxation-controlled diffusion. Plotting ln (Mt/M∞) vs. ln (t), the kinetic parameters n and k can be calculated from the slope and the interception, respectively.

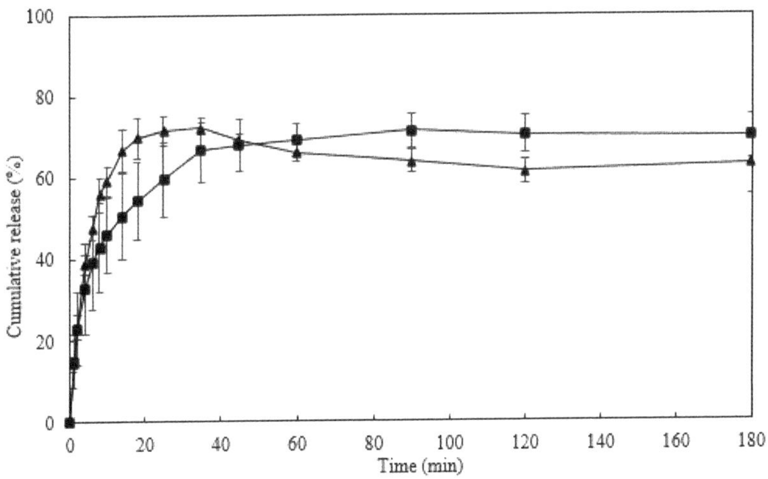

Figure 8. Caffeine release profile of $Na5_1$* hydrogel in PBS (■) and NaCl 0.9% (▲), at 37 °C.

As shown in Figure 9, the kinetic parameters obtained in the PBS solution (n = 0.42 and k = 0.26) revealed a Fickian diffusion which indicates that caffeine diffusion through the hydrogel occurs slower than the relaxation of the polymer chains [55]. On the other hand, the release of caffeine in a NaCl 0.9% solution followed a non-Fickian diffusion, as n = 0.63 and k = 0.24. This mechanism suggests that the drug diffusion rate and the polymer relaxation process are relevant to the drug release rate [33]. The results obtained revealed that drug release from $Na5_1$* hydrogels is not only dependent on the physicochemical properties of the loaded drug, but also on the properties of the medium used to release the drug.

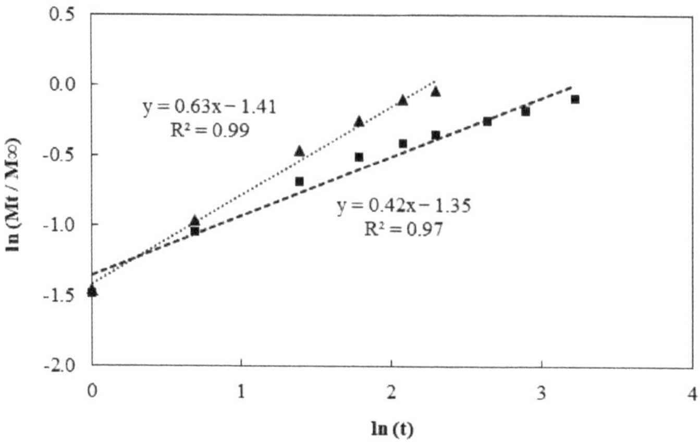

Figure 9. Plot of ln (Mt/M∞) vs. ln (t) for the caffeine release from $Na5_1$* loaded hydrogels in PBS (■) and NaCl 0.9% (▲) solutions, following the Korsmeyer–Peppas model.

4. Conclusions

The procedure for preparing CGC hydrogels comprising polymer dissolution in NaOH by the freeze–thaw method, followed by coagulation by dialysis of the obtained aqueous solution, was optimized by reducing the number of freeze–thaw cycles and the total freezing time. The optimized methodology resulted in CGC hydrogels with improved rheological, mechanical, and swelling properties, which were assayed for their ability for caffeine loading. The loaded hydrogels displayed improved mechanical and rheological properties, with caffeine release profiles following a Fickian diffusion mechanism in the PBS solution and a non-Fickian diffusion in the NaCl 0.9% solution. Therefore, this study demonstrated that CGC can be processed into hydrogels by a simple procedure and the resulting structures possess suitable properties for their use as drug delivery systems.

Author Contributions: Conceptualization, D.A. and T.R.; Methodology, D.A. and T.R.; Investigation, D.A. and T.R.; Writing—original draft preparation, D.A.; Writing—review and editing, F.F. and V.D.A.; Supervision and Funding, F.F. and V.D.A. All authors have read and agreed to the published version of the manuscript.

Funding: This work was financed by national funds from FCT—Fundação para a Ciência e a Tecnologia, I.P., in the scope of the projects UIDP/04378/2020 and UIDB/04378/2020 of the Research Unit on Applied Molecular Biosciences—UCIBIO, LA/P/0140/202019 of the Associate Laboratory Institute for Health and Bioeconomy-i4HB, and UIDP/04129/2020 and UIDB/04129/2020 of the Research Center Linking Landscape, Environment, Agriculture and Food—LEAF. Diana Araújo acknowledges FCT I.P. for PhD fellowship SFRH/BD/140829/2018.

Informed Consent Statement: Not applicable.

Data Availability Statement: Data will be made available upon request.

Conflicts of Interest: The funders had no role in the design of the study; in the collection, analyses, or interpretation of data; in the writing of the manuscript, or in the decision to publish the results.

References

1. Ahsan, A.; Tian, W.-X.; Farooq, M.A.; Khan, D.H. An overview of hydrogels and their role in transdermal drug delivery. *Int. J. Polym. Mater. Polym. Biomater.* **2021**, *70*, 574–584. [CrossRef]
2. Hoffman, A.S. Hydrogels for biomedical applications. *Adv. Drug Deliv. Rev.* **2012**, *64*, 18–23. [CrossRef]
3. Shen, X.; Shamshina, J.L.; Berton, P.; Gurau, G.; Rogers, R.D. Hydrogels based on cellulose and chitin: Fabrication, properties, and applications. *Green Chem.* **2016**, *18*, 53–75. [CrossRef]
4. Klein, M.; Poverenov, E. Natural biopolymer-based hydrogels for use in food and agriculture. *J. Sci. Food Agric.* **2020**, *100*, 2337–2347. [CrossRef]

5. Mitura, S.; Sionkowska, A.; Jaiswal, A.K. Biopolymers for hydrogels in cosmetics: Review. *J. Mater. Sci. Mater. Med.* **2020**, *31*, 1–14. [CrossRef]
6. Chen, G.; Tang, W.; Wang, X.; Zhao, X.; Chen, C.; Zhu, Z. Applications of hydrogels with special physical properties in bio-medicine. *Polymers* **2019**, *11*, 1420. [CrossRef]
7. Xiang, J.; Shen, L.; Hong, Y. Status and future scope of hydrogels in wound healing: Synthesis, materials and evaluation. *Eur. Polym. J.* **2020**, *130*, 109609. [CrossRef]
8. Bashir, S.; Hina, M.; Iqbal, J.; Rajpar, A.H.; Mujtaba, M.A.; Alghamdi, N.A.; Wageh, S.; Ramesh, K.; Ramesh, S. Fundamental Concepts of Hydrogels: Synthesis, Properties, and Their Applications. *Polymers* **2020**, *12*, 2702. [CrossRef]
9. Zhang, H.; Zhang, F.; Wu, J. Physically crosslinked hydrogels from polysaccharides prepared by freeze–thaw technique. *React. Funct. Polym.* **2013**, *73*, 923–928. [CrossRef]
10. Coviello, T.; Matricardi, P.; Marianecci, C.; Alhaique, F. Polysaccharide hydrogels for modified release formulations. *J. Control. Release* **2007**, *119*, 5–24. [CrossRef]
11. Tang, S.; Zhao, L.; Yuan, J.; Chen, Y.; Leng, Y. Physical hydrogels based on natural polymers. In *Hydrogels Based on Natural Polymers*; Chen, Y., Ed.; Elsevier: Amsterdam, The Netherlands, 2019; pp. 51–89. [CrossRef]
12. Farinha, I.; Duarte, P.; Pimentel, A.; Plotnikova, E.; Chagas, B.; Mafra, L.; Grandfils, C.; Freitas, F.; Fortunato, E.; Reis, M.A. Chitin–glucan complex production by Komagataella pastoris: Downstream optimization and product characterization. *Carbohydr. Polym.* **2015**, *130*, 455–464. [CrossRef]
13. Roca, C.; Chagas, B.; Farinha, I.; Freitas, F.; Mafra, L.; Aguiar, F.; Oliveira, R.; Reis, M.A. Production of yeast chitin–glucan complex from biodiesel industry byproduct. *Process Biochem.* **2012**, *47*, 1670–1675. [CrossRef]
14. Meichik, N.R.; Vorob'Ev, D.V. Chitin-glucan complex in cell walls of the Peltigera aphthosa lichen. *Appl. Biochem. Microbiol.* **2012**, *48*, 307–311. [CrossRef]
15. Araújo, D.; Ferreira, I.C.; Torres, C.A.; Neves, L.; Freitas, F. Chitinous polymers: Extraction from fungal sources, characteri-zation and processing towards value-added applications. *J. Chem. Technol. Biotechnol.* **2020**, *95*, 1277–1289. [CrossRef]
16. Singh, A.; Dutta, P.K.; Kumar, H.; Kureel, A.K.; Rai, A.K. Improved antibacterial and antioxidant activities of gallic acid grafted chitin-glucan complex. *J. Polym. Res.* **2019**, *26*, 234. [CrossRef]
17. Kulev, D.; Negrutsa, I. Chitin-glucan complex-food additive with sorbent properties. *J. Hyg. Eng. Des.* **2015**, *11*, 53–56.
18. Marzorati, M.; Maquet, V.; Possemiers, S. Fate of chitin-glucan in the human gastrointestinal tract as studied in a dynamic gut simulator (SHIME®). *J. Funct. Foods* **2017**, *30*, 313–320. [CrossRef]
19. Abdel-Mohsen, A.M.; Jancar, J.; Massoud, D.; Fohlerova, Z.; Elhadidy, H.; Spotz, Z.; Hebeish, A. Novel chitin/chitosan-glucan wound dressing: Isolation, characterization, antibacterial activity and wound healing properties. *Int. J. Pharm.* **2016**, *510*, 86–99. [CrossRef]
20. Araújo, D.; Alves, V.D.; Marques, A.C.; Fortunato, E.; Reis, M.A.M.; Freitas, F. Low Temperature Dissolution of Yeast Chitin-Glucan Complex and Characterization of the Regenerated Polymer. *Bioengineering* **2020**, *7*, 28. [CrossRef]
21. Roy, J.C.; Salaün, F.; Giraud, S.; Ferri, A.; Chen, G.; Guan, J. Solubility of Chitin: Solvents, Solution Behaviors and Their Related Mechanisms. In *Solubility of Polysaccharides*; Xu, Z., Ed.; InTech: Rjeka, Croatia, 2017. [CrossRef]
22. Araújo, D.; Alves, V.D.; Lima, S.A.; Reis, S.; Freitas, F.; Reis, M.A. Novel hydrogels based on yeast chitin-glucan complex: Characterization and safety assessment. *Int. J. Biol. Macromol.* **2020**, *156*, 1104–1111. [CrossRef]
23. Araújo, D.; Freitas, F.; Sevrin, C.; Grandfils, C.; Reis, M.A. Co-production of chitin-glucan complex and xylitol by Koma-gataella pastoris using glucose and xylose mixtures as carbon source. *Carbohydr. Polym.* **2017**, *166*, 24–30. [CrossRef]
24. Pawar, V.; Dhanka, M.; Srivastava, R. Cefuroxime conjugated chitosan hydrogel for treatment of wound infections. *Colloids Surf. B Biointerfaces* **2019**, *173*, 776–787. [CrossRef] [PubMed]
25. Silva, N.H.C.S.; Mota, J.P.; De Almeida, T.S.; Carvalho, J.P.F.; Silvestre, A.J.D.; Vilela, C.; Rosado, C.; Freire, C.S.R. Topical Drug Delivery Systems Based on Bacterial Nanocellulose: Accelerated Stability Testing. *Int. J. Mol. Sci.* **2020**, *21*, 1262. [CrossRef]
26. Zarzycki, R.; Modrzejewska, Z.; Nawrotek, K. Drug release from hydrogel matrices. *Ecol. Chem. Eng. S* **2010**, *17*, 117–136.
27. Ferreira, L.; Figueiredo, M.M.; Gil, M.H.; Ramos, M.A. Structural analysis of dextran-based hydrogels obtained chemoen-zymatically. *J. Biomed. Mater. Res. B: Appl. Biomater.* **2006**, *77*, 55–64. [CrossRef]
28. Chang, C.; Chen, S.; Zhang, L. Novel hydrogels prepared via direct dissolution of chitin at low temperature: Structure and biocompatibility. *J. Mater. Chem.* **2011**, *21*, 3865–3871. [CrossRef]
29. Figueroa-Pizano, M.D.; Vélaz, I.; Peñas, F.J.; Zavala-Rivera, P.; Rosas-Durazo, A.J.; Maldonado-Arce, A.D.; Martínez-Barbosa, M.E. Effect of freeze-thawing conditions for preparation of chitosan-poly (vinyl alcohol) hydrogels and drug release studies. *Carbohydr. Polym.* **2018**, *195*, 476–485. [CrossRef]
30. Hoare, T.R.; Kohane, D.S. Hydrogels in drug delivery: Progress and challenges. *Polymer* **2008**, *49*, 1993–2007. [CrossRef]
31. Song, X.; Zhu, C.; Fan, D.; Mi, Y.; Li, X.; Fu, R.Z.; Duan, Z.; Wang, Y.; Feng, R.R. A Novel Human-Like Collagen Hydrogel Scaffold with Porous Structure and Sponge-Like Properties. *Polymers* **2017**, *9*, 638. [CrossRef]
32. Hurler, J.; Engesland, A.; Poorahmary, B.K.; Škalko-Basnet, N. Improved texture analysis for hydrogel characterization: Gel cohesiveness, adhesiveness, and hardness. *J. Appl. Polym. Sci.* **2012**, *125*, 180–188. [CrossRef]
33. Wong, R.S.H.; Dodou, K. Effect of Drug Loading Method and Drug Physicochemical Properties on the Material and Drug Release Properties of Poly (Ethylene Oxide) Hydrogels for Transdermal Delivery. *Polymers* **2017**, *9*, 286. [CrossRef] [PubMed]

34. Baby, D.K. Rheology of hydrogels. In *Rheology of Polymer Blends and Nanocomposites: Theory, Modelling and Applications*, 1st ed.; Thomas, S., Chandrasekharakurup, S., Chandran, N., Eds.; Elsevier: Amsterdam, The Netherlands, 2019; pp. 193–204.
35. Mushi, N.E.; Kochumalayil, J.; Cervin, N.T.; Zhou, Q.; Berglund, L.A. Nanostructurally Controlled Hydrogel Based on Small-Diameter Native Chitin Nanofibers: Preparation, Structure, and Properties. *ChemSusChem* **2016**, *9*, 989–995. [CrossRef] [PubMed]
36. Ferreira, I.C.; Araújo, D.; Voisin, P.; Alves, V.D.; Rosatella, A.A.; Afonso, C.A.; Freitas, F.; Neves, L.A. Chitin-glucan complex—Based biopolymeric structures using biocompatible ionic liquids. *Carbohydr. Polym.* **2020**, *247*, 116679. [CrossRef]
37. Kim, S.W.; Bae, Y.H.; Okano, T. Hydrogels: Swelling, drug loading, and release. *Pharm. Res.* **1992**, *9*, 283–290. [CrossRef]
38. Kong, B.J.; Kim, A.; Park, S.N. Properties and in vitro drug release of hyaluronic acid-hydroxyethyl cellulose hydrogels for transdermal delivery of isoliquiritigenin. *Carbohydr. Polym.* **2016**, *147*, 473–481. [CrossRef]
39. Wu, S.; Duan, B.; Lu, A.; Wang, Y.; Ye, Q.; Zhang, L. Biocompatible chitin/carbon nanotubes composite hydrogels as neuronal growth substrates. *Carbohydr. Polym.* **2017**, *174*, 830–840. [CrossRef]
40. Udeni Gunathilake, T.M.S.; Ching, Y.C.; Chuah, C.H. Enhancement of curcumin bioavailability using nanocellulose rein-forced chitosan hydrogel. *Polymers* **2017**, *9*, 64. [CrossRef]
41. Raza, H.; Ranjha, N.M.; Razzaq, R.; Ansari, M.; Mahmood, A.; Rashid, Z. Fabrication and in vitro evaluation of 5-florouracil loaded chondroitin sulfate-sodium alginate microspheres for colon specific delivery. *Acta Pol. Pharm.* **2016**, *73*, 495–507.
42. Zhang, S.; Guan, Y.; Fu, G.-Q.; Chen, B.-Y.; Peng, F.; Yao, C.-L.; Sun, R.-C. Organic/Inorganic Superabsorbent Hydrogels Based on Xylan and Montmorillonite. *J. Nanomater.* **2014**, *2014*, 1–11. [CrossRef]
43. Noor, N.; Shah, A.; Gani, A.; Gani, A.; Masoodi, F.A. Microencapsulation of caffeine loaded in polysaccharide based delivery systems. *Food Hydrocoll.* **2018**, *82*, 312–321. [CrossRef]
44. Abosabaa, S.A.; ElMeshad, A.N.; Arafa, M.G. Chitosan Nanocarrier Entrapping Hydrophilic Drugs as Advanced Polymeric System for Dual Pharmaceutical and Cosmeceutical Application: A Comprehensive Analysis Using Box-Behnken Design. *Polymers* **2021**, *13*, 677. [CrossRef] [PubMed]
45. Belščak-Cvitanović, A.; Komes, D.; Karlović, S.; Djaković, S.; Špoljarić, I.; Mršić, G.; Ježek, D. Improving the controlled delivery formulations of caffeine in alginate hydrogel beads combined with pectin, carrageenan, chitosan and psyllium. *Food Chem.* **2015**, *167*, 378–386. [CrossRef] [PubMed]
46. Seyedabadi, M.M.; Rostami, H.; Jafari, S.M.; Fathi, M. Development and characterization of chitosan-coated nanoliposomes for encapsulation of caffeine. *Food Biosci.* **2020**, *40*, 100857. [CrossRef]
47. Alves, V.D.; Costa, N.; Coelhoso, I.M. Barrier properties of biodegradable composite films based on kappa-carrageenan/pectin blends and mica flakes. *Carbohydr. Polym.* **2010**, *79*, 269–276. [CrossRef]
48. Tavagnacco, L.; Engström, O.; Schnupf, U.; Saboungi, M.-L.; Himmel, M.; Widmalm, G.; Cesàro, A.; Brady, J.W. Caffeine and Sugars Interact in Aqueous Solutions: A Simulation and NMR Study. *J. Phys. Chem. B* **2012**, *116*, 11701–11711. [CrossRef]
49. Prince, D.A.; Villamagna, I.J.; Hopkins, C.C.; de Bruyn, J.R.; Gillies, E.R. Effect of drug loading on the properties of tem-perature-responsive polyester–poly(ethylene glycol)–polyester hydrogels. *Polym. Int.* **2019**, *68*, 1074–1083. [CrossRef]
50. Moradi, S.; Barati, A.; Salehi, E.; Tonelli, A.E.; Hamedi, H. Preparation and characterization of chitosan based hydrogels containing cyclodextrin inclusion compounds or nanoemulsions of thyme oil. *Polym. Int.* **2019**, *68*, 1891–1902. [CrossRef]
51. Damiri, F.; Bachra, Y.; Bounacir, C.; Laaraibi, A.; Berrada, M. Synthesis and Characterization of Lyophilized Chitosan-Based Hydrogels Cross-Linked with Benzaldehyde for Controlled Drug Release. *J. Chem.* **2020**, *2020*, 1–10. [CrossRef]
52. Lim, L.Y.; Go, M.-L. Caffeine and nicotinamide enhances the aqueous solubility of the antimalarial agent halofantrine. *Eur. J. Pharm. Sci.* **2000**, *10*, 17–28. [CrossRef]
53. Wang, Q.; Dong, Z.; Du, Y.; Kennedy, J.F. Controlled release of ciprofloxacin hydrochloride from chitosan/polyethylene glycol blend films. *Carbohydr. Polym.* **2007**, *69*, 336–343. [CrossRef]
54. Hu, Y.; Wu, X.Y.; Xu, J.R.; Guo, J. Study on the preparation and drug release property of soybean selenoprotein/carboxymethyl chitosan composite hydrogel. *J. Polym. Eng.* **2018**, *38*, 963–970. [CrossRef]
55. De Stéfano, J.C.Q.; Abundis-Correa, V.; Herrera-Flores, S.D.; Alvarez, A.J. pH-sensitive starch-based hydrogels: Synthesis and effect of molecular components on drug release behavior. *Polymers* **2020**, *12*, 1974. [CrossRef] [PubMed]

polymers

MDPI

Article

Effect of Polymers and Permeation Enhancers in the Release of Quetiapine Fumarate Transdermal Patch through the Dialysis Membrane

Mullaicharam Bhupathyraaj [1,*], K. Reeta Vijaya Rani [2], Sathvik B. Sridhar [3], Javed Shareef [3], Sabin Thomas [4], Nirmala Halligudi [1], Anbazhagan Sockalingam [2], Kiruba Mohandoss [5] and Shyam Sundar [4]

1 College of Pharmacy, National University of Science and Technology, Muscat 130, Oman; nirmalahalligudi@nu.edu.om
2 Surya School of Pharmacy, Vikravandi, Villupuram 605652, Tamilnadu, India; reeta_rani07@yahoo.co.in (K.R.V.R.); sanbubiochem@gmail.com (A.S.)
3 RAK College of Pharmaceutical Sciences, RAK Medical, and Health Sciences University, Ras Al Khaimah P.O. Box 11172, United Arab Emirates; sathvik@rakmhsu.ac.ae (S.B.S.); javedh@rakmhsu.ac.ae (J.S.)
4 School of Pharmacy, College of Pharmacy & Nursing, University of Nizwa, Nizwa 616, Oman; sabin@unizwa.edu.om (S.T.); asundar@unizwa.edu.om (S.S.)
5 Shri Sathya Sai Medical College and Research Institute, Thiruporur, Chengalpet 603108, Tamilnadu, India; mkirupa@gmail.com
* Correspondence: mullaicharam@nu.edu.om

Abstract: Quetiapine Fumarate is potent, and the daily therapeutic dose can be delivered easily across the skin with the help of permeation enhancers. Quetiapine Fumarate-loaded transdermal patches were prepared by solvent evaporation technique. Various formulation parameters, excipients, and their combinations were optimized to get thin, translucent, smooth, stable, and high permeable character patches. A total number of 10 formulations were prepared. All formulations were subjected to various physicochemical evaluations. Three different formulations were prepared and F1, F2, and F3. Various physicochemical studies were carried out and found no significant difference between the three batches. The in vitro release study showed 74.29%, 82.73%, and 77.27%, respectively, up to 24 h. From the results, F2 has been selected as an optimized formulation and evaluated for skin irritation test. The results revealed that there is no irritation produced. The stability study results showed that there is no significant change from its initial nature till the period of three months in both temperatures. Quetiapine Fumarate Transdermal Patch F2 has achieved the goal of extended-release, cost-effectiveness, lowering the dose and frequency of drug administration, and thus may improve patient compliance.

Keywords: schizophrenia; quetiapine fumarate; percutaneous release; transdermal patch

1. Introduction

According to WHO, schizophrenia affects about 24 million people worldwide. It is a treatable disorder, with the treatment being more effective in its initial stages. More than 50% of people with schizophrenia are not receiving appropriate care. A total of 90% of people with untreated schizophrenia are in developing countries [1–3]. Available oral dosage form has severe hepatic first-pass metabolism, resulting in much less oral bioavailability [4]. The major metabolic pathways are sulfoxidation to the sulfoxide metabolite and oxidation to the parent acid metabolite; both metabolites are pharmacologically inactive [5]. Quetiapine is an atypical antipsychotic drug used for the treatment of schizophrenia and bipolar disorder. Its bioavailability is only 9% and it has a short half-life (6 h) due to inherent hepatic metabolism. To address these issues, a matrix-type transdermal patch of Quetiapine Fumarate was developed by using the polymers Poly Vinyl Pyrollidone K30 and Hydroxy Methyl Cellulose K 100. Quetiapine is categorized as a serotonin antagonist, which binds to 5-HT 2A/2C receptors. It has a strong affinity to several dopaminergic receptors but exerts weak antagonism for the D2 receptor, which is responsible for modulating the neuroleptic activity [3]. Quetiapine Fumarate formulations produce dose-dependent side effects like extrapyramidal effect akinesia (inability to initiate movement), akathisia (inability to remain motionless), dyskinesia (disorder resulting in involuntary, repetitive body movements), diatonic symptoms (sustained muscle contractions), and neuroleptic malignant syndrome (muscle rigidity, autonomic instability). The therapeutic equivalent dosage for the transdermal delivery of certain compounds can be significantly less than the corresponding oral dosage, thereby potently reducing the dose-related side effects. Further, a gradual reduction in the dose of the drug could be achieved by simple replacement of the patches containing various doses of the drug [6]. This dosage adjustment may alleviate the side effects that occur after the immediate termination of antipsychotics. The possible overdoses that often occur in the oral route can also be avoided by the transdermal administration of antipsychotics. The patient could be maintained in a well-controlled condition (by maintaining the plasma drug concentration). The transdermal patch can be applied to uncontrolled or calm patients. Quetiapine Fumarate is potent, and the daily therapeutic dose can be delivered easily across the skin with the help of penetration enhancers [7–9]. Concerning the commercial view, the market for transdermal products is ever-increasing (USD 31.5 billion in the year 2015) and approximately 13 active ingredients are approved for use globally [10]. This scenario of the transdermal market shows the commercialization potential of the developed systems in the future.

2. Materials and Methods

2.1. Materials

Quetiapine Fumarate was obtained as a gift sample from Octopus Pharmaceuticals, Chennai, India. HPMC K 100 was obtained from Hi-media laboratories, Mumbai, India. Poly Vinyl Pyrrolidone K 30 was procured from BASF limited, Rosenberg, Germany. Linalool and 1,8-Cineole were procured from Alfa Aesar, Lancaster, UK. DMSO and Glycerin have been purchased from Merck Limited, Vikroli East, Maharashtra, Mumbai. All other chemicals were of analytical grade.

2.2. Methods

2.2.1. Compatibility by FTIR Spectroscopy

Drug-Excipient compatibility was carried out by FT-IR analysis [11]. Initially, the IR spectrum of pure drug, Quetiapine Fumarate, and individual spectrum of PVP, HPMC, glycerin, cineole, linalool, and DMSO were obtained. After that, various admixtures of the drug with other excipients were prepared and the IR Spectra were obtained in KBr pellets using a Perkin Elmer model spectrum (Thane, Maharashtra) in the range 4000–400 cm^{-1}.

2.2.2. Partition coefficient

The partition coefficient was done using n-octanol as the oil phase and phosphate buffer pH 7.4 as the aqueous phase. The two phases were mixed in equal quantity and 10 mg of the weighed amount of drug was added. Then, these were flooded on a mechanical shaker for 2 h. The saturated phases were separated by a separating funnel and an equal volume of both phases n-octanol and phosphate buffer were taken in a conical flask and then analyzed for respective drug controls. The partition coefficient of drug Po/w was calculated by the following formula [12].

$$\text{Partition coefficient [Po/w]} = \text{Coil/Cwater (at equilibrium)}$$

Preparation of transdermal patch of Quetiapine Fumarate

Preliminary trial batches (T1–T10) of Quetiapine Fumarate matrix type transdermal patches were prepared by solvent evaporation method [13] using a Petri plate. The trial batches were composed of PVP K 30 and HPMC K 100 in 1:1, 1:2, 1:3, 1:4, 1:5, 1:6, 1:7, 1:8, 1:9, and 1:10 ratios, respectively (in all formulations the PVP K 30 quantity was kept constant). The polymer solution was prepared using distilled water and was stirred well to get a homogenous mixture using a magnetic stirrer. A specified percentage of glycerin was added as a plasticizer. A weighed quantity of the drug was dissolved in the polymeric solution and finally made up to the required volume using water and kept aside a few hours to remove the air bubbles. Then, the solution was poured into the Petri dish coated with aluminum foil and dried at room temperature for 24 h for solvent evaporation and an inverted funnel was kept over the Petri dish to control the evaporation of the solvent. The patches were removed by peeling and were cut into squares carrying 20 mg of Quetiapine Fumarate/patch area, which were kept in a desiccator for physicochemical evaluations and percentage drug content. Finally, the prepared patches were visually compared with the commercially available transdermal patch of matrix diffusion type (Nu Patch—Zydus Cadila, Ahmedabad, India).

2.3. Evaluation of Preliminary Trial Batches

2.3.1. Water Vapor Transmission (WVT) Rate

A glass vial is used as a transmission cell, in which calcium chloride was placed, acting as a desiccant. A patch that is to be evaluated was placed over the cell. This cell was weighed and placed in a desiccator, which is filled with Potassium Chloride solution (saturated solution) to maintain the 84% RH. Glass vial was removed from the desiccator and reweighed after 24 h for a period of 72 h. The WVT rate was determined using the following formula [14].

$$\text{WVT} = \frac{\text{Final Weight} - \text{Initial Weight} \times 100}{\text{Exposure time} \times \text{Area of patch}},$$

2.3.2. Percentage Moisture Content

The prepared patches were weighed individually and kept in a desiccator containing fused calcium chloride at room temperature for 24 h. After 24 h, the patches were reweighed, and the percentage moisture content was determined using the following formula [15].

$$\text{Percentage moisture content} = \frac{\text{Initial weight} - \text{Final weight} \times 100}{\text{Final weight}}.$$

2.3.3. Percentage Moisture Uptake

The weighed patches were kept in a desiccator at room temperature for 24 h containing a saturated solution of potassium chloride to maintain 84% RH. After 24 h, the patches were reweighed and the percentage moisture uptake was determined using the following formula [16].

$$\text{Percentage moisture content} = \frac{\text{Final weight} - \text{Initial weight} \times 100}{\text{Initial weight}}$$

2.3.4. Thickness

The thickness of the patch was measured at three different points using a digital Vernier caliper and the average thickness was calculated [17].

2.3.5. Weight Uniformity

For each formulation, three randomly selected patches were used. For the weight variation test, three patches from each batch were weighed individually and the average weight was calculated [18].

2.3.6. Folding Endurance

The folding endurance of the patch was determined by repeatedly folding a small strip of the patch (2 cm × 2 cm) at the same place until it broke. The number of times the patch could be folded at the same place without breaking gave the value of folding endurance [19].

2.3.7. Drug Content

The uniformity of drug content of the transdermal patch was determined, based on the dry weight of the drug and polymer used, using a UV/VIS spectrophotometer method [20]. A specified area (2.5 cm^2) of the patch was cut and dissolved in 5 mL of phosphate buffer pH 7.4. Then the solution was transferred to a volumetric flask and the volume was made up to 10 mL. Appropriate dilutions were made using phosphate buffer (pH 7.4), filtered, and analyzed for drug content at 271 nm by using a UV spectrophotometer (Shimadzu-1700, San Francisco, CA, USA).

2.3.8. Incorporation of Permeation Enhancers and Evaluation of Permeation Enhancers Incorporated Trial Batches

After the selection of the best formulation based on the physical properties and drug content evaluations, the selected formulation was again made into three different batches (F1, F2, and F3) with the incorporation of three permeation enhancers (Linalool, 1,8-cineole, and DMSO) in suitable proportions and evaluated for physical parameters such as Weight variation, Percentage Moisture uptake, Percentage moisture content, Thickness, and Percentage drug content.

2.3.9. In Vitro Permeation Study Using Dialysis Membrane

In vitro permeation studies were carried out using Keshary–Chien diffusion cells. The dialysis sac was previously soaked for 24 h in phosphate buffer 7.4. The patches have adhered to the barrier membrane (dialysis membrane) and the sac is tied firmly to the donor compartment of the Keshary–Chien diffusion cell, the receptor compartment of which is filled with 100 mL phosphate buffer 7.4. The total setup was placed on a thermostatically controlled magnetic stirrer set at 37 ± 2 °C. The content of the diffusion cell was stirred at a constant speed (100 rpm). Samples were withdrawn (1 mL) at predetermined time intervals and replaced with the same amount of distilled water to maintain the sink condition. The samples were analyzed for drug content using a UV spectrophotometer at 271 nm. The permeation study was carried out for 24 h [21].

2.3.10. Skin Irritation Study of Quetiapine Fumarate Transdermal Patch

The irritation study was performed on three male healthy rabbits (average weight: 1.43 kg) according to the Draize modified scoring technique [22]. The dorsal surface (50 cm^2) of the rabbit was cleaned and the hair was removed by shaving. The skin was cleaned with rectified spirit. Representative patches were placed over the skin with the use of adhesive tape and were removed after a period of 1 h, 24 h, 48 h, and 72 h and observed for any skin reactions of erythema and edema. The average of 1–72 h readings represents the final score. Ethical clearance was obtained from the institutional animal ethical committee formed for this purpose (IAEC number in Skin irritation study method PCP/IAEC/001/2018).

The Primary Dermal Irritation index (PDI) was calculated from the average scores of the skin reactions. The skin reactions were scored as follows 0—no erythema and edema, 1—very slight erythema and very slight edema, 2—well-defined erythema and slight edema, 3—moderate to severe erythema and moderate edema, 4—severe erythema and severe edema [23–26].

2.3.11. Stability Study

The patch was packed in aluminum foil and tested at accelerated conditions at temperatures of 40 ± 2 °C and 75 ± 5% RH for 3 months [27]. Patches were assessed for visual appearance, color, texture, drug content, and release profile [28].

3. Results

FT-IR spectral analysis of Quetiapine Fumarate and formulations F2 were obtained. The observations concluded that there is no interaction between the drug and other excipients and the results are given below.

The FTIR spectral analysis of Quetiapine Fumarate alone showed the principal peaks at wave numbers of different Functional groups.

Wave Number in cm	Functional groups
3784.09	O-H Stretching
3064.45	C-H Stretching
2086.41	C=C Stretching
1417.06	C-H Deformation
1246.63	C-O Stretching
987.89, 779.05	C-H out of plane bending

The FTIR spectral analysis of formulation F2 alone showed the principal peaks at wave numbers of different Functional groups.

Wave Number in cm^{-1}	Functional groups
3785.37	O-H Stretching
3310.74	=C=H Stretching
2934.92	C-H Stretching
1602.87	C=O Stretching
1416.87	O-H Deformation

The preliminary trial batches, composed of drug Quetiapine Fumarate, polymers PVP K 30, HPMC K 100, plasticizer glycerin, and solvent distilled water, are given in Table 1.

Table 1. Preliminary trial formulations.

S. No.	Quetiapine Fumarate (mg)	PVP K 30 (mg)	HPMC K 100 (mg)	Glycerin (mL)	Water (mL)
T1.	100	50	50	0.4	5
T2.	100	50	100	0.4	5
T3.	100	50	150	0.4	5
T4.	100	50	200	0.4	5
T5.	100	50	250	0.4	5
T6.	100	50	300	0.4	5
T7.	100	50	350	0.4	5
T8.	100	50	400	0.4	5
T9.	100	50	450	0.4	5
T10.	100	50	500	0.4	5

Various physicochemical properties of Preliminary trial formulations such as WVTR, % Moisture uptake, % moisture content, thickness, folding endurance, and % drug content were evaluated for selection of the best formulation and the results are represented in Table 2.

Table 2. Physicochemical properties of preliminary trial formulations.

Batch Code	WVTR 2 (g/c)	% MU	% MC	Thickness (mm)	Folding Endurance (n's)	% Drug Content
T1	0.0182 ± 0.43	12.932 ± 0.94	2.813 ± 0.88	0.38 ± 0.52	192 ± 0.73	98 ± 0.11
T2	0.0291 ± 0.73	6.961 ± 0.77	3.749 ± 0.62	0.42 ± 0.76	145 ± 0.69	97 ± 0.23
T3	0.0149 ± 0.51	10.941 ± 0.82	1.423 ± 0.49	0.29 ± 0.82	162 ± 0.48	98 ± 0.53
T4	0.0281 ± 0.33	9.337 ± 0.61	2.938 ± 0.61	0.31 ± 0.69	202 ± 0.85	98 ± 0.71
T5	0.0441 ± 0.86	14.113 ± 0.06	6.231 ± 0.26	0.51 ± 0.57	214 ± 0.79	99 ± 0.22
T6	0.0301 ± 0.59	11.859 ± 0.27	3.882 ± 0.51	0.21 ± 0.94	154 ± 0.11	98 ± 0.84
T7	0.0221 ± 0.66	10.330 ± 0.84	1.391 ± 0.63	0.38 ± 0.48	198 ± 0.77	98 ± 0.68
T8	0.0182 ± 0.17	9.867 ± 0.49	3.830 ± 0.89	0.34 ± 0.22	172 ± 0.49	98± 0.53
T9	0.0159 ± 0.91	11.769 ± 0.63	2.670 ± 0.44	0.18 ± 0.61	182 ± 0.62	97 ± 0.71
T10	0.0281 ± 0.08	12.891 ± 0.38	1.462 ± 0.74	0.16 ± 0.38	196 ± 0.82	97 ± 0.44

WVTR—Water vapor transmission rate, MU—moisture uptake, MC—moisture content.

Based on the physicochemical properties of the preliminary trial formulations, T5 has been selected as an optimized formulation and selected for incorporation of permeation enhancers. The composition of permeation enhancers incorporated T5 formulation and was divided into three batches, namely F1, F2, and F3. The composition is given in Table 3.

Table 3. Trial batches of optimized formulation with various permeation enhancers.

S. No.	Quetiapine Fumarate (mg)	PVP K 30 (mg)	HPMC K 100 (mg)	Glycerol (mL)	Water (mL)	Cineole (mL)	DMSO (mL)	Linalool (mL)
F1	10	50	250	0.4	5	-	-	0.3
F2	10	50	250	0.4	5	0.3	-	-
F3	10	50	250	0.4	5	-	0.3	-

The physical properties such as weight, % moisture uptake, % moisture content, thickness, and % drug content were evaluated and the results are represented in Table 4.

Table 4. Physicochemical properties of optimized formulation with various permeation enhancers.

Code	Weight	% MU	% MC	Thickness (mm)	% Drug Content
F1	345 ± 3.47	13.543 ± 0.27	6.302 ± 0.15	0.48 ± 0.02	98.86 ± 1.03
F2	380 ± 4.09	14.134 ± 0.31	6.854 ± 0.18	0.53 ± 0.06	99.40 ± 1.76
F3	350 ± 3.12	13.321 ± 0.29	6.451 ± 0.16	0.47 ± 0.04	99.24 ± 1.34

Hence, F2 showed better physical and in vitro permeation results, and was selected as the best formulation. The photo of formulation F2 is represented in Figure 1.

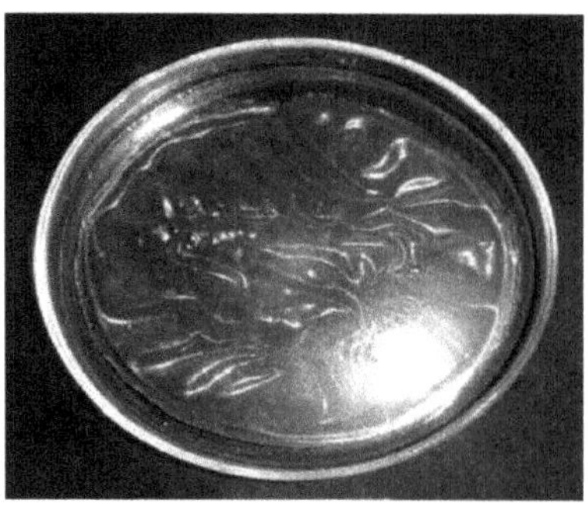

Figure 1. Transdermal patch of F2 formulation.

The comparative in vitro permeation study was carried out for F1, F2, and F3. The formulation F2 shows better release compared to F1 and F2 and the results are explored in Table 5 and Figure 2.

Table 5. Comparative in vitro permeation study results of F1, F2, and F3.

Time (h)	% Drug Release ± S.D. *		
	F1	F2	F3
0	0	0	0
0.5	9.98 ±0.07	9.47 ± 0.02	9.12 ± 0.13
1	12.37 ± 0.04	13.37 ± 0.01	11.84 ± 0.11
2	15.37 ± 0.09	17.98 ± 0.07	15.95 ± 0.13
4	18.26 ± 0.05	23.68 ± 1.21	19.81 ± 0.16
6	22.65 ± 0.12	30.87 ± 0.09	24.71 ± 0.62
8	29.08 ± 0.09	36.62 ± 0.02	30.63 ± 0.17
10	34.74 ± 1.02	43.64 ± 0.08	36.29 ± 0.09
12	40.41 ± 1.05	50.19 ± 0.01	42.78 ± 0.15
14	47.63 ± 1.09	59.59 ± 0.05	49.46 ± 0.16
16	54.83 ± 1.03	65.67 ± 1.01	59.65 ± 0.98
18	60.98 ± 1.20	73.48 ± 0.08	67.35 ± 0.42
20	68.37 ± 1.09	78.93 ± 0.04	73.97 ± 0.50
22	73.92 ± 1.10	83.72 ± 1.21	78.29 ± 0.02
24	74.29 ± 1.12	82.73 ± 1.01	77.27 ± 1.01

* S.D. = standard deviation $n = 3$.

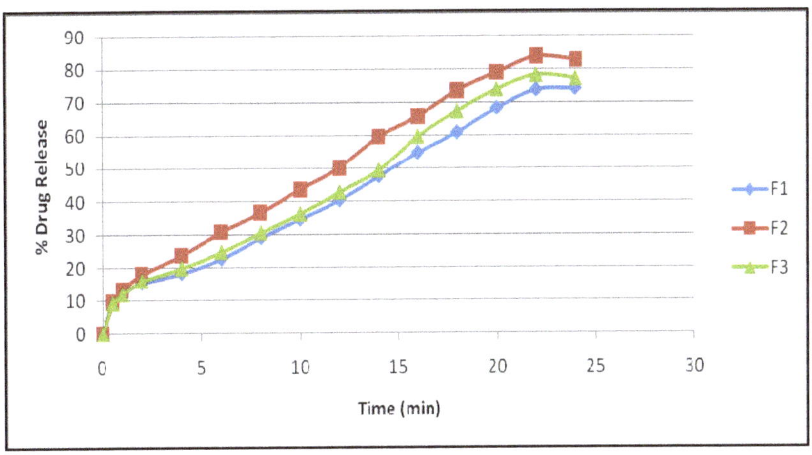

Figure 2. Comparative in vitro permeation profile of F1, F2, and F3.

The optimized formulation F2 has been subjected to a skin irritation study and the skin reactions were scored. The average score and PDI calculated for the patches are given in Tables 6 and 7.

Total: 0/0 = 0

Mean: 0/4 = 0

PDI for 1, 24, 48 and 72 h = 0/4 = 0

Total: 1/0 = 0

Mean: 1/4 = 0.25

PDI for 1, 24, 48 and 72 h = 0.25/4 = 0.0625

Table 6. Erythema and edema of the transdermal placebo patch.

Animal No.	Scores after Treatment with Placebo Patch without Drug after			
	1 h	24 h	48 h	72 h
1	0/0	0/0	0/0	0/0
2	0/0	0/0	0/0	0/0
3	0/0	0/0	0/0	0/0

Table 7. Erythema and edema of the selected (F2) formulation.

Animal No.	Scores after Treatment with a Drug-Loaded Patch after			
	1 h	24 h	48 h	72 h
1	1/0	0/0	0/0	0/0
2	0/0	0/0	0/0	0/0
3	0/0	0/0	0/0	0/0

Further, formulation F2 was subjected to accelerated stability studies. The physicochemical parameters and cumulative in vitro permeability results at the end of the first, second, and third months are interpreted in Tables 8 and 9 and Figure 3.

Table 8. Physicochemical parameters of F2 formulation at 40 ± 2 °C and 75 ± 5% RH at the end of the first, second, and third months.

S. No.	Parameters	Room Temperature	40 ± 2 °C and 75 ± 5% RH
1.	Visual Appearance	Translucent	Translucent
	Initial	No change	No change
	1st month	No change	No change
	2nd month	No change	No change
	3rd month	No change	No change
2	Colour	Dull white	Dull white
	Initial	No change	No change
	1st month	No change	No change
	2nd month	No change	No change
	3rd month	No change	No change
3	Texture	Smooth	Smooth
	Initial	No change	No change
	1st month	No change	No change
	2nd month	No change	No change
	3rd month	No change	No change
4	Drug Content	No change	No change
	Initial	99.48	99.48
	1st month	99.09	98.92
	2nd month	98.48	96.65
	3rd month	97.97	95.34

Table 9. Cumulative percentage of the in vitro permeation study of F2 formulation at 40 ± 2 °C and 75 ± 5% RH.

Time (h)	Percentage of Drugs Released after Storage at the End of		
	1st Month	2nd Month	3rd Month
0	0	0	0
0.5	8.01	8.23	7.89
1	10.3	14.34	10
2	14.4	16.49	12.4
4	15.8	21.43	17.3
6	21.8	27.55	23.7
8	27.3	34.29	28.5
10	32.4	43.87	37.3
12	38.6	50.12	41.8
14	44.9	57.99	49.8
16	53.7	63.27	58.5
18	60.8	70.19	67.4
20	64.6	7.39	71.4
22	69.9	81.32	73.3
24	70.2	80.87	74.7

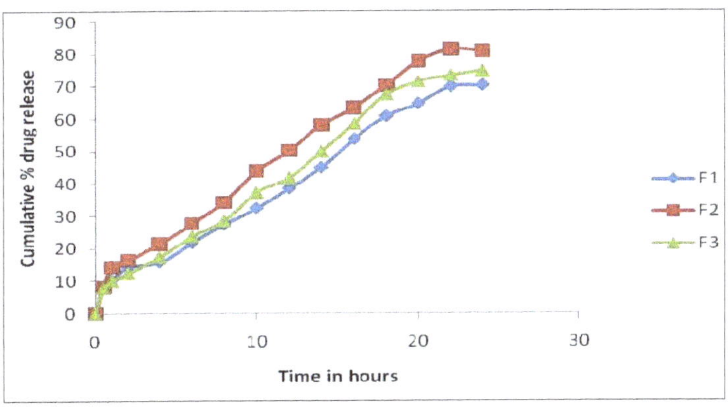

Figure 3. Cumulative percentage of in vitro permeation study of F2 formulation at 40 ± 2 °C and 75 ± 5% RH.

4. Discussion

Solvent evaporation was used to develop a transdermal patch containing quetiapine fumarate. These thin, translucent, smooth, stable, and very permeable patches were created by combining many formulation parameters. Each batch appears to be identical and contains no apparent fractures.

It was evident from the FTIR graphs that the major functional groups of pure Quetiapine Fumarate were not affected by polymer interaction or preparation methods. The drug's partition coefficient was 2.6 in n-octane/phosphate buffer pH 7.4, indicating lipophilicity. Higher valued compounds are extremely lipophilic and stay dissolved in the stratum corneum.

A mixture of PVP K 30 and HPMC K 100 was shown to be the most effective patch-producing polymer. The overall polymer concentration was kept between 0.2 and 0.4 g per patch to provide folding durability, slimness, and an appropriate thickness.

The critical analysis of the plasticizer includes the plasticizer glycerol, which was applied at a concentration of 30% *w*/*w* of the total polymer weight, and was able to generate a flexible patch without having a significant impact on their releasing property.

The patch loses its flexibility and becomes stiff if the amount of plasticizer, glycerol, is larger than 0.1 mL (i.e., 30% *w/w* of polymer), therefore the above two factors were adjusted to make a proper patch. This plasticizer penetrates the polymer particles and softens them. Latex coalescence and patch formation are aided by this softening.

It becomes thick and breaks when the polymer concentration surpasses 0.2–0.4 g/mL. At 50 mg PVP, the patch becomes too thin and no longer fulfills the physicochemical requirements.

The slope of the Korsmeyer–Peppas equation indicates that 'n' is smaller than 0.5. This indicates that the transdermal patch released the medicine via a fickian (diffusion-controlled) transport mechanism, presumably as a result of a polymer that regulates drug diffusion across the polymeric matrix.

The patch's medication release followed the zero-order equation. The correlation between the Hixson–Crowell and motored cube root equations is poor.

The improved formulation's skin irritation was also tested. The mice in the F2 skin irritation investigation showed no erythema or edema after 1 h, 24 h, 48 h, and 72 h. So, it is safe to use.

The formulation T5 had the greatest WVTR, % moisture absorption, % moisture content, thickness, folding endurance, and % drug content of the 10 testing formulations. This may be a result of both polymers' hydrophilicity.

There was no discernible change in the drug content between the patches of formulation T1 to T8. The results suggested that the approach utilized to manufacture a patch with a homogeneous medication content and a small number of patch variables was effective. This refers to the uniform distribution of the medicine throughout the patch production process.

The chosen formulation was compared to a commercial transdermal matrix diffusion patch. Three-month stability tests were conducted. Visual appearance, color, texture, drug content, and drug release were investigated. The results demonstrated no significant change in visual appearance, color, or texture during the period of three months. Accelerated heating lowered the drug content to 95.34%. The accelerated temperatures have no effect on the release kinetics.

Data analysis of the work includes the following:

A natural and synthetic permeation enhancer, linalool 1,8-cineole, was added to the T5 formulation to test its capacity to penetrate the dialysis membrane (three formulations were developed: namely, F1, F2, and F3). Physical factors such as weight fluctuation, thickness, moisture content, moisture absorption, folding endurance, and medication content are also evaluated.

This F1, F2, and F3 weighed 345, 380, and 350 mg. The three formulations were 0.48 mm, 0.53 mm, and 0.47 mm thick. Like the commercial version, this F2 formulation was optimally thick.

F2 had the highest % moisture content and absorption, resulting in a more flexible, easy-to-handle, and smooth patch.

F1–F3 showed a drug content of 98.86%, 99.48%, and 99.24%. There was no change. The 24 h in vitro release profile of F1–F3 exhibited a maximum release of 74.29%, 82.73%, and 77.27%, correspondingly.

The 90% release data were fitted to multiple kinetic models [29–33] to determine the mechanism and rate of drug release. There was an initial burst of the drug followed by a longer sustained release in the F2 formulation. After 24 h, F1 and F3 released little, however, F4 released in an erratic pattern.

This study helped to determine the polymer's efficacy in controlled release testing.

We determined the efficacy of the dialysis membrane permeation enhancers to investigate them further in ex vivo and in vivo in an animal model investigation.

Future studies on the pharmacokinetic characteristics of the novel topical formulation of Quetiapine Fumarate can be done in conjunction with this investigation.

5. Conclusions

Based on the explanation above, it was ascertained that the F2 formulation of Quetiapine Fumarate Transdermal Patch, prepared using a 50:250 proportion of polymers PVP K30 and HPMC K 100, along with the natural permeation enhancer 1,8 cineol, was even more effective in terms of permeation and sustainability. A guideline for developing a non-invasive and sustained release drug delivery system for Quetiapine Fumarate for the successful treatment of schizophrenia is provided in this study endeavor. It is stable under storage settings for three months, with no obvious change in appearance, physical qualities, drug content, or percentage of drug release. The formulation is stable for three months. This formulation may be an alternative to the twice-daily oral conventional pills now available on the market. It has the potential to increase patient compliance, particularly in the case of recalcitrant patients.

Author Contributions: Author contributions: M.B. contributed to the conceptualization and editorial part of the research work. K.R.V.R. contributed to writing the original draft preparation and contributed to the methodology section of the research work. S.B.S. contributed to the software and validation parts of the research work. N.H. contributed to in vitro evaluation studies and reviewed the writing of the article. K.M. contributed to the stability study section of this research work. S.B.S., J.S. and S.T. contributed to the informal analysis of the results. A.S. contributed to the resources and evaluation section of this article. S.S. contributed to the editorial part of research. All authors have read and agreed to the published version of the manuscript.

Funding: This research received no external funding.

Institutional Review Board Statement: Ethical clearance was obtained from the institutional animal ethical committee formed for this purpose. IAEC number in Skin irritation study method PCP/IAEC/001/2018.

Informed Consent Statement: Not applicable.

Data Availability Statement: Not applicable.

Acknowledgments: The authors would like to acknowledge Kiruba Mohandoss for technical support.

Conflicts of Interest: The authors declare no conflict of interest.

References

1. Potu, A.R.; Pujari, N.; Burra, S.; Veerareddy, P.R. Formulation and evaluation of buccoadhesive quetiapine fumarate tablets. *Braz. J. Pharm. Sci.* **2012**, *48*, 335–345. [CrossRef]
2. Kulhara, P.S. What is schizophrenia: A neurodevelopmental and neurodegenerative disorder or a combination of both? A critical analysis. *Indian J. Psychiatry* **2010**, *52*, 21–27. [CrossRef]
3. Ghosh, A.; Chakraborty, K.; Mattoo, S.K. Newer molecules in the treatment of schizophrenia: A clinic update. *Indian J. Pharmacol.* **2011**, *43*, 105–112.
4. Tanner, T.; Marks, R. Delivering drugs by the transdermal route: Review and comment. *Ski. Res. Technol.* **2008**, *14*, 249–260. [CrossRef]
5. Clemency, B.M.; Eggleston, W.; Lindstrom, H.A. Pharmacokinetics and Pharmacodynamics of Naloxone. *Acad. Emerg. Med.* **2019**, *24*, 161–176. [CrossRef]
6. Walker, R.B.; Smith, E.W. The role of percutaneous penetration enhancers. *Adv. Drug Deliver. Rev.* **1996**, *18*, 295–301. [CrossRef]
7. Foldvari, M. Non-invasive administration of drugs through the skin: Challenges in delivery system design. *Pharm. Sci. Technol. Today* **2000**, *3*, 417–425. [CrossRef]
8. Dragicevic-Curic, N.; Scheglmann, D.; Albrecht, V.; Fahr, A. Temoporfin-loaded invasions: Development, characterization and in vitro skin penetration studies. *J. Control. Release* **2008**, *127*, 59–69. [CrossRef]
9. Ritesh, K.; Anil, P. Modified transdermal technologies: Breaking the barriers of drug permeation via the skin. *Trop. J. Pharm. Res.* **2007**, *6*, 633–644.
10. Rowe, R.C. The adhesion of patch coatings to tablet surfaces—The effect of some direct compression excipients and lubricants. *J. Pharm Pharmacol.* **1977**, *29*, 723–726. [CrossRef]
11. Lachman, L.; Lieberman, H.A.; Kanig, J.L. Po/w = Coil/C water (at equilibrium). In *The Theory and Practice of Industrial Pharmacy*, 2nd ed.; Varghese Publishing House: Mumbai, India, 1991; pp. 188–189.

12. Shivaraj, A.; Selvam, R.P.; Mani, T.; Sivakumar, T. Design and evaluation of transdermal drug delivery of ketotifen fumarate. *Int. J. Pharm. Biomed. Res.* **2010**, *1*, 42–47.
13. Mounika, K.; Reddy, B.V.; Reddy, K.N. Formulation and evaluation of carvedilol transdermal patches with hydrophilic polymers. *World J. Pharm. Res.* **2014**, *3*, 815–826.
14. Kumar, J.A.; Pullakandan, N.; Prabhu, S.L.; Gopal, V. Transdermal drug delivery system: An overview. *Int. J. Pharm. Sci. Rev. Research.* **2010**, *3*, 49–54.
15. Kumar, P.D.; Kavitha, K. Development of transdermal patch of nicardipine hydrochloride: An attempt to improve bioavailability. *Int. J. Res. Pharm. Biomed. Sci.* **2011**, *2*, 285–293.
16. Patel, R.P.; Patel, G.; Baria, A. Formulation and evaluation of transdermal patch of Aceclofenac. *Int. J. Drug Deliv.* **2009**, *1*, 41–51. [CrossRef]
17. Trivedi, D.C.; Patel, H.S.; Bhandari, A.K.; Shah, D.A. Preparation and evaluation of transdermal patches of Desloratadine. *Int. J. Biomed. Res.* **2011**, *2*, 359–373. [CrossRef]
18. Garala, K.C.; Shinde, A.J.; Shah, P.H. Formulation and in-vitro characterization of monolith matrix transdermal system using HPMC/Eudragit S100 polymer blends. *Int. J. Pharm. Pharm. Sci.* **2009**, *1*, 108–120.
19. Yadav, V.; Sipai, A.; Mamatha, Y.; Prasanthi, V. Transdermal drug delivery: A technical writeup. *J. Pharm. Sci. Innov.* **2012**, *1*, 5–12.
20. Arabi, H.; Hashemi, S.A.; Ajdari, N. Preparation of a transdermal delivery system and effect of membrane type for scopolamine drug. *Iran. Polym.* **2002**, *11*, 245–250.
21. Draize, J.H.; Woodard, J.; Calvary, H.O. Methods for the Study of Irritation & Toxicity of Substances Applied Topically to the Skin & Mucous Membranes. *J. Pharm. Exp. Thera.* **1944**, *82*, 377–389.
22. Reddy, D.N.; Udupa, N. Formulation and evaluation of oral and transdermal preparations of flurbiprofen and piroxicam incorporated with different carriers. *Drug Dev. Ind. Pharm.* **1993**, *19*, 843–852. [CrossRef]
23. Samuel, Y.; Donbrow, M.; Friedman, M. Sustained release of drugs from ethylcellulose–polyethylene glycol patches and kinetics of drug release. *J. Pharm. Sci.* **1979**, *68*, 325–329. [CrossRef] [PubMed]
24. Chi, S.-C. Transdermal Patch Comprising Paroxetine. WO Patent WO2007011125, 25 January 2007.
25. Shah, H.S.; Tojo, K.; Chie, Y.W. Transdermal controlled delivery of verapamil: Characterization of in vitro skin permeation. *Int. J. Pharm.* **1992**, *86*, 167–173. [CrossRef]
26. European Medicines Agency (EMEA). *Wiley Encyclopedia of Clinical Trials*; Wiley: Hoboken, NJ, USA, 2008; pp. 1–20. Available online: http://doi.wiley.com/10.1002/9780471462422.eoct456 (accessed on 1 January 2022).
27. Banerjee, S.; Chattopadhyay, P.; Ghosh, A.; Bhattacharya, S.S.; Kundu, A.; Veer, V. Accelerated stability testing of a transdermal patch composed of eserine and pralidoxime chloride for prophylaxis against (±)-anatoxin-a poisoning. *J. Food Drug Anal.* **2014**, *22*, 264–270. [CrossRef]
28. Chiang, C. Transdermal Controlled Administration of Indomethacin. Enhancement of Skin Permeability. *Pharm. Res.* **1988**, *2*, 103–106.
29. Singh, J. Transdermal Delivery of Indomethacin: II, Effect of Penetration Enhancers on the In-Vitro Percutaneous Absorption from Patch Formulations. *Pharmazie* **1994**, *8*, 619–620.
30. Mukherjee, B. Design Development Physicochemical and in–vitro Evaluation of Transdermal patches containing Diclofenac Diethyl Ammonium salt. *J. Pharm. Sci.* **2002**, *91*, 2076–2089.
31. Mallikarjuna Setty, C. Chemical Penetration Enhancers for Transdermal Drug Delivery Systems. *Trop. J. Pharm. Res.* **2009**, *8*, 173–179.
32. Adams, B.K.; Ferstl, E.M.; Davis, M.C.; Herold, M.; Kurtkaya, S.; Camalier, R.F.; Hollingshead, M.G.; Kaur, G.; Sausville, E.A.; Rickles, F.R. Synthesis and biological evaluation of novel curcumin analogs as anti-cancer and anti-angiogenesis agents. *Bioorg. Med. Chem.* **2004**, *12*, 3871–3883. [CrossRef]
33. Mohapatra. Design and Characterization of Controlled Release Matrix Tablets of Lamivudine. *Asian J. Pharm. Clin. Res.* **2009**, *2*, 212–220.

polymers

MDPI

Review

Recent Biotechnological Applications of Polyhydroxyalkanoates (PHA) in the Biomedical Sector—A Review

Matheus Silva da Fonseca Diniz *, Murilo Moraes Mourão, Luciana Pereira Xavier and Agenor Valadares Santos *

Laboratory of Biotechnology of Enzymes and Biotransformations, Institute of Biological Sciences, Federal University of Pará, Belém 66075-110, Brazil; mouraomurilo@gmail.com (M.M.M.); lpxavier@ufpa.br (L.P.X.)
* Correspondence: matheusbiotec99@gmail.com (M.S.d.F.D.); avsantos@ufpa.br (A.V.S.)

Abstract: Petroleum-derived plastics are materials of great importance for the contemporary lifestyle, and are widely used commercially because they are low cost, resistant, malleable, and weightless, in addition to their hydrophobic character. However, some factors that confer the qualities of these materials also cause problems, mainly environmental, associated with their use. The COVID-19 pandemic aggravated these impacts due to the high demand for personal protective equipment and the packaging sector. In this scenario, bioplastics are environmentally positive alternatives to these plastics due to their applicability in several areas ranging from packaging, to biomedicine, to agriculture. Polyhydroxyalkanoates (PHAs) are biodegradable biopolymers usually produced by microorganisms as an energy reserve. Their structural variability provides a wide range of applications, making them a viable option to replace polluting materials. PHAs can be applied in various biotechnology sectors, such as producing drug carriers and scaffolds for tissue engineering. This review aimed to survey works published in the last five years on the study and biotechnological application of PHAs in the biomedical sector, exploring the versatility and advantages of their use and helping to understand how to enhance their application.

Keywords: polyhydroxyalkanoate; biodegradable; scaffolds; drug carrier; biomedical; plastics

1. Introduction

One of the most day-to-day materials used is plastic. This material is of great commercial importance since it has applications which range from industrial processes to the packaging of products. Plastic is of great economic importance due to its characteristics of being cheap compared to other materials, light, durable, and malleable, but several economic and social disadvantages are tied to using this material. The emission of greenhouse gases and environmental pollution are two of the problems associated with the use of plastic, mainly due to the linear flow in the value chain, as many applications are single-use products, such as packaging [1].

The versatility of plastics opens up the possibility of various applications, such as in straws, bottles, and plastic bags [2], and the packaging of food, pharmaceuticals, chemicals, and cleaning products [3]. The properties that make plastic an advantageous material are also some of the reasons why it is present in different ecosystems, making it a complex pollutant. The characteristic durability of plastics causes them to remain in the environment for hundreds or even thousands of years. Their hydrophobicity causes them to absorb contaminants, and their low weight allows the locomotion of plastics over long distances. Exposure to environmental factors, such as wave action and the sun, causes the breakdown of these plastics into microplastics (<5 mm), which facilitates their transport to even more environments, and microplastics are often consumed by organisms [4].

The importance of plastics on a day-to-day basis has boosted the increase in production to enable their supply across diverse applications. Thus, the production of 1.5 million tons recorded in the 1950s jumped to 359 million tons in 2018, with an increase in production between 63,000 and 430,000 tons in agricultural areas in Europe and 44,000 and 300,000 tons in North America [5]. From the total of this plastic, it is estimated that 50% of the total weight is composed of plastic materials intended for packaging, which is one of the most worrying residues of this material [6].

During the COVID-19 pandemic, the use of personal protective equipment (PPE), such as gloves and masks, increased considerably, in addition to the increase in other medical hospital waste. These increases overload the treatment facilities for SARS-CoV2 contaminated residues, causing the residues to have inappropriate destinations such as incineration, which is a potential source of greenhouse gases, in addition to other pollutants such as heavy metals. The increased consumption of single-use plastics for packaging and food packaging to avoid contamination was also significant. The increases from the period of social isolation tended to remain after confinement [7].

The pollution caused by these plastics makes strategies for better treatment to reduce the impact of these residues necessary. Circular economy strategies are one of the most common methods of changing the way plastics are disposed of through recycling. However, even this method presents problems, as recycled products can eventually become waste. In this context, chemistry and biotechnology present viable alternatives for treating residues through processes such as pyrolysis, biocatalysts, and the application of microbial cultures that, under appropriate conditions, are capable of degrading these plastics [8]. Another option to reduce the impacts is the replacement of petroleum-derived plastics with degradable biopolymers.

The substitution of these polluting materials contributes to the sustainable objectives established by the United Nations (UN), especially in objective number 12, linked to sustainable consumption and production, reducing the generation of waste, and encouraging companies to employ sustainable practices through the use of biodegradable materials in their production [9]. It also contributes to environmental, social, and governance (ESG) strategies adopted by companies to improve social responsibility, governance, and environmental best practices.

Biopolymers can be classified according to their origin, which includes natural sources, such as keratin, alginate, and collagen; microbial such as PHAs; and synthetic such as polylactic acid (PLA), polyethylene, and polystyrene. Furthermore, they can be classified as biodegradable or nonbiodegradable [10]. Among these polymers, PHAs have drawn attention due to their characteristics which provide products with great versatility, and for being biodegradable, conferring significant advantages over other biopolymers [11].

This review aims to explore the work carried out in the last five years related to the most diverse polyhydroxyalkanoates in the biomedical sector, to better understand the methodologies used and to emphasize the versatility of the biopolymer. A better understanding of how to use PHAs is desired to enhance economic viability and facilitate the optimization and cost reduction of the production process.

1.1. *Polyhydroxyalkanoates*

Polyhydroxyalkanoates (PHAs) are biodegradable biopolymers metabolically produced in nature in small quantities by various prokaryote organisms and some eukaryotes [12] as a form of storage component. They are stored in bacteria in the form of intracellular granules (Figure 1) [13]. Their diverse monomeric structure guarantees these polymers' variable physical characteristics so that they serve different applications [14]. PHA is characterized into three groups according to the number of carbon atoms that make up their monomers: short-chain PHA, with 3–5 carbons in their structure; medium-chain PHA, with 6–14 carbons in their structure; and long-chain PHA, possessing a number of carbons greater than 14 [15].

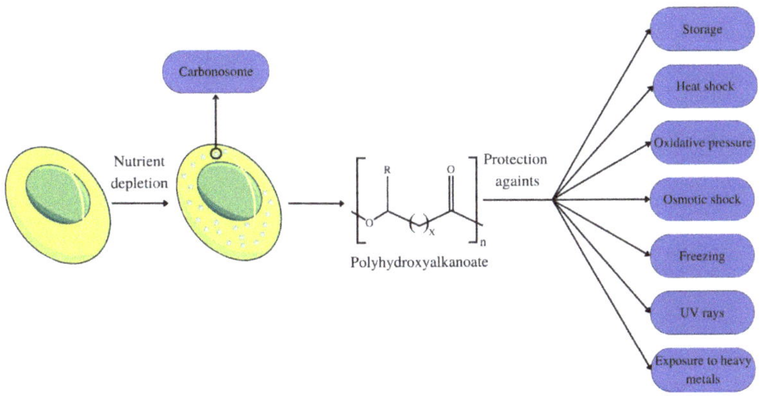

Figure 1. PHA production and cellular functions.

PHAs are a class of polyesters characterized by ester bonds linking the carboxyl group of one monomer with the hydroxyl group of another. The majority of PHAs are produced in the R form [16]. Polyhydroxyalkanoate (PHA) molecules typically consist of a range of 600 to 35,000 monomer units. Approximately 150 distinct monomers have been recognized thus far, and this count is expected to increase as new types of PHAs are developed through chemical or physical modifications of natural PHA or via the utilization of genetically modified organisms [17]. The critical attributes exhibited by PHAs are the piezoelectric effect, nontoxicity, thermoplasticity, gas impermeability, hydrophobicity, enantiomeric purity, and a high degree of polymerization [18]. The multiple mechanical characteristics of polyhydroxyalkanoates (PHAs) arise from the range of monomers that may be included in their polymer structure, resulting in variations in flexibility, strength, and crystallinity. These properties exhibit variation based on factors such as the organism responsible for production, the number of carbons in a monomeric unit, and the specific growth circumstances [19].

Within microbial cells, PHA is present in the form of granules, also referred to as carbonosomes (Figure 1). These granules are complex molecules organized into layers, which are composed of different polypeptides, as well as proteins. Among the proteins present in the layers, phasins stand out, which can fill up to $\frac{3}{4}$ of the granule space, exercising a structural function by maintaining the spherical shape of the granules, in addition to participating in cell metabolism. The size of these granules varies between 100–500 nm, depending on the organism and the composition of the phasins that constitute the granule [20].

This versatility is essential in synthesizing and depolymerizing PHA molecules, promoting polymer accumulation, and determining granules' size, quantity, and distribution. Other examples of phasins include PhaP1 from *Cupriavidus necator*, which can increase the activity of PhaC1 class II and PhaC2 from *Pseudomonas aeruginosa* in vitro, HAF from *Pseudomonas putida*, which regulates PHA synthesis, and ApdA from *Rhodospirillum rubrum* which plays a role in activating the depolymerization of Poly-(3-hydroxybutyrate) [20,21].

Short-chain PHAs (SCL-PHAs) exhibit high crystallinity, hardness, and fragility, but with slow crystallization rates and low impact resistance, along with rapid aging, making their use difficult. Poly-(3-hydroxybutyrate) (PHB), a homopolymer with four carbon subunits of 3-hydroxybutyrate (3HB), is the most frequent SCL-PHA, but other representatives of this group are poly(3-hydroxyoctanoate) (PHO), poly(4-hydroxybutyrate) (P4HB), poly(3-hydroxybutyrate-co-3-hydroxyvalerate) (PHBV), poly(3-hydroxybutyrate-co-4-hydroxybutyrate) (PHB-4HB), and poly(3-hydroxybutyrate-co-3-hydroxyhexanoate) copolymers (PHBHHx), as shown in Figure 2 [22].

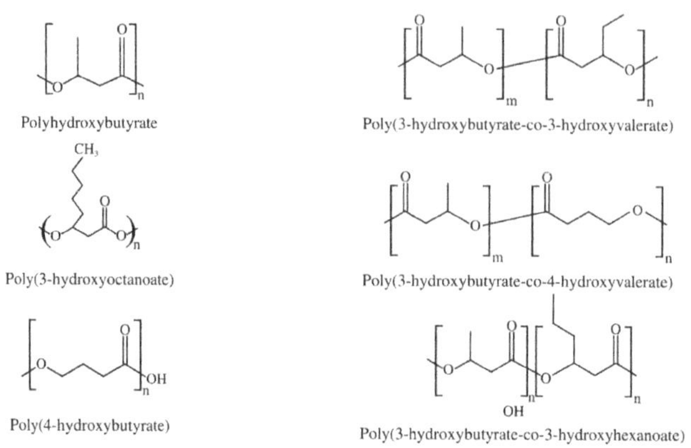

Figure 2. Examples of most common SCL-PHAs.

Medium-chain-length PHA (MCL-PHA) polyhydroxyalkanoates exhibit distinct characteristics in comparison to SCL-PHA. This group is inherently elastomeric, more flexible than SCL-PHA, and has a lower melting point and is less crystalline. The primary producers of MCL-PHA are *Pseudomonas* sp., with β-oxidation as the production pathway in bacteria [20]. Additionally, the MCL-PHA has a lower glass transition temperature, low tensile strength, and excellent elongation until rupture [22].

PHAs are recurrently associated with their primary storage function. Nonetheless, a number of studies indicate that the functions of these polymers within the cell are more complex, including protection against stress factors such as thermal shock, oxidative pressure, osmotic shock, low temperatures, and freezing, as well as protection against the effects of UV rays and exposure to heavy metals [23]. During the growth phase of extremophile microorganisms, PHA accumulates to shield organelles from environmental stress and to support cellular functions [24].

1.2. PHA Synthesis

Some options for production improvement and the synthesis of new PHAs are the control of synthesis through the manipulation of regulatory elements such as enzymes and proteins and the engineering of biosynthesis pathways [25]. Consequently, it is necessary to comprehend these mechanisms in order to visualize the process in its entirety, as well as to use this knowledge to increase the yield and purity of the compounds produced.

Some of the central metabolic pathways, such as the Krebs cycle, glycolytic and pentose–phosphate pathways, and degradation and biosynthesis pathways of fatty acids and amino acids, are directly or indirectly linked to the biosynthetic pathways of PHAs. The 14 pathways of PHA synthesis described so far demonstrate the importance of acetyl-CoA as a precursor to the biosynthesis of several SCL- and MCL-PHAs, such as PHB synthesis (Figure 3) which consists of three reactions catalyzed by acetyl-CoA acetyltransferase (β-ketothiolase; PHAA) catalyzing the condensation of two acetyl-CoA molecules into acetoacetyl-CoA, acetoacetyl-CoA reductase (PhaB) converting acetoacetyl-CoA into (*R*)-3-hydroxybutyl-CoA[(*R*)-3-HB-CoA], and PHA synthase (PhaC) that polymerizes (*R*)-3-HB-CoA in a PHB chain [21]. Under normal growth conditions, β-ketothiolase would be inhibited by the free coenzyme-A from the Krebs cycle. However, during the scarcity of noncarbonic nutrients in the substrate, the entry of acetyl-CoA becomes restricted, and the surplus is directed to the synthesis of PHB [26].

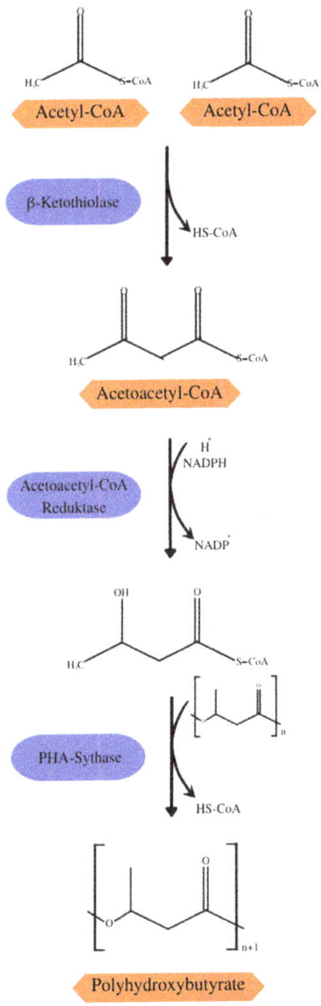

Figure 3. Synthesis of PHB with acetyl-CoA from the Krebs cycle as the precursor.

The synthesis of SCL-PHA occurs in Gram-positive, Gram-negative, and some haloarchaea species, with the majority of SCL-PHA producers able to accumulate PHB, and some species capable of producing the aforementioned copolymers. As stated earlier, acetyl-CoA plays a crucial part in the synthesis of SCL-PHA due to its function as a precursor for the 3HB monomer, which forms the PHB. Succinyl-CoA serves as the precursor for the synthesis of 4-hydroxybutyrate (4HB), an essential monomer. Conversely, 3-hydroxyvalerate (3HV) is derived from propionyl-CoA, acetyl-CoA, and, in certain instances, intermediates of β-oxidation of valerate or other fatty acids. The previously published mechanism for the production of 3HB monomer provides a broad understanding of the biosynthesis of SCL-PHA. Nevertheless, it is important to note that certain modifications may arise, particularly in situations when, in the first stage, condensation takes place between an acetyl-CoA and a propionyl-CoA molecule. This process results in the formation of 3-ketovaleryl-CoA subsequent to the reduction phase facilitated by 3-ketoacyl-CoA reductase, leading to the production of a molecule known as (R)-3-HV-CoA. The conversion of succinyl-CoA to succinate semialdehyde is facilitated by the activity of succinate semialdehyde dehydrogenase, which acts as a catalyst for this reaction. Subsequently, the molecule succinate semialdehyde undergoes oxidation through the action of 4-hydroxybutyrate dehydrogenase, leading to the formation of 4HB. In both cases, the monomers will ultimately be polymerized by the PHA synthases into PHB or one of its copolymers, PHBV, poly(3-hydroxybutyrate-co-4-hydroxybutyrate), and poly(3-hydroxybutyrate-co-3-hydroxyvalerate-co-4-hydroxybutyrate) (PHBV4HB) [25].

MCL-PHA can be synthesized in two ways: β-oxidation and new synthesis of fatty acids. These two pathways work interactively, with β-oxidation being the main route for the synthesis of MCL-PHA from fatty acids through their degradation, while the fatty acid synthesis pathway functions as the main metabolic pathway to obtain MCL-PHA from carbohydrates. The synthesis through related carbon sources (fatty acids) is more accessible than from unrelated carbon sources (carbohydrates), and the addition of fatty acids to the substrate with a structure similar to the monomer of the PHA can improve the synthesis of MCL-PHA [27].

The process of β-oxidation synthesis involves the enzymatic activity of (R)-specific-enoyl-CoA hydratase, acyl-CoA oxidase, and 3-ketoacyl-CoA reductase, which are then polymerized by PHA synthase to produce PHA [28]. One instance involves the operational mechanism of this specific pathway in *Aeromonas* spp., in which acyl-CoA derived from fatty acids undergoes degradation, resulting in the formation of enoyl-CoA intermediates. These intermediates can then be transformed into (R)-3-hydroxyoxycil-CoA through the activity of (R)-specific-enoyl-CoA hydrase. Subsequently, this (R)-3-hydroxyoxycil-CoA can participate in the elongation of the polyester chain by means of PHA synthase [29]. In the pathway of de novo fatty acid synthesis, an intermediate within the fatty acid synthesis pathway undergoes conversion by the transacylase PhaGPp into (R)-3-hydroxydecanoyl-CoA, which exists in its carrier protein form. This conversion process facilitates the utilization of (R)-3-hydroxydecanoyl-CoA as a substrate for the production of PHA [30]. The use of inexpensive substrates, such as glycerol residues, simple sugars, and short-chain volatile fatty acids derived from fermented residues, can be facilitated by this particular approach [31].

A fundamental enzyme in the synthesis pathways is PHA synthase or PhaC en-zymes, which is responsible for the polymerization of hydroxyacyl monomers to obtain PHA polymers. Several classes of PhaC with varying substrate specificities have been isolated from various microorganisms. PhaC of *Ralstonia eutrpha* (PhaCRe) is specific for SCL-PHA. The PHA synthases of *Pseudomonas* sp. are specific for MCL-PHA monomers. An additional example is the PHA synthase of *Aeromonas caviae*, which has a high affinity for units of 3HB and 3HHx and can therefore synthesize the copolymer P(3HB-co-3HHx) [32,33].

PHA is stored as an energy reserve when nutrients such as nitrogen, oxygen, and phosphorus are scarce or when the pH changes. The use of these energy stores for bacterial development occurs when the nutrition supply is restricted, as seen in Figure 1 [34]. This effect is demonstrated in mixed microbial cultures, with the limited nitrogen and phosphorus action resulting in the efficient production of PHA [35], just as the limitation of phosphorus and carbon was advantageous for the production of mcl-PHA [33]. The limitation of sodium nitrate and supplementation with sodium bicarbonate and sodium acetate optimized PHB production in *Stigeoclonium* sp. B23 microalgae [36].

2. PHA in Biotechnology

PHB is recognized for its applications in packaging and medical use (Table 1) due to its thermoplastic properties, but its fragility and high melting temperature limit its industrial use. In the present context, poly-4-hydroxybutyrate (P4HB) has demonstrated considerable potential as a polymer due to its inherent structural strength and high elongation capacity [20].

This class of biopolymers generates considerable interest because they are biomaterials similar to latex or rubber, opening possibilities for applications in several areas including engineering, agriculture, food technology, pharmaceuticals, and, particularly, medicine and pharmacy. For instance, scaffolds have been manufactured with these materials successfully, due to their biocompatibility, which is achieved through the degradation of components that are naturally present in the body, like the polyhydroxybutyrate, that results in R-3-hydroxybutyric acid, a common constituent of blood. The biodegradability of these materials makes them suitable for the development of drug carriers. This is attributed to their low crystallinity and melting point, as well as their degradation through surface erosion. Additionally, they hold promise for applications such as electrocardiograph electrodes and adhesives for dressings [37,38], as illustrated in Figure 4.

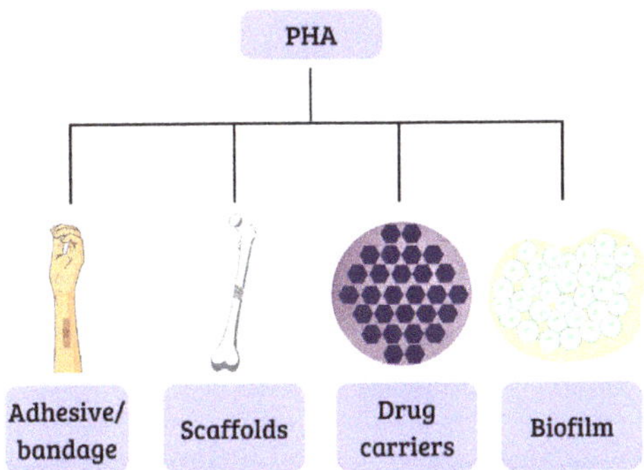

Figure 4. Biomedical applications of PHA.

The commercialization of PHA has been made since the 1980s; however, it has encountered several limitations and conditions that hinder commercial success. These difficulties are mainly associated with instability in thermomechanical properties and high production costs. The elevated costs can be attributed to the significant energy requirements caused by complex sterilization procedures and intensive aeration. Furthermore, the costly downstream processes, slow growth of microorganisms, and slow substrate (carbon) to product (PHA) rate of conversion contribute to overall high costs. Additionally, the discontinuous nature of these processes further adds to the cost of operation. In order to address these challenges, several studies have put forth various alternatives aimed at mitigating manufacturing expenses [39]. These alternatives include utilizing industrial residues as substrates for raw materials [40], chemically modifying PHA, employing genetic engineering techniques to enhance cell growth, manipulating cell morphology, and enhancing characteristics such as altering synthesis pathways [41].

Implementing a circular economy throughout the microbial PHA manufacturing process represents a viable approach to mitigating expenses associated with bioplastic production. By integrating production with wastewater bioremediation, heavy metal and other waste removal, and environmental problem resolution, this approach enables the optimization of output while simultaneously achieving the desired product. An additional element in this approach that aids in the generation of PHA is the incorporation of this substance with other desirable products, such as pigments. Furthermore, residual biomass from cyanobacteria and microalgae used in animal feed production is utilized, and new microbial cultures are established, as illustrated in Figure 5 [42].

2.1. Scaffolds in Fabric Engineering

Tissue engineering offers solutions for repairing or replacing bones, nerves, cartilage, skin, heart valves, bladder, and blood vessels [43]. The multidisciplinary approach involving medicine, cell biology, and biomaterial engineering is part of the principle of tissue engineering. The scaffolds have a crucial role in providing three-dimensional structural support for cellular connections, facilitating the formation of the extracellular matrix (ECM), and serving as a carrier for growth factors and cytokines at the site of repair. The careful choice of a suitable biomaterial is of utmost importance in the development of a scaffold [44].

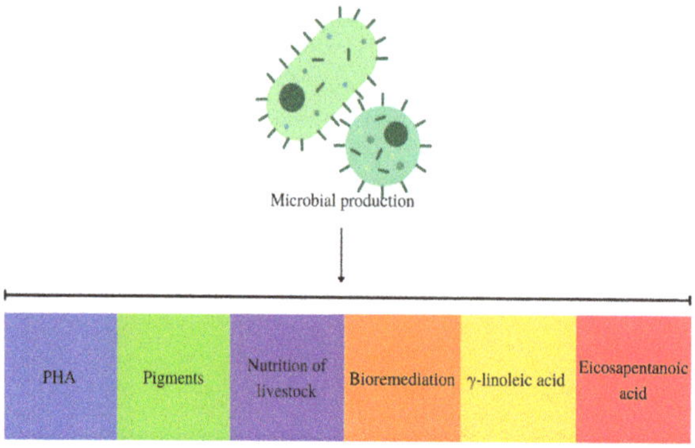

Figure 5. Obtaining PHA with multiproducts in the circular economy.

Scaffolds have the potential for treating bone defects by emulating the role of the extracellular matrix. They offer a three-dimensional environment that facilitates adhesion, proliferation, and differentiation, thus creating favorable physical circumstances for regeneration. Being biodegradable, biocompatible, osteoconductive, osteoinductive, and bioactive are conditions for the ideal scaffold [45].

The field of cardiovascular tissue engineering has made significant contributions to the management of heart diseases. It has played a crucial role in the development of transplant materials and the creation of biological models for preclinical pharmacological testing [46]. Additionally, it has proven to be essential in the structural repair of tissues with limited self-regeneration capabilities, such as tendons, ligaments, cartilage, and meniscus. Furthermore, tissue engineering applications may potentially involve corneas, parts of the urinary system, and bronchi as potential targets [47].

Polymers are one of the most crucial materials in the production of scaffolds due to properties such as hydrophobicity and biodegradability. PHA is used, both alone and combined, for biomedical applications of various types, such as sutures, patches, and dressings, as shown in the PHA produced by *Bacillus cereus* MCCB 281 [48,49].

The suitability of biopolymers P(3HB) and P(3HB-co-3HV) for the production of biocompatible fibrous scaffolds with a favorable shape for cell adherence was demonstrated using the electrospinning technique [50]. The viability of fibrous PHA scaffolds and films in promoting human cell proliferation has been demonstrated by Uribe, Acosta, and Díaz [51]. In addition, it has been demonstrated that these scaffolds and films do not induce toxicity in HEK 293 cells or embryonic renal cells. Furthermore, they have demonstrated resistance to autoclave and ultraviolet (UV) radiation, both of which are frequently utilized techniques for sterilization. The application of fibrous scaffolds has demonstrated improved effectiveness in stimulating cellular proliferation. After 18 days under cellular conditions, these scaffolds degraded, demonstrating their applicability in tissue engineering, where the scaffold degrades while the tissue regenerates.

The combination of PHA with other polymeric materials makes it possible to increase the efficiency of the physical–chemical properties of both polymers. The combination of PHA with additional polymeric substances enables the enhancement of the physical–chemical characteristics exhibited by both polymers. Bacterial cellulose (BC) is a applicable material for use in regenerative medicine due to its exceptional biocompatibility and biodegradability. When combined with PHB, BC demonstrated enhanced effects on osteoblast differentiation and induction of new bone formation, all while exhibiting high biocompatibility and no discernible signs of toxicity [52].

Another example of a combination of polymers is the immobilized PHB with gelatin (gel) and metronidazole (MTZ), a drug with antibacterial, antiprotozoal, and antiamebic capacity, which proved to be efficient in the production of scaffolds with the ability to increase the development of fibroblasts, while also inhibiting the growth of *Escherichia coli* and *Staphylococcus aureus*. The addition of the MTZ allowed increased compatibility with the L-929 fibroblast lineage, also providing good thermal and mechanical properties. The PHB/gel polymer in the proportion of 7:3 (m/m) added to 10% of MTZ showed the best results for skin regeneration, promoting the formation of myofibroblasts, regulating the inflammation of the wound and accelerating its regeneration [53].

The combination of PHB with gelatin and Fe_3O_4 is also viable and has been demonstrated as a feasible approach to produce magnetically-active hybrid scaffolds. These scaffolds have potential applications as dressings, in tissue engineering, and as drug carriers that can be regulated by an external magnetic field. Additionally, they are reported to be nontoxic to cells [54].

Scaffolds produced by the combination of poly(3-hydroxyoctanoate-co-3-hydroxyhexanoate) P(3HO-co-3HHX), an MCL-PHA, with hydroxyapatite, a compound with a structure similar to human bone used in bone implants, proved promising for bone tissue engineering, with excellent osteoconductivity and biocompatibility [55]. The combination of PHA with silver nanoparticles encrusted with graphene (GAg) demonstrated positive results for scaffolds intended to treat chronic injuries and sanitize applications due to their bactericidal potential [56]. Concurrently, positive results were observed with the growth and differentiation of neuronal tissues when PHA was combined with bioactive glass [57].

Three-dimensional printing technology is a viable alternative for producing PHA-based scaffolds, with positive results for potential drug carriers due to the incorporation of the osteogenic growth peptide, which accelerated cell differentiation [58].

Polymeric microspheres are a class of scaffold that offers ample area for cell-tying, enhanced by the 3D porous structure of these scaffolds, and can be kept in suspension. The interconnected porous structure also makes it easier to transport nutrients and metabolites, helping to proliferate and differentiate cells. Poly(3-hydroxylobutyrate-co-3-hydroxyhexanoate-co-3-hydroxyvalerate) (PHBVHHx) porous microspheres were made, and it was noted not only the characteristics mentioned but also a reduction in cell death and apoptosis due to their porous structure, making these microspheres promising for scaffolds and also as drug carriers [59]. The application of PHA in drug carriers will be further discussed next.

2.2. Drug Carriers

The first pharmaceutical–therapeutic application of PHA was probably in drug delivery systems (Figure 4). Studies have shown that the polymer is compatible with mouse fibroblasts, exhibiting no adverse effects on cellular growth and metabolism. Moreover, its ability to be recognized as a degradation product suggests its potential for natural elimination from the human body, hence positioning it as an advantageous option compared to other polymers [60,61]. Impallomeni et al. [62] and Mohandas et al. [49] showed promising results for the development of carriers from MCL-PHA produced by *Pseudomonas aeruginosa* ATCC 27853 and PHA copolymer produced by *Bacillus cereus* MCCB 281 using glycerol as a carbon source, respectively.

Table 1. PHA applications.

Polymer	Producing Body	Application	Author Year	Refs.
PHB	*Stigeoclonium* sp. B3	Drug carrier/biomaterial	Mourão et al., 2020	[36]
PHB	*Stigeoclonium* sp. B3	Drug carrier/biomaterial	Mourão et al., 2021	[40]

Table 1. *Cont.*

Polymer	Producing Body	Application	Author Year	Refs.
PHV	*Bacillus cereus* MCCB 281	Drug carrier	Mohandas et al., 2018	[49]
P(3HB)/P(3HB-co-3HV)/MCL-PHA	*Cupriavidus necator* DSM 428/*Pseudomonas chlororaphis* DSM 19603	Scaffolds	Esmail et al., 2021	[50]
PHB	*Azotobacter vinelandii*	Tissue engineering	Romo-Uribe et al., 2017	[51]
PHB	Unspecified	Scaffolds	Codreanu et al., 2020	[52]
PHB	Unspecified	Scaffolds	El-Shanshory et al., 2022	[53]
PHB	Unspecified	Scaffolds	Pryadko et al., 2022	[54]
P(3HO-co-3HHX)	Unspecified	Scaffolds	Ansari et al., 2017	[55]
PHA	Unspecified	Scaffolds	Lizzarraga-Valderrama et al., 2020	[57]
PHB	Unspecified (commercial)	Tissue engineering	Saska et al., 2018	[58]
PHBVHHx	Unspecified	Scaffolds	Wei et al., 2018	[59]
PHA	*Pseudomonas aeruginosa* ATCC 27853	Tissue/carrier engineering	Impallomeni et al., 2018	[62]
PHB	*Bacillus Cereus* VIT-SSR1	Drug carrier	Evangeline et al., 2019	[63]
PHB	Unspecified (commercial)	Drug carrier	Bini et al., 2017	[64]
PHB	*Pseudomonas aeruginosa* SU-1	Drug carrier	Senthilkumar et al., 2017	[65]
PHBV	Unspecified	Drug carrier	Vardhan et al., 2017	[66]
PHA	Unspecified	Drug carrier	Jiang et al., 2019	[67]
PHB	Unspecified	Drug carrier	Chen et al., 2021	[68]
PHA	Unspecified	Drug carrier	De Freitas E Castro et al., 2021	[69]
PHA	Unspecified	Drug carrier	Huerta-Angeles et al., 2017	[70]
PHA	*Bacillus subtilis* NCDC0671	Drug carrier	Umesh et al., 2018	[71]
P(3HO-co-3HD-co-3HDD)	*Pseudomonas mendocin* CH50	Drug carrier	Owji et al., 2021	[72]
poly(R-3-hydroxybutyrate-co-1,4-butylene adipate)	Unspecified	Drug carrier	Musumeci et al., 2019	[73]
PHBHHx	Unspecified	Drug carrier	Fan et al., 2018	[74]
P3HB	Unspecified	Drug carrier	Shershneva et al., 2018	[75]
PHA	Unspecified	Drug carrier	Canãdas et al., 2021	[76]
PHA	Unspecified	Drug carrier	Pavic et al., 2022	[77]
PHB	Commercial of bacterial origin	Biomedical use	Salama et al., 2018	[78]
P(3HV-co-3HB)	*Halomonas* sp.	Biomaterial	El-Malek et al., 2021	[79]
PHO	*Pseudomonas putida* KT2440	Wound treatment	Balcucho et al., 2023	[80]
P(3HB)/P(3HB-*co*-3HD)	*Pseudomonas mendocina* CH50 and *Bacillus subtilis* OK2	Wound treatment	Kalaoglu-Altan et al., 2021	[81]
PHA	*Pseudomonas chlororaphis* subsp. *aurantiaca*	Adhesive	Pereira et al., 2018	[82]
PHA	*Pseudomonas putida* KT2440	Adhesive and biofilm	Urbina et al., 2018	[83]
P(3HB)/P(3HB/3HV)	Unspecified	Herbicide carrier	Vijayamma et al., 2021	[84]

Mourão et al. [40] showed the potential for adaptive PHB production for drug carriers by the Amazonian microalgae *Stigeoclonium* sp. B23 using hydrolyzed cassava bark as substrate. The tests performed to characterize the PHB showed viability for biomedical application, having thermal, morphological, physicochemical, and biological characteristics similar to PHB and its copolymers produced by bacteria and cyanobacteria. The potential of this microalgae has been demonstrated by Mourão et al. [36].

The combination of PHB with chitosan resulted in a rigid matrix film, making drug carriers effective in releasing curcumin, a curcuminoid with anti-inflammatory, healing, and antioxidant properties widely used in the formulation of drugs [63]. Curcumin was also used in the work carried out by R.A. Bini et al. [64], where PHB nanoparticles were combined with a gelatin matrix to form a nanocomposite, being tested for the loading of curcumin, a hydrophobic drug, and naproxen sodium, a nonsteroidal and hydrophilic anti-inflammatory drug, with curcumin being carried by the PHB. In contrast, naproxen sodium was solubilized in the gelatinous matrix. This combination of the double release of drugs was favorable and has shown that a cheap, sustainable, and straightforward approach is practical for hydrophobic and hydrophilic drugs.

Drug release was reported using PHA produced by *Pseudomonas aeruginosa*, being loaded with curcumin in spherical nanoparticles with sizes between 300 and 500 nm, releasing the drug constantly for more than 5 h [65]. H. Vardhan et al. [66] demonstrated results that point to the nanocomposite composed of PHBV as an alternative to improve breast cancer treatment through the loading of the hydrophobic drug docetaxel. The drug-pumping system applied to fight cancer using docetaxel has also been shown to be effective using a thermogel based on poly[(R)-3-hydroxybutyrate-(R)-3-hydroxyhexanoate] copolymerized with poly(ethylene glycol) and polypropylene glycol. The application of this drug, combined with the carrier, proved to be effective in treating melanomas without causing damage to healthy tissues, proving to be a promising carrier for anticancer drugs [67].

The incorporation of PHA in conjunction with other polymers offers a promising alternative to drug-pumping systems. The junction between PHA and polyvinyl alcohol (PVA) is an example where the specificities of both polymers complement each other, reducing the degradation rate of PVA. Additionally, the PHA membrane supports water-soluble PVA fibers. The efficacy of this combination was demonstrated in the transportation of levofloxacin, resulting in highly efficient bactericidal results against Gram-negative strains of *E. coli* and Gram-positive strains of *S. aureus* [68]. Furthermore, the incorporation of compounds can also serve to modulate the release of the drug by the carrier, as well as the incorporation of superparamagnetic nanoparticles. A carrier with a thermal responsive property regulated by magnetic oscillation was demonstrated to be a promising candidate for veterinary application when PHA copolymers were combined with nanomagnetite to achieve controlled progesterone release [69]. G.H. Angeles et al. [70] also described an effective combination in encapsulating hydrophobic drugs, which is the copolymer formed by combining hyaluronic acid and PHA.

Bacillus subtilis was able to produce PHA from a culture medium containing orange peel as a way to reduce the cost of polymer production. This PHA produced was proved effective in administering levofloxacin, with the release of the drug reaching up to 99.12% [71]. The bacterium *Pseudomonas mendocina* CH50 showed positive results for the production of poly(3-hydroxyoctanoate-co-3-hydroxydecananoate-co-3-hydroxydodecanoate) or P(3HO-co-3HD-co-3HDD). This polymer was used as a film and submitted to a coating process based on the chemical used by mussels to adhere to wet surfaces, thus allowing it to overcome the main barrier for the targeted application of drugs in the oral mucosa. The polydopamine lining showed positive results, increasing film adhesion in in vitro cell proliferation and in in vivo neovascularization [72]. T. Musumeci et al. [73] showed that poly(3-hydroxybutirato-co-butylene adipate) can produce drug nanocarriers.

The formation of PHA nanoparticles can also be performed using the interaction of the polymer with the phasins. PHBHHx was combined with PhaP phasin modified with a targeting peptide that recognizes an epidermal growth factor receptor to develop a targeting system for tumors, proving to be suitable as a drug carrier [74]. The spray-drying technique was used with efficiency in the production of drug carriers. Microparticles of P3HB and P3HB were combined with polyethylene glycol(PEG) for the paclitaxel(PTX) and 5-Fluorouacil(5-FU) drugs carriage. The method showed improvement in yield and greater incorporation of drugs into microparticles, and extended release of drugs led to inhibition of tumor cell growth [75]. PHA was also utilized as an inhaled drug carrier for the lungs; however, additional research is required to determine the viability of these carriers [76].

As drug carriers, the PHA microspheres that were previously mentioned as scaffolds also demonstrated promising results. A. Pavic et al. [77] showed the feasibility of using MCL-PHA in microspheres to transport antifungal drugs to combat candidiasis. These microspheres loaded with polyene demonstrated excellent antifungal activity and decreased the toxic effects of these drugs on zebrafish embryos, being a viable alternative to combat candidiasis; however, further investigation is required.

2.3. Other Applications in Medicine

Polyhydroxyalkanoates have other biomedical applications beyond the production of scaffolds and drug carriers. Salama, Aziz, and Saad [78] demonstrated a process of incorporation of silver nanoparticles into a PHB and chitosan copolymer. Silver nanoparticles' antimicrobial activity led to positive results regarding the Gram-negative bacteria *Escherichia coli* and *Salmonella typhi*, Gram-positive *Streptococcus pneumoniae*, and the fungus *Aspergillus fumigatus*. In addition, chitosan is essential for this outcome, which opens possibilities for the use of this compound in various sectors of the biomedical area. Poly-based nanoparticles (3 HV-co-3HB), produced by *Halomonas pacifica* ASL10 and *Halomonas salifodiane* ASL11 using substrates derived from algae hydrolysate, presented antibacterial characteristics that make them suitable for use in the pharmaceutical industry and food packaging [79].

Another application for PHAs is the treatment of wounds and injuries. Some works report the efficiency of the PHAs, especially blended with other materials. Poly(3-hydroxyoctanoate) showed positive results when incorporated with Ag nanoparticles for antimicrobial wound dressing on the treatment of skin and soft tissue infections, being effective against methicillin-resistant *Staphylococcus aureus*, in addition to nontoxic biodegradation, reducing environmental impact and being suitable for long-term applications [80].

An additional function of PHAs is the treatment is the blend of the SCL-PHA P(3HB) and the MCL-PHA P(3HB-*co*-3HD), combining the properties of both polymers, making it suitable for the confection of wound dressing. These nanofibers were incorporated with AgNPs via a dip-coating method, contributing to cellular metabolism in the wound-healing process [81].

J.R. Pereira et al. [82] describe the production of MCL-PHA from crude glycerol obtained from biodiesel production. The resulting polymer exhibits reduced hydrophobicity and crystallinity, which makes it suitable for the production of elastic and flexible films (Figure 4). Consequently, this material finds potential applications in the biomedical field, particularly in the development of injury treatment materials. This aligns with the findings of Urbina et al. [83], who also investigated MCL-PHA properties. Furthermore, this material demonstrated use in the biomedical sector and the manufacturing of adhesives (Figure 4). Additionally, its elastic, adhesive, and soft properties at room temperature made it suitable for implementation in the field of agriculture.

An application similar to drug delivery was reported by R. Vijayamma et al. [84], where a herbicide delivery system composed of P(3HB) and P(3HB/3HV) was tested. Metribuzin, a herbicide derived from 1,2,4-triazine, demonstrated the best release results with an index of 44–48%, and in the tests carried out during the growth of the plant, *Elsholtzia ciliata* microparticles loaded with metribuzin and tribenuron-methyl demonstrated the best result.

3. Conclusions

Polyhydroxyalkanoates are highly versatile, having applications in several areas, such as in the biomedical sector, and are widely used in scaffolds in tissue engineering, enabling aid in the regeneration of various tissues and adding antimicrobial and anti-inflammatory properties when combined with other materials. Their use in drug carriers is also one of the most common applications due to biocompatibility, due to their degradation resulting in products natural to the body, and biodegradability, which enables the direct application of drugs in the desired tissues, increasing the applicability of the compound and reducing the inherent toxicity of some drugs, and being more attractive than other polymers due to their capacity to be recognized by the human body as degradation product, which is naturally removed from the body. Thus, these polyesters are shown as a viable option for replacing petroleum-derived plastics, reducing the environmental impact caused by them. The economic barrier caused by the high production costs of these polymers is still an obstacle to be faced, as shown in several studies, thus being necessary for further studies to reduce the cost of production and become more commercially attractive by methods like biorefinery, which make it possible to obtain multiple products combined in PHA production, associating the production with bioremediation of residual waters and improving the use of microorganisms that require lower costs in their cultivations, such as microalgae that can be cultivated in residual water and sludge and require fewer nutrients due to their autotrophic behavior.

Author Contributions: Conceptualization, M.S.d.F.D.; methodology, M.S.d.F.D., M.M.M. and A.V.S.; validation, A.V.S., M.M.M. and L.P.X.; formal analysis, M.S.d.F.D., A.V.S., M.M.M. and L.P.X.; investigation, M.S.d.F.D.; resources, A.V.S. and L.P.X.; data curation, M.S.d.F.D.; writing—original draft preparation, M.S.d.F.D.; writing—review and editing, M.S.d.F.D., A.V.S., M.M.M. and L.P.X.; visualization, A.V.S. and L.P.X.; supervision, A.V.S. and L.P.X.; project administration, A.V.S. and L.P.X.; funding acquisition, A.V.S. and L.P.X. All authors have read and agreed to the published version of the manuscript.

Funding: This research was funded by the Coordenação de Aperfeiçoamento de Pessoal de Nível Superior–Brasil (CAPES)-Finance code 001.

Institutional Review Board Statement: Not applicable.

Data Availability Statement: The data presented in this study are available on request from the corresponding author.

Acknowledgments: The authors would like to acknowledge the financial support supplied by Coordenação de Aperfeiçoamento de Pessoal de Nível Superior—Brasil (CAPES), by Pró-Reitoria de Pesquisa e Pós-Graduação da Universidade Federal do Pará (PROPESP/UFPA) and by the National Council for Scientific and Technological Development (CNPq).

Conflicts of Interest: The authors declare no conflict of interest.

References

1. Johansen, M.R.; Christensen, T.B.; Ramos, T.M.; Syberg, K. A Review of the Plastic Value Chain from a Circular Economy Perspective. *J. Environ. Manag.* **2022**, *302*, 113975. [CrossRef] [PubMed]
2. Vimal, K.E.K.; Mathiyazhagan, K.; Agarwal, V.; Luthra, S.; Kirupanandan, S. Analysis of Barriers That Impede the Elimination of Single-Use Plastic in Developing Economy Context. *J. Clean. Prod.* **2020**, *272*, 122629. [CrossRef]
3. Ganesh, K.A.; Anjana, K.; Hinduja, M.; Sujitha, K.; Dharani, G. Review on Plastic Wastes in Marine Environment—Biodegradation and Biotechnological Solutions. *Mar. Pollut. Bull.* **2020**, *150*, 110733. [CrossRef]
4. Bucci, K.; Tulio, M.; Rochman, C.M. What is Known and Unknown about the Effects of Plastic Pollution: A Meta-analysis and Systematic Review. *Ecol. Appl.* **2020**, *30*, e02044. [CrossRef]
5. Okoffo, E.D.; Donner, E.; McGrath, S.P.; Tscharke, B.J.; O'Brien, J.W.; O'Brien, S.; Ribeiro, F.; Burrows, S.D.; Toapanta, T.; Rauert, C.; et al. Plastics in Biosolids from 1950 to 2016: A Function of Global Plastic Production and Consumption. *Water Res.* **2021**, *201*, 117367. [CrossRef]
6. Ncube, L.K.; Ude, A.U.; Ogunmuyiwa, E.N.; Zulkifli, R.; Beas, I.N. An Overview of Plastic Waste Generation and Management in Food Packaging Industries. *Recycling* **2021**, *6*, 12. [CrossRef]
7. Patrício Silva, A.L.; Prata, J.C.; Walker, T.R.; Duarte, A.C.; Ouyang, W.; Barcelò, D.; Rocha-Santos, T. Increased Plastic Pollution Due to COVID-19 Pandemic: Challenges and Recommendations. *Chem. Eng. J.* **2021**, *405*, 126683. [CrossRef]
8. Drzyzga, O.; Prieto, A. Plastic Waste Management, a Matter for the 'Community'. *Microb. Biotechnol.* **2019**, *12*, 66–68. [CrossRef]
9. Objetivos de Desenvolvimento Sustentável | As Nações Unidas No Brasil. Available online: https://brasil.un.org/pt-br/sdgs (accessed on 23 August 2023).
10. Udayakumar, G.P.; Muthusamy, S.; Selvaganesh, B.; Sivarajasekar, N.; Rambabu, K.; Sivamani, S.; Sivakumar, N.; Maran, J.P.; Hosseini-Bandegharaei, A. Ecofriendly Biopolymers and Composites: Preparation and Their Applications in Water-Treatment. *Biotechnol. Adv.* **2021**, *52*, 107815. [CrossRef]

11. Sharma, V.; Sehgal, R.; Gupta, R. Polyhydroxyalkanoate (PHA): Properties and Modifications. *Polymer* **2021**, *212*, 123161. [CrossRef]
12. Kalia, V.C.; Singh Patel, S.K.; Shanmugam, R.; Lee, J.-K. Polyhydroxyalkanoates: Trends and Advances toward Biotechnological Applications. *Bioresour. Technol.* **2021**, *326*, 124737. [CrossRef]
13. Sabapathy, P.C.; Devaraj, S.; Meixner, K.; Anburajan, P.; Kathirvel, P.; Ravikumar, Y.; Zabed, H.M.; Qi, X. Recent Developments in Polyhydroxyalkanoates (PHAs) Production—A Review. *Bioresour. Technol.* **2020**, *306*, 123132. [CrossRef] [PubMed]
14. Taguchi, S.; Matsumoto, K. Evolution of Polyhydroxyalkanoate Synthesizing Systems toward a Sustainable Plastic Industry. *Polym. J.* **2021**, *53*, 67–79. [CrossRef]
15. Ishak, K.A.; Velayutham, T.S.; Annuar, M.S.M.; Sirajudeen, A.A.O. Structure-Property Interpretation of Biological Polyhydroxyalkanoates with Different Monomeric Composition: Dielectric Spectroscopy Investigation. *Int. J. Biol. Macromol.* **2021**, *169*, 311–320. [CrossRef] [PubMed]
16. Raza, Z.A.; Riaz, S.; Banat, I.M. Polyhydroxyalkanoates: Properties and Chemical Modification Approaches for Their Functionalization. *Biotechnol. Prog.* **2018**, *34*, 29–41. [CrossRef] [PubMed]
17. Tan, G.-Y.; Chen, C.-L.; Li, L.; Ge, L.; Wang, L.; Razaad, I.; Li, Y.; Zhao, L.; Mo, Y.; Wang, J.-Y. Start a Research on Biopolymer Polyhydroxyalkanoate (PHA): A Review. *Polymers* **2014**, *6*, 706–754. [CrossRef]
18. Balakrishna Pillai, A.; Jaya Kumar, A.; Thulasi, K.; Kumarapillai, H. Evaluation of Short-Chain-Length Polyhydroxyalkanoate Accumulation in Bacillus Aryabhattai. *Braz. J. Microbiol.* **2017**, *48*, 451–460. [CrossRef]
19. Tarrahi, R.; Fathi, Z.; Seydibeyoğlu, M.Ö.; Doustkhah, E.; Khataee, A. Polyhydroxyalkanoates (PHA): From Production to Nanoarchitecture. *Int. J. Biol. Macromol.* **2020**, *146*, 596–619. [CrossRef]
20. Goswami, M.; Rekhi, P.; Debnath, M.; Ramakrishna, S. Microbial Polyhydroxyalkanoates Granules: An Approach Targeting Biopolymer for Medical Applications and Developing Bone Scaffolds. *Molecules* **2021**, *26*, 860. [CrossRef]
21. Sagong, H.-Y.; Son, H.F.; Choi, S.Y.; Lee, S.Y.; Kim, K.-J. Structural Insights into Polyhydroxyalkanoates Biosynthesis. *Trends Biochem. Sci.* **2018**, *43*, 790–805. [CrossRef]
22. Muthuraj, R.; Valerio, O.; Mekonnen, T.H. Recent Developments in Short- and Medium-Chain- Length Polyhydroxyalkanoates: Production, Properties, and Applications. *Int. J. Biol. Macromol.* **2021**, *187*, 422–440. [CrossRef] [PubMed]
23. Obruca, S.; Sedlacek, P.; Slaninova, E.; Fritz, I.; Daffert, C.; Meixner, K.; Sedrlova, Z.; Koller, M. Novel Unexpected Functions of PHA Granules. *Appl. Microbiol. Biotechnol.* **2020**, *104*, 4795–4810. [CrossRef]
24. Obulisamy, P.K.; Mehariya, S. Polyhydroxyalkanoates from Extremophiles: A Review. *Bioresour. Technol.* **2021**, *325*, 124653. [CrossRef] [PubMed]
25. Mitra, R.; Xu, T.; Chen, G.; Xiang, H.; Han, J. An Updated Overview on the Regulatory Circuits of Polyhydroxyalkanoates Synthesis. *Microb. Biotechnol.* **2022**, *15*, 1446–1470. [CrossRef]
26. Verlinden RA, J.; Hill, D.J.; Kenward, M.A.; Williams, C.D.; Radecka, I. Bacterial Synthesis of Biodegradable Polyhydroxyalkanoates. *J. Appl. Microbiol.* **2007**, *102*, 1437–1449. [CrossRef] [PubMed]
27. Ai, M.; Zhu, Y.; Jia, X. Recent Advances in Constructing Artificial Microbial Consortia for the Production of Medium-Chain-Length Polyhydroxyalkanoates. *World J. Microbiol. Biotechnol.* **2021**, *37*, 2. [CrossRef]
28. Guleria, S.; Singh, H.; Sharma, V.; Bhardwaj, N.; Arya, S.K.; Puri, S.; Khatri, M. Polyhydroxyalkanoates Production from Domestic Waste Feedstock: A Sustainable Approach towards Bio-Economy. *J. Clean. Prod.* **2022**, *340*, 130661. [CrossRef]
29. Możejko-Ciesielska, J.; Marciniak, P.; Szacherska, K. Polyhydroxyalkanoates Synthesized by Aeromonas Species: Trends and Challenges. *Polymers* **2019**, *11*, 1328. [CrossRef]
30. Rehm, B.H.A.; Mitsky, T.A.; Steinbüchel, A. Role of Fatty Acid De Novo Biosynthesis in Polyhydroxyalkanoic Acid (PHA) and Rhamnolipid Synthesis by Pseudomonads: Establishment of the Transacylase (PhaG)-Mediated Pathway for PHA Biosynthesis in *Escherichia coli*. *Appl. Environ. Microbiol.* **2001**, *67*, 3102–3109. [CrossRef]
31. Blunt, W.; Lagassé, A.; Jin, Z.; Dartiailh, C.; Sparling, R.; Gapes, D.J.; Levin, D.B.; Cicek, N. Efficacy of Medium Chain-Length Polyhydroxyalkanoate Biosynthesis from Different Biochemical Pathways under Oxygen-Limited Conditions Using Pseudomonas Putida LS46. *Process Biochem.* **2019**, *82*, 19–31. [CrossRef]
32. Lu, J.; Tappel, R.C.; Nomura, C.T. Mini-Review: Biosynthesis of Poly(Hydroxyalkanoates). *Polym. Rev.* **2009**, *49*, 226–248. [CrossRef]
33. Możejko-Ciesielska, J.; Serafim, L.S. Proteomic Response of Pseudomonas Putida KT2440 to Dual Carbon-Phosphorus Limitation during Mcl-PHAs Synthesis. *Biomolecules* **2019**, *9*, 796. [CrossRef] [PubMed]
34. Urtuvia, V.; Villegas, P.; González, M.; Seeger, M. Bacterial Production of the Biodegradable Plastics Polyhydroxyalkanoates. *Int. J. Biol. Macromol.* **2014**, *70*, 208–213. [CrossRef] [PubMed]
35. Valentino, F.; Karabegovic, L.; Majone, M.; Morgan-Sagastume, F.; Werker, A. Polyhydroxyalkanoate (PHA) Storage within a Mixed-Culture Biomass with Simultaneous Growth as a Function of Accumulation Substrate Nitrogen and Phosphorus Levels. *Water Res.* **2015**, *77*, 49–63. [CrossRef] [PubMed]
36. Mourão, M.M.; Gradíssimo, D.G.; Santos, A.V.; Schneider, M.P.C.; Faustino, S.M.M.; Vasconcelos, V.; Xavier, L.P. Optimization of Polyhydroxybutyrate Production by Amazonian Microalga Stigeoclonium Sp. B23. *Biomolecules* **2020**, *10*, 1628. [CrossRef] [PubMed]
37. Silva, J.B.; Pereira, J.R.; Marreiros, B.C.; Reis, M.A.M.; Freitas, F. Microbial Production of Medium-Chain Length Polyhydroxyalkanoates. *Process Biochem.* **2021**, *102*, 393–407. [CrossRef]

38. Muhammadi; Shabina; Afzal, M.; Hameed, S. Bacterial Polyhydroxyalkanoates-Eco-Friendly next Generation Plastic: Production, Biocompatibility, Biodegradation, Physical Properties and Applications. *Green Chem. Lett. Rev.* **2015**, *8*, 56–77. [CrossRef]
39. Chen, G.-Q.; Jiang, X.-R. Engineering Bacteria for Enhanced Polyhydroxyalkanoates (PHA) Biosynthesis. *Synth. Syst. Biotechnol.* **2017**, *2*, 192–197. [CrossRef]
40. Mourão, M.M.; Xavier, L.P.; Urbatzka, R.; Figueiroa, L.B.; da Costa, C.E.F.; Dias, C.G.B.T.; Schneider, M.P.C.; Vasconcelos, V.; Santos, A.V. Characterization and Biotechnological Potential of Intracellular Polyhydroxybutyrate by Stigeoclonium Sp. B23 Using Cassava Peel as Carbon Source. *Polymers* **2021**, *13*, 687. [CrossRef]
41. Chen, G.-Q.; Chen, X.-Y.; Wu, F.-Q.; Chen, J.-C. Polyhydroxyalkanoates (PHA) toward Cost Competitiveness and Functionality. *Adv. Ind. Eng. Polym. Res.* **2020**, *3*, 1–7. [CrossRef]
42. Gomes Gradíssimo, D.; Pereira Xavier, L.; Valadares Santos, A. Cyanobacterial Polyhydroxyalkanoates: A Sustainable Alternative in Circular Economy. *Molecules* **2020**, *25*, 4331. [CrossRef]
43. Klimek, K.; Ginalska, G. Proteins and Peptides as Important Modifiers of the Polymer Scaffolds for Tissue Engineering Applications—A Review. *Polymers* **2020**, *12*, 844. [CrossRef]
44. Jafari, M.; Paknejad, Z.; Rad, M.R.; Motamedian, S.R.; Eghbal, M.J.; Nadjmi, N.; Khojasteh, A. Polymeric Scaffolds in Tissue Engineering: A Literature Review: Polymeric Scaffolds in Tissue Engineering. *J. Biomed. Mater. Res.* **2017**, *105*, 431–459. [CrossRef] [PubMed]
45. Qu, H.; Fu, H.; Han, Z.; Sun, Y. Biomaterials for Bone Tissue Engineering Scaffolds: A Review. *RSC Adv.* **2019**, *9*, 26252–26262. [CrossRef] [PubMed]
46. Scott, L.; Jurewicz, I.; Jeevaratnam, K.; Lewis, R. Carbon Nanotube-Based Scaffolds for Cardiac Tissue Engineering—Systematic Review and Narrative Synthesis. *Bioengineering* **2021**, *8*, 80. [CrossRef] [PubMed]
47. Mafi, P.; Hindocha, S.; Mafi, R.; Khan, W.S. Evaluation of Biological Protein-Based Collagen Scaffolds in Cartilage and Musculoskeletal Tissue Engineering—A Systematic Review of the Literature. *CSCR* **2012**, *7*, 302–309. [CrossRef] [PubMed]
48. Soleymani Eil Bakhtiari, S.; Karbasi, S.; Toloue, E.B. Modified Poly(3-Hydroxybutyrate)-Based Scaffolds in Tissue Engineering Applications: A Review. *Int. J. Biol. Macromol.* **2021**, *166*, 986–998. [CrossRef]
49. Mohandas, S.P.; Balan, L.; Jayanath, G.; Anoop, B.S.; Philip, R.; Cubelio, S.S.; Bright Singh, I.S. Biosynthesis and Characterization of Polyhydroxyalkanoate from Marine Bacillus Cereus MCCB 281 Utilizing Glycerol as Carbon Source. *Int. J. Biol. Macromol.* **2018**, *119*, 380–392. [CrossRef] [PubMed]
50. Esmail, A.; Pereira, J.R.; Zoio, P.; Silvestre, S.; Menda, U.D.; Sevrin, C.; Grandfils, C.; Fortunato, E.; Reis, M.A.M.; Henriques, C.; et al. Oxygen Plasma Treated-Electrospun Polyhydroxyalkanoate Scaffolds for Hydrophilicity Improvement and Cell Adhesion. *Polymers* **2021**, *13*, 1056. [CrossRef]
51. Romo-Uribe, A.; Meneses-Acosta, A.; Domínguez-Díaz, M. Viability of HEK 293 Cells on Poly-β-Hydroxybutyrate (PHB) Biosynthesized from a Mutant Azotobacter Vinelandii Strain. Cast Film and Electrospun Scaffolds. *Mater. Sci. Eng. C* **2017**, *81*, 236–246. [CrossRef]
52. Codreanu, A.; Balta, C.; Herman, H.; Cotoraci, C.; Mihali, C.V.; Zurbau, N.; Zaharia, C.; Rapa, M.; Stanescu, P.; Radu, I.-C.; et al. Bacterial Cellulose-Modified Polyhydroxyalkanoates Scaffolds Promotes Bone Formation in Critical Size Calvarial Defects in Mice. *Materials* **2020**, *13*, 1433. [CrossRef]
53. El-Shanshory, A.A.; Agwa, M.M.; Abd-Elhamid, A.I.; Soliman, H.M.A.; Mo, X.; Kenawy, E.-R. Metronidazole Topically Immobilized Electrospun Nanofibrous Scaffold: Novel Secondary Intention Wound Healing Accelerator. *Polymers* **2022**, *14*, 454. [CrossRef] [PubMed]
54. Pryadko, A.S.; Botvin, V.V.; Mukhortova, Y.R.; Pariy, I.; Wagner, D.V.; Laktionov, P.P.; Chernonosova, V.S.; Chelobanov, B.P.; Chernozem, R.V.; Surmeneva, M.A.; et al. Core-Shell Magnetoactive PHB/Gelatin/Magnetite Composite Electrospun Scaffolds for Biomedical Applications. *Polymers* **2022**, *14*, 529. [CrossRef] [PubMed]
55. Ansari, N.F.; Annuar, M.S.M.; Murphy, B.P. A Porous Medium-chain-length Poly(3-hydroxyalkanoates)/Hydroxyapatite Composite as Scaffold for Bone Tissue Engineering. *Eng. Life Sci.* **2016**, *17*, 420–429. [CrossRef]
56. Mukheem, A.; Muthoosamy, K.; Manickam, S.; Sudesh, K.; Shahabuddin, S.; Saidur, R.; Akbar, N.; Sridewi, N. Fabrication and Characterization of an Electrospun PHA/Graphene Silver Nanocomposite Scaffold for Antibacterial Applications. *Materials* **2018**, *11*, 1673. [CrossRef] [PubMed]
57. Lizarraga-Valderrama, L.R.; Nigmatullin, R.; Ladino, B.; Taylor, C.S.; Boccaccini, A.R.; Knowles, J.C.; Claeyssens, F.; Haycock, J.W.; Roy, I. Modulation of Neuronal Cell Affinity of Composite Scaffolds Based on Polyhydroxyalkanoates and Bioactive Glasses. *Biomed. Mater.* **2020**, *15*, 045024. [CrossRef]
58. Saska, S.; Pires, L.C.; Cominotte, M.A.; Mendes, L.S.; De Oliveira, M.F.; Maia, I.A.; Da Silva, J.V.L.; Ribeiro, S.J.L.; Cirelli, J.A. Three-Dimensional Printing and in Vitro Evaluation of Poly(3-Hydroxybutyrate) Scaffolds Functionalized with Osteogenic Growth Peptide for Tissue Engineering. *Mater. Sci. Eng. C* **2018**, *89*, 265–273. [CrossRef]
59. Wei, D.-X.; Dao, J.-W.; Liu, H.-W.; Chen, G.-Q. Suspended Polyhydroxyalkanoate Microspheres as 3D Carriers for Mammalian Cell Growth. *Artif. Cells Nanomed. Biotechnol.* **2018**, *46*, 473–483. [CrossRef]
60. Prakash, P.; Lee, W.-H.; Loo, C.-Y.; Wong, H.S.J.; Parumasivam, T. Advances in Polyhydroxyalkanoate Nanocarriers for Effective Drug Delivery: An Overview and Challenges. *Nanomaterials* **2022**, *12*, 175. [CrossRef]

61. Koller, M. Biodegradable and Biocompatible Polyhydroxy-Alkanoates (PHA): Auspicious Microbial Macromolecules for Pharmaceutical and Therapeutic Applications. *Molecules* **2018**, *23*, 362. [CrossRef]
62. Impallomeni, G.; Ballistreri, A.; Carnemolla, G.M.; Rizzo, M.G.; Nicolò, M.S.; Guglielmino, S.P.P. Biosynthesis and Structural Characterization of Polyhydroxyalkanoates Produced by Pseudomonas Aeruginosa ATCC 27853 from Long Odd-Chain Fatty Acids. *Int. J. Biol. Macromol.* **2018**, *108*, 608–614. [CrossRef]
63. Evangeline, S.; Sridharan, T.B. Biosynthesis and Statistical Optimization of Polyhydroxyalkanoate (PHA) Produced by Bacillus Cereus VIT-SSR1 and Fabrication of Biopolymer Films for Sustained Drug Release. *Int. J. Biol. Macromol.* **2019**, *135*, 945–958. [CrossRef] [PubMed]
64. Bini, R.A.; Silva, M.F.; Varanda, L.C.; Da Silva, M.A.; Dreiss, C.A. Soft Nanocomposites of Gelatin and Poly(3-Hydroxybutyrate) Nanoparticles for Dual Drug Release. *Colloids Surf. B Biointerfaces* **2017**, *157*, 191–198. [CrossRef] [PubMed]
65. Senthilkumar, P.; Dawn, S.S.; Sree Samanvitha, K.; Sanjay Kumar, S.; Narendra Kumar, G.; Samrot, A.V. Optimization and Characterization of Poly[R]Hydroxyalkanoate of Pseudomonas Aeruginosa SU-1 to Utilize in Nanoparticle Synthesis for Curcumin Delivery. *Biocatal. Agric. Biotechnol.* **2017**, *12*, 292–298. [CrossRef]
66. Vardhan, H.; Mittal, P.; Adena, S.K.R.; Mishra, B. Long-Circulating Polyhydroxybutyrate-Co-Hydroxyvalerate Nanoparticles for Tumor Targeted Docetaxel Delivery: Formulation, Optimization and in Vitro Characterization. *Eur. J. Pharm. Sci.* **2017**, *99*, 85–94. [CrossRef]
67. Jiang, L.; Luo, Z.; Loh, X.J.; Wu, Y.-L.; Li, Z. PHA-Based Thermogel as a Controlled Zero-Order Chemotherapeutic Delivery System for the Effective Treatment of Melanoma. *ACS Appl. Bio. Mater.* **2019**, *2*, 3591–3600. [CrossRef] [PubMed]
68. Chen, S.; Gao, J.; Yan, E.; Wang, Y.; Li, Y.; Lu, H.; Fan, L.; Wang, D.; An, Q. A Novel Porous Composite Membrane of PHA/PVA via Coupling of Electrospinning and Spin Coating for Antibacterial Applications. *Mater. Lett.* **2021**, *301*, 130279. [CrossRef]
69. De Freitas E Castro, M.; Mendonça, T.T.; Da Silva, L.F.; Gomez, J.G.C.; Sanchez Rodriguez, R.J. Carriers Based on Poly-3-Hydroxyalkanoates Containing Nanomagnetite to Trigger Hormone Release. *Int. J. Biol. Macromol.* **2021**, *166*, 448–458. [CrossRef]
70. Huerta-Angeles, G.; Brandejsová, M.; Nigmatullin, R.; Kopecká, K.; Vágnerová, H.; Šmejkalová, D.; Roy, I.; Velebný, V. Synthesis of Graft Copolymers Based on Hyaluronan and Poly(3-Hydroxyalkanoates). *Carbohydr. Polym.* **2017**, *171*, 220–228. [CrossRef]
71. Umesh, M.; Priyanka, K.; Thazeem, B.; Preethi, K. Biogenic PHA Nanoparticle Synthesis and Characterization from *Bacillus subtilis* NCDC0671 Using Orange Peel Medium. *Int. J. Polym. Mater. Polym. Biomater.* **2018**, *67*, 996–1004. [CrossRef]
72. Owji, N.; Mandakhbayar, N.; Gregory, D.A.; Marcello, E.; Kim, H.; Roy, I.; Knowles, J.C. Mussel Inspired Chemistry and Bacteria Derived Polymers for Oral Mucosal Adhesion and Drug Delivery. *Front. Bioeng. Biotechnol.* **2021**, *9*, 663764. [CrossRef] [PubMed]
73. Musumeci, T.; Cupri, S.; Bonaccorso, A.; Impallomeni, G.; Ballistreri, A.; Puglisi, G.; Pignatello, R. Technology Assessment of New Biodegradable Poly(R-3-Hydroxybutyrate-Co-1,4-Butylene Adipate) Copolymers for Drug Delivery. *J. Appl. Polym. Sci.* **2019**, *136*, 47233. [CrossRef]
74. Fan, F.; Wang, L.; Ouyang, Z.; Wen, Y.; Lu, X. Development and Optimization of a Tumor Targeting System Based on Microbial Synthesized PHA Biopolymers and PhaP Mediated Functional Modification. *Appl. Microbiol. Biotechnol.* **2018**, *102*, 3229–3241. [CrossRef]
75. Shershneva, A.; Murueva, A.; Nikolaeva, E.; Shishatskaya, E.; Volova, T. Novel Spray-Dried PHA Microparticles for Antitumor Drug Release. *Dry. Technol.* **2018**, *36*, 1387–1398. [CrossRef]
76. Cañadas, O.; García-García, A.; Prieto, M.; Pérez-Gil, J. Polyhydroxyalkanoate Nanoparticles for Pulmonary Drug Delivery: Interaction with Lung Surfactant. *Nanomaterials* **2021**, *11*, 1482. [CrossRef]
77. Pavic, A.; Stojanovic, Z.; Pekmezovic, M.; Veljović, Đ.; O'Connor, K.; Malagurski, I.; Nikodinovic-Runic, J. Polyenes in Medium Chain Length Polyhydroxyalkanoate (Mcl-PHA) Biopolymer Microspheres with Reduced Toxicity and Improved Therapeutic Effect against Candida Infection in Zebrafish Model. *Pharmaceutics* **2022**, *14*, 696. [CrossRef]
78. Salama, H.E.; Aziz, M.S.A.; Saad, G.R. Thermal Properties, Crystallization and Antimicrobial Activity of Chitosan Biguanidine Grafted Poly(3-Hydroxybutyrate) Containing Silver Nanoparticles. *Int. J. Biol. Macromol.* **2018**, *111*, 19–27. [CrossRef]
79. El-malek, F.A.; Rofeal, M.; Farag, A.; Omar, S.; Khairy, H. Polyhydroxyalkanoate Nanoparticles Produced by Marine Bacteria Cultivated on Cost Effective Mediterranean Algal Hydrolysate Media. *J. Biotechnol.* **2021**, *328*, 95–105. [CrossRef]
80. Balcucho, J.; Narváez, D.M.; Tarazona, N.A.; Castro-Mayorga, J.L. Microbially Synthesized Polymer-Metal Nanoparticles Composites as Promising Wound Dressings to Overcome Methicillin-Resistance Staphylococcus Aureus Infections. *Polymers* **2023**, *15*, 920. [CrossRef]
81. Kalaoglu-Altan, O.I.; Baskan, H.; Meireman, T.; Basnett, P.; Azimi, B.; Fusco, A.; Funel, N.; Donnarumma, G.; Lazzeri, A.; Roy, I.; et al. Silver Nanoparticle-Coated Polyhydroxyalkanoate Based Electrospun Fibers for Wound Dressing Applications. *Materials* **2021**, *14*, 4907. [CrossRef]
82. Pereira, J.R.; Araújo, D.; Marques, A.C.; Neves, L.A.; Grandfils, C.; Sevrin, C.; Alves, V.D.; Fortunato, E.; Reis, M.A.M.; Freitas, F. Demonstration of the Adhesive Properties of the Medium-Chain-Length Polyhydroxyalkanoate Produced by Pseudomonas Chlororaphis Subsp. Aurantiaca from Glycerol. *Int. J. Biol. Macromol.* **2019**, *122*, 1144–1151. [CrossRef]

83. Urbina, L.; Wongsirichot, P.; Corcuera, M.Á.; Gabilondo, N.; Eceiza, A.; Winterburn, J.; Retegi, A. Application of Cider By-Products for Medium Chain Length Polyhydroxyalkanoate Production by Pseudomonas Putida KT2440. *Eur. Polym. J.* **2018**, *108*, 1–9. [CrossRef]
84. Vijayamma, R.; Maria, H.J.; Thomas, S.; Shishatskaya, E.I.; Kiselev, E.G.; Nemtsev, I.V.; Sukhanova, A.A.; Volova, T.G. A Study of the Properties and Efficacy of Microparticles Based on P(3HB) and P(3HB/3HV) Loaded with Herbicides. *J. Appl. Polym. Sci.* **2022**, *139*, 51756. [CrossRef]

polymers

MDPI

Article

Chitosan Microparticles Loaded with New Non-Cytotoxic Isoniazid Derivatives for the Treatment of Tuberculosis: In Vitro and In Vivo Studies

Ionut Dragostin [1,†], Oana-Maria Dragostin [1,*,†], Andreea Teodora Iacob [2,†], Maria Dragan [2,†], Carmen Lidia Chitescu [1,†], Luminita Confederat [3], Alexandra-Simona Zamfir [4], Rodica Tatia [5], Catalina Daniela Stan [2] and Carmen Lacramioara Zamfir [6]

1 Research Centre in the Medical-Pharmaceutical Field, Faculty of Medicine and Pharmacy, Dunarea de Jos University, 35 Al. I. Cuza Str., 800017 Galati, Romania; ionut.dragostin@yahoo.com (I.D.); camen.chitescu@ugal.ro (C.L.C.)

2 Faculty of Pharmacy, University of Medicine and Pharmacy Grigore T. Popa, 16 Universitatii Str., 700115 Iasi, Romania; andreea.panzariu@umfiasi.ro (A.T.I.); maria.wolszleger@umfiasi.ro (M.D.); catalina.stan@umfiasi.ro (C.D.S.)

3 Faculty of Medicine, University of Medicine and Pharmacy Grigore T. Popa, 16 Universitatii Str., 700115 Iasi, Romania; luminita.confederat@yahoo.com

4 Department of Pneumology, Faculty of Medicine, University of Medicine and Pharmacy Grigore T. Popa, 16 Universitatii Str., 700115 Iasi, Romania; simona-zamfir@umfiasi.ro

5 Department of Cellular and Molecular Biology, National Institute of Research and Development for Biological Sciences, 296 Splaiul Independentei, 060031 Bucharest, Romania; rodica.tatia@gmail.com

6 Department of Histology, Faculty of Medicine, University of Medicine and Pharmacy Grigore T. Popa, 16 Universitatii Str., 700115 Iasi, Romania; carmen.zamfir@umfiasi.ro

* Correspondence: oana.dragostin@ugal.ro

† These authors contributed equally to this work.

Abstract: Lately, in the world of medicine, the use of polymers for the development of innovative therapies seems to be a major concern among researchers. In our case, as a continuation of the research that has been developed so far regarding obtaining new isoniazid (INH) derivatives for tuberculosis treatment, this work aimed to test the ability of the encapsulation method to reduce the toxicity of the drug, isoniazid and its new derivatives. To achieve this goal, the following methods were applied: a structural confirmation of isoniazid derivatives using LC-HRMS/MS; the obtaining of microparticles based on polymeric support; the determination of their loading and biodegradation capacities; in vitro biocompatibility using MTT cell viability assays; and, last but not least, in vivo toxicological screening for the determination of chronic toxicity in laboratory mice, including the performance of a histopathological study and testing for liver enzymes. The results showed a significant reduction in tissue alterations, the disappearance of cell necrosis and microvesicular steatosis areas and lower values of the liver enzymes TGO, TGP and alkaline phosphatase when using encapsulated forms of drugs. In conclusion, the encapsulation of INH and INH derivatives with chitosan had beneficial effects, suggesting a reduction in hepatotoxicity and, therefore, the achievement of the aim of this paper.

Keywords: encapsulation; microparticles; isoniazid derivatives; chitosan; drug toxicity; biocompatibility; tissue alterations

1. Introduction

In one of our previous research works [1], isoniazid was subjected to a structural modulation process that aimed to synthesize new isonicotinoyl hydrazones, which proved to have an improved pharmacotoxicological profile. As part of an ongoing research process, the present paper focused on the effects of the encapsulation of isoniazid derivatives in polymer matrices (chitosan microparticles).

The benefits of the microencapsulation process include the separation of components, which ensures the prevention of drug incompatibility, and the protection of the substances from acids, alkalis, heat, ultraviolet and oxidant agents, which extends their validity [2]. In this regard, the encapsulation method forms a physical barrier around a sensitive compound to decrease its reactivity with extrinsic factors, exert an important impact on the drug delivery profile and provide new features according to the support material used [3]. This technique can modify the drug release profile as it uses dual pH-responsive, porous, double-layered polymers with opposite charges for controlled release applications, which allows the encapsulated active ingredients to remain active, despite the gastric acidity levels, and ready to be released in the intestines [4]. Among the many benefits, the role of microencapsulation as a tool for increasing the low bioavailability of different compounds must be highlighted, such as polyphenols that are used in the treatment of diabetes [5].

Over time, chitosan has become an interesting material for medical applications and has attracted considerable interest due to its unique combination of properties, such as biocompatibility, biodegradability and low toxicity [6]. Due to these characteristics, chitosan has been assigned a number of applications, either by itself or in combination with other natural polymers, in the food, pharmaceutical, textile, agriculture and cosmetics industries [7]. In addition, the following intrinsic therapeutic effects of chitosan are also known: anticholesteremic, antidiabetic, antimicrobial, antioxidant, antidepressant, antitumor and hemostatic [8]. Microencapsulation based on chitosan is usually used to improve the bioavailability of drugs and obtain their sustained release [9]. The process of obtaining microparticles based on biopolymers, such as chitosan, is different depending on the type of active substance that is embedded; for example, the encapsulation of essential oils through the spray-drying technique is used for *Candida albicans* biofilms [10] while the self-assembled chitosan microparticle formation through the depolymerization of chitosan (using potassium persulphate at 60 °C followed by cooling) is used as a tumor angiogenesis inhibitor [11].

In addition to the existing results that have been presented in the literature, the novelty of this study was the testing of the ability of the encapsulation method to reduce the drug toxicity of isoniazid and its new derivatives, as described in our previous paper [1].

2. Materials and Methods

2.1. Materials

Chitosan with medium molecular weight (CS MMW, 425 kDa, deacetylation degree of 85%), isoniazid (INH), acetic acid, sodium hydroxide, sodium tripolyphosphate (TPP) and organic solvents (p.a.) were purchased from Sigma-Aldrich, Tokyo, Japan. All solvents and reagents were used without preliminary purification. The synthesis of isoniazid derivatives (INH-a, INH-b and INH-c) in the laboratory was described in a previous study by the same authors [1].

Cell and material culture: A stabilized line of NCTC mouse fibroblasts (929 clones) was used for the biocompatibility analysis and cell morphology analysis.

The assayed cell cultures from well plates, after 72 h of experimentation, were stained with hematoxylin and eosin (Sigma Corporation, Tokyo, Japan) dyes. The cell morphology was examined and pictures were taken using an inverted microscope (Nikon, Tokyo, Japan) and a photo camera (Zeiss AxioCam MRc 5, Göttingen, Germany).

2.2. Methods

2.2.1. Structural Confirmation of Isoniazid Derivatives

As a continuation of the ^{1}H-NMR and ATR-FTIR analysis from our previous study [1], in the present work, Q Exactive high-resolution quadrupole Orbitrap LC-HRMS/MS was applied for targeted ion fragmentation (t-MS^2) for the additional structural confirmation of the compounds.

The analysis was carried out using a Thermo Scientific Dionex Ultimate 3000 UHPLC system, consisting of a pump (Series RS) that was coupled with a column compartment

(Series TCC-3000RS) and an autosampler (Series WPS-3000RS). The UHPLC system was controlled by Chromeleon 7.2 software (Thermo Fisher Scientific, Waltham, MA, USA and Dionex Softron GMbH, part of Thermo Fisher Scientific, Bremen, Germany). An ultra-performance Accucore UHPLC Column C18 (150 × 2.1 mm, 2.6 μm; Thermo Scientific) was used. The mobile phase consisted of a mixture of ultra-pure water containing 500 μL L^{-1} of formic acid (pH 2.5) and methanol with 500 μL L^{-1} of formic acid.

A HESI (heated electrospray) ion source was used for the ionization in the positive mode. The ion source parameters were optimized. The full scan HRMS analysis of the compounds was performed using a Q Exactive mass spectrometer. The full scan data in positive mode were acquired at a resolving power of 70,000 FWHM at m/z 200. A scan range of m/z 100–1000 Da was chosen, the automatic gain control (AGC) was set at 3e6 and the injection time was set to 200 ms. The scan rate was set at 2 scan s^{-1}.

For structural information, the fragmentation events were performed in selective reaction monitoring (SRM) mode and were successively applied to each of the selected compounds at a resolving power of 35,000 FWHM and increasing collision energy (15, 20, 30, 60 and 80 NCE (normalized collision energy)).

The data were processed by the Quan/Qual Browser Xcalibur 2.3 (Thermo Fisher). The mass tolerance window was set to 5 ppm. Mass Frontier 8.0 software (Thermo Scientific, Waltham, MA, USA) was used to generate the fragmentation patterns of the selected compounds for a comparison analysis.

2.2.2. Preparation of Chitosan Microparticles Loaded with Isoniazid Derivatives

The isoniazid derivatives (INH-a, INH-b and INH-c) were loaded into chitosan microparticles for oral administration using the ionic gelation method, according to an adapted procedure from the literature [9] that was previously applied by us [12]. For each batch, 0.5 g of active substance (INH, INH-a, INH-b or INH-c) was suspended in 50 mL of 2% w/v viscous chitosan solution. Then, 5 mL of the medium viscosity suspension was added dropwise through a syringe needle (18 G) to 5 mL of 2% w/v TPP solution using gentle stirring. The formed microparticles were left in contact with the crosslinking agent solution for 24 h at room temperature to achieve efficient crosslinking and were subjected to gentle shaking at 300 rpm to prevent their adhesion. Subsequently, the microparticles were extracted from the TPP solution, washed three times with distilled water (to remove excess crosslinking agent) and dried by exposure to the open air (Figure 1).

2.2.3. Microscopic Characterization of the Obtained Microparticles

The size and morphology of the obtained microparticles were analyzed using the Fei Quanta 200F scanning electron microscope (SEM), which allowed for the examination of details with a great magnitude and resolution.

2.2.4. Loading Efficiency (LE)

The loading efficiency of INH derivatives in the chitosan microparticles was evaluated using the UV spectrophotometric method. The content of INH and its derivatives in the TPP solution was evaluated spectrophotometrically at 280 nm, after removing the chitosan microparticles. The loading efficiency (%) was calculated using a calibration curve for each INH derivative and the following formula [13]:

$$\% \text{ LE (loading efficiency)} = (C_0 - C_1/C_0) \times 100 \quad (1)$$

where C_0 is 5 mg/mL (the initial concentration of INH derivative in the chitosan solution that was dripped into the TPP solution) and C_1 is the concentration of INH derivative in the TPP solution (mg/mL) after removing the formed microparticles.

2.2.5. Assessment of In Vitro Biodegradation Capacity

The biodegradation capacity of the microparticles was studied according to the literature and using techniques that were applied by us in a previous study [14], using lysozyme 152,975 IU/mg as the biodegradation agent after pre-solubilization at 37 °C in a phosphate buffer (pH 7.4) at a concentration of 10,000 IU/mL. Until the balance of swelling was established, the microparticles were left in the buffer solution (pH 7.4) and were then transferred to the lysozyme buffer and kept in the incubator at 37 °C for 7 days. The solution was changed daily. On Days 1, 4 and 7, the microparticles were removed from the buffer medium and weighed, following a decrease in mass due to biodegradation. The evaluation of the biodegradation percentage (D %) was calculated using the following formula:

$$D\,\% = \frac{W_0 - W_x}{W_0 \times 100} \tag{2}$$

where W_0 is the mass of the microparticle before incubation and at the equilibrium of its swelling capacity in the buffer solution and W_x is the mass of the microparticle after incubation (on Days 1, 4 and 7).

2.2.6. Evaluation of In Vitro Biocompatibility Using MTT Cell Viability Analysis

An evaluation of the cytotoxicity of the microencapsulated samples was performed by directly exposing cultured cells to samples, followed by a thiazolyl tetrazolium bromide MTT cell proliferation assay.

For this test, the microparticles were sealed in polyethylene foil and sterilized by exposure to UV radiation for 8 h. Cells in 24-well plates were incubated for 24 h at 37 °C and 95% relative humidity in air containing 5% CO_2 to induce cell adhesion. After 24 h of cell incubation, the medium was replaced with 500 μL of fresh culture medium containing microparticles from each sample. In parallel with the samples, cell cultures treated with H_2O_2 hydrogen peroxide (0.03%) were used as a positive control and untreated cells were used as a control culture. The plates were incubated at 37 °C and the quantitative assessment of cytotoxicity was performed after the desired exposure time (24, 48 and 72 h) using MTT thiazolyl tetrazolium bromide. Cell morphology was assessed after 72 h using an inverted microscope. All tests were performed in triplicate.

Cell viability was performed and calculated according to the formula described in the literature [15]:

$$\text{Cell viability (\%)} = \text{OD test}/\text{OD control} \times 100 \tag{3}$$

where OD test is the optical density of the microparticle sample and OD control is the optical density of the untreated control.

For this, the culture medium was replaced with fresh medium containing an MTT solution in a 10:1 (v/v) ratio and the plates were incubated at 37 °C for 3 h. Further, a volume of 500 μL of isopropanol was added to each well to dissolve the formazan crystals by gentle shaking on a platform for 3 h. As previously described, the optical density (OD) of the colored solution was read at 570 nm using a Mithras LB940 microplate reader (Berthold Technologies, Bad Wildbad, Germany).

2.2.7. Biological Evaluation

Chronic toxicity studies were performed according to the guidelines for the deontology and ethics of laboratory animal studies (national law no. 206/27 May 2004; directives 2010/63/EU and CE86/609/EEC) and after obtaining approval from the research ethics commission at the University of Medicine and Pharmacy, Grigore T. Popa, Iasi, (17 April 2018). The experimental protocol included the housing and handling of animals and the administration of test compounds in encapsulated form, as well as their scarification after anesthetic administration (ketamine i.p. 100 mg/kg body weight).

In Vivo Toxicological Screening: The Determination of Chronic Toxicity

Based on the results obtained from the acute toxicity test (as detailed in our previous study [1]), the chronic toxicity of the isoniazid derivative compounds was studied by the oral administration of the microencapsulated form in doses of 1/10 from DL 50, expressed in mg/kg body weight (1/10 from 175.2 for INH, 352.54 for INH-a, 1778.8 for INH-b and 1251.5 for INH-c).

Thus, the doses calculated based on acute toxicity and used for chronic toxicity were as follows:

- 17.52 mg/kg body weight for isoniazid (INH) in Group 1;
- 35.234 mg/kg body weight for INH-a in Group 2;
- 177.88 mg/kg body weight for INH-b in Group 3;
- 125.15 mg/kg body weight for INH-c in Group 4.

All encapsulated compounds were orally administered daily, in a single dose, for 30 days. The empty chitosan microparticles were used as a control in Group 5, while Group 6 was used as the untreated control batch.

At the end of the experiment, biological products from the blood and liver fragments were collected for biochemical and histopathological analyses. The liver fragments were subsequently fixed in 10% buffered formalin for histopathological examination.

Histopathological Examination

In order to evaluate the effects of the microencapsulation of selected compounds on chronic tissue toxicity, a histopathological study was performed on liver tissue fragments to identify possible morphological changes that were produced by the chronic administration of the encapsulated substances. The liver fragments fixed in the 10% buffered formalin were subsequently processed following the specific steps of the histopathological technique [16,17].

The microscopic examination was performed using a Nikon Eclipse 50i microscope to see whether there were any tissue alterations that demonstrated the impact of the administration of the substances incorporated in the chitosan microparticles.

Evaluation of Biochemical Parameters

The biochemical parameters of serum aminotransferases (GPT and GOT) and alkaline phosphatase were determined as markers of hepatotoxicity using the ABX Pentra 400 automated biochemical analyzer (Horiba, Kyoto, Japan). The kits were from Diamedix and the reagents were in boxes that ensured their increased stability and the extended linearity of the results.

2.2.8. Statistical Analysis

Data were analyzed using an analysis of variance (ANOVA; $p < 0.05$). All determinations were made in triplicate and the results were expressed as the mean $\pm$ standard deviation (SD). Graphical representations were made for all results that referred to the evaluation of the biochemical parameters, which took into account the mean of all animals in each group and showed the standard deviation as error bars.

3. Results

3.1. *Mass Spectrometric Structural Confirmation of the Active Compounds*

Recently, high-resolution mass spectrometry (HR-MS/MS) has been applied to the elucidation of the structures of various known and unknown compounds [18,19]. High mass spectral resolving power (RP), which reduces interference, and high mass accuracy, which allows the prediction of molecular formula for spectral peaks, have been proven to be powerful tools for the identification of new chemicals. Structural proposals that are only based on mass spectral evidence are now commonly reported in peer-reviewed literature [18,20,21].

In the present study, a high-resolution full scan analysis and fragmentation experiments for protonated molecules using increasing collision energy were used for compound identification (Figure 1) and the structural characterization of the isoniazid derivatives. Mass Frontier 8.0 software (Thermos Scientific, Waltham, MA, USA) was used to compare the identification of the fragmentation patterns. The fragment ions of isoniazid that were observed in the MS^2 spectrum of isoniazid and its derivatives are shown in Table 1.

Table 1. The target compounds, deprotonated precursors and fragment ions. The common fragments for the parent compound and its derivatives are in bold.

Compound	R.T. (min)	Formula	Exact Mass	Error (ppm)	Adduct Ion (*m/z*)	MS^2 Fragments (*m/z*)
INH	4.25	$C_6H_7N_3O$	137.0589	1.47	138.0667	122.0712; 120.0556; 108.04429; 106.0287; 96.0443; 59.0239
INH-a	11.17	$C_{13}H_{11}N_3O$	225.0902	1.25	226.0980	208.08692; 148.0505; 147.05527; 106.0287; 96.0443
INH-b	11.27	$C_{13}H_{10}N_4O_3$	270.0753	0.86	271.0831	255.0876; 253.0720; 224.0818; 192.0430; 148.0505; 136.0393; 106.0287; 96.0443
INH-c	13.40	$C_{13}H_{10}N_3OBr$	303.0007	0.5	304.0085	288.0130; 285.9974;224.9658; 224.0818; 168.9647; 148.0505; 106.0287; 96.0443

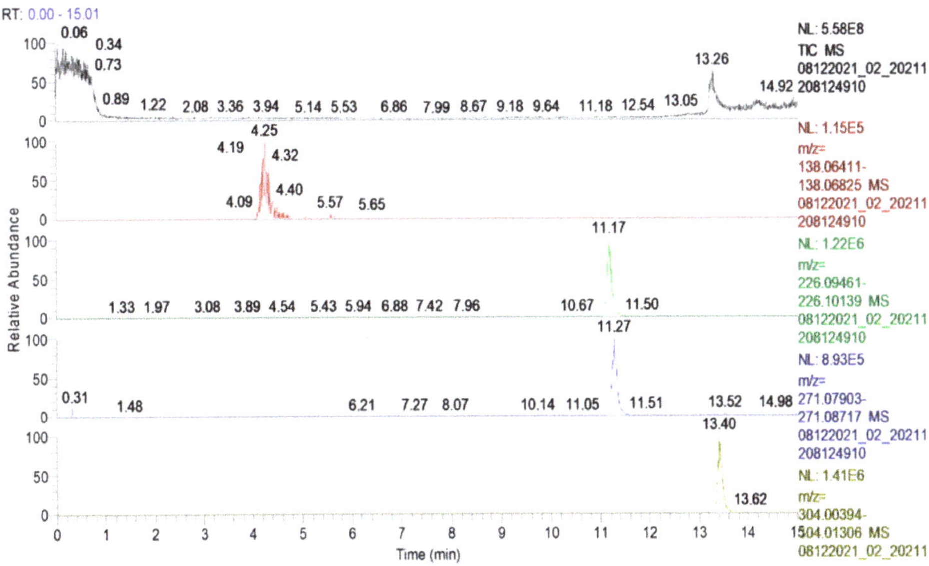

Figure 1. UHPL-HRMS full scan extracted chromatograms of the target compounds (isoniazid and its derivatives). From top to bottom: INH *m/z* 138.0667; INH-a *m/z* 226.0980; INH-b *m/z* 271.0831; INH-c *m/z* 304.0085.

The fragment ions *m/z* 106.087 and 96.0443, which are both generated by the cleavage of the C–N bond and result in the loss of the [-NH-NH_2] group, were common for all compounds, which showed the connection to the same parent compound (isoniazid) (Table 1).

For structure confirmation, other fragment ions were identified and compared to the fragmentation patterns that were generated by the Mass Frontier software for all derivatives. In Figure 2, the fragmentation pattern that was identified for INH-c is compared to the parent compound INH as an example.

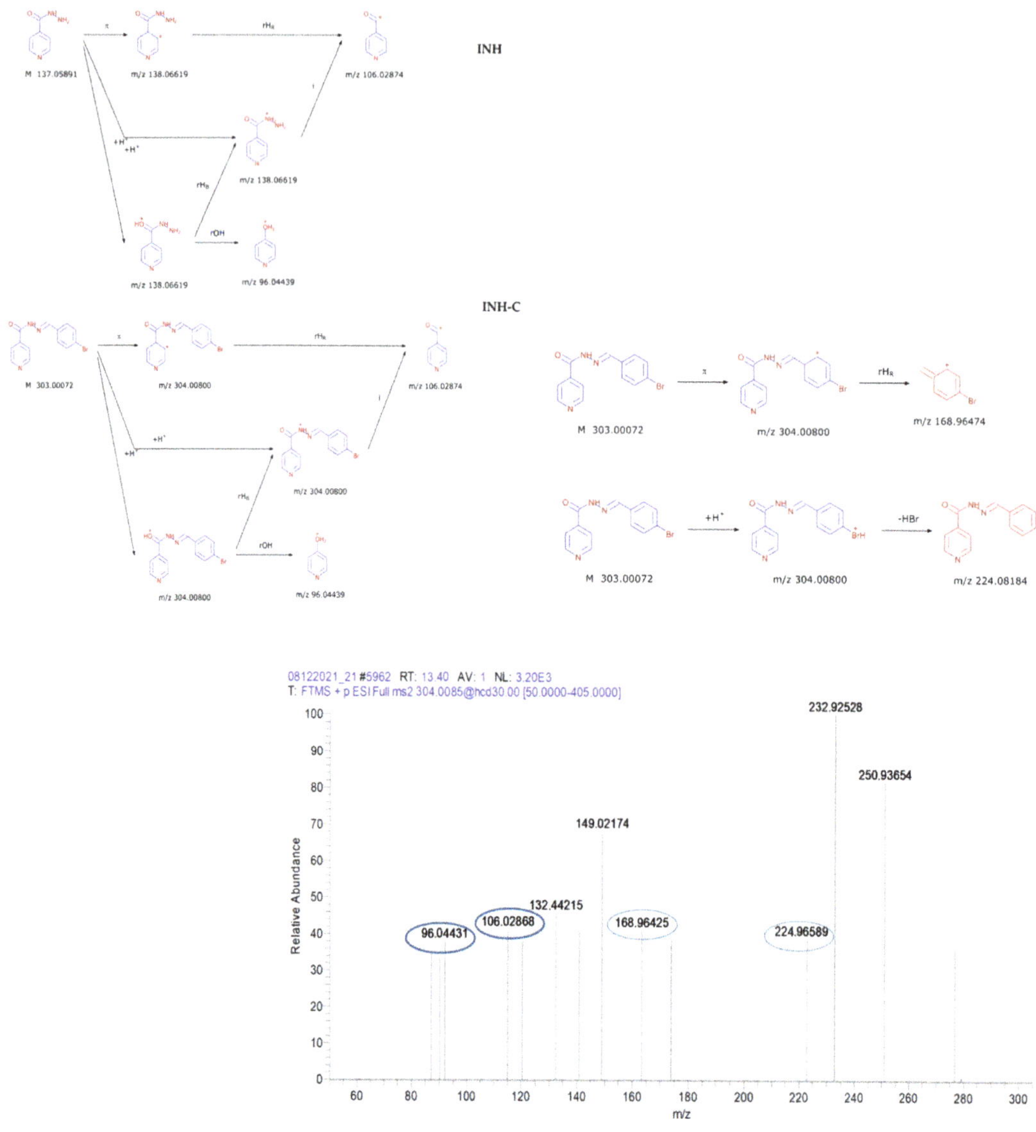

Figure 2. Common fragmentation pathways of INH-c and the parent compound INH and the characteristic fragments of the derivative compound identified in the MS^2 spectra.

3.2. Macroscopic and Microscopic Characterization of the Obtained Microparticles

The color of the microparticles that contained encapsulated active substances varied from white to yellow, according to substance (Figure 3).

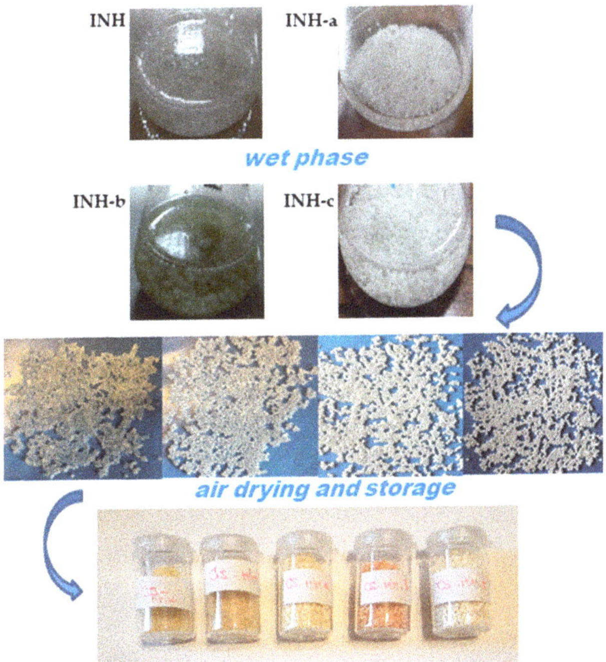

Figure 3. Obtaining chitosan microparticles containing isoniazid (INH) and its condensation products (INH-a, INH-b and INH-c): the wet phase and air drying.

The morphological study showed that the (empty) chitosan microparticles had a regular contour, a spherical and slightly rough surface and a size of about 250 μm in the dry state, showing relative transparency (Figure 4).

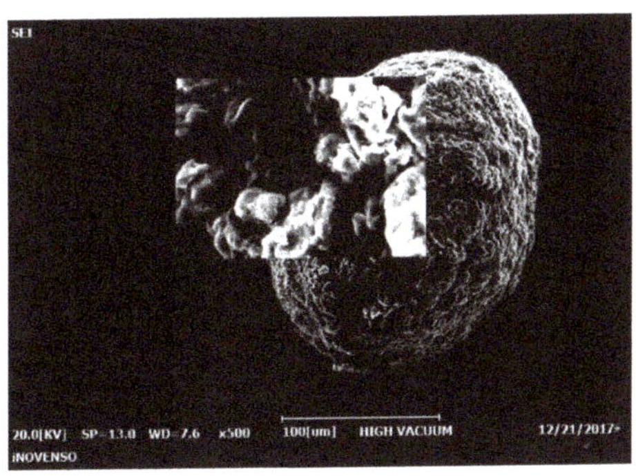

Figure 4. SEM image of an empty chitosan microparticle.

The most important SEM characteristics for chitosan microparticles containing active substances were an irregular contour and a smooth, matte and cracked surface, which could be explained by the hydrophobic nature of the encapsulated substances and by their accentuated crystallinity (which was described in our previous study [1]). Their size was in the range of 500–800 μm in the dry state, which demonstrated a significant increase in diameter compared to the empty microparticles that did not contain an active substance (Figure 5).

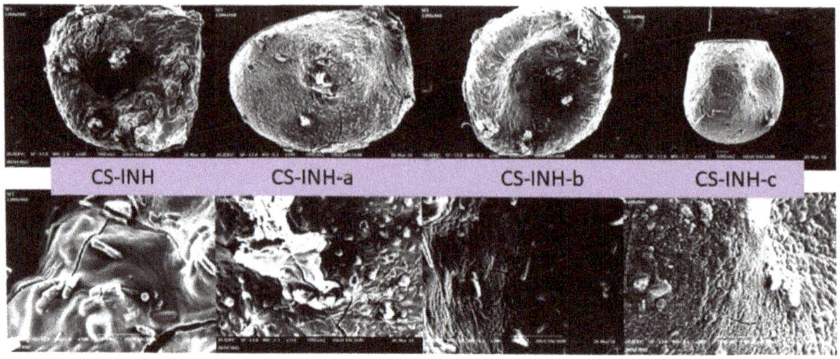

Figure 5. SEM image of chitosan microparticles containing active substances (CS–INH, CS–INH-a, CS–INH-b and CS–INH-c).

3.3. Loading Efficiency (LE)

The loading efficiency of INH and its derivatives in the chitosan microparticles are shown in Table 2.

Table 2. Loading efficiency (LE) of chitosan microparticles.

Compound	C_0 (mg/mL)	C_1 (mg/mL)	LE (%), $n = 3$
CS–INH	5	0.282	94.36 ± 0.7
CS–INH-a	5	0.52	89.60 ± 0.9
CS–INH-b	5	0.39	92.17 ± 0.8
CS–INH-c	5	0.45	90.89 ± 0.7

As can be observed, the amount of INH derivative that was loaded into the matrix of chitosan was slightly lower than the encapsulated isoniazid (between 89.6 and 92.17% compared to 94.36% for pure isoniazid).

3.4. In Vitro Biodegradation Capacity

The results of the in vitro biodegradation using lysozyme, which was monitored over 7 days, are shown in Figure 6.

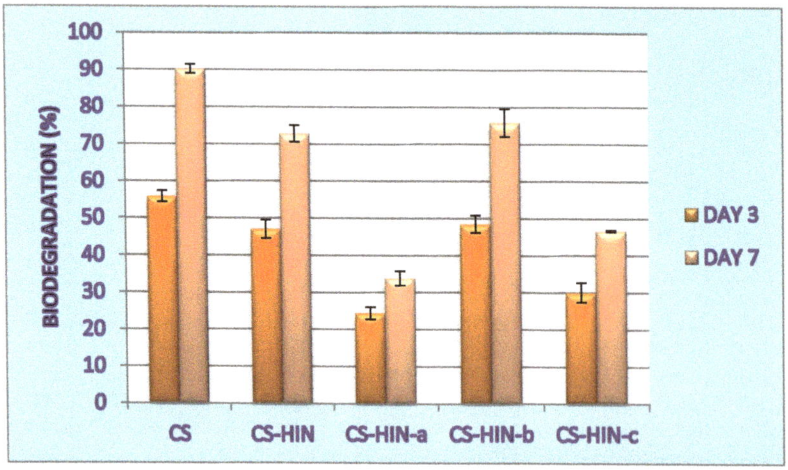

Figure 6. In vitro biodegradability of microparticles containing active substances compared to that of empty microparticles (CS).

From Day 3 of incubation, biodegradation occurred more than 20%, ranging from 24.34% (for CS–INH-a microparticles) to 48.49% (for CS–INH-b). Thus, there was a decrease in the biodegradation capacity of the corresponding microparticles due to the increase in the degree of lipophilicity and the occurrence of condensation reactions between isoniazid and different aromatic benzaldehydes. Compared to these microparticles, the empty chitosan microparticles displayed the highest biodegradation ratio of 55.83% (on Day 3). On the last day of incubation (Day 7), the biodegradation of empty microparticles increased to 90.15%, while it increased to a maximum of 75.87% for microparticles containing isoniazid derivatives (for CS–INH-b).

3.5. *In Vitro Biocompatibility Using MTT Cell Viability Analysis*

The cell viability test results are shown in Table 3.

Table 3. The cell viability values (%) for the tested samples of powder compounds and microparticles after 24 h, 48 h and 72 h.

Samples	**Cell Viability 24 h**	**(%) 48 h**	**72 h**
CS–INH-a	83.92 ± 0.7	71.08 ± 0.9	63.99 ± 0.6
CS–INH-b	106.54 ± 0.9	99.09 ± 1.2	89.64 ± 0.8
CS–INH-c	94.96 ± 0.5	80.22 ± 0.9	94.14 ± 0.7
CS–INH	100.27 ± 1.1	98.37 ± 0.7	95.53 ± 0.9
CS (Negative Control)	98.23 ± 0.8	90.92 ± 0.4	97.10 ± 0.6
H_2O_2 0.03% (Positive Control)	11.99 ± 0.8	5.02 ± 0.5	2.90 ± 1.4
Untreated Control	100.00	100.00	100.00

80–100%, non-cytotoxic compounds; 50–80%, mildly cytotoxic compounds; 30–50%, moderately cytotoxic compounds; <30%, severely cytotoxic compounds (ISO 10993-5, Geneva 2003).

The MTT testing results for the biocompatibility evaluation of the CS–INH microparticles on NCTC cells revealed that microparticles that were based on INH-b and INH-c compounds had manifested biocompatibility with fibroblast cells at all three intervals with viability values of 80.22–106.54% (CS–INH-b at 24 h), which were close to the values that were registered by the negative control, CS–INH and CS. The sample of CS–INH-a microparticles after 24 h was non-cytotoxic, while an increased cytotoxic effect was induced after 48 and 72 h.

The values recorded for the encapsulated derivatives were close to those of the empty chitosan microparticles (CS), i.e., 90.92–98.32%, which were used as the negative control (Table 3).

The influence of the tested microparticles on the cell culture presented in the images indicated a good biocompatibility of samples with NCTC cells, with exception of CS–INH-a, which induced a slightly cytotoxic effect with slow cell proliferation and the modification of the cell shape to an elongated or spindly form (Figure 7).

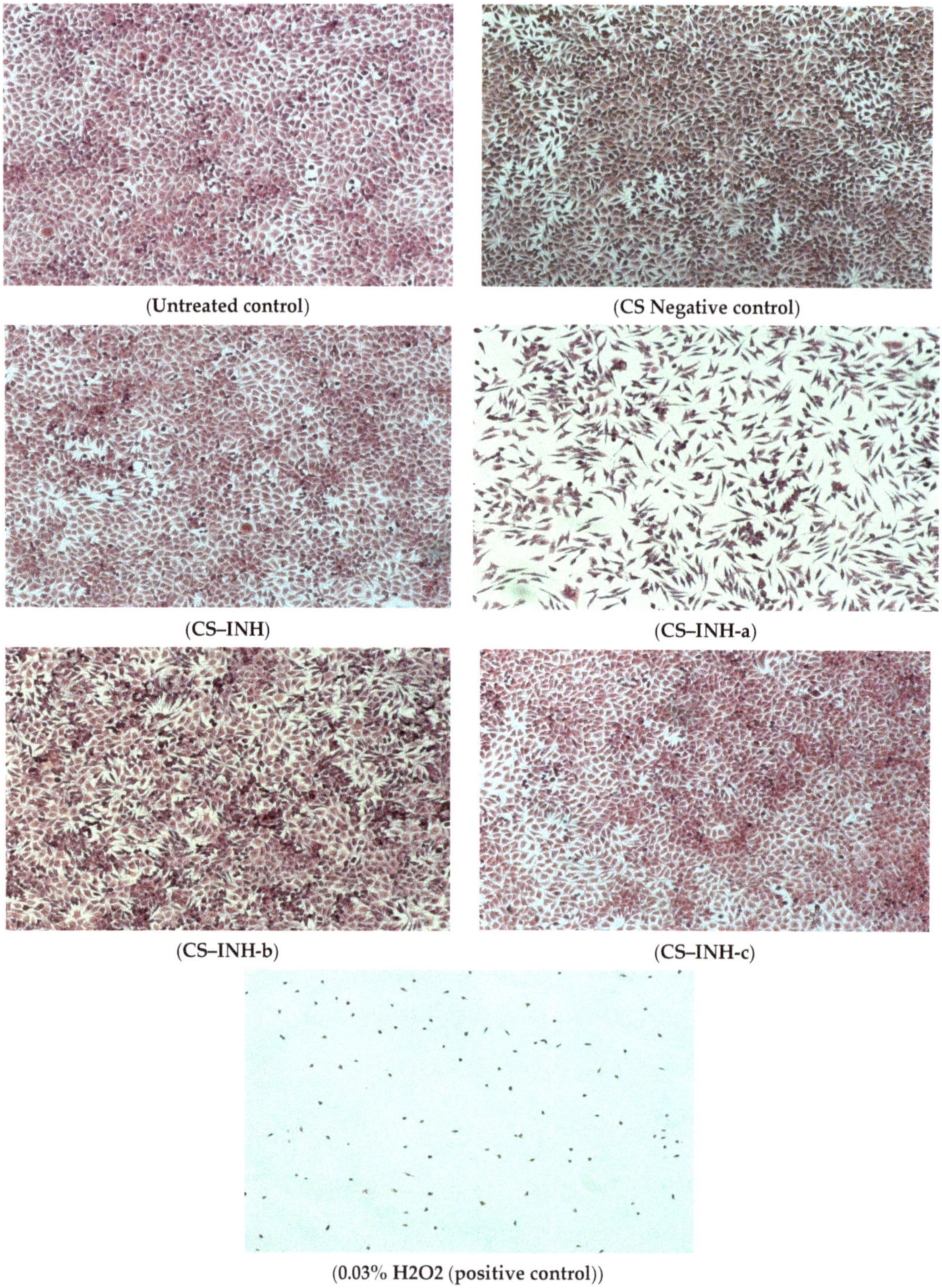

Figure 7. Cell morphology stained with hematoxylin–eosin after 72 h of incubation with microparticles of CS–INH derivatives, CS and CS–INH negative controls.

3.6. *Biological Evaluation*

3.6.1. In Vivo Toxicological Screening: The Determination of Chronic Toxicity

In order to evaluate the impact of the administered encapsulated compounds, on the tissues, liver fragments were examined and the following aspects were highlighted:

Group 1, which received INH microparticles (CS–INH), showed fewer signs of liver damage, manifested only by sinusoidal dilatations and rare inflammatory reactions. Comparing this group to the non-encapsulated INH [1], it could be seen that the areas of microvesicular steatosis disappeared through the encapsulation process, probably due to the hypolipidemic effects of chitosan (as mentioned in the introduction of this paper). Group 2, which received INH-a microparticles (CS–INH-a), also exhibited a pronounced dilation of the sinusoidal capillary system without necrosis but with rare inflammatory infiltrates, while Group 3 (INH-b microparticles) revealed extensive vascular congestion. For Group 4, which received INH-c microparticles, some of the hepatocytes exhibited finely vacuolar cytoplasm. In addition, due to the non-toxic effects of chitosan (which were referred to in the introduction section), Group 5 (CS, non-active empty microparticles) showed normal-looking hepatocytes. The untreated control batch (Group 6) revealed a normal liver morphology, with hepatocyte cords radiating from the centrilobular vein and a normal configuration of liver parenchyma (Figure 8).

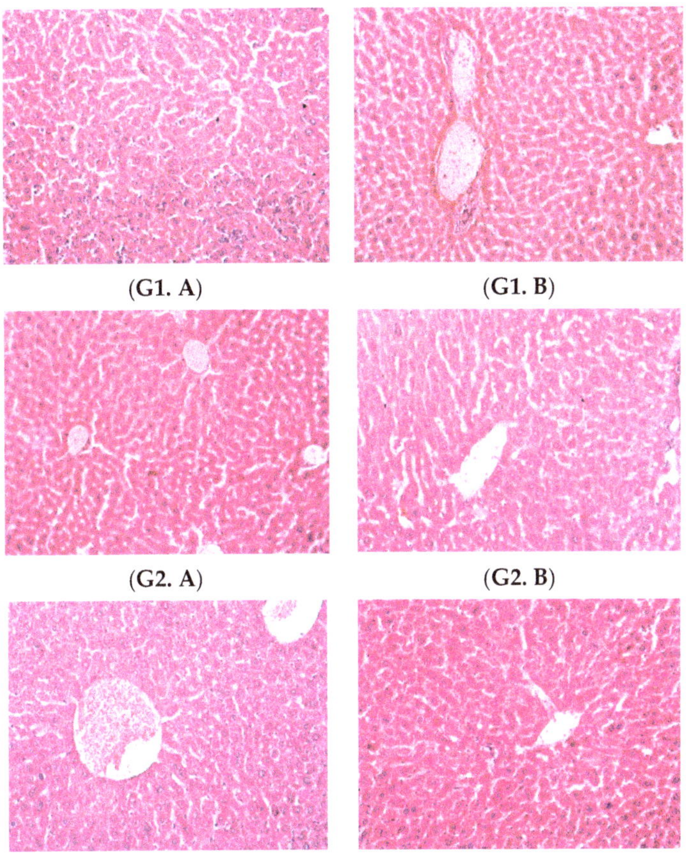

Figure 8. *Cont.*

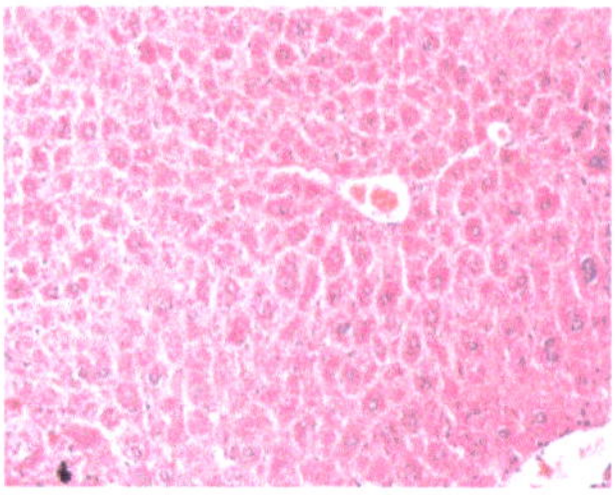

(G4)

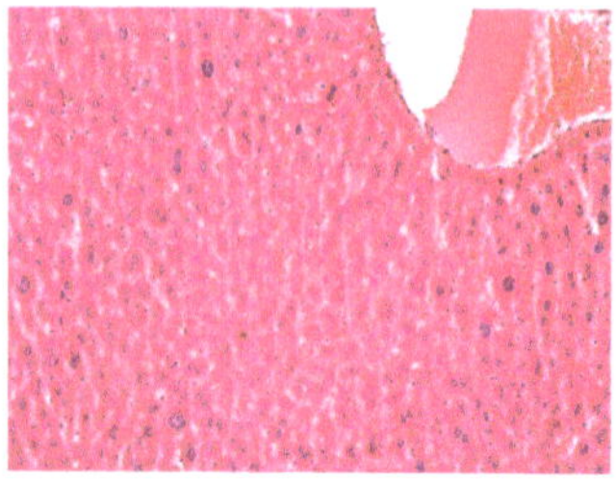

(G5)

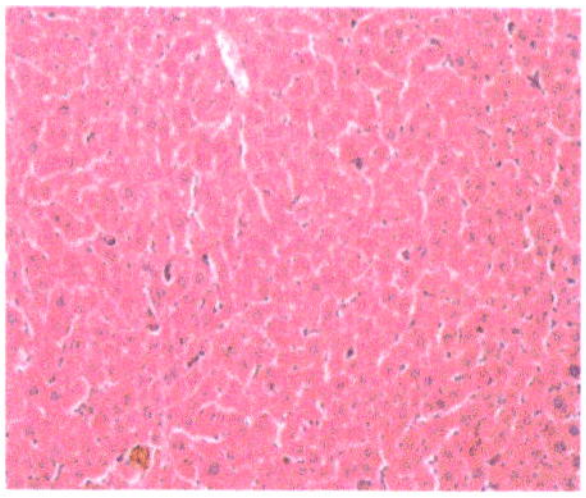

(G6)

Figure 8. Liver, hematoxylin and eosin staining × 20: Group 1, (**G1. A**) sinusoidal dilatations, capillary mononuclear cell infiltrations and the onset of necrosis and (**G1. B**) small areas of perivascular inflammation; Group 2, (**G2. A**) areas of vascular congestion and (**G2. B**) sinusoidal dilatations and reduced areas of inflammation; Group 3, (**G3. A**) extensive vascular congestion and (**G3. B**) slight sinusoidal dilatations; Group 4 (**G4**), hepatocellular vacuolation and reduced vascular congestion; Group 5 (**G5**), normal liver morphology; Group 6 (**G6**), normal liver morphology.

3.6.2. Evaluation of Biochemical Parameters

Alanine aminotransferase (ALT)/glutamic pyruvate transaminase (GPT).

A significant reduction in GPT values was observed in the chitosan–INH group (85.75 U/L) and the three encapsulated derivatives: chitosan–INH-a (78.46 U/L) and chitosan–INH-b (76.20 U/L). The lowest values of the GPT were recorded in the group that received INH-c–chitosan (75.15 U/L) and the control group that received empty chitosan microparticles (22.97 U/L), as well as the untreated control group (23.70 U/L) (Figure 9).

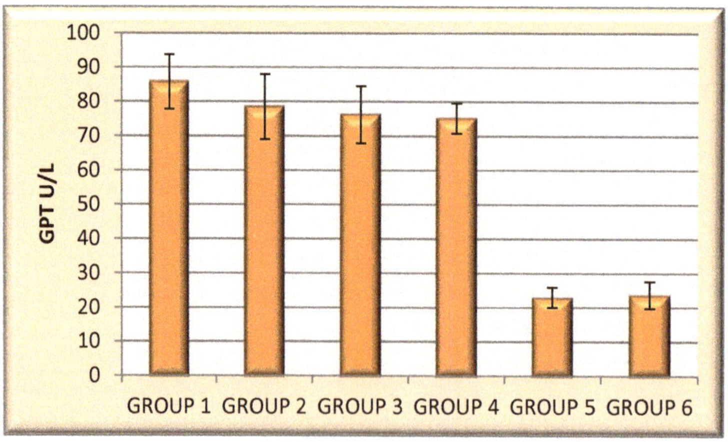

Figure 9. Mean GPT values for the encapsulated samples from the six groups of animals.

Aspartataminotransferase (ASAT)/glutamic oxaloacetic transaminase (TOG).

A slight reduction in GOT values was observed for the encapsulated isoniazid (chitosan–INH; 159.58 U/L) and the three encapsulated derivatives: chitosan–INH-a (147.13 U/L), chitosan–INH-b (145.06 U/L) and chitosan–INH-c (145.18 U/L). The lowest values of the GOT indicator were found in the control group that received empty chitosan microparticles (76.28 U/L) and the untreated control group (74.76 U/L) (Figure 10).

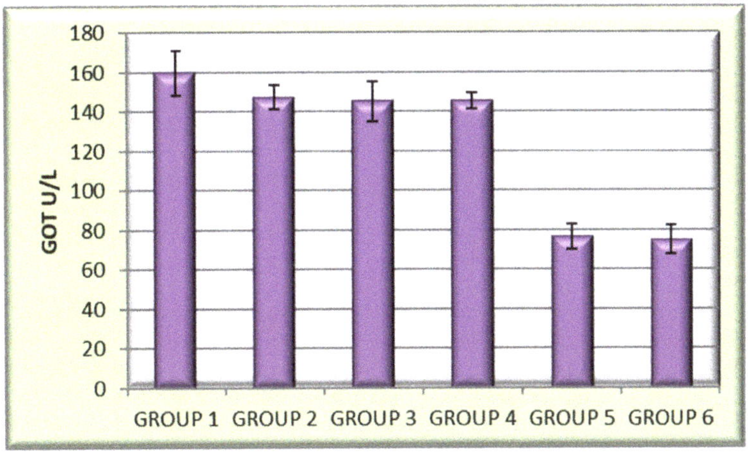

Figure 10. Mean GOT values for the encapsulated samples from the six groups of animals.

Alkaline phosphatase (ALP).

Alkaline phosphatase values ranged from 175.44 U/L to 182.06 U/L for INH and its encapsulated derivatives. Those values were significantly higher than those recorded in the control group (98.16 U/L) and untreated control group (97.21 U/L) (Figure 11).

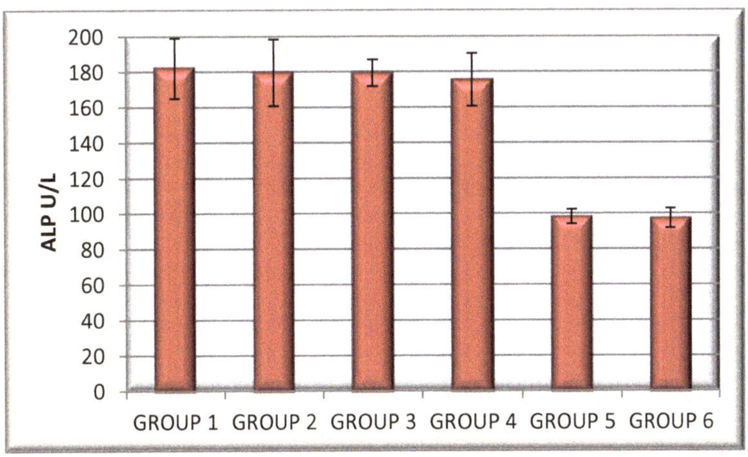

Figure 11. Mean alkaline phosphatase values for the encapsulated samples from the six groups of animals.

4. Discussion

Hepatotoxicity may be induced by drugs such as isoniazid, paracetamol, halothane, hydralazine, diclofenac and carbamazepine, which cause hepatotoxic reactions within 1 to 8 weeks in combination with a rash, fever and eosinophilia [22]. One of the main side effects of isoniazid is hepatotoxicity, which is a major cause of treatment discontinuation among tuberculosis patients. Human genetic studies have shown that cytochrome P450 2E1 (CYP2E1) is involved in the metabolism and hepatotoxicity of INH.

As demonstrated in our previous study [1], even when antimicrobial activity decreased with the introduction of attractive electron groups at positions 2 and 4 in the structure of benzaldehyde, higher MIC values in the case of new derivatives (0.84–4.16 μg/mL) were recorded. This happened simultaneously with a significant reduction in the toxicity of the studied compounds of the three isoniazid derivatives. Moreover, their activity was similar or even higher than that of other isoniazid derivatives that have been mentioned in the literature: new thioamides as isoniazid derivatives, with MIC values between 0.391 and 6.25 μg/mL [23]; new isoniazid–azole hybrids, with MIC values between 0.195 and 1.56 μM; and the new isoniazid–pyrrole hybrid, which was found to be active with a MIC value of 3.2 μg/mL [24].

Hepatic toxicity caused by INH can be manifested by the occurrence of cell necrosis, steatosis or both due to the metabolites of the drug, including hydrazine and acetyl hydrazine, having destructive effects on liver cells. In animal experiments, a positive and significant correlation has been reported between plasma hydrazine levels and the severity of liver cell destruction caused by isoniazid treatment [25].

In our previous study [1], we concluded that the least hepatic damage was exhibited by the group of animals that received the INH-c derivative, which showed only slight vascular congestion. Thus, the condensation of isoniazid with bromobenzaldehyde had the most favorable influence on reducing the toxicity of the parent compound at the tissue level.

Microvesicular steatosis was only reported in the group of animals that received isoniazid in suspension, while cell necrosis was completely absent in the case of INH-c but still present in the INH-a and INH-b derivatives cases. Although condensation with bromobenzaldehyde has been proven to be the most favorable, the use of benzaldehyde and nitrobenzaldehyde for the chemical modification of isoniazid presented positive effects in both the reduction in cell necrosis and the disappearance of microvesicular steatosis, while maintaining increased therapeutic potential.

On the other hand, microparticles have been studied as a method for the controlled release of drugs for the pulmonary administration of active substances used to treat various lung diseases, such as asthma, chronic obstructive pulmonary disease or various infectious pathologies [26]. Microencapsulation offers the advantages of protecting the drug against pulmonary metabolism, while ensuring the sustained and prolonged release of the drug. These benefits could be all the more beneficial in the treatment of tuberculosis by increasing patient compliance with long-term therapies, thus allowing for less frequent dosing. Another advantage could be the reduction in the long-term side effects that are associated with systemic or oral tuberculosis therapies [27]. In addition, chitosan has beneficial effects (through its biocompatibility, bioactivity and non-toxicity) on the toxicological and pharmacokinetic profiles of the active substances [28].

Furthermore, during the microencapsulation process, the size of the microparticles is easily controllable by setting the necessary parameters, which is a plus in that the resulting particles can efficiently penetrate into the alveoli due to their aerodynamic shape. Thus, higher intracellular drug concentrations in macrophages could be obtained [29].

There are various studies in the literature that have aimed to encapsulate active principles that are used in the treatment of tuberculosis using biodegradable polymers, such as polylactic acid (PLA) [30], poly (lactic-co-glycolic acid) (PLGA) [31], hyaluronic acid [32] or chitosan [33,34]. Among the substances included in the microencapsulation process, both first-line drugs that are used in the treatment of tuberculosis (such as rifampicin, isoniazid and pyrazinamide) and second- or third-line drugs (such as para-aminosalicylic acid (PAS), capreomycin, amikacin and moxifloxacin) have been considered [35–37].

In one study [38], a dose–effect relationship was observed between poly (lactic-co-glycolic acid) microspheres that contained rifampicin and pulmonary bacterial loading, after either single or repeated administration. In addition, guinea pigs that were treated with these microspheres had a significantly lower number of viable bacteria but reduced inflammation and lung damage compared to control animals.

In another study presenting the encapsulation of rifampicin in chitosan, the resulting microparticles showed a significant increase in the release of the active substance compared to other polymers [39]. The same authors encapsulated isoniazid in chitosan microparticles and found that their size allowed release to the lower airways, but they did not include animal studies to test the benefits to the treatment of tuberculosis.

In our study, the influence of microencapsulation on cell biocompatibility was tested in vitro and the influence on hepatotoxicity was tested in vivo. An improvement in cytotoxicity was observed for the drugs that were encapsulated in chitosan compared to the non-encapsulated versions, which was confirmed by higher values of cell viability percentage.

For example, INH-c, the derivative with the highest cell viability after 72 h, had a cell viability percentage of 94.14% in its encapsulated form, while the percentage was 72.47% in its non-encapsulated form [1].

The encapsulation of isoniazid derivatives using chitosan as a microencapsulation polymer displayed beneficial effects on chronic toxicity reduction. Hepatocellular lesions caused by necrosis and microvesicular steatosis were only observed in animals that received pure isoniazid during the 30-day experiment [1].

These abnormalities were reduced by administering microencapsulated treatment, both in terms of the reduction in areas of cell necrosis and the disappearance of microvesicular steatosis. Similar effects were observed for all isoniazid derivatives in encapsulated form, in contrast to non-encapsulated compounds.

Thus, the importance of using chitosan to reduce isoniazid hepatotoxicity was particularly emphasized. Unlike non-encapsulated isoniazid INH [1], the encapsulated form CS–INH resulted in reduced areas of advanced cell necrosis as well as the disappearance of microvesicular steatosis. The lipid-lowering effects of chitosan, which are known in the literature [40], had a positive influence on microvesicular stasis. Moreover, the hepatoprotective effects of chitosan could be explained by its antioxidant action, as mentioned in various studies [41–44].

Regarding the encapsulated form of INH-a (in contrast to the non-encapsulated version for which areas of the onset of necrosis were described), the disappearance of areas of cell necrosis was noticed, which was similar to the administration of INH-b microcapsules. In the case of INH-c, the non-encapsulated form did not produce significant liver damage and no changes in tissue morphology were reported for the encapsulated form.

The hepatotoxicity analysis of laboratory animals was also evaluated by the determination of liver enzyme activity. Thus, serum GPT and GOT levels (markers for liver damage) increased significantly in animals that received parental isoniazid, indicating hepatotoxicity due to the oxidative destruction of hepatocytes, which resulted in the release of enzymes into the vascular compartments. The hepatotoxicity was significantly diminished by the administration of drugs in a microencapsulated form due to the intrinsic properties of chitosan and the structural changes that were made to the isoniazid molecule, which were addressed in the introduction section. Thus, in the case of groups that received encapsulated substances, there was a decrease in the values expressing the activity of liver enzymes compared to the groups that received non-encapsulated substances. This was closely related to the disappearance of cell necrosis areas and microvesicular steatosis that was highlighted during the histopathological examination, which confirmed the correlation of the data obtained from the two types of analysis.

In this context, the use of chitosan microencapsulation for tuberculostatic therapies could represent an important strategy for reducing liver enzyme levels, removing microvesicular steatosis and reducing cell necrosis areas. All of this would have a very important contribution to reducing the risk of fulminant liver failure.

5. Conclusions

Hepatotoxicity induced by tuberculostatic therapy is a significant adverse effect that interferes with the effective administration of tuberculosis treatment. In this regard, our study presented the beneficial effects of the microencapsulation of specific drugs on liver damage induced by anti-TB therapy. The chitosan microparticles loaded with isoniazid derivatives that were subjected to our research proved to be an advantageous formulation for reducing the hepatotoxicity of the studied compounds and ensuring the disappearance of cell necrosis and microvesicular steatosis compared to their administration in suspension. In conclusion, the microparticles obtained in innovative pharmaceutical forms have potential biological application as antimicrobial agents in the treatment of tuberculosis, for which both a favorable antimicrobial action and a low incidence of adverse reactions are required given the problem of high toxicity caused by current tuberculostatic medication.

Author Contributions: Conceptualization, O.-M.D. and C.L.Z.; methodology, O.-M.D., R.T. and C.L.Z.; software, C.L.C.; validation, O.-M.D., R.T. and C.L.Z.; formal analysis, I.D., A.T.I., M.D., C.L.C., A.-S.Z. and C.D.S.; investigation, R.T., M.D., L.C. and C.L.C.; resources, O.-M.D. and C.L.Z.; data curation, I.D., A.T.I., M.D. and L.C.; writing—original draft preparation, I.D., A.T.I. and A.-S.Z.; writing—review and editing, O.-M.D. and C.L.C.; visualization, C.D.S. and C.L.Z.; supervision, C.L.Z. All authors have read and agreed to the published version of the manuscript.

Funding: This research received no external funding.

Institutional Review Board Statement: The animal study protocol was approved by the Institutional Ethics Committee—University of Medicine and Pharmacy, Grigore T. Popa, Iasi (17 April 2018).

Informed Consent Statement: Not applicable.

Data Availability Statement: Not applicable.

Conflicts of Interest: The authors declare no conflict of interest.

References

1. Dragostin, I.; Dragostin, O.M.; Samal, S.K.; Dash, S.; Tatia, R.; Dragan, M.; Confederat, L.; Ghiciuc, C.M.; Diculencu, D.; Lupușoru, C.E.; et al. New isoniazid derivatives with improved pharmaco-toxicological profile: Obtaining, characterization and biological evaluation. *Eur. J. Pharm. Sci.* **2019**, *137*, 104974. [CrossRef] [PubMed]
2. Dragostin, I.; Dragostin, O.; Pelin, A.-M.; Grigore, C.; Zamfir, C.L. The importance of polymers for encapsulation process and for enhanced cellular functions. *J. Macromol. Sci.—Pure Appl. Chem.* **2017**, *54*, 489–493. [CrossRef]
3. Baltrusch, K.L.; Torres, M.D.; Domínguez, H.; Florez-Fernandez, N. Spray-drying microencapsulation of tea extracts using green starch, alginate or carrageenan as carrier materials. *Int. J. Biol. Macromol.* **2022**, *203*, 417–429. [CrossRef]
4. Thuekeaw, S.; Angkanaporn, K.; Chirachanchai, S.; Nuengjamnong, C. Dual pH responsive via double—Layered microencapsulation for controlled release of active ingredients in simulated gastrointestinal tract: A model case of chitosan-alginate microcapsules containing basil oil (*Ocimum basilicum* Linn.). *Polym. Degrad. Stab.* **2021**, *191*, 109660. [CrossRef]
5. Annunziata, G.; Jiménez-García, M.; Capó, X.; Moranta, D.; Arnone, A.; Tenore, G.C.; Sureda, A.; Tejada, S. Microencapsulation as a tool to counteract the typical low bioavailability of polyphenols in the management of diabetes. *Food Chem. Toxicol.* **2020**, *139*, 111248. [CrossRef] [PubMed]
6. Feng, Y.; Xia, W. Preparation, characterization and antibacterial activity of water-soluble O-fumaryl-chitosan. *Carbohydr. Polym.* **2011**, *83*, 1169–1173. [CrossRef]
7. Kong, M.; Chen, X.G.; Xing, K.; Park, H.J. Antimicrobial properties of chitosan and mode of action: A state of the art review. *Int. J. Food Microbiol.* **2010**, *144*, 51–63. [CrossRef] [PubMed]
8. Dragostin, O.M.; Samal, S.K.; Lupascu, F.; Pânzariu, A.; Dubruel, P.; Lupascu, D.; Tuchilus, C.; Vasile, C.; Profire, L. Development and Characterization of Novel Films Based on Sulfonamide-Chitosan Derivatives for Potential Wound Dressing. *Int. J. Mol. Sci.* **2015**, *16*, 29843–29855. [CrossRef]
9. Lupascu, F.G.; Dash, M.; Samal, S.K.; Dubruel, P.; Lupusoru, C.E.; Lupusoru, R.V.; Dragostin, O.; Profire, L. Development, optimization and biological evaluation of chitosan scaffold formulations of new xanthine derivatives for treatment of type-2 diabetes mellitus. *Eur. J. Pharm. Sci.* **2015**, *77*, 122–134. [CrossRef] [PubMed]
10. Garcia, L.G.S.; da Rocha, M.G.; Lima, L.R.; Cunha, A.P.; de Oliveira, J.S.; de Andrade, A.R.C.; Ricardo, N.M.P.S.; Pereira-Neto, W.A.; Sidrim, J.J.C.; Rocha, M.F.G.; et al. Essential oils encapsulated in chitosan microparticles against *Candida albicans* biofilms. *Int. J. Biol. Macromol.* **2021**, *166*, 621–632. [CrossRef]
11. Punarvasu, T.P.; Prashanth, K.V.H. Self-assembled chitosan derived microparticles inhibit tumor angiogenesis and induce apoptosis in Ehrlich-ascites-tumor bearing mice. *Carbohydr. Polym.* **2022**, *278*, 118941. [CrossRef]

12. Dragostin, O.M.; Tatia, R.; Samal, S.K.; Oancea, A.; Zamfir, A.S.; Dragostin, I.; Lisa, E.-L.; Apetrei, C.; Zamfir, C.L. Designing of Chitosan Derivatives Nanoparticles with Antiangiogenic Efect for Cancer Therapy. *Nanomaterials* **2020**, *10*, 698. [CrossRef]
13. Bai, J.; Zhanga, Y.; Chen, L.; Yan, H.; Zhang, C.; Liu, L.; Xu, X. Synthesis and characterization of paclitaxel-imprinted microparticles for controlled release of an anticancer drug. *Mater. Sci. Eng. C* **2018**, *92*, 338–348. [CrossRef]
14. Dragostin, O.M.; Samal, S.K.; Dash, M.; Lupascu, F.; Pânzariu, A.; Tuchilus, C.; Ghetu, N.; Danciu, M.; Dubruel, P.; Pieptu, D.; et al. New antimicrobial chitosan derivatives for wound dressing applications. *Carbohydr. Polym.* **2016**, *141*, 28–40. [CrossRef] [PubMed]
15. Batir-Marin, D.; Mircea, C.; Boev, M.; Burlec, A.F.; Corciova, A.; Fifere, A.; Iacobescu, A.; Cioanca, O.; Verestiuc, L.; Hancianu, M. In Vitro Antioxidant, Antitumor and Photocatalytic Activities of Silver Nanoparticles Synthesized Using Equisetum Species: A Green Approach. *Molecules* **2021**, *26*, 7325. [CrossRef]
16. Avwioro, G. Histochemical Uses of Haematoxylin—A Review Godwin Avwioro. *J. Phys. Conf. Ser.* **2011**, *1*, 24–34.
17. Mogoanta, L. *Ghid de Tehnici de Histologie si Imunohistochimie*; Medicala Universitara: Craiova, Romania, 2007.
18. Géhin, C.; Holman, S.W. Advances in high-resolution mass spectrometry applied to pharmaceuticals in 2020: A whole new age of information. *Anal. Sci. Adv.* **2021**, *2*, 142–156. [CrossRef]
19. Morariu, D.I.; Avasilcăi, L.; Vieriu, M.; Lungu, I.I.; Morariu, B.; Robu, S.; Tiutunaru, D.; Cioancă, O.; Hăncianu, M. Estimation of Quinolones, Ceftiofur and Thiamphenicol Residues Levels In Honey. *Farmacia* **2021**, *69*, 3. [CrossRef]
20. Floros, D.J.; Petras, D.; Kapono, C.A.; Melnik, A.V.; Ling, T.J.; Knight, R. Mass spectrometry based molecular 3D-cartography of plant metabolites. *Front. Plant Sci.* **2017**, *8*, 429. [CrossRef]
21. Carrano, L.; Naggi, A.; Urso, E. High Resolution Mass Spectrometry for the Recognition and Structural Characterization of a New Antimicrobial Compound. *Pharmacol. Pharm.* **2018**, *9*, 135. [CrossRef]
22. Daly, A.K.; Day, C.P. Genetic association studies in drug-induced liver injury. *Drug Metab. Rev.* **2012**, *44*, 116–126. [CrossRef]
23. Matei, L.; Bleotu, C.; Baciu, I.; Draghici, C.; Ionita, P.; Paun, A.; Chifiriuc, M.C.; Sbarcea, A.; Zarafu, I. Synthesis and bioevaluation of some new isoniazid derivatives. *Bioorg. Med. Chem.* **2013**, *21*, 5355–5361. [CrossRef]
24. Hu, Y.-Q.; Zhang, S.; Zhao, F.; Gao, C.; Feng, L.-S.; Lv, Z.-S.; Xu, Z.; Wu, X. Isoniazid derivatives and their anti-tubercular activity. *Eur. J. Med. Chem.* **2017**, *133*, 255–267. [CrossRef]
25. Jahromi, H.K.; Pourahmad, M.; Abedi, H.A.; Karimi, M.; Jahromi, Z.K. Protective effects of salep against isoniazid liver toxicity in wistar rats. *J. Tradit. Complement Med.* **2017**, *8*, 239–240. [CrossRef]
26. Panchal, R.; Patel, H.; Patel, V.; Joshi, P.; Parikh, A. Formulation and evalution of montelukast sodium—Chitosan based spray dried microspheres for pulmonary drug delivery. *J. Pharm. Bioallied Sci.* **2012**, *4*, S110–S111. [CrossRef] [PubMed]
27. Pham, D.-D.; Fattal, E.; Tsapis, N. Pulmonary drug delivery systems for tuberculosis treatment. *Int. J. Pharm.* **2015**, *478*, 517–529. [CrossRef]
28. Ranganathan, P.; Mutharani, B.; Chen, S.-M.; Sireesha, P. Biocompatible chitosan-pectin polyelectrolyte complex for simultaneous electrochemical determination of metronidazole and metribuzin. *Carbohydr. Polym.* **2019**, *214*, 317–327. [CrossRef] [PubMed]
29. Sharma, R.; Muttil, P.; Yadav, A.B.; Rath, S.K.; Bajpai, V.K.; Mani, U.; Misra, A. Uptake of inhalable microparticles affects defence responses of macrophages infected with *Mycobacterium tuberculosis* H37Ra. *J. Antimicrob. Chemother.* **2007**, *59*, 499–506. [CrossRef]
30. Liu, S.; Qin, S.; He, M.; Zhou, D.; Qin, Q.; Wang, H. Current applications of poly (lactic acid) composites in tissue engineering and drug delivery. *Compos. Part B* **2020**, *199*, 108238. [CrossRef]
31. O'Connora, G.; Krishnane, N.; Fagan-Murphy, A.; Cassidy, J.; O'Learyd, S.; Robertsone, B.D.; Keaned, J.; O'Sullivand, M.P.; Cryan, S.-A. Inhalable poly (lactic-co-glycolic acid) (PLGA) microparticles encapsulating all-trans-Retinoic acid (ATRA) as a host-directed, adjunctive treatment for *Mycobacterium tuberculosis* infection. *Eur. J. Pharm. Biopharm.* **2019**, *134*, 153–165. [CrossRef]
32. Mukhtar, M.; Szakonyi, Z.; ´Farkas, A.; Burian, K.; Kokai, D.; Ambrus, R. Freeze-dried vs spray-dried nanoplex DPIs based on chitosan and its derivatives conjugated with hyaluronic acid for tuberculosis: In vitro aerodynamic and in silico deposition profiles. *Eur. Polym. J.* **2021**, *160*, 110775. [CrossRef]
33. Pandey, R.P.; Kumar, S.; Ahmad, S.; Vibhuti, A.; Raj, V.S.; Verma, A.K.; Sharma, P.; Leal, E. Use Chou's 5-steps rule to evaluate protective efficacy induced by antigenic proteins of *Mycobacterium tuberculosis* encapsulated in chitosan nanoparticles. *Life Sci.* **2020**, *256*, 117961. [CrossRef] [PubMed]
34. Abdel-Aziz, M.M.; Abu Elella, M.H.; Mohamed, R.R. Green synthesis of quaternized chitosan/silver nanocomposites for targeting *Mycobacterium tuberculosis* and lung carcinoma cells (A-549). *Int. J. Biol. Macromol.* **2020**, *142*, 244–253. [CrossRef]
35. Bhat, A.R.; Wani, F.A.; Behera, K.; Khan, A.B.; Patel, R. Formulation of biocompatible microemulsions for encapsulation of anti-TB drug rifampicin: A physicochemical and spectroscopic study. *Colloids Surf. A Physicochem. Eng. Asp.* **2022**, *645*, 128846. [CrossRef]
36. Manning, T.; Mikula, R.; Lee, H.; Calvin, A.; Darrah, J.; Wylie, G.; Phillips, D.; Bythell, B.J. The copper (II) ion as a carrier for the antibiotic capreomycin against *Mycobacterium tuberculosis*. *Bioorg. Med. Chem. Lett.* **2014**, *24*, 976–982. [CrossRef]
37. Abdelghany, S.; Parumasivam, T.; Pang, A.; Roediger, B.; Tang, P.; Jahn, K.; Britton, W.J.; Chan, H.-K. Alginate modified-PLGA nanoparticles entrapping amikacin and moxifloxacin as a novel host-directed therapy for multidrug-resistant tuberculosis. *J. Drug Deliv. Sci. Technol.* **2019**, *52*, 642–651. [CrossRef]
38. Suarez, S.; O'Hara, P.; Kazantseva, M.; Newcomer, C.E.; Hopfer, R.; McMurray, D.N.; Hickey, A.J. Airways delivery of rifampicin microparticles for the treatment of tuberculosis. *J. Antimicrob. Chemother.* **2001**, *48*, 431–434. [CrossRef] [PubMed]
39. Kundawala, A.J. Treating tuberculosis with chitosan microparticles loaded with rifampicin as respirable powder for pulmonary delivery. *Indian J. Nov. Drug Deliv.* **2012**, *4*, 57–65.

40. Zhao, X.-H.; Qian, L.; Yin, D.-L.; Zhou, Y. Hypolipidemic effect of the polysaccharides extracted from pumpkin by cellulase-assisted method on mice. *Int. J. Biol. Macromol.* **2014**, *64*, 137–138. [CrossRef] [PubMed]
41. Tan, W.; Zhang, J.; Mi, Y.; Dong, F.; Li, Q.; Guo, Z. Synthesis, characterization, and evaluation of antifungal and antioxidant properties of cationic chitosan derivative via azide-alkyne click reaction. *Int. J. Biol. Macromol.* **2018**, *120*, 318–324. [CrossRef] [PubMed]
42. Afonso, C.R.; Hirano, R.S.; Gaspar, A.L.; Chagas, E.G.L.; Carvalho, R.A.; Silva, F.V.; Leonardi, G.R.; Lopes, P.S.; Silva, C.F.; Yoshida, C.M.P. Biodegradable antioxidant chitosan films useful as an anti-aging skin mask. *Int. J. Biol. Macromol.* **2019**, *132*, 1262–1273. [CrossRef] [PubMed]
43. Yingjun, J.; Jianghao, H.; Xueqing, X. Maintenance of the antioxidant capacity of fresh-cut pineapple by procyanidin-grafted chitosan. *Postharvest Biol. Technol.* **2019**, *154*, 79–86.
44. Savic, I.M.; Savic Gajic, I.M.; Milovanovic, M.G.; Zerajic, S.; Gajic, D.G. Optimization of Ultrasound-Assisted Extraction and Encapsulation of Antioxidants from Orange Peels in Alginate-Chitosan Microparticles. *Antioxidants* **2022**, *11*, 297. [CrossRef]

polymers

MDPI

Article

Modification of the Polymer of a Bone Cement with Biodegradable Microspheres of PLGA and Loading with Daptomycin and Vancomycin Improve the Response to Bone Tissue Infection

Joaquin García-García [1,‡], Galo Azuara [2,‡], Oscar Fraile-Martinez [3,4], Cielo García-Montero [3,4], Miguel Angel Álvarez-Mon [3,4], Sara Ruíz-Díez [3], Melchor Álvarez-Mon [3,4,5], Julia Buján [3,4], Natalio García-Honduvilla [3,4], Miguel A. Ortega [3,4,*,†] and Basilio De la Torre [6,7,†]

1 Service of Orthopedic Surgery of University Hospital Principe de Asturias, 28805 Alcalá de Henares, Madrid, Spain; joaquingarciagarcia@gmail.com
2 Service of Traumatology of University Hospital of Guadalajara, 19002 Guadalajara, Spain; galoazu@hotmail.com
3 Departments of Medicine and Medical Specialities, Faculty of Medicine and Health Sciences, University of Alcalá, Alcalá de Henares, 28801 Madrid, Spain; oscarfra.7@hotmail.com (O.F.-M.); cielo.gmontero@gmail.com (C.G.-M.); maalvarezdemon@icloud.com (M.A.Á.-M.); sruizdiez16@gmail.com (S.R.-D.); mademons@gmail.com (M.Á.-M.); mjulia.bujan@uah.es (J.B.); natalio.garcia@uah.es (N.G.-H.)
4 Ramón y Cajal Institute of Sanitary Research (IRYCIS), 28034 Madrid, Spain
5 Immune System Diseases-Rheumatology, Oncology Service an Internal Medicine, University Hospital Príncipe de Asturias, (CIBEREHD), 28806 Alcala de Henares, Spain
6 Department of Surgery, Medical and Social Sciences, Faculty of Medicine and Health Sciences, University of Alcalá, Alcalá de Henares, 28801 Madrid, Spain; bjtorre@gmail.com
7 Service of Traumatology of University Hospital Ramón y Cajal, 28034 Madrid, Spain
* Correspondence: miguel.angel.ortega92@gmail.com
† These authors shared senior authorship in this work.
‡ These authors contributed equally to this work.

Abstract: Chronic infections are one of the most serious adverse outcomes of prosthetic surgery. Prosthetic revision surgery using a bone cement loaded with antibiotics between the two stages of the surgery is commonly performed. However, this method often fails to reach the minimum inhibitory concentration and promotes antibiotic resistance, thus emphasizing the need for improving the current available therapies. Materials and methods: In this study, we performed a study of the in vivo response of a polymer-based construct of poly (lactic-*co*-glycolic acid) (PLGA) in the solid phase of Palacos R® in combination with vancomycin, daptomycin, and/or linezolid. To test its effectiveness, we applied an in vivo model, using both histological and immunohistochemical analyses to study the bone tissue. Results: The presence of PLGA in the combination of vancomycin with daptomycin showed the most promising results regarding the preservation of bone cytoarchitecture and *S. aureus* elimination. Conversely, the combination of vancomycin plus linezolid was associated with a loss of bone cytoarchitecture, probably related to an increased macrophage response and inefficient antimicrobial activity. Conclusions: The modification of Palacos R® bone cement with PLGA microspheres and its doping with the antibiotic daptomycin in combination with vancomycin improve the tissue response to bone infection.

Keywords: preclinical model; bone; polymer-based construct of poly (lactic-*co*-glycolic acid) (PLGA); infection

1. Introduction

Chronic infection is one of the most serious complications of prosthetic surgery, with a prevalence of 0.3–2.2% in primary surgery [1] and 3–4% in revision surgery [2]. Prosthetic infection is a therapeutic challenge due to the appearance of biofilms, made by microorganisms that are enveloped and protected in them, which strongly adhere to the surface of the implant [3]. Several strategies can be applied to treat prosthetic infections, mainly depending on the time of onset of the infection. For chronic infections, prosthetic revision surgery performed in two stages has been the most commonly used method, using a bone cement spacer between the first and second stages [4,5]. This bone cement spacer is loaded with antibiotics, whose release results in much higher local concentrations than can be achieved through intravenous injection. This release is higher in the first few days but persists up to several weeks, in many cases at concentrations below the minimum inhibitory concentration (MIC) of the microbe, which has been associated with the development of antibiotic resistance, the colonization of spacers, and, therefore, therapeutic failure [6,7].

Approximately 20% of *Staphylococcus aureus* and 80% of *Staphylococcus epidermidis* isolated from prosthetic infections are resistant to methicillin [8]. The antibiotic of choice for the treatment of methicillin-resistant staphylococcal infections is vancomycin [9]. The growing abundance of multidrug-resistant bacteria makes it necessary to incorporate new antibiotics into bone cement. Daptomycin and linezolid are the two main alternatives to vancomycin in cases of bacterial resistance [10], having demonstrated good release kinetics and antibacterial properties in in vitro cement [11]. Daptomycin is a calcium-dependent cyclic lipopeptide. It was the first in class of a new group of calcium-dependent membrane-binding lipopeptides [12]. Linezolid is the first member of the class of oxazolidinones. The compound is a synthetic antibiotic that inhibits bacterial protein synthesis by binding to rRNA [13].

Prosthetic infection affects the surrounding bone tissue and its remodeling capacity, which are difficult to evaluate in clinical practice. Experimental animal models are necessary to investigate these issues. There are widely used histological scales created by different authors [14,15], but they evaluate very specific aspects of the bone histoarchitecture and exhibit great variability depending on the strain of the bacteria, the inoculum, and the selected animal model. For this reason, we developed a new histological staging system that can assess bone tissue in a complete way, providing a global picture of the conservation and remodeling of bone histoarchitecture in the presence of infection and under the influence of bone cement [16,17].

Our study group managed to increase the release of vancomycin, daptomycin, and linezolid in vitro, significantly improving the kinetics of their temporary release by incorporating biodegradable microparticles of poly(lactic-*co*-glycolic acid) (PLGA) in the solid phase of Palacos R cement ® [11]. Using these developed cements and applying the new histological staging system that we developed [16], in this paper, we conducted the present study to (1) assess the histological behavior of the modified cement loaded with vancomycin, daptomycin, and/or linezolid in a rabbit model of osteomyelitis; (2) assess the tissue distribution of bacteria; and (3) analyze the presence of macrophages around the different formulations.

2. Materials and Methods

2.1. Formulations of Bone Cement

The formulations of acrylic bone cements were loaded with two new-generation antibiotics, linezolid (Pfizer, Peapack, NJ, USA) and daptomycin (Cubist Pharmaceutical Incorporation, Kenilworth, NJ, USA), individually or in combination with vancomycin. The composition of the solid phase of commercial bone cement Palacos R® (Heraeus Medical, Wehrheim, Germany) was modified by incorporating microspheres of a biodegradable copolymer, PLGA, as described in our previous study [11].

2.2. Titanium Rods Coated with Hydroxyapatite Contaminated with S. aureus

The rods used in the experimental model were made of grade 4 titanium and measured 3 mm in diameter and 10 mm in length. They had a coating with a thickness less than

150 microns, made with a solution of poly(methyl methacrylate) (PMMA) in acetone (10%) and crude hydroxyapatite (HA) in suspension at a ratio of PMMA:HA of 1:1.2 [16].

A clinical strain of *Staphylococcus aureus* was selected, which was sensitive to vancomycin (MIC: 1 μg/mL), linezolid (MIC: 2 μg/mL), daptomycin (MIC: 0.5 μg/mL), gentamicin (MIC: 0.5 μg/mL), rifampicin (MIC: 0.5 μg/mL), and clindamycin (MIC: 0.25 μg/mL) and resistant to other antibiotics, including cloxacillin, cephalosporins, and quinolones. A suspension was prepared with an inoculum of the germ equivalent to 1.2×10^9 colony-forming units/mL. In each tube with this suspension, a titanium rod coated with PMMA:HA was introduced for 24 h to achieve contamination. The presence of many cocci adhering to the rod after extraction from the suspension was observed by scanning electron microscopy [16]. This rod became the implant contaminated with methicillin-resistant *S. aureus* (MRSA) that we introduced into the bone tissue of the rabbit [16,17].

2.3. Animal Experimental Model

As the experimental animal, we used rabbits *(Oryctolagus cuniculus)* of the New Zealand variety, male, with a weight between 2.5 and 3.2 kg and an age of 12 weeks at the beginning of the study. They had physeal cartilage in the distal femur, as they were not skeletally mature [17]. The animals were divided into six study groups (1–6) according to the use of the commercial cement Palacos R® or the experimental cement Palacos R + PLGA and the presence of the antibiotics under study (vancomycin, daptomycin, and/or linezolid). Each group had five rabbits (Table 1).

Table 1. Compositions of the experimental formulations.

Group	Cement	Antibiotic
1	Palacos R	Daptomycin
2	Palacos R + PLGA	Daptomycin
3	Palacos R	VancomycinDaptomycin
4	Palacos R	VancomycinLinezolid
5	Palacos R+PLGA	VancomycinDaptomycin
6	Palacos R+PLGA	VancomycinLinezolid

Two surgeries were performed, namely, the initial surgery and the final surgery, separated by 3 weeks (20–22 days). The surgical technique is described in an article by our study group, Azuara et al. [17]

Animals were managed in accordance with the current International Regulations on Experimental Animals (609/86/EEC and ETS 123) at the Animal Research Center of the University of Alcala. The study protocol received approval from the Committee on the Ethics of Animal Experiments of the University of Alcala (CEI UAH 2011017). The diet of the animals was available ad libitum.

2.4. Sample Processing: Histological and Immunohistochemical Techniques

After fixation and decalcification of the samples by Osteosoft®, the inclusion and cutting process was performed. Different histological techniques were performed: hematoxylin–eosin (HE) for the staining of bone and cartilage tissue, Gram staining for the staining of bacteria, and immunohistochemistry. The monoclonal antibody used to detect macrophages in rabbits was RAM-11 (DAKO ref. M633 dilution 1:50). Immunohistochemical detection of the antigen of interest was performed using the avidin–biotin complex method, using alkaline phosphatase as a tracer and Fast Red as a revealing solution. In all cases, the same biological material without primary antibody was used as a negative control. Staining was visualized under a Zeiss Axiophot optical microscope (Carl Zeiss, Oberkochen, Germany).

2.5. Histological Staging Methods

2.5.1. Destructuring of Bone Histoarchitecture

To quantify the histological results, a new histological staging method was used, as described in the previous work by Ibarra et al. [16]. In this preliminary study, the experimental cement Palacos R + PLGA was evaluated in vivo in comparison with the commercial cement Palacos R® and in the presence or absence of bacteria.

2.5.2. Presence of Bacteria and the Macrophage Response

To measure the presence of bacteria and the macrophage response, we used a semi-quantitative measurement method similar to that described by Remmele et al. [18]. This grading method is based on the assignment of a value estimating the percentage of the area occupied by the cells under study out of the total tissue sample. In our case, we measured the area occupied by bacteria stained with Gram stain and macrophages labeled with the monoclonal antibody RAM-11. With these percentages, we formed categories that covered the different ranges: 0 = negative, 1 = up to 25%, 2 = from 25 to 75%, and 3 = more than 75% of the total number of cells.

2.6. Statistical Analysis

In the evaluation of bone structure, two blinded evaluators applied the histoarchitectural staging model to five samples per rabbit, obtaining means per rabbit ($n = 5$) and per group ($n = 6$). Lin's agreement index (Lin's rho = 0.91, with a confidence interval of 95% (0.8–1.0)) was analyzed between the two evaluators. The semiquantitative assessments were performed in 15 microscopic fields per sample, obtaining the means per animal and per group. The normality of the averages obtained was verified by statistical tests and graphs. Given that they did not have a normal distribution, bivariate analyses were performed using the Mann–Whitney U test. All these analyses were carried out in the statistical program Prism® 6.0 (GraphPad Software Incorporation, San Diego, CA, USA). The significance values are represented in the graphs as follows: * $p < 0.05$, ** $p < 0.01$, and *** $p < 0.001$.

3. Results

3.1. Evaluation of Bone Histoarchitecture: Quantitative Analysis of the Degree of Destruction

In the comparative analysis between groups with Palacos R® cement and with Palacos R cement + PLGA, the presence of PLGA significantly improved the preservation of bone architecture when the combination of antibiotics used was vancomycin with daptomycin (Group 5). However, Palacos R + PLGA bone cement loaded with vancomycin and linezolid (Group 6) induced severe destruction of the bone architecture, so treatment with linezolid in combination with vancomycin did not preserve the bone histoarchitecture (Figures 1 and 2).

3.2. Analysis of the Tissue Distribution of Bacteria

The distribution of *S. aureus* followed a fairly homogeneous pattern across all groups studied. The bacteria colonized the peripheral soft tissues and moved from the implant site to the physeal cartilage, subsequently occupying the epiphyseal bone and articular cartilage.

The distribution of *S. aureus* was not modified by the use of Palacos R® bone cement, regardless of the combination of antibiotics used (Groups 1, 3, and 4). *S. aureus* was less abundant in the presence of PLGA in all groups studied except the vancomycin + linezolid group (Group 6). This reduction was very significant when the combination of antibiotics was vancomycin + daptomycin (Group 5) (Figures 3 and 4). Treatment with linezolid + vancomycin failed to control bacterial dispersion (Figure 5).

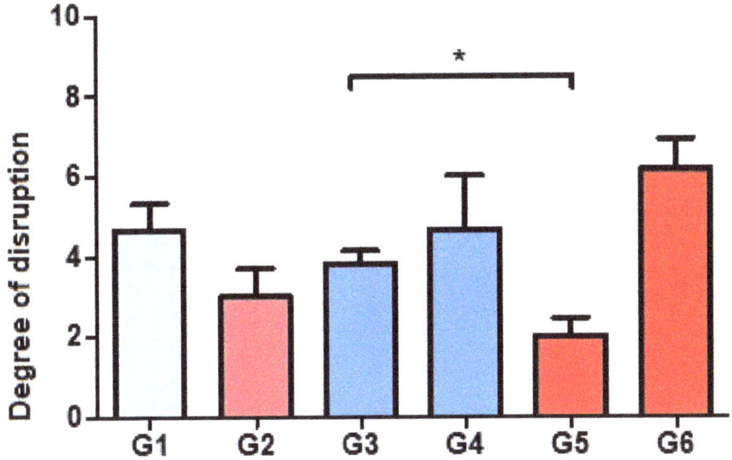

Figure 1. Histograms corresponding to the means and standard deviations of Groups 1 to 6. * $p < 0.05$. Degree of disruption: light (from 1 to 3), moderate (from 4 to 6), severe (from 7 to 9).

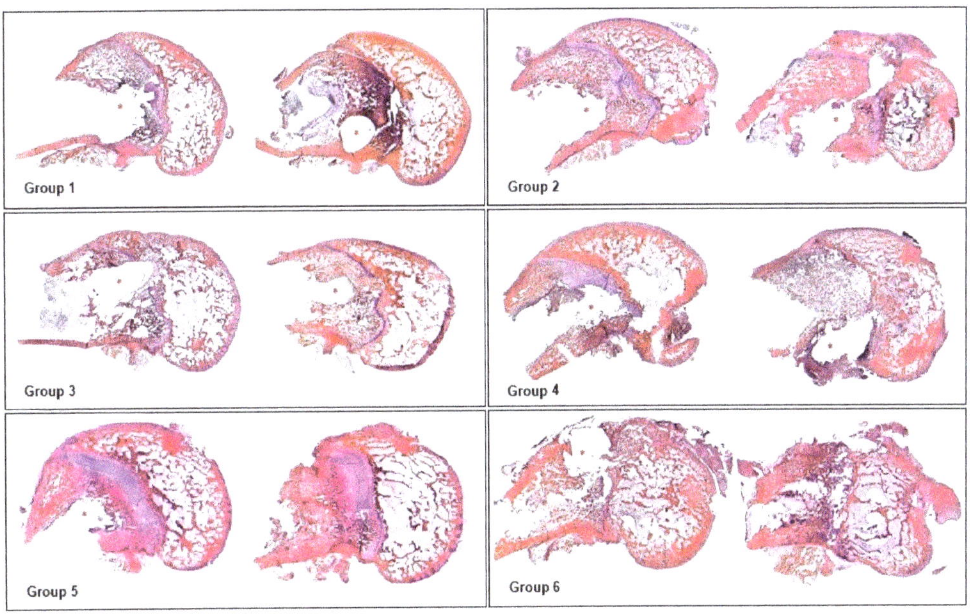

Figure 2. Representative panoramic images of groups of rabbits carrying contaminated rods. HE (5×). Group 1: Palacos R® and daptomycin. Group 2: Palacos R + PLGA and daptomycin. Group 3: Palacos R®, vancomycin, and daptomycin. Group 4: Palacos R®, vancomycin, and linezolid. Group 5: Palacos R + PLGA, vancomycin, and daptomycin. Group 6: Palacos R + PLGA, vancomycin, and linezolid. * Location of the rod.

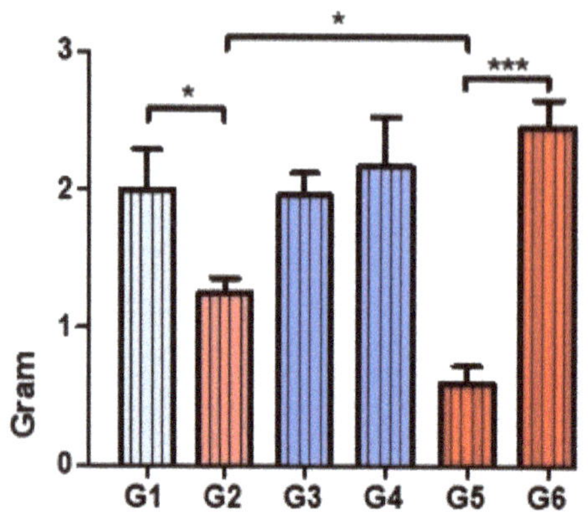

Figure 3. Gram staining. Histograms corresponding to the means and standard deviations of the Groups 1 to 6. * $p < 0.05$. *** $p < 0.001$.

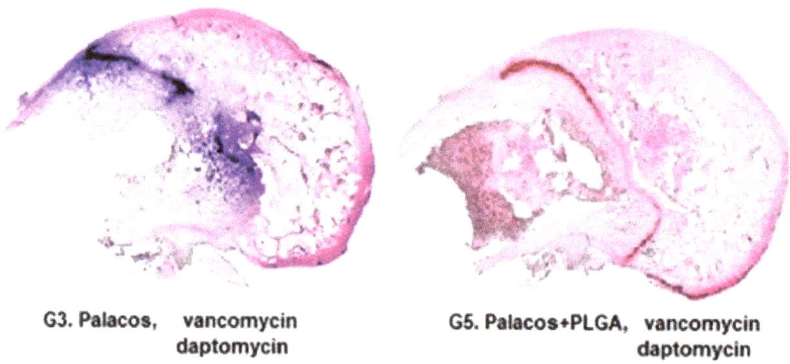

Figure 4. Gram-stained samples (dyed purple) of Groups 3 and 5 (5×). Note the smaller amount and smaller distribution area of bacteria in the sample of Group 5 with respect to Group 3.

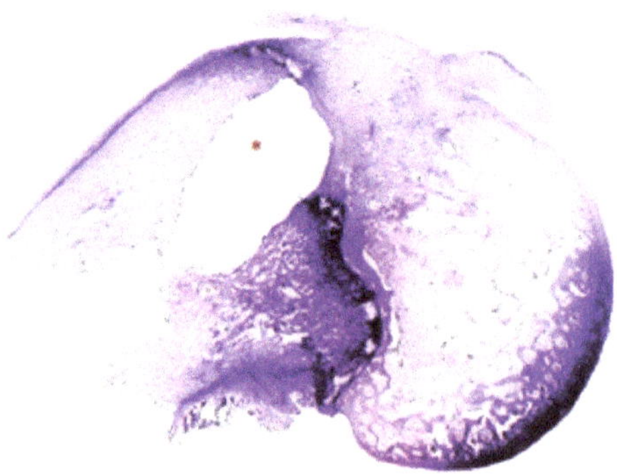

Figure 5. Gram staining shows the intense presence of *S. aureus* in the entire region studied in the presence of linezolid. Image corresponding to the control group (commercial cement and only linezolid). Magnification: 5×. * Imprint of contaminated rod.

3.3. Assessment of the Presence of Macrophages

We found macrophagic cells in all groups in the capsular areas surrounding the implant, as well as in the areas with more bacteria (Figure 6). In the Palacos R + PLGA with vancomycin + linezolid group (Group 6), a greater macrophage response was associated with a greater degree of bone destruction and a high presence of bacteria (Figure 6).

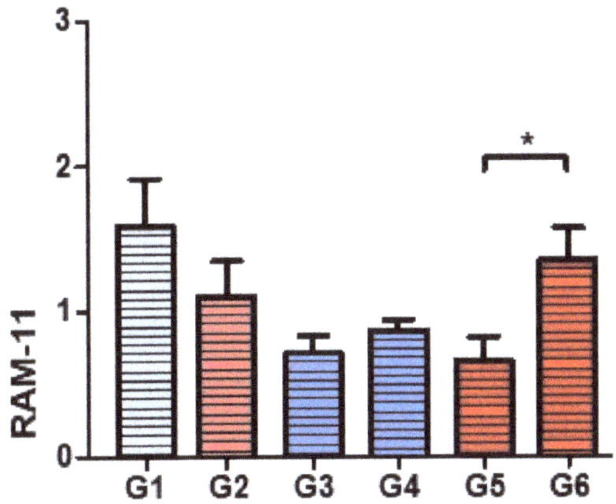

Figure 6. RAM 11 staining. Histograms corresponding to the means and standard deviations of Groups 1 to 6. * $p < 0.05$.

4. Discussion

The increase in prosthetic infections caused by multiresistant bacteria has made it necessary to use new antibiotics, such as daptomycin and linezolid [7,19]. Their incorporation into bone cement has been studied in vitro [11], where they have demonstrated similar efficacy to the available gentamicin–vancomycin spacers [20]. However, there are no in vivo studies that show the behavior of the combination of these new antibiotics against bone tissue. We have published in vivo experimental studies in which we have studied the response of bone tissue to the presence of cements loaded only with linezolid [17]. In vitro studies with linezolid seem to show good antibacterial activity [11,20]. In our in vivo model [16,17], linezolid did not control bacterial growth or dispersion, showing large colonies of bacteria in the sheet extending from the implantation site to the articular cartilage under the microscope. This phenomenon was correlated with the loss of the structure of bone tissue observed in the groups that incorporated linezolid, where the tissue was so affected that, in many cases, it did not resist the action of the usual decalcifiers and routine manipulation during the processing of tissues to obtain histological samples. This observed response is in line with the toxicity that linezolid seems to exert locally and with the results of previous in vitro studies, since linezolid has been the antibiotic with the highest cytotoxicity of those studied [11]. This bone toxicity may be in line with the adverse effects observed in systemic treatment with linezolid, which include thrombocytopenia, anemia, lactic acidosis, and peripheral neuropathy [21].

However, it has been shown in vitro that daptomycin retains antibacterial properties against *S. aureus* after 5 days of incorporation into PMMA [22]. Its release is significantly higher in combination with gentamicin and tobramycin due to a synergistic effect [23]. Our working group has also previously demonstrated the synergy of daptomycin with vancomycin added to modified experimental cement (Palacos R + PLGA), where there is greater release of the combination of antibiotics than of the isolated antibiotics, as well as a biphasic release [11].

Taking into account these results, we considered it necessary to investigate the in vivo behavior of daptomycin and linezolid in combination with vancomycin, as well as the behavior of daptomycin individually loaded both in the commercial cement Palacos R® and in our modified experimental cement (Palacos R + PLGA). The good bacterial control results obtained with daptomycin in our experimental model seem to be related to its ability to inhibit the formation of new biofilms [24], although it has no activity against already established biofilms [25]. In our model, the presence of *S. aureus* in the groups treated with Palacos R + PLGA cement and daptomycin was reduced to small colonies most often located in the physeal cartilage or in the peripheral soft tissues near the cartilage. Unlike the bone treated with the formulations loaded with linezolid, the bone treated with daptomycin had a well-preserved structure, and the macrophages showed a wide dispersion throughout the bone tissue, helping to reduce the presence of bacteria. This behavior may be explained by the lipid nature of daptomycin, which, as has been proven in recent studies, facilitates tissue penetration [26,27], in contrast to vancomycin, which exerts its action more locally because it has greater difficulty diffusing through bone tissue [28].

We found only one experimental study of animals treated with cement loaded with daptomycin. Rouse et al. [29] published a study in 2006 in which they treated a rat model of tibial osteomyelitis with 7.5% daptomycin, among other options, concluding that daptomycin loaded in cement is a valid option for the treatment of osteomyelitis caused by MRSA.

Our results show that Palacos R® cement modified with PLGA microspheres and loaded with daptomycin allows significant preservation of the tissue structure against *S. aureus* infection. Daptomycin controls bacterial dispersion such that, in our study, at 3 weeks, there were barely any bacteria in the rabbits' soft tissues or bone tissues, only small colonies in the physeal cartilage. Similar results were obtained with PLGA-modified Palacos R® cement loaded with the synergistic combination of daptomycin and vancomycin. One of the advantages of our experimental model is that it revealed the existence of a common pattern of bacterial dispersion, which had not been described before. In the absence of bone cement loaded with antibiotics or when the antibiotic used was linezolid, at 3 weeks, the bacteria occupied the entire femoral condyle, producing extreme degradation of the bone histoarchitecture [17]. As the bacterial control by the antibiotic increased, as happened with daptomycin, the bacterial colonies were reduced to small clusters in the dispersion routes through the peripheral soft tissues to the physeal cartilage. This experimental model confirms the great affinity of *S. aureus* for peripheral soft tissues and for physeal cartilage, areas that have a concentration of many bacteria in cases of uncontrolled infection.

Our experimental model also allowed us to assess the immune response at 3 weeks after implant placement by visualizing the location of macrophages. This is because, in recent years, specific cell markers have been developed to count specific cell types in histological preparations, such as the monoclonal antibody RAM-11 for rabbit macrophages, as we have used in our model [30]. The incorporation of PLGA microspheres in commercial Palacos R® cement induces changes in the behavior of macrophages. The results suggest that, although the quantitative assessments may be similar, the macrophages present a different localization depending on the type of cement used. Thus, in the groups with PLGA, these cells appeared to accumulate in the fibrous capsule without forming granulomas, remaining as macrophages, while in the groups with commercial cement Palacos R®, these cells, in addition to being found in the vascular lagoons, which are their natural niche, were observed in cartilage territories, a preferred place for the accumulation of bacteria. This fact is related to the lesser control of bacterial dispersion in the groups that had the commercial cement Palacos R®. In the presence of vancomycin, macrophages preferentially accumulated in the capsular territories around the contaminated implant and in the peripheral soft tissues, through which the bacterial migration routes are established. In contrast, in the presence of daptomycin, macrophages were dispersed throughout all tissues studied, including bone marrow and trabecular bone. This observation could be explained by the greater ease with which daptomycin penetrates tissues, given its lipophilic nature, allowing it to achieve greater bacterial control [26]. This reasoning could explain the good results obtained by the Palacos R + PLGA cement + daptomycin group, in which the two mechanisms of antibacterial action described complemented each other [11].

Due to the use of these new antibiotics, one of the concerns in the scientific community is the emergence of resistance. Although there is a clear link between the use of linezolid and the appearance of resistance despite its reduced use [31], after more than 10 years of clinical use, more than 99% of bacterial strains are still susceptible to linezolid [32]. Similarly, a recent international daptomycin surveillance program reported susceptibility rates of methicillin-sensitive *S. aureus* and MRSA of 99.9% [33].

Therefore, we can state that formulations of bone cements with the introduction of PLGA microspheres can improve the release of antibiotics, especially daptomycin, individually or in combination with vancomycin, and that these formulations loaded with daptomycin preserve the bone structure that surrounds them to a greater extent, as demonstrated in our experimental model. This may have clinical implications for the treatment of prosthetic infections caused by bacteria resistant to conventional antibiotics through the use of spacers made with these formulations of bone cement loaded with daptomycin.

5. Conclusions

The modification of Palacos R® bone cement with PLGA microspheres and its doping with the antibiotic daptomycin in combination with vancomycin improve the tissue response to bone infection. However, the antibiotic linezolid in combination with vancomycin did not obtain good results in our model, regardless of the type of cement used.

Author Contributions: Conceptualization, J.G.-G., G.A. and B.D.l.T.; investigation, J.G.-G., G.A., O.F.-M., C.G.-M., M.A.Á.-M., S.R.-D., M.Á.-M., J.B., N.G.-H., M.A.O. and B.D.l.T.; data curation, J.G.-G., G.A., M.A.O. and B.D.l.T.; writing—original draft preparation, J.G.-G., G.A., O.F.-M., C.G.-M., M.A.Á.-M., S.R.-D., M.Á.-M., J.B., N.G.-H., M.A.O. and B.D.l.T.; writing—review and editing, J.G.-G., G.A., O.F.-M., C.G.-M., M.A.Á.-M., S.R.-D., M.Á.-M., J.B., N.G.-H., M.A.O. and B.D.l.T.; supervision, M.A.O. and B.D.l.T.; project administration, M.A.O. and B.D.l.T.; funding acquisition, M.Á.-M., M.A.O. and B.D.l.T. All authors have read and agreed to the published version of the manuscript.

Funding: This research was funded by the SECOT (Spanish Society of Orthopedic Surgery and Traumatology); 2016/0055, MINECO (Spanish Ministry of Economy and Competitiveness); MAT2014-51918-C2-1-R, MAT-201784277-R, MITIC-CM (Community of Madrid); B2017/BMD-3804 and Halekulani S.L.

Institutional Review Board Statement: Animals were managed in accordance with the current International Regulations on Experimental Animals (609/86/EEC and ETS 123) at the Animal Research Center of the University of Alcala. The study protocol received approval from the Committee on the Ethics of Animal Experiments of the University of Alcala (CEI UAH 2011017).

Informed Consent Statement: Not applicable.

Data Availability Statement: The data used to support the findings of the present study are available from the corresponding author upon request.

Acknowledgments: The authors thank F. J. Parra-Ruiz and A. González-Gómez for assistance in the work.

Conflicts of Interest: The authors declare no conflict of interest.

References

1. Lindeque, B.; Hartman, Z.; Noshchenko, A.; Cruse, P. Infection After Primary Total Hip Arthroplasty. *Orthopedics* **2014**, *37*, 257–265. [CrossRef] [PubMed]
2. Chen, K.-H.; Tsai, S.-W.; Wu, P.-K.; Chen, C.-F.; Wang, H.-Y.; Chen, W.-M. Partial component-retained two-stage reconstruction for chronic infection after uncemented total hip arthroplasty: Results of sixteen cases after five years of follow-up. *Int. Orthop.* **2017**, *41*, 2479–2486. [CrossRef] [PubMed]
3. Yarwood, J.M.; Paquette, K.M.; Tikh, I.B.; Volper, E.M.; Greenberg, E.P. Generation of Virulence Factor Variants in Staphylococcus aureus Biofilms. *J. Bacteriol.* **2007**, *189*, 7961–7967. [CrossRef]
4. Zimmerli, W.; Trampuz, A.; Ochsner, P.E. Prosthetic-joint infections. *N. Engl. J. Med.* **2004**, *351*, 1645–1654. [CrossRef]

5. Cabrita, H.B.; Croci, A.T.; de Camargo, O.P.; de Lima, A.L.L.M. Prospective study of the treatment of infected hip arthroplasties with or without the use of an antibiotic-loaded cement spacer. *Clin. Sao Paulo Braz.* **2007**, *62*, 99–108. [CrossRef] [PubMed]
6. Anagnostakos, K. Therapeutic Use of Antibiotic-loaded Bone Cement in the Treatment of Hip and Knee Joint Infections. *J. Bone Jt. Infect.* **2017**, *2*, 29–37. [CrossRef] [PubMed]
7. Tande, A.J.; Patel, R. Prosthetic joint infection. *Clin. Microbiol. Rev.* **2014**, *27*, 302–345. [CrossRef]
8. Lamagni, T. Epidemiology and burden of prosthetic joint infections. *J. Antimicrob. Chemother.* **2014**, *69*, i5–i10. [CrossRef]
9. Liu, C.; Bayer, A.; Cosgrove, S.E.; Daum, R.S.; Fridkin, S.K.; Gorwitz, R.J.; Kaplan, S.L.; Karchmer, A.W.; Levine, D.P.; Murray, B.E.; et al. Clinical Practice Guidelines by the Infectious Diseases Society of America for the Treatment of Methicillin-Resistant Staphylococcus aureus Infections in Adults and Children: Executive Summary. *Clin. Infect. Dis.* **2011**, *52*, 285–292. [CrossRef]
10. Pillai, S.K.; Wennersten, C.; Venkataraman, L.; Eliopoulos, G.M.; Moellering, J.R.C.; Karchmer, A.W. Development of Reduced Vancomycin Susceptibility in Methicillin-SusceptibleStaphylococcus aureus. *Clin. Infect. Dis.* **2009**, *49*, 1169–1174. [CrossRef]
11. Parra-Ruíz, F.; González-Gómez, A.; Fernández-Gutiérrez, M.; Parra, J.; García-García, J.; Azuara, G.; De la Torre, B.; Buján, J.; Ibarra, B.; Duocastella-Codina, L.; et al. Development of advanced biantibiotic loaded bone cement spacers for arthroplasty associated infections. *Int. J. Pharm.* **2017**, *522*, 11–20. [CrossRef]
12. Hashemian, S.M.; Farhadi, T.; Ganjparvar, M. Linezolid: A review of its properties, function, and use in critical care. *Drug Des. Dev. Ther.* **2018**, *12*, 1759–1767. [CrossRef]
13. Gray, D.A.; Wenzel, M. More Than a Pore: A Current Perspective on the In Vivo Mode of Action of the Lipopeptide Antibiotic Daptomycin. *Antibiotics* **2020**, *9*, 17. [CrossRef]
14. Petty, W.; Spanier, S.; Shuster, J.J.; Silverthorne, C. The influence of skeletal implants on incidence of infection. Experiments in a canine model. *J. Bone Jt. Surg. Am.* **1985**, *67*, 1236–1244. [CrossRef]
15. Smeltzer, M.S.; Thomas, J.R.; Hickmon, S.G.; Skinner, R.A.; Nelson, C.L.; Griffith, D.; Parr, T.R., Jr.; Evans, R.P. Characterization of a rabbit model of staphylococcal osteomyelitis. *J. Orthop. Res. Off Publ. Orthop. Res. Soc.* **1997**, *15*, 414–421. [CrossRef]
16. Ibarra, B.; García-García, J.; Azuara, G.; Vázquez-Lasa, B.; Ortega, M.A.; Asúnsolo, Á.; Román, J.S.; Buján, J.; Honduvilla, N.G.; De La Torre, B. Polylactic-co-glycolic acid microspheres added to fixative cements and its role on bone infected architecture. *J. Biomed. Mater. Res. Part B Appl. Biomater.* **2019**, *107*, 2517–2526. [CrossRef]
17. Azuara, G.; García-García, J.; Ibarra, B.; Parra-Ruiz, F.J.; Asúnsolo, A.; Ortega, M.A.; Vázquez-Lasa, B.; Buján, J.; San Román, J.; de la Torre, B. Experimental study of the application of a new bone cement loaded with broad spectrum antibiotics for the treatment of bone infection. *Rev. Espanola Cirugia Ortop. Traumatol.* **2019**, *63*, 95–103. [CrossRef]
18. Remmele, W.; Schicketanz, K.-H. Immunohistochemical determination of estrogen and progesterone receptor content in human breast cancer: Computer-assisted image analysis (QIC score) vs. subjective grading (IRS). *Pathol. Res. Pract.* **1993**, *189*, 862–866. [CrossRef]
19. Ariza, J.; Cobo, J.; Baraia-Etxaburu, J.; de Benito, N.; Bori, G.; Cabo, J.; Corona, P.; Esteban, J.; Horcajada, J.P.; Lora-Tamayo, J.; et al. Executive summary of management of prosthetic joint infections. Clinical practice guidelines by the Spanish Society of Infectious Diseases and Clinical Microbiology (SEIMC). *Enferm. Infecc. Microbiol. Clin.* **2017**, *35*, 189–195. [CrossRef]
20. Anagnostakos, K.; Kelm, J.; Grün, S.; Schmitt, E.; Jung, W.; Swoboda, S. Antimicrobial properties and elution kinetics of linezolid-loaded hip spacers in vitro. *J. Biomed. Mater Res. B Appl. Biomater.* **2008**, *87*, 173–178. [CrossRef]
21. Beibei, L.; Yun, C.; Mengli, C.; Nan, B.; Xuhong, Y.; Rui, W. Linezolid versus vancomycin for the treatment of Gram-positive bacterial infections: Meta-analysis of randomised controlled trials. *Int. J. Antimicrob. Agents* **2010**, *35*, 3–12. [CrossRef] [PubMed]
22. Kuechle, D.K.; Landon, G.C.; Musher, D.M.; Noble, P.C. Elution of Vancomycin, Daptomycin, and Amikacin From Acrylic Bone Cement. *Clin. Orthop. Relat. Res.* **1991**, *264*, 302–308. [CrossRef]
23. Kaplan, L.; Kurdziel, M.; Baker, K.; Verner, J. Characterization of Daptomycin-loaded Antibiotic Cement. *Orthopedics* **2012**, *35*, e503–e509. [CrossRef] [PubMed]
24. Eick, S.; Hofpeter, K.; Sculean, A.; Ender, C.; Klimas, S.; Vogt, S.; Nietzsche, S. Activity of Fosfomycin- and Daptomycin-Containing Bone Cement on Selected Bacterial Species Being Associated with Orthopedic Infections. *BioMed Res. Int.* **2017**, *2017*, 2318174. [CrossRef]
25. Stefani, S.; Campanile, F.; Santagati, M.; Mezzatesta, M.L.; Cafiso, V.; Pacini, G. Insights and clinical perspectives of daptomycin resistance in Staphylococcus aureus: A review of the available evidence. *Int. J. Antimicrob. Agents* **2015**, *46*, 278–289. [CrossRef]
26. Grillon, A.; Argemi, X.; Gaudias, J.; Ronde-Ousteau, C.; Boeri, C.; Jenny, J.-Y.; Hansmann, Y.; Lefebvre, N.; Jehl, F. Bone penetration of daptomycin in diabetic patients with bacterial foot infections. *Int. J. Infect. Dis.* **2019**, *85*, 127–131. [CrossRef]
27. Mescola, A.; Ragazzini, G.; Alessandrini, A. Daptomycin Strongly Affects the Phase Behavior of Model Lipid Bilayers. *J. Phys. Chem. B* **2020**, *124*, 8562–8571. [CrossRef]
28. Bue, M.; Tøttrup, M.; Hanberg, P.; Langhoff, O.; Birke-Sørensen, H.; Thillemann, T.M.; Søballe, K. Bone and subcutaneous adipose tissue pharmacokinetics of vancomycin in total knee replacement patients. *Acta Orthop.* **2018**, *89*, 95–100. [CrossRef]
29. Rouse, M.S.; Piper, K.E.; Jacobson, M.; Jacofsky, D.J.; Steckelberg, J.M.; Patel, R. Daptomycin treatment of Staphylococcus aureus experimental chronic osteomyelitis. *J. Antimicrob. Chemother.* **2006**, *57*, 301–305. [CrossRef]
30. Kanazawa, T.; Soejima, T.; Murakami, H.; Inoue, T.; Katouda, M.; Nagata, K.; Wang, J.-W.; Fong, C.-Y.; Su, Y.-S.; Yu, H.-N. An immunohistological study of the integration at the bone-tendon interface after reconstruction of the anterior cruciate ligament in rabbits. *J. Bone Jt. Surg. Br. Vol.* **2006**, *88*, 682–687. [CrossRef]

31. Baos, E.; Candel, F.J.; Merino, P.; Pena, I.; Picazo, J.J. Characterization and monitoring of linezolid-resistant clinical isolates of Staphylococcus epidermidis in an intensive care unit 4 years after an outbreak of infection by cfr-mediated linezolid-resistant Staphylococcus aureus. *Diagn. Microbiol. Infect. Dis.* **2013**, *76*, 325–329. [CrossRef]
32. Flamm, R.K.; Mendes, R.; Ross, J.E.; Sader, H.; Jones, R.N. An international activity and spectrum analysis of linezolid: ZAAPS Program results for Diagn. *Microbiol. Infect. Dis.* **2013**, *76*, 206–213. [CrossRef]
33. Sader, H.S.; Flamm, R.K.; Jones, R.N. Antimicrobial activity of daptomycin tested against Gram-positive pathogens collected in Europe, Latin America, and selected countries in the Asia-Pacific Region (2011). *Diagn. Microbiol. Infect. Dis.* **2013**, *75*, 417–422. [CrossRef]

polymers MDPI

Article

Layer-By-Layer Self-Assembled Dip Coating for Antifouling Functionalized Finishing of Cotton Textile

Sana Javaid [1,2], Azhar Mahmood [1], Habib Nasir [1], Mudassir Iqbal [1], Naveed Ahmed [3] and Nasir M. Ahmad [4,*]

1 School of Natural Science (SNS), National University of Science and Technology (NUST), Islamabad 44000, Pakistan; sana.javaid@sns.nust.edu.pk (S.J.); dr.azhar@sns.nust.edu.pk (A.M.); habibnasir@sns.nust.edu.pk (H.N.); mudassir.iqbal@sns.nust.edu.pk (M.I.)
2 Department of Chemistry, University of Wah, Wah Cantt 47040, Pakistan
3 Department of Pharmacy, Quaid-i-Azam University, Islamabad 44000, Pakistan; natanoli@qau.edu.pk
4 Polymer Research Lab, School of Chemical and Materials Engineering (SCME), National University of Sciences and Technology (NUST), Islamabad 44000, Pakistan
* Correspondence: nasir.ahmad@scme.nust.edu.pk; Tel.: +92-519-085-5213

Abstract: The fouling of surfaces such as textiles is a major health challenge, and there is a continuous effort to develop materials and processes to overcome it. In consideration of this, this study regards the development of antifouling functional nanoencapsulated finishing for the cotton textile fabric by employing a layer-by-layer dip coating technique. Antifouling textile finishing was formulated by inducing the nanoencapsulation of the antifouling functional group inside the hydrophobic polymeric shell. Cotton fabric was taken as a substrate to incorporate antibacterial functionality by alternatively fabricating multilayers of antifouling polymeric formulation (APF) and polyelectrolyte solution. The surface morphology of nanoencapsulated finished textile fabric was characterized through scanning electron microscopy to confirm the uniform distribution of nanoparticles on the cotton textile fabric. Optical profilometry and atomic force microscopy studies indicated increased surface roughness in the coated textile substrate as compared to the uncoated textile. The surface thickness of the fabricated textile increased with the number of deposited bilayers on the textile substrate. Surface hydrophobicity increased with number of coating bilayers with θ values of x for single layer, up to y for 20 bilayers. The antibacterial activity of the uncoated and layer-by-layer coated finished textile was also evaluated. It was significant and exhibited a significant zone of inhibition against microbial strains Gram-positive *S. aureus* and Gram-negative *E. coli*. The bilayer coating exhibited water repellency, hydrophobicity, and antibacterial activity. Thus, the fabricated textile could be highly useful for many industrial and biomedical applications.

Keywords: antifouling polymeric formulation (APF); finishing; nanoencapsulation; cotton textile; layer-by-layer technique

1. Introduction

Textile substrates are susceptible to the growth of bacteria, fungi, and various other microbes [1]. The proliferation rate of bacteria and other microbes is aggravated under suitable conditions of optimal temperature, pH, humidity, and moisture. The deterioration and biofouling of textile material in industrial and medical or health sectors pose serious threats. The spread of pathogenic microbes through contaminated textiles is a major source of nosocomial infections [2]. Gram-negative and Gram-positive bacteria species such as *Escherichia coli* and *Staphylococcus aureus* are among those bacterial strains responsible for hospital-acquired infections [3–5]. In consideration of such concerns, the global demand for antifouling textiles has tremendously increased due to consumer awareness regarding hygiene, health, and safety [6]. The annual production of medical textiles is predicted to significantly increase to up to 3500 metric tons by 2025 [7].

Textile functionalization through antifouling agents has allowed for many technological applications in biomedical fields, such as bandages, wound dressing, artificial sutures, body-implanted and nonimplanted devices, and tissue scaffolding [8,9]. Nanotechnology plays a significant role in the functionalization and modification of textiles by incorporating immobilized nanoparticles on the textile surface by encapsulating active agents [10]. Advancements in nanoencapsulation for functional finishes impart dyeing, anti-UV, flame-retardant, insect- and moth-repellent, antibacterial, self-healing, and aroma-producing properties to textiles [11–14].

Polymeric natural and synthetic nanomaterial is an efficient carrier of active agents to specific sites owing to its biodegradability, biocompatibility, and nontoxicity [15]. Polycaprolactone fulfills the requirements of bioapplications with suitable features for encapsulating antifouling functionalities [16,17]. Because of the above, the development and modification of biodegradable PCL nanoparticles for the slow release of an active agent through nanoencapsulation are very important, particularly in skin applications [18–20]. Different approaches for achieving nanoencapsulation by using biodegradable polymeric shells and bioactive core material include emulsification through coacervation, diffusion, or evaporation, and interfacial and in situ polymerization [21–24]. Both physical and chemical methods for nanoencapsulation have been studied. Natural cotton textiles incorporating nanocapsules for imparting antibacterial properties require attention for hygienic wound dressing [25]. Eugenol-loaded human serum albumin/silk fibroin (HSA/SF) nanocapsules with antibacterial properties were synthesized for the functionalization of cotton/polyethylene terephthalate blends. Functionalization was achieved for cotton wound dressing with significant antibacterial properties against *Escherichia coli* and *Staphylococcus aureus*. The in vitro release of pH-sensitive antimicrobial agents from nanocapsules reached the textile surface to inhibit bacteria [26]. Antibacterial formulations are generally coated on textiles with various techniques such as pad–dry–cure (dipping method) or padding, spray coating, and the sol-gel technique [27]. Padding is the most conventional treatment for functionalized, but it is difficult to manipulate the effect of shearing forces acting on textile surfaces of due to the large number of process parameters. Among various coating techniques, the layer-by-layer (LBL) deposition of polyelectrolytes for the antifouling functional finishing of cotton is the simplest, easiest, and environmentally friendly approach [28]. Regarding this, chitosan/pentasodium tripolyphosphate as antibacterial functionality was coated on ionized cotton using the LBL deposition of oppositely charged polyelectrolytes. The subsequent deposition of multiple bilayers consequently increased the antibacterial efficacy of cotton against *E. coli* and *S. aureus* [29].

Considering the significance of antibacterial functional textiles for biomedical and industrial applications, we employed nanoencapsulation for antifouling finishing. Nanocapsules were prepared with biodegradable polycaprolactone (PCL)-encapsulated cefotaxime as antifouling moiety, and they were immobilized on a cotton textile. PCL is semicrystalline and biocompatible, achieving significant performance in the slow release of active molecules on fabric surfaces [30]. Cefotaxime, a third-generation semisynthetic cephalosporine, was chosen as an antifouling functional moiety having a broad range of antibacterial activity, particularly against *E. coli* and *S. aureus*. The efficacy of cefotaxime in bactericidal action by inhibiting the synthesis of peptidoglycan is associated with the high stability of its beta-lactamase. The toxicity of cefotaxime was significantly reduced by encapsulation, and caused the slow and sustained release that is particularly suitable for fabricating functional textiles. The molecular structure of cefotaxime is presented in Figure 1.

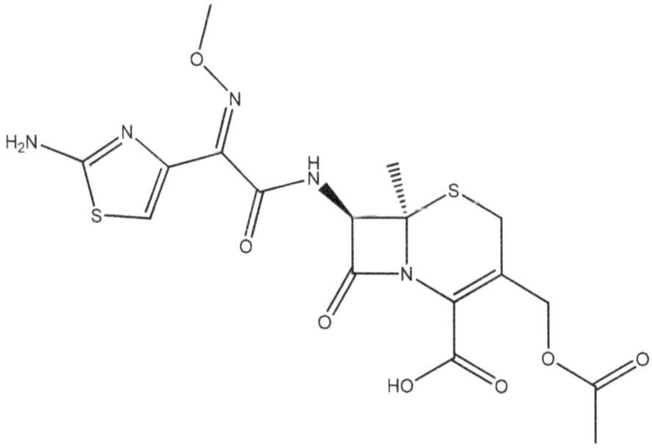

Figure 1. Molecular structure of cefotaxime.

Antifouling polymeric formulations (APFs) were previously developed through nanoprecipitation and employed to fabricate antifouling textiles. Thus, the layer-by-layer technique was used to develop self-assembled multilayers through the dip coating of an antifouling formulation to fabricate cotton textiles for the in vitro and gradual release of antifouling activity. A cotton substrate was selected as the model textile fabric for antifouling due to its hydrophilicity and rapid deterioration in moist environments to allow for microbial degradation. The layer-by-layer coating of a polymeric formulation on cotton textile fabric induced significant hydrophobicity. The fabricated textile was analyzed through various characterization techniques. Surface wettability was analyzed by measuring the contact angle (θ), and surface roughness, thickness, and morphology were studied through optical profilometry, atomic force microscopy (AFM), and scanning electron microscopy (SEM). In vitro antibacterial tests were conducted to assess antifouling properties. The efficacy of antifouling activity of the developed cotton textile was dependent on the concentration of APF, deposited via self-assembled bilayers. The fabricated textile showed antibacterial activity against Gram-positive *Staphylococcus aureus* and Gram-negative *Escherichia coli* with a significant reduction in bacterial colony numbers. The novel nanoencapsulation finishing could be used in developing woven and nonwoven textiles with antibacterial functionalities. It could be utilized to fabricate various types of industrial smart textiles, including cotton wound dressing for healing skin irritations and skin allergies.

2. Materials and Methods

Polycaprolactone (PCL) (Mw-1400 g/mol), polyvinyl alcohol (PVA) (Mw 3100 g/mol), poly (diallyl dimethyl ammonium chloride) (PDAC) (average Mw 200,000–350,000, 20 wt% solution), and dichloromethane (DCM) (Mw-84.93) with 99.9% purity were purchased from Sigma Aldrich, Germany. Deionized ultrapure water with total dissolved solids (TDS~0.00) was used for solution preparation and washing. Cefotaxime (Mw 455.47 g/mol in powder form) as an antifouling functional moiety was obtained from Nectar Life sciences, India. Mercerized bleached (100%) whitish pure cotton with areal density of 1.31 g/m^3 was obtained from the National Textile University (NTU), Faisalabad, Pakistan. Sodium chloride (NaCl), purchased from Sigma Aldrich (Germany), was used as a buffer saline solution. Mueller-Hinton II Agar (MHA) (Biolab, Budapest, Hungry) was utilized as the nutrient broth for the bacterial culture. In vitro antibacterial assay was performed against clinical bacterial strains of Gram-positive *Staphylococcus aureus* (ATCC 6538) and Gram-negative *Escherichia coli* (ATCC 8739).

2.1. Synthesis of Antifouling Polymeric Formulation

The antifouling polymeric formulation (APF) was synthesized and characterized of as described in previous work [31]. The nanoencapsulation of antifouling functional cefotaxime was conducted by using polycaprolactone as the shell material through nano-precipitation. The organic phase containing polycaprolactone (PCL) was dissolved in dichloromethane (DCM), while the aqueous phase containing polyvinyl alcohol (PVA) was used as a stabilizing agent. To optimize the blank formulation, an organic phase was slowly injected into an aqueous phase. After achieving a stable formulation by varying the process parameters, cefotaxime was dissolved in the organic phase and slowly injected into an aqueous phase. The subsequent addition of the organic phase containing cefotaxime into an aqueous phase formed the antifouling polymeric nanoformulation.

2.2. Characterization of Antifouling Polymeric Formulation

The prepared antifouling polymeric nanoformulation was analyzed with different techniques. Surface morphology was studied with scanning electron microscopy (SEM) (model JEOL JSM 6490 LA, Tokyo, Japan); applied voltage was 10 KV. The encapsulation of antifouling functionality in the polymeric nanoparticles was evidenced with Fourier transform infrared (FTIR) using an FTIR spectrophotometer (Modal Shimadzu 8400, Markham, ON, Canada). Dynamic light scattering (DLS) (Nano ZS, Malvern Instruments, Worcestershire, UK) was used to observe the average size, surface charge, and distribution of nanoparticles.

2.3. Fabrication of Antifouling Polymeric Coating on Textile

Pretreated and mercerized cotton textile fabric was received as plain whitish woven with an areal density of 1.31 g/m^3. The antifouling polymeric coating on the cotton textile was fabricated through a layer-by-layer technique. Cotton textile fabric was alternatively dipped in a polycation solution of 10^{-3} M poly(diallyl dimethyl ammonium chloride) (PDAC) solution and 100 mL of diluted antifouling polymeric formulation (APF) comprising 25 mg of PCL in a series of steps. Antibacterial tests were conducted against clinical isolates of Gram-positive *Staphylococcus aureus* (ATCC 6538) and Gram-negative *Escherichia coli* (ATCC 8739).

The pretreated cotton textile was cut into swatches of 75 by 25 mm. Negatively charged cotton swatches were first dipped in PDAC (a polycation solution maintained at neutral pH) for 10 min, resulting in positive charge deposition (single ion layer) followed by subsequent washing with deionized water twice for 5 min in two separate beakers to achieve the uniform deposition of charges, and to remove unbound or loosely attached molecules from the textile surface. For the opposite charge of another single ion layer, cotton swatches were dipped in an antifouling polymeric formulation (APF maintained at neutral pH) for 10 min, followed by washing with deionized water twice for 5 min in two separate beakers. Thus, a bilayer (1bL) of oppositely charged ions was formed due to electrostatic interaction on the cotton textile between the positively charged polyelectrolyte and negatively charged antifouling polymeric formulation (APF). The whole process of fabricating a single bilayer was accomplished in 40 min. The same procedure was repeated to cyclically fabricate 5, 10, 15, and 20 bilayers without the textile drying in each step. Fabricated textiles with top layers of negatively charged APF were considered for the slow release of antifouling functionality and characterized for an in vitro antibacterial test. The schematic representation of the layer-by-layer coating of the cotton textile is given in Figure 2.

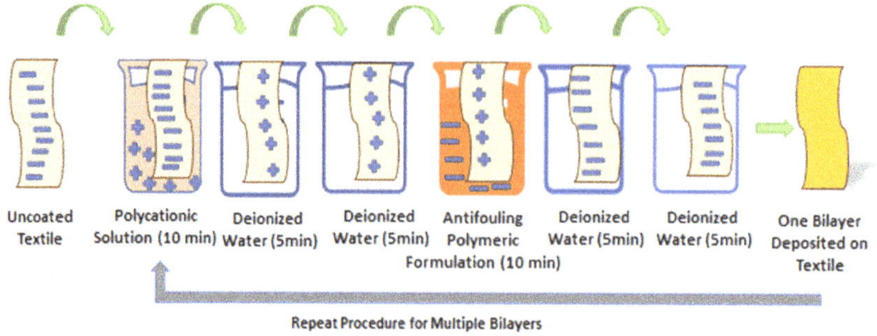

Figure 2. Schematic representation of fabrication of antifouling functional cotton textile through layer-by-layer technique.

2.4. Characterization of Fabricated Cotton Textile Fabric

The layer-by-layer coated textile fabric was analyzed by different techniques. The antifouling finished cotton textile fabric was analyzed for surface morphology, thickness, and roughness. The surface morphology of the coated and uncoated textile samples was observed through SEM (Modal JSM 6490LA, JEOL, Tokyo, Japan). Textile swatches of equal dimensions were fixed on the glass slide, followed by sputter-coating with gold for analysis. Optical profilometry was used for measuring the difference between the average roughness and thickness of the uncoated and coated textiles in micrometers. We fixed 20 × 5 mm textile swatches at the center of the glass slide with double tape, and analyzed them through a 2D noncontact profilometer (model NANOVEA PS-50, USA). The subsequent roughness due to the layer-by-layer deposition of the nanoformulation inducing nanoencapsulated finished cotton textile fabric was further confirmed through scanning probe microscopy (Modal JSPM-5200, JEOL, Japan). Then, 20 × 5 mm textile swatches were cut and fixed on the glass slides with the double tape, and were analyzed in tapping mode. Uncoated and coated textile samples were topographically analyzed through 3D plots for detailed visualization. The surface wettability and hydrophobicity of the fabricated cotton textile were analyzed by measuring the contact angle (θ). Textile swatches (20 × 5 mm dimension) were clipped at the center of the glass slide and placed under a drop-shape analyzer (Model Kruss Gmbh 2014–2020, Hamburg, Germany).

2.5. In Vitro Antibacterial Bioassay

Standard qualitative textile testing with AATCC TM 100 with slight modification was performed to assess antibacterial activity over the course of 24 h [32]. Clinical isolates of *Staphylococcus aureus* (ATCC 6538) and *Escherichia coli* (ATCC 8739) were utilized to check the antibacterial activity of the fabricated cotton textile by agar disk diffusion or Kirby–Bauer assay. For this purpose, a Muller Hinton Agar (MHA) solution was prepared as commercially recommended and autoclaved overnight. The freshly prepared nutrient broth was streaked with microbial cells and incubated overnight at 37 °C. Then, a 10 mL saline solution was poured into a 50 mL sterilized falcon tube and inoculated with pathogenic colonies, followed by vortexed mixing for constant distribution. Afterwards, 0.5 McFarland was utilized to set the optical density and turbidity [33]. MHA (25 mL) solution was poured in sterilized Petri dishes and it was allowed to solidify the agar; then, it was placed in an incubator at 37 °C overnight. Solid agar plates were streaked with sterile cotton swabs by adding the bacterial culture (about 50 microliters). Streaking was conducted by rotating the Petri plate about 60 degrees thrice for uniform inoculation. Uncoated and coated (1, 5, 10, 15, and 20 bl) textiles of 40 × 10 mm dimensions were placed onto solid agar plates by gently pressing the textile at the center of each Petri dish. Plates were placed in an incubator at 37 °C overnight, and antibacterial activity was observed after 24 h. The zone of inhibition around the textile sample was measured as mean standard deviation.

3. Results and Discussion

3.1. Characterization of Antifouling Polymeric Formulation (AFP)

Cefotaxime was nanoencapsulated as a core antifouling functionality inside the polymeric shell for the slow release of in vitro antibacterial activity as described previously [31]. Polycaprolactone nanoparticles were characterized for average size, surface charge, and morphology. The average size and zeta potential of the drug-loaded polymeric nanoparticles were 216 nm and −11.1 mV, respectively, as shown in Figure 3. Uniformity in particle size distribution was observed at a PDI of 0.4 (less than 0.5 shows uniformity in distribution). The negatively charged zeta potential of PCL nanoparticles was attributed to the hydrolysis of the carboxyl and hydroxyl groups present at the surface.

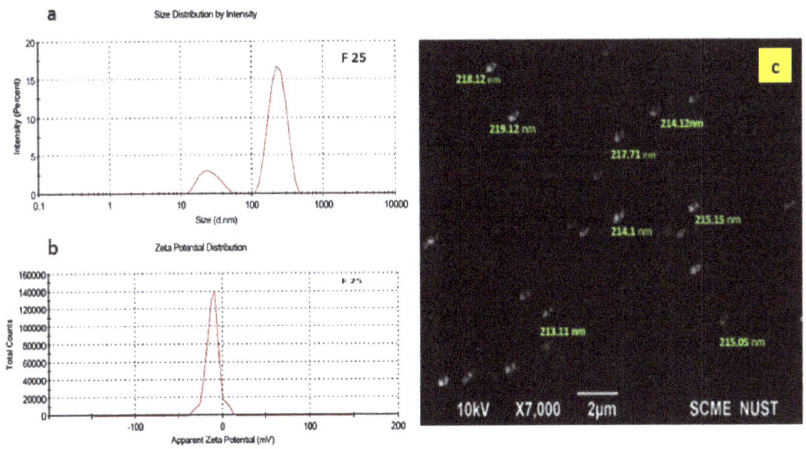

Figure 3. (**a**) Zeta size, (**b**) zeta potential, and (**c**) scanning electron microscope images of drug loaded nanoformulation.

3.2. Characterization of Antifouling Fabricated Cotton Textile

3.2.1. Surface Morphology Studies

The surface morphology of layer-by-layer-deposited thin films of antifouling polymeric (APF) coating was studied by observing the SEM micrographs. Cross-sectional views of SEM micrographs showed the surface morphology of the uncoated and coated cotton textiles, as given in Figure 4. The smooth and uniform surface of the uncoated mercerized cotton textile fabric is shown in Figure 4a, while a significant increase in roughness in the morphology of the coated textile appeared after coating 1 bilayer up to 20 bilayers as shown in Figure 4b–f. The multilayer coating of polyelectrolytes over the cotton textile surface produced an agglomeration [34]. The density of agglomerates increased with the sequential adsorption of the number of bilayers. SEM micrographs of the coated textile fabric showed the uniform and homogeneous deposition of nanoparticles over the surface. There were fewer agglomerates with the coating of 1 and 5 bilayers, as shown in Figure 4b,c, as compared with the coated textile fabric with 10, 15, and 20 bilayers, as shown in Figure 4d–f, with a subsequent increase in the population of agglomerates. The less and dense agglomeration of charged particles over the textile surface depends upon various factors, including the nature and concentration of the polyelectrolyte solution (cationic or anionic polyelectrolyte), immersion time in deionized water, polyelectrolytes, and nanoformulations, besides the cyclical drying process in each step [35]. The roughened surface of the fabricate textile fabric after subsequent bilayer coating potentially created hydrophobicity and antifouling activity [36].

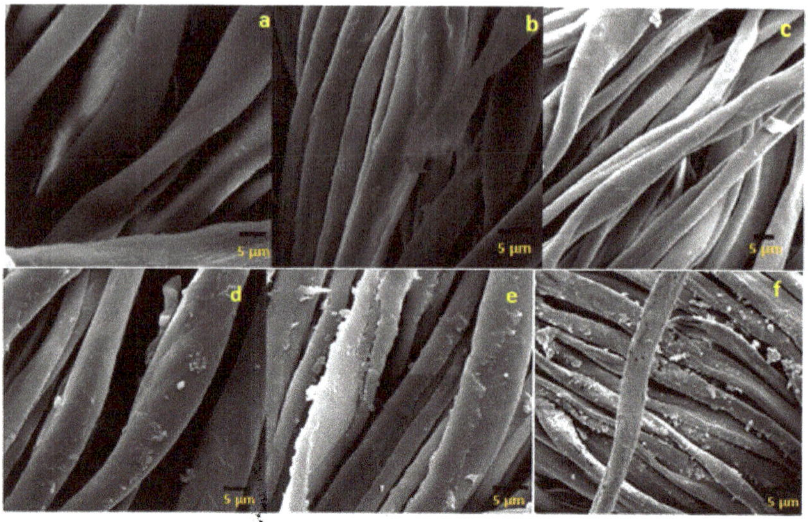

Figure 4. (**a**) EM images of uncoated cotton textile fabric; coated cotton textile with (**b**) 1 BL, (**c**) 5 BL, (**d**) 10 BL, (**e**) 15 BL, and (**f**) 20 BL.

The surface roughness of the functionalized cotton textile from the layer-by-layer dip coating of polyelectrolyte solution and APF formulation was observed through optical profilometry. Mercerized/bleached cotton was smooth, shiny, and whitish before coating. There was significantly increased average roughness (Ra) in the deposited thin film over cotton textile [37]. The average roughness (Ra) of the uncoated (0 BL) and coated textiles (1, 5, 10, 15, and 20 BL) was calculated in µm with a 2D noncontact profilometer. A marked difference in the roughness profile of the uncoated and coated textiles was attributed to the incremental deposition of antifouling polymeric formulation (APF) and poly(diallyl dimethyl ammonium chloride) (PDAC) solution. The comparatively smooth surface of uncoated textile and increasing trend in average roughness of the coated textiles with a deposited number of bilayers are shown in Figure 5. This was also evidenced by the scanning electron micrographs of the uncoated and coated textiles [38].

The surface topography of the uncoated and coated textile samples was further studied by using atomic force microscopy (AFM). The topography of the uncoated and coated textiles was better visualized by scanning a small area and taking 3D plots, as shown in Figure 6. Tapping mode AFM was used for the topographical analysis of the textile samples by scanning an area in µm. The marked difference in the roughness profile of the uncoated and coated textiles was quantitively analyzed in terms of root mean square roughness (Rq) by plotting a graph between Rq values and the number of deposited bilayers on the cotton textile fabrics, as shown in Figure 7. Rq values were evaluated in terms of height or thickness as measured from the mean level of surface [39]. The value of Rq increased with the increase in the number of deposited bilayers, confirming that the Rq value varies with the composition and density of the sample [40]. Topographical AFM 3D plots showed characteristic bright and dark portions, with brighter groves indicating the deposition of nanoparticles. As the number of deposited bilayers increased, the bright portion of the peaks and valleys groves became more significant in the case of the coated textiles with 5, 10, 15, and 20 bilayers. These bright groves and thickness in peak and valleys resulted due to the electrostatic interaction between the tip and surface of the textile. The denser the textile surface with the increased number of bilayers was, the more significant the portion of bright groves with characteristic roughness that appeared was. The resulting variation depended upon the alternative deposition of the antifouling polymeric formulation (APF) and PDAC solution in the modified textile samples [41]. The mercerized cotton was smooth and shiny, while the nanoencapsulated finished cotton textile induced the aggregation of nanoparticles with the addition of some bilayers on the textile surface, thus enhancing micro- and nanoscale roughness [38,42]. The acquired roughness (Rq) of the nanoencapsulated finished cotton textile fabric was evidenced by contact angle measurement for water repellency and the desirable antifouling properties.

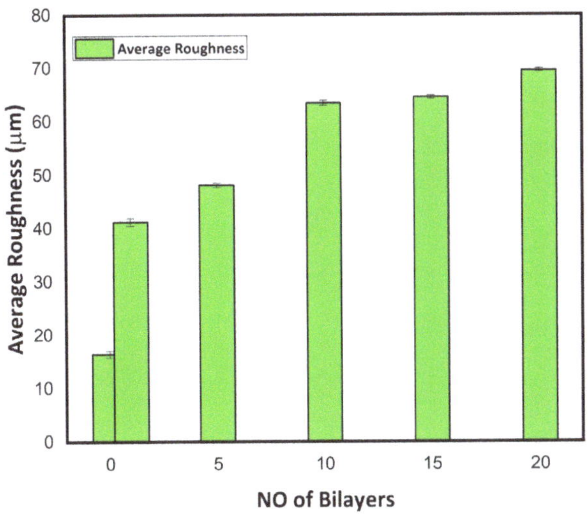

Figure 5. Average roughness (Ra) profile in μm of uncoated and coated cotton textile fabrics (1, 5, 10, 15, and 20 BL).

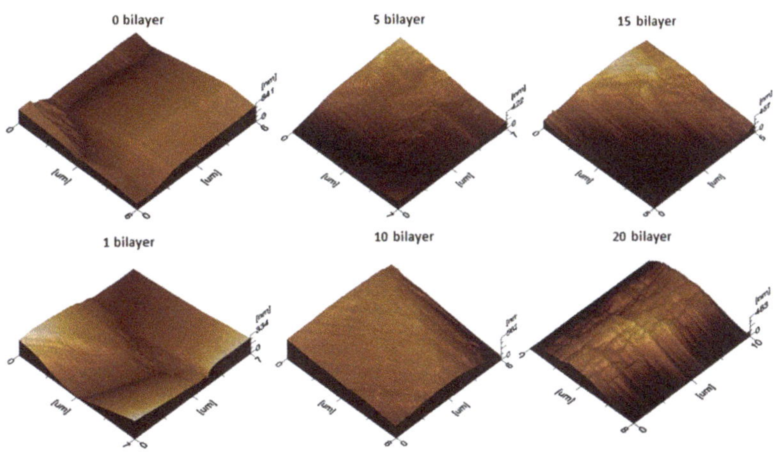

Figure 6. Atomic force microscopy (AFM) images of uncoated (0 BL) and coated cotton textile (1, 5, 10, 15, and 20 BL).

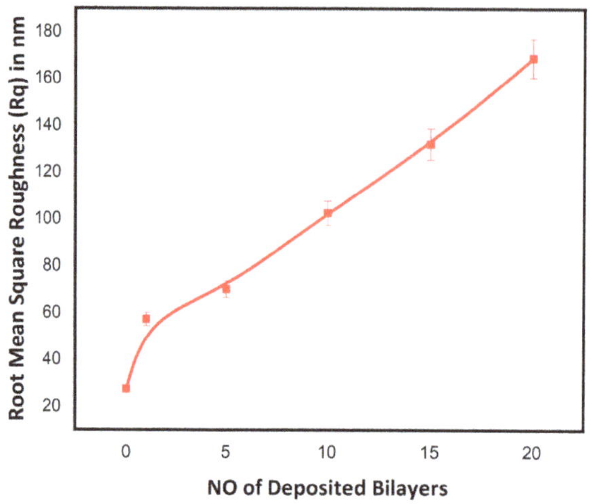

Figure 7. Root mean square roughness (Rq) in nm of uncoated and coated textile with deposited number of bilayers (1, 5, 10, 15, and 20 BL).

3.2.2. Surface Wettability Analyses

The surface wettability of thenanoencapsulated finished cotton textile fabric was studied by measuring the contact angle (θ) as shown in Figure 8. The hydrophilicity and hydrophobicity of the fabricated textile was quantitatively analyzed by observing the values of the contact angle (θ) and surface energy (γ_s). The surface energy of the finished textile substrates was calculated from the contact angle measurement by considering the interaction at the interface of solid and liquid surfaces [43,44]. Variation in contact angle (θ) and surface energy (γ_s) with the number of deposited bilayers in the nanoencapsulated finished textile is shown in Figure 9a,b respectively. The uncoated cotton textile with a low contact angle (θ) had a mercerized fine cotton texture and the presence of hydroxyl group that lead to maximal hydrophilicity and high surface energy (γ_s). Moreover, the increased value of surface energy was associated with pronounced wettability and maximal surface coverage, and this could lead to a poor antifouling effect [36]. The water contact angle (θ) of the 1 bilayer coated textile was greater than that of the uncoated textile substrate due to the hydrophobic nature of PCL. The increased value of the contact angle (θ) resulted from the deposition of 5 bilayers, and was associated with the roughness and uneven protrusions on the textile substrate [45]. Further coating with 10, 15 and 20 bilayers showed an increased value of θ with differences among them as compared to the 1- and 5-bilayer-coated textile due to the interpenetration of the next bilayer with the previously adsorbed bilayer [46]. Thus, a greater number of bilayers coating led to nonuniform deposition and inhomogeneous surface of the textiles. Therefore, water repellency and comparative hydrophobicity were a result of the deposited bilayers in the nanoencapsulated finished cotton textile fabric.

The surface energy of the fabricated cotton textile was evaluated through the results of contact angle (θ) [47]. Thus, Young's model summarizes the theoretical description in Young's equation ($\gamma_s = \gamma_{s-l} + \gamma_l \cos\theta$), a best fit for smooth surfaces before coating applications. γ_s stands for surface energy of solids, γ_{s-l} is the surface energy of the solid–liquid interface, γ_l is the surface energy of liquids, and θ is the contact angle of a liquid droplet. The developed layer-by-layer dip coating of antifouling polymeric formulation (APF) induces significant water repellency associated with the hydrophobicity of PCL [48].

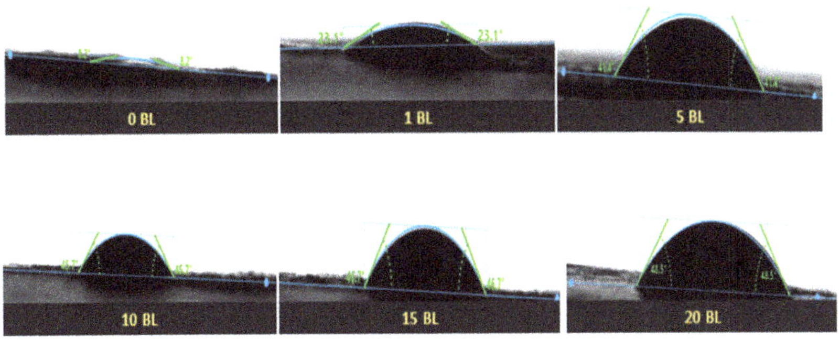

Figure 8. Contact angle measurement of 0, 1, 5, 10, 15, and 20 BL coated cotton textile.

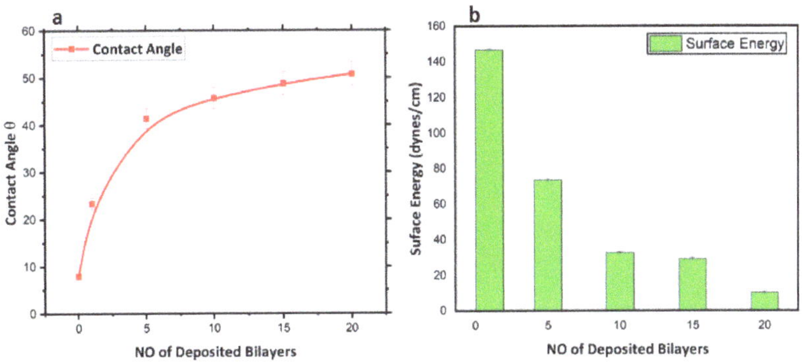

Figure 9. (**a**) Contact angle θ of fabricated cotton textile plot versus number of deposited bilayers; (**b**) surface energy of fabricated cotton textile plot versus number of deposited bilayers.

3.2.3. Antibacterial Assay of Textile Fabric

The antibacterial activity of the uncoated and nanoencapsulated finished cotton textiles was established against two bacterial strains, Gram-positive *S. aureus* and Gram-negative *E. coli*, by using a qualitative agar disk diffusion assay [49]. The antibacterial test results of the uncoated and coated textile fabrics against *E. coli* and *S. aureus* are shown in Figures 10 and 11 respectively. The uncoated textile was taken as a control against both strains to compare the antibacterial efficacy of the nanoencapsulated finished cotton textile fabricated with the layer-by-layer coating of the antifouling polymeric formulation (APF). The antibacterial efficacy of fabricated finished textile was analyzed by placing the textile swatches (40 × 10 mm) in intimate contact on agar plates prestreaked with bacterial inoculum [50]. A clear zone of inhibition was observed around the modified textiles coated with 1, 5, 10, 15, and 20 bilayers. The zone of inhibition against both strains progressively increased around the textile swatches fabricated with a progressive coating of bilayers. These results were attributed to the significant release of antifouling functional drugs from the polymeric nanoencapsulated finished textile in a concentration-dependent manner [31,51]. The release of antifouling drug cefotaxime from APF was responsible for the antifouling activity of the fabricated textiles. The antibacterial activity of the fabricated textiles was further compared against both strains with a greater zone of inhibition against *E. coli* as compared to *S. aureus*, as shown in Figure 12. Thin-walled Gram-negative *E. coli* was more susceptible to bactericidal action than Gram-positive *S. aureus* was. Moreover, there was electrostatic repulsion between Gram-negative *E. coli* and the negatively charged layer of the antifouling polymeric formulation (APF) coated on the textile [52]. The antifouling activity of the nanoencapsulated finished textile fabric was attributed to the slow release of cefotaxime encapsulated inside the PCL shell. As the number of deposited bilayers increased, antifouling activity increased due to the increased release of cefotaxime form PCL nanoparticles. Thus, the antifouling polymeric formulation (APF) encapsulated nanoparticles of cefotaxime, as a notable antibacterial moiety has significant bactericidal action attributed to the high stability of beta-lactamase enzymes, thus inhibiting the synthesis of bacterial cell walls. It has significant bactericidal action against Gram-negative bacterial strains, while it is associated with adequate action towards Gram-positive bacteria [53].

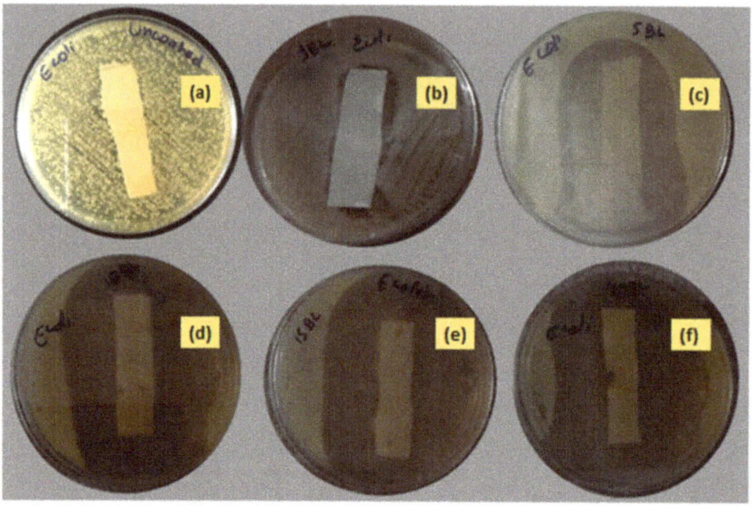

Figure 10. Antibacterial activity of (**a**) uncoated and coated in (**b**) 1 BL, (**c**) 5 BL, (**d**) 10 BL, (**e**) 15 BL, and (**f**) 20 BL against *Escherichia coli*.

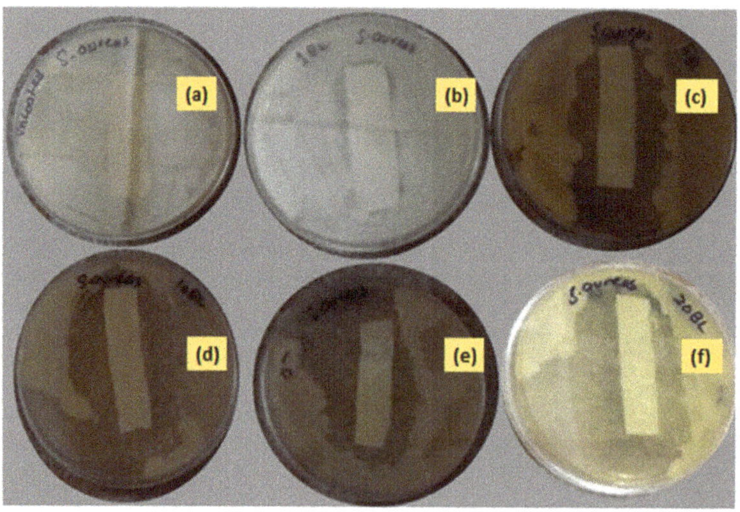

Figure 11. Antibacterial activity of (**a**) uncoated and coated in (**b**) 1 BL, (**c**) 5 BL, (**d**) 10 BL, (**e**) 15 BL, and (**f**) 20 BL textiles against *Staphylococcus aureus*.

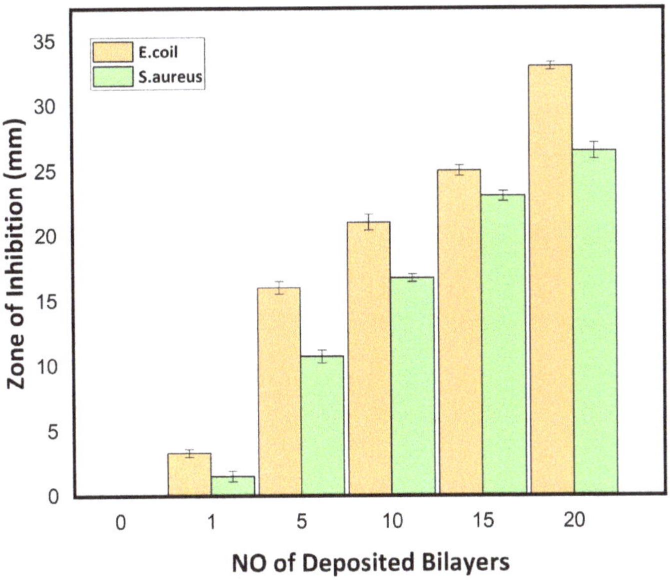

Figure 12. Zone of inhibition of uncoated (0 BL) and coated textiles (1, 5, 10, 15, and 20 BL) against *Escherichia coli* and *Staphylococcus aureus*.

4. Conclusions

The growing demand for health and hygiene regarding textile products has led to industrial concerns about antifouling textiles. Thus, the present work illustrates a novel approach to fabricating nanoencapsulated finished cotton textile fabrics by incorporating the antifouling functional group finishing of cotton textile achieved by employing a layer-by-layer technique through the alternate dip coating of oppositely charged polyelectrolyte solution and antifouling polymeric formulation (APF). SEM and AFM images confirmed the significant deposition of polymeric nanoparticles with the increasing number of coating bilayers. The optical profilometry analysis of the functionalized textiles showed that the multilayer deposition of thin films resulting in the subsequent coating of 1 to 20 bilayers increased surface roughness from 16.4 to 69.4 μm. Further, the surface wettability of the functionalized textiles was evaluated by measuring the contact angle (θ), which increased from 23° to 48.5°, and surface energy decreased from 145 to 18 dynes/cm, which showed the decreased wettability and increased surface hydrophobicity associated with the deposited bilayers. The nanoencapsulated finished textiles were also analyzed through an in vitro antibacterial assay against Gram-negative *E. coli* and Gram-positive *S. aureus*, with a significant reduction in the bacterial cell colonies of both strains. Moreover, the bactericidal efficacy of the fabricated textiles was evaluated in a concentration-dependent manner, and showed an exponential zone of inhibition with the number of coating bilayers on the textiles. The zone of inhibition against both strains increased with the number of deposited bilayers due to the increased release of cefotaxime from the polymeric nanoparticles. The resulting antifouling functional finishing, achieved through the nanoencapsulation of the antifouling functional group, could provide a substantial route in fulfilling the growing global needs for antifouling textiles. The nanoencapsulated finished antifouling cotton textile fabric could be used in wound dressing, for skin infections, and in various other industrial applications.

Author Contributions: S.J. conducted the experimental work and wrote the original manuscript. N.M.A. designed and evaluated this work as the principal investigator. A.M., formal analysis.

H.N., M.I. and N.A., final editing. All authors have read and agreed to the published version of the manuscript.

Funding: This research received no external funding.

Institutional Review Board Statement: Not applicable.

Informed Consent Statement: Not applicable.

Data Availability Statement: All data are available to the readers.

Acknowledgments: Sana Javaid acknowledges the financial support of NUST Research Directorate. Nasir M. Ahmad acknowledges the support of Higher Education Commission (HEC), Pakistan through NRPU grant no. 3526 and 6020.

Conflicts of Interest: The authors declare no conflict of interest.

References

1. Mahall, K. Microbiological Damage to Fibers. In *Quality Assessment of Textiles*; Springer: Berlin/Heidelberg, Germany, 1993; pp. 195–213. [CrossRef]
2. Morais, D.S.; Guedes, R.M.; Lopes, M.A. Antimicrobial Approaches for Textiles: From Research to Market. *Materials* **2016**, *9*, 498. [CrossRef] [PubMed]
3. Khan, H.A.; Ahmad, A.; Mehboob, R. Nosocomial infections and their control strategies. *Asian Pac. J. Trop. Biomed.* **2015**, *5*, 509–514. [CrossRef]
4. Abla, H.-H.; Chafia, B.; Abdesselam, L.; Houcine, L.; Kaddour, B.; Farida, S. Multidrug-resistant bacteria isolated from patients hospitalized in Intensive Care Unit in University Hospital of Constantine, Algeria (2011–2015). *Afr. J. Microbiol. Res.* **2016**, *10*, 1328–1336. [CrossRef]
5. Paszkiewicz, M.; Gołąbiewska, A.; Rajski, Ł.; Kowal, E.; Sajdak, A.; Zaleska-Medynska, A. The Antibacterial and Antifungal Textile Properties Functionalized by Bimetallic Nanoparticles of Ag/Cu with Different Structures. *J. Nanomater.* **2016**, *2016*, 6056980. [CrossRef]
6. Windler, L.; Height, M.; Nowack, B. Comparative evaluation of antimicrobials for textile applications. *Environ. Int.* **2013**, *53*, 62–73. [CrossRef]
7. Gao, Y.; Cranston, R. Recent Advances in Antimicrobial Treatments of Textiles. *Text. Res. J.* **2008**, *78*, 60–72.
8. Höfer, D. Antimicrobial Textiles—Evaluation of Their Effectiveness and Safety. *Curr. Probl. Dermatol.* **2006**, *33*, 42–50. [CrossRef]
9. Mahltig, B.; Zhang, J.; Wu, L.; Darko, D.; Wendt, M.; Lempa, E.; Rabe, M.; Haase, H. Effect pigments for textile coating: A review of the broad range of advantageous functionalization. *J. Coat. Technol. Res.* **2016**, *14*, 35–55. [CrossRef]
10. Martí, M.; Martínez, V.; Rubio, L.; Coderch, L.; Parra, J.L. Biofunctional textiles prepared with liposomes: In vivo and in vitro assessment. *J. Microencapsul.* **2011**, *28*, 799–806. [CrossRef]
11. Nelson, G. Application of microencapsulation in textiles. *Int. J. Pharm.* **2002**, *242*, 55–62. [CrossRef]
12. Cheng, S.Y.; Yuen, M.C.W.; Kan, C.W.; Cheuk, K.K.L.; Chui, C.H.; Lam, K.H. Cosmetic textiles with biological benefits: Gelatin microcapsules containing vitamin C. *Int. J. Mol. Med.* **2009**, *24*, 411–419.
13. Monllor, P.; Bonet-Aracil, M.; Cases, F. Characterization of the behaviour of flavour microcapsules in cotton fabrics. *Eur. Polym. J.* **2007**, *43*, 2481–2490. [CrossRef]
14. Xiao, Z.; Liu, W.; Zhu, G.; Zhou, R.; Niu, Y. A review of the preparation and application of flavour and essential oils microcapsules based on complex coacervation technology. *J. Sci. Food Agric.* **2013**, *94*, 1482–1494. [CrossRef]
15. Kumari, A.; Yadav, S.K.; Yadav, S.C. Biodegradable polymeric nanoparticles based drug delivery systems. *Colloids. Surf. B Biointerfaces* **2010**, *75*, 1–18. [CrossRef]
16. Cesca, F.; Limongi, T.; Accardo, A.; Rocchi, A.; Orlando, M.; Shalabaeva, V.; Di Fabrizio, E.; Benfenati, F. Fabrication of biocompatible free-standing nanopatterned films for primary neuronal cultures. *RSC Adv.* **2014**, *4*, 45696–45702. [CrossRef]
17. Limongi, T.; Lizzul, L.; Giugni, A.; Tirinato, L.; Pagliari, F.; Tan, H.; Das, G.; Moretti, M.; Marini, M.; Brusatin, G.; et al. Laboratory injection molder for the fabrication of polymeric porous poly-epsilon-caprolactone scaffolds for preliminary mesenchymal stem cells tissue engineering applications. *Microelectron. Eng.* **2017**, *175*, 12–16. [CrossRef]
18. Sinha, V.R.; Bansal, K.; Kaushik, R.; Kumria, R.; Trehan, A. Poly-ϵ-caprolactone microspheres and nanospheres: An overview. *Int. J. Pharm.* **2004**, *278*, 1–23. [CrossRef]
19. Madhaiyan, K.; Sridhar, R.; Sundarrajan, S.; Venugopal, J.R.; Ramakrishna, S. Vitamin B12 loaded polycaprolactone nanofibers: A novel transdermal route for the water soluble energy supplement delivery. *Int. J. Pharm.* **2013**, *444*, 70–76. [CrossRef]
20. Alonso, C.; Martí, M.; Barba, C.; Lis, M.; Rubio, L.; Coderch, L. Skin penetration and antioxidant effect of cosmeto-textiles with gallic acid. *J. Photochem. Photobiol. B Biol.* **2016**, *156*, 50–55. [CrossRef]
21. Reis, C.; Neufeld, R.J.; Ribeiro, A.; Veiga, F. Nanoencapsulation I. Methods for preparation of drug-loaded polymeric nanoparticles. *Nanomed. Nanotechnol. Biol. Med.* **2006**, *2*, 8–21. [CrossRef]
22. Xing, F.; Cheng, G.; Yi, K.; Ma, L. Nanoencapsulation of capsaicin by complex coacervation of gelatin, acacia, and tannins. *J. Appl. Polym. Sci.* **2005**, *96*, 2225–2229. [CrossRef]

23. Moinard-Chécot, D.; Chevalier, Y.; Briançon, S.; Beney, L.; Fessi, H. Mechanism of nanocapsules formation by the emulsion–diffusion process. *J. Colloid Interface Sci.* **2008**, *317*, 458–468. [CrossRef] [PubMed]
24. Quaglia, F.; Ostacolo, L.; Mazzaglia, A.; Villari, V.; Zaccaria, D.; Sciortino, M.T. The intracellular effects of non-ionic amphiphilic cyclodextrin nanoparticles in the delivery of anticancer drugs. *Biomaterials* **2009**, *30*, 374–382. [CrossRef] [PubMed]
25. Ghayempour, S.; Montazer, M. Micro/nanoencapsulation of essential oils and fragrances: Focus on perfumed, antimicrobial, mosquito-repellent and medical textiles. *J. Microencapsul.* **2016**, *33*, 497–510. [CrossRef]
26. Quartinello, F.; Tallian, C.; Auer, J.; Schön, H.; Vielnascher, R.; Weinberger, S.; Wieland, K.; Weihs, A.M.; Herrero-Rollett, A.; Lendl, B.; et al. Smart textiles in wound care: Functionalization of cotton/PET blends with antimicrobial nanocap-sules. *J. Mater. Chem. B* **2019**, *7*, 6592–6603. [CrossRef]
27. Singha, K. A Review on Coating & Lamination in Textiles: Processes and Applications. *Am. J. Polym. Sci.* **2012**, *2*, 39–49. [CrossRef]
28. Aubert, V.F.; Mogrovejo, V.A.; Tabary, N.; Maton, M.; Chai, F.; Neut, C.; Martel, B.; Blanchemain, N. Evaluation of antibacterial textile covered by layer-by-layer coating and loaded with chlorhexidine for wound dressing application. *Mater. Sci. Eng. C Mater. Biol. Appl.* **2019**, *100*, 554–563. [CrossRef]
29. Shirvan, A.R.; Nejad, N.H.; Bashari, A. Antibacterial finishing of cotton fabric via the chitosan/TPP self-assembled nano layers. *Fibers Polym.* **2014**, *15*, 1908–1914. [CrossRef]
30. Massella, D.; Leone, F.; Peila, R.; Barresi, A.A.; Ferri, A. Functionalization of Cotton Fabrics with Polycaprolactone Nanoparticles for Transdermal Release of Melatonin. *J. Funct. Biomater.* **2017**, *9*, 1. [CrossRef]
31. Javaid, S.; Ahmad, N.; Mahmood, A.; Nasir, H.; Iqbal, M.; Ahmad, N.; Irshad, S. Cefotaxime Loaded Polycaprolactone Based Polymeric Nanoparticles with Antifouling Properties for In-Vitro Drug Release Applications. *Polymers* **2021**, *13*, 2180. [CrossRef]
32. Pinho, E.; Magalhães, L.; Henriques, M.; Oliveira, R. Antimicrobial activity assessment of textiles: Standard methods comparison. *Ann. Microbiol.* **2010**, *61*, 493–498. [CrossRef]
33. Andrews, J.M. Testing, BSAC standardized disc susceptibility testing method (version 5). *J. Antimicrob. Chemother.* **2006**, *58*, 511–529. [CrossRef]
34. Ostrander, J.W.; Mamedov, A.A.; Kotov, N.A. Two modes of linear layer-by-layer growth of nanoparticle–polylectrolyte multilayers and different interactions in the layer-by-layer deposition. *J. Am. Chem. Soc.* **2001**, *123*, 1101–1110. [CrossRef]
35. Rivero, P.J.; Urrutia, A.; Goicoechea, J.; Arregui, F.J. Nanomaterials for Functional Textiles and Fibers. *Nanoscale Res. Lett.* **2015**, *10*, 501. [CrossRef]
36. Ouadil, B.; Amadine, O.; Essamlali, Y.; Cherkaoui, O.; Zahouily, M. A new route for the preparation of hydrophobic and antibacterial textiles fabrics using Ag-loaded graphene nanocomposite. *Colloids Surf. A Physicochem. Eng. Asp.* **2019**, *579*, 123713. [CrossRef]
37. Bhushan, B. *Modern Tribology Handbook*, 1st ed.; Two Volume Set; CRC Press: Boca Raton, FL, USA, 2000. [CrossRef]
38. Cai, R.; Glinel, K.; De Smet, D.; Vanneste, M.; Mannu, N.; Kartheuser, B.; Nysten, B.; Jonas, A.M. Environmentally Friendly Super-Water-Repellent Fabrics Prepared from Water-Based Suspensions. *ACS Appl. Mater. Interfaces* **2018**, *10*, 15346–15351. [CrossRef]
39. Simpson, G.J.; Sedin, D.L.; Rowlen, K.L. Surface Roughness by Contact versus Tapping Mode Atomic Force Microscopy. *Langmuir* **1999**, *15*, 1429–1434. [CrossRef]
40. Koc, S.K. Applications of Atomic Force Microscopy in Textiles. *J. Eng. Fibers Fabr.* **2015**, *10*, 155892501501000118. [CrossRef]
41. Dong, X.; Gao, S.; Huang, J.; Li, S.; Zhu, T.; Cheng, Y.; Zhao, Y.; Chen, Z.; Lai, Y. A self-roughened and biodegradable superhydrophobic coating with UV shielding, solar-induced self-healing and versatile oil–water separation ability. *J. Mater. Chem. A* **2018**, *7*, 2122–2128. [CrossRef]
42. Jonas, A.M.; Cai, R.; Vermeyen, R.; Nysten, B.; Vanneste, M.; De Smet, D.; Glinel, K. How roughness controls the water repellency of woven fabrics. *Mater. Des.* **2019**, *187*, 108389. [CrossRef]
43. Li, S.; Huang, J.; Chen, Z.; Chen, G.; Lai, Y. A review on special wettability textiles: Theoretical models, fabrication technologies and multifunctional applications. *J. Mater. Chem. A* **2016**, *5*, 31–55. [CrossRef]
44. Żenkiewicz, M. Methods for the calculation of surface free energy of solids. *J. Achiev. Mater. Manuf. Eng.* **2007**, *24*, 137–145.
45. Deng, H.; Zhou, X.; Wang, X.; Zhang, C.; Ding, B.; Zhang, Q.; Du, Y. Layer-by-layer structured polysaccharides film-coated cellulose nanofibrous mats for cell culture. *Carbohydr. Polym.* **2010**, *80*, 474–479. [CrossRef]
46. Deng, H.; Wang, X.; Liu, P.; Ding, B.; Du, Y.; Li, G.; Hu, X.; Yang, J. Enhanced bacterial inhibition activity of layer-by-layer structured polysaccharide film-coated cellulose nanofibrous mats via addition of layered silicate. *Carbohydr. Polym.* **2011**, *83*, 239–245. [CrossRef]
47. Calvimontes, A. The measurement of the surface energy of solids using a laboratory drop tower. *NPJ Microgravity* **2017**, *3*, 25. [CrossRef]
48. Siddig, E.A.; Zhang, Y.; Yang, B.; Wang, T.; Shi, J.; Guo, Y.; Xu, Y.; Zhang, J. Plasma-exposed TiO_2 nanoparticles on polyethylene terephthalate matrix surface and its effects on the durable hydrophobic coating. *Text. Res. J.* **2021**. [CrossRef]
49. Haase, H.; Jordan, L.; Keitel, L.; Keil, C.; Mahltig, B. Comparison of methods for determining the effectiveness of antibacterial functionalized textiles. *PLoS ONE* **2017**, *12*, e0188304. [CrossRef] [PubMed]
50. Lalitha, P.; Jayanthi, P.; Sujitha, R. Antimicrobial Activity of Perspiration Pads and Cotton Cloth Fabricated with the Ethyl Acetate Extract of Eichhornia crassipes (Mart.) Solms. *J. Text.* **2014**, *2014*, 943287. [CrossRef]

51. Hipler, U.-C.; Elsner, P.; Fluhr, J.W. Antifungal and antibacterial properties of a silver-loaded cellulosic fiber. *J. Biomed. Mater. Res. Part B Appl. Biomater.* **2006**, *77*, 156–163. [CrossRef] [PubMed]
52. Niño-Martínez, N.; Salas Orozco, M.F.; Martínez-Castañón, G.-A.; Torres Méndez, F.; Ruiz, F. Molecular Mechanisms of Bacterial Resistance to Metal and Metal Oxide Nanoparticles. *Int. J. Mol. Sci.* **2019**, *20*, 2808. [CrossRef]
53. Lefrock, J.L.; Prince, R.A.; Left, R.D. Mechanism of Action, Antimicrobial Activity, Pharmacology, Adverse Effects, and Clinical Efficacy of Cefotaxime. *Pharmacother. J. Hum. Pharmacol. Drug Ther.* **1982**, *2*, 174–184. [CrossRef]

polymers

MDPI

Review

Cutting-Edge Progress in Stimuli-Responsive Bioadhesives: From Synthesis to Clinical Applications

Elham Khadem [1], Mahshid Kharaziha [1,*], Hamid Reza Bakhsheshi-Rad [2], Oisik Das [3] and Filippo Berto [4,*]

[1] Department of Materials Engineering, Isfahan University of Technology, Isfahan 84156-83111, Iran; khadem_77@yahoo.com

[2] Advanced Materials Research Center, Department of Materials Engineering, Najafabad Branch, Islamic Azad University, Najafabad, Iran; rezabakhsheshi@pmt.iaun.ac.ir

[3] Structural and Fire Engineering Division, Department of Civil, Environmental and Natural Resources Engineering, Luleå University of Technology, 97187 Luleå, Sweden; oisik.das@ltu.se

[4] Department of Mechanical and Industrial Engineering, Norwegian University of Science and Technology, 7491 Trondheim, Norway

* Correspondence: ma.kharaziha@gmail.com (M.K.); filippo.berto@ntnu.no (F.B.)

Abstract: With the advent of "intelligent" materials, the design of smart bioadhesives responding to chemical, physical, or biological stimuli has been widely developed in biomedical applications to minimize the risk of wounds reopening, chronic pain, and inflammation. Intelligent bioadhesives are free-flowing liquid solutions passing through a phase shift in the physiological environment due to stimuli such as light, temperature, pH, and electric field. They possess great merits, such as ease to access and the ability to sustained release as well as the spatial transfer of a biomolecule with reduced side effects. Tissue engineering, wound healing, drug delivery, regenerative biomedicine, cancer therapy, and other fields have benefited from smart bioadhesives. Recently, many disciplinary attempts have been performed to promote the functionality of smart bioadhesives and discover innovative compositions. However, according to our knowledge, the development of multifunctional bioadhesives for various biomedical applications has not been adequately explored. This review aims to summarize the most recent cutting-edge strategies (years 2015–2021) developed for stimuli-sensitive bioadhesives responding to external stimuli. We first focus on five primary categories of stimuli-responsive bioadhesive systems (pH, thermal, light, electric field, and biomolecules), their properties, and limitations. Following the introduction of principal criteria for smart bioadhesives, their performances are discussed, and certain smart polymeric materials employed in their creation in 2015 are studied. Finally, advantages, disadvantages, and future directions regarding smart bioadhesives for biomedical applications are surveyed.

Keywords: bioadhesive; stimuli-responsive materials; wound healing; drug delivery

1. Introduction

An increase in the cost of healthcare and the age of the population has resulted in a rising request for bioadhesives and drug delivery systems [1]. The term "bioadhesion" was introduced for the first time in the 1970s. Bioadhesion is defined as the phenomenon in which two materials, one of which should be biological, are attached for a long time by interfacial tailoring [2]. Due to biodegradability, biocompatibility, and large molecular weight, bioadhesives can be applied in many hard-and soft-tissue applications, drug delivery, reinforcing fragile tissues in therapy, and helping with hemostasis [3]. Despite the potential benefits of bioadhesives, the functionality of commercially available bioadhesives is limited [4]. For example, existing adhesives are designedto support the injured tissues mechanically. To further explore the potential of bioadhesives in medicine, scientists have developed multifunctional bioadhesives with antimicrobial ability to limit microbial adherence and growth [5,6]. Additionally, some bioadhesives with self-healing capabilities can prolong the adhesive's available time [7]. However, nearly all common materials are in a static condition when utilized as bioadhesives; "smart", "intelligent", or "stimuli-responsive" bioadhesives have been introduced [8]. Smart bioadhesives are generally defined as materials that can sense and react to different stimuli, including photoradiation (UV, visible light), temperature, pH, specific ions, solvents, electric and magnetic fields, redox conditions, mechanical stress, and biomolecules. Smart bioadhesives are of great interest for treatment systems where they can be used to control the release of drugs, close wounds, and fix devices on/in the body through non-invasive methods without damage until they have reached their desired aim. They are not only inexpensive, but they are also easy to control. Therefore, smart bioadhesives with this potential can change their performance and adhesive behavior in response to changing physiological conditions and promote treatment processes [9].

Advances in mono bioadhesives have been reviewed elsewhere [3,10],and are not the focus of this review article. Moreover, the history and classification of bioadhesives [5], various types of polymers and composites used as bioadhesives [11], as well as the application of bioadhesive hydrogels for drug delivery (via buccal, transdermal, gastrointestinal, parenteral, vaginal, and rectal routes) [4], wound healing (wound closure, sealing leakage, and immobilization) [12], and biomedical engineering [13] have been discussed by various research groups. In another review, hydrogels with multiple stimulus-responsive mechanisms were introduced, and their applications in emerging biomedical applications were examined [14]. Furthermore, basic background knowledge in designing environment-sensitive hydrogels [15], chemical force microscopy of stimuli-responsive adhesive copolymers [16], and their use as an intelligent carrier in the clinical field have been addressed [17]. In 2018, El-Sherbiny et al. [9] described some synthesis approaches, unique features, and different structures of stimuli-responsive polymers in thin films and nanostructures. Over the years, many disciplinary endeavors have been reported to optimize the functionality of smart bioadhesives and explore new and innovative applications. Nevertheless, investigations are ongoing in the field of smart bioadhesives. Despite extensive studies on stimulus-responsive polymers [18] and hydrogels [19] and many disciplinary attempts to optimize the functionality of smart bioadhesives, according to our knowledge, the development of multifunctional bioadhesives for various biomedical applications has not been adequately explored. In addition, some published review papers have directly or indirectly focused on the biomedical applications of adhesive and non-adhesive hydrogels [20–24], for example, Hwang et al. describe fundamental adhesion mechanisms in the development of multifunctional advanced skin adhesive patches. In comparison with previous review articles, this article provides a comprehensive overview of developed smart adhesives, their limitations, and future directions and challenges for the next generation of smart bioadhesives. In this regard, we will summarize the most recent cutting-edge strategies (years 2015–2021) used for stimuli-sensitive bioadhesives that can respond as external stimuli, self-heal, and remold shapes. First, the principal criteria for bioadhesives and types of smart bioadhesives will be discussed. Hereafter, the applicants of smart bioadhesives in various applications, including tissue engineering, wound healing, and drug delivery, are being studied. Figure 1 schematically presents an overview of these technologies and their applications. Finally, the limitations and challenges of current studies and future directions of smart bioadhesives will be discussed.

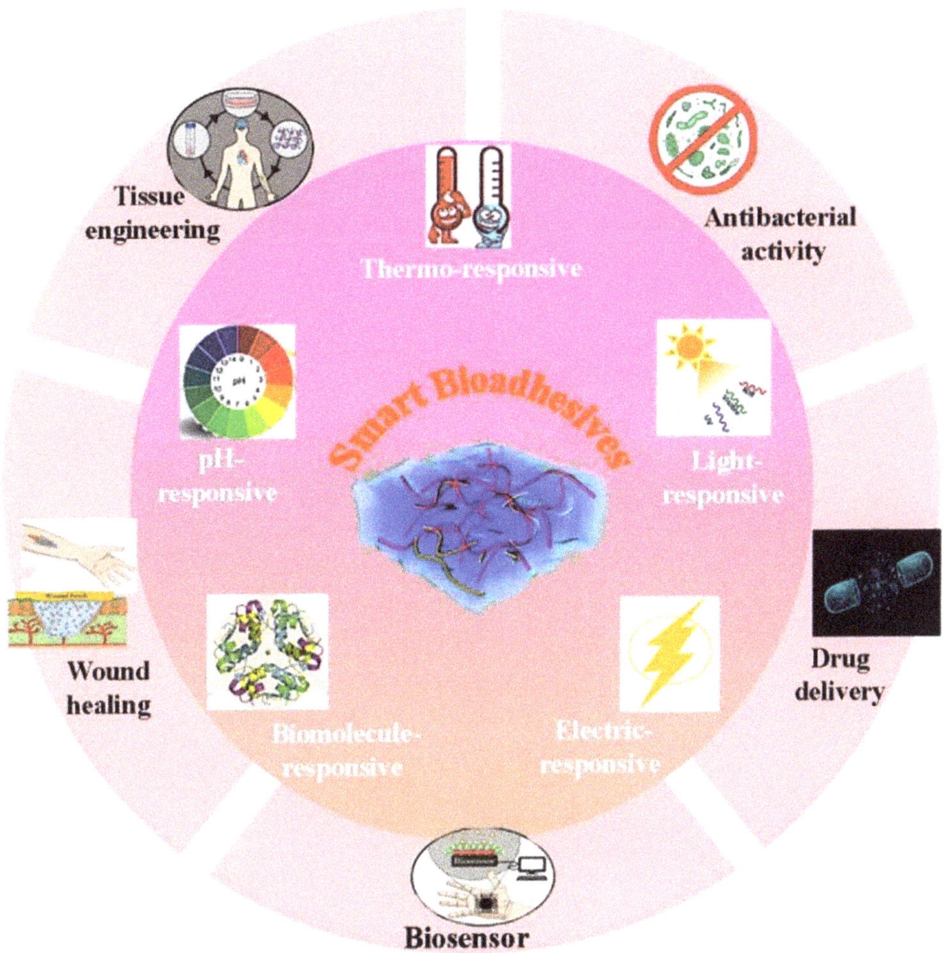

Figure 1. Schematic representation of various types of intelligent bioadhesives applied for different biomedical applications.

2. Principal Criteria in Bioadhesive

The adhesion term is defined as the potency of the adhesive to flow, wet the entire surface, and develop specific physicochemical intermolecular forces between the substrates and adhesive [25]. The phenomenon that two materials, one of which should be biological in character, are linked for a long time by interface tailoring is known as bioadhesion [3]. A bioadhesive system aims to make tight contact with the biologic substrate for a long time by interfacial forces. Bioadhesion in biological systems can be divided into class one with attachment between two biological phases, such as platelet aggregation and tissue repair; class two, adhesion between two biological phases (e.g., platelet aggregation and wound healing); and class three, adhesion of a biological degree to an artificial substrate (e.g., cell adhesion to culture dishes and biofilm formation on artificial devices), and adhesion of artificial material to a biological substrate (e.g., adhesion of synthetic hydrogels to soft tissues and adhesion of sealants to dental enamel), which are all examples of bioadhesion in biological systems [26,27]. Irrespective of the type and application, three factors are essential in the design of bioadhesives, including the capability to create powerful interfacial interactions, the ability to maintain cohesive character after curing, and the need to remain biocompatible during bioadhesive service life [5]. In addition to the mentioned characteristics, bioadhesives should have additional properties for specific applications. Bioadhesives should form strong interfacial interactions with tissue. These interfacial bonds can comprise hydrogen bonds, hydrophobic interactions, electrostatic interaction, diffusion, physical interlocking, and chemical cross-linking (such as ester, isocyanate, and aldehyde groups to primary amine) [28]. Nevertheless, the physical interactions poorly act within a high humidity environment because water interferes and may not be impressive in a biological system [29]. In some situations, the incorporation of cross-linking agents into adhesive systems leads to the formation of covalent interactions, such as disulfide cross-linking, Schiff-based chemistry, and enzyme-mediated cross-linking [30]. Therefore, these techniques can be applied to fabricate two and three-dimensional (2D and 3D) networks. However, the development of strong interfacial bonds is not sufficient, and bioadhesives should provide cohesive strength and remain stable for specific times to have adequate support for wound healing [16]. On the other hand, controlling bioadhesives' hydrophilic and swelling ability is essential for delivering therapeutically active agents in various ways, such as oral and transdermal [4,31,32]. In addition, bioadhesives should be biocompatible to achieve favorable in-vivo results. This means that they do not provide any unpleasant systemic or local effects during deployment and throughout the lifetime of the bioadhesive. In this regard, bioadhesives should not be cytotoxic, allergenic, irritants, or carcinogenic and need to maintain their mechanical integrity with surrounding tissues [33].

Tissue sealants, tissue adhesives, and hemostatic agents are examples of biomedical adhesives. Bioadhesives are categorized into internal and external, depending on the function and compatibility. Internal (intra-corporal) bioadhesives are often applied to repair chronic organ leakages and reduce bleed complications [34]. On the contrary, external bioadhesives are commonly utilized for wound closure and epidermal grafting. Based on the interaction with various organs and tissues, internal bioadhesives are likely to show better biocompatibility, adhesiveness, and strength properties on wet surfaces/environments than external bioadhesives. External bioadhesives are also expected to have a shorter closure time and higher durability than internal bioadhesives [35]. On the other hand, intelligent bioadhesives can also be categorized into three main classes depending on the type of stimuli: chemical (pH, oxidant, and glucose), physical (temperature, light, ultrasound, and pressure), and biological (enzymes, antigen, and ligand) responsive [36,37]. Smart bioadhesives are discussed in the following section.

3. Smart Bioadhesives and Their Applications

Smart bioadhesives are attributed to stimuli-responsive compounds with high performances that demonstrate reversible transitions in properties, including solubility, shape, molecular assembly, and surface characteristics in response to a stimulus [38]. Responsive compounds with dynamic properties, including wettability switch, mass transport, and mechanical actuation to inert materials, can have tremendous effects on smart bioadhesives [39]. In general, reversibility in bond association and dissociation aids in the reconstruction of polymeric networks in bioadhesives with stimulus-responsive characteristics. A structurally dynamic material is used in a distinct method for reversible adhesion. The presence of an active bond in structurally dynamic materials allows the material to change one or more properties [40] reversibly. When subjected to a suitable stimulus, the dynamic bond will undergo constant reversible exchange/cleavage, resulting in changes in the material's properties, such as modulus and viscosity. The bonds stop exchanging when the trigger is removed, and the material returns to its previous state [41]. For example, poly(*N*-isopropylacrylamide) (PNIPAM) is a promising macromolecule in thermo-responsive bioadhesives with a low critical solution temperature (LCST) of about 32 °C. According to the reports, hydrophilic PNIPAM reversibly alters to a hydrophobic state by increasing the external temperature until LCST. So, adhesiveness occurs between the room temperature and the body temperature [42]. Poly (acrylic acid) (PAA) is also a pH-responsive bioadhesive, which can protonate or deprotonate with pH changes. PAA can be swelled through electrostatic repulsion and experience high sorption and release in drug delivery systems [43]. Impressive self-assembly in the liquid state and mass transport in the solid-state can be attained in light-responsive polymers such as azobenzenes that isomerize quickly from one state to another and change size under UV light [44]. Electrochemical-responsive polymers are another smart adhesive group that responds to electric fields by changing their size or shape [45]. More details on the types of smart bioadhesives and their applications are provided in the following section.

3.1. Light-Responsive Bioadhesives

Light-responsive smart adhesives as noninvasive tools to regulate cell adhesion can be applied to tissue engineering, cell diagnostics, and medicine. The physicochemical behaviors of photosensitive molecules are altered or degraded in response to light irradiation with suitable wavelength and intensity [46,47]. Table 1 summarizes light-responsive bioadhesives trouped by stimulus responses and contains information about the inspiration and application. The light-responsiveness of O-nitrobenzyl was first cited by Ciamician and Silber about a century ago [48]. UV radiation activates most photochemical processes, such as acrylate polymerization, thiol-ene reaction, nitrobenzyl, and spiropyran groups. However, UV light-induced injury to biological specimens and live organs may restrict its use in-vitro and in-vivo because of intrinsic cytotoxicity and poor tissue penetration [49]. Photo-activation with near-infrared (NIR) light due to neglectable phototoxicity, easy access, clean, inexpensive, and sufficient penetration into the tissue can be considerable [50]. Li et al. [51] utilized spiropyran (SP) conjugated multi-shell upconversion nanoparticles (UCNPs) for adjusting cell adhesion/detachment reversibly and noninvasively. The UCNPs are ceramic lattices incorporated with trivalent lanthanide ions that could convert NIR light to UV radiation and activate photochemical processes on request. High-power and low-power NIR treatments were used to activate ring-opening and ring-closing procedures, respectively (Figure 2A). Such conversions caused the relation between SP and the cellular protein surface to be replaceable, resulting in reversible cell adhesion and detachment. Bian et al. [52] synthesized a reversible visible-light-responsive biofunctional surface by interacting the host–guest of azobenzene derivatized polycation/polyanion on a cyclodextrin (CD)-terminated substrate for switching from antibacterial to bioadhesion. They showed that the polyanions with COO^- groups provided bioadhesive properties, while the azobenzene functionalized polycations with quaternary ammonium groups had vigorous antibacterial activity. They could be switched by alternate assembly when exposed to visible light. Light-responsive bio-inspired MnO_2 hybrid (BMH) bioadhesives were employed in a research study for efficient melanoma photo-thermo-chemotherapy and multidrug-resistant (MDR) bacteria-contaminated healing of wounds(Figure 2B) [53]. As one of the "light-responsive" materials, MnO_2 nanosheet was produced to induce spatial and temporal controlled hyperthermia for further photothermal therapy. Furthermore, the two-dimensional nanosheets could be perfect bioactive molecule delivery carriers, allowing potent synergistic treatments for cancer and wound healing due to their high surface area and high binding energy via electrostatic and polar interactions. Based on the results, enhancing the local access to oxygen increased the cellular toxicity of doxorubicin (DOX) versus melanoma.

Table 1. Different types of light-responsive bioadhesives and their applications.

Compounds	Stimulus-Response Agents	Application	Summary	Role of Stimuli	Ref.
Spiropyran, multishellupconversion nanoparticles	Multishell Upconversion nanoparticles	–	The interactions between spiropyran and cell surface protein fibronectin were switchable even after 10 cycles.	By simply decreasing/increasing the excitation power density of the same 980 nm laser, cell adhesion/detachment can be switched quickly.	[51]
Catechol functionalized chitosan	MnO_2 nanosheets	–	BMH hydrogel successfully eliminated cancer cells in vitro giant solid tumors in vivo and had effective antibacterial properties without antibiotics.	By NIR irradiation, BMH hydrogel reduced the hypoxic tumor microenvironment by degrading internal hydrogen peroxide into oxygen and simultaneously releasing the anticancer doxorubicin hydrochloride.	[53]
Chitosan–polyvinyl alcohol-loaded tannic acid-TiO_2	–	Artificial electronic skin	Irradiation causes a change in surface wettability from hydrophobic to hydrophilic, leading to increases in electrical characteristics, mechanical strength, and adhesive properties.	Controllable swelling ratio upon irradiation with UV and visible light.	[54]
Thiol–PEG/ maleimide	Upconverting nanoparticles	Tissue engineering	Preparing light-sensitive adhesive hydrogels with spatiotemporally regulated biological functions for cell culture without causing significant photodamage to the cells	Photochemical processes are activated by converting NIR light (974 nm) into local UV emission.	[55]
PNIPAM/graphene oxide (GO)	Graphene (808 nm)	Cell capture	The bioadhesives efficiently captured cells via the adhesive oligopeptide and released a NIR light stimulus, suitable for cell preservation and therapeutic cell delivery.	NIR light efficiently triggered cell release; continuous NIR irradiation efficiently released the cells from adhesive hydrogel.	[56]
Dodecyl, chitosan	WS_2 nanosheets	Wound healing	Bioadhesive hydrogels with a positive charge, macropores, and alkyl chains could catch and limit microorganisms.	WS_2 nanosheets produced heat when exposed to NIR, and the antibiotic was triggered to release at the wound site.	[57]
PDA and PNIPAM	PDA	Wound healing	The coating of PDA–NPs onto hydrogel surfaces was effective in cell affinity, tissue adhesiveness, and growth factor/protein immobilization ability.	Pulsatile release of drugs and quick healing (1 min) after unfavorable damage with the assistance of NIR laserirradiation.	[58]

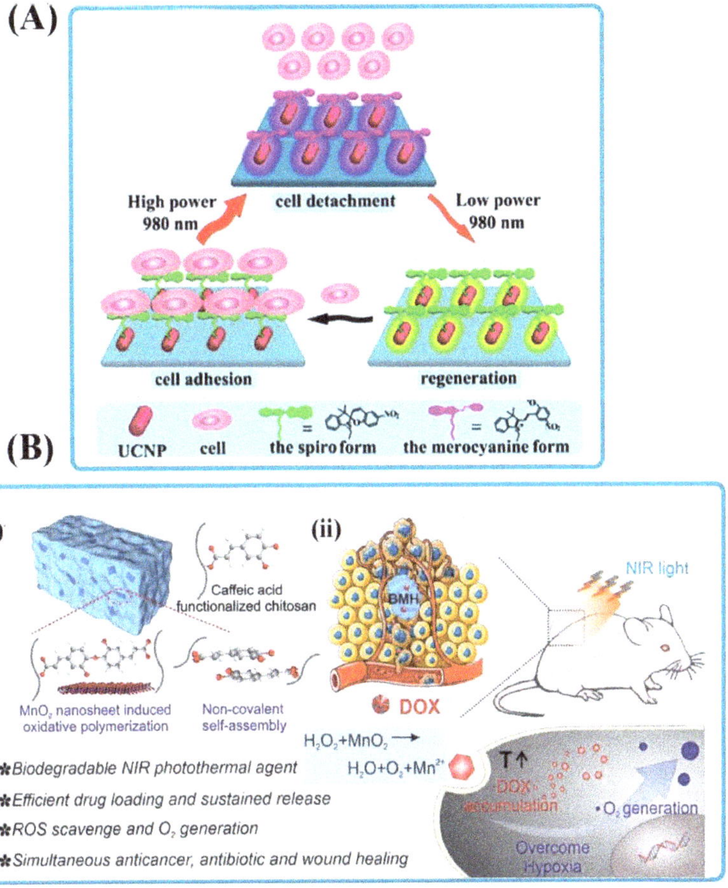

Figure 2. Light-responsive bioadhesives. (**A**) An illustration of SP–UCNP usage as a NIR-triggered photo-switch to modulate cell adhesion/detachment in a non-invasive and reversible manner by adjusting the power density of a laser. Reprinted with permission from Ref. [51]. Copyright 2015, ACS Publications. (**B**) (**i**) Schematic illustration of BMH hydrogel's composition and structure for simultaneous anti-cancer treatment and MDR bacteria-infected scar tissue. (**ii**) The nanostructure of BMH hydrogel effectively increased chemotherapy by enhancing O_2 generation via breaking endogenous H_2O_2 and enhancing intracellular buildup of DOX by PTT Reprinted with permission from Ref. [53]. Copyright 2020, Elsevier.

Although light-responsive bioadhesives possess many advantages, such as minor damage to cells, remote modulation, and high controllability of stimulus, they still have several limitations to being converted into medical products, including the incapacity of light sources to penetrate tissue [59], the use of UV light as a non-biofriendly source [60], and weak mechanical strength [61]. Many light-responsive bioadhesives require complex synthesis techniques that restrict their potential to be developed. Though NIR-triggered agents are more popular than UV-triggered agents, no investigation studied the effect of NIR-triggered agents on deep tissues. Only a few studies have been conducted on superficial disease models. In addition, due to their lower efficiency, these systems require a longer exposure time to have a therapeutic impact [62]. The unwanted extreme warmth may injure the surrounding healthy cells as a result of the unwanted extreme warmth.

Furthermore, most of these studies were conducted in-vitro, and it is essential to continue development and confirm results in-vivo.

3.2. Thermo-Responsive Bioadhesives

The temperature is of great interest among external stimuli because of its broad application, effortless control, and capability to use in-vitro and in-vivo states [63]. As a minimally invasive technique, thermo-sensitive bioadhesives show conformational changes in response to temperature stimuli, particularly near-physiological human body temperature, to generate interim polymer chain cross-linking via multiple physical interactions [64]. An ideal thermo-responsive system is a polymer solution with low viscosity at ambient temperature that, after injecting into target sites, changes into a gel at body temperature. Some studies focused on thermo-responsive bioadhesives are listed in Table 2.

Table 2. Different types of thermo-responsive bioadhesives and their applications.

Compounds	Stimulus-Response Agents	Application	Summary	Role of Stimuli	Ref.
Pluronic® 127 hydroxypropyl methylcellulose (HPMC)	Pluronic® F127	Wound infections	Ex vivo and in vivo studies showed bioadhesives with suitable antibacterial therapy of burn wound infections and anti-inflammatory activities. HPMC adhesive increased gel and bioadhesive strength	Formation of a stiff gel by increasing temperature from 4 to 32–37 °C.	[65]
Poly(acrylic acid)(PAA)/PNIPAM-co-dopamine methacrylamide (PDA)	PNIPAM	Epidermal sensors	The hydrogel with adhesive strength and self-healing ability demonstrated unusual fatigue and crack resistance properties.	Temperature-sensitive hydrogels, the lowest adhesion strength of hydrogel was at 25 °C.	[66]
Gelatin and chondroitin sulfate	Chondroitin sulfate	Surgical adhesive for sealing	In vivo and ex vivo, the injectable self-healing bioadhesive is used as a multifunctional tissue adhesive/sealant for closing bleeding wounds.	Exceptional tissue adherence at 37 °C diminished at low temperatures (20 °C), allowing it to detach from tissue easily.	[67]
Polydopamine-coated Tetronics–tyramine	Tetronic, tyramine (37 to 4 °C)	Tissue engineering	Adhesive hydrogels promoted human dermal fibroblast attachment, controlled by serum protein adsorption, creating a cell sheet after growth.	Cell sheet translocation process by changing temperature from 37 °C to 4 °C.	[68]
Hyaluronic acid (HA), methylcellulose, polyethylene glycol (PEG)	Methylcellulose	Surgical adhesive for sealing	Free-flowing, injectable at ambient temperature, gelation point about 40 ± 2 s, and lack of cellular toxicity	The transition of bioadhesive from sol at four °C to gel state at 37 °C.	[69]

Table 2. *Cont.*

Compounds	Stimulus-Response Agents	Application	Summary	Role of Stimuli	Ref.
Catechol modified quaternized chitosan, poly(d,l-lactide)-poly(ethylene glycol)-poly(d,l-lactide) (PLEL)	PLEL	Wound healing	The injectable thermo-sensitive adhesive hydrogel offered excellent properties as a wound dressing for promoting wound healing (only in 7 days), biocompatibility, and bioactivity through in vivo degradation, stimulated endothelial cells migration, and angiogenesis.	The temperature-triggered reversible sol (25 °C)–gel (37 °C) transition of PLEL solution.	[70]
Galactose modified xyloglucan (mXG) and hydroxybutyl chitosan	Galactose modified xyloglucan	Wound healing	According to in vivo findings, bioadhesive was an excellent anti-adhesion system for avoiding repeated adhesion following adhesiolysis, promoting wound healing and reducing scar formation.	Gelation temperature and time depended on the total solid content of bioadhesive hydrogels.	[71]
PIPAAm, butyl methacrylate (BMA)	PNIPAAm	Regenerative medicine and tissue engineering	Increasing BMA concentration improved the cell adhesion, owing to increased cellular protein adsorption.	Celladhesion and detachment from hydrophobized thermos-responsive brushes.	[72]
PNIPAAm-g-chitosan	PNIPAAm	Tissue engineering	Hydrogels showed outstanding biocompatibility to MSCs, fibroblasts, and osteoblasts, allowing cell encapsulation without toxicity.	LCST at around 30.71–32.02 °C indicated hydrogels had potential for in situ injection.	[73]
Pluronics, hyaluronic acid, corn silk extract, and nanosilver	Pluronics	Wound healing	From a biological point of view, hydrogels had good biocompatibility and exhibited antibacterial activity toward gram-positive and gram-negative bacteria.	Viscoelastic parameters changed in the temperature ranging from 25 to 40 °C.	[74]
Collagen, chitosan, and bioactive glass	Chitosan	Bone tissue engineering	The addition of collagen to the system resulted in larger pore size and enough interconnectivity, making it suitable for use as biomaterials for bone tissue engineering.	Gelation temperature at 37 °C.	[75]

Polymers are classified into two categories [68]. The first case is LCST, which is insoluble above its critical temperature. The second case is the upper critical solution temperature (UCST), which precipitated and underwent a phase shift in its critical temperature, the temperature at which the polymers keep miscible in solution. At the same time, phase separation occurs when the temperature rises over the critical value, which is called "negative temperature-sensitive polymers" in LCST materials (e.g., PNIPAM, gelatin, and carrageenan) [76]. In contrast, UCST materials are known as "positive temperature-sensitive polymers". They are miscible at room temperature, while their solubility diminishes when the temperature drops below the critical value, causing phase separation. Examples include (acrylamide-co-butyl methacrylate), PAA, and polyacrylamide (PAAm) [77]. The thermo-sensitive microstructure changes of a supramolecular hydrogel bioadhesive containing ureidopyrimidinone (UPy) and PNIPAM (Figure 3A) were investigated by scanning electron microscope (SEM) images (Figure 3B) [78]. This bioadhesive displayed large pores at the 25 °C (approximately 3.4 μm) while the size of the pores decreased to around 0.82 μm at 37 °C. Interestingly, the pores could regain their primary size (about 4.0 μm) by turning back the temperature to 25 °C. This suggested that the macromolecule chains became hydrophilic at a lower temperature than the LCST and may generate bigger pores during the lyophilization process. In contrast, the polymer chains were dehydrated and collapsed at a temperature higher than LCST, and the size of pores became smaller. This bioadhesive was a good candidate for drug delivery application. Zheng et al. [70] designed a bioadhesive based on quaternized chitosan (QCS-C) embedded into poly(d,l lactide)-poly(ethylene glycol)-poly(d,l-lactide) (PLEL) for wound healing. In the below LCST, the PLEL polymer was a random coil unimer (Figure 3C). With the increase in temperature above LCST, the PLEL structure changed to a micelle because of the hydrophilic poly(ethylene glycol) chain (outer shell) and the hydrophobic poly(d,l-lactide) chain (inner core). The presence of QCS-C could be effective in decreasing sol–gel transition temperature. In addition, the rheological property of the bioadhesive indicated that it could flow freely below the gel point and was fully suitable for in situ injection. As the temperature increased from 33 to 40 °C, the storage modulus approached the loss modulus, indicating a semisolid property. As a result, the human body's temperature may be ideal for therapeutic wound management. Zhang et al. [71] prepared a thermos-responsive bioadhesive by simply combining galactose modified xyloglucan (mXG) and hydroxybutyl chitosan (HBC). The obtained bioadhesive as a cytocompatible and hemocompatible hydrogel prevented repeated adhesion after adhesiolysis, enhanced wound healing, and reduced tissue injury.

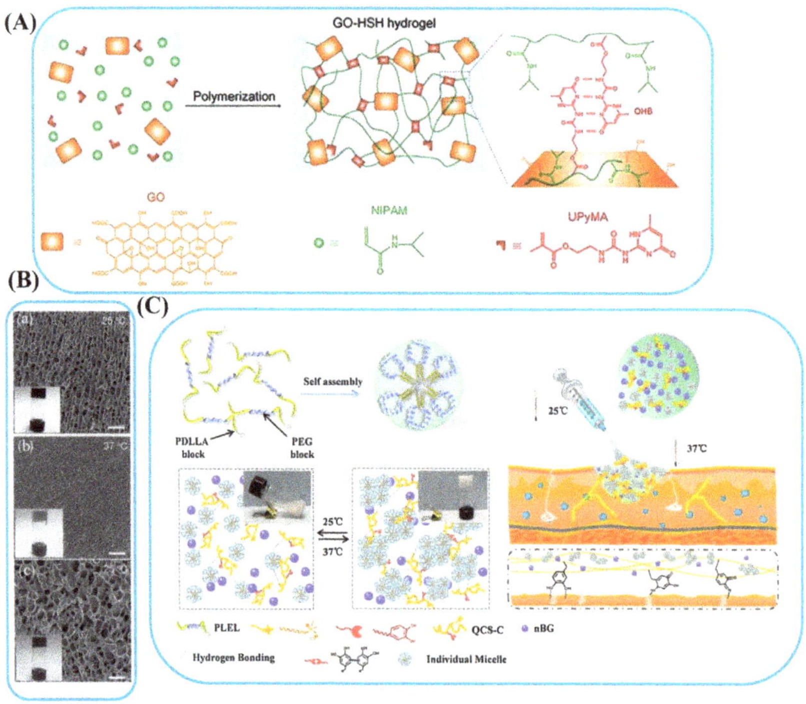

Figure 3. Thermal-responsive bioadhesives. (**A**) Schematic models of the synthesis of supramolecular

hydrogel (HSH) containing GO, NIPAM, and UPy ethyl methacrylate monomer. (**B**) SEM images of a bioadhesive in both states of hydration and dehydration: (**a**) at 25 °C, (**b**) dehydrated at 37 °C above the LCST, and (**c**) restored to the hydrate conditionat 25 °C. Reprinted with permission from Ref. [78]. Copyright 2017, Wiley-VCH. (**C**) Schematic diagram of thermo-sensitive injectable PLEL-nano bioactive glass-QCS-C composite hydrogel for wound healing. Reprinted with permission from Ref. [70]. Copyright 2020, Elsevier.

Comprehensive studies have focused on applying thermo-sensitive bioadhesives for tissue engineering and in vitro transplantable tissues. In tissue engineering and regenerative medicine, intelligent bioadhesives can be employed as injection systems to transfer growth factors and cell stimuli-responsive surfaces to regulate cell adherence or penetration [79]. For instance, Moreira et al. created a bioactive thermogelling chitosan-based injection of bioadhesive hydrogel for bone regeneration. Recently, regenerative medicines with the ability for cell culture to remedy the lost functions of organs and tissue have been becoming promising treatments. To form transplantable tissues invitro, selecting a cell separation method with enough purity and function after dissociation is interesting. In this regard, the separated cells using the thermo-responsive adhesive brush have shown high function [80]. Even though this process requires a relatively long time, cell purification is not needed for constructing tissues. Moreover, separated cells using this polymer show good function without correction of the cell surface, which is significant for manufactured tissue transplantation.Furthermore, the separation can be accomplished simply by changing the external temperature of the adhesive brush surfaces that have been created [81]. Polymer brushes are unique macromolecular structures with a dense array of polymer chains immobilized on a surface or interface by one of their end chains. These structures have promising applications for stimuli-responsive and cell adhesive surfaces [82]. In the study by Nagase et al. [72], thermo-responsive copolymer bioadhesives were developed by copolymerizing butyl methacrylate (BMA) into PIPAAm. The adhesion characteristics of copolymer brush surfaces at 37 °C and detachment at 20 or 10 °C were confirmed for human umbilical vein endothelial cells (HUVECs) and normal human dermal fibroblasts (NHDFs), respectively (Figure 4A,B).

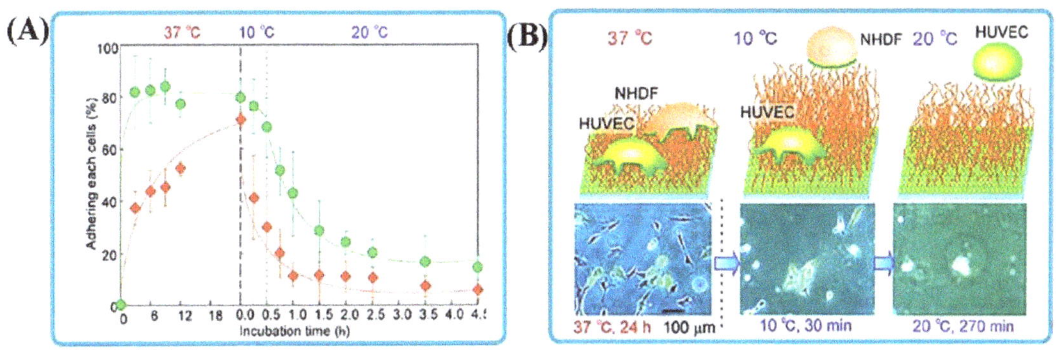

Figure 4. Smart bioadhesives for tissue engineering. (**A**) Diagram of adhesion on and detachment of green fluorescent protein (GFP)-HUVECs and NHDFs from IPB-5 in culture medium. NHDFs and GFP-HUVECs are represented as orange squares and green circles, respectively. Cell adhesion was carried out at 37 °C for 24 h; after which the cells were incubated for 30 min at 10 °C, followed by a recovery period at 20 °C. (**B**) Morphology of GFP-HUVECs and NHDFs on and detachment from IPB-5. Reprinted with permission from Ref. [72]. Copyright 2013, ACS Publications.

To investigate bone tissue repair, Saravanan et al. [83] developed a thermosensitive chitosan/glycerophosphate adhesive hydrogel containing graphene oxide (GO) with the applicability of injectable. They found that the inclusion of GO into the matrix significantly improved swelling and protein adsorption ability. They further concluded that a GO-containing chitosan/GP hydrogel possessed the ability to produce osteogenic differentiation of mesenchymal stem cells (MSCs), making it appropriate for bone tissue engineering.

The induction of antibacterial properties is another promising application of thermal-responsive adhesives. The mechanism of antibacterial surfaces depending on bactericidal agents is classified into releasing-based and contacting-based bactericidal agents [84]. In the first case, biocides are usually incorporated or preloaded into a matrix and then released into the surroundings to kill the bacteria (e.g., releasing drugs to decrease infection in sores). Bioadhesives containing quaternary ammonium are extensively utilized for contacting-based mechanisms for the second group. It is related to the low toxicity, excellent cell membrane infiltration character, extended residence time, environmental constancy, and biological activity of ammonium [85]. For example, dopamine (DA) as an anchoring site was loaded into the polyethersulfone (PES) membrane surface to develop an adhesive layer, then the adhesive was stuck onto the membrane by using photoinduced cross-linking copolymerization of methacryloxyethyltrimethyl ammonium chloride (DMC) and NIPAAm (Figure 5A) [84]. The results showed that the quaternary ammonium salts in the hydrogel film could lead to the destruction of the adhering bacteria. On the other hand, the dead bacteria detached from the surface by decreasing the temperature below the LCST of PNIPAM. In addition, the clotting test revealed that the changed surfaces improved blood compatibility and prevented hemolysis. In other work, 3D printable thermo-responsive PNIPAM/cellulose nanofibrils (CNFs) were developed to provide a new platform for regulating LCST properties and tuning bioadhesive behaviors [86]. In response to temperature, the hydrogel system containing 2% CNF had exchangeable bioadhesion. Above the LCST, the adhesion of the PNIPAm/CNF hydrogels to bacteria was stronger. It could be related to the wholly extended CNF, which made a semi-interpenetrating polymer with PNIPAm. The bioadhesive was severely weakened at 40 °C. The CNF chains were divided into small separate sections, considerably reducing the bacteria-CNF contact area. Therefore, temperature control might be utilized to keep or release bacteria that have developed on the hydrogel surface (Figure 5B).

Despite advantages such as easy accessibility, low side effects, stability of drugs, etc., thermo-responsive materials have disadvantages such as poor mechanical strength, limitation in drug loading ability [87,88], and uncontrollable on/off state of actuation [89], which need more studies.

3.3. *pH-Responsive Bioadhesives*

Ionic polymers are commonly used in pH-responsive adhesives. Protonation or deprotonation of ionic side chains in these polymers can cause swelling of polymer backbones due to electrostatic repulsion [90]. pH-responsive materials can be divided into anionic and cationic bioadhesives based on the pendant group in the polymer chains [91]. When the adhesive's acid dissociation constant (pK_a) is lower than the pH of the surrounding aqueous solution, the swelling/deswelling behavior of anionic materials is triggered by osmotic pressure. In contrast, the cationic materials that contain donor electron groups such as amine become protonated and swelled in an aqueous solution with a lower pH ($<pK_a$) [14]. For a better understanding, some polymers' swelling and shrinking behavior, such as chitosan, is referred to as external and environmental pH [92]. At lower pH, the protonation of the amine group of chitosan generated electrostatic repulsion, allowing polymer chains to extend and interact with water molecules more efficiently, thereby enabling water solubility. The amine group is deprotonated when the pH rises, implying no net charge [93]. Further, the amine groups destroy the chitosan structure and decrease the water solubility. As a result, the pK_a value is effective in the water solubility of some pH-responsive polymers. Pores are typically produced in other polymers with pH-dependent solubility for more

specialized pH-sensitive uses. The waterinsolubility of these polymers at low pH (e.g., in the stomach) and their solubility at higher pH (e.g., in the small and large intestines) causes materials to leach out and form a porous and permeable film. In bioadhesives, synthetic polymers such as PAA, proteins, and polysaccharides are commonly classified as pH-sensitive polymers [5,94]. According to studies, diseasedcaries in people's mouths fluctuate between 4.5 and 6. Therefore, dual adhesive membranes and oral drugs can be applied to protect from oral infections and oral tissue regeneration. Table 3 summarizes different types of pH-responsive bioadhesives and their applications.

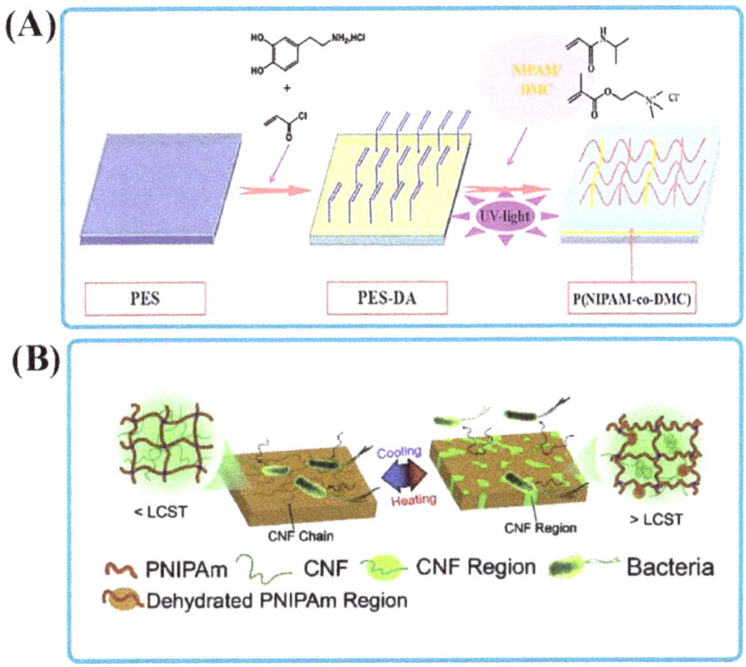

Figure 5. Smart bioadhesives for antibacterial activity. (**A**) A schematic illustration of modification procedure of PES membranes by ene-functionalized dopamine to form an adhesive layer and then attaching it onto the membranes via photo-induced surface cross-linking copolymerization. Reprinted with permission from Ref. [84]. Copyright 2018, Wiley-VCH. (**B**) Schematic of the effect of temperature on microorganism's growth and adhesion for PNIPAm/CNF hydrogels. Reprinted with permission from Ref. [86]. Copyright 2020, Elsevier.

Table 3. Different types of pH-responsive bioadhesives and their applications.

Compounds	Stimulus-Response Agents	Application	Summary	Role of Stimuli	Ref.
PAA/Zinc (II) ion	PAA and dopamine	-	Coacervate bioadhesive with good mechanical and self-healing properties.	Oxidation of catechol groups at basic pH favored the formation of strong adhesion.	[95]

Table 3. *Cont.*

Compounds	Stimulus-Response Agents	Application	Summary	Role of Stimuli	Ref.
Carbon nanotubes and GO/tectomers	Tectomer	Tissue engineering	The hybrid materials can be used as pH-switchable bioadhesive coatings and scaffolds for tumor models in ex vivo studying.	Controlled release from a pH-dependent peptidic coating.	[96]
Chitosan-grafted-dihydrocaffeicacid/ oxidized pullulan	Chitosan-g-dihydrocaffeic acid	Drug delivery	Good injectability, a decent gelation duration, and pH-dependent equilibrated swelling ratios, morphologies, and rheological properties were observed by bioadhesive hydrogels.	At acidic conditions, the hydrogels had a larger swelling ratio and pore size than at pH 7.4.	[97]
D-α-tocopheryl PEG 1000 succinate conjugated chitosan.	Chitosan	Drug delivery	Invivo pharmacokinetic results demonstrated the relative bioavailability of bioadhesive micelles was effective beneficial for brain cancer therapies with the prolonged release.	A pH decrease triggered the drug release.	[98]
Dopamine-conjugated HA/ mesoporous silica	Dopamine	Drug Delivery	In vivo studies confirmed the injection of bioadhesives could achieve high therapeutic efficiency against tumor growth while avoiding significant damage to healthy organs.	The faster release rate of the drug at pH 5.0 than at pH 7.4.	[99]
Collagen and PEG	Collagen	Diabetic wound repair	Bioadhesive loaded stem cell factor as an anti-inflammatory and biocompatibility dressing was used for tissue regeneration.	Effective in drug release rate.	[100]
Chitosan and pectin	Chitosan	Drug delivery and tissue regeneration	Based on ex vivo testing, membranes loaded with antimicrobial peptides had simultaneous antibacterial effectiveness against oral streptococci as well as cytocompatibility with both soft and hard tissue.	Temporary preventive and therapeutic distribution in the oral cavity with a 'supply on demand' release behavior in a pH-controlled manner	[101]

Table 3. *Cont.*

Compounds	Stimulus-Response Agents	Application	Summary	Role of Stimuli	Ref.
PAA and PAAm	PAA and PAAm	Drug delivery	In vitro findings showed dual pH-responsive bioadhesive hydrogel can release lipophilic or hydrophilic pharmaceuticals based on the pH of the environment while preventing drug metabolism, degradation, and excretion.	In alkaline or acid conditions, the bioadhesive can conduct programmable and bidirectional bending by shrinking anionic and cationic networks and asymmetric swelling.	[102]
PAA	PAA	Sensor	Bacterial detachment is caused by increasing brush thickness, disparity, and solution pH.	Tuning the attachment and detachment of bacteria in various pH values.	[103]

For pH-controlled delivery of antimicrobial peptides (AMP) into the oral cavity, Boda and coworkers [101] prepared bioadhesive membranes combined with chitosan and pectin derivatives with dual adherence to soft and hard tissue surfaces (Figure 6A). Pure chitosan membranes indicated suitable adhesion to enamel tissue/hard, whereas the presence of oxidized pectin can be an effective way to increase mucoadhesion. One of the drawbacks of this work was that the effect of pH on the adhesive qualities of membranes was not studied. Yadav et al. [103] developed pH-sensitive adhesive for antibacterial applications. They learned how to brush dispersity and thickness affected the initial attachment and future detachment of *Staphylococcus epidermidis* bacteria to a pH-responsive PAA brush system. With increasing pH value, the instinct properties of PAA changed from neutral, hydrophobic, and dried up to negatively charged, hydrophilic, and swollen. Switching from pH=4 to 9 also removed microorganisms from the brush surface. Based on the results, an optimal thickness of 13 up to 18 nm was recognized for maximizing microorganism detachment on the PAA brushes at pH 4. The brush dispersion did not affect bacterial adhesion. Recently, a series of injectable pH-responsive self-healing bioadhesives have been produced by radical polymerization of acryloyl-6-aminocaproic acid (AA) and AA-g-N-hydroxysuccinimide (AA-NHS) for wound healing applications [104]. The good hemostatic performance, histomorphological evaluations, and wound healing results demonstrated the therapeutic efficacy of the AA/AA-NHS hydrogel in a swine gastric hemorrhage/wound model (Figure 6B).

Despite the reality that pH-responsive bioadhesives have demonstrated wide applications in medicine, there are still issues with the evolution of a bioadhesive, which can behave in favorable procedures under basic and acidic situations. The swelling property is necessary for bioadhesives because liquid absorption is critical during tissue regeneration [105]. However, some adhesives lose their mechanical strength due to solution uptake. On the other hand, the pH value of the media pH may differ depending on the intensity of the complaint or the type of damaged tissues, making it difficult to maintain the bioadhesive's adhesion capabilities during the therapy process [106]. Furthermore, the actions of pH-sensitive bioadhesives might be initiated during administration or use, rendering these systems susceptible to off-target distribution.

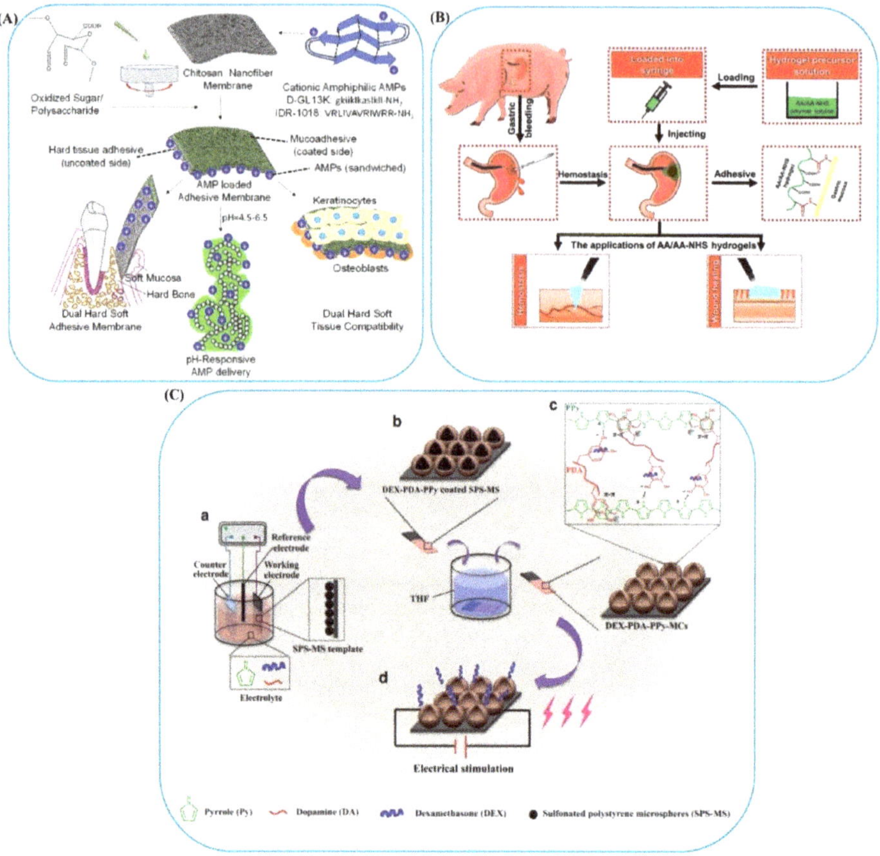

Figure 6. (**A**) pH-responsive bioadhesives. A scheme of the preparation stages of oral tissue adhesive membranes coated with AMP and pH-responsive release of AMP to acidogenic oral biofilm. Reprinted with permission from Ref. [101]. Copyright 2020, ACS Publications. (**B**) The use of AA/AA-NHS bioadhesive hydrogels for wound healing and blood clotting. Reprinted with permission from Ref. [104]. Copyright 2021, Springer Nature. (**C**) Electro-responsive bioadhesives: An illustration of the procedure for preparation of dexamethasone-loaded PDA–PPyMCs, (**a**) process of electrochemical deposition, (**b**) eliminating the sulfonated polystyrene microspheres template by tetrahydrofuran etching, (**c**) hydrogen bonding, and π–π interactions between PDA and PPy, and (**d**) drug delivery by electrical stimulation. Reprinted with permission from Ref. [107]. Copyright 2017, Springer Nature.

3.4. Electromagnetic-Responsive Bioadhesives

Field-responsive polymers can be exploited in applying sonic, magnetic, electric, and electromagnetic fields. An electric field's changing geometrical shapes and sizes can be depicted as a synergy of coulombic, electrophoretic, and electroosmotic interactions [47]. Electro-responsive polymers are classified into two groups: ionic and dielectric [108]. Various types of electro-responsive bioadhesives and their applications are given in Table 4. The first group is known as conducting polymers, in which response to an electric field leads to the mobility of free ions and a change in the local concentration of ions in solution or within the material [109]. The migration of ions in an electric field can cause the asymmetric distribution of charged ions, the formation of a concentration gradient of ions, the generation of an osmotic pressure difference, and finally, the production of a swollen hydrogel. On the other hand, the second group includes dielectric elastomers and electrostrictive polymers created by electrostatic (coulombic) forces [110]. Polydopamine–polypyrrolemicrocapsules (PDA–PPyMCs), as electro-responsive and conductive polymers, have been synthesized on titanium electrodes that can release medications locally and accurately [107]. The preparation steps of the electro-responsive system are described in Figure 6C. Based on the results, the adhesion strength of the PDA–PPy for reacting with various substrates was enhanced by increasing the PDA amount. Also, PDA–PPyMCs presented an excellent ability to attach to cells and drug-loading due to strong cell affinity, porous form, electro-responsivity, and good conductivity. Thismeans they could be used as a conductive substrate to transfer electrical impulses to stimulate cell action. However, precise control over the magnitude and duration of electric current provided a unique advantage to electro-responsive bioadhesives. Moreover, their applications have drawbacks, such as wired and bulky instruments [111]. Furthermore, they need to implant electrodes in the bioadhesive matrix, which determines their applications for topical or subdermal implants.

Table 4. Different types of electromagnetic-responsive bioadhesives and their applications.

Compounds	Stimulus-Response Agents	Application	Summary	Role of Stimuli	Ref.
Poly [anilineTetramer-methacrylamide]-co-[dopamine methacrylamide]-co-[poly(ethylene glycol) methyl ether methacrylate]}	-	Bone tissue engineering	A conductive bioadhesive with biocompatibility and strong adhesion was prepared for regeneration of comminuted bone fracture; the adhesive strength of hydrogel was less than that of the cortical bone and showed in in vivo cytotoxicity.	Electrical conductivity of bioadhesive enhanced with the increase of AT, which improved cellular activities.	[112]
AA and PEG dimethacry-late/GO/gelatin	Graphene oxide	Wound healing	Adhesive hydrogel with good thermal and mechanical stability indicated viability of more than 94% for human fibroblasts, while curcumin-loaded samples showed a reduction of bacteria of 90%.	At 0 and V, the slow and fast release was achieved, while intermediate kinetics was found at 12 and V.	[113]
Xanthan gum, chitosan, and iron oxide magnetic	-	Muscle, skin, cartilage, and connective tissue engineering	In vitro studies showed that bioadhesive hydrogels improved fibroblasts' growth and adherence in an external magnetic field compared to the pristine hydrogel.	In a magnetic field, adhesion and proliferation of fibroblasts were enhanced in hydrogels containing magnetic nanoparticles.	[114]
PAA grafted gum ghatti (GGH)	Gum ghatti	Drugs delivery by the skin	A histopathology examination demonstrated reversible changes in skin structure.	The release was observed over a two-fold increase in the drug after applying an electric stimulus.	[115]

Table 4. *Cont.*

Compounds	Stimulus-Response Agents	Application	Summary	Role of Stimuli	Ref.
Nanoclay (laponite), multiwalled carbon nanotubes (CNTs), and NIPAM	CNTs	Human motion sensing	Multifunctional conductive flexible hydrogels with self-healing, sticky, and 3D printable properties without any toxicity for the L929 cells.	Conductive bioadhesive hydrogels for wearable electronic devices revealed good electrical stability and multifunctional stretchability.	[116]
Chitosan-aniline oligomer/polyvinyl alcohol	Polyaniline	Tissue engineering	Biocompatibility testing demonstrated the conductive substrate offered the platform with more cellular activity than non-conductive materials.	Rising in drug release after electrical stimulation in comparison with non-stimulated webs.	[117]
GO-PAA	Graphene oxide	Artificial muscle and tissue engineering scaffold	Bioadhesive hydrogel showed good compatibility with bone marrow-derived mesenchymal stem cells.	Under the circumstance of electrical stimulation, the morphology of adherent cells was changed, and the differentiation of neural stem cells was promoted.	[118]

3.5. Biomolecule-Responsive Bioadhesive

Sensitive bioadhesives to biomolecule amounts can activate the localized therapeutic drug release to mimic the short- and long-term molecular regulatory processes seen in tissues at the cellular level. Biomolecule responsive bioadhesives have been highly regarded for their structural transition in response to the main target biomolecule [119]. For instance, glucose-responsive adhesives can indicate structural changes in reaction to the glucose amount in diabetes disease. Insulin injections may possess various problems, such as a long treatment period and diet restrictions, which present an alternative therapy way with the capability of immediate responses to blood glucose levels; safe and continual administration seems essential [17]. Types of electro-responsive bioadhesives and their applications are listed in Table 5. Zhou et al. [120] developed a stimulus-sensitive turnover method using a bioadhesive oral delivery nanoparticle system coupled with glucose oxidase (GOx) and insulin as an intelligent glucose-responsive switch. The L-cysteine–alginate in glucose-responsive nanoparticles with a suitable weight ratio of 2:1 showed good encapsulation efficiency, bioadhesion, and pH stability, which are favorable for oral delivery. In-vitro studies revealed that glucose-responsive nanoparticles switch insulin release behavior "ON" in response to a hyperglycemic condition by catalysis of GOx and "OFF" in response to normal blood glucose levels. Despite promising features, obstacles such as poor stability in various environmental conditions, unfavorable behavior at physiological pH, poor glucose selectivity, and slow response rate raise concerns about the clinical usage of glucose-responsive bioadhesives [121].

Table 5. Different types of electro-responsive bioadhesives and their applications.

Compounds	Stimulus-Response Agents	Application	Summary	Role of Stimuli	Ref.
Thioglycolic acid, chitosan, gold nanoparticle	Thioglycolic acid	-	Ultra-low concentrations of thrombin, as well as low molecular weight anatoxin, are detected selectively and reproducibly.	Detect early biomarkers in complex body fluid.	[122]
Phenylboronic acid and cis -diol modified PEG	Modified PEG	Drug delivery	The injectable, self-healing and adhesive hydrogel could have applications in 3D cell culture substrates for tissue engineering and controlled macromolecule release.	Size-dependent controlled release of proteins encapsulated within the network and the glucose-responsive release of larger proteins.	[123]
Hyaluronic acid cross-linked with divinyl sulfone.	Hyaluronic acid	Diabetic patients	The released insulin from glucose-responsive nanocarriers displayed a practical hypoglycemic effect for a longer time after oral administration to diabetic rats than insulin-loaded nanocarriers.	Regulation of insulin.	[124]
2-nitroimidazole–l-cysteine–alginate	2-nitroimidazole	Diabetic patients	Invivo experiments on type I diabetic rats showed that the hyperglycemia risk was reduced following oral administration, and a standard glucose range was maintained for a long time.	Blood glucose regulation via glucose catalysis by glucose-responsive adhesives.	[120]

3.6. *Multi-Responsive Bioadhesives*

Multi-responsive bioadhesives have been created to develop multifunctional bioadhesives for different biomedical applications [125]. These bioadhesives are specifically attractive for drug delivery applications. Recent advances in intelligent drug delivery adhesive carriers have great promise. They provide a way to promote formulations tailored to drug delivery systems and the release of drug control based on stimuli responses [126]. An intelligent drug delivery system can release an active chemical at the proper place and at a rate that adjusts in response to disease progression [127].Some of the multi-responsive bioadhesives and their applications are provided in Table 6. Le et al. [128] synthesized pH and temperature-sensitive injectable bioadhesives of poly (sulfamethazine-ester-urethane) (PSMEU) and poly(ethylene glycol) (PEG) by in-situ developing injectable hydrogelators (Figure 7A). Although PEG–PSMEU bioadhesive was free-flowing at ambient temperature, it quickly became a gel when exposed to body physiological conditions (pH 7.4 and 37 °C). These bioadhesives could promote skin wound repair due to their superior adhesive, bioresorbable, and mechanical characteristics. In addition, hypodermic implantation of PEG–PSMEU repaired the damaged skin and led to wound healing without an inflammatory response. In their research, Lee et al. [129] indicated that alginate–boronic acid hydrogel could be orally administrated in drug delivery systems because of its tolerance to high acidic conditions in the stomach (Figure 7B). In addition, the adhesive nature of alginate-BA could increase the residual time in the body. For the first 30 min, a clear fluorescent signal was observed in the esophagus region for alginate-BA (Figure 7C, right panel, yellow circle). It was related to alginate-BA's gelation in the esophagus region (pH upshifts to 7.4). On the other hand, alginate was wiped off, and no fluorescence was noticed when administered orally (top, left panel). The alginate-BA hydrogel led to strong fluorescent signals (e.g., red spots) in the colon after 24 h. However, alginate alone had weak dispersive signals of residual fluorescence (bottom, left panel). Abebe et al. [130] synthesized a self-adhesive hydrogel based on modified alginate with gallic acid (GA) and in-situ polymerization of polyacrylic acid for pH and strain-responsive transdermal delivery. According to the findings, stretching led to an increase in the release rate, but strain percentage had the opposite impact. As demonstrated in Figure 7D, strain percentage of 100% resulted in increased release within the first 10 min, but the releasing pattern reversed after 20 min at pH = 5.5. In 2019, a triple stimuli-responsive hydrogel with self-adhesive, self-healing capabilities was developed using *N*, *N*-diethylacrylamide (as a thermo-sensitive part), PDA (as a NIR light-sensitive part), and acrylic acid (as a pH-sensitive part) [131]. The resulting bioadhesive was extensively employed in wound dressings and wearable technology. Furthermore, the hydrogel with an optimum mass fraction of 0.4 wt.% of PDA was for the removal of methylene blue, with a maximum adsorption capacity of 305.4 mg/g.

Table 6. Different types of multi-responsive bioadhesives and their applications.

Compounds	Stimulus/Stimulus-Response Agents	Application	Summary	Role of Stimuli	Ref.
PEG, PSMEU	pH and thermal/PSMEU	Wound healing	Bioadhesive hydrogels were used in vivo to seal cutaneous wounds, absorb wound exudates, and promote tissue regeneration in the injured area.	Free-flowing PEG–PSMEU copolymer sols (pH 8.5, 23 °C) were converted into stable gels in the body (pH 7.4, 37 °C).	[128]
Alginate–boronic acid conjugate	pH- and glucose/boronic acid-diol complexation	Drug delivery systems	Alginate-BA hydrogels showed great promise in various applications, including pressure-sensitive biological glues to biomedical substrates requiring stretchability, self-healing, and multiresponsiveness.	Effect on the viscoelastic and mechanical properties of bioadhesive hydrogels.	[129]
Dopamine functionalized 4-armed PEG (4-arm-PEG-DA) and phenylboronic acid	pH, glucose, and dopamine triple-responsive/Dopamine and modified PEG	Drug delivery, Tissue engineering	Bioadhesive showed good adherence to tissues, and in vitro cytotoxicity experiments showed hydrogels were very cytocompatible.	The disintegration rate of hydrogel increased by decreasing pH value from 9 to 3.	[132]
PNIPAM/PDA/clay	Light-and thermos/PDA, PNIPAM	Electronic skin	In vitro cytotoxicity results indicated that hydrogel with high adhesiveness and biocompatibility suggested good cell affinity and biocompatibility.	Locally controllable deformation of the hydrogel by remote NIR irradiation.	[133]

Table 6. *Cont.*

Compounds	Stimulus/Stimulus-Response Agents	Application	Summary	Role of Stimuli	Ref.
Thiolated chitosan and thiolated chondroitin sulfate	pH and redox/Amino groups, carboxyl and sulfate groups	Wound healing and tissue engineering	Multilayer systems with disulfide bonds aided tuning cell contact, film degradation, and controlled release of bioactive compounds.	Cross-linking in alkaline pH or reduction of disulfide bonds changed mechanical and surface properties and cell function.	[134]
Collagen (COL), guar gum (GG), PNIPAM, GO	Light and thermal/PNIPAM and GO	Wound healing, wearable electronic devices, and sensors.	A bioadhesive hydrogel with many functions was synthesized, including quick wound healing, super-ductility, injectability, remoldability, conductive, thermo-sensitive, NIR-responsive, and accelerated wound healing.	Phase change occurs shortly after touches the human body.	[135]
PAA, oligo(ethylene glycol) methacrylate, 2-(2-methoxyethoxy) ethyl methacrylate, chitosan	pH and thermal/PAA (pH-sensitive) and oligo(ethylene glycol) methacrylate and 2-(2-methoxyethoxy) ethyl methacrylate (Thermal sensitive)	Drug delivery	In vitro cytotoxicity studies confirmed that hydrogels had excellent cell compatibility, with 5-Fu-loaded hydrogels having a lower cell growth inhibition efficiency for normal LO2 cells but a higher cell growth inhibition efficiency for cancer HepG2 cells than pure 5-Fu at the same drug concentration.	The value of medication released was low in an acidic environment (pH 1.2) but high in a neutral environment.	[136]
Poly (1-butyl-3-vinylimidazolium bis(trifluoromethane-sulfonyl)imide) ([PBVIm] [TFSI])	Strain and electric	Utilized in clothing to monitor various body movements	Membranes possessed washable, comfortable, good mechanical properties and satisfactory moisture proof sensing performance.	-	[137]
1-vinyl-3-butylimidazolium bromide ([VBIM+] Br-) ionic liquid, vinyl-modified lignin (v-lignin), acrylamide (AM), borax, ammonium persulfate	Strain and thermoresponsive	Electronic skin, human–machine interface, and remote medical healthcare	Hydrogel showed high stretchability, excellent toughness, and impressive stress loading-unloading cyclic stability.	Motion capture and gesture identification by the hydrogel strain sensor.	[138]
Lignin/poly(ionic liquids)/ 3-butyl-1-isopropyl-1H-imidazol-3-ium bromide/1-vinylimidazole and bromobutane	pH and temperature responsive	Drug delivery	The hybrid hydrogel was more successful at killing malignant cells in an invitro cytotoxicity and drug release testing.	Drug release occurred at intracellular acidic pH.	[139]

Table 6. *Cont.*

Compounds	Stimulus/Stimulus-Response Agents	Application	Summary	Role of Stimuli	Ref.
1-vinyl-3-butylimidazolium tetrafluoroborate/1-butyl-3-methylimidazolium tetrafluoroborate	Strain and light	Reusable wearable electronics	Ionogel integrated excellent mechanical properties, ultra-strong adhesive, self-healing ability, and recyclability.	Detection of physical motion and physiological signals of human body.	[140]
Cetylpyridinium salicy-late/cetpylpyridinium chloride	pH and temperature	Drug delivery	The preparation of hybrid pharmaceutical ionogels through encapsulation of the chemotherapeutic drug imatinib mesylate within the ionogel matrix.	The maximum release drug was conducted at an acidic pH at 37 °C.	[141]
Dual-cross-linked ionohydrogel	Temperature and strain	Wearable ionotronic devices	The bioadhesives possessed excellent mechanical properties, transparency, high ionic conductivity, and robust adhesion, along with the advantages of superior antifreezing and long-term antidehydration properties.		[142]
1-methyl-3-(oxiran-2-ylmethyl)-1H-imidazol-3-ium chloride/methoxy polyethylenglycol-aldehyde/chitosan	Magnetic, pH responsive	Drug delivery	The findings of the cytotoxicity assay demonstrated that medications loaded nanocarriers have a higher cytotoxicity effect than free drugs.	pH-responsive branched nanocarrier for co-delivery of DOX and MTX.	[143]

Ionic liquids (ILs) are a type of organic salt that is made up of cation–anion pairs of an organic ion and an inorganic counterion, in which the cationic or anionic part is a relatively large organic moiety [144]. These cations and anions affect their properties in these compositions, including polarity, electroconductivity, viscosity, and volume. Consequently, ILs are ionresponsive making them attractive for various applications, including multi-responsive adhesives [145]. For example, dual-cross-linked ionohydrogelhas been developed using IL binary solvent system [142]. The IL incorporated adhesive hydrogel revealed admirable mechanical characteristics, transparency, high ionic conductivity making it promising for flexible ionotronic adhesive devices. Kuddushi et al. [141] also developed a stimuli-responsive and self-healable bioadhesive based on an ester-functionalized IL. Results demonstrated that the hydrogel was responsive to intracellular biological stimuli, including acidic pH of cancerous cells and temperature, making it promising for the controlled release of anticancer drugs. In addition, the morphology of hydrogel was changed by changing the shape and size of the gelator. In another interesting study, microwave-responsive adhesives were developed using simple mixing of acrylic adhesives with ionic liquids [146]. Fast response to microwave irradiation was reported via local heating of the IL. This response resulted in adhesive failure in less than 30 s. One of the main applications of ILs based adhesives is sensing [140]. In a recent study, a self-healable and ultrastrong adhesive ionogel was developed for multifunctional strain sensors. Li et al. [147] prepared a polysiloxane-supported ionogel by locking ionic 1-ethyl-3-methylimidazolium bis(trifluoromethylsulfonyl)imide ([EMIM][Tf2N]), into poly(aminopropylmethylsiloxane) (PAPMS) grafted with [2-(methacryloyloxy)ethyl] trimethylammonium chloride (METAC). Due to its adhesive behavior and high ionic conductivity, the obtained ionogelwas promising for flexible electronic devices such as sensors. In addition, Yu et al. [148] fabricated an adhesive hydrogel by multiple cross-linking between a PIL and k-carrageenan. The results indicate that the addition of PIL has a great influence on the adhesion strength of hydrogels. It can be due to interactions with charged groups or polar groups through ion–dipole and dipole–dipole interactions. Due to having suitable electrochemical performance, high mechanical stability, and strain sensitivity, these conductive adhesives can be appropriate for wearable strain sensors and the monitoring of human health [149].

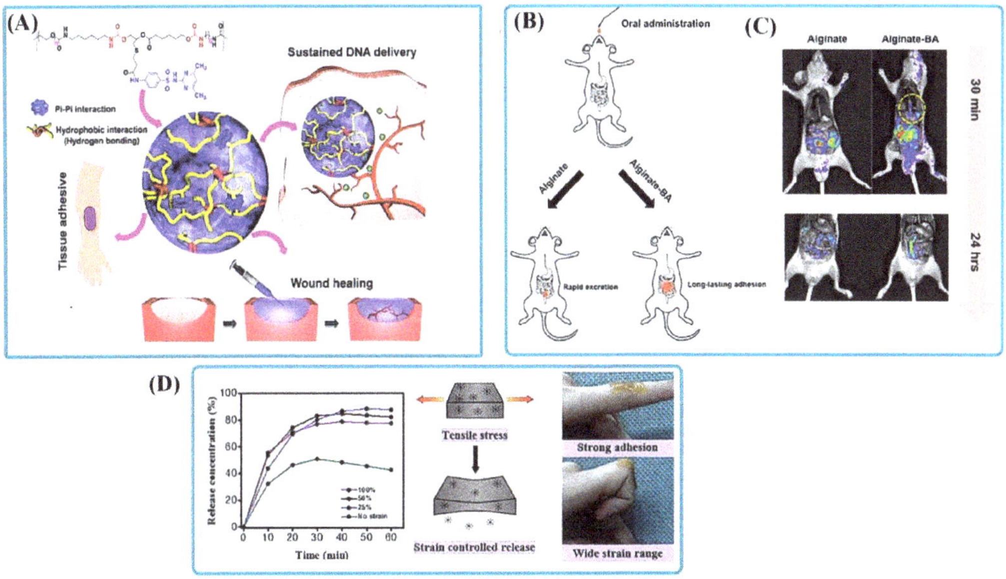

Figure 7. Multi-responsive bioadhesives. (**A**) Schematic of transition of sol-to-gel phase in PEG–PSMEU bioadhesives and their biomedical application in wound healing. Reprinted with permission from Ref. [128]. Copyright 2018, ACS Publications. (**B**) Schematic of oral delivery of alginate and alginate-BA to mice. (**C**) BALB/c mice were given rhodamine B isothiocyanate–dextran plus alginate (left) and alginate-BA (right) solutions and killed after 30 min and 24 h. Reprinted with permission from Ref. [129]. Copyright 2018, ACS Publications. (**D**) Pattern of strain-controlled release related to GA hydrogel adhesive at different strain percentages and strong adhesion on human skin with stretching. Reprinted with permission from Ref. [130]. Copyright 2020, Elsevier.

4. Clinical Applications of Stimuli-Responsive Bioadhesives

Intelligent adhesives have been widely applied in biomedical applications, including tissue engineering, drug delivery, epidermal sensors, tissue sealants, and wound healing. The stimuli-responsive bioadhesives can be developed to regulate their adhesiveness based on applied stimulations. After the wound is healed, the bioadhesive must be removed from the wound closure process [150,151]. However, there is a risk of injuring tissue and triggering pain for patients during the removal process. The ability to modulate adhesiveness on-demand with stimuli-responsive bioadhesives allows the bioadhesive to be removed from the wound site without causing any damage or pain [13]. When the intelligent bioadhesive is used to adhere to dynamic tissues, it is subjected to periodic external stresses. The bioadhesive may suffer irreversible physical damage as a result. Self-healing functionality is used as an efficient technique to assure the stability of bioadhesives in dynamic tissue settings. After the physical injury, self-healing bioadhesives repair their mechanical structure while keeping their original characteristics [152]. The stimuli-responsive bioadhesive can be programmed to deliver therapeutic medicines on demand. Due to the dynamic nature of the tissue healing process, temporal management of tissue is required for more effective tissue repair with fewer adverse outcomes. The stimuli-responsive bioadhesives allow for control by releasing antibiotics and therapeutic medicines at the right time to promote cellular differentiation, tissue-specific gene expression, and native tissue healing [153].

Internal body tissues, including bone, heart, nerve, kidney, nerve, and muscle, can also benefit from stimuli-responsive bioadhesives [34]. These interior tissues are moister and rougher compared to the skin tissue environment. The internal tissue environment is intricate and dynamic according to numerous biochemical parameters. As a result, implementing controlled responsiveness can be difficult. Another critical problem is creating stimuli-responsive bioadhesives that are stable in vivo for an extended time [154]. Furthermore, because internal tissue procedures inevitably necessitate incisions, a protracted adhesion process increases the risk of bacterial infection, inflammatory reactions, and tissue damage. Biocompatible bioadhesives with low cytotoxicity have demonstrated an essential role in this field.

Contact between electrodes and tissues is required to obtain electrical biosignals from the body. The touched electrode will experience dynamic motions in epidermal and interior tissues [155]. This might result in mechanical deformations such as stretching, compression, and bending during contact. Internal tissues have a stricter environment for electrode–tissue contact due to muscle contractions from gastrointestinal peristalsis, pulmonary cycles, and heart muscles. The electrodes may now be strongly retained on the tissue surface due to the invention of conductive smart bioadhesives. Furthermore, flexible bioadhesives provide for conformal contact between the electrode and the tissue, allowing direct electrical signal delivery [156]. The clinical applications of smart bioadhesives can be summarized and grouped in Table 7 as follows.

Table 7. Clinical applications of smart bioadhesives.

Compounds	Stimuli	Application	Summary	Ref
Poly(glycerol sebacate)-co-poly(ethylene glycol)-g-catechol	Photothermal	Wound closure	Bioadhesives perform superior wound closure and healing of skin incisions than medical glue and surgical suture, with good hemostasis and a high killing ratio of bacteria.	[157]
Ferric ion, protocatechualdehyde containing catechol and aldehyde groups and quaternized chitosan	NIR responsiveness	Wound closure	Bioadhesives presents good biocompatibility, hemostasis, antibacterial activity, injectability, and multifunctional adhesiveness.	[158]
Hyaluronic acid-graft-dopamine andreduced graphene oxide	NIR responsiveness	Drug delivery	Bioadhesive hemostatic antioxidative conductive hydrogels with sustained drug release properties are an ideal wound dressing for promoting full-thickness skin regeneration.	[159]
Poly(*N*-isopropylacrylamide) terminated with catechols/ polypyrrole nanoparticles	pH, temperature, and NIR light–responsive	Drug delivery	Bioadhesive with multi-responsive behavior, especially NIR light response, can be profitable in removable sealant materials and remotely controlled release systems.	[160]

Table 7. *Cont.*

Compounds	Stimuli	Application	Summary	Ref
Graphene aerogel/poly(*N*-isopropylacrylamide) hydrogel/polydopamine nanoparticles	Thermo- and NIR responsiveness	Drug delivery	Correlation between the drug release and the resistance allowed the drug-release behavior of the bioadhesive hydrogels to be monitored using electrical signals	[161]
Alginate/PNIPAm	Thermoresponsive	Sealing leakage and wound healing	Inspired by embryonic wound contraction, bioadhesive can support skin wound healing with stretchability, toughness, tissue adhesion, and antimicrobial function.	[162]
Gelatin Methacryloyl (GelMA)/PhenylIsothio-cyanate-Modified Gelatin	Light-responsive	Hemostasis	The produced bioadhesive with injectability and immediate hemostatic effect can be used as a fast cross-linkable hemostatic agent for irregular wounds in oral/dental surgical procedures.	[163]
Hemocoagulase/GelMA	Visible light-responsive	Hemostasis	The bioadhesives resulted in fast hemostasis and tissue sealing through the activation and aggregation of platelets as well as the effective transformation of fibrinogen into fibrin.	[164]
GO/poly(vinylalcohol)/PAA grafted with *N*-hydroxysuccinimide ester	Electro-responsive	Bioelectronic	The obtained bioadhesive with biocompatibility, applicability, mechanical and electrical stability, and recording and stimulation functionalities can be used to improve tissue–device integration and enhance the performance of biointegrated electronic devices.	[155]
Gelatin/PAAm/Clay hydrogel	Salt ions, pH, and stress	BioSensor	A capacitive pressure sensor with ability of high conductivity, high self-healing efficiency, and robust adhesion has been designed for monitoring human motions.	[165]

4.1. Wound Healing of Soft and Hard Tissues

One of the most common uses of smart bioadhesives is wound healing. For many years, wound closure has been carried out with wires, sutures, and staples [166]. Nevertheless, concerns about the sign of scar, secondary damage, wicking-induced infection, slowed wound healing, and complicated postoperative care has limited their applications. Smart bioadhesives have become more popular because people are more concerned with their physical appearance [167]. Zhao et al. [157] created a new stimulus-responsive bioadhesive made up of a prepolymer of poly(glycerol sebacate)-co-poly(ethylene glycol)-g-catechol. Three incisions (2 cm) were made on the rats' backs to evaluate their capacity to close wounds. The smart bioadhesive-treated group had higher fibroblast recruitment and proliferation, as well as less inflammatory infiltration, on the seventh day after surgery. Liang et al. [158] fabricated a smart bioadhesive through dual-dynamic bond cross-linking between Fe, protocatechualdehyde containing catechol and aldehyde groups and quaternized chitosan. A full-thickness incision model was used to examine the wound closure's efficacy. On day 7 post-surgery, the sealed incision treated with the bioadhesive exhibited complete epidermis and dermis structures and higher collagen deposition levels than the control group, and the incision closed with surgical sutures. Treatment for wounds of brittle and hard tissues is another type of wound healing where smart bioadhesives are beneficial [168]. Bioadhesives, especially for small fragments of bone, are a quick and easy way to repair damaged portions of hard tissues. The lack of fixation of the small fragments of bone typically results in bone resorption, which can lead to deformation of the bone union, bone movement, and nonunion [169]. Accordingly, Yan et al. [112] developed an electrically conducting bioadhesive to attach tiny bone fragments in comminuted bone fractures. Aniline tetramer and dopamine were added to the system to enhance the cell adhesion, proliferation, and osteogenic differentiation of MC3T3-E1 cells.

4.2. Drug Delivery

The recent innovations in smart drug carrier systems seem promising, as they supply a means to promote formulations of targeted drug delivery systems, and drug release control based on stimuli response [170]. Smart bioadhesives in delivery have an advantage over typical hydrogel delivery systems in that they can fix delivered objects on the site. Mucoadhesion is effective in enhancing the bioavailability of poorly absorbed drugs by lengthening their residence time in the gastrointestinal tract, resulting in lower doses and dosage frequency [159,171]. Yan et al. [160] prepared adhesive hydrogel with pH, temperature, and NIR light-responsive behavior for use in controlled release systems. Zhu et al. fabricated a stimuli-responsive bioadhesive by incorporating graphene aerogel into a PNIPAM network with incorporated PDA nanoparticles. The NIR controllability of the bioadhesive for the DOX release was excellent. When the hydrogels were exposed to a NIR laser for one minute, DOX was released, and the amount released rose dramatically. After the laser was turned off, no more drug release was recorded. Smart bioadhesives can also be an important topic in both agricultural and environmental chemistry. For example, the Ca–alginate/PNIPAm-based photothermal adhesive was designed to control the release of imidacloprid (IMI) by sunlight [172]. Researchers showed that the accumulative release percentage of IMI was about 29.8% at 15 °C and increased to about 60.4% at 40 °C. Smart bioadhesives can also be loaded with cells and growth factors. Using a suitable scaffold biomaterial as a cell transport vehicle can create a favorable microenvironment for extending cell survival [173,174].

4.3. Leak Sealants in Medical

A common complication of surgeries and injuries is leakage. Headaches, meningitis, and seizures can result from cerebrospinal fluid leaks caused by traumas or brain and sinus surgery [175]. Gastric fluid leakage, common during surgical operations, can result in infection and significant tissue destruction [176]. As a result, leakage control is critical in lowering operation risks, complications, and costs. Tissue sealants, also known as smart bioadhesives for leakage prevention, have piqued the interest of researchers and have showed tremendous promise in the clinic. Blacklow et al. created a thermo-responsive bioadhesive to speed wound healing, contracted at body temperature. Bioadhesive dressings could help heal wounds in other epithelial tissues such as the gut, lung, and liver. Bleeding is one of the most common side effects caused by surgical procedures, injury, diseases, and medications [177]. Hemostasis sealants are widely accessible on the market. However, they have separate limits. Chang et al. presented a hemostatic photo-responsive bioadhesive based on gelatin methacryloyl that was able to prevent bleeding following oral/dental surgical procedures. According to the findings, the bioadhesive could be immediately extruded into the bleeding site and shortened blood clotting time by 45%.

Furthermore, it may be easily removed from the bleeding site after clotting and prevent subsequent wound harm. Guo et al. [164] prepared a hemostatic smart bioadhesive composed of hemocoagulase (the same as reptilase) and GelMA, inspired by the coagulation function of snake venom. Blood clotting time with visible light-responsive bioadhesive was about 45 s compared with 5 to 6 min without bioadhesive. Hemostatic bioadhesives achieved hemostasis in 45 s on a liver incision and 34 s on a cut rat tail, reducing blood loss by 79 and 78%, respectively.

4.4. Wearable Medical Devices

Nowadays, implantable and wearable medical devices such as tissue scaffolds, biodetectors, and biosensors are attracting a substantial amount of attention [178–204]. However, conformal and stable contact between such devices and the target tissue needs to be established. This fixation requires the use of sutures and wires, raising the risk of infection, scaffold deterioration, and subsequent injury. Smart bioadhesives have the potential to replace invasive fixing procedures with a noninvasive adhesion method. Deng et al. designed an electrical bioadhesive based on a thin layer of graphene nanocomposite that can provide rapid and on-demand detachable integration of bioelectronic devices onto a variety of wet tissues. They then successfully recorded an epicardial electrocardiogram using the synthesized bioadhesive on-site and electrically stimulated a sciatic nerve in a rat model. Zhu et al. prepared a smart ionic gelatin/PAAm/clay bioadhesive with high conductivity and high self-healing efficiency, which can be used as a capacitive pressure sensor for human motion monitoring.

5. Conclusions and Future Perspectives

Smart bioadhesives have become a subject of interest recently, not only because they increase the environmental sustainability and the bioadhesive's mechanical and biological features but also the reliability of adhesion when compared with synthetic adhesives. Smart bioadhesives are stimuli-responsive materials that undergo phase and morphology variations in response to environmental stimuli (e.g., temperature, pH, electricity, light, and magnetic fields) and establish a link between therapeutic aims and drug delivery. Wound cover, tissue engineering, skin sensors, and medication delivery systems are just some applications for smart bioadhesives. Once assessing recently established smart bioadhesives, we motivated the creation, ideologies, and applications to accomplish an organized review and offer broad support for outlook materials design that show great potential in treatment areas. Though noteworthy advancement has been achieved in developing smart bioadhesives, several unsolved issues and significant obstacles in materials manufacturing and efficiency evaluation impede their useful application and industrialization. In detail: (i) the first and foremost difficulty is finding low-cost and straightforward approaches to manufacture bioinspired adhesive structures or integrating stimuli-responsive materials into adhesive; due their time-consuming nature or severe conditions, many present synthetic processes are challenging to scale up, limiting these advancements for use only in laboratories. (ii) On the other hand, one inherent problem of smart bioadhesive hydrogels is the dependence on an aqueous system, as the stimulation event is carried out by the transfer of water between the environment and the adhesive [179]. Although some studies have confirmed that a smart bioadhesive can be agitated by moisture in the air, the stimulation performance in the open air will be less than in the water environment [180]. For example, this obstacle may prevent the reconstruction of injured muscles with artificial tissue. (iii) Another challenge is the compromise between degradation and structural and functional stability of the bioadhesive after a long period for different applications. For example, we require bioadhesives that self-destruct after drug delivery to the target tissue [181]. (iv) Another difficulty that must be addressed before smart bioadhesives can be used commercially is drug delivery monitoring [182]. (v) Recently, smart bioadhesives have been utilized for several disease treatments using the transdermal drug method. However, TDD's fundamental problems, such as the differences in drug dosing amounts between humans, parts of the skin, and gender, remain unresolved. The severe problem is that TDD does not accurately control dose absorption through injured and irritated skins, which have unpredictable drug permeability. Therefore, these groups of bioadhesives need more evaluation [183–197]. (vi) In contrast to various stimuli (temperature, pH, light, etc.) that have been introduced for drug delivery, a mechanical stimulus (e.g., compressive, tensile, and shear stress) can be created by the skin itself during body movement without the requirement for any external ambient stimulus. It thus makes mechanical stimulus an inexpensive drug delivery monitoring design [130]. (vii) Eventually, despite the progress in bioengineering methods for innovative bioadhesive preparation, numerous factors, such as reaction time, degradability, inflammatory, and immunological response of these materials, must be carefully evaluated to fabricate more cytocompatible bioadhesives for tissue engineering and drug delivery.

In the end, stimuli-responsive materials have broadened the scope of smart bioadhesives by boosting the accuracy in modifying therapeutic molecules' efficacy and decreasing their off-target toxicity. Nevertheless, it is challenging to develop intelligent systems for responding to several physiological signals or external stimuli at nanoscale level. Furthermore, efforts to maintain the payload in place until desired stimulation, the ability to reach deeper layers of tissue, and minimizing unwanted tissue injury are important issues that require further progress. We hope that this study will appeal to increasing notice from research groups performing interdisciplinary research in medical science, polymer science, and engineering and which more collaborative endeavors will be devoted to the progress of intelligent bioadhesives.

Author Contributions: E.K.: investigation, writing—original draft preparation, M.K.: supervision, writing—review and editing, H.R.B.-R.: writing—review and editing, O.D.: writing—review and editing, F.B.: writing—review and editing and funding acquisition. All authors have read and agreed to the published version of the manuscript.

Funding: This research received no external funding.

Institutional Review Board Statement: Not applicable.

Informed Consent Statement: Not applicable.

Acknowledgments: This study received financial support from the Research Affairs Division of Isfahan University of Technology (IUT), Isfahan. I. R. Iran, the National Elite Foundation (NEF) Tehran, I. R. Iran, and Iran National Science Foundation (INSF), Tehran, I. R. Iran (Grant Number (99003685).

Conflicts of Interest: The authors declare that they have no competing/financial conflict of interests in this paper.

References

1. Li, X.; Su, X. Multifunctional smart hydrogels: Potential in tissue engineering and cancer therapy. *J. Mater. Chem. B* **2018**, *6*, 4714–4730. [CrossRef] [PubMed]
2. Khanlari, S.; Dubé, M.A. Bioadhesives: A review. *Macromol. React. Eng.* **2013**, *7*, 573–587. [CrossRef]
3. Ramesh, M.; Kumar, L.R. *Bioadhesives, Green Adhesives: Preparation, Properties and Applications*; John Wiley & Sons: Hoboken, NJ, USA, 2020; pp. 145–164.
4. Chopra, H.; Kumar, S.; Singh, I. Bioadhesive Hydrogels and Their Applications. In *Bioadhesives in Drug Delivery*; Scrivener Publishing LLC: Beverly, MA, USA, 2020; pp. 147–170.
5. Pinnaratip, R.; Bhuiyan, M.S.A.; Meyers, K.; Rajachar, R.M.; Lee, B.P. Multifunctional biomedical adhesives. *Adv. Healthc. Mater.* **2019**, *8*, 1801568. [CrossRef] [PubMed]
6. Tavakoli, S.; Mokhtari, H.; Kharaziha, M.; Kermanpur, A.; Talebi, A.; Moshtaghian, J. A multifunctional nanocomposite spray dressing of Kappa-carrageenan-polydopamine modified ZnO/L-glutamic acid for diabetic wounds. *Mater. Sci. Eng. C* **2020**, *111*, 110837. [CrossRef]
7. Rajabi, N.; Kharaziha, M.; Emadi, R.; Zarrabi, A.; Mokhtari, H.; Salehi, S. An adhesive and injectable nanocomposite hydrogel of thiolated gelatin/gelatin methacrylate/Laponite® as a potential surgical sealant. *J. Colloid Interface Sci.* **2020**, *564*, 155–169. [CrossRef]
8. Cabane, E.; Zhang, X.; Langowska, K.; Palivan, C.G.; Meier, W. Stimuli-responsive polymers and their applications in nanomedicine. *Biointerphases* **2012**, *7*, 9. [CrossRef]

9. El-Sherbiny, I.M.; Khalil, I.A.; Ali, I.H. Updates on stimuli-responsive polymers: Synthesis approaches and features. In *Polymer Gels*; Springer: New York, NY, USA, 2018; pp. 129–146.
10. Bruschi, M.L.; Borghi-Pangoni, F.B.; Junqueira, M.V.; de Souza Ferreira, S.B. Nanostructured therapeutic systems with bioadhesive and thermoresponsive properties. In *Nanostructures for Novel Therapy*; Elsevier: Amsterdam, The Netherland, 2017; pp. 313–342.
11. Dey, A.; Bhattacharya, P.; Neogi, S. Bioadhesives in Biomedical Applications: A Critical Review. *Rev. Adhes. Adhes.* **2020**, *8*, 130–152. [CrossRef]
12. Duan, W.; Bian, X.; Bu, Y. Applications of bioadhesives: A mini review. *Front. Bioeng. Biotechnol.* **2021**, *9*, 716035. [CrossRef]
13. Park, J.; Kim, Y.; Chun, B.; Seo, J. Rational engineering and applications of functional bioadhesives in biomedical engineering. *Biotechnol. J.* **2021**, *16*, 2100231. [CrossRef]
14. Shi, Q.; Liu, H.; Tang, D.; Li, Y.; Li, X.; Xu, F. Bioactuators based on stimulus-responsive hydrogels and their emerging biomedical applications. *NPG Asia Mater.* **2019**, *11*, 1–21. [CrossRef]
15. Bratek-Skicki, A. Towards a new class of stimuli-responsive polymer-based materials–Recent advances and challenges. *Appl. Surf. Sci. Adv.* **2021**, *4*, 100068. [CrossRef]
16. Beaussart, A.; Ngo, T.C.; Derclaye, S.; Kalinova, R.; Mincheva, R.; Dubois, P.; Leclère, P.; Dufrêne, Y.F. Chemical force microscopy of stimuli-responsive adhesive copolymers. *Nanoscale* **2014**, *6*, 565–571. [CrossRef] [PubMed]
17. Sharifzadeh, G.; Hosseinkhani, H. Biomolecule-responsive hydrogels in medicine. *Adv. Healthc. Mater.* **2017**, *6*, 1700801. [CrossRef] [PubMed]
18. Aguilar, M.R.; San Román, J. Introduction to smart polymers and their applications. In *Smart Polymers and Their Applications*; Elsevier: Amsterdam, The Netherland, 2019; pp. 1–11.
19. Shoukat, H.; Buksh, K.; Noreen, S.; Pervaiz, F.; Maqbool, I. Hydrogels as potential drug-delivery systems: Network design and applications. *Ther. Deliv.* **2021**, *12*, 375–396. [CrossRef]
20. El-Husseiny, H.M.; Mady, E.A.; Hamabe, L.; Abugomaa, A.; Shimada, K.; Yoshida, T.; Tanaka, T.; Yokoi, A.; Elbadawy, M.; Tanaka, R. Smart/stimuli-responsive hydrogels: Cutting-edge platforms for tissue engineering and other biomedical applications. *Mater. Today Bio* **2022**, *13*, 100186. [CrossRef]
21. Pourjavadi, A.; Heydarpour, R.; Tehrani, Z.M. Multi-stimuli-responsive hydrogels and their medical applications. *New J. Chem.* **2021**, *45*, 15705–15717. [CrossRef]
22. Ding, M.; Jing, L.; Yang, H.; Machnicki, C.; Fu, X.; Li, K.; Wong, I.; Chen, P.-Y. Multifunctional soft machines based on stimuli-responsive hydrogels: From freestanding hydrogels to smart integrated systems. *Mater. Today Adv.* **2020**, *8*, 100088. [CrossRef]
23. Hwang, I.; Kim, H.N.; Seong, M.; Lee, S.H.; Kang, M.; Yi, H.; Bae, W.G.; Kwak, M.K.; Jeong, H.E. Multifunctional smart skin adhesive patches for advanced health care. *Adv. Healthc. Mater.* **2018**, *7*, 1800275. [CrossRef]
24. Guillon, E.; Das, D.; Jülich, D.; Hassan, A.-R.; Geller, H.; Holley, S. Fibronectin is a smart adhesive that both influences and responds to the mechanics of early spinal column development. *Elife* **2020**, *9*, e48964. [CrossRef]
25. Pathak, K.; Malviya, R. Introduction, Theories and Mechanisms of Bioadhesion. In *Bioadhesives in Drug Delivery*; Scrivener Publishing LLC: Beverly, MA, USA, 2020; pp. 1–27.
26. Lenaerts, V.M.; Gurny, R. *Bioadhesive Drug Delivery Systems*; CRC Press: Boca Raton, FL, USA, 1989.
27. Kaurav, H.; HariKumar, S.; Kaur, A. Mucoadhesive microspheres as carriers in drug delivery: A review. *Int. J. Drug Dev. Res.* **2012**, *4*, 21–34.
28. Masareddy, R.S.; Patil, A.S.; Gadad, A.P. Bioadhesive Nanoparticulate Drug Delivery System. In *Nanopharmaceutical Advanced Delivery Systems*; Scrivener Publishing LLC: Beverly, MA, USA, 2021; pp. 309–331.
29. Bouten, P.J.; Zonjee, M.; Bender, J.; Yauw, S.T.; van Goor, H.; van Hest, J.C.; Hoogenboom, R. The chemistry of tissue adhesive materials. *Prog. Polym. Sci.* **2014**, *39*, 1375–1405. [CrossRef]
30. Huang, S.; Kong, X.; Xiong, Y.; Zhang, X.; Chen, H.; Jiang, W.; Niu, Y.; Xu, W.; Ren, C. An overview of dynamic covalent bonds in polymer material and their applications. *Eur. Polym. J.* **2020**, *141*, 110094. [CrossRef]
31. Zhang, H.; Zhao, T.; Newland, B.; Duffy, P.; Annaidh, A.N.; O'Cearbhaill, E.D.; Wang, W. On-demand and negative-thermo-swelling tissue adhesive based on highly branched ambivalent PEG–catechol copolymers. *J. Mater. Chem. B* **2015**, *3*, 6420–6428. [CrossRef] [PubMed]
32. Jin, S.G.; Kim, K.S.; Kim, D.W.; Kim, D.S.; Seo, Y.G.; Go, T.G.; Youn, Y.S.; Kim, J.O.; Yong, C.S.; Choi, H.-G. Development of a novel sodium fusidate-loaded triple polymer hydrogel wound dressing: Mechanical properties and effects on wound repair. *Int. J. Pharm.* **2016**, *497*, 114–122. [CrossRef] [PubMed]
33. Korde, J.M.; Kandasubramanian, B. Biocompatible alkyl cyanoacrylates and their derivatives as bio-adhesives. *Biomater. Sci.* **2018**, *6*, 1691–1711. [CrossRef] [PubMed]
34. Zhu, W.; Chuah, Y.J.; Wang, D.-A. Bioadhesives for internal medical applications: A review. *Acta Biomater.* **2018**, *74*, 1–16. [CrossRef] [PubMed]
35. Schricker, S.R.; Palacio, M.L.; Bhushan, B. Designing nanostructured block copolymer surfaces to control protein adhesion. *Philos. Trans. Royal Soc. A* **2012**, *370*, 2348–2380. [CrossRef]
36. Rizwan, M.; Yahya, R.; Hassan, A.; Yar, M.; Azzahari, A.D.; Selvanathan, V.; Sonsudin, F.; Abouloula, C.N. pH sensitive hydrogels in drug delivery: Brief history, properties, swelling, and release mechanism, material selection and applications. *Polymers* **2017**, *9*, 137. [CrossRef]

37. Karolewicz, B. A review of polymers as multifunctional excipients in drug dosage form technology. *Saudi Pharm. J.* **2016**, *24*, 525–536. [CrossRef]
38. Mahinroosta, M.; Farsangi, Z.J.; Allahverdi, A.; Shakoori, Z. Hydrogels as intelligent materials: A brief review of synthesis, properties and applications. *Mater. Today Chem.* **2018**, *8*, 42–55. [CrossRef]
39. Ebara, M.; Kotsuchibashi, Y.; Narain, R.; Idota, N.; Kim, Y.-J.; Hoffman, J.M.; Uto, K.; Aoyagi, T. *Smart Biomaterials*; Springer: New York, NY, USA, 2014.
40. Lu, W.; Le, X.; Zhang, J.; Huang, Y.; Chen, T. Supramolecular shape memory hydrogels: A new bridge between stimuli-responsive polymers and supramolecular chemistry. *Chem. Soc. Rev.* **2017**, *46*, 1284–1294. [CrossRef] [PubMed]
41. Michal, B.T.; Spencer, E.J.; Rowan, S.J. Stimuli-responsive reversible two-level adhesion from a structurally dynamic shape-memory polymer. *ACS Appl. Mater. Interfaces* **2016**, *8*, 11041–11049. [CrossRef] [PubMed]
42. Zhang, J.; Cui, Z.; Field, R.; Moloney, M.G.; Rimmer, S.; Ye, H. Thermo-responsive microcarriers based on poly (*N*-isopropylacrylamide). *Eur. Polym. J.* **2015**, *67*, 346–364. [CrossRef]
43. Swift, T.; Swanson, L.; Geoghegan, M.; Rimmer, S. The pH-responsive behaviour of poly (acrylic acid) in aqueous solution is dependent on molar mass. *Soft Matter* **2016**, *12*, 2542–2549. [CrossRef] [PubMed]
44. Martella, D.; Nocentini, S.; Micheletti, F.; Wiersma, D.S.; Parmeggiani, C. Polarization-dependent deformation in light responsive polymers doped by dichroic dyes. *Soft Matter* **2019**, *15*, 1312–1318. [CrossRef] [PubMed]
45. Sénéchal, V.; Saadaoui, H.; Rodriguez-Hernandez, J.; Drummond, C. Electro-responsive polyelectrolyte-coated surfaces. *Faraday Discuss.* **2017**, *199*, 335–347. [CrossRef] [PubMed]
46. Hao, Y.; Meng, J.; Wang, S. Photo-responsive polymer materials for biological applications. *Chin. Chem. Lett.* **2017**, *28*, 2085–2091. [CrossRef]
47. Manouras, T.; Vamvakaki, M. Field responsive materials: Photo-, electro-, magnetic-and ultrasound-sensitive polymers. *Polym. Chem.* **2017**, *8*, 74–96. [CrossRef]
48. Hao, Y.; Cui, H.; Meng, J.; Wang, S. Photo-responsive smart surfaces with controllable cell adhesion. *J. Photochem. Photobiol. A Chem.* **2018**, *355*, 202–211. [CrossRef]
49. He, F.; Yang, G.; Yang, P.; Yu, Y.; Lv, R.; Li, C.; Dai, Y.; Gai, S.; Lin, J. A new single 808 nm NIR light-induced imaging-guided multifunctional cancer therapy platform. *Adv. Funct. Mater.* **2015**, *25*, 3966–3976. [CrossRef]
50. Kasiński, A.; Zielińska-Pisklak, M.; Oledzka, E.; Sobczak, M. Smart hydrogels–synthetic stimuli-responsive antitumor drug release systems. *Int. J. Nanomed.* **2020**, *15*, 4541. [CrossRef] [PubMed]
51. Li, W.; Chen, Z.; Zhou, L.; Li, Z.; Ren, J.; Qu, X. Noninvasive and reversible cell adhesion and detachment via single-wavelength near-infrared laser mediated photoisomerization. *J. Am. Chem. Soc.* **2015**, *137*, 8199–8205. [CrossRef] [PubMed]
52. Bian, Q.; Chen, S.; Xing, Y.; Yuan, D.; Lv, L.; Wang, G. Host-guest self-assembly toward reversible visible-light-responsive switching for bacterial adhesion. *Acta Biomater.* **2018**, *76*, 39–45. [CrossRef]
53. Wang, S.; Zheng, H.; Zhou, L.; Cheng, F.; Liu, Z.; Zhang, H.; Zhang, Q. Injectable redox and light responsive MnO_2 hybrid hydrogel for simultaneous melanoma therapy and multidrug-resistant bacteria-infected wound healing. *Biomaterials* **2020**, *260*, 120314. [CrossRef] [PubMed]
54. Ryplida, B.; Lee, K.D.; In, I.; Park, S.Y. Light-Induced Swelling-Responsive Conductive, Adhesive, and Stretchable Wireless Film Hydrogel as Electronic Artificial Skin. *Adv. Funct. Mater.* **2019**, *29*, 1903209. [CrossRef]
55. Zheng, Y.; Chen, Z.; Jiang, Q.; Feng, J.; Wu, S.; Del Campo, A. Near-infrared-light regulated angiogenesis in a 4D hydrogel. *Nanoscale* **2020**, *12*, 13654–13661. [CrossRef] [PubMed]
56. Li, W.; Wang, J.; Ren, J.; Qu, X. 3D graphene oxide–polymer hydrogel: Near-infrared light-triggered active scaffold for reversible cell capture and on-demand release. *Adv. Mater.* **2013**, *25*, 6737–6743. [CrossRef]
57. Yang, N.; Zhu, M.; Xu, G.; Liu, N.; Yu, C. A near-infrared light-responsive multifunctional nanocomposite hydrogel for efficient and synergistic antibacterial wound therapy and healing promotion. *J. Mater. Chem. B* **2020**, *8*, 3908–3917. [CrossRef]
58. Han, L.; Zhang, Y.; Lu, X.; Wang, K.; Wang, Z.; Zhang, H. Polydopamine nanoparticles modulating stimuli-responsive PNIPAM hydrogels with cell/tissue adhesiveness. *ACS Appl. Mater. Interfaces* **2016**, *8*, 29088–29100. [CrossRef]
59. Abueva, C.D.; Chung, P.-S.; Ryu, H.-S.; Park, S.-Y.; Woo, S.H. Photoresponsive Hydrogels as Drug Delivery Systems. *Med. Lasers Eng. Basic Res. Clin. Appl.* **2020**, *9*, 6–11. [CrossRef]
60. Lv, S.-W.; Liu, Y.; Xie, M.; Wang, J.; Yan, X.-W.; Li, Z.; Dong, W.-G.; Huang, W.-H. Near-infrared light-responsive hydrogel for specific recognition and photothermal site-release of circulating tumor cells. *ACS Nano* **2016**, *10*, 6201–6210. [CrossRef] [PubMed]
61. Jiang, Z.; Tan, M.L.; Taheri, M.; Yan, Q.; Tsuzuki, T.; Gardiner, M.G.; Diggle, B.; Connal, L.A. Strong, Self-Healable, and Recyclable Visible-Light-Responsive Hydrogel Actuators. *Angew. Chem.* **2020**, *132*, 7115–7122. [CrossRef]
62. Lv, Z.; He, S.; Wang, Y.; Zhu, X. Noble Metal Nanomaterials for NIR-Triggered Photothermal Therapy in Cancer. *Adv. Healthc. Mater.* **2021**, *10*, 2001806. [CrossRef]
63. Bi, J.; Song, K.; Wu, S.; Zhang, Y.; Wang, Y.; Liu, T. Effect of thermal-responsive surfaces based on PNIPAAm on cell adsorption/desorption. *Int. J. Polym. Mater. Polym. Biomater.* **2019**, *68*, 145–151. [CrossRef]
64. Abuwatfa, W.H.; Awad, N.S.; Pitt, W.G.; Husseini, G.A. Thermosensitive Polymers and Thermo-Responsive Liposomal Drug Delivery Systems. *Polymers* **2022**, *14*, 925. [CrossRef]
65. Arafa, M.G.; El-Kased, R.F.; Elmazar, M. Thermoresponsive gels containing gold nanoparticles as smart antibacterial and wound healing agents. *Sci. Rep.* **2018**, *8*, 1–16. [CrossRef]

66. Zhang, R.; Ruan, H.; Fu, Q.; Zhu, X.; Yao, Y. A high strain, adhesive, self-healable poly (acrylic acid) hydrogel with temperature sensitivity as an epidermal sensor. *Mater. Adv.* **2020**, *1*, 329–333. [CrossRef]
67. Zhou, L.; Dai, C.; Fan, L.; Jiang, Y.; Liu, C.; Zhou, Z.; Guan, P.; Tian, Y.; Xing, J.; Li, X. Injectable Self-Healing Natural Biopolymer-Based Hydrogel Adhesive with Thermoresponsive Reversible Adhesion for Minimally Invasive Surgery. *Adv. Funct. Mater.* **2021**, *31*, 2007457. [CrossRef]
68. Lee, Y.B.; Shin, Y.M.; Kim, E.M.; Lee, J.-Y.; Lim, J.; Kwon, S.K.; Shin, H. Mussel adhesive protein inspired coatings on temperature-responsive hydrogels for cell sheet engineering. *J. Mater. Chem. B* **2016**, *4*, 6012–6022. [CrossRef]
69. Sultana, T.; Gwon, J.-G.; Lee, B.-T. Thermal stimuli-responsive hyaluronic acid loaded cellulose based physical hydrogel for post-surgical de novo peritoneal adhesion prevention. *Mater. Sci. Eng. C* **2020**, *110*, 110661. [CrossRef]
70. Zheng, Z.; Bian, S.; Li, Z.; Zhang, Z.; Liu, Y.; Zhai, X.; Pan, H.; Zhao, X. Catechol modified quaternized chitosan enhanced wet adhesive and antibacterial properties of injectable thermo-sensitive hydrogel for wound healing. *Carbohydr. Polym.* **2020**, *249*, 116826. [CrossRef] [PubMed]
71. Zhang, E.; Guo, Q.; Ji, F.; Tian, X.; Cui, J.; Song, Y.; Sun, H.; Li, J.; Yao, F. Thermoresponsive polysaccharide-based composite hydrogel with antibacterial and healing-promoting activities for preventing recurrent adhesion after adhesiolysis. *Acta Biomater.* **2018**, *74*, 439–453. [CrossRef] [PubMed]
72. Nagase, K.; Hatakeyama, Y.; Shimizu, T.; Matsuura, K.; Yamato, M.; Takeda, N.; Okano, T. Hydrophobized thermoresponsive copolymer brushes for cell separation by multistep temperature change. *Biomacromolecules* **2013**, *14*, 3423–3433. [CrossRef] [PubMed]
73. Wu, S.-W.; Liu, X.; Miller, A.L., II; Cheng, Y.-S.; Yeh, M.-L.; Lu, L. Strengthening injectable thermo-sensitive NIPAAm-g-chitosan hydrogels using chemical cross-linking of disulfide bonds as scaffolds for tissue engineering. *Carbohydr. Polym.* **2018**, *192*, 308–316. [CrossRef] [PubMed]
74. Makvandi, P.; Ali, G.W.; Della Sala, F.; Abdel-Fattah, W.I.; Borzacchiello, A. Biosynthesis and characterization of antibacterial thermosensitive hydrogels based on corn silk extract, hyaluronic acid and nanosilver for potential wound healing. *Carbohydr. Polym.* **2019**, *223*, 115023. [CrossRef] [PubMed]
75. Moreira, C.D.; Carvalho, S.M.; Mansur, H.S.; Pereira, M.M. Thermogelling chitosan–collagen–bioactive glass nanoparticle hybrids as potential injectable systems for tissue engineering. *Mater. Sci. Eng. C* **2016**, *58*, 1207–1216. [CrossRef]
76. Umapathi, R.; Reddy, P.M.; Rani, A.; Venkatesu, P. Influence of additives on thermoresponsive polymers in aqueous media: A case study of poly (*N*-isopropylacrylamide). *Phys. Chem. Chem. Phys.* **2018**, *20*, 9717–9744. [CrossRef]
77. Kim, Y.-J.; Matsunaga, Y.T. Thermo-responsive polymers and their application as smart biomaterials. *J. Mater. Chem. B* **2017**, *5*, 4307–4321. [CrossRef]
78. Chen, Y.; Cheng, W.; Teng, L.; Jin, M.; Lu, B.; Ren, L.; Wang, Y. Graphene oxide hybrid supramolecular hydrogels with self-healable, bioadhesive and stimuli-responsive properties and drug delivery application. *Macromol. Mater. Eng.* **2018**, *303*, 1700660. [CrossRef]
79. Mantha, S.; Pillai, S.; Khayambashi, P.; Upadhyay, A.; Zhang, Y.; Tao, O.; Pham, H.M.; Tran, S.D. Smart hydrogels in tissue engineering and regenerative medicine. *Materials* **2019**, *12*, 3323. [CrossRef]
80. Xue, X.; Thiagarajan, L.; Braim, S.; Saunders, B.R.; Shakesheff, K.M.; Alexander, C. Upper critical solution temperature thermo-responsive polymer brushes and a mechanism for controlled cell attachment. *J. Mater. Chem. B* **2017**, *5*, 4926–4933. [CrossRef] [PubMed]
81. Nagase, K.; Uchikawa, N.; Hirotani, T.; Akimoto, A.M.; Kanazawa, H. Thermoresponsive anionic copolymer brush-grafted surfaces for cell separation. *Colloids Surf. B Biointerfaces* **2020**, *185*, 110565. [CrossRef] [PubMed]
82. Shen, Y.; Li, G.; Ma, Y.; Yu, D.; Sun, J.; Li, Z. Smart surfaces based on thermo-responsive polymer brushes prepared from L-alanine derivatives for cell capture and release. *Soft Matter* **2015**, *11*, 7502–7506. [CrossRef] [PubMed]
83. Saravanan, S.; Vimalraj, S.; Anuradha, D. Chitosan based thermoresponsive hydrogel containing graphene oxide for bone tissue repair. *Biomed. Pharmacother.* **2018**, *107*, 908–917. [CrossRef]
84. Wang, Q.; Feng, Y.; He, M.; Huang, Y.; Zhao, W.; Zhao, C. Thermoresponsive Antibacterial Surfaces Switching from Bacterial Adhesion to Bacterial Repulsion. *Macromol. Mater. Eng.* **2018**, *303*, 1700590. [CrossRef]
85. Sutton, A.; Shirman, T.; Timonen, J.V.; England, G.T.; Kim, P.; Kolle, M.; Ferrante, T.; Zarzar, L.D.; Strong, E.; Aizenberg, J. Photothermally triggered actuation of hybrid materials as a new platform for In Vitro cell manipulation. *Nat. Commun.* **2017**, *8*, 1–13. [CrossRef]
86. Sun, X.; Tyagi, P.; Agate, S.; McCord, M.G.; Lucia, L.A.; Pal, L. Highly tunable bioadhesion and optics of 3D printable PNIPAm/cellulose nanofibrils hydrogels. *Carbohydr. Polym.* **2020**, *234*, 115898. [CrossRef]
87. Ferber, S.; Behrens, A.M.; McHugh, K.J.; Rosenberg, E.M.; Linehan, A.R.; Sugarman, J.L.; Jayawardena, H.S.N.; Langer, R.; Jaklenec, A. Evaporative cooling hydrogel packaging for storing biologics outside of the cold chain. *Adv. Healthc. Mater.* **2018**, *7*, 1800220. [CrossRef]
88. Xu, X.; Liu, Y.; Fu, W.; Yao, M.; Ding, Z.; Xuan, J.; Li, D.; Wang, S.; Xia, Y.; Cao, M. Poly (*N*-isopropylacrylamide)-based thermoresponsive composite hydrogels for biomedical applications. *Polymers* **2020**, *12*, 580. [CrossRef]
89. Luckanagul, J.A.; Pitakchatwong, C.; Bhuket, P.R.N.; Muangnoi, C.; Rojsitthisak, P.; Chirachanchai, S.; Wang, Q.; Rojsitthisak, P. Chitosan-based polymer hybrids for thermo-responsive nanogel delivery of curcumin. *Carbohydr. Polym.* **2018**, *181*, 1119–1127. [CrossRef]

90. Kumar, T.M.; Paul, W.; Sharma, C.P.; Kuriachan, M. Bioadhesive, pH responsive micromatrix for oral delivery of insulin. *Trends Biomater. Artif. Organs* **2005**, *18*, 198–202.
91. Dai, S.; Ravi, P.; Tam, K.C. pH-Responsive polymers: Synthesis, properties and applications. *Soft Matter* **2008**, *4*, 435–449. [CrossRef] [PubMed]
92. Yang, J.; Dahlström, C.; Edlund, H.; Lindman, B.; Norgren, M. pH-responsive cellulose—Chitosan nanocomposite films with slow release of chitosan. *Cellulose* **2019**, *26*, 3763–3776. [CrossRef]
93. Desbrières, J.; Guibal, E. Chitosan for wastewater treatment. *Polym. Int.* **2018**, *67*, 7–14. [CrossRef]
94. Ofridam, F.; Tarhini, M.; Lebaz, N.; Gagniere, E.; Mangin, D.; Elaïssari, A. pH-sensitive polymers: Classification and some fine potential applications. *Polym. Adv. Technol.* **2021**, *32*, 1455–1484. [CrossRef]
95. Wang, W.; Xu, Y.; Li, A.; Li, T.; Liu, M.; von Klitzing, R.; Ober, C.K.; Kayitmazer, A.B.; Li, L.; Guo, X. Zinc induced polyelectrolyte coacervate bioadhesive and its transition to a self-healing hydrogel. *RSC Adv.* **2015**, *5*, 66871–66878. [CrossRef]
96. Garriga, R.; Jurewicz, I.; Seyedin, S.; Bardi, N.; Totti, S.; Matta-Domjan, B.; Velliou, E.G.; Alkhorayef, M.A.; Cebolla, V.L.; Razal, J.M. Multifunctional, biocompatible and pH-responsive carbon nanotube-and graphene oxide/tectomer hybrid composites and coatings. *Nanoscale* **2017**, *9*, 7791–7804. [CrossRef]
97. Liang, Y.; Zhao, X.; Ma, P.X.; Guo, B.; Du, Y.; Han, X. pH-responsive injectable hydrogels with mucosal adhesiveness based on chitosan-grafted-dihydrocaffeic acid and oxidized pullulan for localized drug delivery. *J. Colloid Interface Sci.* **2019**, *536*, 224–234. [CrossRef]
98. Agrawal, P.; Singh, R.P.; Sharma, G.; Mehata, A.K.; Singh, S.; Rajesh, C.V.; Pandey, B.L.; Koch, B.; Muthu, M.S. Bioadhesive micelles of d-α-tocopherol polyethylene glycol succinate 1000: Synergism of chitosan and transferrin in targeted drug delivery. *Colloids Surf. B Biointerfaces* **2017**, *152*, 277–288. [CrossRef]
99. Wu, D.; Shi, X.; Zhao, F.; Chilengue, S.T.F.; Deng, L.; Dong, A.; Kong, D.; Wang, W.; Zhang, J. An injectable and tumor-specific responsive hydrogel with tissue-adhesive and nanomedicine-releasing abilities for precise locoregional chemotherapy. *Acta Biomater.* **2019**, *96*, 123–136. [CrossRef]
100. Zhang, L.; Zhou, Y.; Su, D.; Wu, S.; Zhou, J.; Chen, J. Injectable, Self-healing and pH Responsive Stem Cell Factor Loaded Collagen Hydrogel as Dynamic Bioadhesive Dressing for Diabetic Wound Repair. *J. Mater. Chem. B* **2021**, *9*, 5887–5897. [CrossRef] [PubMed]
101. Boda, S.K.; Fischer, N.G.; Ye, Z.; Aparicio, C. Dual Oral Tissue Adhesive Nanofiber Membranes for pH-Responsive Delivery of Antimicrobial Peptides. *Biomacromolecules* **2020**, *21*, 4945–4961. [CrossRef] [PubMed]
102. Han, Z.; Wang, P.; Mao, G.; Yin, T.; Zhong, D.; Yiming, B.; Hu, X.; Jia, Z.; Nian, G.; Qu, S. Dual pH-responsive hydrogel actuator for lipophilic drug delivery. *ACS Appl. Mater. Interfaces* **2020**, *12*, 12010–12017. [CrossRef]
103. Yadav, V.; Jaimes-Lizcano, Y.A.; Dewangan, N.K.; Park, N.; Li, T.-H.; Robertson, M.L.; Conrad, J.C. Tuning bacterial attachment and detachment via the thickness and dispersity of a pH-responsive polymer brush. *ACS Appl. Mater. Interfaces* **2017**, *9*, 44900–44910. [CrossRef] [PubMed]
104. He, J.; Zhang, Z.; Yang, Y.; Ren, F.; Li, J.; Zhu, S.; Ma, F.; Wu, R.; Lv, Y.; He, G. Injectable Self-Healing Adhesive pH-Responsive Hydrogels Accelerate Gastric Hemostasis and Wound Healing. *Nano-Micro Lett.* **2021**, *13*, 1–17. [CrossRef] [PubMed]
105. Baloğlu, E.; Özyazıcı, M.; Hızarcıoğlu, S.Y.; Karavana, H.A. An in vitro investigation for vaginal bioadhesive formulations: Bioadhesive properties and swelling states of polymer mixtures. *Il Farmaco* **2003**, *58*, 391–396. [CrossRef]
106. Pan, G.; Li, F.; He, S.; Li, W.; Wu, Q.; He, J.; Ruan, R.; Xiao, Z.; Zhang, J.; Yang, H. Mussel-and Barnacle Cement Proteins-Inspired Dual-Bionic Bioadhesive with Repeatable Wet-Tissue Adhesion, Multimodal Self-Healing, and Antibacterial Capability for Nonpressing Hemostasis and Promoted Wound Healing. *Adv. Funct. Mater.* **2022**, 2200908. [CrossRef]
107. Xie, C.; Li, P.; Han, L.; Wang, Z.; Zhou, T.; Deng, W.; Wang, K.; Lu, X. Electroresponsive and cell-affinitive polydopamine/polypyrrole composite microcapsules with a dual-function of on-demand drug delivery and cell stimulation for electrical therapy. *NPG Asia Mater.* **2017**, *9*, e358. [CrossRef]
108. Ali, I.; Xudong, L.; Xiaoqing, C.; Zhiwei, J.; Pervaiz, M.; Weimin, Y.; Haoyi, L.; Sain, M. A review of electro-stimulated gels and their applications: Present state and future perspectives. *Mater. Sci. Eng. C* **2019**, *103*, 109852. [CrossRef]
109. Palza, H.; Zapata, P.A.; Angulo-Pineda, C. Electroactive smart polymers for biomedical applications. *Materials* **2019**, *12*, 277. [CrossRef]
110. Romasanta, L.J.; López-Manchado, M.A.; Verdejo, R. Increasing the performance of dielectric elastomer actuators: A review from the materials perspective. *Prog. Polym. Sci.* **2015**, *51*, 188–211. [CrossRef]
111. Adesanya, K.; Vanderleyden, E.; Embrechts, A.; Glazer, P.; Mendes, E.; Dubruel, P. Properties of electrically responsive hydrogels as a potential dynamic tool for biomedical applications. *J. Appl. Polym. Sci.* **2014**, *131*, 1–9. [CrossRef]
112. Yan, H.; Li, L.; Wang, Z.; Wang, Y.; Guo, M.; Shi, X.; Yeh, J.-M.; Zhang, P. Mussel-inspired conducting copolymer with aniline tetramer as intelligent biological adhesive for bone tissue engineering. *ACS Biomater. Sci. Eng.* **2019**, *6*, 634–646. [CrossRef]
113. di Luca, M.; Vittorio, O.; Cirillo, G.; Curcio, M.; Czuban, M.; Farfalla, A.; Hampel, S.; Nicoletta, F.P.; Iemma, F. Electro-responsive graphene oxide hydrogels for skin bandages: The outcome of gelatin and trypsin immobilization. *Int. J. Pharm.* **2018**, *546*, 50–60. [CrossRef] [PubMed]
114. Rao, K.M.; Kumar, A.; Han, S.S. Polysaccharide-based magnetically responsive polyelectrolyte hydrogels for tissue engineering applications. *J. Mater. Sci. Technol.* **2018**, *34*, 1371–1377. [CrossRef]

115. Birajdar, R.P.; Patil, S.B.; Alange, V.V.; Kulkarni, R.V. Electro-responsive polyacrylamide-grafted-gum ghatti copolymer for transdermal drug delivery application. *J. Macromol. Sci. Part A* **2019**, *56*, 306–315. [CrossRef]
116. Deng, Z.; Hu, T.; Lei, Q.; He, J.; Ma, P.X.; Guo, B. Stimuli-responsive conductive nanocomposite hydrogels with high stretchability, self-healing, adhesiveness, and 3D printability for human motion sensing. *ACS Appl. Mater. Interfaces* **2019**, *11*, 6796–6808. [CrossRef]
117. Bagheri, B.; Zarrintaj, P.; Samadi, A.; Zarrintaj, R.; Ganjali, M.R.; Saeb, M.R.; Mozafari, M.; Park, O.O.; Kim, Y.C. Tissue engineering with electrospun electro-responsive chitosan-aniline oligomer/polyvinyl alcohol. *Int. J. Biol. Macromol.* **2020**, *147*, 160–169. [CrossRef]
118. Qiao, K.; Guo, S.; Zheng, Y.; Xu, X.; Meng, H.; Peng, J.; Fang, Z.; Xie, Y. Effects of graphene on the structure, properties, electro-response behaviors of GO/PAA composite hydrogels and influence of electro-mechanical coupling on BMSC differentiation. *Mater. Sci. Eng. C* **2018**, *93*, 853–863. [CrossRef]
119. Vázquez-González, M.; Willner, I. Stimuli-Responsive Biomolecule-Based Hydrogels and Their Applications. *Angew. Chem. Int. Ed.* **2020**, *59*, 15342–15377. [CrossRef]
120. Zhou, X.; Wu, H.; Long, R.; Wang, S.; Huang, H.; Xia, Y.; Wang, P.; Lei, Y.; Cai, Y.; Cai, D. Oral delivery of insulin with intelligent glucose-responsive switch for blood glucose regulation. *J. Nanobiotech.* **2020**, *18*, 1–16. [CrossRef] [PubMed]
121. Turner, J.G.; White, L.R.; Estrela, P.; Leese, H.S. Hydrogel-Forming Microneedles: Current Advancements and Future Trends. *Macromol. Biosci.* **2021**, *21*, 2000307. [CrossRef] [PubMed]
122. Lin, Z.T.; Gu, J.; Li, C.H.; Lee, T.R.; Xie, L.; Chen, S.; Cao, P.Y.; Jiang, S.; Yuan, Y.; Hong, X. A nanoparticle-decorated biomolecule-responsive polymer enables robust signaling cascade for biosensing. *Adv. Mater.* **2017**, *29*, 1702090. [CrossRef] [PubMed]
123. Yesilyurt, V.; Webber, M.J.; Appel, E.A.; Godwin, C.; Langer, R.; Anderson, D.G. Injectable self-healing glucose-responsive hydrogels with pH-regulated mechanical properties. *Adv. Mater.* **2016**, *28*, 86–91. [CrossRef]
124. Li, L.; Jiang, G.; Yu, W.; Liu, D.; Chen, H.; Liu, Y.; Huang, Q.; Tong, Z.; Yao, J.; Kong, X. A composite hydrogel system containing glucose-responsive nanocarriers for oral delivery of insulin. *Mater. Sci. Eng. C* **2016**, *69*, 37–45. [CrossRef]
125. Sigen, A.; Xu, Q.; Johnson, M.; Creagh-Flynn, J.; Venet, M.; Zhou, D.; Lara-Sáez, I.; Tai, H.; Wang, W. An injectable multi-responsive hydrogel as self-healable and on-demand dissolution tissue adhesive. *Appl. Mater. Today* **2021**, *22*, 100967.
126. Aminu, N.; Toh, S. Applicability of nanoparticles-hydrogel composite in treating periodontal diseases and beyond. *Asian J. Pharm. Clin. Res.* **2017**, *10*, 65–70. [CrossRef]
127. Alvarez-Lorenzo, C.; Bromberg, L.; Concheiro, A. Light-sensitive intelligent drug delivery systems. *Photochem. Photobiol.* **2009**, *85*, 848–860. [CrossRef]
128. Le, T.M.D.; Duong, H.T.T.; Thambi, T.; Giang Phan, V.; Jeong, J.H.; Lee, D.S. Bioinspired pH-and temperature-responsive injectable adhesive hydrogels with polyplexes promotes skin wound healing. *Biomacromolecules* **2018**, *19*, 3536–3548. [CrossRef]
129. Hong, S.H.; Kim, S.; Park, J.P.; Shin, M.; Kim, K.; Ryu, J.H.; Lee, H. Dynamic bonds between boronic acid and alginate: Hydrogels with stretchable, self-healing, stimuli-responsive, remoldable, and adhesive properties. *Biomacromolecules* **2018**, *19*, 2053–2061. [CrossRef]
130. Abebe, M.W.; Appiah-Ntiamoah, R.; Kim, H. Gallic acid modified alginate self-adhesive hydrogel for strain responsive transdermal delivery. *Int. J. Biol. Macromol.* **2020**, *163*, 147–155. [CrossRef] [PubMed]
131. Li, S.; Xu, J.; Yao, G.; Liu, H. Self-adhesive, self-healable, and triple-responsive hydrogel doped with polydopamine as an adsorbent toward methylene blue. *Ind. Eng. Chem. Res.* **2019**, *58*, 17075–17087. [CrossRef]
132. Shan, M.; Gong, C.; Li, B.; Wu, G. A pH, glucose, and dopamine triple-responsive, self-healable adhesive hydrogel formed by phenylborate–catechol complexation. *Polym. Chem.* **2017**, *8*, 2997–3005. [CrossRef]
133. Di, X.; Kang, Y.; Li, F.; Yao, R.; Chen, Q.; Hang, C.; Xu, Y.; Wang, Y.; Sun, P.; Wu, G. Poly (*N*-isopropylacrylamide)/ polydopamine/clay nanocomposite hydrogels with stretchability, conductivity, and dual light-and thermo-responsive bending and adhesive properties. *Colloids Surf. B Biointerfaces* **2019**, *177*, 149–159. [CrossRef] [PubMed]
134. Esmaeilzadeh, P.; Köwitsch, A.; Liedmann, A.; Menzel, M.; Fuhrmann, B.; Schmidt, G.; Klehm, J.; Groth, T. Stimuli-responsive multilayers based on thiolated polysaccharides that affect fibroblast cell adhesion. *ACS Appl. Mater. Interfaces* **2018**, *10*, 8507–8518. [CrossRef] [PubMed]
135. Zhang, M.; Deng, F.; Tang, L.; Wu, H.; Ni, Y.; Chen, L.; Huang, L.; Hu, X.; Lin, S.; Ding, C. Super-ductile, injectable, fast self-healing collagen-based hydrogels with multi-responsive and accelerated wound-repair properties. *Chem. Eng. J.* **2021**, *405*, 126756. [CrossRef]
136. Che, Y.; Li, D.; Liu, Y.; Yue, Z.; Zhao, J.; Ma, Q.; Zhang, Q.; Tan, Y.; Yue, Q.; Meng, F. Design and fabrication of a triple-responsive chitosan-based hydrogel with excellent mechanical properties for controlled drug delivery. *J. Polym. Res.* **2018**, *25*, 1–17. [CrossRef]
137. Wang, Z.; Si, Y.; Zhao, C.; Yu, D.; Wang, W.; Sun, G. Flexible and washable poly (ionic liquid) nanofibrous membrane with moisture proof pressure sensing for real-life wearable electronics. *ACS Appl. Mater. Interfaces* **2019**, *11*, 27200–27209. [CrossRef]
138. Qu, X.; Zhao, Y.; Chen, Z.A.; Wang, S.; Ren, Y.; Wang, Q.; Shao, J.; Wang, W.; Dong, X. Thermoresponsive Lignin-Reinforced Poly (Ionic Liquid) Hydrogel Wireless Strain Sensor. *Research* **2021**, *2021*, 9845482. [CrossRef]
139. Kuddushi, M.; Ray, D.; Aswal, V.; Hoskins, C.; Malek, N. Poly (vinyl alcohol) and functionalized ionic liquid-based smart hydrogels for doxorubicin release. *ACS Appl. Bio Mater.* **2020**, *3*, 4883–4894. [CrossRef]
140. Xiang, S.; Zheng, F.; Chen, S.; Lu, Q. Self-healable, recyclable, and ultrastrong adhesive ionogel for multifunctional strain sensor. *ACS Appl. Mater. Interfaces* **2021**, *13*, 20653–20661. [CrossRef] [PubMed]

141. Kuddushi, M.; Patel, N.K.; Rajput, S.; El Seoud, O.A.; Mata, J.P.; Malek, N.I. Temperature-Responsive Low Molecular Weight Ionic Liquid Based Gelator: An Approach to Fabricate an Anti-Cancer Drug-Loaded Hybrid Ionogel. *ChemSystemsChem* **2020**, *2*, e1900053. [CrossRef]
142. Zhang, X.; Cui, C.; Chen, S.; Meng, L.; Zhao, H.; Xu, F.; Yang, J. Adhesive Ionohydrogels Based on Ionic Liquid/Water Binary Solvents with Freezing Tolerance for Flexible Ionotronic Devices. *Chem. Mater.* **2022**, *34*, 1065–1077. [CrossRef]
143. Rahimi, M.; Shafiei-Irannejad, V.; Safa, K.D.; Salehi, R. Multi-branched ionic liquid-chitosan as a smart and biocompatible nano-vehicle for combination chemotherapy with stealth and targeted properties. *Carbohydr. Polym.* **2018**, *196*, 299–312. [CrossRef] [PubMed]
144. Correia, D.M.; Fernandes, L.C.; Martins, P.M.; García-Astrain, C.; Costa, C.M.; Reguera, J.; Lanceros-Méndez, S. Ionic liquid—Polymer composites: A new platform for multifunctional applications. *Adv. Funct. Mater.* **2020**, *30*, 1909736. [CrossRef]
145. Hu, X.; Huang, J.; Zhang, W.; Li, M.; Tao, C.; Li, G. Photonic Ionic Liquids Polymer for Naked-Eye Detection of Anions. *Adv. Mater.* **2008**, *20*, 4074–4078. [CrossRef]
146. Usuba, M.; Hongo, C.; Matsumoto, T.; Nishino, T. On-demand easy peeling of acrylic adhesives containing ionic liquids through a microwave irradiation stimulus. *Polym. J.* **2018**, *50*, 1051–1056. [CrossRef]
147. Li, Z.; Wang, J.; Hu, R.; Lv, C.; Zheng, J. A highly ionic conductive, healable, and adhesive polysiloxane—Supported ionogel. *Macromol. Rapid Commun.* **2019**, *40*, 1800776. [CrossRef]
148. Yu, Y.; Xie, F.; Gao, X.; Zheng, L. Double-network hydrogels with adjustable surface morphology and multifunctional integration for flexible strain sensors. *Soft Matter* **2021**, *17*, 4352–4362. [CrossRef]
149. Zhang, D.; Tang, Y.; Zhang, Y.; Yang, F.; Liu, Y.; Wang, X.; Yang, X.; Zheng, J. Highly stretchable, self-adhesive, biocompatible, conductive hydrogels as fully polymeric strain sensors. *J. Mater. Chem. A* **2020**, *8*, 20474–20485. [CrossRef]
150. Pagano, C.; Marinozzi, M.; Baiocchi, C.; Beccari, T.; Calarco, P.; Ceccarini, M.R.; Chielli, M.; Orabona, C.; Orecchini, E.; Ortenzi, R.; et al. Bioadhesive Polymeric Films Based on Red Onion Skins Extract for Wound Treatment: An Innovative and Eco-Friendly Formulation. *Molecules* **2020**, *25*, 318. [CrossRef] [PubMed]
151. Zheng, K.; Gu, Q.; Zhou, D.; Zhou, M.; Zhang, L. Recent progress in surgical adhesives for biomedical applications. *Smart Mater. Med.* **2021**, *3*, 41–65. [CrossRef]
152. Xiong, Y.; Zhang, X.; Ma, X.; Wang, W.; Yan, F.; Zhao, X.; Chu, X.; Xu, W.; Sun, C. A review of the properties and applications of bioadhesive hydrogels. *Polym. Chem.* **2021**, *12*, 3721–3739. [CrossRef]
153. Brahmbhatt, D. Bioadhesive drug delivery systems: Overview and recent advances. *Int. J. Chem. Life Sci.* **2017**, *6*, 2016. [CrossRef]
154. Jiao, Y.; Pang, X.; Liu, M.; Zhang, B.; Li, L.; Zhai, G. Recent progresses in bioadhesive microspheres via transmucosal administration. *Colloids Surf. B Biointerfaces* **2016**, *140*, 361–372. [CrossRef]
155. Deng, J.; Yuk, H.; Wu, J.; Varela, C.E.; Chen, X.; Roche, E.T.; Guo, C.F.; Zhao, X. Electrical bioadhesive interface for bioelectronics. *Nat. Mater.* **2020**, *20*, 229–236. [CrossRef]
156. Wang, X.; Sun, X.; Gan, D.; Soubrier, M.; Chiang, H.-Y.; Yan, L.; Li, Y.; Li, J.; Yu, S.; Xia, Y.; et al. Bioadhesive and conductive hydrogel-integrated brain-machine interfaces for conformal and immune-evasive contact with brain tissue. *Matter* **2022**, *5*, 1204–1223. [CrossRef]
157. Zhao, X.; Liang, Y.; Huang, Y.; He, J.; Han, Y.; Guo, B. Physical double—Network hydrogel adhesives with rapid shape adaptability, fast self—Healing, antioxidant and NIR/pH stimulus—Responsiveness for multidrug—Resistant bacterial infection and removable wound dressing. *Adv. Funct. Mater.* **2020**, *30*, 1910748. [CrossRef]
158. Liang, Y.; Li, Z.; Huang, Y.; Yu, R.; Guo, B. Dual-Dynamic-Bond Cross-Linked Antibacterial Adhesive Hydrogel Sealants with On-Demand Removability for Post-Wound-Closure and Infected Wound Healing. *ACS Nano* **2021**, *15*, 7078–7093. [CrossRef]
159. Liang, Y.; Zhao, X.; Hu, T.; Chen, B.; Yin, Z.; Ma, P.X.; Guo, B. Adhesive Hemostatic Conducting Injectable Composite Hydrogels with Sustained Drug Release and Photothermal Antibacterial Activity to Promote Full—Thickness Skin Regeneration During Wound Healing. *Small* **2019**, *15*, e1900046. [CrossRef]
160. Yan, Y.H.; Rong, L.H.; Ge, J.; Tiu, B.D.B.; Cao, P.F.; Advincula, R.C. Mussel—Inspired hydrogel composite with multi—Stimuli responsive behavior. *Macromol. Mater. Eng.* **2019**, *304*, 1800720. [CrossRef]
161. Zhu, Y.; Zeng, Q.; Zhang, Q.; Li, K.; Shi, X.; Liang, F.; Han, D. Temperature/near-infrared light-responsive conductive hydrogels for controlled drug release and real-time monitoring. *Nanoscale* **2020**, *12*, 8679–8686. [CrossRef] [PubMed]
162. Blacklow, S.O.; Li, J.; Freedman, B.R.; Zeidi, M.; Chen, C.; Mooney, D.J. Bioinspired mechanically active adhesive dressings to accelerate wound closure. *Sci. Adv.* **2019**, *5*, eaaw3963. [CrossRef] [PubMed]
163. Chang, W.C.; Tai, A.Z.; Tsai, N.Y.; Li, Y.C.E. An Injectable Hybrid Gelatin Methacryloyl (GelMA)/Phenyl Isothiocyanate-Modified Gelatin (Gel-Phe) Bioadhesive for Oral/Dental Hemostasis Applications. *Polymers* **2021**, *13*, 2386. [CrossRef]
164. Guo, Y.; Wang, Y.; Zhao, X.; Li, X.; Wang, Q.; Zhong, W.; Mequanint, K.; Zhan, R.; Xing, M.; Luo, G. Snake extract–laden hemostatic bioadhesive gel cross-linked by visible light. *Sci. Adv.* **2021**, *7*, eabf9635. [CrossRef]
165. Zhu, Y.; Lin, L.; Chen, Y.; Song, Y.; Lu, W.P.; Guo, Y. A self-healing, robust adhesion, multiple stimuli-response hydrogel for flexible sensors. *Soft Matter* **2020**, *16*, 2238–2248. [CrossRef]
166. Schnabel, B.; Scharf, M.; Schwieger, K.; Windolf, M.; van der Pol, B.; Braunstein, V.; Appelt, A. Biomechanical comparison of a new staple technique with tension band wiring for transverse patella fractures. *Clin. Biomech* **2009**, *24*, 855–859. [CrossRef]
167. Rathi, S.; Saka, R.; Domb, A.J.; Khan, W. Protein—Based bioadhesives and bioglues. *Polym. Adv. Technol.* **2019**, *30*, 217–234. [CrossRef]

168. Rahimnejad, M.; Zhong, W. Mussel-inspired hydrogel tissue adhesives for wound closure. *RSC Adv.* **2017**, *7*, 47380–47396. [CrossRef]
169. Elliott, D.S.; Newman, K.J.H.; Forward, D.P.; Hahn, D.M.; Ollivere, B.; Kojima, K.; Handley, R.; Rossiter, N.; Wixted, J.; Moran, C.G. A unified theory of bone healing and nonunion: BHN theory. *Bone Jt. J.* **2016**, *98*, 884–891. [CrossRef]
170. Shirvan, A.R.; Bashari, A.; Hemmatinejad, N. New insight into the fabrication of smart mucoadhesive buccal patches as a novel controlled-drug delivery system. *Eur. Polym. J.* **2019**, *119*, 541–550. [CrossRef]
171. Hanafy, N.A.N.; Leporatti, S.; El-Kemary, M. Mucoadhesive hydrogel nanoparticles as smart biomedical drug delivery system. *Appl. Sci.* **2019**, *9*, 825. [CrossRef]
172. Zheng, D.; Bai, B.; Zhao, H.; Xu, X.; Hu, N.; Wang, H. Stimuli-responsive Ca-alginate-based photothermal system with enhanced foliar adhesion for controlled pesticide release. *Colloids Surf. B Biointerfaces* **2021**, *207*, 112004. [CrossRef] [PubMed]
173. Qu, M.; Jiang, X.; Zhou, X.; Wang, C.; Wu, Q.; Ren, L.; Zhu, J.; Zhu, S.; Tebon, P.; Sun, W.; et al. Stimuli-Responsive Delivery of Growth Factors for Tissue Engineering. *Adv. Healthc. Mater.* **2020**, *9*, e1901714. [CrossRef] [PubMed]
174. Abdollahiyan, P.; Baradaran, B.; de la Guardia, M.; Oroojalian, F.; Mokhtarzadeh, A. Cutting-edge progress and challenges in stimuli responsive hydrogel microenvironment for success in tissue engineering today. *J. Control. Release* **2020**, *328*, 514–531. [CrossRef] [PubMed]
175. Epstein, N. Hemostasis and other benefits of fibrin sealants/glues in spine surgery beyond cerebrospinal fluid leak repairs. *Surg. Neurol. Int.* **2014**, *5*, 304–314. [CrossRef] [PubMed]
176. Wu, J.; Yuk, H.; Sarrafian, T.L.; Guo, C.F.; Griffiths, L.G.; Nabzdyk, C.S.; Zhao, X. An off-the-shelf bioadhesive patch for sutureless repair of gastrointestinal defects. *Sci. Transl. Med.* **2022**, *14*. [CrossRef]
177. Kiratli, H.; Tarlan, B. Subconjunctival hemorrhage: Risk factors and potential indicators. *Clin. Ophthalmol.* **2013**, *7*, 1163–1170. [CrossRef]
178. Lee, G.-H.; Moon, H.; Kim, H.; Lee, G.H.; Kwon, W.; Yoo, S.; Myung, D.; Yun, S.H.; Bao, Z.; Hahn, S.K. Multifunctional materials for implantable and wearable photonic healthcare devices. *Nat. Rev. Mater.* **2020**, *5*, 149–165. [CrossRef]
179. Chen, X.; Yuk, H.; Wu, J.; Nabzdyk, C.S.; Zhao, X. Instant tough bioadhesive with triggerable benign detachment. *Proc. Natl. Acad. Sci. USA* **2020**, *117*, 15497–15503. [CrossRef]
180. Ionov, L. Hydrogel-based actuators: Possibilities and limitations. *Mater. Today* **2014**, *17*, 494–503. [CrossRef]
181. Wang, M.; Zhai, Y.; Ye, H.; Lv, Q.; Sun, B.; Luo, C.; Jiang, Q.; Zhang, H.; Xu, Y.; Jing, Y.; et al. High Co-loading Capacity and Stimuli-Responsive Release Based on Cascade Reaction of Self-Destructive Polymer for Improved Chemo-Photodynamic Therapy. *ACS Nano* **2019**, *13*, 7010–7023. [CrossRef] [PubMed]
182. Zheng, F.; Xiong, W.; Sun, S.; Zhang, P.; Zhu, J.J. Recent advances in drug release monitoring. *Nanophotonics* **2019**, *8*, 391–413. [CrossRef]
183. Naseem, F.; Shah, S.U.; Rashid, S.A.; Farid, A.; Almehmadi, M.; Alghamdi, S. Metronidazole Based Floating Bioadhesive Drug Delivery System for Potential Eradication of H. pylori: Preparation and In Vitro Characterization. *Polymers* **2022**, *14*, 519. [CrossRef]
184. Céspedes-Valenzuela, D.N.; Sánchez-Rentería, S.; Cifuentes, J.; Gantiva-Diaz, M.; Serna, J.A.; Reyes, L.H.; Ostos, C.; la Portilla, C.C.-D.; Muñoz-Camargo, C.; Cruz, J.C. Preparation and Characterization of an Injectable and Photo-Responsive Chitosan Methacrylate/Graphene Oxide Hydrogel: Potential Applications in Bone Tissue Adhesion and Repair. *Polymers* **2021**, *14*, 126. [CrossRef]
185. Wong, Y.L.; Pandey, M.; Choudhury, H.; Lim, W.M.; Bhattamisra, S.K.; Gorain, B. Development of In-Situ Spray for Local Delivery of Antibacterial Drug for Hidradenitis Suppurativa: Investigation of Alternative Formulation. *Polymers* **2021**, *13*, 2770. [CrossRef]
186. Spiridon, I.; Andrei, I.-M.; Anghel, N.; Dinu, M.; Ciubotaru, B.-I. Development and Characterization of Novel Cellulose Composites Obtained in 1-Ethyl-3-methylimidazolium Chloride Used as Drug Delivery Systems. *Polymers* **2021**, *13*, 2176. [CrossRef]
187. Ageitos, J.; Robla, S.; Valverde-Fraga, L.; Garcia-Fuentes, M.; Csaba, N. Purification of Hollow Sporopollenin Microcapsules from Sunflower and Chamomile Pollen Grains. *Polymers* **2021**, *13*, 2094. [CrossRef]
188. Anghel, N.; Dinu, V.; Verestiuc, L.; Spiridon, I. Transcutaneous Drug Delivery Systems Based on Collagen/Polyurethane Composites Reinforced with Cellulose. *Polymers* **2021**, *13*, 1845. [CrossRef]
189. Antov, P.; Krišt'ák, L.; Réh, R.; Savov, V.; Papadopoulos, A.N. Eco-Friendly Fiberboard Panels from Recycled Fibers Bonded with Calcium Lignosulfonate. *Polymers* **2021**, *13*, 639. [CrossRef]
190. Nifant'ev, I.; Shlyakhtin, A.; Komarov, P.; Tavtorkin, A.; Kananykhina, E.; Elchaninov, A.; Vishnyakova, P.; Fatkhudinov, P.; Ivchenko, P. In Vitro and In Vivo Studies of Biodegradability and Biocompatibility of Poly (εCL)-b-Poly (EtOEP)-Based Films. *Polymers* **2020**, *12*, 3039. [CrossRef] [PubMed]
191. Saha, N.; Saha, N.; Sáha, T.; Öner, E.T.; Brodnjak, U.; Redl, H.; Von Byern, J.; Sáha, P. Polymer Based Bioadhesive Biomaterials for Medical Application—A Perspective of Redefining Healthcare System Management. *Polymers* **2020**, *12*, 3015. [CrossRef] [PubMed]
192. Rusu, L.-C.; Ardelean, L.C.; Jitariu, A.-A.; Miu, C.A.; Streian, C.G. An Insight into the Structural Diversity and Clinical Applicability of Polyurethanes in Biomedicine. *Polymers* **2020**, *12*, 1197. [CrossRef] [PubMed]
193. Spiridon, I.; Anghel, N.; Dinu, M.V.; Vlad, S.; Bele, A.; Ciubotaru, B.-I.; Verestiuc, L.; Pamfil, D. Development and Performance of Bioactive Compounds-Loaded Cellulose/Collagen/Polyurethane Materials. *Polymers* **2020**, *12*, 1191. [CrossRef]

194. Wu, Y.; Rashidpour, A.; Almajano, M.P.; Metón, I. Chitosan-Based Drug Delivery System: Applications in Fish Biotechnology. *Polymers* **2020**, *12*, 1177. [CrossRef]
195. Krabicová, I.; Appleton, S.L.; Tannous, M.; Hoti, G.; Caldera, F.; Pedrazzo, A.R.; Cecone, C.; Cavalli, R.; Trotta, F. History of Cyclodextrin Nanosponges. *Polymers* **2020**, *12*, 1122. [CrossRef]
196. Wei, S.-M.; Pei, M.-Y.; Pan, W.-L.; Thissen, H.; Tsai, S.-W. Gelatin Hydrogels Reinforced by Absorbable Nanoparticles and Fibrils Cured In Situ by Visible Light for Tissue Adhesive Applications. *Polymers* **2020**, *12*, 1113. [CrossRef]
197. Pahlevanzadeh, F.; Emadi, R.; Valiani, A.; Kharaziha, M.; Poursamar, S.A.; Bakhsheshi-Rad, H.R.; Ismail, A.F.; RamaKrishna, S.; Berto, F. Three-dimensional printing constructs based on the chitosan for tissue regeneration: State of the art, developing directions and prospect trends. *Materials* **2020**, *13*, 2663. [CrossRef]
198. Elahpour, N.; Pahlevanzadeh, F.; Kharaziha, M.; Bakhsheshi-Rad, H.R.; Ramakrishna, S.; Berto, F. 3D printed microneedles for transdermal drug delivery: A brief review of two decades. *Int. J. Pharm.* **2021**, *597*, 120301. [CrossRef]
199. Pahlevanzadeh, F.; Mokhtari, H.; Bakhsheshi-Rad, H.R.; Emadi, R.; Kharaziha, M.; Valiani, A.; Poursamar, S.A.; Ismail, A.F.; RamaKrishna, S.; Berto, F. Recent trends in three-dimensional bioinks based on alginate for biomedical applications. *Materials* **2020**, *13*, 3980. [CrossRef]
200. Pahlevanzadeh, F.; Bakhsheshi-Rad, H.R.; Kharaziha, M.; Kasiri-Asgarani, M.; Omidi, M.; Razzaghi, M.; Ismail, A.F.; Sharif, S.; RamaKrishna, S.; Berto, F. CNT and rGO reinforced PMMA based bone cement for fixation of load bearing implants: Mechanical property and biological response. *J. Mech. Behav. Biomed. Mater.* **2021**, *116*, 104320. [CrossRef] [PubMed]
201. Pahlevanzadeh, F.; Bakhsheshi-Rad, H.R.; Brabazon, D.; Kharaziha, M.; Ismail, A.F.; Sharif, S.; Razzaghi, M.; Berto, F. *Additive Manufacturing of Polymer Matrix Composites*; Elsevier: Amsterdam, The Netherlands, 2021.
202. Mokhtari, H.; Tavakoli, S.; Safarpour, F.; Kharaziha, M.; Bakhsheshi-Rad, H.R.; Ramakrishna, S.; Berto, F. Recent advances in chemically-modified and hybrid carrageenan-based platforms for drug delivery, wound healing, and tissue engineering. *Polymers* **2021**, *13*, 1744. [CrossRef] [PubMed]
203. Rajabi, N.; Rezaei, A.; Kharaziha, M.; Bakhsheshi-Rad, H.R.; Luo, H.; RamaKrishna, S.; Berto, F. Recent advances on bioprinted gelatin methacrylate-based hydrogels for tissue repair. *Tissue Eng. Part A* **2021**, *27*, 679–702. [CrossRef] [PubMed]
204. Bai, Y.; Liu, X.; Shi, S.Q.; Li, J. A Tough and Mildew-Proof Soybean-Based Adhesive Inspired by Mussel and Algae. *Polymers* **2020**, *12*, 756. [CrossRef] [PubMed]

polymers

MDPI

Review

New Opportunities for Organic Semiconducting Polymers in Biomedical Applications

Kyunghun Kim [1], Hocheon Yoo [2,*] and Eun Kwang Lee [3,*]

1 Samsung Advanced Institute of Technology (SAIT), Suwon 16678, Korea; kkhspecial@gmail.com
2 Department of Electronic Engineering, Gachon University, 1342 Seongnam-daero, Seongnam 13120, Korea
3 Department of Chemical Engineering, Pukyong National University, Busan 48513, Korea
* Correspondence: hyoo@gachon.ac.kr (H.Y.); eklee@pknu.ac.kr (E.K.L.)

Abstract: The life expectancy of humans has been significantly elevated due to advancements in medical knowledge and skills over the past few decades. Although a lot of knowledge and skills are disseminated to the general public, electronic devices that quantitatively diagnose one's own body condition still require specialized semiconductor devices which are huge and not portable. In this regard, semiconductor materials that are lightweight and have low power consumption and high performance should be developed with low cost for mass production. Organic semiconductors are one of the promising materials in biomedical applications due to their functionalities, solution-processability and excellent mechanical properties in terms of flexibility. In this review, we discuss organic semiconductor materials that are widely utilized in biomedical devices. Some advantageous and unique properties of organic semiconductors compared to inorganic semiconductors are reviewed. By critically assessing the fabrication process and device structures in organic-based biomedical devices, the potential merits and future aspects of the organic biomedical devices are pinpointed compared to inorganic devices.

Keywords: organic semiconductors; biomedical engineering; chemical sensors; biosensors; field-effect transistors

1. Introduction

Organic semiconductors are materials that can replace inorganic silicon semiconductors owing to their flexibility and lightness, tunable optoelectronic properties [1] through sophisticated molecular structure control, and low-cost processes. On the other hand, the commercialization of organic transistors and organic biosensors as next-generation electronic devices has been impeded by their lower charge carrier mobility than that of inorganic transistors and their intrinsically operational instability. This low charge carrier mobility has been attributed to poor transfer integral and significant energy loss when charge carriers transport through molecules [2]. Recently, research on the development and the improvement of high-performance organic electronics based on organic molecules and polymer semiconductors has been very active, and considerable advances in their electrical properties have been made. Scientific reports on improving the charge carrier mobility of organic semiconductors have revealed a variety of fundamental and technical approaches that exceeded the mobility of 10 $cm^2 \cdot V^{-1} \cdot s^{-1}$, or even higher than that of amorphous silicon [3]. For example, high-performance organic/polymer semiconductors are realized by modifying their π-conjugation length and degree at the molecular level. In addition, the energy loss in organic/polymer semiconductors is reduced further by inducing a high degree of molecular packing and minimizing their grain boundaries in fabrication steps. Thus, organic semiconductors are a key material for implementing next-generation flexible/wearable displays, smart cards, and chemical and biosensors.

One of the key advantages of using organic semiconductors in biomedical engineering applications is their better chemical compatibility in various synthesis steps compared to inorganic material-based semiconductors [4–6]. Therefore, electrical properties can be modified through synthetic routes depending on the selection of organic solvents, biodegradability with a proper molecular design, and the selection of functional groups in a molecular structure with a large degree of freedom (Figure 1).

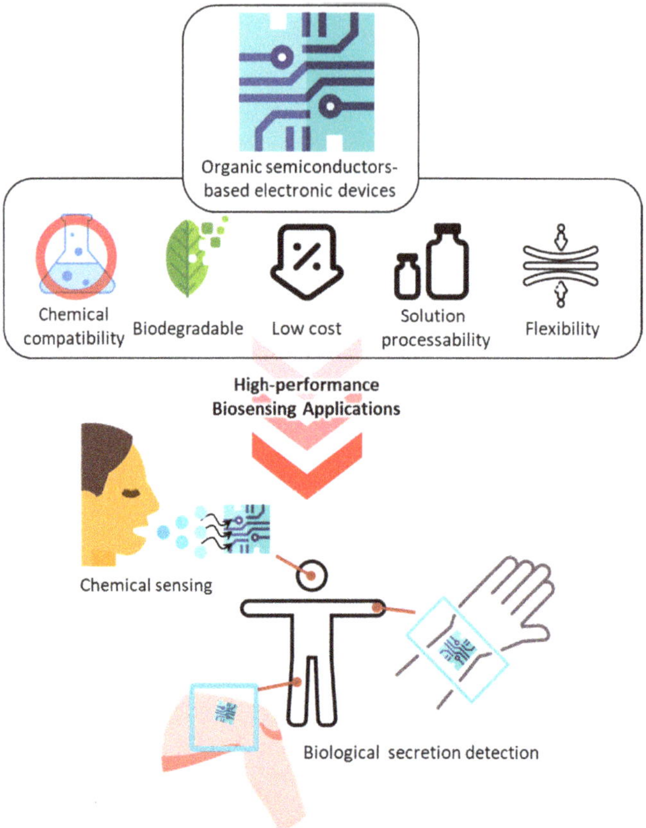

Figure 1. Schematic diagram of the use of organic semiconductors in biomedical applications.

Low-cost manufacturing techniques are an ongoing and ever-present research topic in the modern electronics industry. Conventional Si-based electronic devices require high-tech resources and time-consuming fabrication processes for high performance. Normally, high temperatures, around over 500 °C, are used in the film formation process (such as physical/chemical vapor deposition) to form inorganic electronic component layers on the top of a heterogeneous component. In contrast, the film formation conditions of organic semiconductors are less harsh than those used for inorganic semiconductors. A low temperature (under 200 °C) and simple fabrication processes are utilized widely for organic thin film formation, e.g., drop casting, spin coating, and dip coating. A low temperature allows the fabrication of flexible/wearable devices using polymeric substrates prone to damage by high temperature. Thus, the low-cost mass production of flexible/wearable devices for biomedical applications could be realized.

In contrast to inorganic semiconductors, organic/polymer semiconductors are intrinsically flexible or stretchable, depending on their chemical structure. The Young's modulus of organic/polymer semiconductors ranges from 0.1 to 1 GPa [7]. On the other hand, the Young's modulus of inorganic semiconductors ranges from several tens to hundreds of GPa, depending on their mechanical structure. Organic/polymer semiconductors are promising materials in biomedical applications in which flexibility and stretchability are prerequisites. In terms of the mechanical properties, organic/polymer semiconductors are much lighter than inorganic ones. The lightness allows devices based on the organic/polymer semiconductors to become mechanically imperceptible when fabricated into thin-film-based wearable devices [8].

One of the important advantages of organic/polymer semiconductors is their chem/biosensing characteristics in electrochemical applications to detect biological analytes from a human body. A variety of electrochemical sensing is performed in aqueous systems, in which reduction and oxidation (redox) reactions generate electrical signals. Electrical signal characteristics are influenced significantly by morphological structure, the molecular design of organic semiconductors, and the interaction between the semiconductor and the biomolecular analytes [9]. The main electrical signal detection mechanisms can be categorized into direct and indirect detections. An electrical signal can be generated by either a direct or indirect interaction between an organic semiconductor and a biomolecular analyte. The former can measure a clear electrical signal well because of direct sensing and low cost (no further functionalization is needed on the active channel). The latter has the advantage of the chemical modifiability of the organic semiconductor and long-term use, because it can avoid direct contact between the organic semiconductor and the biomolecular analyte.

Conventional Si-based biomedical devices show poor biocompatibility owing to their toxicity and short reliability in an in vivo system [10]. The most distinguishable merit of organic/polymer semiconductor devices in biomedical applications is their implantation in a human body without rejection. For example, functionalizing organic semiconductors with amine groups improves biocompatibility because of cationic charging around the amine group [10–13]. In addition, the biocompatibility of a variety of organic/polymer semiconductors and conductors (such as poly(3-hexylthiophene-2,5-diyl) (P3HT), 6,13-Bis(triisopropylsilylethynyl)pentacene (TIPS-pentacene), 3,6-bis(5-(benzofuran-2-yl)thiophen-2-yl)-2,5-bis(2-ethylhexyl)pyrrolo [3,4-c]pyrrole-1,4-dione (DPP(TBFu)$_2$), and poly(3,4-ethylenedioxythiophene) polystyrene sulfonate (PEDOT:PSS)) was evaluated in terms of stability and cell adhesion [14]. All the materials were electrically stable in an aqueous medium with sufficient protection. The cell adhesion ability for each organic semiconductor depends on chemical structure and composition. Although the cell adhesion of organic semiconductors is intrinsically not as good as that of standard culture plastics owing to their hydrophobicity, the cell adhesion property can be improved by surface functionalization with a self-assembled deposition of collagen IV.

Although organic/polymer semiconductors are still far from commercialization for high-end electronic devices because of their poorer electrical performance than their inorganic counterparts, research on organic/polymer semiconductors is growing because of their tailorable charge transport characteristics according to their surrounding environment, which means that they have potential for chem/bio sensors and logic devices. Many different types of bioanalytes can be detected by changing the major charge carrier polarity (hole or electron) of organic semiconductors. Various external stimulations, such as light, heat, pressure, and chemicals, plus the fine control of the interfacial conditions, can change the polarity of organic semiconductors. For example, self-assembled monolayers (SAMs) can be used as an interfacial layer between the organic semiconductor and metal electrodes [15]. A thiol-terminated SAM can functionalize an Ag electrode in a solution-processable way. The functionalized Ag electrode alters the injection property of the organic semiconductor, which results in the polarity change.

The practical utilization of organic semiconductors is still in its infancy. Especially, the detection of biosignals using an organic semiconductor as an active component is not comparable to the use of inorganic counterparts [4,16]. However, the electrical performance of organic semiconductors can be enhanced by applying the three-terminal electrode system, e.g., transistors. This review outlines the recent development of various organic polymer semiconductors that are mainly used in field-effect transistors in biomedical applications compared to previous papers. Section 2 presents the organic semiconductors used in biomedical applications using *p*- and *n*-type charge transport. In addition, ways to enhance the electrical performance of organic semiconductors are also presented. Some fabrication techniques for organic semiconductors are introduced in terms of low-cost solution processes and large-area fabrication in Section 3. Section 4 reports a variety of device structures of organic field-effect transistors (OFETs) and organic semiconductor-based chemical sensors. Organic semiconductors can transduce several external stimuli into electrical signals. In Section 5, various device platforms, including OFETs on unconventional substrates and OFET-based biomedical sensors, will be presented. Finally, the critical issues of organic semiconductors in biomedical engineering applications will be discussed in the Conclusion.

2. Materials: Organic Semiconductors in Biomedical Engineering

Research on organic semiconductors in biomedical engineering has focused on synthesizing high-performance organic semiconductor materials. Recently, improved device performance in biomedical engineering has been reported via various thin film formation technologies, the development of an insulating layer, the optimization of the device structure, and improvement of the synthetic material routes.

The electrical properties of organic semiconductors can vary significantly depending on their molecular structures and molecular arrangement in the active channel. In particular, the polarity (e.g., *p*-type, *n*-type, and ambipolar) of organic semiconductors can be modulated by how they are stacked at the molecular level. In addition, using one-dimensional (1D) organic semiconductor structures formed by supramolecular self-assembly (SA) can enhance electrical performance by increasing the molecular crystallinity in which a misaligned molecular packing is minimized within the structure.

The research trends and prospects on various organic semiconductor materials used widely in biomedical electronic devices are discussed in this section. This section is mainly divided into the introduction of *p*-type organic semiconductors and *n*-type organic semiconductors. Their fundamental chemical structure design according to their polarity and functional groups constituting the molecule will be discussed at the molecular level. Owing to the maturity of organic semiconductors in FETs and biosensors, this paper introduces the technological trends of organic semiconductors in high-performance OFETs and biosensors.

2.1. *p*-Type Organic Semiconductors in Biomedical Engineering

Organic semiconductor materials are classified into *p*-type and *n*-type semiconductors according to the type of charge carriers (polarity) that contribute to the flow of current, and unipolar and ambipolar semiconductors are classified according to the combination of carriers. In addition, organic semiconductor materials are classified into molecular (Figure 2) and polymeric (Figure 3) semiconductors according to the repeated units contained.

In *p*-type semiconductors, a hole is used as a major charge carrier. The hole injected from the source electrode moves through the molecular orbital with the highest occupied molecular orbital (HOMO) energy level, allowing the current to flow into the electronic device. In general, the HOMO level, ranging from −4.5 to −5.5 eV, of organic semiconductor materials, is similar to the work function (WF) value of many commercially available metals (e.g., Ag, Au, Cu, Ti, Cr, Ni, and Pt), so hole injection from these metal electrodes is reliable. Therefore, stable charge transfer is possible compared to an electron in air. For this reason, in the development of organic semiconductor materials thus far, *p*-type semiconductors have outperformed *n*-type semiconductors, and relatively higher hole mobility has been reported compared to that of its counterparts. Representative *p*-type organic semiconductors include pentacene, fused aromatic compounds, oligothiophene, and rubrene. The thin-film formation of *p*-type organic semiconductors or the fabrication of their single crystals through vacuum deposition, solution processing, or physical vapor deposition have been used to employ these *p*-type organic semiconductors in devices as an active layer.

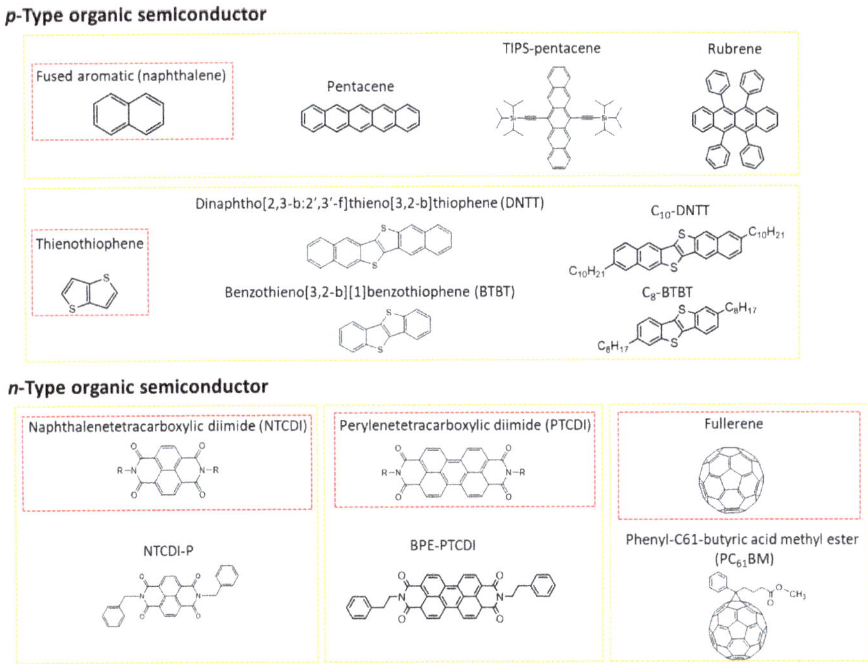

Figure 2. Organic semiconductors (small molecule-based) with respect to polarity (*p*- and *n*-type) in biomedical applications. A molecule in red box is a unit molecule in yellow box.

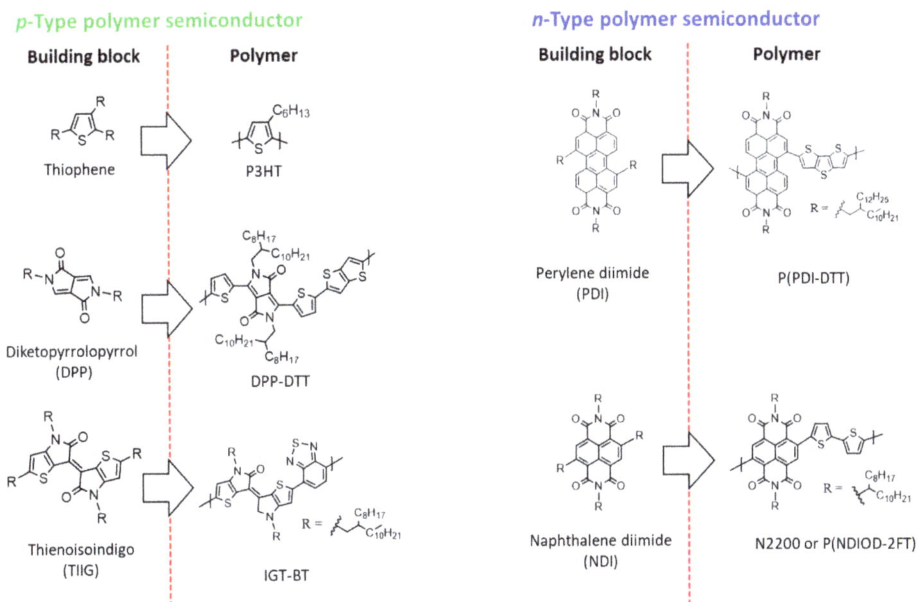

Figure 3. Polymer semiconductors with respect to polarity (*p*- and *n*-type) in biomedical applications.

Research into designing the molecular structure, the morphology control of semiconductor thin films, modification at the dielectric–semiconductor interface, and the formation of microstructures and process optimization improves the electrical performance of *p*-type organic semiconductor-based transistors and biosensors. In particular, in the 2000s, high-performance organic semiconductors surpassing the charge mobility of amorphous silicon ($1\ cm^2 \cdot V^{-1} \cdot s^{-1}$) were reported. For example, pentacene, a representative example of a *p*-type organic semiconductor, exhibits a hole mobility ranging from 1 to 10 $cm^2 \cdot V^{-1} \cdot s^{-1}$ depending on the deposition and dielectric–semiconductor conditions [17].

A research team at Pohang University of Science and Technology introduced a small molecule called m-bis(triphenylsilyl)benzene (TSB3) as a molecular insulator to improve the interfacial properties between the insulating layer and pentacene in semiconductors [18]. A high hole mobility of 6.3 $cm^2 \cdot V^{-1} \cdot s^{-1}$ was realized in the TSB3/pentacene heterojunction structure in the active channel. A low glass transition temperature (T_g, 33 °C) of TSB3 induces a soft rubbery phase at the interface between TSB3 and pentacene. The soft rubbery phase helps reduce the number of crystal grains, improving the electrical performance. In particular, the surface phase separation formed a spontaneous nanoporous pentacene thin film, which can be applied as a high-performance chemical sensor. In a similar study, a record hole mobility of 15–40 $cm^2 \cdot V^{-1} \cdot s^{-1}$ was reported in a pentacene single-crystal thin film grown through physical vapor deposition (PVD) by introducing 6,13-pentacenequinone with a molecular structure similar to pentacene as an interface control material [19].

Fused aromatic compounds of organic semiconductors offer strengthened air stability by improving the degree of crystal packing and molecular orientation. Typically, a thienothiophene-based compound fused with benzene and naphthalene shows excellent electrical performance because of the improved crystal packing [20]. In addition, thin films can be manufactured via a solution process because they have an alkyl chain substituent as a solubilizing group at the benzene end group. Unlike pentacene, the lateral CH-π intermolecular interaction is disturbed by the influence of the alkyl chain in the molecules, resulting in a two-dimensional π-stack structure. Therefore, thus, more improved mobility can be expected than that of pentacene derivatives.

A solution-processable *p*-type organic semiconductor, 2,7-Dioctyl [1]benzothieno [3,2-b][1]benzothiophene (C_8-BTBT), can be fabricated into a single-crystal thin film by inkjet printing. The maximum hole mobility from a single crystal is 31.3 $cm^2 \cdot V^{-1} \cdot s^{-1}$ [21]. Professor J. Huang and Z. Bao's research team reported very high hole mobility of 43 $cm^2 \cdot V^{-1} \cdot s^{-1}$ in C_8-BTBT thin films using a novel off-center spin-coating method [22]. They induced molecular orientation using centrifugal force generated when the organic semiconductor solution was dropped outside the spin-coater chuck on a rotating substrate. Furthermore, improved electrical performance was obtained by reducing the electrical trapping sites at their interface by introducing a vertical phase separation of a solution blended with C_8-BTBT and polystyrene (PS). In a similar study, a research team led by Professor K. Takimiya and J. Takeya in Japan developed an air-stable *p*-type organic semiconductor derivative of dinaphtho [2,3-b:2′,3′-f]thieno [3,2-b]thiophene (DNTT). They fabricated a patterned crystalline thin film with a hole mobility of 11 $cm^2 \cdot V^{-1} \cdot s^{-1}$, in which the gradual movement of the liquid–air boundary when the solvent evaporated could simultaneously control the direction and location of crystal growth. The hole mobility was improved by a factor of four compared to the spin-coated one. A rubrene single crystal was used as a *p*-type organic semiconductor to understand the anisotropic property of charge transfer according to the lattice orientation of the crystal. The dielectric contact-free intrinsic mobility reached 40 $cm^2 \cdot V^{-1} \cdot s^{-1}$ [23]. In one biosensing application, an electrolyte gate FET was fabricated using C_8-BTBT/polystyrene as a semiconductor and a blend of BTBT-biotin and a siloxane dimer of BTBT as an influenza virus-detecting layer, which was modified with streptavidin and the DNA aptamer RHA0385 [24].

Although polymer semiconductors have relatively low crystallinity compared to organic molecules, large-area printing is possible through solution processing, such as inkjet printing. Their excellent mechanical properties make them highly attractive as next-generation electronic materials. Poly(3-hexylthiophene) (P3HT), a representative *p*-type polymer semiconductor, is used widely in transistors and various applications, such as solar cells and sensors, because of its excellent optical and electrical properties [5,6]. In general, P3HT shows a low hole mobility of 0.001–0.1 $cm^2 \cdot V^{-1} \cdot s^{-1}$. Many studies have examined the molecular weight, regioregularity, solvent, thin-film morphology, thickness, manufacturing processes, humidity control, and design of side chain groups to improve the fundamental charge mobility. As a result, many new high-performance conjugated structures in polymer semiconductors have been developed.

Organic dye-based diketopyrrolopyrrole (DPP) and thienoisoindigo (TIIG) structures are based on a conjugated bicyclic lactam structure. Hence, their planar structures facilitate backbone alignment and induce strong intermolecular π-bonding. They were mainly introduced as backbone units of polymer semiconductors. Most polymer semiconductors are developed through donor(D)–acceptor(A)-type molecular design; electron-deficient DPP and IIG structures have been used as effective acceptors. A donor–acceptor polymer in the alternating structure can form a partial charge transfer state in the ground state and induce a very small π–interplanar distance. Furthermore, a thienyl-DPP structure in which thiophenes are linked covalently on both sides of the DPP structure has been designed, further inducing the flattening of the main chain in the molecule and improving intermolecular packing through the intra-interaction between oxygen in the carbonyl group and sulfur in the thiophene.

On the other hand, a commercially available conducting polymer, PEDOT:PSS (Figure 4), has been utilized in organic electrochemical transistors for the application of biomedical engineering.

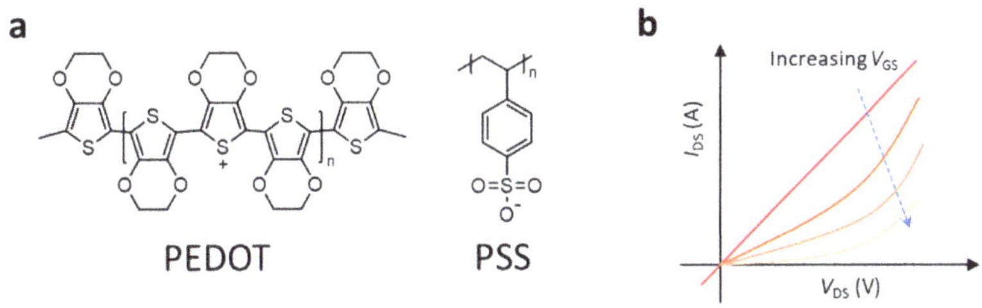

Figure 4. (**a**) Chemical structure of PEDOT:PSS. (**b**) The typical output curve of PEDOT:PSS depending on the gate voltage. The electrical conductance ($G = I\,V^{-1}$) of PEDOT:PSS is decreased with increasing gate voltage due to a gating effect in the electrochemical transistor.

The PEDOT:PSS is solution-processable with water and is a blended system of PEDOT as a conducting polymer and PSS as a dopant. The PEDOT:PSS, originally, is in an oxidized state with high conductivity, but its conductivity can be modulated (decreased) by a gating effect in the electrochemical transistor. Below is a typical reaction equation of PEDOT:PSS in an electrochemical transistor depending on the gate voltage condition:

$$\text{PEDOT:PSS} + M^+ + e^- \rightleftharpoons \text{PEDOT} + \text{M:PSS}$$

where M is a cation and e^- is an electron in the electrolyte of the electrochemical transistor. Under the zero-gate voltage, a high concentration of M^+ in the electrolyte is presented, which results in high conductivity in PEDOT:PSS. On the other hand, applying (increasing) the gate voltage, PSSs are oxidized with Ms, and PEDOT:PSS becomes less conducting compared to the one without a gate voltage applied, as shown in Figure 4b. During the operation of the PEDOT:PSS-based organic electrochemical transistor, a steady state of redox reaction occurs throughout the interfaces of the PEDOT–electrolyte and the source–PEDOT–drain region. In particular, both electronic species (electrons and holes) and ionic species are transported to the region of the PEDOT–electrolyte. On the other hand, only electronic species are transported to the region of the source–PEDOT–drain. Achieving a high on/off ratio in a PEDOT:PSS-based organic electrochemical transistor is troublesome. Intrinsically, PEDOT:PSS is an organic conductor material that can flow an electrical current under a zero-gate bias. Thus, the on/off ratio with respect to the gate voltage is not as high as that of a typical organic semiconductor material. Increasing the width of the gate channel while maintaining the length of the gate channel would improve the on/off ratio [25]. The highest on/off ratio obtained was up to 10^3 when a width-to-length ratio of the channel was larger than 10 times.

A recent research paper reported that crosslinked PEDOT:PSS-based organic electrochemical sensors are chemically robust in various organic solvents, such as dimethyl sulfoxide (DMSO), n-methyl-2-pyrrolidone (NMP), and acetone [26,27]. 3-Glycidyloxypropyl trimethoxysilane (GPOS) was mixed as a crosslinker in the solution of PEDOT:PSS to crosslink the PEDOT:PSS. The crosslinked PEDOT:PSS film as the active channel in an organic electrochemical transistor was used to detect methylene blue (MB), a biomarker for a certain redox activation in response to a change in the potential of the electrolyte with respect to the gate voltage. Moreover, during patterning a thin film of crosslinked PEDOT:PSS conventional photolithography techniques can be applied without damaging the film, providing a degree of freedom in device fabrication.

2.2. *n*-Type Organic Semiconductors in Biomedical Engineering

Research on *n*-type organic semiconductors for biomedical engineering applications has lagged far behind that on *p*-type organic semiconductors. An *n*-type organic semiconductor in which electrons are major charge carriers is oxidized easily by oxygen, moisture, and ozone molecules under ambient conditions, and its performance tends to deteriorate significantly, making it relatively difficult to develop. On the other hand, since the development of *n*-type organic semiconductors is essential for manufacturing *p*-*n* junction solar cells and high-performance complementary circuits for medical use, many studies have reported improved molecular designs, device performance, and the air stability of *n*-type organic semiconductors.

The electrons in an *n*-type semiconductor device are the major charge carriers that flow through the lowest unoccupied molecular orbital (LUMO) energy level of an organic semiconductor, so the energy level engineering of the LUMO level of the organic semiconductor and the proper selection of metal electrodes is important. In general, an electron-deficient group typically has low reduction potential, resulting in a low LUMO level in *n*-type organic semiconductors. In particular, perylene diimides (PDIs) and naphthalene diimides (NDIs), representative *n*-type semiconductors, have relatively high electron affinity, high electron mobility, and chemical and thermal stability. Hence, they are being applied as the most promising electron-deficient group unit thus far. In addition, air stability can be increased by lowering the LUMO level further through the functionalization of the core or end group, or by producing high-performance *n*-type semiconductors that are stable in air by enhancing molecular packing.

The first PDI-based *n*-type low-molecular organic semiconductor-based transistor was announced by Horowitz et al. [28]. Based on vacuum-deposited PDI thin films, in which the phenyl group was substituted at the *N*,*N*′ position, the electrical performance in FETs was only 10^{-4} $cm^2 \cdot V^{-1} \cdot s^{-1}$ in electron mobility. On the other hand, *N*,*N*′-bis(2-phenylethyl)-perylene-3,4:9,10-tetracarboxylic diimide (BPE-PTCDI), in which the center of the PDI was replaced with an ethyl group material, exhibited a highly π-conjugated structure, reaching an electron mobility up to 0.1 $cm^2 \cdot V^{-1} \cdot s^{-1}$ in FETs. Furthermore, the electron mobility was improved to 1.4 $cm^2 \cdot V^{-1} \cdot s^{-1}$ by fabricating single-crystal nanowires by recrystallization [28].

The alkyl chains are substituted at the *N*,*N*′ position of PDI to be soluble in organic solvents and enable the solution processing of low-molecular-weight PDI derivatives. In this case, an electron mobility up to 2.1 $cm^2 \cdot V^{-1} \cdot s^{-1}$ was reported by improving the molecular crystallinity and optimizing the thin-film morphology. In addition, the air stability of the *n*-type organic semiconductor device was significantly improved by introducing a fluorocarbon substituent at the *N,N′* position or an electron-withdrawing substituent with a high electronegativity, such as –CN, –F, or –Cl at the center position [29].

Facchetti and Morpurgo et al. synthesized N,N′-bis(*n*-alkyl)-(1,7 and 1,6)-dicyanoperylene-3,4:9,10-bis(dicarboximide)s (PDIF-CN2), which achieved high electron affinity and excellent air stability by substituting an electron-withdrawing group, –CN, to an aromatic center in a PDI derivative substituted with a fluorinated alkyl chain. A record-high electron mobility of 10.8 $cm^2 \cdot V^{-1} \cdot s^{-1}$ was reported in the low-temperature region (230K) in a device structure using a vacuum gap as a dielectric layer after PDIF-CN2 was grown as a single crystal using the physical vapor transport method [30].

The first NDI molecule-based transistor with an electron mobility of 10^{-4} $cm^2 \cdot V^{-1} \cdot s^{-1}$ has been reported [31]. Subsequently, the NDI-based molecular semiconductor N,N′-bis(cyclohexyl) naphthalene-1,4,5,8-bis (dicarboximide) showed a high electron mobility of 7.5 $cm^2 \cdot V^{-1} \cdot s^{-1}$ under an argon atmosphere [32]. Typically, the molecular packing and electromagnetic movement characteristics in the solid phase of the organic semiconductor can be controlled by substituting the N,N′ position and the aromatic π–central region of the imide group.

Würthner et al. developed an *n*-type semiconductor that introduced an electron-withdrawing group, i.e., a fluorinated alkyl chain, in the NDI main chain that enhanced the air stability. The observation of the optical anisotropy of the fluorinated NDI demonstrated the self-aggregation of the fluorinate group in the sheared thin film, which resulted in enhanced electrical properties. An NDI-Cl_2 single crystal was synthesized by solvent evaporation. The crystal growth direction of NDI-Cl_2 affected the charge transfer and charge injection directions. As a result, in transistor applications, it was possible to achieve an electron mobility (8.6 $cm^2 \cdot V^{-1} \cdot s^{-1}$) that was more than twice that of a thin-film transistor [33].

On the other hand, the electron mobility of a typical *n*-type organic molecular semiconductor, fullerene (C_{60}), was low (0.08 $cm^2 \cdot V^{-1} \cdot s^{-1}$). On the other hand, a single crystal of C_{60} was fabricated using a droplet-pinned crystallization (DPC) method developed by Bao et al. The maximum electron mobility of a single crystal of C_{60} using this method was 11 $cm^2 \cdot V^{-1} \cdot s^{-1}$ [34]. In the DPC process, a small piece of a Si substrate was placed on another silicon substrate onto which was polymer insulator was coated. A droplet of the fullerene solution evaporated slowly by controlling the annealing temperature to form a single crystal bundle around the small piece of Si substrate. The solvent was comprised of either a single or dual solvent to handle the morphology of the single crystal. In particular, in the case of the dual-solvent-based system, in which the boiling point and surface tension were different in both solvents, solvent circulation occurred inside of the solution, and the concentration distribution became uniform inside the droplet, resulting in a ribbon-like single crystal of C_{60}. This single crystal exhibited an excellent electron mobility up to 11 $cm^2 \cdot V^{-1} \cdot s^{-1}$ in transistors. Similar to the DPC, one-dimensional self-assembly induced by a π–π interaction of the *n*-type organic molecular semiconductor is a promising methodology to enhance electrical performance.

Many reports on high-performance PDI and NDI-based *n*-type polymer semiconductors have been published in the form of transistors and biosensors with excellent air and operational stability and electron mobility. As an example of PDI-based polymers, Marder et al. and Lee et al. synthesized poly{[N,N′-dioctylperylene-3,4,9,10-bis(dicarboximide)-1,7(6)-diyl]-alt-[(2,5-bis(2) -ethyl-hexyl)-1,4-phenylene)bis(ethyn-2,1-diyl]} (PDIC8-EB) by coupling brominated PDI with diethynylbenzene. A nanowire suspension of PDIC8-EB can be prepared after complete dissolution, filtration, and recrystallization. An electron mobility of 0.1 $cm^2 \cdot V^{-1} \cdot s^{-1}$ was reported in a PDIC8-EB nanowire-based transistor [35].

The electrical performances of NDI- and PDI-based polymer semiconductors were compared using a typical NDI-based polymer semiconductor, poly([N,N′-bis(2-octyldodecyl)-naphthalene-1,4,5,8-bis(dicarboximide)-2,6-diyl]-alt-5,5′-(2,2′-bithiophene)) (P(NDI2OD-T2)), that shows a higher electron mobility than P(PDI2OD-T2), a PDI-based polymer semiconductor. This is due to the extended conjugated structure of NDI with a higher electron affinity than PDI, and the high stereoregularity of the main chain. In particular, Kim and Noh et al. reported the major charge carrier polarity modification of the P(NDI2OD-T2) ambipolar polymer and the enhanced electron mobility by blending with a small amount of an organic dopant, bis(cyclopentadienyl)–cobalt(II) (cobaltocene, CoCp2) [36].

In addition, an NDI-derived semiconductor copolymerized with an electron-donating group, poly{N,N″-bis(2-octyl-dodecyl)-1,4,5,8-naphthalenedicarboximide-2,6-diyl]-alt-5,5′-(2,2′-thienylenevinylene-thienylene)} (PNDI-TVT), was developed. A high charge mobility up to 1.8 $cm^2 \cdot V^{-1} \cdot s^{-1}$ was obtained [37]. In particular, although the PNDI-TVT exhibited ambipolar transport behavior with Au electrodes, *n*-type transport behavior with cesium carbonate-treated Au electrodes were obtained. This was attributed to the slightly increased polymer crystallinity under electrode regions, improved electron injection, and hole-blocking properties with cesium carbonate-treated metal electrodes.

Many synthetic routes for *n*-type polymer semiconductors have been reported by introducing amide or ester groups to the polymer core. For example, in the case of a benzodifurandione-based polymer, poly(p-phenylene vinylene) (BDPPV), with the electron-accepting group in the core, its high electron mobility of 1.1 $cm^2 \cdot V^{-1} \cdot s^{-1}$ has been reported with a top-gated transistor [38].

Electron-deficient units, including IIG and DPP, have been utilized as effective electron-accepting units. In particular, IIG can be copolymerized with electron-deficient cores to form acceptor–acceptor-type copolymers. An electron mobility of 0.22 $cm^2 \cdot V^{-1} \cdot s^{-1}$ was reported in a copolymer with benzothiazole [39]. A nitrile (CN) group was introduced to the *p*-type polymer semiconductor (PDPP-TVT) to synthesize a high-performance *n*-type polymer semiconductor (PDPP-CNTVT). The CN group is an electron-withdrawing group that enhances *n*-type charge transport behavior [39]. Therefore, the highest electron mobility of 7 $cm^2 \cdot V^{-1} \cdot s^{-1}$ was achieved by controlling thin-film thickness and the concentration of the solution of the transistor.

This insightful point suggests the importance of systematic molecular design and the optimization of thin-film formation conditions for realizing high-performance n-type organic transistors and sensors [40].

In summary, we have included a comparison table of key advantages (pros) and current challenges (cons) for the representative organic semiconductor and inorganic semiconductors that are utilized in biomedical applications (Table 1). The overall electrical performance of organic semiconductors, including conductivity and charge carrier mobility, etc., lags behind inorganic alternatives. However, their optical and synthetic properties can satisfy the demands of biomedical devices.

Table 1. A comparison table of key advantages and challenges for representative organic semiconductors and inorganic semiconductors.

Materials	Examples	Key Advantages	Challenges
Organic semiconductor	TIPS-pentacene P3HT PEDOT:PSS C_{10}-DNTT C_8-BTBT	Low cost, low temperature, large area solution process, light weight, flexibility and stretchability, tunable optical and electrical properties by synthetic routes.	Generally lower conductivity, lower field-effect mobility, lower thermal stability, lower lifetime.
Inorganic semiconductor	Si, Ge Oxide (e.g., In-Ga-Zn-O) III-V (e.g., GaAs, GaN, InN, AlN) II-VI (e.g., CdSe, CdS, ZnSe, ZnS, ZnTe)	Better conductivity, better field-effect mobility, thermal stability, long lifetime.	Hard, heavy, high-cost vacuum process.

3. Fabrication

In terms of optimizing thin-film formation conditions, aligning the semiconducting molecules in a certain direction with respect to source/drain electrodes in a FET architecture is important because the molecules have intrinsically anisotropic structures and the main charge transport pathways are in the π–π stacking direction [41]. In addition, the organic semiconductor films need to be patterned to reduce leakage current and prevent crosstalk between the neighboring FETs [42,43]. Several efforts have been made to align and pattern the organic semiconducting crystal films using various solution printing techniques, such as solution shearing, slot-die coating, soft lithography, and direct writing techniques [44–48]. This section introduces how these techniques work to fabricate organic semiconductor arrays for possible application to flexible biomedical devices.

3.1. Soft Lithography

Soft lithography is a patterning method that utilizes elastomeric stamps, such as polydimethylsiloxane (PDMS) and polyurethane acrylate (PUA) molds, to replicate structures. This technique has attracted considerable attention and has been investigated widely for decades because it enables high-resolution patterns that range from nanometer to micrometer precision; it also has low cost and high-throughput fabrication. The patterning of organic semiconductor films using soft lithography is performed by contacting PDMS stamps onto the film and increasing the substrate temperature so that the crystals in contact with the stamps can be absorbed into the stamps [49,50]. When the stamps are removed from the substrate, the film regions that were not in contact with the stamps remain on the substrate, and high-resolution patterns can be obtained. On the other hand, this method has a disadvantage in that the size or orientation of the remaining semiconductor patterns cannot be controlled because the molecules are already crystallized before the patterning process. When the molecules are not oriented in the desired direction, the OFET performance is degraded. The performance variation over the OFET array increases, and the further application of the OFET array into the biomedical application is restricted.

Methods in which organic semiconductors can be patterned and oriented in the desired direction over a large area have been devised to solve this problem (Figure 5a) [51]. First, soft lithography capable of patterning and crystallization simultaneously was introduced for organic semiconductors using solvent vapor annealing. The PDMS stamps were immersed in the solvent reservoirs to allow the stamps to absorb the solvent, and then put into contact with the as-cast amorphous semiconductor film. At this time, the film in contact with the stamps was dissolved in the solvent, and then absorbed into the stamp while the remaining patterns were crystallized by the solvent vapor in the stamps. Uniformly oriented semiconductor patterns could be induced by crystallizing each linear pattern divided into micrometer-scale lengths rather than random nucleation and crystallization for an unpatterned film over a large area. The crystal size and uniformity of the crystal orientation were improved when the aspect ratio of the patterns increased (i.e., a decrease in pattern widths), as confirmed by polarized optical microscopy and grazing-incidence wide-angle X-ray scattering (Figure 5b). An OFET array comprising the patterned semiconducting crystals exhibited a higher average field-effect mobility (μ_{FET}) compared to the unpatterned one. In addition, the coefficient of variation decreased over an entire array, indicating that the electrical performance of the array was improved, and a reliable OFET array was achieved by virtue of uniform organic semiconductor patterns (Figure 5c,d).

As an alternative to absorbing organic semiconductor molecules into elastomeric stamps, there is another soft lithography method that uses capillary force. The concept of capillary force lithography is that pattern replication is achieved by annealing the constituent polymer film beyond the glass transition temperature (T_g) while placing the elastomeric stamps onto the polymer film [52]. Wetting the stamp wall with the polymer melt lowers the total free energy. Thus, the polymer melt can rise to fill up the voids between the polymer and the mold. This patterning method using capillary force can also be applied to a small molecule semiconductor/solvent system so that the organic semiconductor solution sandwiched between the substrate and the stamp rises up the mold wall. As the solvent evaporates, the semiconductor molecules self-assemble and crystallize.

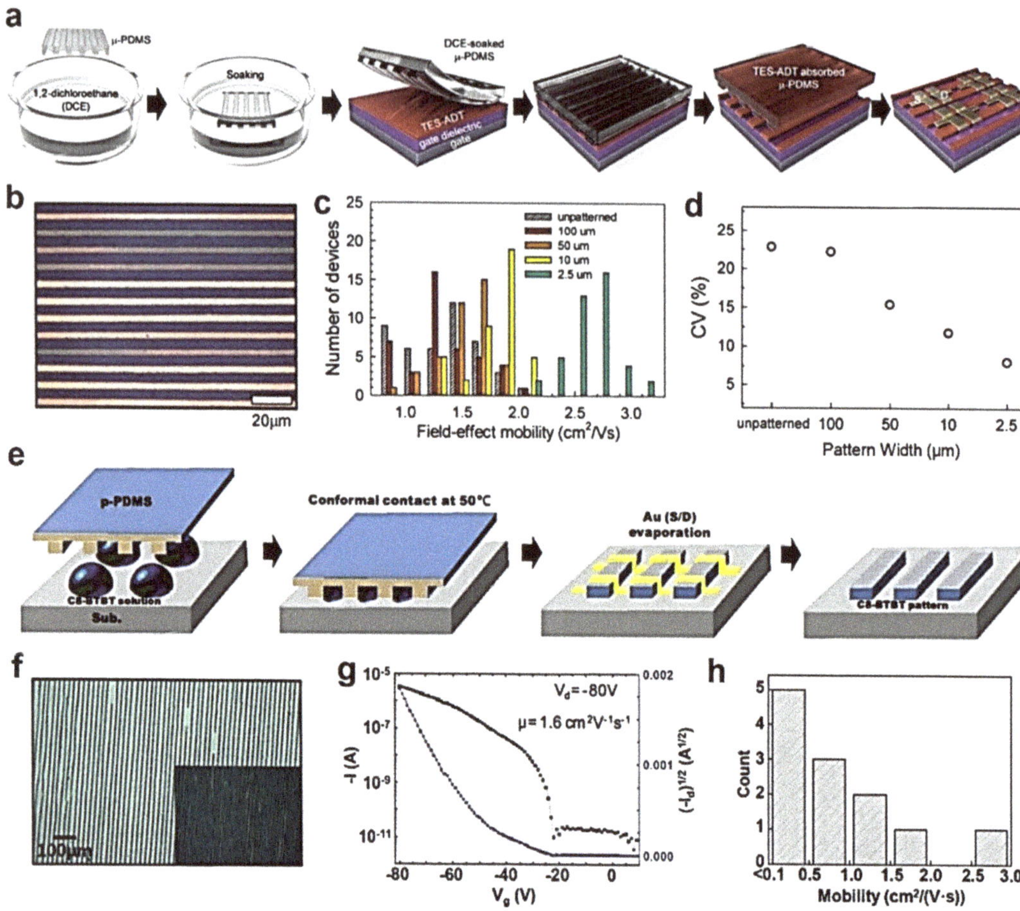

Figure 5. (**a**) Schematic illustration of the patterning process using PDMS stamp with 1,2-dichloroethane solvent. (**b**) An optical microscope image of the patterned TES-ADT semiconductor. (**c**) Distribution of μ_{FET}s of TES-ADT-based OFETs and (**d**) the corresponding coefficient of variation (CV) values. (**e**) Schematic illustration of the patterning process using solvent-assisted capillary lithography with a PDMS stamp. (**f**) Optical microscope images of the patterned C_8-BTBT semiconductor. (**g**) Transfer characteristic of the OFETs with C_8-BTBT patterns. (**h**) Distribution of μ_{FET}s of C_8-BTBT-based OFETs. Reprinted from [51], Copyright 2016 with permission from Royal Society of Chemistry.

For example, C_8-BTBT, which is a well-known solution-processable small molecule semiconductor, was dissolved in 1,2,4-trichlorobenzene, and the solution was confined between the target substrate and pre-patterned PDMS stamp (Figure 5e) [53]. As the small molecule did not have T_g, the authors applied capillary force lithography to the C_8-BTBT solution by heating the sample to 50 °C, giving sufficient mobility to the C_8-BTBT molecules during solidification within the trenches. The confined solution formed a meniscus and rose up the mold. After solvent evaporation, the width of the resulting C_8-BTBT line patterns ranged from 5 to 20 μm (Figure 5f). In-plane X-ray diffraction and selected area electron diffraction analysis of the patterns showed that C_8-BTBT crystals nucleated at the walls of the PDMS stamps, which induced the directional growth of the organic crystals. At the bottom gate, top-contact OFETs with C_8-BTBT patterns and a 200 nm thick SiO_2 dielectric layer exhibited an average μ_{FET} of 0.9 $cm^2 \cdot V^{-1} \cdot s^{-1}$ and a highest μ_{FET} of 2.6 $cm^2 \cdot V^{-1} \cdot s^{-1}$, indicating moderately high and uniform electrical performance (Figure 5g,h). Hence, the crystalline orientation could be controlled over a large area during the patterning of the organic semiconductor films.

In addition, various capillary force lithography techniques employing PUA molds have been introduced. The PUA mold provides more accurate and high-resolution patterning because of its higher hardness than that of PDMS, and it does not swell in the presence of a solvent (i.e., impermeable) [54–56].

3.2. Direct Writing Techniques

Although soft lithography has many advantages, as described above, it requires high-cost photo- or e-beam lithography to produce a master mold. This process can also leave low-molecular-weight oligomers on the surface after making contact with the elastomeric stamp with the substrate [57]. Such contamination may alter surface roughness, surface energy, or both, affecting the subsequent process. Therefore, in recent years, direct writing techniques that print organic semiconductors in a non-contact manner without complicated and high-cost patterning processes over a large area have been investigated extensively. Among the various direct writing techniques, aerosol jet printing, inkjet printing, capillary pen printing, and electrohydrodynamic jet printing have been used to print an array of organic semiconductors.

A capillary pen writing paradigm offers an efficient patterning approach for the printing of organic semiconductors because it is simple, and dispensing organic ink is unaffected by the printing environment (e.g., ambient temperature, humidity) (Figure 6a) [58]. The capillary pen is combined with a three-axis motorized position controller for the automated and programmed printing of organic semiconductor crystal arrays. Two hundred and fifty-three dots of TIPS-pentacene and poly(dimethyl-triarylamine) (PTAA) blended semiconductor crystals and five hundred and six dots of PEDOT:PSS electrodes were printed on a 2×2 cm^2 flexible polymer substrate to fabricate bottom-contact OFETs (Figure 6b). The average μ_{FET} was 0.025 cm^2·V^{-1}·s^{-1} with a standard deviation of 0.01 cm^2·V^{-1}·s^{-1} and an overall device yield of more than 80% (Figure 6c,d). Hence, the direct writing of organic semiconducting crystals using the capillary pen is reproducible.

In aerosol jet printing, a functional ink is aerosolized and ejected to the substrate by a carrier gas, providing high resolution as narrow as 10 μm (Figure 6e) [59]. P3HT polymer semiconductor, ion gel electrolyte gate dielectric, and PEDOT:PSS electrode materials have been printed on SiO_2 and plastic substrates by aerosol jet printing to fabricate 45 transistors. By optimizing the printing conditions and post-annealing process for each material (e.g., thickness and width), the P3HT-based transistors showed a high average μ_{FET} of 1.38 and 1.35 cm^2·V^{-1}·s^{-1} on SiO_2 and plastic substrates, respectively. Furthermore, the printed transistors exhibited stability upon gate bias stress and bending stress on a plastic substrate, suggesting that aerosol jet printing can print high-quality organic semiconductors (Figure 6f,g).

The electrohydrodynamic jet (E-jet) printing technique has attracted considerable attention recently because of its high-resolution printing. E-jet printing ejects a functional ink from the nozzle by applying a strong electric field between the nozzle and the substrate [60]. When the ink is subjected to an electric field, electric charges accumulate on the meniscus. At the critical limit of the electric field, surface charge repulsion exceeds the surface tension, changing the circular meniscus to a Taylor cone. Thus, a droplet of fluid is ejected towards the substrate. The formation of a Taylor cone makes it possible to print fine-line patterns with widths narrower than the nozzle size. For example, although the nozzle diameter is on the micrometer scale, printing hundreds of nm wide organic semiconducting nanowires has been reported.

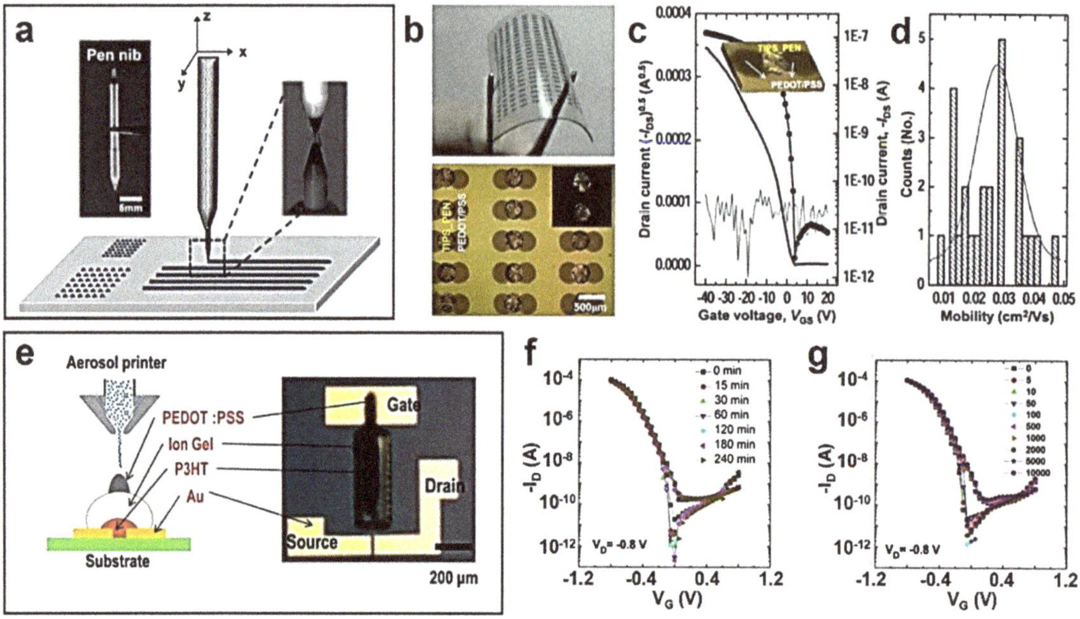

Figure 6. (**a**) Schematic diagram of a capillary pen writing system. (**b**) Optical microscopy images of OFETs with TIPS-pentacene dots on a flexible substrate. (**c**) Transfer characteristics of the OFETs with TIPS-pentacene dots and (**d**) distribution of μ_{FET}s. (**e**) Schematic illustration of an aerosol jet printing system. Transfer characteristics of the printed P3HT-based OFETs according to (**f**) gate bias stress and (**g**) bending stress. Reprinted from [58], Copyright 2013 with permission from John Wiley and Sons.

To print organic semiconductors such as poly(9-vinyl carbazole) (PVK), P3HT, and poly{[N,N0-bis(2-octyldodecyl)-naphthalene-1,4,5,8-bis(dicarboximide)-2,6-diyl]-alt-5,50-(2,20-bithiophene)} (N2200) using E-jet printing, Lee et al. set the nozzle-to-substrate distance at less than 1 cm to suppress the chaotic whipping motion of the ejected ink commonly observed when performing electrospinning. Organic semiconducting nanowires were printed and oriented along the desired directions using computerized x-y stage movement. Printing 3.96 wt.% PVK solution in styrene using an E-jet printer resulted in 289 nm wide PVK nanowires with a regular spacing of 50 μm. Printing approximately 15 nm long nanowires took only 2 min. *p*-Type P3HT and *n*-type N2200 semiconducting nanowires were printed by mixing their solutions with poly(ethylene oxide) and PVK, respectively, to increase the viscosity. P3HT and N2200 nanowires, 780 and 248 nm wide, respectively, were printed on SiO_2 (100 nm)/Si substrates, and their bottom-gate FETs exhibited a μ_{FET} of 0.015 and 0.03 $cm^2 \cdot V^{-1} \cdot s^{-1}$, respectively, which were comparable to those of pure P3HT and N2200-based FETs. These FETs with well-aligned *p*- and *n*-type polymer semiconducting nanowires highlight the promising applications of E-jet printing in large-area electronic applications.

4. Device Structure

Organic semiconductor-based devices are used in various biomedical electronic systems through structural modifications. This section revisits the recent advances in the development of biomedical applications through the structural engineering of nanoscale device geometry. Rather than using the conventional film structure of organic semiconductors, various dimensional modifications to materials have been attempted. Nanotubes, nanomeshes, nanopores, and nanofillers have allowed organic semiconductors to detect biosensing signals with enhanced sensitivity. In 2020, Park et al. presented carboxylated polypyrrole nanotubes (CPNTs) and dopamine-specific aptamers for improved biosensors (Figure 7a) [61].

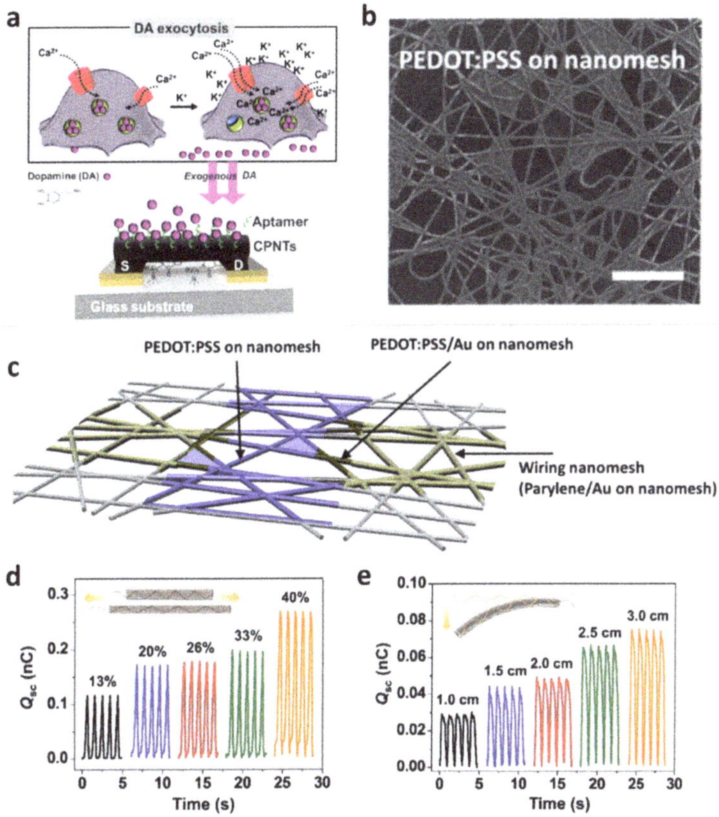

Figure 7. Nanotube and nanomesh structure devices. (**a**) Schematic of liquid-ion-gating for aptasensor-based oncarboxylated polypyrrole nanotubes. Reprinted from [61], copyright (2020) with permission from Springer Nature. (**b**) Enlarged image of PEDOT:PSS on nanomesh. (**c**) Nanomesh organic electrochemical transistor structure. Reprinted from [62], copyright (2020) with permission from American Chemical Society. (**d**,**e**) Measured Q_{sc} of nanocomposite polymer electrolyte triboelectric nanogenerator under different mechanical energy harvesting patterns and bending. Reprinted from [63], copyright (2021) with permission from John Wiley and Sons.

Through their experiments, they showed that the smaller diameter CPNTs with a diameter size of 120 nm exhibited 100 times higher sensitivity and selectivity than the counterparts in the wider CPNTs with a diameter size of 200 nm. They implemented an electrical-detection liquid-ion-gated biosensor based on exogenous dopamine-specific aptamer-release detection where cellular dopamine-specific aptamers are released by Ca^{2+} transport through the calcium ion channels. Someya et al. presented another structure engineered for biomedical organic devices, a nanomesh PEDOT:PSS structure on polyurethane nanofibers (Figure 7b,c) [62]. Electrophysiological detection, such as electrocardiogram (ECG) signals, was demonstrated using the proposed nanomesh PEDOT:PSS electrochemical transistors, enabling on-skin electrodes with local amplification to collect high-quality physiological signals [62]. In 2021, a nanofiller-based triboelectric nanogenerator for polymer electrolytes was reported as another healthcare monitoring application using nanostructured organic devices [63]. Li et al. presented polycation-modified carbon dot-assisted polyvinyl alcohol nanocomposite polymer electrolytes, providing triboelectric effects that responded to different mechanical stimuli (Figure 7d,e) [63]. An impressive demonstration in this study was the monitoring of physiological signals and the full joint range of motion in a fast, real-time, and non-invasive manner.

A porous structure has been proposed in recent years to increase the ability of biosensing remarkably. Salleo et al. presented molecularly selective membrane-based electrochemical transistors in a nanoporous structure to allow the real-time monitoring of the human stress hormone cortisol. The device detected the cortisol hormone selectively with the analysis of real body samples [64]. In 2021, Moon et al. also proposed a porous structure with a porous ion gel composed of poly(ethyl acrylate-ran-styrene-randivinylbenzene) and the ionic liquid (IL) of 1-ethyl-3 methylimidazolium bis(trifluoromethylsulfonyl) imide ((EMI)(TFSI)), formed by a crosslinking polymerization. Owing to pressure-dependent electrochemical properties in the proposed porous ion gel, the monitoring of various human motions, such as finger bending, was demonstrated (Figure 8) [65].

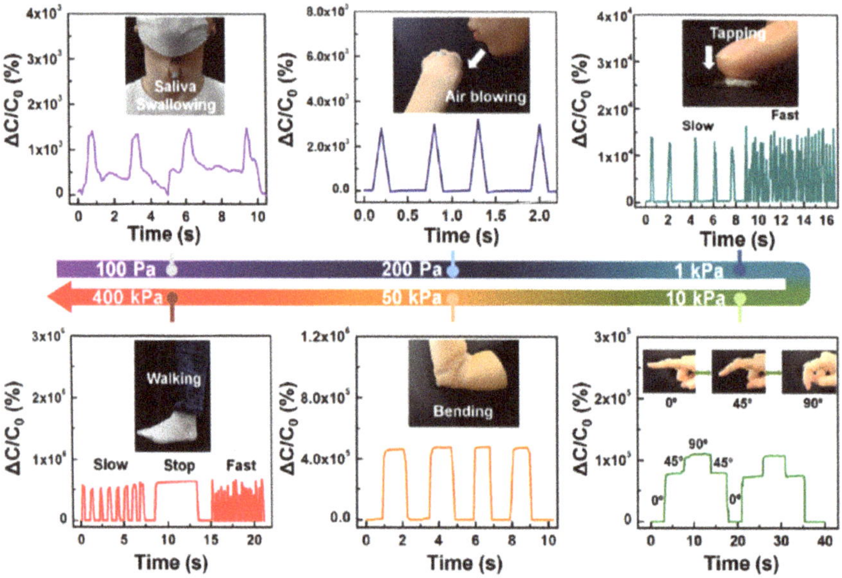

Figure 8. Real-time monitoring of human motions in different magnitudes of pressures based on nanopore ion gel. Reprinted from [65], copyright (2021) with permission from American Chemical Society.

Rather than the above-revisited nanostructures in organic materials, efforts to perform structural engineering at a device level were also made. In 2022, Minami et al. reported oxytocin detection at pg·mL^{-1} (part per trillion level) using an extended-gate structure with OFETs [66]. An anti-oxytocin antibody-attached self-assembled monolayer (SAM) was used in their devices; modified streptavidin on SAM was applied to immobilize biotinylated anti-oxytocin antibodies via biotin-avidin interaction. A floating-gate-structured organic device has also been proposed as another biomedical application in which temperature- and pressure-sensing are available (Figure 9a) [67].

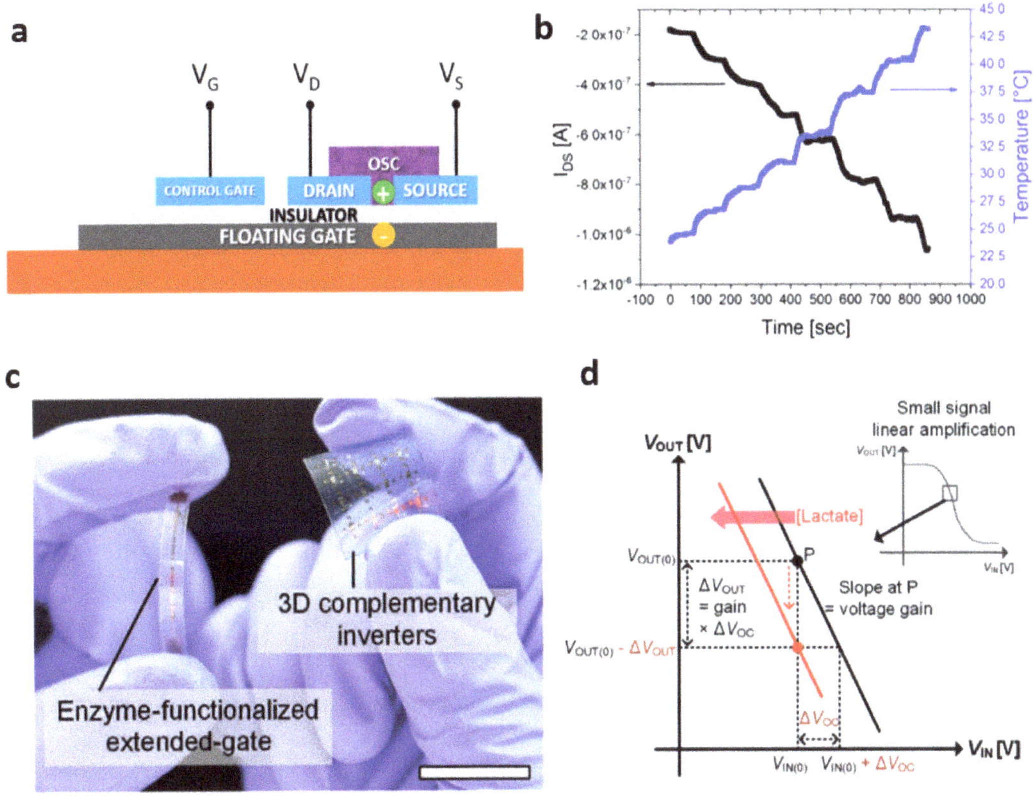

Figure 9. Extended gate/floating gate/3D-stacked. (**a**) Schematic of organic charge-modulated transistors. (**b**) Measured current variation recorded in real time as a function of the temperature. Reprinted from [67], copyright (2018) with permission from MDPI. (**c**) Measured image of the 3D inverter and extended-gated structured lactate detector. (**d**) Schematic of the extended-gate-type lactate detector using the complementary inverter. Reprinted from [68], Copyright (2020) with permission from John Wiley and Sons.

Depending on the variation in temperature in the range of 18.5–50 °C and pressure in the range of 10^2–10^3 Pa, the measured current was linearly changed in the presented floating-gate organic device (Figure 9b). In contrast to the conventional planar-type organic device structure, a vertically integrated biosensor based on organic semiconductors was reported. In 2020, Jung et al. presented a vertically integrated inverter circuit compromising a bottom *n*-type poly{[N,N′-bis(2-octyldodecyl)-naphthalene-1,4,5,8-bis-(dicarboximide)-2,6-diyl]-alt-5,5′-(2,2′-bithiophene)} (P[NDI2OD-T2])-based transistor and a top *p*-type poly(N-alkyl diketopyrrolo-pyrrole dithienylthieno-[3,2-b]thiophene (DPP-DTT)-based transistor [68]. As an interesting approach, this study used the shared gate, located between the two transistors, for both *p*-type and *n*-type transistors (Figure 9c). The shared gate was connected to an extended gate where the detection of the lactate was carried out (Figure 9d).

5. Applications

The techniques for synthesizing high-mobility organic semiconductors and the large-area printing of uniformly oriented organic semiconductors have matured. Various applications using organic semiconductor-based electronics, such as driving circuits of flexible/rollable display, photodetector array, radio frequency identification tags, and wearable biomedical sensors, have been demonstrated. Among them, considerable research is being conducted on wearable biomedical sensors because of the recent increase in medical spending and the interest in personal healthcare. Devices on various substrates, such as flexible/stretchable substrates, cylindrical metal wires, or fibers, have been used as wearable biomedical sensors.

5.1. OFETs on a Flexible Substrate

OFETs printed on a flexible substrate have been developed as pressure sensors and for sensing the pulse of the wrist. Monolithic OFETs that directly possess pressure-sensitive components can work actively as pressure sensors. This active matrix sensor array is advantageous in reducing power consumption. Lim et al. proposed a unique OFET structure that combined centro-apically self-organized organic semiconductors on top of printed hemispheric microstructures and an elastomeric PDMS top-gate dielectric (Figure 10a) [69].

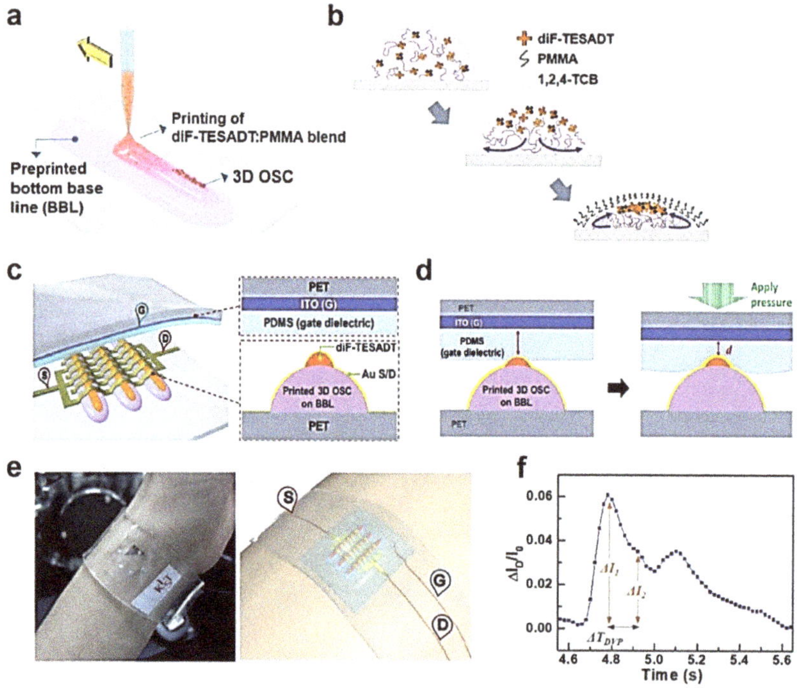

Figure 10. Schematic illustration of (**a**) the fabrication process of the pressure sensor, (**b**) the self-organization of the organic semiconductor, (**c**) the cross-sectional structure of the OFETs, and (**d**) the operation mechanism of OFETs as pressure sensors. (**e**) A photograph and a schematic illustration of the wrist wearing the pressure sensor using a PDMS band. (**f**) Current response to real-time pulse at constant source–drain (−60 V) and source–gate (−40 V) voltages. Three distinguishable peaks were detected. Reprinted from [69], copyright 2017 with permission from the American Chemical Society.

A 2,8-Difluoro-5,11-bis(triethylsilylethynyl)-anthradithiophene (diF-TESADT) semiconductor and poly(methyl methacrylate) (PMMA) insulator blended solution was line-printed using a dispenser printer. During solidification, diF-TESADT molecules were segregated vertically on top of PMMA, and self-assembled to form semiconducting crystals (Figure 10b). The three-dimensional (3D) semiconductor array was prepared on a bottom polyethylene terephthalate (PET) flexible substrate directly contacted with the PDMS gate dielectric on indium-tin-oxide (ITO)-coated top PET substrates (Figure 10c). In this structure, the OFETs operated as pressure sensors. The capacitance increased due to the decrease in elastomeric PDMS gate dielectric thickness when external pressure was applied to the OFETs. As a result, the drain current of the OFETs increased (Figure 10d). The combination of 3D microstructured organic semiconductors with a flat elastomeric dielectric layer would improve the sensitivity of the sensors, as reported previously. The proposed pressure sensors exhibited a high sensitivity of 1.07 kPa^{-1}, a rapid response and a short relaxation time of 18 ms, and good stability over 1000 cycles, which satisfied the conditions for the real-time monitoring of the pulse of the wrist artery. Fabrication on the flexible substrate enabled the wearing of sensors on the wrist (Figure 10e). When observing the current modulation of the sensors integrated on the wrist, a typical pulse shape was obtained with three prominent peaks, which can be used to assess the subjects' health (Figure 10f). Therefore, the sensors introduced in this study have demonstrated the low-cost, large-area fabrication of FET-based pressure sensors.

5.2. OFETs on Unconventional Substrates

The intrinsic electronic and mechanical properties of organic materials enable their integration onto a cylindrical fiber-shaped substrate, a key component of the newly emerging electronic textiles (e-textile) for wearable electronics. The fabrication of OFETs on the cylindrical substrates can be obtained using: (1) the fabrication of OFETs at the intersection of two fibers using an electrolyte gate dielectric; and (2) the fabrication of OFETs on a single-fiber substrate.

Jang et al. fabricated cylindrical OFETs using an Al single metallic fiber as a cylindrical gate substrate, cross-linkable insulating poly(4-vinylphenol) (PVP), poly(vinyl cinnamate) (PVCN) polymers as gate dielectrics, vacuum-deposited pentacene as a semiconductor, and a thermally evaporated Au as a source/drain electrode [70]. Al metallic fiber was electropolished to a lower surface roughness before being used as a gate substrate. The surface roughness was smoothed gradually for up to 10 min and increased afterward. PVP and PVCN polymer gate dielectrics were coated on the Al wire surface by a dip-coating process and crosslinked by thermal annealing and UV illumination. Dip-coated dielectric films showed a surface roughness of approximately 0.3 nm, which was low enough to grow high-quality pentacene semiconductor crystals on the surface without disturbance. Pentacene crystals vacuum-sublimed on the wire showed similar grain sizes of 1 μm compared to the ones deposited on the flat substrates. The thermal evaporation of Au with a stencil mask defined the source/drain electrodes, and the resulting OFETs showed typical hysteresis-free transfer and output characteristics with average μ_{FET}s of 0.24 (with PVP) and 0.53 $cm^2 \cdot V^{-1} \cdot s^{-1}$ (with PVCN). Under bending stress, the OFETs with PVCN dielectric exhibited excellent stability compared to the OFETs with the PVP dielectric. OFETs with PVCN dielectric maintained their electrical performance even at the smallest bending radius of 1.0 cm. The superior bending stability of the OFETs with the PVCN dielectric was attributed to the more rigid and stiff thermally crosslinked PVP than photo-cured PVCN, and the main chain of PVCN may be more flexible than that of PVP.

Kim et al. also reported fibriform OFETs using Au microfibers as a gate electrode (Figure 11a) [71].

The Au microfibers were treated with polydopamine to promote adhesion. An organic semiconductor/dielectric blend solution comprising 2,8-difluoro-5,11-bis(triethylsilylethynyl) anthradithiophene (diF-TES-ADT) semiconductor and PMMA dielectric was coated onto the polydopamine-coated Au fiber. Upon solvent vapor annealing, vertical phase separation between dif-TES-ADT and PMMA and the crystallization of dif-TES-ADT molecules occurred. Source/drain electrodes were defined by depositing 100 nm thick Au dots onto the metal insulator semiconductor microfiber through a transmission electron microscopy grid mask (Figure 11b). The average μ_{FET} and on/off ratio of 30 fibriform OFETs were 0.19 $cm^2 \cdot V^{-1} \cdot s^{-1}$ and ~10^4, respectively. The device performance was maintained up to 80% of the original values when the microfiber was bent with a bending radius down to 3.0 mm. Finally, the authors showed fibriform OFET-embedded textiles by weaving the OFETs with cotton (Figure 11c). PEDOT:PSS/graphene oxide-coated conducting threads worked as source/drain electrodes, and their fibriform OFET showed a reasonable μ_{FET} and on/off ratio of 0.17 $cm^2 \cdot V^{-1} \cdot s^{-1}$ and ~10^3, respectively (Figure 11d).

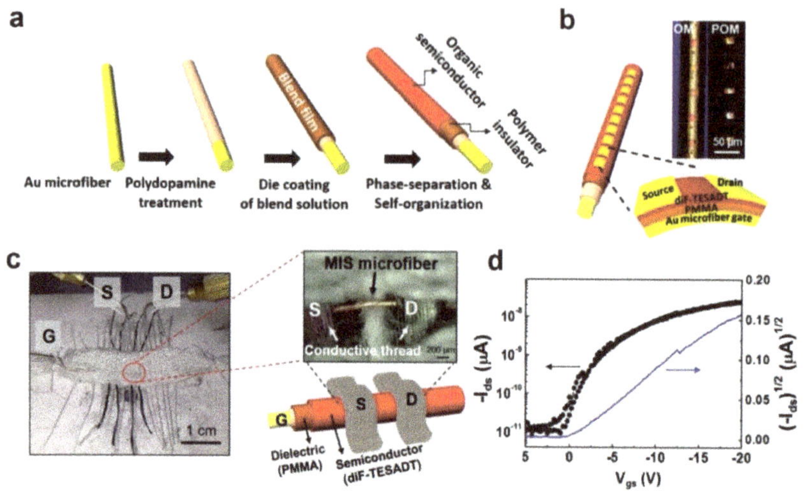

Figure 11. (**a**) Schematic illustration for the fabrication of fibriform OFETs using Au microfibers. (**b**) Schematic illustration and optical microscopy images of the fibriform OFETs. (**c**) Photograph, optical microscopy image, and schematic illustration of the fibriform OFETs embedded in the textiles. (**d**) Transfer characteristics of the fibriform OFETs at a source–drain voltage of −20 V. Reprinted from [71], copyright 2016 with permission from American Chemical Society.

Recently, as a more advanced concept of OFETs with an unconventional substrate, a fiber OFET was introduced. Kim et al. reported a spirally wrapped carbon nanotube (CNT) microelectrode with the desired dimension using inkjet printing and an agarose-gel-assisted transfer technique to overcome the previously reported limitations in fabricating fiber OFETs (e.g., the use of high-cost vacuum-assisted technologies or difficulty in the fine control of channel length) (Figure 12a) [72].

The dimension of the CNT microelectrode was controlled by the inkjet printing conditions (e.g., the number of printing layers and the interval between patterns). With the printed CNT microelectrode, the authors fabricated OFETs and organic photodiode. In particular, the photodiode comprising poly([2,6′-4,8-di(5-ethylhexylthienyl)benzo [1,2-b;3,3-b]-dithiophene]{3-fluoro-2[(2-ethylhexyl)carbonyl]thieno [3,4-b]thiophenediyl})(PTB7-Th) and [6,6]-phenyl-C71-butyric-acid methyl ester ($PC_{71}BM$) organic semiconductors was utilized in wearable biomedical devices measuring photoplethysmography (PPG) (Figure 12b). The photodiode measured the real-time PPG signals with discernible systolic and diastolic peaks when integrated into the textile PPG bandage (Figure 12c).

The research group also demonstrated fiber OFETs with the double-stranded assembly of electrode microfibers (Figure 13a) [73].

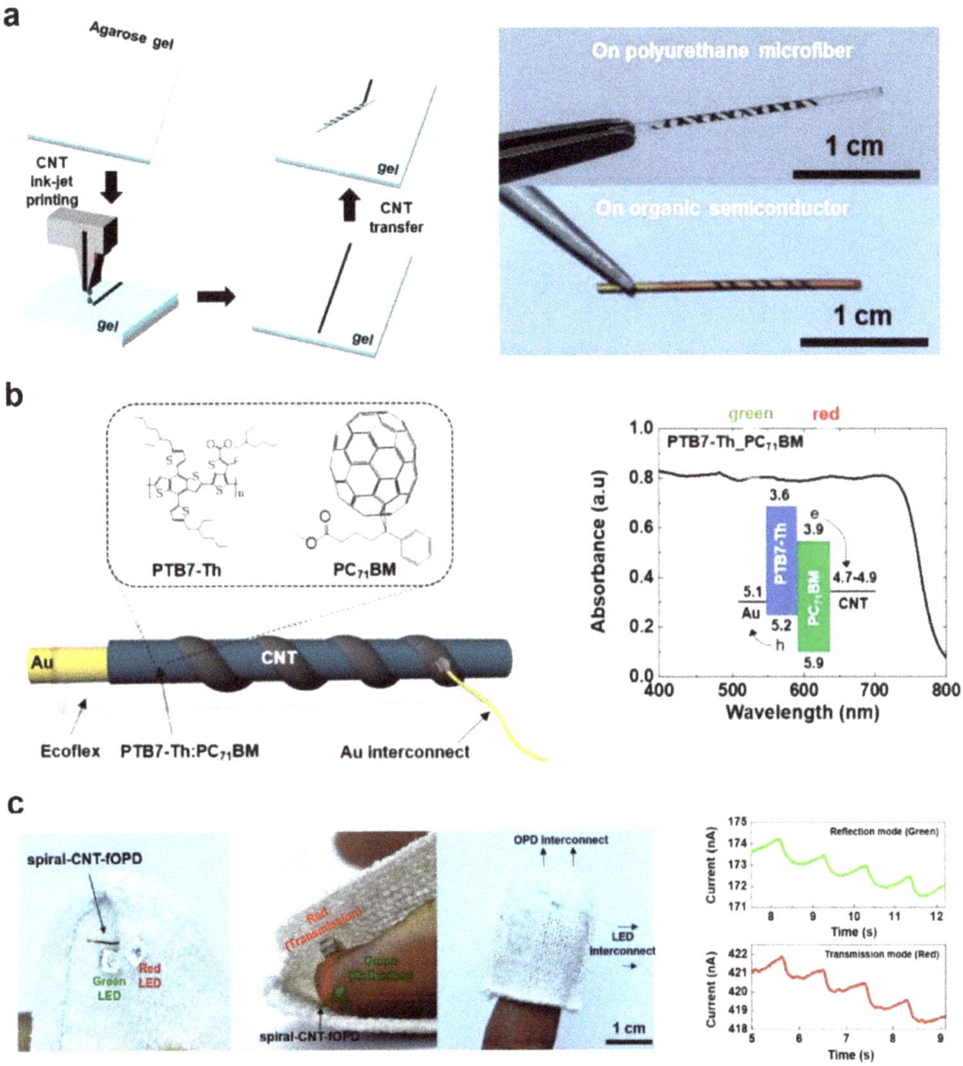

Figure 12. (**a**) Schematic illustration of the fabrication of CNT microelectrodes using inkjet and rolling-transfer techniques. (**b**) Schematic illustration of the device structure of an organic photodiode, chemical structure of PTB7-Th and $PC_{71}BM$, and absorption spectrum of the PTB7-Th:$PC_{71}BM$ photoactive layer. (**c**) Photographs of green and red light-emitting diodes and photodiodes embedded in the textiles for the measurement of PPG signals and PPG signals from the photodiodes in the reflection and transmission modes. Reprinted from [72], copyright 2022 with permission from American Chemical Society.

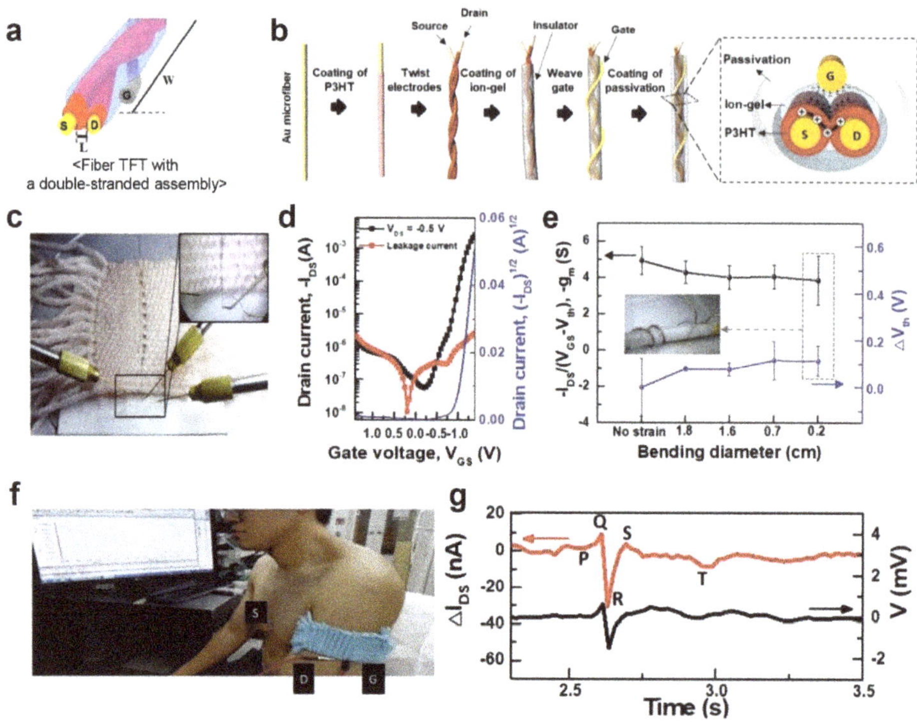

Figure 13. (**a**) Schematic illustration of a fiber OFET with a double-stranded assembly. (**b**) Schematic illustration of the fabrication process of the fiber OFET. (**c**) Photograph of the fiber OFETs embedded in the textile. (**d**) Transfer characteristic of the OFET. (**e**) Change in transconductance and threshold voltage according to bending diameter. (**f**) A photograph of a human subject wearing the fiber OFETs embedded in the textile for the measurement of ECG signal. (**g**) Single ECG trace from potentiometric recording (black line) and drain current (red line) from the fiber OFETs. Reprinted from [73], copyright 2019 with permission from John Wiley and Sons.

Two Au microfibers, 100 µm in diameter for use as source/drain electrodes, were coated with a P3HT semiconductor using a home-built die-coating system (Figure 13b). Subsequently, they were twisted so that the P3HT channel could be located between the source/drain fibers. In addition, they used a poly(vinylidene fluoride-*co*-hexafluoropropylene) (P(VDF-HFP)) and 1-ethyl-3-methylimidazolium bis(trifluoromethylsulfonyl)imide ((EMIM)(TFSI)) ion–gel gate dielectric material as a solid electrolyte in which an IL was dispersed in a polymer matrix. The resulting fiber OFETs exhibited typical transfer characteristics with low voltage operation below 1.3 V and an on/off ratio of 10^5. After a successful demonstration, the fiber OFETs were embedded into a textile by weaving (Figure 13c). The embedded OFETs showed negligible changes in electrical properties because the device could endure bending deformation (Figure 13d). During bending more than 1000 times, the electrical properties, including the transconductance and threshold voltage of the fiber OFETs, were maintained up to 80% with a bending radius of 2 mm (Figure 13e). Finally, the application of the fiber OFETs for the real-time monitoring of human body signals was demonstrated (Figure 13f). Among various electrophysiological signals from the human body, a recording ECG was tested. The contraction and relaxation of the heart muscle usually results in ECG signals with an amplitude of a few hundred microvolts. Therefore, the gate and source electrodes of the fiber OFETs were attached to the wrist and chest of the human subject, respectively, so that the ECG signals could be delivered to the gate electrode and amplified by the fiber OFETs. When the fiber OFETs were set to the subthreshold swing regime, where the drain current changed abruptly, the typical ECG signals containing typical P-Q-R-S-T subwaves were amplified (Figure 13g). This suggests that the fiber OFET possessed sufficient resolution to record ECG signals; thus, it can be applied to e-textiles for wearable healthcare devices.

6. Conclusions

This review paper has given an overview of organic semiconductors in biomedical applications. Table 2 is a summary table to compare the materials, device structures, fabrication methods, and applications in previously reported organic material-based healthcare devices.

Table 2. Comparison of materials, device structures, fabrication methods, and applications in previously reported organic material-based healthcare devices.

Material	Device Structure	Representative Method	Application	Refs.
Carbon dots/ polyvinyl alcohol	Fiber triboelectric nanogenerator	Microwave-assisted pyrolytic reaction	Monitoring of physiological signals	[63]
Carboxylated polypyrrole nanotubes	Liquid-ion-gated FET	Reverse microemulsion polymerization	Dopamine detection	[61]
PEDOT:PSS on nanomesh	Organic electrochemical transistor	Spray coating of PEDOT:PSS	On-skin ECG signal detection	[62]
Porous PEA-r-PS-r-PDVB	Porous ion gel	Use of sugar template	Monitoring of human motions	[65]
PEDOT:PSS	Organic electrochemical transistor	Laser-patterned microcapillary	Cortisol sensing	[64]
PEDOT:PSS/ TIPS-Pentacene	Floating-gate transistor	Inkjet printing	Temperature sensing	[67]
DPP-DTT/ P[NDI2OD-T2]	Shared-gate structure	Spin-coating	Lactate sensing	[68]
C6-DNT-VW	Extended-gate transistor	Printing	Oxytocin sensing	[66]
diF-TES-ADT	Vertical transistor	Printing	Monitoring of the radial artery pulse	[69]

Organic semiconductors have several advantages in mechanical and electrical aspects over the inorganic counter parts. There are apparent limitations when it comes to electrical conductivity and charge carrier mobility compared to inorganic counterparts. They are molecularly bulky, which hinders the materialization of perfect crystalline structure at a molecular level, whereas inorganic semiconductors typically consist of atomic crystalline structures, in which the packing scale is smaller than that of organic semiconductors. This irresistible material's nature should differentiate the direction between organic and inorganic research and development. The development of inorganic semiconductors might focus on the high performance and large capacity of the final device, whereas the development of organic semiconductors might focus on the mass production and multifunctional properties. Thus, inorganic and organic semiconductors could complement each other.

In addition, recent fabrication methods under the spotlight for organic semiconductors, including molecules and polymers, have been overviewed in this review. Soft lithography and direct writing techniques are promising methodologies based on solution processibility. These enable a precise and fine patterning of organic semiconductors on unconventional substrates. Especially, soft lithography utilizes flexible and bendable elastomeric stamps to transfer certain patterns of organic semiconductors. Flexible and bendable elastomeric stamps provide the direct printing of organic semiconductors on three-dimensional curvatures. The unique processability of those methodologies reduces tremendous fabrication costs compared to those based on conventional fabrication processes such as chemical and physical vapor deposition, and thermal and e-beam evaporators, which require high-vacuum systems.

Through the survey on material properties and fabrication advantages, organic semiconductors have great potential in biomedical applications in terms of flexible and wearable medical devices. Owing to the excellent mechanical properties of organic semiconductors, intrinsically flexible electronic devices can be realized. However, several practical challenges of organic semiconductors have blocked their commercialization.

First, their unsatisfactory robustness hinders their commercialization. Organic semiconductors are physically and chemically vulnerable. The hardness—the resistance of a material to allow plastic deformation—of organic semiconductors is too low to withstand the scratches formed during daily activities. Thus, organic semiconductors with intrinsically physical robustness should be developed. Several approaches are proposed, for example, making composite materials, mixing inorganic dielectric materials, and the modification of functional groups in organic semiconductors. On the other hand, applying an encapsulation layer on organic semiconductors is a good approach to protect the underlying layers, but additional processes are needed.

Second, the reliability of organic semiconductors has fallen far behind for long-term operation and fabrication uniformity. Owing to the intrinsic degradation of organic semiconductors over continuous operation, organic biomedical devices could deliver wrong information and sensing signals over a long period of time. This is a critical flaw of organic semiconductors in biomedical applications. In addition, fabrication uniformity should be achieved for the reliability of large-area and multi-array devices. Especially, organic semiconductors in biomedical applications can be utilized in active layers of multi-sensing components by applying functionalization to organic semiconductors. In this respect, an array of unit devices are inevitable.

Lastly, biodegradability and biocompatibility are important challenges that must be managed. With increasing demand for the mass production of organic biomedical devices, technologies for the natural disposal of organic biomedical devices should be developed. One of the approaches to solve this issue is natural dissolution on the human skin surface by environmentally friendly solvents such as water. In addition, the devices are normally attached to the human body to record biosignals. Imperceptibility and aesthetic points are of great significance. In order to achieve this, reducing the device thickness and improving transparency should be studied. By focusing on the uniqueness of organic semiconductors rather than their disadvantages, organic semiconductor-based biomedical devices could be expanded for more advanced applications.

Author Contributions: K.K., H.Y. and E.K.L. performed the literature research and analysis, and wrote the paper. All authors have read and agreed to the published version of the manuscript.

Funding: This research was supported by a National Research Foundation of Korea (NRF) grant funded by Ministry of Science and ICT (Grant No. NRF-2022R1F1A1076383).

Institutional Review Board Statement: Not applicable.

Informed Consent Statement: Not applicable.

Data Availability Statement: Not applicable.

Conflicts of Interest: The authors declare no conflict of interest.

References

1. Hussain, S.; Zhao, H.; Zhou, L.; Zhou, X.; Iyer, P.K.; Lv, F.; Liu, L.; Wang, S. An Optoelectronic Device for Rapid Monitoring of Creatine Kinase Using Cationic Conjugated Polyelectrolyte. *Adv. Mater. Technol.* **2019**, *4*, 1900361. [CrossRef]
2. Giannini, S.; Blumberger, J. Charge Transport in Organic Semiconductors: The Perspective from Nonadiabatic Molecular Dynamics. *Acc. Chem. Res.* **2022**, *55*, 819–830. [CrossRef] [PubMed]
3. Paterson, A.F.; Singh, S.; Fallon, K.J.; Hodsden, T.; Han, Y.; Schroeder, B.C.; Bronstein, H.; Heeney, M.; McCulloch, I.; Anthopoulos, T.D. Recent Progress in High-Mobility Organic Transistors: A Reality Check. *Adv. Mater.* **2018**, *30*, 1801079. [CrossRef] [PubMed]
4. Ohayon, D.; Inal, S. Organic Bioelectronics: From Functional Materials to Next-Generation Devices and Power Sources. *Adv. Mater.* **2020**, *32*, 2001439. [CrossRef]
5. Chen, X.; Hussain, S.; Hao, Y.; Tian, X.; Gao, R. Review—Recent Advances of Signal Amplified Smart Conjugated Polymers for Optical Detection on Solid Support. *ECS J. Solid State Sci. Technol.* **2021**, *10*, 037006. [CrossRef]
6. Chen, X.; Hussain, S.; Abbas, A.; Hao, Y.; Malik, A.H.; Tian, X.; Song, H.; Gao, R. Conjugated polymer nanoparticles and their nanohybrids as smart photoluminescent and photoresponsive material for biosensing, imaging, and theranostics. *Microchim. Acta* **2022**, *189*, 83. [CrossRef]
7. Roth, B.; Savagatrup, S.; de los Santos, N.V.; Hagemann, O.; Carlé, J.E.; Helgesen, M.; Livi, F.; Bundgaard, E.; Søndergaard, R.R.; Krebs, F.C.; et al. Mechanical Properties of a Library of Low-Band-Gap Polymers. *Chem. Mater.* **2016**, *28*, 2363–2373. [CrossRef]
8. Root, S.E.; Savagatrup, S.; Printz, A.D.; Rodriquez, D.; Lipomi, D.J. Mechanical Properties of Organic Semiconductors for Stretchable, Highly Flexible, and Mechanically Robust Electronics. *Chem. Rev.* **2017**, *117*, 6467–6499. [CrossRef]
9. Hopkins, J.; Fidanovski, K.; Lauto, A.; Mawad, D. All-Organic Semiconductors for Electrochemical Biosensors: An Overview of Recent Progress in Material Design. *Front. Bioeng. Biotechnol.* **2019**, *7*, 237. [CrossRef]
10. Bonaventura, G.; Iemmolo, R.; La Cognata, V.; Zimbone, M.; La Via, F.; Fragalà, M.E.; Barcellona, M.L.; Pellitteri, R.; Cavallaro, S. Biocompatibility between Silicon or Silicon Carbide surface and Neural Stem Cells. *Sci. Rep.* **2019**, *9*, 11540. [CrossRef]
11. Yu, T.; Malugin, A.; Ghandehari, H. Impact of silica nanoparticle design on cellular toxicity and hemolytic activity. *ACS Nano* **2011**, *5*, 5717–5728. [CrossRef] [PubMed]
12. Hsiao, I.L.; Fritsch-Decker, S.; Leidner, A.; Al-Rawi, M.; Hug, V.; Diabaté, S.; Grage, S.L.; Meffert, M.; Stoeger, T.; Gerthsen, D.; et al. Biocompatibility of Amine-Functionalized Silica Nanoparticles: The Role of Surface Coverage. *Small* **2019**, *15*, 1805400. [CrossRef] [PubMed]
13. Morris, A.S.; Adamcakova-Dodd, A.; Lehman, S.E.; Wongrakpanich, A.; Thorne, P.S.; Larsen, S.C.; Salem, A.K. Amine modification of nonporous silica nanoparticles reduces inflammatory response following intratracheal instillation in murine lungs. *Toxicol. Lett.* **2015**, *241*, 207–215. [CrossRef] [PubMed]
14. Šafaříková, E.; Šindlerová, L.Š.; Stříteský, S.; Kubala, L.; Vala, M.; Weiter, M.; Víteček, J. Evaluation and improvement of organic semiconductors' biocompatibility towards fibroblasts and cardiomyocytes. *Sens. Actuators B Chem.* **2018**, *260*, 418–425. [CrossRef]
15. Roh, J.; Lee, T.; Kang, C.-M.; Kwak, J.; Lang, P.; Horowitz, G.; Kim, H.; Lee, C. Injection-modulated polarity conversion by charge carrier density control via a self-assembled monolayer for all-solution-processed organic field-effect transistors. *Sci. Rep.* **2017**, *7*, 46365. [CrossRef]
16. Zeglio, E.; Rutz, A.L.; Winkler, T.E.; Malliaras, G.G.; Herland, A. Conjugated Polymers for Assessing and Controlling Biological Functions. *Adv. Mater.* **2019**, *31*, 1806712. [CrossRef]
17. Dong, H.; Fu, X.; Liu, J.; Wang, Z.; Hu, W. 25th anniversary article: Key points for high-mobility organic field-effect transistors. *Adv. Mater.* **2013**, *25*, 6158–6183. [CrossRef]
18. Kang, B.; Jang, M.; Chung, Y.; Kim, H.; Kwak, S.K.; Oh, J.H.; Cho, K. Enhancing 2D growth of organic semiconductor thin films with macroporous structures via a small-molecule heterointerface. *Nat. Commun.* **2014**, *5*, 4752. [CrossRef]
19. Jurchescu, O.D.; Popinciuc, M.; van Wees, B.J.; Palstra, T. Interface-Controlled, High-Mobility Organic Transistors. *Adv. Mater.* **2007**, *19*, 688–692. [CrossRef]
20. Ebata, H.; Izawa, T.; Miyazaki, E.; Takimiya, K.; Ikeda, M.; Kuwabara, H.; Yui, T. Highly soluble [1]benzothieno[3,2-b]benzothiophene (BTBT) derivatives for high-performance, solution-processed organic field-effect transistors. *J. Am. Chem. Soc.* **2007**, *129*, 15732–15733. [CrossRef]
21. Minemawari, H.; Yamada, T.; Matsui, H.; Tsutsumi, J.Y.; Haas, S.; Chiba, R.; Kumai, R.; Hasegawa, T. Inkjet printing of single-crystal films. *Nature* **2011**, *475*, 364–367. [CrossRef] [PubMed]
22. Yuan, Y.; Giri, G.; Ayzner, A.L.; Zoombelt, A.P.; Mannsfeld, S.C.B.; Chen, J.; Nordlund, D.; Toney, M.F.; Huang, J.; Bao, Z. Ultra-high mobility transparent organic thin film transistors grown by an off-centre spin-coating method. *Nat. Commun.* **2014**, *5*, 3005. [CrossRef] [PubMed]
23. Takeya, J.; Yamagishi, M.; Tominari, Y.; Hirahara, R.; Nakazawa, Y.; Nishikawa, T.; Kawase, T.; Shimoda, T.; Ogawa, S. Very high-mobility organic single-crystal transistors with in-crystal conduction channels. *Appl. Phys. Lett.* **2007**, *90*, 102120. [CrossRef]
24. Poimanova, E.Y.; Shaposhnik, P.A.; Anisimov, D.S.; Zavyalova, E.G.; Trul, A.A.; Skorotetcky, M.S.; Borshchev, O.V.; Vinnitskiy, D.Z.; Polinskaya, M.S.; Krylov, V.B.; et al. Biorecognition Layer Based On Biotin-Containing [1]Benzothieno[3,2-b][1]benzothiophene Derivative for Biosensing by Electrolyte-Gated Organic Field-Effect Transistors. *ACS Appl. Mater. Interfaces* **2022**, *14*, 16462–16476. [CrossRef] [PubMed]
25. Khodagholy, D.; Rivnay, J.; Sessolo, M.; Gurfinkel, M.; Leleux, P.; Jimison, L.H.; Stavrinidou, E.; Hervé, T.; Sanaur, S.; Owens, R.M.; et al. High transconductance organic electrochemical transistors. *Nat. Commun.* **2013**, *4*, 2133. [CrossRef]

26. Nguyen-Dang, T.; Chae, S.; Harrison, K.; Llanes, L.C.; Yi, A.; Kim, H.J.; Biswas, S.; Visell, Y.; Bazan, G.C.; Nguyen, T.-Q. Efficient Fabrication of Organic Electrochemical Transistors via Wet Chemical Processing. *ACS Appl. Mater. Interfaces* **2022**, *14*, 12469–12478. [CrossRef]
27. Tang, K.; Miao, W.; Guo, S. Crosslinked PEDOT:PSS Organic Electrochemical Transistors on Interdigitated Electrodes with Improved Stability. *ACS Appl. Polym. Mater.* **2021**, *3*, 1436–1444. [CrossRef]
28. Oh, J.H.; Lee, H.W.; Mannsfeld, S.C.B.; Stoltenberg, R.M.; Jung, E.; Jin, Y.W.; Kim, J.M.; Yoo, J.-B.; Bao, Z. Solution-processed, high-performance n-channel organic microwire transistors. *Proc. Natl. Acad. Sci. USA* **2009**, *106*, 6065–6070. [CrossRef]
29. Schmidt, R.; Oh, J.H.; Sun, Y.S.; Deppisch, M.; Krause, A.M.; Radacki, K.; Braunschweig, H.; Könemann, M.; Erk, P.; Bao, Z.; et al. High-Performance Air-Stable n-Channel Organic Thin Film Transistors Based on Halogenated Perylene Bisimide Semiconductors. *J. Am. Chem. Soc.* **2009**, *131*, 6215–6228. [CrossRef]
30. Minder, N.A.; Ono, S.; Chen, Z.; Facchetti, A.; Morpurgo, A.F. Band-Like Electron Transport in Organic Transistors and Implication of the Molecular Structure for Performance Optimization. *Adv. Mater.* **2011**, *24*, 503–508. [CrossRef]
31. Laquindanum, J.G.; Katz, H.E.; Dodabalapur, A.; Lovinger, A.J. n-Channel Organic Transistor Materials Based on Naphthalene Frameworks. *J. Am. Chem. Soc.* **1996**, *118*, 11331–11332. [CrossRef]
32. Shukla, D.; Nelson, S.F.; Freeman, D.C.; Rajeswaran, M.; Ahearn, W.G.; Meyer, D.M.; Carey, J.T. Thin-Film Morphology Control in Naphthalene-Diimide-Based Semiconductors: High Mobility n-Type Semiconductor for Organic Thin-Film Transistors. *Chem. Mater.* **2008**, *20*, 7486–7491. [CrossRef]
33. He, T.; Stolte, M.; Würthner, F. Air-Stable n-Channel Organic Single Crystal Field-Effect Transistors Based on Microribbons of Core-Chlorinated Naphthalene Diimide. *Adv. Mater.* **2013**, *25*, 6951–6955. [CrossRef] [PubMed]
34. Li, H.; Tee, B.C.K.; Cha, J.J.; Cui, Y.; Chung, J.W.; Lee, S.Y.; Bao, Z. High-Mobility Field-Effect Transistors from Large-Area Solution-Grown Aligned C60 Single Crystals. *J. Am. Chem. Soc.* **2012**, *134*, 2760–2765. [CrossRef] [PubMed]
35. Hahm, S.G.; Rho, Y.; Jung, J.; Kim, H.; Sajoto, T.; Kim, F.S.; Barlow, S.; Park, C.E.; Jenekhe, S.A.; Marder, S.R.; et al. High-Performance n-Channel Thin-Film Field-Effect Transistors Based on a Nanowire-Forming Polymer. *Adv. Funct. Mater.* **2012**, *23*, 2060–2071. [CrossRef]
36. Liu, C.; Jang, J.; Xu, Y.; Kim, H.J.; Khim, D.; Park, W.-T.; Noh, Y.-Y.; Kim, J.-J. Effect of Doping Concentration on Microstructure of Conjugated Polymers and Characteristics in N-Type Polymer Field-Effect Transistors. *Adv. Funct. Mater.* **2014**, *25*, 758–767. [CrossRef]
37. Kim, R.; Amegadze, P.S.K.; Kang, I.; Yun, H.-J.; Noh, Y.-Y.; Kwon, S.-K.; Kim, Y.-H. High-Mobility Air-Stable Naphthalene Diimide-Based Copolymer Containing Extended π-Conjugation for n-Channel Organic Field Effect Transistors. *Adv. Funct. Mater.* **2013**, *23*, 5719–5727. [CrossRef]
38. Ding, L.; Wang, Z.-Y.; Wang, J.-Y.; Pei, J. Organic Semiconducting Materials Based on BDOPV: Structures, Properties, and Applications. *Chin. J. Chem.* **2019**, *38*, 13–24. [CrossRef]
39. Kim, G.; Han, A.R.; Lee, H.R.; Lee, J.; Oh, J.H.; Yang, C. Acceptor–acceptor type isoindigo-based copolymers for high-performance n-channel field-effect transistors. *Chem. Commun.* **2014**, *50*, 2180–2183. [CrossRef]
40. Tang, M.L.; Oh, J.H.; Reichardt, A.D.; Bao, Z. Chlorination: A general route toward electron transport in organic semiconductors. *J. Am. Chem. Soc.* **2009**, *131*, 3733–3740. [CrossRef]
41. Lee, J.Y.; Roth, S.; Park, Y.W. Anisotropic field effect mobility in single crystal pentacene. *Appl. Phys. Lett.* **2006**, *88*, 252106. [CrossRef]
42. Kim, J.; Kim, M.-G.; Kim, J.; Jo, S.; Kang, J.; Jo, J.-W.; Lee, W.; Hwang, C.; Moon, J.; Yang, L.; et al. Scalable Sub-micron Patterning of Organic Materials Toward High Density Soft Electronics. *Sci. Rep.* **2015**, *5*, 14520. [CrossRef] [PubMed]
43. Khim, D.; Baeg, K.-J.; Kim, J.; Kang, M.; Lee, S.-H.; Chen, Z.; Facchetti, A.; Kim, D.-Y.; Noh, Y.-Y. High performance and stable N-channel organic field-effect transistors by patterned solvent-vapor annealing. *ACS Appl. Mater. Interfaces* **2013**, *5*, 10745–10752. [CrossRef]
44. He, Z.; Chen, J.; Sun, Z.; Szulczewski, G.; Li, D. Air-flow navigated crystal growth for TIPS pentacene-based organic thin-film transistors. *Org. Electron.* **2012**, *13*, 1819–1826. [CrossRef]
45. Sakamoto, K.; Ueno, J.; Bulgarevich, K.D.; Miki, K. Anisotropic charge transport and contact resistance of 6,13-bis(triisopropylsilylethynyl) pentacene field-effect transistors fabricated by a modified flow-coating method. *Appl. Phys. Lett.* **2012**, *100*, 123301. [CrossRef]
46. Giri, G.; Park, S.; Vosgueritchian, M.; Shulaker, M.M.; Bao, Z. High-mobility, aligned crystalline domains of TIPS-pentacene with metastable polymorphs through lateral confinement of crystal growth. *Adv. Mater.* **2013**, *26*, 487–493. [CrossRef]
47. Chang, J.; Chi, C.; Zhang, J.; Wu, J. Controlled growth of large-area high-performance small-molecule organic single-crystalline transistors by slot-die coating using a mixed solvent system. *Adv. Mater.* **2013**, *25*, 6442–6447. [CrossRef] [PubMed]
48. Diao, Y.; Tee, B.C.K.; Giri, G.; Xu, J.; Kim, H.; Becerril, H.A.; Stoltenberg, R.M.; Lee, T.H.; Xue, G.; Mannsfeld, S.C.B.; et al. Solution coating of large-area organic semiconductor thin films with aligned single-crystalline domains. *Nat. Mater.* **2013**, *12*, 665–671. [CrossRef]
49. Bae, I.; Kang, S.J.; Shin, Y.J.; Park, Y.J.; Kim, R.H.; Mathevet, F.; Park, C. Tailored Single Crystals of Triisopropylsilylethynyl Pentacene by Selective Contact Evaporation Printing. *Adv. Mater.* **2011**, *23*, 3398–3402. [CrossRef]
50. Dickey, K.C.; Subramanian, S.; Anthony, J.E.; Han, L.-H.; Chen, S.; Loo, Y.-L. Large-area patterning of a solution-processable organic semiconductor to reduce parasitic leakage and off currents in thin-film transistors. *Appl. Phys. Lett.* **2007**, *90*, 244103. [CrossRef]

51. Kim, K.; Jang, M.; Lee, M.; An, T.K.; Anthony, J.E.; Kim, H.; Yang, H.; Park, C.E. Unified film patterning and annealing of an organic semiconductor with micro-grooved wet stamps. *J. Mater. Chem. C* **2016**, *4*, 6996–7003. [CrossRef]
52. Suh, K.Y.; Kim, Y.S.; Lee, H. Capillary Force Lithography. *Adv. Mater.* **2001**, *13*, 1386–1389. [CrossRef]
53. Jo, P.S.; Vailionis, A.; Park, Y.M.; Salleo, A. Scalable fabrication of strongly textured organic semiconductor micropatterns by capillary force lithography. *Adv. Mater.* **2012**, *24*, 3269–3274. [CrossRef] [PubMed]
54. Park, K.S.; Cho, B.; Baek, J.; Hwang, J.K.; Lee, H.; Sung, M.M. Single-Crystal Organic Nanowire Electronics by Direct Printing from Molecular Solutions. *Adv. Funct. Mater.* **2013**, *23*, 4776–4784. [CrossRef]
55. Hwang, J.K.; Cho, S.; Dang, J.M.; Kwak, E.B.; Song, K.-K.; Moon, J.; Sung, M.M. Direct nanoprinting by liquid-bridge-mediated nanotransfer moulding. *Nat. Nanotechnol.* **2010**, *5*, 742–748. [CrossRef] [PubMed]
56. Kim, K.; Rho, Y.; Kim, Y.; Kim, H.; Hahm, S.G.; Park, C.E. A Lattice-Strained Organic Single-Crystal Nanowire Array Fabricated via Solution-Phase Nanograting-Assisted Pattern Transfer for Use in High-Mobility Organic Field-Effect Transistors. *Adv. Mater.* **2016**, *28*, 3209–3215. [CrossRef]
57. Felmet, K.; Loo, Y.-L.; Sun, Y. Patterning conductive copper by nanotransfer printing. *Appl. Phys. Lett.* **2004**, *85*, 3316–3318. [CrossRef]
58. Kang, B.; Min, H.; Seo, U.; Lee, J.; Park, N.; Cho, K.; Lee, H.S. Directly Drawn Organic Transistors by Capillary Pen: A New Facile Patterning Method using Capillary Action for Soluble Organic Materials. *Adv. Mater.* **2013**, *25*, 4117–4122. [CrossRef]
59. Kim, H.; Hong, K.; Lee, K.H.; Frisbie, C.D. Performance and stability of aerosol-jet-printed electrolyte-gated transistors based on poly(3-hexylthiophene). *ACS Appl. Mater. Interfaces* **2013**, *5*, 6580–6585. [CrossRef]
60. Min, S.-Y.; Kim, T.S.; Kim, B.J.; Cho, H.; Noh, Y.-Y.; Yang, H.; Cho, J.H.; Lee, T.-W. Large-scale organic nanowire lithography and electronics. *Nat. Commun.* **2013**, *4*, 1773. [CrossRef]
61. Park, S.J.; Lee, J.; Seo, S.E.; Kim, K.H.; Park, C.S.; Lee, S.H.; Ban, H.S.; Lee, B.D.; Song, H.S.; Kim, J.; et al. High-Performance Conducting Polymer Nanotube-based Liquid-Ion Gated Field-Effect Transistor Aptasensor for Dopamine Exocytosis. *Sci. Rep.* **2020**, *10*, 3772. [CrossRef] [PubMed]
62. Wang, J.; Lee, S.; Yokota, T.; Jimbo, Y.; Wang, Y.; Nayeem, O.G.; Nishinaka, M.; Someya, T. Nanomesh Organic Electrochemical Transistor for Comfortable On-Skin Electrodes with Local Amplifying Function. *ACS Appl. Electron. Mater.* **2020**, *2*, 3601–3609. [CrossRef]
63. Li, Z.; Xu, B.; Han, J.; Huang, J.; Fu, H. A Polycation-Modified Nanofillers Tailored Polymer Electrolytes Fiber for Versatile Biomechanical Energy Harvesting and Full-Range Personal Healthcare Sensing. *Adv. Funct. Mater.* **2021**, *32*, 2106731. [CrossRef]
64. Parlak, O.; Keene, S.T.; Marais, A.; Curto, V.F.; Salleo, A. Molecularly selective nanoporous membrane-based wearable organic electrochemical device for noninvasive cortisol sensing. *Sci. Adv.* **2018**, *4*, eaar2904. [CrossRef] [PubMed]
65. Kwon, J.H.; Kim, Y.M.; Moon, H.C. Porous Ion Gel: A Versatile Ionotronic Sensory Platform for High-Performance, Wearable Ionoskins with Electrical and Optical Dual Output. *ACS Nano* **2021**, *15*, 15132–15141. [CrossRef] [PubMed]
66. Ohshiro, K.; Sasaki, Y.; Zhou, Q.; Lyu, X.; Yamanashi, Y.; Nakahara, K.; Nagaoka, H.; Minami, T. Oxytocin detection at ppt level in human saliva by an extended-gate-type organic field-effect transistor. *Analyst* **2022**, *147*, 1055–1059. [CrossRef]
67. Lai, S.; Viola, F.A.; Cosseddu, P.; Bonfiglio, A. Floating Gate, Organic Field-Effect Transistor-Based Sensors towards Biomedical Applications Fabricated with Large-Area Processes over Flexible Substrates. *Sensors* **2018**, *18*, 688. [CrossRef]
68. Baek, S.; Kwon, J.; Mano, T.; Tokito, S.; Jung, S. A Flexible 3D Organic Preamplifier for a Lactate Sensor. *Macromol. Biosci.* **2020**, *20*, 2000144. [CrossRef]
69. Yeo, S.Y.; Park, S.; Yi, Y.; Kim, H.; Lim, J.A. Highly Sensitive Flexible Pressure Sensors Based on Printed Organic Transistors with Centro-Apically Self-Organized Organic Semiconductor Microstructures. *ACS Appl. Mater. Interfaces* **2017**, *9*, 42996–43003. [CrossRef]
70. Jang, J.; Nam, S.; Hwang, J.; Park, J.-J.; Im, J.; Park, C.-E.; Kim, J.M. Photocurable polymer gate dielectrics for cylindrical organic field-effect transistors with high bending stability. *J. Mater. Chem.* **2012**, *22*, 1054–1060. [CrossRef]
71. Kim, H.M.; Kang, H.W.; Hwang, K.; Lim, H.S.; Ju, B.K.; Lim, J.A. Metal–Insulator–Semiconductor Coaxial Microfibers Based on Self-Organization of Organic Semiconductor:Polymer Blend for Weavable, Fibriform Organic Field-Effect Transistors. *Adv. Funct. Mater.* **2016**, *26*, 2706–2714. [CrossRef]
72. Kim, H.; Kang, T.-H.; Ahn, J.; Han, H.; Park, S.; Kim, S.J.; Park, M.-C.; Paik, S.-H.; Hwang, K.; Yi, H.; et al. Spirally Wrapped Carbon Nanotube Microelectrodes for Fiber Optoelectronic Devices beyond Geometrical Limitations toward Smart Wearable E-Textile Applications. *ACS Nano* **2020**, *14*, 17213–17223. [CrossRef] [PubMed]
73. Kim, S.J.; Kim, H.; Ahn, J.; Hwang, K.; Ju, H.; Park, M.-C.; Yang, H.; Kim, H.; Jang, H.W.; Lim, J.A. A New Architecture for Fibrous Organic Transistors Based on a Double-Stranded Assembly of Electrode Microfibers for Electronic Textile Applications. *Adv. Mater.* **2019**, *31*, 1900564. [CrossRef] [PubMed]

polymers

MDPI

Article

Microscopic and Spectroscopic Imaging and Thermal Analysis of Acrylates, Silicones and Active Pharmaceutical Ingredients in Adhesive Transdermal Patches

Barbara Mikolaszek [1], Marzena Jamrógiewicz [2], Krystyna Mojsiewicz-Pieńkowska [2] and Małgorzata Sznitowska [1,*]

1 Department of Pharmaceutical Technology, Faculty of Pharmacy, Medical University of Gdańsk, 80-416 Gdańsk, Poland; barbara.mikolaszek@gumed.edu.pl

2 Department of Physical Chemistry, Faculty of Pharmacy, Medical University of Gdańsk, 80-416 Gdańsk, Poland; marzena.jamrogiewicz@gumed.edu.pl (M.J.); krystyna.pienkowska@gumed.edu.pl (K.M.-P.)

* Correspondence: malgorzata.sznitowska@gumed.edu.pl; Tel.: +48-58-349-1080; Fax: +48-58-349-1090

Abstract: Dermal or transdermal patches are increasingly becoming a noteworthy alternative as carriers for active pharmaceutical ingredients (APIs), which makes their detailed physicochemical evaluation essential for pharmaceutical development. This paper demonstrates mid-infrared (FTIR) and Raman spectroscopy with complementary microscopic methods (SEM, optical and confocal Raman microscopy) and differential scanning calorimetry (DSC) as tools for the identification of the state of model API (testosterone TST, cytisine CYT or indomethacin IND) in selected adhesive matrices. Among the employed spectroscopic techniques, FTIR and Raman may be used not only as standard methods for API identification in the matrix, but also as a means of distinguishing commercially available polymeric materials of a similar chemical structures. A novel approach for the preparation of adhesive polymers for the FTIR analysis was introduced. In silicone matrices, all three APIs were suspended, whereas in the case of the acrylic PSA, Raman microscopy confirmed that only IND was dissolved in all three acrylic matrices, and the dissolved fraction of the CYT differed depending on the matrix type. Moreover, the recrystallization of TST was observed in one of the acrylates. Interestingly, a DSC analysis of the acrylic patches did not confirm the presence of the API even if the microscopic images showed suspended particles.

Keywords: adhesive patches; silicones; acrylates; indomethacin; testosterone; cytisine; infrared spectroscopy (FTIR); Raman spectroscopy; microscopy; differential scanning calorimetry (DSC)

1. Introduction

Over the last 30 years, polymeric patches have been used for the transdermal delivery of a variety of active substances with the aim of achieving a local or systemic effect. Active pharmaceutical ingredients (APIs) such as nicotine, testosterone or diclofenac sodium are presented in such application forms [1,2].

Sufficient skin adhesion is assured by pressure-sensitive adhesive polymers (PSAs) which adhere easily to the skin surface under light pressure and are capable of staying in the area of application for up to several days, while at the same time, the patch can be easily and painlessly removed without damaging the skin [3,4]. Among the most commonly used PSA polymers, one should mention polyisobutylene and acrylates, with silicone matrices being increasingly investigated due to their excellent biocompatibility and low skin irritation potential. A quite broad selection of commercially available products can be found within all three chemical groups, but detailed information on the polymer structure and the excipients present in the material are either insufficient or undisclosed. Limited research focusing on the correlation between the polymer structure and its interaction with APIs is available, which, even in the case of controlled release formulations, makes the primary selection of the optimal polymer based on a lucky guess rather than on an evidence-based methodology.

Currently, the most popular are transdermal patches wherein the API is incorporated in the adhesive matrix, since the coating of the non-adhesive matrix with an adhesive layer involves a more complex technology [1]. API may be present in the polymeric matrix in a crystalline or amorphous form, or it can be dissolved. The dissolved fraction is critical to enable the diffusion of the API onto the skin surface from where it can be absorbed [5]. The high release and absorption rate may be retained due to the saturation of the matrix, which occurs in the presence of the coexisting suspended and undissolved fraction of the API, because the solute flux can be proportionally enhanced by increasing the thermodynamic activity of the drug in the vehicle according to the Higuchi equation [6]. The prediction of an API solubility in a polymer without experimental work is very difficult or impossible, especially taking into consideration the diversity of the polymer chain constructions, even within a single chemical group [7,8].

Considering the above, the evaluation of transdermal patches should consist of an investigation into the APIs' capability to dissolve in the matrix, as well as towards the determination of the particle size and distribution of the non-dissolved fraction, combined with detailed observations of drug–polymer interactions [9]. Among others, microscopic techniques were reported to be used in order to estimate the solubility of APIs [10–13]. The observation of the physical state of an API in the patch should enable determining the changes which might occur during storage or during its relatively long (up to several days) application on the skin. Additionally, tools for confirming the identity of the API in medicinal products are especially valuable nowadays, when incidents of counterfeit products are becoming more common.

This paper describes a comparative analysis of PSA polymers, namely silicone and acrylate matrices, intended for transdermal patch formulations. A number of spectroscopic and microscopic techniques were selected to visualize the effect of the matrix type on the fraction of the dissolved API. Previous works have been mostly limited to the evaluation of one type of polymer with a selected active substance, while a more comprehensive approach was hardly ever presented. For this work, two silicones and three polyacrylates were chosen due to their versatile physicochemical properties (the differences in the structure and properties described in Table 1). Three APIs of different lipophilicity and potential for transdermal delivery were investigated (Table 2).

The following microscopic techniques were employed: optical microscopy, scanning electron microscopy and microscopy with a Raman spectrometer. The microscopic observations were supported by DSC studies performed in order to determine the solubility of the API and the potential interactions of the drug with the polymer. In the choice of the matrix-forming polymer, the possibility of a simple identification technique of the polymer type and active substance as part of the product control must also be considered. The spectral evaluation of the selected PSA matrices, with or without an API, was carried out to demonstrate the possibility of using the FTIR and Raman spectrometry for the identification of medicated patches.

Table 1. Characteristics of the commercial polymers used for the preparation of adhesive matrices [14–16].

PSA Type	Acrylate			Silicone	
Symbol	A1	A2	A3	S1	S2
Brand name	DuroTak® 387-2287	DuroTak® 87-4098	DuroTak® 87-2852	Bio-PSA MD7-4502	Soft Skin Adhesive MG 7-9850
Film formation	Solvent evaporation				Polymerization
Structure and chemical name	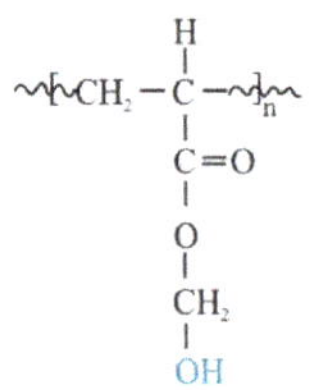 Acrylate-vinylacetate	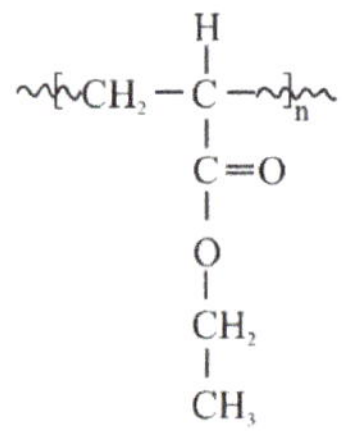 Acrylate-vinylacetate	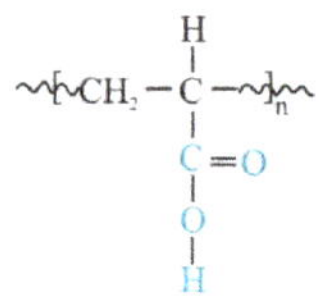 Acrylic acid	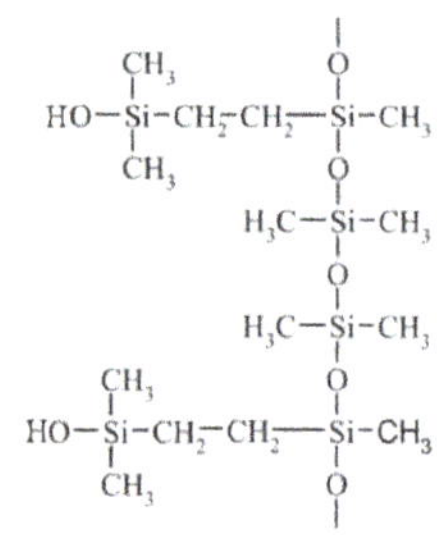	
Additives	Vinyl acetate	Vinyl acetate	Crosslinker	No data	Crosslinker
Solvent	Ethyl acetate	Ethyl acetate	Ethyl acetate, n-heptan, isopropanol, toluen, acetylacetone	Ethyl acetate	–
Content of solids (%)	50.5	38.5	33.5	60–65	100
Viscosity * (mPa*s)	18,000	6500	2500	1500	2900 **

* Viscosity of a polymer solution. ** Viscosity of part A and part B before the curing process.

Table 2. Characteristics of the investigated drug substances (APIs) [17].

	Indomethacin (IND)	Cytisine (CYT)	Testosterone (TST)
Chemical formula	$C_{19}H_{16}ClNO_4$	$C_{11}H_{14}N_2O$	$C_{19}H_{28}O_2$
log P/clog P	3.4–4.25	0.6–1.06	2.99–3.37
Molecular weight (g/mol)	357.8	190.2	288.4
Melting point T_m (°C)	158–162	152–153	155
Polymorphism	$\alpha, \beta, \gamma, \delta$, new forms: ε, ζ, η	No data	4 polymorphic forms
Solubility in water (mg/mL)	0.0024 (pH dependent)	8.14	0.0333
Particle size * (µm)	3.1 ± 2.6	6.6 ± 3.9	5.8 ± 2.6

*—average size (±sd) in the drug powder used in the experiments (experimental data).

2. Materials and Methods

2.1. Materials

The materials used for patch formulation were as follows. DuroTak® 387-2287 (A1), 87-4098 (A2), 87-2852 (A3) acrylic adhesives solutions were gifted by the manufacturer (Henkel, Brussels, Belgium). Standard Silicone Adhesive solution Bio-PSA MD7-4502 (S1) and two-part (A&B) silicone elastomer Liveo™ Soft Skin Adhesive MG 7-9850 SSA (S2) were provided gratis by DuPont (Brussels, Belgium). Polyethylene (PE) membrane (Esselte, Warsaw, Poland) was used as a backing layer and fluoropolymer-coated Scotchpak® 1020 as a release liner (3M, Neuss, Germany). APIs were: indomethacin IND (Sigma–Aldrich, Steinheim, Germany); testosterone TST (Ipca Laboratories, Mumbai, India); and cytisine CYT (Xieli Pharmaceutical, Sichuan, China). Characteristics of the polymers used for the adhesive matrices are presented in Table 1 and APIs are characterized in Table 2.

2.2. Preparation of the Patches

IND, CYT or TST were added to the adhesive polymer solutions A1, A2, A3 and S1 in a concentration of 5% (*w*/*w*) based on the polymer dry weight. Each blend was then mixed in a planetary mixer (ARE-250 Thinky, Tokyo, Japan) for 4 min at 2000 rpm and defoamed for 2 min at 2200 rpm, before being transferred to high vacuum for 5 min to remove the remaining air bubbles. In the case of the S2 matrix, part A was preliminarily mixed with the APIs at 500 rpm for 2 min, and then part B containing a crosslinker was added, mixed and de-aired as described above. The blends were then casted on the PE membranes using a lab coater (Camag, Berlin, Germany) with the gap adjusted to obtain a dry patch thickness of approximately 150 µm. The step of solvent evaporation for A1, A2, A3 and S1 films was carried out in a drying oven at 40 °C for 3 h and then for 12 h at 23 ± 1 °C. The curing process (cross-linking of S2) was conducted at 23 ± 1 °C for 12 h in a high vacuum. The prepared films were covered with a release liner, packed into polyethylene bags with a tight zipper closure, stored in a controlled environment (25 ± 1 °C, RH 60%) and analyzed within 7–30 days. The thickness of the obtained films was accurately measured with an infrared gauge MiniTest 730 (Electro Physik, Cologne, Germany).

2.3. Raman Spectroscopy and Confocal Raman Microscopy

A WITec Alfa300 Access microscope with Raman spectrometer (WITec, Ulm, Germany) equipped with a 785 nm laser was used to acquire spectra of the polymeric adhesive patches and APIs. Mainly Zeiss objectives (×10, ×50, ×100) were used for the measurements, with a laser power of 60–80 mW. Data analysis was performed using WITec Project Plus software.

Raman spectra of the drug-loaded acrylate and siloxane polymeric films were collected and compared with the spectra of reference APIs and pure polymer films, taking into account non-overlapping peaks.

A Raman imaging mode was employed to visualize the API distribution in polymeric matrices. Zeiss 50×/NA 0.75 was used for Raman spectroscopy acquisition. The laser was focused on the adhesive patch surface to avoid backing layer interference. In each sample, three randomly chosen 50 µm × 50 µm areas were scanned with a 1 µm step size with an exposure time of 1 s. The Raman maps were processed to reduce non-chemical effects.

For a given API-loaded patch, spectra were normalized to the maximal intensity of the API's characteristic/signature peak. To generate intensity maps, the signature peaks for each API were selected as non-overlapping with the peaks of the polymer matrix, and a center point of a 10 cm^{-1} range was used. For CYT, a signature peak at 1210 cm^{-1} was chosen for all investigated polymers. In the case of IND, 745 cm^{-1} was found to be suitable for polyacrylates, while the 1700 cm^{-1} band allowed to identify API in silicone polymers. The TST spectra showed the 950 cm^{-1} peak as a signature in all polymers except for A2, where only the 1615 cm^{-1} peak did not interfere with the polymer spectrum.

2.4. Scanning Electron Microscopy (SEM) and Optical Microscopy

To assess the uniformity of particle distribution, crystallization and morphology of the particles, the microscopic images of the films surface were obtained using an optical microscope (Nikon Eclipse i50, Nikon Instruments, Tokyo, Japan). Zeiss objectives 10× were used for a general overview of the sample and the images within an in-depth range of 15 µm were obtained with Ph2 DLL 40× magnification lens. The images that were captured in a different axial (z) dimension were processed with NIS Elements Advanced Research 3.20 software (Nikon Instruments, Tokyo, Japan).

Details of the patches surface were observed with a scanning electron microscope (Phenom Pro Generation 5, Thermo Fisher, Eindhoven, The Netherlands) using an in-line detection mode at 5–10 kV, with a backscattered (BSD) or secondary electron detector (SED). Samples of approximately 5 × 5 mm were coated with a thin layer of gold in an ion-sputtering device (thickness 5 or 10 nm).

2.5. Fourier Transform Infrared Spectroscopy (FTIR)

FTIR spectra were obtained using Jasco-4700 instrument (4000–400 cm^{-1} with 32 scans, 4 cm^{-1} resolution; Jasco Company, Tokyo, Japan). Thin film method using salt (KBr) plates was applied by dissolving 10–20 mg of sample in 1–2 drops of solvent mixture (ethyl acetate 65%, isopropanol 19%, hexane 12%, toluene 3% and benzene 1% *v*/*v*) and placing one drop of the solution on one salt plate. After evaporation, the analysis was performed. The spectrum from a clean plate was recorded as a background and the analysis of spectra was performed using Spectra Analysis software (Jasco Company, Tokyo, Japan).

2.6. Differential Scanning Calorimetry (DSC)

DSC thermograms were obtained using the DSC-1 STARe System (Mettler Toledo AG, Schwerzenbach, Switzerland) combined with the intercooler system (HUBER TC 100) and the program STARe Evaluation Software version 16.30. All patches samples (10–13 mg) or pure APIs (0.5 mg) were sealed in flat-bottomed aluminum pans (40 µL). The indium calibration standard was used to calibrate the DSC instrument. A 75 mL/min nitrogen flow was used and the heating rate was 5°C/min. The analytical temperature range was from −40 °C to +350 °C. The thermogram recorded for an empty pan was treated as a baseline. The melting temperature (T_m) and glass transition temperature (T_g) values were determined as the midpoint of the endotherm and inflection in the DSC thermograms, respectively.

In an additional experiment, thermograms were recorded for a sample prepared by layering 0.5 mg of API next to the placebo acrylate or silicone patch, without mixing. The enthalpy (ΔH) was compared with enthalpy determined for the drug-loaded patches and for pure API (0.5 mg).

3. Results

3.1. FTIR Spectroscopy

Figure 1 shows the IR spectra of the tested placebo patches: three acrylic polymers and two siloxane polymers [18]. All characteristic bands according to chemical structure of the material were identified and presented in Table 3. Spectra of all tested polyacrylates are characterized by a wide band (wavenumber at 3700–3000 cm^{-1}) corresponding to the stretching vibrations of the hydroxyl group (free OH group from water and hydrogen bonds). In this region, both acrylic or silicone matrices present typical FTIR spectra bands between 3000 and 2800 cm^{-1}, corresponding to asymmetric C-H (CH_3); specific ranges for the carbonyl acrylic group C=O are observed at 1736 cm^{-1}; while vinyl group from silicone derivative structure CH_2 is visible at wavenumber 1409 cm^{-1}. Characteristic $-CH_2$, $-CH_3$ and −CH deformation vibrations (1450–1372 cm^{-1}) and ester group vibrations O-C for acrylate polymers are observed at 1121 and 1168 cm^{-1}. Silicone matrices present typical FTIR spectra corresponding to a C-H band for Si-CH_3 at 1255 cm^{-1}, Si-O-Si band at 1092 and 1023 cm^{-1}, Si-C from Si-CH_3 at 864 and C-H (CH_3) at 800 cm^{-1}. Other bands are not specific and their presence results from a solvent or other additives.

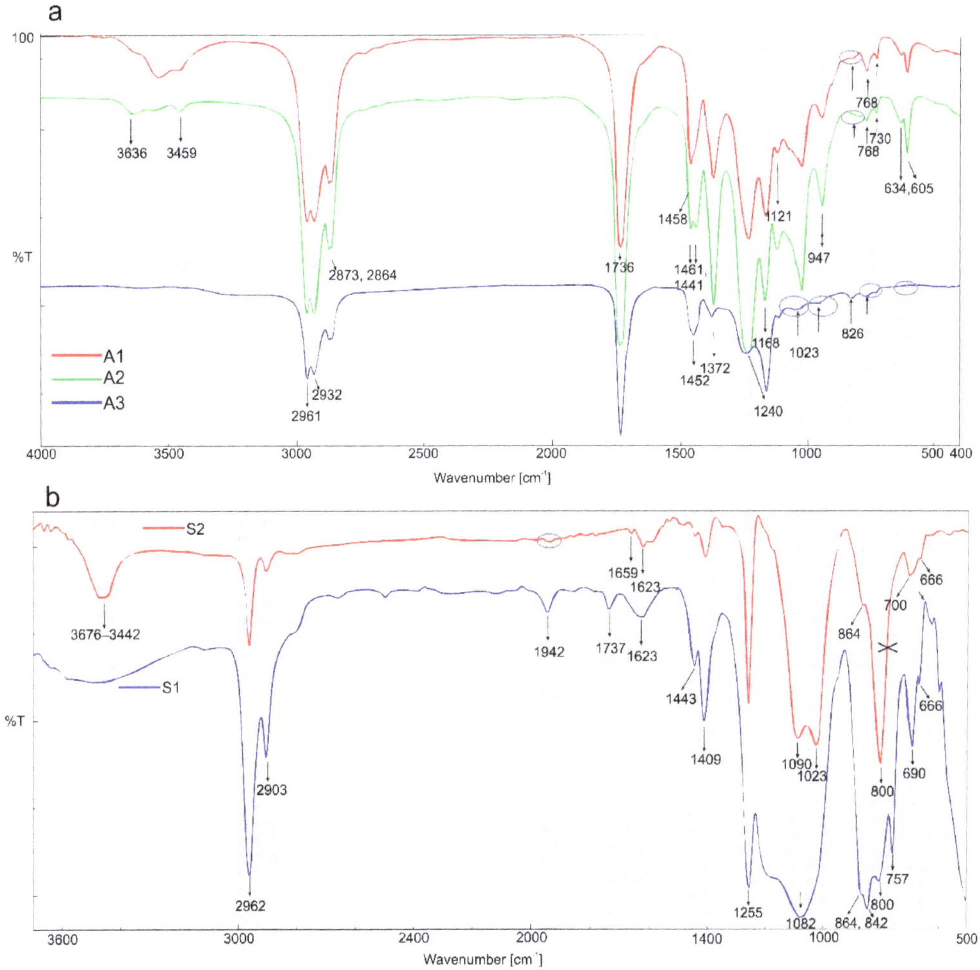

Figure 1. FTIR spectra of the investigated placebo patches: (**a**) polyacrylates; and (**b**) polysiloxanes. The grey bands mark differences or similarities between the polymers' spectra.

Table 3. FTIR identification of IND (a), CYT (b) and TST (c) in the polymeric matrices (bands or wavenumber ranges underlined are characteristic of API identification in respective patches and not present in placebo polymer film) [19–22].

(a)

Chemical Bond	IND 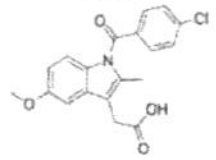	Acrylates						Silicones			
		A1	A1 + API	A2	A2 + API	A3	A3 + API	S1	S1 + API	S2	S2 + API
O-H intermolecular bonded	-	3538, 3459	3541, 3454	-	-	3448	3541, 3454	3485	3349	3474	-
C-H (CH_2) stretching alkane	3088, 2993, 2960, 2933, 2833	2956, 2927, 2875–2859	2956, 2927, 2875–2859	-	-	2956, 2927, 2867	2956, 2927, 2856	2961, 2902	2961, 2902	-	3088
C=O stretching vinylphenyl ester	-	1738	-	-	-	1735	-	1737	1728	-	-
Asymmetric acid –C=O ketone	1709	-	-	-	-	-	-	-	1709	-	1710
(benzoilo)-C=O amide	1685, 1590 1611–1593	-	1686, 1618–1592	-	1686, 1591	-	1602	1644, 1632, 1622	1687–1678, 1609–1590	-	1680, 1610–1592
C-H bending alkane methyl group	-	1462	1459–1414	1463, 1435	-	1465–1448	1454	-	-	-	-
–C=C-	1478, 1458, 1437	-	-	-	-	-	-	1442	1479, 1453–1442	1443	1479, 1458, 1436
C-H (CH_3)	1400	-	-	-	-	-	-	1414	1413–1401	1409	1411
O-H bending alcohol	1371,1357	1373	1373	1373	-	1380	1378–1353	-	1370, 1356	-	1370, 1357
–C-O- acidic group	1321	-	1321	-	1320	-	1321	-	1316	-	1319
–C-O-C-	1292, 1261, 1224	1238	-	1237	-	1260–1238	1258	1254	1254	1260	1260, 1224
–C-N- or/and C-O stretching ester	1178, 1148	-	-	-	-	-	-	-	-	1180	1175
	-	1166	-	1166	-	1164	1164	-	-	-	1145
O-H alcohol	-	1123	1123	1123	-	1116	116	-	-	-	-
–C- Cl-	1089, 1067	-	1086, 1068	-	1089, 1069	small 1079, broad 1043	1089, 1069	1076	1076	1093	1088, 1069
–C-O-C-	1036, 1015	-	-	-	-	1035	1035, 1015	-	-	-	1036
C-O stretching primary alcohol	-	1023	1023	1023	-	-	-	-	-	1024	1018
C-O vinyl acetate	995	-	-	-	-	-	-	-	-	-	-
	968	944	-	945	945	broad 964	-	-	-	-	-
–C-H-	926, 803, 692	-	-	-	-	-	926	946, 692	926, 692	660	926, 692, 662, 483
strong C-H bending	-	-	835	-	835	-	835	-	-	-	-

Table 3. *Cont.*

(b)												
Chemical bond	CYT 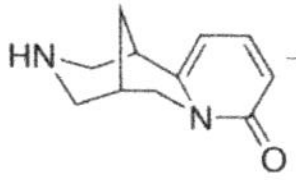	Acrylates						Silicones				
		A1	A1 + API	A2	A2 + API	A3	A3 + API	S1	S1 + API	S2	S2 + API
–C-Cl- and strong C-H bending	854, 833, 734	-	856, 803	800	853, 800	-	853	864, 842, 803	864, 842, 803	865, 800	866, 846, 800
C-H bending	754, 733	768, 728	755, 728	770	770, 754	772	770, 755	757	757	-	754, 560, 481
ν (N-H) intermolecular bonded/O-H	3420	3533, 3450	3533, 3449	3628, 3542, 3454	3628, 3542, 3454	3448	3450	3485	3349	3473	3405
C-H (CH_2) stretching alkane	2944, 2736	2960, 2927, 2875, 2857	2960, 2927, 2875, 2857	2953, 2935, 2879, 2863	2953, 2935, 2879, 2863	2956, 2927, 2867	2956, 2927, 2867, 2736	2961, 2902	2961, 2902	2964, 2902	2964, 2902
C=O stretching vinyl/phenyl ether	-	1737	1737–1726	1745–1725	1745–1725	1737	1737–1726	1736, 1625	-	-	-
N-C=O	1644	-	1649	-	1655	-	1644	1648	1641	-	1643
–C=C- cyclic alkene	1546	-	1546	-	1546	-	1546	1547	1547	-	1544
C-H bending alkane methyl group	1469, 1449, 1425, 1410	1460	1460	1462	1462	1465–1448	1465–1448	1447	1447	1442	1446, 1470
O-H bending alcohol	-	1380	1380	1372	1372	1380	1380	1411	-	-	-
–C-N- stretching	1358, 1348	-	-	-	-	-	-	-	1360	-	1357
C-O stretching ester	-	-	-	-	-	-	-	-	-	-	1307
–C-O-C stretching vinyl ether/–C-N-	1263, 1229	1236	1236	1255–1229	1255–1229	1260–1238	1260–1238	1259	1259	1262	1262, 1227
C-N stretching amine	1166, 1154, 1114	1167	1168	1171	1171	1164, 1116	1164, 1116	-	-	-	1151
C-O stretching ester	-	1123	1123	1123	1123	1117	1117	-	-	-	-
C-O stretching ester or C-N	1061	1025	1025	1023	1023	Small broad 1062–1026	1065–1027	1078	1078	1092, 1024	1092, 1062, 1025
C-O-	992	-	-	944	944	Broad 964	978	947	-	-	943
–C=C-	944, 805, 740	947, 728	947, 802, 728	794, 770, 727	770, 731, 802	-	806, 729	803, 757	803, 757	799	799, 740
–C-H bending	916, 865, 827	827	827	-	864	863, 827	865, 827	865, 690	865, 690	864	917, 864
-C=C-	613, 577, 543, 488	629, 606	606, 566, 486	605, 496	605, 570, 486	491	613, 581, 545, 488	599	599	662	662, 613, 575, 544, 487

Table 3. *Cont.*

(c)											
Chemical bond	TST 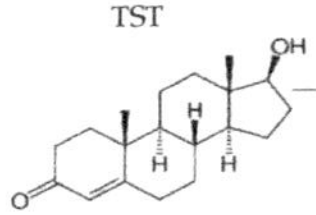	Acrylates						Silicones			
		A1	A1 + API	A2	A2 + API	A3	A3 + API	S1	S1 + API	S2	S2 + API
ν (O-H) intermolecular bonded	3411	3533, 3450	3533, 3449	3628, 3542, 3454	3628, 3542, 3454	3454	3454	3485	3610–3250	3473	3428
C-H (CH_2) stretching alkane	2940, 2871	2960, 2927, 2875, 2857	2960, 2927, 2875, 2857	2953, 2935, 2879, 2863	2953, 2935, 2879, 2863	2958, 2933, 2876	2956, 2927, 2867, 2736	2961, 2902	2961, 2902	2964, 2902	2964, 2902
C=O stretching vinyl/phenyl ether	1662	1737, 1678	1743–1725, 1678	1745–1725	1738, 1676	1737	1737	1853, 1736, 1619	1853, 1768, 1662	-	1661
–C=C- cyclic alkene	1614	-	1617	-	1617	-	1619	-	1613	-	1614
C-H bending alkane methyl group	1467, 1449, 1433, 1417	1460	1460	1462, 1437	1462, 1437	1452	1465–1448	1444	1447	1442	1446
O-H bending alcohol	1375	1373	1373	1372	1372	1378	1378	1412	1412	1412	1416
–C-H- stretching	1351, 1333	-	-	-	-	1334	1334	-	-	1378, 1350, 1330	1375, 1350, 1331
–C-O-C stretching vinyl ether/–C-N-	1272, 1230, 1131, 1056	1238	1244	1255–1229	1239	1260–1234	1260–1234	1254	1254	1259	1259–1230
C-O stretching ester	1189, 1113	1169, 1122	1169, 1122	1171, 1124	1171, 1124	1165, 1116	1116	1189	1189	-	-
–C=C- cyclic alkene	1018, 953, 941, 866, 831	1023, 827	1023, 867, 830	1024, 827	1024, 867	1051, 828,	1051, 867, 828	1076, 843	1076, 843	1089, 1024	1054, 1023
C-O-	956, 941	945	945	944	944	Broad 964	978	948	-	-	956, 941
–C-H bending	778	769, 729	769, 729	770, 728	770, 728	769, 729	769, 739	867, 690, 799, 757	867, 690, 799, 757	926, 864, 797, 702	864, 797, 702
–C=C-	684	630, 605	684, 630, 605	697, 630, 605	684, 630, 605	694	684	690, 599,	690, 599,	685	635
–C-C-	566, 512, 459, 438	491	512, 566	-	512, 566	568, 443	566, 512, 438	-	-	-	-

Infrared spectra of the tested patches containing 5% *w/w* of API are presented in Figure 2A–C. Each spectra of placebo polymeric matrices (acrylic A1–A3) compared to patches with APIs spectra are collected and presented in Supplementary Materials Figures S1–S5. In Table 3, all bands identified in the patches loaded with three different APIs are listed. Generally, all investigated polyacrylic patches allow one to observe the presence of the incorporated API. In the spectra of IND patches (Figure 2A), the following bands were characteristic for API: near 1618–1591 cm^{-1} (C=O benzoilo amide), at 1321 cm^{-1} corresponding to C-N vibrations, at 1089, 1068 and 835 cm^{-1} for C-Cl bond, and 754 cm^{-1} for C-H bending. For CYT in the patch, only three characteristic bands were observed in the spectra (Figure 2B), namely N=C=O at 1644 cm^{-1}, cyclic -C=C- at 1546 cm^{-1} as well as in wavenumber the range of 806–802 cm^{-1}. TST can be identified in acrylic matrices only based on the characteristic -C=C- cyclic alkene bands at wavenumbers: 1619–1617, 867, 684 and 566 cm^{-1} (Figure 2C).

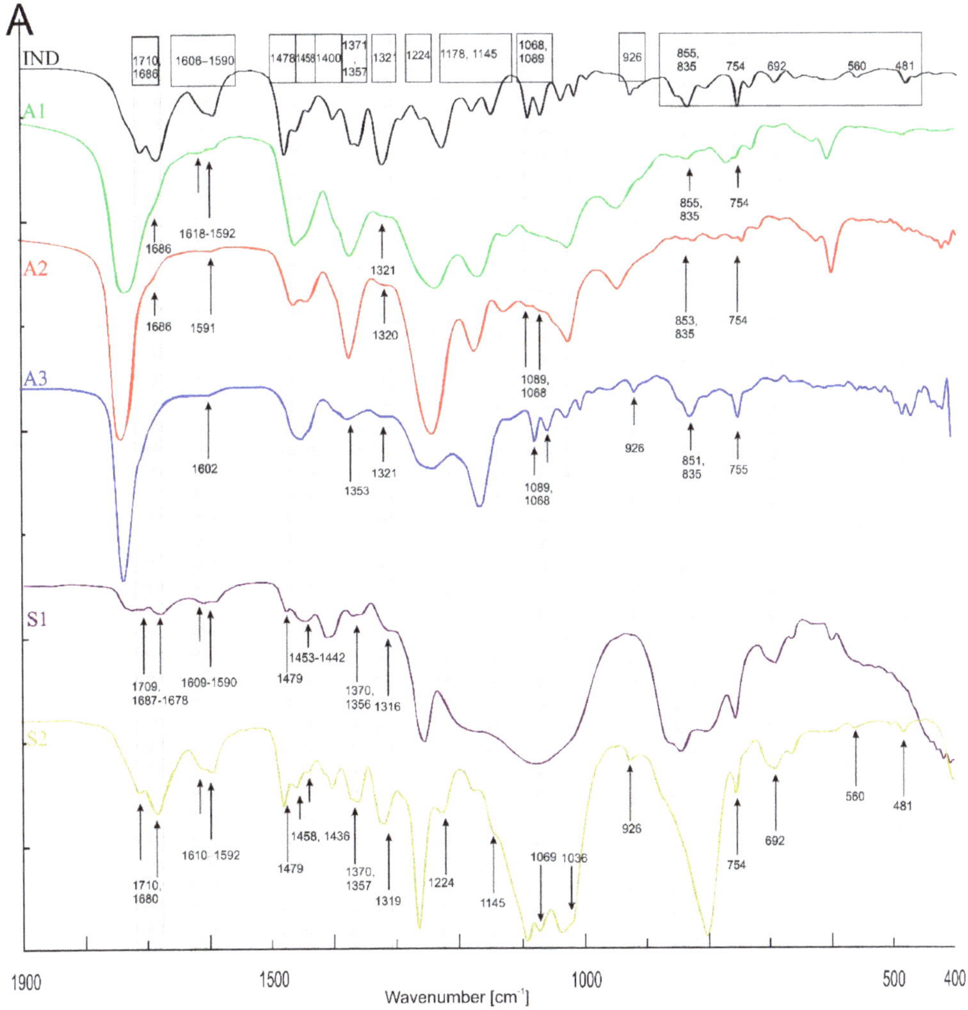

Figure 2. *Cont.*

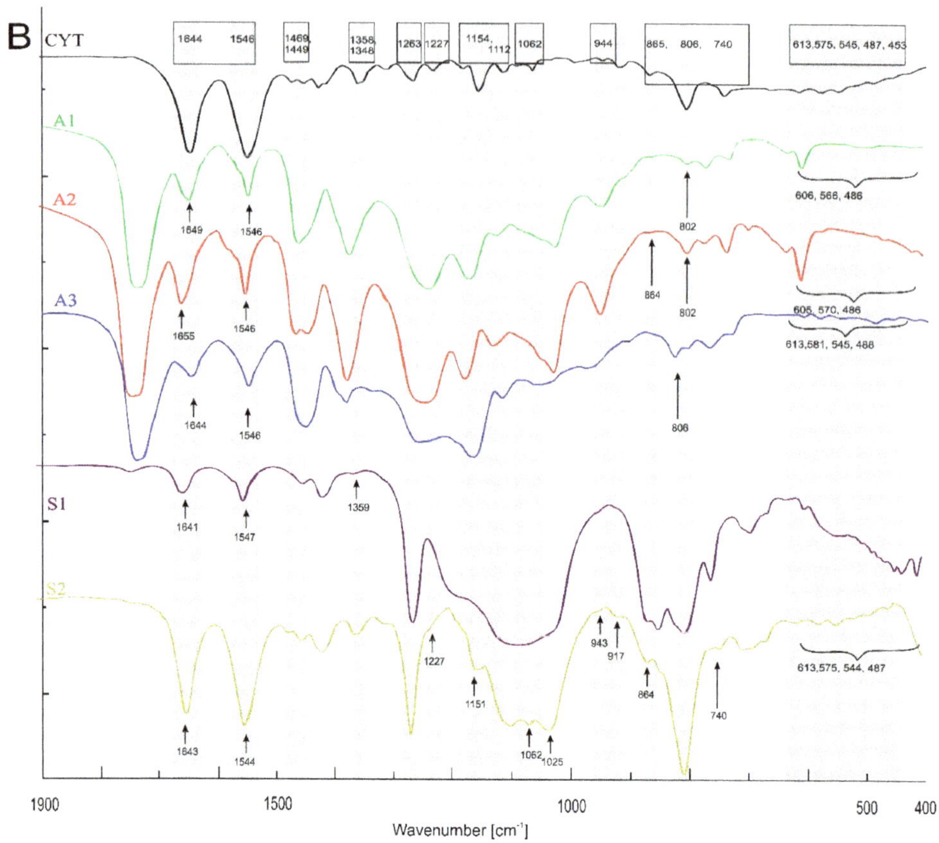

Figure 2. *Cont.*

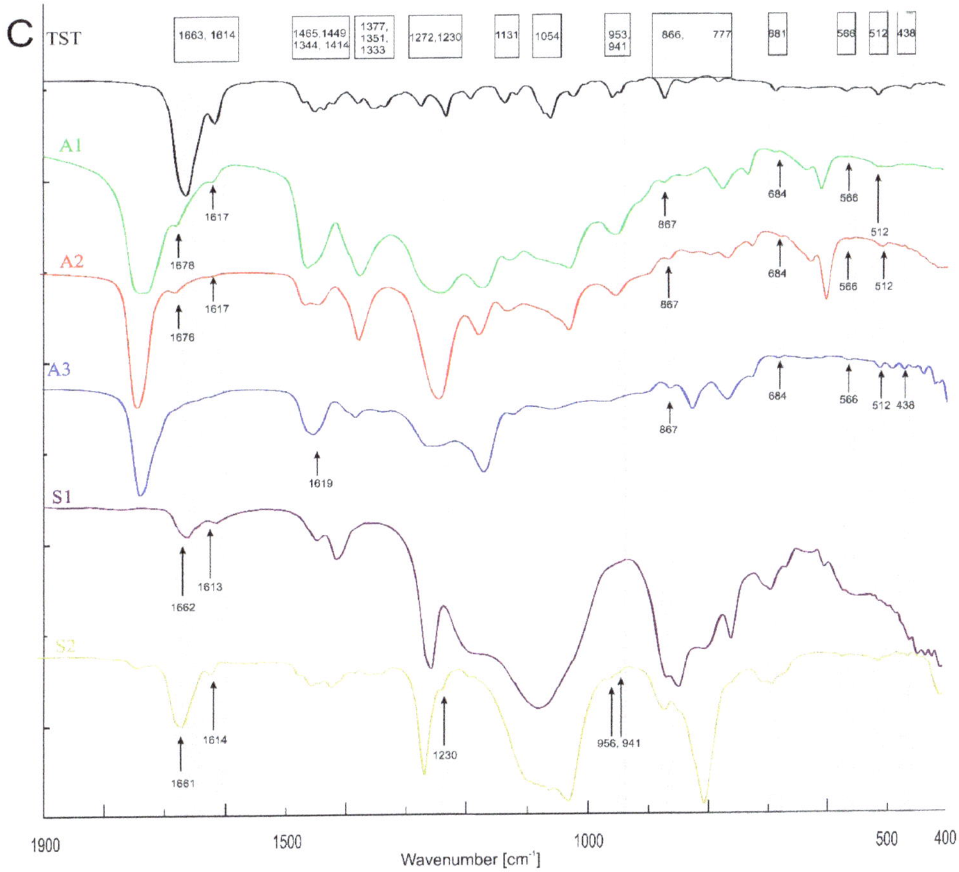

Figure 2. FTIR spectra of polymeric patches with API: (**A**) indomethacin, (**B**) cytisine, (**C**) testosterone.

3.2. Raman Spectroscopy

The Raman spectra are complementary to the FTIR spectra described above [23]. In Figure 3, spectra of placebo polymer matrices are presented. Generally, the stretching vibrations C=O at 1740 cm^{-1}, $-CH_2$, $-CH_3$ and $-CH$ deformation vibrations at 1446–1456 cm^{-1} as well as at 830, 633 (A2), 526 (A1) are observed in acrylates. Vibration of "additional" CH_2 group corresponds to the band near 633 cm^{-1} for A2 and S2 at 614 cm^{-1} observed also in the FTIR spectra.

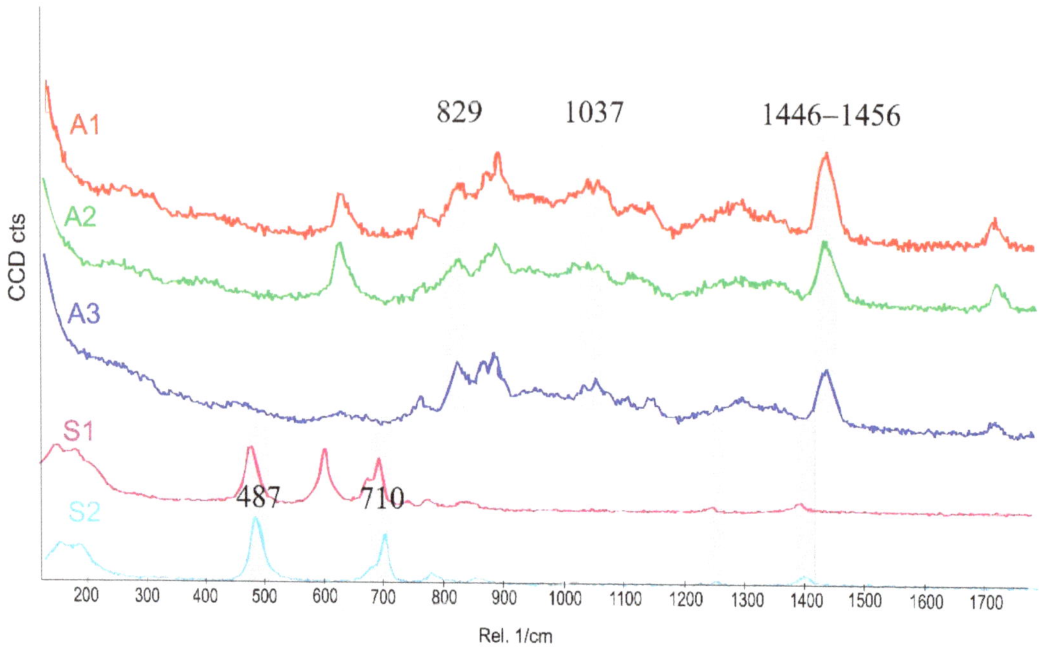

Figure 3. Raman spectra of placebo patches: polyacrylates (A1–A3) and polysiloxanes (S1, S2). The grey bands mark differences or similarities between the spectra of the polymers.

The data compiled in Figure 3 and Table 4 show that the spectra of polysiloxanes are less complex and have more potential space for the observation of the incorporated APIs. Practically, for all three APIs, all of the most characteristic bands are visible in the Raman spectra. Moreover, in the case of IND, an additional signal at 1698 cm^{-1} (characteristic for a benzoilo −C=O)—absent in polyacrylates—was identified. Notably, the Raman band at 487 cm^{-1} in S1 and S2 is derived from nanosilica particles embedded in the silicone matrix.

The characteristic bands of the investigated APIs in Raman spectra are also listed in Table 4a–c, together with the bands identified in the API-loaded patches (the spectra are presented as Supplementary Materials—Figure S6). In all acrylic patches, the IND signals were identified at 734–737 cm^{-1} for C-H bending. In the spectra of IND in A3 and A2, additional specific bands at 1358–1361 cm^{-1} (O-H group) and at 1220–1223 cm^{-1} corresponding to C-O-C vibrations were shown. For CYT in acrylates, there is -C-O-C stretching vinyl ether or −C-N- at 1207–1222 cm^{-1} and −C=C- near 613 and 659 cm^{-1}. TST is observed in the acrylic matrices due to its C-H- methyl group at 1611 cm^{-1}. This is the only signal of TST in A3. Additional signals in A1 and A2 with TST appeared near 1663 and 1000, 946 cm^{-1} corresponding to −C=C- cyclic alkene and C=C bending, respectively.

Table 4. Raman identification of IND (b), CYT (a) and TST (c) in the polymeric matrices (peaks or wavenumber ranges underlined are characteristic for API identification in respective patches and not present in placebo polymer film) [19–22].

(a)

Chemical bond	CYT 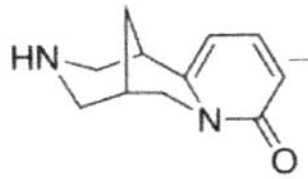	Acrylates						Silicones			
		A3	A3 + API	A1	A1 + API	A2	A2 + API	S1	S1 + API	S2	S2 + API
C=O stretching vinyl/phenyl ether	-	1731	-	1733	-	1734	-	-	-	-	-
N-C=O	1644	-	-	-	-	-	1644	-	-	-	
–C=C- cyclic alkene	1540, 1561	-	-	-	1543	-	1540, 1561	-	1546	-	1561, 1541
C-H bending alkane methyl group	1419, 1468	1450	1457	1449	1460	1445	1419, 1468	-	1470, 1445	1407	1468
O-H bending alcohol	-	1312	1380	1200–1380	1311	1200–1380	-	1412	1416	-	1415
–C-N- stretching	1344	1250–1400	1358	1200–1380	1274	1200–1380	1344	-	-	-	-
-C-O-C stretching vinyl ether/–C-N-	1263, 1223, 1207	-	1218	1200–1380	1210	1200–1380	1263, 1223, 1207	1257	1266, 1226, 1207	1262	1262, 1223, 1210
C-N stretching amine	1153	1154	1165, 1151	1146	1159, 1142	-	1153	-	1152	-	1155
C-O stretching ester	1110	-	-	1115	1107	1123	1110	-	1110,	-	-
C-O stretching ester or C-N	1011, 1035, 1056, 1077, 1092	1060	1035	1041, 1061	1094, 1080, 1062, 1042, 1015	1061, 1020	1011, 1035, 1056, 1077, 1092	-	1087, 1057, 1039, 1011	-	1094, 1078, 1056, 1036, 1014
C-O-	992	-	978	-	983	-	992	-	-	-	978
–C=C-	977, 735	890, 772	806	-	797, 741, 717, 739	895, 877	977, 735	790, 709,	978	793	793
–C-H bending	905, 860, 818	875, 832	838	821, 891	908, 868, 821,	830	905, 860, 818	854	908,	865	905, 864, 818, 790, 737,
–C=C-	716, 655, 613, 574, 539, 511, 485, 455, 376, 352	-	454, 525	625, 766	659, 616, 578, 540, 489, 455, 381, 357	630, 430–330	716, 655, 613, 577, 539, 511, 485, 455, 376, 352	691, 613, 486	714, 656, 615, 575, 485, 377	707,688, 488	715, 614, 575, 529, 488, 378, 354

Table 4. *Cont.*

(b)											
Chemical bond	IND 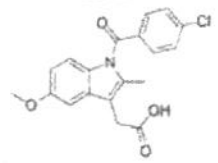	Acrylates						Silicones			
		A3	A3 + API	A1	A1 + API	A2	A2 + API	S1	S1 + API	S2	S2 + API
C=O stretching vinyl/phenyl ester	-	1731	-	-	-	1731	1731	-	1728	-	-
Asymmetric acid -C=O ketone	1698	-	-	-	-	-	-	-	1697	-	1699
(benzoilo) –C=O amide	1620, 1589	-	-	-	-	-	1586	-	1619, 1587	-	1619, 1589
C-H bending alkane methyl group	1467	1450	1456	1457	1447	1446	1443	-	-	-	1469
–C=C-	1438	-	-	1434	-	-	-	-	1438	-	1439
C-H (CH_3)	1396	-	-	-	-	-	-	1409	1410	1410	1394
O-H bending alcohol	1360	-	1358	1357	1359	1380–1240	1355	-	1360	-	1360
–C-O- acidic group	1311	-	1327	1327	1306	1310	1310	-	1310	-	1310
-C-O-C-	1263, 1221	-	1220	1217	-	-	1223	1265	1265, 1220	1265	1265, 1222
–C-N- or/and C-O stretching ester	1172	1160	1150	-	-	-	-	-	1170	-	1173
	1145	1149	-	1164, 1149	1148	-	-	-	1145	-	1148
O-H alcohol	-	1114	-	-	-	1116, 1120	1125	-	-	-	-
-C-Cl-	1068, 1087	1062	-	-	1060	1064, 1045	1064, 1045, 1086	-	1088	-	1089, 1069
–C-O-C-	1023	1045	1040	1033	-	1022	1022	-	1022	-	1023
C-O stretching primary alcohol	-	997	997	-	-	-	-	-	-	-	1018
C-O vinyl acetate	966	940–980	970	997, 972	-	-	-	-	966	-	967
–C-H-	926, 907	-	926	939, 897	-	-	-	-	908	-	907
strong C-H bending	-	879, 896	835, 896	840, 896	892	892	892	-	-	-	-
–C-Cl- strong C-H bending	840	829	807	805, 838	827	830	830	870–840	839	864	841
C-H bending	756, 738, 701, 630, 399, 271	772	736, 399	526, 452, 395	734, 628	633	630, 698, 737	786 754, 706, 687, 614, 490	755, 738, 700, 615, 487, 431, 411, 397, 271	790, 708, 490	756, 738, 700, 630, 617, 490, 432, 413, 397, 275

Table 4. *Cont.*

(c)										
Chemical bond	TST 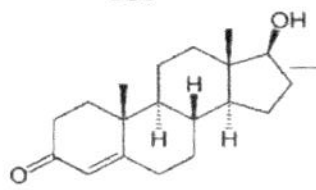	Acrylates						Silicones		
		A3	A3 + API	A1	A1 + API	A2	A2 + API	S1	S1 + API	S2	S2 + API
C=O stretching vinyl/phenyl ether	-	1735	-	1731	-	1734	1731	-	-	-	-
−C=C- cyclic alkene	1654	-	-	-	1663	-	1664	-	1655	-	1652
C-H bending alkane methyl group	1610	-	1611	-	1613	-	1612	-	1612	-	1611
O-H bending alcohol	1450	1454	-	1446	1446	1446	1447	-	-	-	1446
−C-O-C stretching vinyl ether/-C-N-	1349	-	-	-	-	-	-	1409	1410	1411	-
C-O stretching ester	-	-	-	1347	1326	-	-	-	1330	-	1331
−C=C- cyclic alkene	-	1310	-	1308	-	1308	-	-	-	-	1310
C-O-	1281, 1233	-	-	1238	1233	-	1231	1261	1283, 1232	1263	1280, 1233
−C-H bending −C=C-	1205, 1192	1159	-	-	-	-	1195	-	1151	-	1151
	1134	-	-	1143	1153	1120	1129	-	1131	-	1131
−C-C-	1113	1117	-	1118	1115	-	-	-	-	-	-
C-O stretching	1091, 1032	1047	-	1062	-	1063, 1020	1061, 1020, 1086	-	-	-	1091
C=C bending	998, 949, 912, 901, 868	966	-	892, 873, 826,	999, 949, 903, 870, 939	892	1001, 946, 894,	-	997, 947, 902, 869	-	998, 948, 902, 867
C-H bending	839, 746, 717, 681, 662, 541	896, 879, 835, 771	-	766, 628	748, 720, 684, 635, 544	831, 765, 631	829	852, 788, 707, 614, 487	837, 746	863, 790, 709, 490	837, 745, 712, 683

3.3. Optical Microscopy, the SEM and Raman Mapping

The microstructure of all five polymers mixed with three API was investigated by optical microscopy and compared with the API-free films, as shown in Figure 4. Each optical image was taken within 15 μm depth of focus, starting at the sample surface. The API-free films were transparent, and free from solid particles. Due to the roughness of the patch surface, the backscattered light effect could be observed as irregular spots on the optical microscope images. Acrylate patches demonstrated a much rougher surface, with multiple parallelly orientated "furrows", while silicones were smooth, as visible under a SEM.

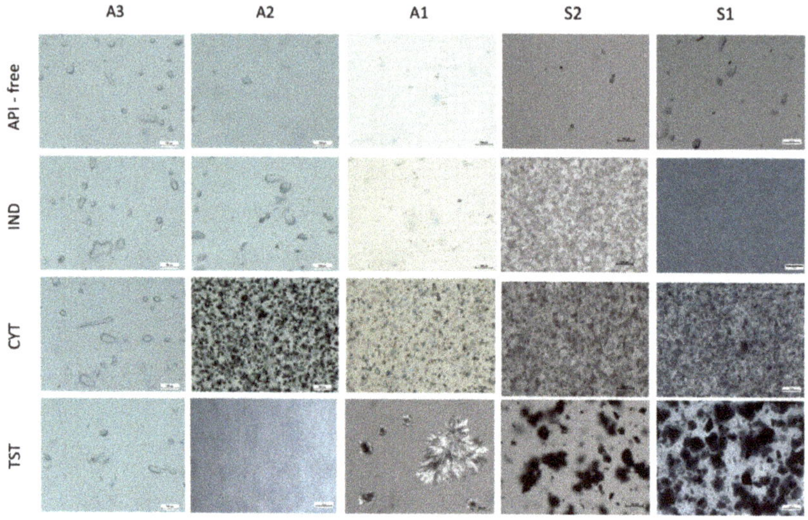

Figure 4. Optical microscope images of acrylate and silicone patches (scale 100 μm).

In the silicone matrices, the shape and size (below 10 μm) of the API particles did not change when compared with the particles of the raw material used for the preparation of the patches. It was only in the case of silicone patches with TST that the agglomerates were observed.

By contrast, the optical images of polyacrylic polymers showed diverse interactions, mainly dependent on the type of API. Only IND patches prepared with all three polyacrylates (A1–A3) were transparent, without visible particles. The material of patch A3 was the one which also enabled the dissolution of TST and CYT producing clear patches. In the case of CYT the patches, A1 and A2 showed the homogenous distribution of solid particles; however, the number of particles in CYT-A1 images appeared to be significantly smaller. No re-crystallization of CYT was observed. In contrast, during the first 24 h of storage, the TST re-crystallized in A1 and A2 patches: needle-shaped crystals were homogenously distributed in the A2 patch, while in A1, it formed large crystal aggregates.

A comparative analysis of the SEM micrographs with optical images allowed to obtain additional information on the location of the crystalline API. Only in the case of A2 with CYT, a considerable number of drug particles was located superficially, whereas the TST crystals present in A1 and A2 were not observed using the SEM (Figure 5) [24].

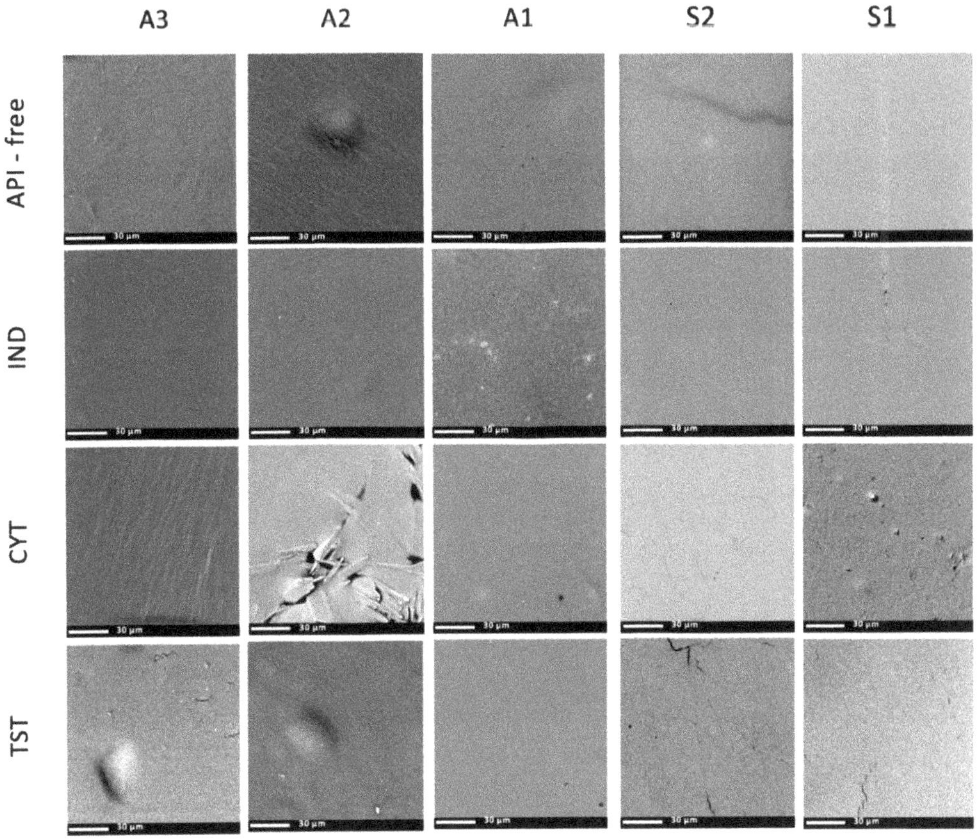

Figure 5. SEM images of acrylate and silicone patches (scale 30 μm).

The detailed study of the microstructure of the adhesive patches continued using a SEM, where the area of interest was the patch surface. As shown in Figure 5, the microstructure of a polymer patch in the presence of an API undergoes subtle but noticeable changes. Acrylate patches presented a much rougher surface, with multiple parallelly arranged "furrows", while silicones were smooth. On the other hand, pin-hole-like structures were only present in the patches prepared by solvent evaporation and were not spotted in the S2 matrix.

Detailed Raman maps were generated (Figure 6), where the intensity of the Raman signatures of IND, CYT or TST enabled the location of the regions with the API. In probing with a confocal system of a depth-of-focus of ca. 1–2 μm, an image of the thin surface layer of the patch was created.

In the case of polyacrylates, IND was found to be present on the patch surface forming a homogenously distributed layer of low intensity; a similar image was recorded for A3/CYT. Single particles, their diameters below 1 μm, unnoticed under the imaging techniques described above (SEM and optical microscopy), were occasionally identified. The CYT distribution demonstrated the most pronounced variation. CYT in the A2 resembled elevated particles observed under the previous imaging techniques, while for A1/CYT, the particles were found near the surface, probably still covered by the polymer, as proven by plain SEM micrographs (Figure 5). A noticeable decrease in the TST concentration in the polymeric matrix was found in the case of A1 and A2, which was due to crystal formation in the deeper patch layers. Especially in the A2, some regions devoid of TST presence in the surface layer were identified. Surprisingly, in the A3 patches, a considerable number of evenly distributed TST-rich areas were noted, which indicated the presence of TST particles sized below 1 µm, undetected by the other imaging techniques used.

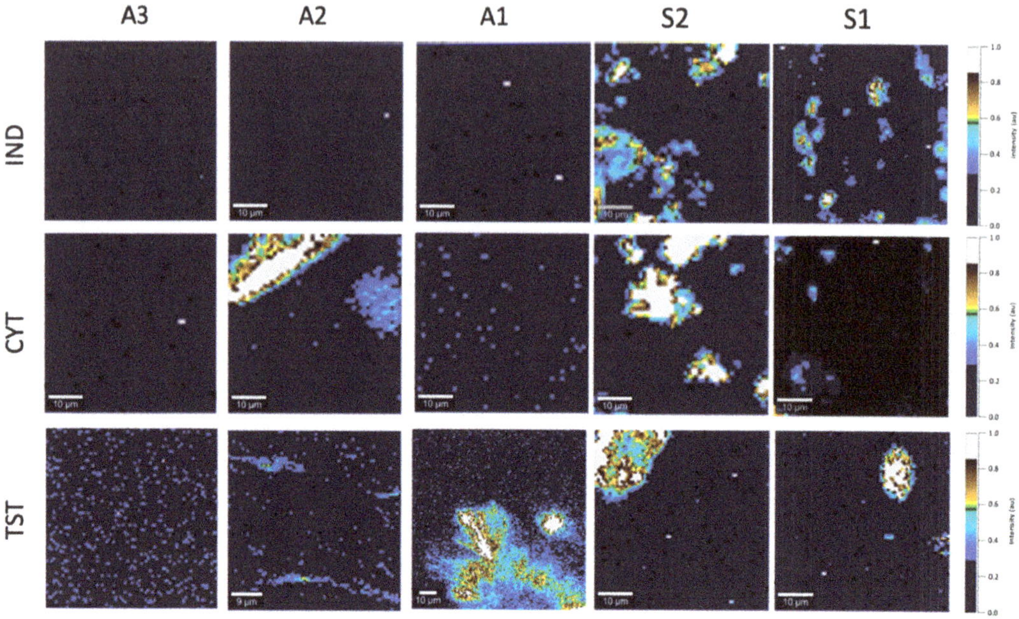

Figure 6. Raman image of indomethacin (IND), cytisine (CYT) and testosterone (TST) distribution on the surface of acrylate (A) and silicone (S) patches (scale 10 µm).

Silicone polymers exhibited the most noticeable fluctuations of the spectra: alongside stronger signals, from the sites corresponding to the API particles, a gradual drop in signal intensity was recorded in the areas surrounding the particles (the 'hallo' effect). Due to the dominating fraction of solid drug particles in the S1 and S2 silicone patches, numerous black spots representing drug-free areas were observed in the distribution maps (Figure 6). In the case of S1/CYT, an undetectable amount of dissolved API was observed superficially, with only single particles submerged in the polymeric matrix.

3.4. DSC

Clear endotherms for all investigated pure APIs were observed. The determined melting temperatures, T_m (Table 5), corresponded to the values reported in the literature [25]. Thermograms were also recorded for drug-loaded silicone patches (S1, S2) and endothermic peaks with T_m corresponding to T_m of pure API were obtained confirming the presence of IND, TST and CYT in the matrices. Sample thermograms for the S2 silicone patches are presented in Figure 7a, where a lack of any significant shift of the peaks compared to pure API indicated the absence of interaction between the incorporated API and the silicone polymers (Table 5).

Table 5. Melting temperatures of API recorded in thermograms of the tested silicone patches and the effect of API on glass temperatures recorded for acrylic patches.

	Placebo Patch	IND	CYT	TST
	Melting temperature T_m (°C)			
Pure API		160.9	156.1	154.4
A1, A2, A3		n.d.	n.d.	n.d.
S1		160.0	151.0	152.7
S2		160.1	155.9	153.4
	Glass transition temperature T_g (°C)			
A1	−26.0	−25.7	−26.7	−27.8
A2	−13.8	−17.6	−17.9	−15.3
A3	−14.3	−18.1	−17.6	−19.6

n.d.—not detected.

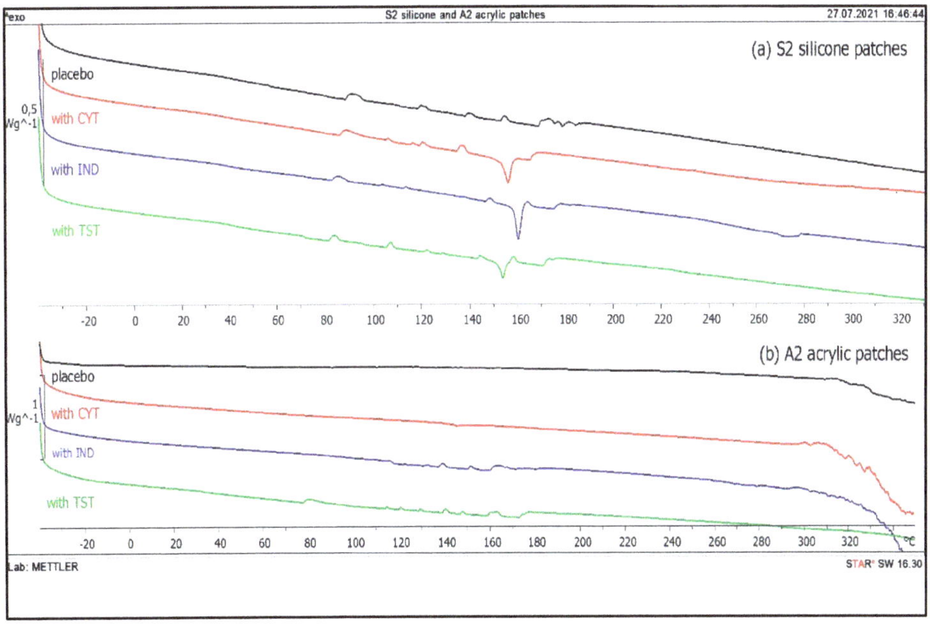

Figure 7. DSC thermograms of S2 silicone patches (**a**) and A2 acrylic patches (**b**): placebo (black), with CYT (red), IND (blue), TST (green).

Contrary to the above, melting endotherms associated with the investigated APIs were absent in the acrylate patches (Figure 7b and Table 5). The effect might have been predicted in the case of a matrix where the API was dissolved and microscopic observations did not reveal any visible particles (all A3 patches, and IND in A1 and A2 patches). Interestingly, although solid API particles were present in the acrylic patches with CYT and TST (Figure 4), endotherms were missing in their recorded DSC thermograms. In order to explain the reason of the low endotherm intensity of the API in silicone matrices and the lack of endotherm noted for acrylate patches, further investigation was performed, whereby 10 mg of API-free polymer was placed in aluminum pans with 0.5 mg of CYT (to simulate 5% w/w API content in the patch). The measured enthalpy of CYT (ΔH) was very low (ranging from −4.0 to −5.2 J/g) compared to the value measured for the same amount of pure CYT (−73.2 J/g). Moreover, no differences were observed linked to the presence of a particular polymer. In the Supplementary Materials (Table S1), more detailed results are presented, demonstrating that, by increasing the amount of API (CYT or IND) in the test pan, proportionally higher enthalpy was observed.

Further investigation consisted in the glass transition (T_g) determination of the acrylates, as presented in Table 5. Due to equipment-related limitations (operating temperature range: −85 °C–350 °C), the T_g of silicone matrices could not be determined (T_g below −120 °C [26]). A significant decrease in T_g of the A2 and A3 polymers was noted in the presence of all investigated APIs. Interestingly, the T_g of the A1 was found to be much lower compared to A2 and A3; moreover, no significant shift was observed in the presence of the API.

4. Discussion

Despite the fact that adhesive matrices have been used as a drug carrier in the transdermal delivery of APIs for almost four decades, surprisingly limited research has been dedicated to evaluating the advanced physicochemical parameters of those polymeric materials when combined with the API.

Various physicochemical properties of either the polymer or the API may influence the structure or appearance of the patches. At some technological stages, such as the evaporation of the solvent during the thin layer formation, uncontrolled re-crystallization may occur. Some effects may also result in changes in the microstructure of the polymer matrix. Three complementary methods: optical microscopy, scanning electron microscopy and Raman mapping were employed to study the appearance of adhesive patches with the emphasis laid on the physical state of the API and its distribution in the matrix.

The matrix PSA materials under investigation differed not only in the chemical structure, but also in the process of forming the adhesive films. The investigated acrylic PSA polymers (A1–A3) contained organic solvents (Table 1), which were evaporated in order to form an adhesive patch. Similarly, one of the silicone PSAs (S1) contained a solvent, while another one (S2) formed a matrix in a cross-linking reaction between its two components. It was expected that these differences might influence both the microscopic organization of the polymeric matrix and the solid state of the API in the final patch formulation.

When the patches were observed under optical microscopy, it was found that silicone patches were smooth and acrylate patches presented a much rougher surface. Such differences between the acrylic and silicone matrices may result from the different plasticity of the materials and the microdeformation of the surface of the acrylate patches occurring during solvent evaporation in the course of the preparation process. Acrylate may be more susceptible to stretching and shrinking due to polymer chain mobility. On the other hand, pin-hole-like structures were only present in the patches prepared by solvent evaporation.

4.1. Silicone Patches

In regard to the data obtained with the imaging techniques, all model APIs, independent of their lipophilicity or the silicone matrix type, were present in the S1 and S2 as suspended particles. The SEM revealed the location of the particles in the inner part of the polymeric matrix. From the shape and size of the particles, it was concluded that neither significant dissolution, nor other interference from the silicone matrix occurred. This also indicates that in the silicone patches, the dissolved fraction of the API capable of diffusing was inconsiderable. The residue solvent in the S1 was considered irrelevant, since neither API dissolution nor crystallization occurred, even though the latter effect has been suggested by other authors [27]. The distribution of the particles was noticeably homogenous with sufficient incorporation into the matrix, a fact confirmed by SEM, since no particles in the images of the patch surface were noted, even for TST, where the formation of agglomerates had been confirmed in optical micrographs. The mentioned agglomerates pose a problem in terms of the product quality and indicates the necessity of adding a levigating agent or a surfactant to the TST formulations, a step not needed for less lipophilic APIs.

Silicone matrices give IR and Raman spectra containing relatively few bands since the functional groups in this type of polymer are rather predictable. This makes it possible to easily identify the active substances present in the patch. It is noteworthy that, in contrast to the Raman spectroscopy, a direct FTIR analysis of APIs in the adhesive patches proved unfeasible. Hence, a method of preparing samples was proposed: before recording the spectra, the patch samples were dissolved in a mixture of organic solvents, the solution was placed on a salt plate, and a FTIR analysis was performed after evaporation.

Numerous bands were identified in the drug-loaded patches spectra that can be used for routine patch quality control. This is especially the case for S2, where almost all bands specific for the three APIs are visible (Figure 2A–C and Figures 3 and 4). For the S1 and S2 matrices, the wavenumber region between 1900 and 1300 cm^{-1} is the most suitable for analyzing the added substances. In this region, the IND bands at 1709 cm^{-1} (characteristic for the symmetric benzoilo amide −C=O group) and at 1478 cm^{-1} (the olefin group -C=C-) are observed. An interesting result of a comparison between all polymeric matrices is noticed for the −C-O-C- group of IND, because only S1 allows to observe it at 1224 cm^{-1}. Inversely, it does not allow seeing the −C-Cl- at 1068/1089 cm^{-1}. For CYT, the most characteristic bands of the N-C=O and −C=C- cyclic alkene group vibration are visible at 1644 and 1546 cm^{-1}, and CYT lends itself to much clearer identification in the S2, then in S1, as the following bands are sharply visible in the FTIR spectrum: at 1644 cm^{-1} (amine), 1546 (pyridone −C=O), 1166 (C-N), 1151 (C-N stretching amine), 943, as well as a group of bands: 613, 575, 544, 487 (−C=C- aromatic), and 806 (amine). On the other hand, TST is easily observed only at 1663 cm^{-1} (C=O stretching vinyl/phenyl ether) and 1617 cm^{-1} (−C=C- cyclic alkene) in both polysiloxanes patches, while other bands (from −C-O-C- or -C-O at 1230, 953 and 941 cm^{-1}) are hidden in the S1 but strongly manifested in S2.

Similarly to the FTIR analysis, in the Raman spectra, practically all characteristic bands are visible for all three APIs in the silicone matrices.

Raman mapping is considered especially useful for identifying APIs and excipients in solid state mixtures [28]. In this study, the patches were analyzed not only for the homogeneous distribution of the undissolved particles, but also with the aim of identifying regions with dissolved API. For example, the absence of CYT signals apart from the particles identified in the S1, with no such effect in the S2, would indicate that both silicone matrices differed significantly in terms of solubility of this hydrophilic drug substance. Silicone polymers exhibited noticeable fluctuations of the API spectra in microscopic images: in addition to stronger signals from the sites corresponding to the API particles, a gradual drop in intensity in the areas surrounding the particles was recorded. Such a 'hallo' effect can result from lower API concentration and may be attributed to the saturation of the matrix with dissolved API at the particle/matrix interphase; however, it could also represent the 'iceberg' effect.

The use of silicone matrices in which drug substances form a suspension also allows their easy identification using the DSC method. The endotherms of all three APIs in the silicone patches were visible on the thermograms and the lack of shifts signaled that there was no API interaction with the silicone matrix during the preparation of the patches. Moreover, the lack of marked T_g for the API also indicates the absence of amorphization due to the contact of the API with the matrix components.

4.2. Acrylic Patches

In contrast to the silicone matrices, greater differences were observed in acrylate matrices, both as a result of the differences in the properties of the active substances, and in the varieties of the polymer materials used. It can be concluded that, especially in the case of these materials, it is difficult to predict what physical system will be created as an effect of introducing the active substance. It was only in the case of IND that a solution system was formed in each of the three investigated matrices. It was demonstrated by the

DSC that in the case of the A2 and A3 matrices, the dissolved IND plasticized the matrix reducing the glass transition temperature, which is evidence of the API interacting with the matrix polymer.

In the case of the acrylic patches, where an API is dissolved in the matrix, the applicability of the microscopic techniques is limited to the evaluation of the re-crystallization of the API, as was noted for TST in the A1 (Figure 4). Although the only structural variation claimed for the A1 polyacrylate is the presence of the hydroxyl group in the polymer chain, it was considered insufficient to explain the stable dissolved state of the TST in the case of the A2 and A3 matrices. Since other additives are present in the polymer solutions, undisclosed by the manufacturer in terms of their quantity or type, those variations in composition may partially influence the solubility of the API in the final matrix. Optical microscopy allowed for not only detecting the undissolved TST particles, but also precisely demonstrating the shape of the crystals, which indicated the re-crystallization of the originally dissolved TST. Interestingly, a further detailed comparative analysis of the SEM micrographs with optical images allowed obtaining additional data on the location of the crystalline API. The examination of the patch surface using an SEM excluded the presence of the TST (Figure 5), which suggested the location of the crystals in the deeper layers of the matrix and proves that the re-crystallization of TST occurred not on the air/patch contact surface but in deeper layers of the polymeric matrix. The observation leads to the conclusion that re-crystallization was not initiated by external conditions (e.g., contact with air or mechanical stress) as previously observed for fentanyl patches [24]. One cannot rule out that the observed phenomena for TST in the acrylate patches indicate that the dissolved fraction is present in the concentration close to the saturated state, which should result in high thermodynamic activity and thus an optimal release rate [13].

The differences in the potential capability of the acrylate polymers to dissolve the APIs were most profoundly shown for CYT (Figures 4–6). All microscopic images showed that CYT was undissolved in the A2 and A1 matrices, but the number of particles in the CYT-A1 images appeared to be significantly smaller. This suggests the increasing solubility of CYT in polyacrylic matrices in the following order: A2 (undissolved) < A1 (partially dissolved) < A3 (dissolved). The complete dissolution of CYT was confirmed only in the case of the A3 matrix, which differs from A1 and A2 not only due to the presence of the COOH functional group in the polymeric structure, but also in that it contains a mixture of solvents (Table 1). CYT is the most hydrophilic substance among the examined APIs (Table 2) and in this case, the choice of proper polymeric material for the acrylic patch formulation is crucial for the development of a product with the API in an expected physical state.

The Raman maps provided some additional information on the presence of the API particles of the diameter below 1 μm, unnoticed by the SEM or optical microscopy. Such particles were occasionally identified even in the patches with IND which, on the basis of other microscopic techniques, were considered to contain the API in the dissolved state.

Compared to silicones, the acrylic polymers contain more functional groups, and the placebo patches give IR and Raman spectra with numerous specific bands, which makes the identification of the API in the patch by spectral methods more difficult. Nevertheless, FTIR and Raman spectra were obtained for each of the active substances tested, enabling API identification. The spectral analysis also distinguished three placebo matrices. In this case, the samples obtained after dissolving the patch and evaporating the solvent were also needed to record the FTIR spectra.

In the case of IND, the characteristic bands identified in the silicone patches at 1709 cm^{-1} and at 1478 cm^{-1} were absent for polyacrylates. Notably, the band near 1686 cm^{-1} was absent in the A3 matrix. Acrylic polymers also hide the vibration of IND's O-H bending alcohol due to its own vibration in this area of the spectrum. For CYT, the vibration of the most characteristic bands of the N-C=O and −C=C- cyclic alkene group was visible at 1644 and 1546 cm^{-1}, both in silicones and acrylates; however, the band at 1360 cm^{-1} (−C-N- stretching) present in the S1 silicone patch was fully covered in the acrylic patches. Similarly to the silicone patches, the TST band is easily observed at 1663 cm^{-1}.

Like in the FTIR analysis, all polyacrylic matrices show the presence of added APIs in the Raman spectra.

The data obtained using the DSC indicate a vast difference in the state in which API occurs in silicone and acrylic polymers. Despite the same low API content in the investigated matrices (5% *w/w*), crystalline form was confirmed only in silicone matrices. Even if the particles of suspended API were present, as in polymer A2 with CYT, or when the re-crystallization of TST in the A1 and A2 occurred, no API endotherm was noted. The lack of API peaks in the thermograms of the acrylic patches leads to the conclusion that a substantial amount of the API was dissolved in the acrylic matrices and the signals from undissolved particles were too weak.

It may also be concluded that due to the insufficient sensitivity of the DSC method and the excessively low API concentration, the particles visualized by microscopic techniques in the A1 and A2 were not identified in the form of endotherms. In similar cases of a decrease in peak intensity and enthalpy, only few researchers to date have addressed the issue and the suggested relevance of low API content [29], the loss of API crystallinity due to its incorporation in dissolved form [30], API dissolution in the matrix during the melting in the DSC analysis [31], or phase transition (e.g., amorphization) resulting from interaction with the polymer [32]. However, considering the above, the amorphization of the API was neglected in the case of all matrices, since no glass transition, T_g was noted on the thermograms for the investigated APIs.

In spite of the failure to identify the API in acrylic patches, the DSC technique appeared to be useful to observe the interaction between the API and the polymer. The effect of API on the physical state of the acrylic matrix was demonstrated by the changes in T_g of the polymers. A significant decrease in T_g in the A2 and A3 polymers was noted in the presence of all investigated APIs, which indicates the increased polymer chain mobility caused by the API plasticization effect [33–35]. Moreover, it was assumed that the lowest T_g of the A1 polymer ($T_g = -26$ °C) initially indicates its most elastic properties among the analyzed viscoelastic materials, and in this case, none of the APIs could further increase mobility of the polymer chain.

5. Conclusions

The present study indicates that the applied techniques are valuable for identifying APIs in transdermal adhesive patches, and potentially useful for the quality control of the medicated patches.

The modern microscopic and spectroscopic techniques enable the broad characterization of the adhesive matrices and the observation of the active substance distribution within their structure. Among the employed spectroscopic techniques, FTIR and Raman may be used not only as standard methods for API identification in the matrix, but also to distinguish commercially available polymeric materials of a similar chemical structure. The spectra of the silicone and polyacrylate polymers presented in this work may be especially useful as a reference whenever the information provided by the manufacturers of the materials is insufficient. The presented methodology is considered as an essential tool for the precise determination of the polymer identity. Moreover, the proposed procedure for the sample preparation for FTIR analysis was developed and optimized. The results of the performed spectroscopic analysis prove that it is useful to introduce the IR and Raman methods in the routine quality control of the patches because of the simplicity of the identification they offer and their uncomplicated methodology.

Although the evaluation of the API state in the matrix or the distribution, morphology and particle size of the undissolved fraction can be defined by optical microscopy, the surface-focused SEM can provide additional information on the three-dimensional location of the API, which can be helpful, e.g., in interpreting the release kinetics. As a single microscopic technique, the Raman microscopy not only allows producing images of the undissolved particles, but is especially valuable in the evaluation of the dissolved fraction distribution within the matrix.

The DSC was the only tested technique which failed to identify the API in the acrylic patches, even if undissolved API was present. However, in the case of silicone patches, where the API endotherms were recorded, it enabled the confirmation that the API was practically undissolved, while measurements of the T_g values of the acrylic matrix polymers indicated that dissolved API could act as a plasticizer in some types of matrices.

Supplementary Materials: The following supporting information can be downloaded at: https://www.mdpi.com/article/10.3390/polym14142888/s1, Figure S1. Infrared spectra of the acrylic patches A1—placebo and drug loaded: (a) IND; (b) CYT; and (c) TST. The grey bands mark the differences or similarities between the spectra of the placebo polymers and API's characteristic band. Figure S2. Infrared spectra of the acrylic patches A2—placebo and drug loaded: (a) IND; (b) CYT; and (c) TST. The grey bands mark the differences or similarities between the spectra of the placebo polymers and API's characteristic band. Figure S3. Infrared spectra of the acrylic patches A3—placebo and drug loaded: (a) IND; (b) CYT; and (c) TST. The grey bands mark the differences or similarities between the spectra of the placebo polymers and API's characteristic band. Figure S4. Infrared spectra of the silicone patches S1—placebo and drug loaded: (a) IND; (b) CYT; and (c) TST. The grey bands mark the differences or similarities between the spectra of the placebo polymers and API's characteristic band. Figure S5. Infrared spectra of the silicone patches S2—placebo and drug loaded: (a) IND; (b) CYT; and (c) TST. The grey bands mark the differences or similarities between the spectra of the placebo polymers and API's characteristic band. Figure S6. Raman spectra of polymeric patches with (A) indomethacin; (B) cytisine; and (C) testosterone. Arrows indicate the characteristic API bands also described by rectangular gray stripe. Table S1. The effect of placebo patches on the enthalpy values of endotherms of API.

Author Contributions: B.M.—Raman mapping analysis, SEM, optical microscope, review of literature, data assessment, cooperation in editing and preparation of the manuscript and preparation of figures and tables (30%); M.J.—work concept, FTIR, Raman analysis, review of literature, data assessment, cooperation in editing and preparation of the manuscript and preparation of figures and tables (30%); K.M.-P.—DSC analysis, data assessment, cooperation in editing and preparation of the manuscript and preparation of DSC figures and tables (20%); M.S.—substantive supervision over the concept of work, review of literature, editing and preparing the manuscript and interpretation (20%). All authors have read and agreed to the published version of the manuscript.

Funding: This research was supported by the Polish National Science Centre (NCN, grant number 2018/29/B/NZ7/01111).

Institutional Review Board Statement: Not applicable.

Informed Consent Statement: Not applicable.

Data Availability Statement: Not applicable.

Acknowledgments: The authors gratefully thank the following postgraduate students: Daria Butkiewicz, Anna Strzelec and Damian Rosłonowski for the valuable contribution to the film preparation. DuPont, Belgium, and Henkel, USA, are gratefully acknowledged for providing the authors with polymers.

Conflicts of Interest: The authors declare no conflict of interest.

References

1. Pastore, M.N.; Kalia, Y.N.; Horstmann, M.; Roberts, M.S. Transdermal patches: History, development and pharmacology. *Br. J. Pharmacol.* **2015**, *172*, 2179–2209. [CrossRef] [PubMed]
2. Davis, D.A.; Martins, P.P.; Zamloot, M.S.; Kucera, S.A.; Williams, R.O.; Smyth, H.D.; Warnken, Z.N. Complex drug delivery systems: Controlling transdermal permeation rates with multiple active pharmaceutical ingredients. *AAPS PharmSciTech* **2020**, *21*, 1–11. [CrossRef] [PubMed]
3. Cilurzo, F.; Gennari, C.G.M.; Minghetti, P. Adhesive properties: A critical issue in transdermal patch development. *Expert Opin. Drug Deliv.* **2012**, *9*, 33–45. [CrossRef] [PubMed]
4. Dasht Bozorg, B.; Banga, A.K. Effect of different pressure-sensitive adhesives on performance parameters of matrix-type transdermal delivery systems. *Pharmaceutics* **2020**, *12*, 209. [CrossRef]
5. Haq, A.; Chandler, M.; Michniak-Kohn, B. Solubility-physicochemical-thermodynamic theory of penetration enhancer mechanism of action. *Int. J. Pharm.* **2020**, *575*, 118920. [CrossRef] [PubMed]

6. Siepmann, J.; Peppas, N.A. Higuchi equation: Derivation, applications, use and misuse. *Int. J. Pharm.* **2011**, *418*, 6–12. [CrossRef] [PubMed]
7. Knopp, M.M.; Gannon, N.; Porsch, I.; Rask, M.B.; Olesen, N.E.; Langguth, P.; Holm, R.; Rades, T. A Promising New Method to Estimate Drug-Polymer Solubility at Room Temperature. *J. Pharm. Sci.* **2016**, *105*, 2621–2624. [CrossRef] [PubMed]
8. Kim, E.S.; Song, D.B.; Choi, K.H.; Lee, J.H.; Suh, D.H.; Choi, W.J. Robust and recoverable dual cross-linking networks in pressure-sensitive adhesives. *J. Polym. Sci.* **2020**, *58*, 3358–3369. [CrossRef]
9. Chadha, R.; Bhandari, S. Drug-excipient compatibility screening-role of thermoanalytical and spectroscopic techniques. *J. Pharm. Biomed. Anal.* **2014**, *87*, 82–97. [CrossRef]
10. Hassan, S.; Adam, F.; Bakar, M.A.; Mudalip, S.A. Evaluation of solvents' effect on solubility, intermolecular interaction energies and habit of ascorbic acid crystals. *J. Saudi Chem. Soc.* **2019**, *23*, 239–248. [CrossRef]
11. Medarević, D.; Djuriš, J.; Barmpalexis, P.; Kachrimanis, K.; Ibrić, S. Analytical and computational methods for the estimation of drug-polymer solubility and miscibility in solid dispersions development. *Pharmaceutics* **2019**, *11*, 372. [CrossRef] [PubMed]
12. Mikolaszek, B.; Kazlauske, J.; Larsson, A.; Sznitowska, M. Controlled Drug Release by the Pore Structure in Polydimethylsiloxane Transdermal Patches. *Polymers* **2020**, *12*, 1520. [CrossRef]
13. Kumar, A.; Singh, P.; Nanda, A. Hot stage microscopy and its applications in pharmaceutical characterization. *Appl. Microsc.* **2020**, *50*, 12. [CrossRef] [PubMed]
14. Available online: https://www.dupont.com/liveo/wearable-medical-devices-skin-contact-devices.html (accessed on 1 July 2022).
15. Czech, Z.; Kurzawa, R. Acrylic pressure-sensitive adhesive for transdermal drug delivery systems. *J. Appl. Polym. Sci.* **2007**, *106*, 2398–2404. [CrossRef]
16. Benedek, I.; Feldstein, M.M. (Eds.) *Handbook of Pressure-Sensitive Adhesives and Products: Three Volume Set*; CRC Press: Boca Raton, FL, USA, 2009.
17. MedicinesComplete Database, Martindale Complete Drug Reference. Available online: https://www.medicinescomplete.com (accessed on 1 July 2022).
18. Available online: https://www.henkel-adhesives.com/content/dam/uai/AIH/master/images/drug-delivery-polymers/durotak-gelva-production-guide-US-4pager-160920.pdf (accessed on 1 July 2022).
19. Ramírez-Hernández, A.; Aguilar-Flores, C.; Aparicio-Saguilán, A. Fingerprint analysis of FTIR spectra of polymers containing vinyl acetate. *Dyna* **2019**, *86*, 198–205. [CrossRef]
20. Parhi, R.; Padilam, S. In vitro permeation and stability studies on developed drug-in-adhesive transdermal patch of simvastatin. *Bull. Fac. Pharm. Cairo Univ.* **2018**, *56*, 26–33. [CrossRef]
21. Minaeva, V.A.; Minaev, B.F.; Hovorun, D.N. Vibrational spectra of the steroid hormones, estradiol and estriol, calculated by density functional theory. The role of low-frequency vibrations. *Ukr. Biokhim. Zh.* **2008**, *80*, 82–95.
22. Available online: https://www.sigmaaldrich.com/PL/pl/technical-documents/technical-article/analytical-chemistry/photometry-and-reflectometry/ir-spectrum-table (accessed on 1 July 2022).
23. Lee, D.C.; Webb, M.L. *Pharmaceutical Analysis*; CRC Press: Boca Raton, FL, USA, 2003; pp. 254–259.
24. Xu, T.; Yilmaz, H.; Willett, D.R.; Strasinger, C.; Rodriguez, J.D.; Keire, D.A.; Wokovich, A.M. Raman mapping of fentanyl transdermal delivery systems with off-label modifications. *Analyst* **2020**, *145*, 953–962. [CrossRef]
25. Available online: https://pubchem.ncbi.nlm.nih.gov/ (accessed on 1 July 2022).
26. Benedek, I.; Heymans, L.J. *Pressure-Sensitive Adhesives and Applications*; Marcel Dekker, Inc.: New York, NJ, USA, 1996; p. 145.
27. Kunst, A.; Geoffrey, L. Effect of residual solvent in polymer adhesive matrix on release and skin permeation of scopolamine. *Int. J. Pharm.* **2015**, *491*, 42–48. [CrossRef]
28. Kann, B.; Windbergs, M. Chemical imaging of drug delivery systems with structured surfaces—A combined analytical approach of confocal Raman microscopy and optical profilometry. *AAPS J.* **2013**, *15*, 505–510. [CrossRef]
29. Jirapornchai, S.; Siripornpinyo, P.; Chaiprasit, S. Formulation, Characterization, and In Vitro Evaluation of Transdermal Patches for Inhibiting Crystallization of Mefenamic Acid. *J. Drug Deliv.* **2017**, 1–7. [CrossRef]
30. Jain, P.; Banga, A.K. Induction and Inhibition of Crystallization in Drug-in-Adhesive-Type Transdermal Patches. *Pharm. Res.* **2013**, *30*, 562–571. [CrossRef] [PubMed]
31. Alhijjaj, M.; Yassin, S.; Reading, M.; Axel Zeitler, J.; Belton, P.; Qi, S. Characterization of Heterogeneity and Spatial Distribution of Phases in Complex Solid Dispersions by Thermal Analysis by Structural Characterization and X-ray Micro Computed Tomography. *Pharm. Res.* **2017**, *34*, 971–989. [CrossRef] [PubMed]
32. Rouf Akram, M.; Ahmad, M.; Abrar, A.; Sarfraz, R.M.; Mahmood, A. Formulation design and development of matrix diffusion controlled transdermal drug delivery of glimepiride. *Drug Des. Dev. Ther.* **2018**, *12*, 349–364. [CrossRef]
33. Budd, P.M.; McKeown, N.B.; Fritsch, D. Free volume and intrinsic microporosity in polymers. *J. Mater. Chem.* **2005**, *15*, 1977–1986. [CrossRef]
34. Liu, C.; Hui, M.; Quan, P.; Fang, L. Drug in adhesive patch of palonosetron: Effect of pressure sensitive adhesive on drug skin permeation and in vitro-in vivo correlation. *Int. J. Pharm.* **2016**, *511*, 1088–1097. [CrossRef] [PubMed]
35. Siepmann, F.; Le Brun, V.; Siepmann, J. Drugs acting as plasticizers in polymeric systems: A quantitative treatment. *J. Control. Release* **2006**, *115*, 298–306. [CrossRef] [PubMed]

polymers

MDPI

Article

Antiseptic Polymer–Surfactant Complexes with Long-Lasting Activity against SARS-CoV-2

Vyacheslav S. Molchanov [1], Andrey V. Shibaev [1], Eduard V. Karamov [2,3], Viktor F. Larichev [2], Galina V. Kornilaeva [2], Irina T. Fedyakina [2], Ali S. Turgiev [2,3], Olga E. Philippova [1,*] and Alexei R. Khokhlov [1]

1 Physics Department, Lomonosov Moscow State University, 119991 Moscow, Russia; molchan@polly.phys.msu.ru (V.S.M.); shibaev@polly.phys.msu.ru (A.V.S.); khokhlov@polly.phys.msu.ru (A.R.K.)

2 Gamaleya National Research Center for Epidemiology and Microbiology of the Russian Ministry of Health, 123098 Moscow, Russia; karamov2004@yandex.ru (E.V.K.); vlaritchev@mail.ru (V.F.L.); kornilaeva@yandex.ru (G.V.K.); irfed2@mail.ru (I.T.F.); turgiev@ld.ru (A.S.T.)

3 National Medical Research Center of Phthisiopulmonology and Infectious Diseases of the Russian Ministry of Health, 127473 Moscow, Russia

* Correspondence: phil@polly.phys.msu.ru

Abstract: Antiseptic polymer gel–surfactant complexes were prepared by incorporating the low-molecular-weight cationic disinfectant cetylpyridinium chloride into the oppositely charged, slightly cross-linked polymer matrices. Three types of polymers were used: copolymers of acrylamide and sodium 2-acrylamido-2-methylpropane sulfonate; copolymers of acrylamide and sodium methacrylate; copolymers of vinylpyrrolidone and sodium methacrylate. It was shown that the rate of the release of the cationic disinfectant from the oppositely charged polymer gels could be tuned in a fairly broad range by varying the concentration of the disinfectant, the degree of swelling, and degree of cross-linking of the gel and the content/type of anionic repeat units in the polymer matrix. Polymer–surfactant complexes were demonstrated to reduce SARS-CoV-2 titer by seven orders of magnitude in as little as 5 s. The complexes retained strong virucidal activity against SARS-CoV-2 for at least one week.

Keywords: polyelectrolyte; hydrogel; surfactant; COVID-19; SARS-CoV-2; disinfectant; polyacrylamide

1. Introduction

The coronavirus disease (COVID-19), caused by severe acute respiratory syndrome coronavirus 2 (SARS-CoV-2), appeared in China at the end of 2019 [1]. By March 2020, the disease was raging throughout most countries of the world, and the World Health Organization (WHO) declared COVID-19 a pandemic [1,2]. By 24 May 2022, the pandemic had caused more than 6.3 million deaths, according to Worldometers [3].

The fight against the infection is complicated by the high mutation rate of the virus. This requires introducing changes into the composition of the vaccines, which cannot be achieved fast enough. In addition, no specific therapy was proposed that can address all variants and mutants of SARS-CoV-2 at once. These factors make early prevention of the infection crucial to control its spreading. The preventive measures should comprise, in particular, the use of effective and human-friendly antiseptic coatings that restrict contact transmission through surfaces contaminated by respiratory droplets of the infected people [4]. It was reported that, at room temperature, SARS-CoV-2 could persist on different surfaces (such as plastic, stainless steel, metal, and glass) for up to 3–4 days [5–8]. Moreover, the virus was found on various surfaces in the cabins of the Diamond Princess cruise ship 17 days after infected passengers vacated [9]. Commonly used disinfectants, including alcohols and quaternary ammonium compounds, can efficiently inactivate the virus [10–15], but their action does not last long because they are easily washed away [16]. To provide continuous action, they should be reused frequently, which can cause undesirable toxicological effects [2,17,18]. An alternative possibility for ensuring prolonged antiseptic activity would be to use polymers as carriers for disinfectant molecules [2,19]. Polymer matrices can provide a prolonged release of the disinfectant in a controlled fashion [19].

Of particular interest is to use polymer matrices composed of anionic polyelectrolyte gel that is able to form complexes with oppositely charged disinfectants—quaternary ammonium compounds. These compounds are believed to exert antiviral effects by disrupting the protective lipid coating [20,21] of enveloped viruses (all coronaviruses, including SARS-CoV-2, belong to enveloped viruses [22]). The disruption may proceed through adsorption of quaternary ammonium compounds to the negatively charged surface of the virus via electrostatic attraction, followed by the penetration of hydrophobic alkyl chain into the lipid bilayer of the envelope. An intact viral envelope is a prerequisite to virus interaction with its receptors on the target cells, which makes possible penetration therein and initiation of the infection [23]. Thus, envelope perturbation/disruption is a way to inactivate SARS-CoV-2 and prevent infection [24].

Many quaternary ammonium compounds were shown to be quite efficient against SARS-CoV-2; those include cetylpyridinium chloride (CPC) [24–26], benzalkonium chloride (BAC) [6,13,15], di-N-didodecyldimethylammonium bromide, and di-N-decyldimethylamm onium chloride [14]. For instance, CPC that has already been FDA approved for use in commercial healthcare products [27] demonstrated virucidal activity against SARS-CoV-2 both in vitro [24,25] and in vivo [26]. BAC was used for the fabrication of face masks [28] and face shields [29], capable of inactivating SARS-CoV-2. Note that the above-mentioned quaternary ammonium compounds are cationic surfactants since each contains long hydrophobic tail(s) in addition to a cationic group.

Being introduced into a polymer–gel matrix, the molecules of a disinfectant can be gradually released over extended periods, providing a long-term antiseptic activity [19,30–34]. It was shown that the presence in the gel of units with a charge opposite to that of the disinfectant increases its load significantly and favors prolonged release [30].

Many efforts are directed toward the development of antiseptic systems based on polyelectrolyte gels and oppositely charged disinfectants [19,30–34]. For example, Ref. [19] describes gels based on poly(4-vinylbenzyl chloride-co-acrylic acid) and poly(sodium 4-styrenesulfonate-co-glycidyl methacrylate) copolymers modified by covalent or electrostatic binding of the biocides (4-vinyl benzyl dimethylhexadecylammonium chloride and cetyltrimethylammonium 4-styrenesulfonate, respectively). It was shown [19] that the system with the electrostatically bound cationic surfactant is more active against *S. aureus* and *P. aeruginosa*. In another paper [31], cross-linked multilayer films with a pH-triggered release of an antibacterial agent were prepared from poly(methacrylic acid) and an oppositely charged antimicrobial peptide derived from lactoferrin. In a recent study [34], microgels based on an interpenetrating network of poly-N-isopropylacrylamide/polyacrylic acid, loaded with the oppositely charged disinfectant BAC were elaborated. The microgels show antiseptic efficacy against *Bacillus subtilis* and *S. aureus*, which is approximately equal to that of commercial antibiotics.

Most of the reports on polymer gel/disinfectant systems have focused on antibacterial properties. Studies of antiviral activity of such materials are scarce [35,36], most of them dealing with uncross-linked polyelectrolytes. For instance, in Ref. [37], a water-soluble antiseptic complex of poly(N-vinylpyrrolidone-co-crotonic acid) with BAC was proposed, which was shown to effectively destroy nonenveloped and enveloped viruses (coronaviruses, rotaviruses, adenoviruses, AH1N1 influenza virus). In a recent report [38], a water-insoluble complex of polystyrene sulfonate with BAC exhibiting anti-SARS-CoV-2 activity was elaborated. To the best of our knowledge, no complexes based on cross-linked polyelectrolyte gels were tested for the virucidal activity against SARS-CoV-2. At the same time, cross-linking provides many advantages for producing carriers for antiseptic agents. For instance, cross-linked materials do not lose polymer during use (e.g., as a result of dissolution) and, therefore, can be fully rejuvenated by reloading with the disinfectant.

The aim of the present study is to develop antiseptic coatings with long-lasting activity against SARS-CoV-2, which are based on negatively charged polyelectrolyte gels with an embedded model antiseptic—the quaternary ammonium compound CPC [39–41] (Chart 1). Three types of negatively charged polyelectrolyte gels were used (Chart 1): poly(acrylamide-co-sodium 2-acrylamido-2-methylpropane sulfonate) (PAAm-AMPSA), poly(acrylamide-co-sodium methacrylate) (PAAm-MA), and poly(vinylpyrrolidone-co-sodium methacrylate) (PVP-MA). Such systems offer promise as coatings for various surfaces in public areas (hospitals, shopping malls, public transport, etc.).

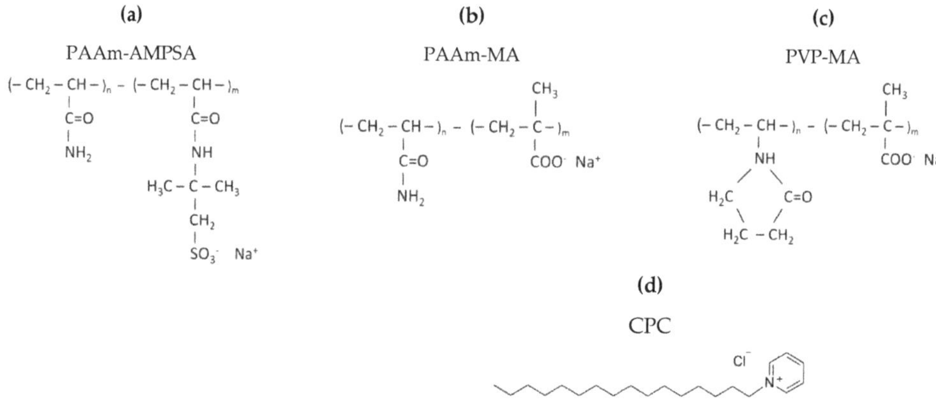

Chart 1. Chemical structures of anionic polyelectrolytes: (**a**) poly(acrylamide-co-sodium 2-acrylamido-2-methylpropane sulfonate) (PAAm-AMPSA), (**b**) poly(acrylamide-co-sodium methacrylate) (PAAm-MA), (**c**) poly(vinylpyrrolidone-co-sodium methacrylate) (PVP-MA), and (**d**) chemical structure of cationic surfactant cetylpyridinium chloride (CPC).

2. Materials and Methods

2.1. Materials

Acrylamide (AAm) (purity > 98%), 2-acrylamido-2-methyl-1-propanesulfonic acid (AMPSA) (purity > 99%), methacrylic acid (MA) (purity > 99%), ammonium persulfate (purity > 98%), N,N,N′,N′-tetramethylethylenediamine (purity > 99%), 1-vinyl-2-pyrrolidone (VP) (purity > 99%), and CPC (purity > 98%, critical micelle concentration (CMC) in water 0.9 mM [42–44]) were provided by Sigma Aldrich (St. Louis, MO, USA). N,N′-methylenebisacrylamide (purity > 99%), azobisisobutyronitrile (purity > 98%), and polyethylene glycol (PEG) with a molar mass of 1000 g/mol were obtained from Fluka (New York, NY, USA). Isopropanol was purchased from Acros Organics (Geel, Belgium). All chemicals were used as received. The solutions were prepared using distilled deionized water obtained by MilliQ system (Millipore, Burlington, MA, USA).

Vero E6 cells (ATCC, Manassas, VA, USA; catalog number CRL-1586) were cultured in high glucose Dulbecco's modified Eagle's medium (DMEM; Sigma-Aldrich, St. Louis, MO, USA) supplemented with 5% fetal calf serum (FCS), 2 mM L-glutamine, and antibiotics (150 u/mL penicillin and 150 u/mL streptomycin) at 37 °C in 5% CO_2.

The stock of SARS-CoV-2 (strain HCoV-19/Russia/Moscow-PMVL-12/2020 (EPI_ISL_572398) isolated from a patient) was the culture liquid withdrawn from cultures of the infected Vero E6 cells.

2.2. Preparation of the Hydrogels

Hydrogels were synthesized by free-radical copolymerization of uncharged and charged comonomers and a cross-linker (N,N′-methylenebisacrylamide).

Synthesis of PAAm-AMPSA and PAAm-MA hydrogels was carried out in water. Monomers, cross-linker, initiator (ammonium persulfate), and PEG were dissolved in water in appropriate quantities. After that, the solution was degassed by purging argon for 5 min, and a catalyst (N,N,N′,N′-tetramethylethylenediamine) was added. Polymerization was carried out for 8 h at room temperature.

Synthesis of PVP-MA gels was performed in isopropanol. After the dissolution of monomers, cross-linker, and initiator (azobisisobutyronitrile), polymerization was carried out for 2 h at 70 °C under constant purging of argon through the reaction medium.

After synthesis, the gels were put into a large volume of 10^{-4} M NaOH and then distilled water for 12 h to remove unreacted chemicals; the outer solution was changed a few times. In the case of PEG-containing gels, the outer solution had the same PEG concentration as the gels in order to prevent PEG release from the gel. For PAAm-AMPSA gels, the sol fraction was less than 0.5%, as determined by weighing the dried samples. For PVP-MA gels, the sol fraction was larger (~30%), which may be due to the lower reactivity of VP monomers as compared to other vinyl monomers [45].

The degree of swelling of the gels was determined according to the formula [46]: $\beta = (m_{sw} - m_0)/m_0$, where m_{sw} is the mass of the swollen gel and m_0 is the mass of the dry gel.

2.3. Loading of the Hydrogels with Disinfectants

To load the hydrogels with CPC, two methods were used (Figure 1). Method 1: the gels were immersed in an aqueous solution of CPC (the concentration of CPC was equal to 0.028 M or 0.14 M for the gels containing 2 and 10 mol% of charged units, respectively). Method 2: 0.1–0.5 mL 0.244 M aqueous CPC was poured onto the surface of the hydrogel. The hydrogels prepared using methods 1 and 2 were designated by letters v and s, respectively.

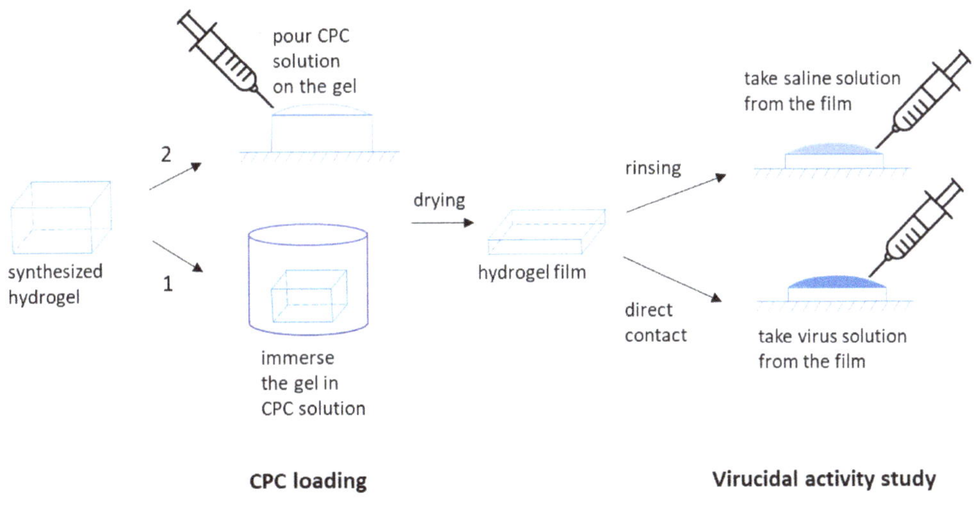

Figure 1. Schematic representation of the experiments on CPC loading and virucidal activity study.

Note that, in these experiments, we tried to avoid the formation of dense shells of polymer–surfactant complex, which are observed when the surface region collapses because of complex formation, whereas the central part of the gel remains in the swollen state [47–50]. Such collapsed shells release a considerable amount of CPC, so that the disinfectant is quickly exhausted, shortening the antiseptic activity of the material. We observed that the formation of collapsed shells could be avoided if the concentration of CPC solution used for loading is rather low (for instance, lower than 0.25 M for PAAm-AMPSA-4s hydrogel).

To reload the hydrogels with CPC after their full consumption, an appropriate volume of solution containing 7 wt% CPC and 25 wt% PEG was poured on their surface.

2.4. UV-Spectroscopy

UV-spectroscopy was used to follow the release of CPC from the hydrogel films. The experiments were performed with the SF-2000 spectrophotometer (OKB Spektr, Saint-Petersburg, Russia). CPC concentration was determined from the intensity of the absorbance peak of the pyridinium ring at 259 nm (Figure S1, Supplementary Materials). The intensity–concentration calibration curve was obtained by using CPC water solutions with known concentrations in the range 0.01–0.3 mM (Figure S2, Supplementary Materials). It gave the extinction coefficient of 4070 L/mol·cm, which is consistent with the literature data [43]. The solutions with CPC released from the polymer matrix were diluted 10–200 times with distilled water before the measurements in order to obtain the optical density D in the range from 0.1 to 1.

2.5. NMR Spectroscopy

^{1}H NMR measurements were carried out on a Bruker AV-600 spectrometer (Billerica, MA, USA) at 30 °C in D_2O as a solvent. The gel samples were degraded for 10 min by ultrasonic homogenizer Cole-Parmer CPX500 in order to break the gel, and then were put in standard 5 mm NMR tubes (Norell). The acquisition time of 2 s was used and 256 scans were performed. The data were processed using MestReNova software, including baseline, reference, and phase correction. ^{1}H chemical shifts were referenced to the HOD signal.

2.6. Mechanical Tests

Indentation load–unload tests of the hydrogel coatings were performed on an LS5 testing machine (Lloyd Instruments Ltd., Segensworth, Hampshire, UK) at room temperature. Load–depth curves were obtained on disk-like samples (diameter 30 mm, height 4 mm) with a 5 kN load cell and cylindrical indenter (diameter 8 mm) at the constant load–unload speed of 0.5 mm/min.

2.7. Release Studies

To study CPC release from hydrogel samples, two main approaches were used. In the first approach, a small volume of saline solution was poured onto the gel surface. In the second approach, the gels were immersed in a large volume of saline (exceeding about 100 times that of the polymer gel).

2.8. Antiviral Activity Tests

2.8.1. Virucidal Activity of CPC Solutions

Solutions of CPC (1 mL) were incubated at room temperature with an equal volume of the virus stock for 1 h. To avoid possible toxic effects of CPC on the cells (Table S1, Supplementary Materials), the samples (CPC + virus) were centrifuged at 27,000 rpm for 1 h. A positive control (the virus stock without CPC) was used in every run. Viral pellets were resuspended in 300 μL of support medium (DMEM, 1% FCS), and serial 10-fold dilutions in support medium were prepared for titer determination. The dilutions thus obtained were used to infect confluent Vero-E6 monolayers in 96-well plates; following 2 h of incubation, the inoculum was removed, and the plates were washed twice with FCS-free DMEM. Thereafter, the plates were filled with another type of support medium (DMEM, 2% FCS) and further incubated at 37 °C in 5% CO_2 for 96 h.

Virus-induced cytopathic effects (structural changes in the cells caused by the viral invasion) were assessed by microscopic examination of the wells; the percentage of the infected cells was used for virus quantification. The amount of the active virus was determined by endpoint dilution (titration) and expressed in fifty-percent tissue culture infective doses ($TCID_{50}$). The titer was calculated using the Spearman–Kärber method and presented as lg $TCID_{50}$/0.1 mL [51,52].

The virucidal activity of CPC was assessed by the difference in the virus titers (A) between control (A_c) and experimental (A_e) samples:

$$A = A_c - A_e$$

The protection index, or inhibition coefficient (IC), was calculated using the formula:

$$IC = [(A_c - A_e)/A_c] \times 100\%$$

2.8.2. Virucidal Activity of CPC Formulated into Gels

Two experimental setups were used to assess the efficacy of CPC formulated into gels (Figure 1).

(i) Washes from the surface of the gel: phosphate-buffered saline, pH 7.2 (450 μL), was applied dropwise onto pieces of CPC-containing gels, and placed in Petri dishes; after 5 s, the flowing liquid was collected and used for incubation with the virus (300 μL) and spectrophotometric measurement of the amount of CPC released from the gels as a result of the washing (residual volume, less than 150 μL). The virucidal activity of CPC washed off the surface of the gels was determined (as described above) 24 h (on day 1) and 168 h (on day 7) after the gels were prepared.

(ii) Direct contact of the virus with the surface of CPC-containing gels: 150 μL of support medium containing the known amount of the virus was applied onto the surface of the gel for 5 s, after which the flowing liquid was collected and the titer of the virus, determined as described above (in order to establish changes in the titer caused by the contact with the surface of the CPC-containing gel).

3. Results and Discussion

3.1. Preparation and Characterization of Antiseptic Polymer Coatings

To prepare antiseptic polymer coatings, three types of negatively charged copolymer gels were used: PAAm-AMPSA, PAAm-MA, and PVP-MA. They were prepared by free-radical copolymerization of uncharged and charged monomers, using N,N′-methylenebisacrylamide as a cross-linker. The degree of cross-linking was very low (Table 1), which facilitates the formation of coatings from these gels while avoiding the loss of polymer when in contact with water. The composition of the gels was confirmed by ^{1}H NMR (Figure 2 and Table 1).

Table 1. Composition of copolymer gels used for film preparation.

Samples	Total Concentration of Monomers, M	Fraction of Charged Monomer Units		Concentration of Cross-Linker, M	Degree of Cross-Linking	Concentration of PEG, wt%
		Feed	^{1}NMR			
PAAm-AMPSA-1v	1.4	0.02	0.019	0.00175	1:800	10
PAAm-AMPSA-1s	1.4	0.02	0.020	0.00175	1:800	10
PAAm-AMPSA-2s	1.4	0.02	0.018	0.00175	1:800	20
PAAm-AMPSA-3v	2.8	0.02	0.020	0.00280	1:1000	0
PAAm-AMPSA-4v	2.8	0.10	0.091	0.00280	1:1000	0
PAAm-AMPSA-5v	2.8	0.02	0.018	0.00560	1:500	0
PAAm-MA	2.8	0.02	0.017	0.00560	1:500	0
PVP-MA	2.8	0.10	0.30	0.014	1:200	0

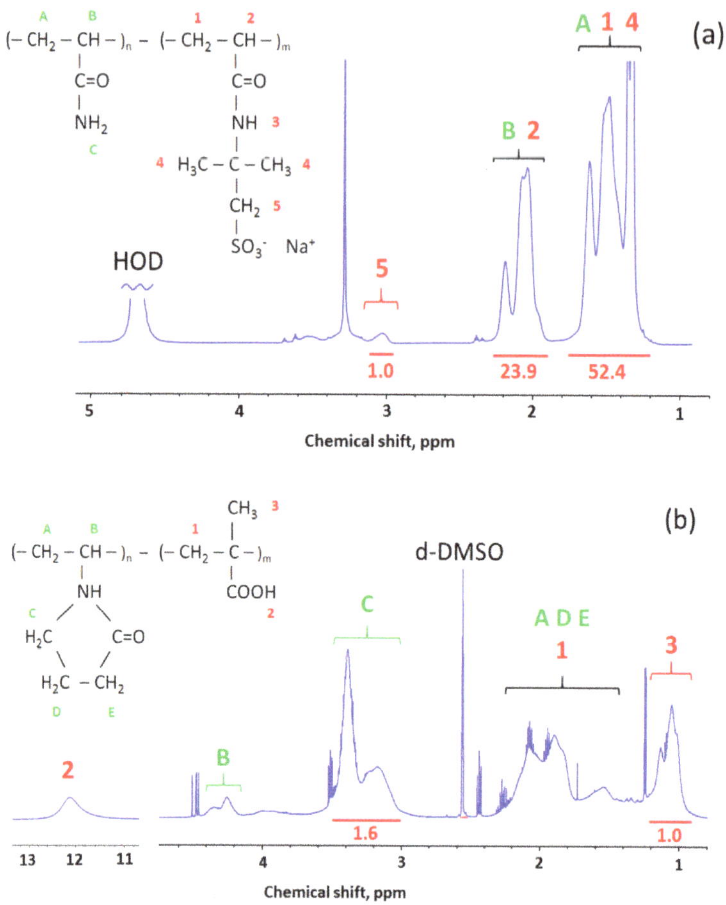

Figure 2. ^1H NMR spectra of synthesized copolymer gels: (**a**) PAAm-AMPSA in D_2O, (**b**) PVP-MA in d-DMSO. Peak assignments were made according to Refs. [53–55]. At the top, the peaks are labeled by letters (A, B, C, etc.) and numbers (1, 2, 3, etc.) corresponding to different protons in the polymers' chemical structure drawn in the Figure. At the bottom, peak integrals used for quantification of the polymers' composition are written under horizontal red lines.

The polymer gels were further modified by electrostatic binding of the cationic surfactant CPC acting as a disinfectant. The disinfectant loading was performed in two ways. In the first way, the whole gel sample was immersed in CPC solution. In the second way, CPC solution was applied only onto one surface of the gel (Figure 1).

The interaction of ionic surfactants with oppositely charged polyelectrolyte gels has been the subject of extensive studies, both experimental [42,43,47–49,56–58] and theoretical [49,59]. It was shown that surfactant ions penetrate the gel as a result of an ion exchange reaction with polymer counterions. If the initial charge density on the gel subchains is high, the within-the-gel concentration of the surfactant may considerably exceed its concentration in the outer solution. As a consequence, micelle formation will start within the gel in the absence of micelles in the outer solution. Note that the critical association concentration of surfactant within the oppositely charged gel is always 1–2 orders of magnitude lower than the CMC of the same surfactant in water [58]. Theoretically [59], the reason for this phenomenon was understood as described below. When the micelles are formed in the solution, some parts of counterions become condensed near their highly charged surface, losing their entropy of independent translational motion. By contrast, within the gel, the micelle charge is neutralized by ions that are initially immobilized on polymer chains so that no additional loss of translational entropy occurs.

When the amount of the bound surfactant ions is close to that of the charged polymer units, the volume of the initially highly swollen hydrogel decreases considerably, i.e., the gel collapse occurs [60,61]. The collapse is due to the decrease in the osmotic pressure within the gel, resulting from surfactant aggregation into micelles. The collapsed gel can further absorb the surfactant over the equimolar surfactant/gel charged units ratio corresponding to electroneutrality due to hydrophobic interactions. In this case, surfactant ions penetrate the gel together with their counterions, which causes the osmotic pressure to increase, leading to the reswelling of the gel [62].

The collapsed gels seem to be not suitable as antiseptic coatings. To avoid gel collapsing, we used CPC either in deficiency (CPC/charged gel units ratio 0.5) or in excess (CPC/charged gel units ratio 10).

3.2. Kinetics of the Disinfectant Release

The kinetics of CPC release from polymer coatings was studied under different conditions, using PAAm-AMPSA and PAAm-MA gels.

It is important for various practical applications to explore the release of disinfectant into small droplets of aqueous solution appearing on the surface, e.g., as a result of coughing/sneezing or touching by wet hands; therefore, to estimate the amount of CPC released, the gel samples were rinsed by a very small volume (0.2 mL) of 0.9 wt% aqueous NaCl solution. The amount of disinfectant passed into the solution at different contact times was determined by UV-spectroscopy using the absorption band of the pyridine ring of CPC at 259 nm.

Figure 3 shows CPC release from PAAm-AMPSA-4 gels loaded with the disinfectant by whole gel immersion (PAAm-AMPSA-4v) and surface application (PAAm-AMPSA-4s). In both samples, the CPC/AMPSA molar ratio was the same (0.5). One can see that the release is much faster from PAAm-AMPSA-4s than PAAm-AMPSA-4v: the amount of CPC released in 30 s equaled, respectively, 28 μg/mL and 100 μg/mL; therefore, 'surface incorporation' of CPC makes it possible to increase its release from the gel significantly. This is due to the accumulation of the surfactant in the vicinity of the gel surface that was exposed to water droplets. Furthermore, the lower CPC content on the opposite surface of the hydrogel sample improves its adhesion to various surfaces, e.g., glass or polypropylene; therefore, this way of loading provides two advantages: faster CPC release and better adhesion of the polymer coating to different surfaces.

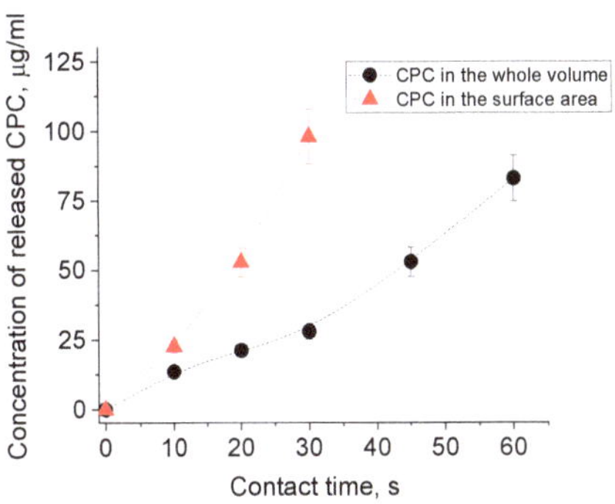

Figure 3. CPC release into a small volume (0.2 mL) of physiological saline from dried PAAm-AMPSA hydrogels with 10 mol% of AMPSA, which were filled with CPC by immersing them in CPC solution (PAAm-AMPSA-4v, black circles) or by applying CPC solution onto the surface of the gel (PAAm-AMPSA-4s, red triangles) at the CPC/AMPSA molar ratio equal to 0.5.

Further acceleration of CPC release can be achieved by increasing the content of CPC in the gel. All the above-described experiments were performed at the CPC/AMPSA molar ratio equal to 0.5, i.e., under the conditions of CPC deficiency relative to the number of oppositely charged polymer network units. In further studies, the CPC/AMPSA molar ratio was increased to 10. In this case, already in 2 s, the concentration of CPC in the water droplet reached 300 μg/mL, which far exceeds the concentration sufficient for complete SARS-CoV-2 inactivation (see Section 3.4). The amount of the surfactant released did not change on subsequent pouring of water onto the same place for at least 30 times during a period of 3–4 weeks; therefore, such polymer/surfactant complexes are very promising for the production of long-lasting antiseptic coatings.

One can expect that the amount of the released CPC can be finely tuned by varying other factors in addition to the initial CPC content in the gel. Those factors include characteristics of the polymer gel matrix such as the degree of swelling and the degree of cross-linking determining the gel pore size as well as the content/type of anionic groups responsible for the interaction with the disinfectant. To explore their effects on CPC release, the gel samples were immersed in a large volume of water (exceeding about 100 times that of the polymer gel) since it is difficult to measure the release profile accurately by using small water droplets.

Let us first examine the effect of the initial degree of swelling of the gel on the release of the disinfectant. The data for initially water-swollen and dried PAAm-AMPSA hydrogels are shown in Figures 4 and 5, respectively. The kinetic curves of CPC release from the water-swollen hydrogel (Figure 4b) demonstrate a $t^{1/2}$ time dependence of the fraction of the released surfactant that is characteristic of Fickian diffusion [63,64]. The approximation is valid up to approximately 60% of the released solute, which is in accordance with previously reported data for polymer systems with Fickian release behavior [63]. In Fickian behavior, the relaxation processes of the polymer are very slow in comparison to the diffusion rate of the solute inside the hydrogel [65]. Figure 5b demonstrates that CPC release from the initially dried PAAm-AMPSA hydrogel also obeys Fickian law at the early stage (up to approximately 50% of the released solute).

A comparison of Figures 4 and 5 indicates that CPC release proceeds faster from the swollen as compared to the initially dried PAAm-AMPSA hydrogel. For instance, in the case of PAAm-AMPSA-3v, 50% CPC was released in 25 min and 2.2 h from the swollen and initially dried gels, respectively. This behavior is expected because the pores of the swollen gel are open for release, whereas the dried gel should first swell to make possible CPC diffusion into the outer solution. A similar effect was observed for PAAm-MA gels (Figure S3, Supplementary Materials).

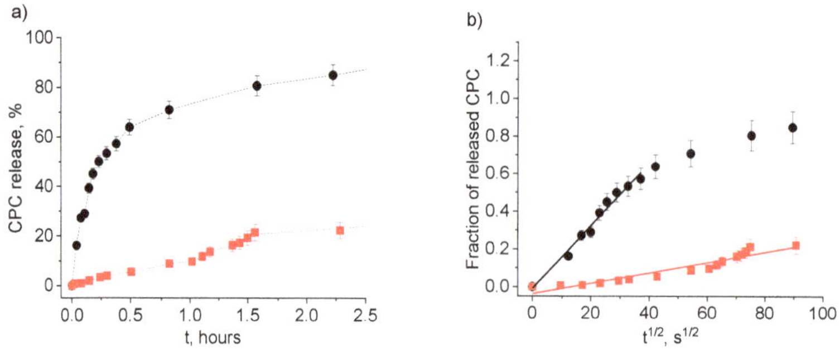

Figure 4. Release of CPC into physiological saline from water-swollen PAAm-AMPSA hydrogels containing 2 wt% AMPSA (PAAm-AMPSA-3v; degree of swelling, 16.0) (black circles) and 10 wt% AMPSA (PAAm-AMPSA-4v; degree of swelling, 17.1) (red squares) as a function of time t (**a**) and $t^{1/2}$ (**b**). In both samples, the CPC/AMPSA molar ratio is 0.5, and the degree of cross-linking is 1:1000.

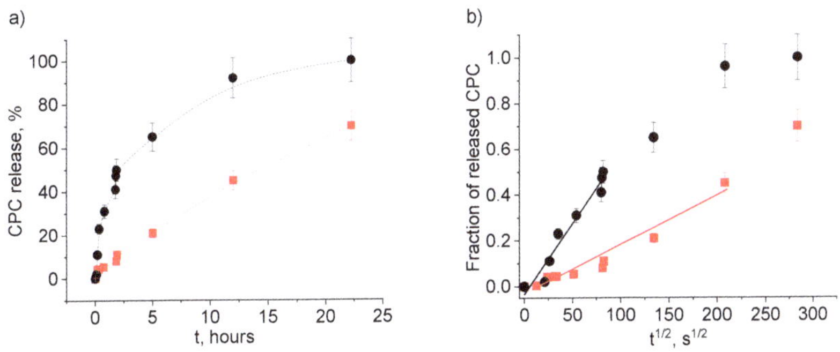

Figure 5. Release of CPC into physiological saline from dried PAAm-AMPSA hydrogels containing 2 wt% AMPSA (PAAm-AMPSA-3v, Table 1) (black circles) and 10 wt% AMPSA (PAAm-AMPSA-4v, Table 1) (red squares) as a function of time t (**a**) and $t^{1/2}$ (**b**). In both samples, the CPC/AMPSA molar ratio is 0.5, and the degree of cross-linking is 1:1000.

To accelerate the diffusion from the dried gel, one can add hydrophilic additives (e.g., PEG) that help to retain some water in the gel matrix (Figure 6). Drying of the PAAm-MA gels with no added PEG results in an almost complete loss of water after 4 h; however, in the presence of PEG, the gels retain 25% and 20% of water after 8 h of drying at 20 °C and 50 °C, respectively. The mechanical properties of dried gel coatings were studied using a cylindrical indenter with a diameter of 8 mm, imitating finger touches (Figure S4, Supplementary Materials). It was shown that in the presence of PEG, the coating is much softer: the indentation penetration depth of 30% is achieved only at 1 N load (instead of 54 N load in the absence of PEG). The load–unload test shows an 80% reversible (elastic) deformation for PEG-containing samples and only a 5% reversible deformation for neat samples without PEG (almost fully plastic deformation). High elastic deformation of the PEG-containing coating ensures a tighter contact between the human finger and touch surface, which should increase the effectiveness of the virucidal efficacy.

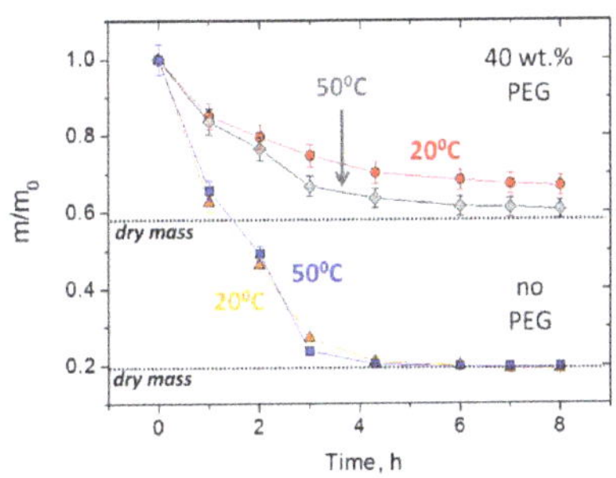

Figure 6. Loss of mass upon drying of PAAm-MA (10 mol%) gels containing no PEG (triangles—drying at 20 °C, squares—50 °C) or 40 wt% PEG400 (circles—drying at 20 °C, diamonds—50 °C). The initial gel samples in the form of disks with 0.5 mm thickness and 10 mm diameter were deposited on a glass substrate during drying.

Another parameter, which is expected to affect the rate of release, is the gel cross-linking density [66,67]. Figure 7a (full and dashed black curves) shows that increasing the degree of cross-linking slows down the rate of CPC release, as expected. Such behavior, observed previously for many hydrogels [65,67], was attributed to a smaller network mesh size and a less flexible hydrogel structure.

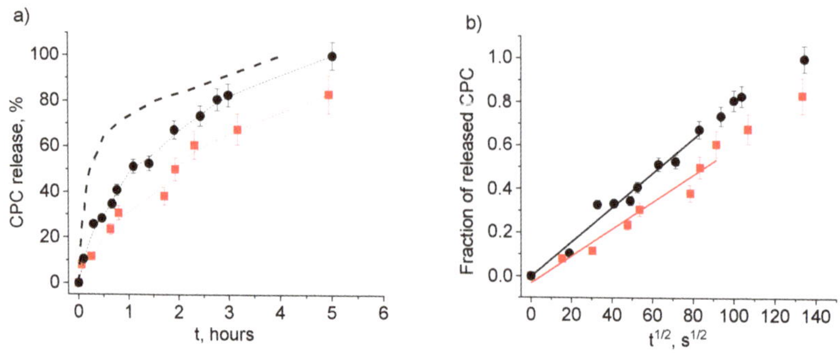

Figure 7. Release of CPC into physiological saline from water-swollen PAAm hydrogels with 2 wt% anionic units of different types: AMPSA (PAAm-AMPSA-6v, Table 1; degree of swelling, 5.1) (black circles) and MA (PAAm-MA, Table 1; degree of swelling, 5.0) (red squares) as a function of time t (**a**) and $t^{1/2}$ (**b**). In both samples, the CPC/charged units molar ratio is 0.5; the degree of cross-linking is 1:500. The black dashed line describes CPC release from water-swollen PAAm-AMPSA hydrogel with 2 wt% AMPSA (PAAm-AMPSA-3v, Table 1; degree of swelling, 16.0), having a lower degree of cross-linking (1:1000).

Let us now examine the effect of the content of anionic groups responsible for the interaction with CPC on its release. Figures 4 and 5 demonstrate that, for both swollen and dried samples, the release rate depends significantly on the degree of charging of the gel (at constant CPC/AMPSA molar ratios). The greater the gel charge, the slower the release. For instance, CPC ions are completely released from more charged swollen PAAm-AMPSA gel (10 mol% AMPSA) in 12 h, whereas less charged swollen gel releases all CPC ions in 4 h; therefore, charged monomer units impede the release of oppositely charged surfactant ions from the gel. This behavior can be exploited for tuning the release time of the disinfectant in order to obtain a long-acting disinfectant coating.

Figure 7 allows us to compare CPC release profiles for hydrogels with two different types of negatively charged units (AMPSA and MA). It is seen that the release rate is somewhat lower for hydrogels with MA groups. This may indicate to stronger interaction of carboxylic groups with CPC as compared to sulfonate groups, which is consistent with a higher charge density of alkyl carbonates headgroups as compared to alkyl sulfates [68]. Previously, clear differences in the affinity of carboxylic and sulfonate groups to various cations were demonstrated both by experimental and computational results [68,69]. They were attributed to a competition for a given cation between hydrating water molecules and the anion, which is qualitatively related to charge distributions, polarizabilities, and effective field strengths of the ions [70].

Thus, the rate of the release of cationic disinfectant from oppositely charged polymer gel can be tuned in a fairly broad range by varying the concentration of CPC, the degree of swelling and the degree of cross-linking of the gel and the content/type of anionic repeat units in the polymer matrix.

3.3. Reloading

The polyelectrolyte gel/disinfectant complexes belong to released-based [71] antiseptic materials. The lifetime of their antiseptic activity is limited by the amount of disinfectant embedded in the polymer matrix, which will exhaust over time. One can estimate that the complexes under study will keep their activity during ca. 10^3 treatments with 0.2 mL of saline solution. Further, the polymer matrix can be refilled with the disinfectant and reused. This is illustrated in Figure 8. Since it is very time consuming to make 1000 separate treatments with 0.2 mL water aliquots to remove most of the CPC, we first removed 98% of CPC by immersing hydrogel in 200 mL of physiological saline solution, dried the hydrogel and then studied the release of the remaining disinfectant by adding many times 0.2 mL portions of saline. The kinetics of the release of disinfectant to these final volumes of salt solution is illustrated in Figure 8. When the CPC was completely removed, the polymer matrix was reloaded with the disinfectant. Figure 8 shows that the reloaded gel effectively releases the disinfectant. In this case, the first portions of released CPC contain a large amount of disinfectant, probably, because, upon reloading, some CPC molecules did not penetrate deep into the polymer matrix. To avoid this, the dried hydrogel could be treated preliminarily with water to induce gel swelling and only then treated with a CPC solution.

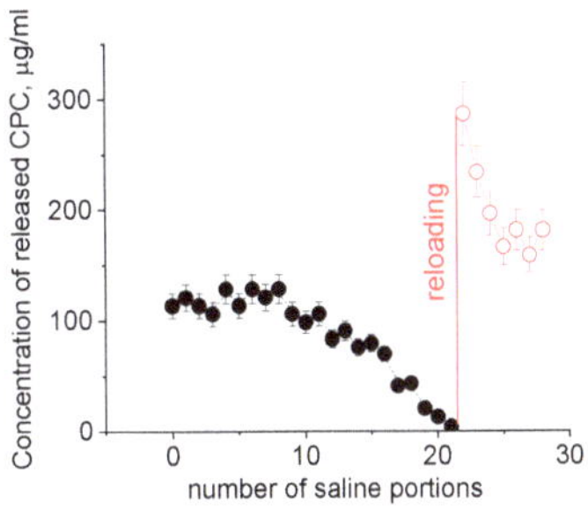

Figure 8. Release of CPC from dried PAAm-AMPSA-4s hydrogels to a small volume (0.2 mL) of physiological saline for original hydrogel preliminarily washed with a large volume of saline (black circles) and for reloaded hydrogel (red circles). Initial CPC/AMPSA molar ratio in the hydrogel matrix is 0.5; the CPC/AMPSA molar ratio in the hydrogel matrix upon reloading is 0.35; the solution used for reloading contained 7 wt% CPC and 25 wt% PEG.

Thus, the polyelectrolyte gels can be reloaded with disinfectant by spraying or pouring the disinfectant solution on the gel surface.

3.4. Virucidal Properties

We first studied the virucidal activity of CPC in the absence of the polymer gel. Table 2 shows that CPC completely inactivates SARS-CoV-2 (IC = 100%) at concentrations equal to or higher than 0.07 mM (or 0.0024 wt%).

Table 2. SARS-CoV-2 inactivation by CPC solution (contact time, 1h).

Concentration		Virus Titer			Inhibition Coefficient IC, %
mM	wt%	Control A_c	Experiment A_e	Log10 Reduction A	
0.28	0.0095	7.0	0	7.0	100
0.14	0.0048	7.0	0	7.0	100
0.07	0.0024	7.0	0	7.0	100
0.025	0.0009	0	7.0	0	0

Table 3 shows the results of SARS-CoV-2 inactivation by CPC-containing polymer coatings. It is seen that all samples studied completely inactivate the virus (IC = 100%), reducing its titer by seven orders of magnitude. For all samples, the concentration of the released CPC exceeds that required for complete inactivation of the virus. An important point that we should pay attention to is the very short duration of the contact between the buffer solution or the virus-containing liquid with the gel coating (5 s), which was sufficient to release from the polymers an amount of CPC that ensures complete (100%) inactivation of high-titer SARS-CoV-2. There is still no exact information in the available publications regarding how many viral particles correspond to one $TCID_{50}$. Even if we assume 10 $TCID_{50}$ to be equivalent to as little as 2 to 4 viral particles [72], our data indicate that the surfactant released from the gel in 5 s inactivates 2 to 4 million viral particles.

Table 3. SARS-CoV-2 inactivation by antiseptic polymer gels (contact time, 5 s).

Sample	Experimental Conditions	Initial CPC Concentration in the Gel, wt%	Concentration of Released CPC, wt%	Virus Titer			Inhibition Coefficient IC, %
				Control A_c	Experiment A_e	Log10 Reduction A	
PAAm-AMPSA-1v	Rinsing *, day 1	8.3	0.057	7.00	0.00	7.00	100
PAAm-AMPSA-1s	Rinsing *, day 1	2.0	0.100	7.00	0.00	7.00	100
PAAm-AMPSA-2s	Rinsing *, day 1	2.0	0.145	7.00	0.00	7.00	100
PVP-MA-v	Rinsing *, day 1	2.0	0.078	7.00	0.00	7.00	100
PAAm-AMPSA-2s	Rinsing *, day 7	2.0	0.100	7.00	0.00	7.00	100
PVP-MA-v	Rinsing *, day 7	2.0	0.016	7.00	0.00	7.00	100
PAAm-AMPSA-1v	Direct contact **	8.3	-	7.00	0.00	7.00	100
PVP-MA-v	Direct contact **	2.0	-	7.00	0.00	7.00	100

* Polymer gels were rinsed by applying dropwise 450 μL physiological saline onto their surface and collecting the flowing liquid in 5 s. Rinsing was performed on day 1 or day 7 after preparation of the gel sample. ** Direct contact of polymer gels with virus-containing liquid for 5 s.

Table 3 shows that the coating retains high antiviral activity for at least 7 days. Moreover, the studies of CPC release show that the amount of the surfactant released remains high enough to completely inactivate SARS-CoV-2 even in 3–4 weeks. Thus, the proposed antiseptic coatings offer significant promise as means of long-term disinfection of surfaces that can get recontaminated many times, remaining a potential source of infection.

Note that the concentrations of released CPC (Table 3) that are effective against SARS-CoV-2 are close to CPC concentrations (0.025–0.1 wt%) considered by the US Food and Drug Administration as safe when formulated in mouth rinses [73]. Several CPC toxicity studies have shown that LD_{50} of CPC is 250 mg/kg subcutaneously [73,74] and >500 mg/kg orally [75]. Percutaneous absorption of CPC is not believed to be significant [76].

Polymer matrices make it possible to tune both the concentration of the released cationic disinfectant and the rate of its release, which paves the way to finding conditions under which CPC will be delivered in desired amounts. This is very important since the use of excessive amounts of disinfectants may have a negative impact on the environment.

4. Conclusions

Polyelectrolyte gels with electrostatically bound disinfectant ions were demonstrated to be capable of complete and fast elimination of infectious SARS-CoV-2. The gels exhibit long-lasting antiviral activity and can be easily reloaded when the disinfectant is exhausted. When compared to common low-molecular-weight disinfectants, such materials can be regarded as safer and environmentally friendly sustainable biocides. They can deliver a high disinfectant agent concentration locally, thus minimizing the toxic effects. The coatings made from these polymer/disinfectant systems can serve to inactivate viruses in respiratory droplets appearing on surfaces (e.g., as a result of coughing or sneezing of the infected people). Such coatings can be used to cover various surfaces in public areas, such as door handles, light switches, elevator buttons, and so on. Considering that the mechanism of action of quaternary ammonium disinfectants is most likely universal, it is conceivable that the materials under study will be efficient in inactivating other enveloped viruses. Note that the proposed approach is general, and CPC should be viewed as a model disinfectant. In further studies, the same approach can be applied to other cationic disinfectants (such as BAC or double-chain surfactants such as di-N-didodecyldimethylammonium bromide [77]).

Supplementary Materials: The following supporting information can be downloaded at: https://www.mdpi.com/article/10.3390/polym14122444/s1, Figure S1: Absorption spectra of CPC solutions at 0.24 mmol/L and 0.075 mmol/L; Figure S2: Dependence of the intensity of the pyridinium ring absorbance peak at 259 nm on CPC concentration; Figure S3: Release of CPC into physiological saline from water-swollen and dried PAAm-MA hydrogels containing 2 mol% MA units (sample PAAm-MA, Table 1) as a function of time t (a) and $t^{1/2}$ (b); Figure S4: Panel (a): Load-unload test (a force as a function of indentation depth) for PAAm-AMPSA-4s hydrogels without PEG and with 22 wt% PEG. Panel (b): Part of the graph, shown as a dashed rectangle in panel (a), enlarged for clarity; Table S1: Cytotoxicity of cetylpyridinium chloride (CPC) and benzalkonium chloride (BAC) to Vero E6 cells. References [78–81] are cited in the Supplementary Materials.

Author Contributions: Conceptualization, O.E.P., E.V.K. and A.R.K.; investigation, V.S.M., A.V.S., V.F.L., I.T.F. and G.V.K.; resources, A.R.K.; writing—original draft preparation, V.S.M., A.V.S. and O.E.P.; writing—review and editing, V.S.M., A.V.S., A.S.T. and O.E.P. All authors have read and agreed to the published version of the manuscript.

Funding: The reported study was funded by Russian Foundation for Basic Research according to research project No. 20-04-60302.

Institutional Review Board Statement: Not applicable.

Informed Consent Statement: Not applicable.

Data Availability Statement: The data presented in this study are openly available.

Conflicts of Interest: The authors declare no conflict of interest.

References

1. Martellucci, C.A.; Flacco, M.E.; Cappadona, R.; Bravi, F.; Mantovani, L.; Manzoli, L. SARS-CoV-2 pandemic: An overview. *Adv. Biol. Regul.* **2020**, *77*, 100736. [CrossRef] [PubMed]
2. Kunduru, K.R.; Kutner, N.; Nassar-Marjiya, E.; Shaheen-Mualim, M.; Rizik, L.; Farah, S. Disinfectants role in the prevention of spreading the COVID-19 and other infectious diseases: The need for functional polymers! *Polym. Adv. Technol.* **2022**, 1–9. [CrossRef] [PubMed]
3. Worldometers. Coronavirus. Available online: https://www.worldometers.info/coronavirus/#countries (accessed on 24 May 2022).
4. Stadnytskyi, V.; Anfinrud, P.; Bax, A. Breathing, speaking, coughing or sneezing: What drives transmission of SARS-CoV-2? *J. Intern. Med.* **2021**, *290*, 949–951. [CrossRef] [PubMed]
5. Aboubakr, H.A.; Sharafeldin, T.A.; Goyal, S.M. Stability of SARS-CoV-2 and other coronaviruses in the environment and on common touch surfaces and the influence of climatic conditions: A review. *Transbound. Emerg. Dis.* **2021**, *68*, 296–312. [CrossRef]
6. Chin, A.W.H.; Chu, J.T.S.; Perera, M.R.A.; Hui, K.P.Y.; Yen, H.-L.; Chan, M.C.W.; Peiris, M.; Poon, L.L.M. Stability of SARS-CoV-2 in different environmental conditions. *Lancet Microbe* **2020**, *1*, E10. [CrossRef]
7. Van Doremalen, N.; Bushmaker, T.; Morris, D.H.; Holbrook, M.G.; Gamble, A.; Williamson, B.N.; Tamin, A.; Harcourt, J.L.; Thornburg, N.J.; Gerber, S.I.; et al. Aerosol and Surface Stability of SARS-CoV-2 as Compared with SARS-CoV-1. *N. Engl. J. Med.* **2020**, *382*, 1564–1567. [CrossRef]
8. Kampf, G.; Todt, D.; Pfaender, S.; Steinmann, E. Persistence of cor-onaviruses on inanimate surfaces and their inactivation with biocidalagents. *J. Hosp. Infect.* **2020**, *104*, 246–251. [CrossRef]
9. Moriarty, L.F.; Plucinski, M.M.; Marston, B.J.; Kurbatova, E.V.; Knust, B.; Murray, E.L.; Pesik, N.; Rose, D.; Fitter, D.; Kobayashi, M.; et al. Public health responsesto COVID-19 outbreaks on cruise ships—worldwide, February–March 2020. *MMWR Morb. Mortal. Wkly. Rep.* **2020**, *69*, 347–352. [CrossRef]
10. Kratzel, A.; Todt, D.; V'kovski, P.; Steiner, S.; Gultom, M.; Thao, T.T.N.; Ebert, N.; Holwerda, M.; Steinmann, J.; Niemeyer, D.; et al. Inactivation of severe acute respiratory syndrome coronavirus 2 by WHO-recommended hand rub formulations and alcohols. *Emerg. Infect. Dis.* **2020**, *26*, 1592–1595. [CrossRef]
11. Leslie, R.A.; Zhou, S.S.; Macinga, D.R. Inactivation of SARS-CoV-2 by commercially available alcohol-based hand sanitizers. *Am. J. Infec. Control* **2021**, *49*, 401–402. [CrossRef]
12. Basak, D.; Deb, S. Sensitivity of SARS-CoV-2 towards alcohols: Potential for alcohol-related toxicity in humans. *Life* **2021**, *11*, 1334. [CrossRef] [PubMed]
13. Hirose, R.; Bandou, R.; Ikegaya, H.; Watanabe, N.; Yoshida, T.; Daidoji, T.; Naito, Y.; Itoh, Y.; Nakaya, T. Disinfectant effectiveness against SARS-CoV-2 and influenza viruses present on human skin: Model-based evaluation. *Clin. Microbiol. Infect.* **2021**, *27*, 1042.e1–1042.e4. [CrossRef] [PubMed]
14. Guo, X.; Chen, Y.; Wang, L.; Wu, X.; Fan, J.; Li, F.; Zeng, X.; Ge, Y.; Chi, Y.; Zhang, L.; et al. In vitro inactivation of SARS-CoV-2 by commonly used disinfection products and methods. *Sci. Rep.* **2021**, *11*, 2418. [CrossRef]
15. Ogilvie, B.H.; Solis-Leal, A.; Lopez, J.B.; Poole, B.D.; Robinson, R.A.; Berges, B.K. Alcohol-free hand sanitizer and other quaternary ammonium disinfectants quickly and effectively inactivate SARS-CoV-2. *J. Hosp. Infect.* **2021**, *108*, 142–145. [CrossRef]
16. Tamimi, A.H.; Carlino, S.; Gerba, C.P. Long-term efficacy of a self-disinfecting coating in an intensive care unit. *Am. J. Infect. Control* **2014**, *42*, 1178–1181. [CrossRef]
17. Nabi, G.; Wang, Y.; Hao, Y.; Khan, S.; Wu, Y.; Li, D. Massive use of disinfectants against COVID-19 poses potential risks to urban wildlife. *Environ. Res.* **2020**, *188*, 109916. [CrossRef]

18. Hora, P.I.; Pati, S.G.; McNamara, P.J.; Arnold, W.A. Increased use of quaternary ammonium compounds during the SARS-CoV-2 pandemic and beyond: Consideration of environmental implications. *Environ. Sci. Technol. Lett.* **2020**, *7*, 622–631. [CrossRef]
19. Druvari, D.; Koromilas, N.D.; Lainioti, G.C.; Bokias, G.; Vasilopoulos, G.; Vantarakis, A.; Baras, I.; Dourala, N.; Kallitsis, J.K. Polymeric quaternary ammonium-containing coatings with potential dual contact-based and release-based antimicrobial activity. *ACS Appl. Mater. Interfaces* **2016**, *8*, 35593–35605. [CrossRef]
20. Springthorpe, V.S.; Sattar, S.A. Chemical disinfection of virus-contaminated surfaces. *Crit. Rev. Environ. Sci. Control* **1990**, *20*, 169–229. [CrossRef]
21. McDonnell, G.; Russell, A.D. Antiseptics and disinfectants: Activity, action, and resistance. *Clin. Microbiol. Rev.* **1999**, *12*, 147–179. [CrossRef]
22. Gorbalenya, A.E.; Baker, S.C.; Baric, R.S.; de Groot, R.J.; Drosten, C.; Gulyaeva, A.A.; Haagmans, B.L.; Lauber, C.; Leontovich, A.M.; Neuman, B.W.; et al. The species Severe acute respiratory syndrome-related coronavirus: Classifying 2019-nCoV and naming it SARS-CoV-2. *Nat. Microbiol.* **2000**, *5*, 536–544. [CrossRef]
23. Prince, D.L.; Prince, H.N.; Thraenhart, O.; Muchmore, E.; Bonder, E.; Pugh, J. Methodological approaches to disinfection of human hepatitis B virus. *J. Clin. Microbiol.* **1993**, *31*, 3296–3304. [CrossRef]
24. Baker, N.; Williams, A.J.; Tropsha, A.; Ekins, S. Repurposing quaternary ammonium compounds as potential treatments for COVID-19. *Pharm. Res.* **2020**, *37*, 104. [CrossRef] [PubMed]
25. Munos-Basagoiti, J.; Perez-Zsolt, D.; Leon, R.; Blanc, V.; Raich-Regué, D.; Cano-Sarabia, M.; Trinité, B.; Pradenas, E.; Blanco, J.; Gispert, J.; et al. Mouthwashes with CPC reduce the infectivity of SARS-CoV-2 variants in vitro. *J. Dental Res.* **2021**, *100*, 1265–1272. [CrossRef] [PubMed]
26. Mezarina Mendoza, J.; Trelles Ubillús, B.P.; Salcedo Bolívar, G.T.; Castañeda Palacios, R.; Herrera Lopez, P.; Padilla Rodríguez, D.A.; Uchima Koecklin, K.H. Antiviral effect of mouthwashes against SARS-COV-2: A systematic review. *Saudi Dent. J.* **2022**, *34*, 167–193. [CrossRef]
27. Wong, P.T.; Leroueil, P.R.; Smith, D.M.; Ciotti, S.; Bielinska, A.U.; Janczak, K.W.; Mullen, C.H.; Ii, J.V.G.; Taylor, E.M.; Passmore, C.; et al. Formulation, high throughput in vitro screening and in vivo functional characterization of nanoemulsion-based intranasal vaccine adjuvants. *PLoS ONE* **2015**, *10*, e0126120. [CrossRef]
28. Martí, M.; Tuñón-Molina, A.; Aachmann, F.L.; Muramoto, Y.; Noda, T.; Takayama, K.; Serrano-Aroca, Á. Protective face mask filter capable of inactivating SARS-CoV-2, and methicillin-resistant *Staphylococcus aureus* and *Staphylococcus epidermidis*. *Polymers* **2021**, *13*, 207. [CrossRef]
29. Tuñón-Molina, A.; Martí, M.; Muramoto, Y.; Noda, T.; Takayama, K.; Serrano-Aroca, Á. Antimicrobial face shield: Next generation of facial protective equipment against SARS-CoV-2 and multidrug-resistant bacteria. *Int. J. Mol. Sci.* **2021**, *22*, 9518. [CrossRef]
30. Tiller, J.C.; Sprich, C.; Hartmann, L. Amphiphilic conetworks as regenerative controlled releasing antimicrobial coatings. *J. Control. Release* **2005**, *103*, 355–367. [CrossRef]
31. Pavlukhina, S.; Lu, Y.; Patimetha, A.; Libera, M.; Sukhishvili, S. Polymer multilayers with pH-triggered release of antibacterial agents. *Biomacromolecules* **2010**, *11*, 3448–3456. [CrossRef]
32. Liakos, I.; Rizzello, L.; Bayer, I.S.; Pompa, P.P.; Cingolani, R.; Athanassiou, A. Controlled antiseptic release by alginate polymer films and beads. *Carbohydr. Polym.* **2013**, *92*, 176–183. [CrossRef] [PubMed]
33. Mohan, A.; Al-Sayah, M.H.; Ahmed, A.; El-Kadri, O.M. Triazine-based porous organic polymers for reversible capture of iodine and utilization in antibacterial application. *Sci. Rep.* **2022**, *12*, 2638. [CrossRef] [PubMed]
34. Kozhunova, E.Y.; Komarova, G.A.; Vyshivannaya, O.V.; Nasimova, I.R.; Kuvarina, A.E.; Sadykova, V.S. Antiseptic materials on the base of polymer interpenetrating networks microgels and benzalkonium chloride. *Int. J. Mol. Sci.* **2022**, *23*, 4394. [CrossRef] [PubMed]
35. Imani, S.M.; Ladouceur, L.; Marshall, T.; Maclachlan, R.; Soleymani, L.; Didar, T.F. Antimicrobial nanomaterials and coatings: Current mechanisms and future perspectives to control the spread of viruses including SARS-CoV-2. *ACS Nano* **2020**, *14*, 12341–12369. [CrossRef]
36. Nasri, N.; Rusli, A.; Teramoto, N.; Jaafar, M.; Ku Ishak, K.M.; Shafiq, M.D.; Abdul Hamid, Z.A. Past and current progress in the development of antiviral/antimicrobial polymer coating towards COVID-19 prevention: A review. *Polymers* **2021**, *13*, 4234. [CrossRef]
37. Panarin, E.F. Biologically active polymer nanosystems. *Russ. Chem. Bull.* **2017**, *66*, 1812–1820. [CrossRef]
38. Gentili, V.; Pazzi, D.; Rizzo, S.; Schiuma, G.; Marchini, E.; Papadia, S.; Sartorel, A.; Di Luca, D.; Caccuri, F.; Bignozzi, C.A.; et al. Transparent polymeric formulations effective against SARS-CoV-2 infection. *ACS Appl. Mater. Interfaces* **2021**, *13*, 54648–54655. [CrossRef]
39. Pitten, F.A.; Kramer, A. Efficacy of cetylpyridinium chloride used as oropharyngeal antiseptic. *Arzneimittelforschung* **2001**, *51*, 588–595. [CrossRef]
40. Fromm-Dornieden, C.; Rembe, J.-D.; Schäfer, N.; Böhm, J.; Stuermer, E.K. Cetylpyridinium chloride and miramistin as antiseptic substances in chronic wound management—Prospects and limitations. *J. Med. Microbiol.* **2015**, *64*, 407–414. [CrossRef]
41. Hoang, T.P.N.; Ghori, M.U.; Conway, B.R. Topical antiseptic formulations for skin and soft tissue infections. *Pharmaceutics* **2021**, *13*, 558. [CrossRef]
42. Philippova, O.E.; Hourdet, D.; Audebert, R.; Khokhlov, A.R. Interaction of hydrophobically modified poly(acrylic acid) hydrogels with ionic surfactants. *Macromolecules* **1996**, *29*, 2822–2830. [CrossRef]

43. Philippova, O.E.; Chtcheglova, L.A.; Karybiants, N.S.; Khokhlov, A.R. Two mechanisms of gel/surfactant binding. *Polym. Gels Netw.* **1998**, *6*, 409–421. [CrossRef]
44. Mukhim, T.; Dey, J.; Das, S.; Ismail, K. Aggregation and adsorption behavior of cetylpyridinium chloride in aqueous sodium salicylate and sodium benzoate solutions. *J. Colloid Interface Sci.* **2010**, *350*, 511–515. [CrossRef] [PubMed]
45. Pound, G. Reversible Addition Fragmentation Chain Transfer (RAFT) Mediated Polymerization of N-Vinylpyrrolidone. Ph.D. Dissertation, Stellenbosch University, Stellenbosch, South Africa, 2008.
46. Andreeva, A.S.; Philippova, O.E.; Khokhlov, A.R.; Islamov, A.K.; Kuklin, A.I. Effect of the mobility of charged units on the microphase separation in amphiphilic polyelectrolyte hydrogels. *Langmuir* **2005**, *21*, 1216–1222. [CrossRef] [PubMed]
47. Khandurina, Y.V.; Rogacheva, V.B.; Zezin, A.B.; Kabanov, V.A. Interaction of cross-linked polyelectrolytes with oppositely charged surfactants. *Polym. Sci.* **1994**, *36*, 184–188.
48. Hansson, P.; Schneider, S.; Lindman, B. Phase separation in polyelectrolyte gels interacting with surfactants of opposite charge. *J. Phys. Chem. B* **2002**, *106*, 9777–9793. [CrossRef]
49. Tararyshkin, D.; Kramarenko, E.; Khokhlov, A. Two-phase structure of polyelectrolyte gel/surfactant complexes. *J. Chem. Phys.* **2007**, *127*, 164905. [CrossRef]
50. Hansson, P. Volume transition and phase coexistence in polyelectrolyte gels interacting with amphiphiles and proteins. *Gels* **2020**, *6*, 24. [CrossRef]
51. Flint, S.J.; Racaniello, V.R.; Rall, G.F.; Skalka, A.M.; Enquist, L.W. The infectious cycle. In *Principles of Virology*, 4th ed.; ASM Press: Washington, DC, USA, 2015; Volume 1, pp. 24–52.
52. Kärber, G. Beitrag zur kollektiven Behandlung pharmakologischer Reihenversuche. *Naunyn-Schmiedebergs Arch. Exp. Pathol. Pharmakol.* **1931**, *162*, 480–483. [CrossRef]
53. Nandy, K.; Srivastava, A.; Afgan, S.; Kumar, R.; Yadav, D.K.; Ganesan, V. Trithiocarbonate-mediated RAFT synthesis of a block copolymer: Silver nanoparticles integration and sensitive recognition of Hg^{2+}. *Polym. Bull.* **2022**. [CrossRef]
54. Martínez-Cornejo, V.; Velázquez-Roblero, J.; Rosiles-González, V.; Correa-Duran, M.; Avila-Ortega, A.; Hernández-Núñez, E.; Le Lagadec, R.; González-Díaz, M.O. Synthesis of poly(2-acrylamido-2-methylpropane sulfonic acid) and its block copolymers with methyl methacrylate and 2-hydroxyethyl methacrylate by quasiliving radical polymerization catalyzed by a cyclometalated ruthenium(II) complex. *Polymers* **2020**, *12*, 1663. [CrossRef] [PubMed]
55. Loría-Bastarrachea, M.I.; Herrera-Kao, W.; Cauich-Rodríguez, J.V.; Cervantes-Uc, J.M.; Vázquez-Torres, H.; Ávila-Ortega, A. A TG/FTIR study on the thermal degradation of poly(vinyl pyrrolidone). *J. Therm. Anal. Calorim.* **2011**, *104*, 737–742. [CrossRef]
56. Starodubtsev, S.G. Influence of topological structure of polyelectrolyte networks on their interaction with oppositely charged micelle-forming surfactants. *Polym. Sci. Ser. B* **1990**, *32*, 925–930.
57. Khokhlov, A.R.; Kramarenko, E.Y.; Makhaeva, E.E.; Starodubtzev, S.G. Collapse of polyelectrolyte networks induced by their interaction with oppositely charged surfactants. *Macromolecules* **1992**, *25*, 4779–4783. [CrossRef]
58. Philippova, O.E.; Starodoubtzev, S.G. Interaction of slightly cross-linked gels of poly(diallyldimethylammonium bromide) with sodium dodecyl sulfate: Diffusion of surfactant ions in gel. *J. Polym. Sci. Part B Polym. Phys.* **1993**, *31*, 1471–1476. [CrossRef]
59. Khokhlov, A.R.; Kramarenko, E.Y.; Makhaeva, E.E.; Starodubtzev, S.G. Collapse of polyelectrolyte networks induced by their interaction with an oppositely charged surfactant. Theory. *Makromol. Chem. Theory Simul.* **1992**, *1*, 105–118. [CrossRef]
60. Kramarenko, E.Y.; Philippova, O.E.; Khokhlov, A.R. Polyelectrolyte networks as highly sensitive polymers. *Polym. Sci. Ser. C* **2006**, *48*, 1. [CrossRef]
61. Zaroslov, Y.D.; Philippova, O.E.; Khokhlov, A.R. Change of elastic modulus of strongly charged hydrogels at the collapse transition. *Macromolecules* **1999**, *32*, 1508–1513. [CrossRef]
62. Piculell, L.; Sjöström, J.; Lynch, I. Swelling isotherms of surfactant-responsive polymer gels. *Progr. Colloid Polym. Sci.* **2003**, *122*, 103–112. [CrossRef]
63. Ritger, P.L.; Peppas, N.A. A simple equation for description of solute release. I. Fickian and non-Fickian release from non-swellable devices in the form of slabs, spheres, cylinders or discs. *J. Control. Release* **1987**, *5*, 23–36. [CrossRef]
64. Ritger, P.L.; Peppas, N.A. A simple equation for description of solute release. II. Fickian and anomalous release from swellable devices. *J. Control. Release* **1987**, *5*, 37–42. [CrossRef]
65. Serra, L.; Domenech, J.; Peppas, N.A. Drug transport mechanisms and release kinetics from molecularly designed poly(acrylic acid-g-ethylene glycol) hydrogels. *Biomaterials* **2006**, *27*, 5440–5451. [CrossRef] [PubMed]
66. Karybiants, N.S.; Philippova, O.E.; Starodoubtsev, S.G.; Khokhlov, A.R. Conformational transitions in poly(methacrylic acid) gel/poly(ethylene glycol) complexes. Effect of the gel cross-linking density. *Macromol. Chem. Phys.* **1996**, *197*, 2373–2378. [CrossRef]
67. Khan, S.; Ranjha, N.M. Effect of degree of cross-linking on swelling and on drug release of low viscous chitosan/poly(vinyl alcohol) hydrogels. *Polym. Bull.* **2014**, *71*, 2133–2158. [CrossRef]
68. Vlachy, N.; Jaroda-Cwiklik, B.; Vácha, R.; Touraud, D.; Jungwirth, P.; Kunz, W. Hofmeister series and specific interactions of charged headgroups with aqueous ions. *Adv. Colloid Interface Sci.* **2009**, *146*, 42–47. [CrossRef]
69. Strauss, U.P.; Leung, Y.P. Volume changes as a criterion for site binding of counterions by polyelectrolytes. *J. Am. Chem. Soc.* **1965**, *87*, 1476–1480. [CrossRef]
70. Bungenberg de Jong, H.G. *Colloid Science*; Elsevier Publishing Co., Inc.: New York, NY, USA, 1949; Volume II, Chapter 9.

71. Cloutier, M.; Mantovani, D.; Rosei, F. Antibacterial coatings: Challenges, perspectives, and opportunities. *Trends Biotechnol.* **2015**, *33*, 637–652. [CrossRef]
72. Shrivastava, S.; Patil, H.P.; Mhaske, S.T.; Palkar, S.; Lalwani, S.; Mishra, A.C.; Arankalle, V.A. Isolation and genetic characterization of SARS-CoV-2 from Indian patients in a single family without H/O travel abroad. *Virus Genes* **2021**, *57*, 245–249. [CrossRef]
73. Food and Drug Administration. 21 CFR Part 356. Oral health care drug products for over-the-counter human use; Tentative final monograph for oral antiseptic drug products. *Fed. Regist.* **1994**, *59*, 6094.
74. Nelson, J.W.; Lyster, S.C. The toxicity of myristyl-gamma-picolinium chloride. *J. Am. Pharm. Assoc.* **1946**, *35*, 89–94. [CrossRef]
75. Updegraff, D.M.; Kvam, D.C.; Robertson, J.E. N-(1,1-dihydroperfluorooctyl)pyridinium trifluoromethanesulfonate, a new quaternary ammonium antiseptic. *J. Pharm. Sci.* **1970**, *59*, 188–192. [CrossRef] [PubMed]
76. Wilson, J.T.; Burr, I.M. Benzalkonium chloride poisoning in infant twins. *Arch. Pediatr. Adolesc. Med.* **1975**, *129*, 1208–1209. [CrossRef] [PubMed]
77. Karamov, E.V.; Larichev, V.F.; Kornilaeva, G.V.; Fedyakina, I.T.; Turgiev, A.S.; Shibaev, A.V.; Molchanov, V.S.; Philippova, O.E.; Khokhlov, A.R. Cationic Surfactants as Disinfectants against SARS-CoV-2. *Int. J. Mol. Sci.* **2022**, *23*, 6645. [CrossRef]
78. Barltrop, J.A.; Owen, T.C.; Cory, A.H.; Cory, J.G. 5-(3-carboxymethoxyphenyl)-2-(4,5-dimethylthiazolyl)-3-(4-sulfophenyl)tetrazolium, inner salt (MTS) and related analogs of 3-(4,5-dimethylthiazolyl)-2,5-diphenyltetrazolium bromide (MTT) reducing to purple water-soluble formazans as cell-viability indicators. *Bioorg. Med. Chem. Lett.* **1991**, *1*, 611–614. [CrossRef]
79. Berridge, M.V.; Tan, A.S. Characterization of the cellular reduction of 3-(4,5-dimethylthiazol-2-yl)-2,5-diphenyltetrazolium bromide (MTT): Subcellular localization, substrate dependence, and involvement of mitochondrial electron transport in MTT reduction. *Arch. Biochem. Biophys.* **1993**, *303*, 474–482. [CrossRef] [PubMed]
80. Khorolsuren, Z.; Lang, O.; Vag, J.; Kohidai, L. Effect of dental antiseptic agents on the viability of human periodontal ligament cells. *Saudi Dent. J.* **2021**, *33*, 904–911. [CrossRef]
81. Mukherjee, P.K.; Esper, F.; Buchheit, K.; Arters, K.; Adkins, I.; Ghannoum, M.A.; Salata, R.A. Randomized, double-blind, placebo-controlled clinical trial to assess the safety and effectiveness of a novel dual-action oral topical formulation against upper respiratory infections. *BMC Infect. Dis.* **2017**, *17*, 74. [CrossRef]

polymers

MDPI

Article

Biodegradable and Biocompatible Thermoplastic Poly(Ester-Urethane)s Based on Poly(ε-Caprolactone) and Novel 1,3-Propanediol Bis(4-Isocyanatobenzoate) Diisocyanate: Synthesis and Characterization

Alejandra Rubio Hernández-Sampelayo [1,2], Rodrigo Navarro [1,*], Dulce María González-García [3,4], Luis García-Fernández [1,5], Rosa Ana Ramírez-Jiménez [1,5], María Rosa Aguilar [1,5] and Ángel Marcos-Fernández [1,*]

1 Institute of Polymer Science and Technology (CSIC), Juan de la Cierva, 3, 28006 Madrid, Spain; alejandra@ictp.csic.es (A.R.H.-S.); luis.garcia@csic.es (L.G.-F.); raramirez@ictp.csic.es (R.A.R.-J.); mraguilar@ictp.csic.es (M.R.A.)
2 Universidad Nacional de Educación a Distancia (UNED), Facultad de Ciencias, C/Bravo Murillo, 38, 28015 Madrid, Spain
3 Instituto Politécnico Nacional, Escuela Superior de Ingeniería Química e Industrias Extractivas, UPALM-Zacatenco, Col Lindavista, Mexico City 07738, Mexico; dgonzalezg@ipn.mx
4 Universidad de Guanajuato, Departamento de Química, Noria Alta s/n, Guanajuato 36050, Mexico
5 Biomedical Research Networking Center in the Subject Area of Bioengineering, Biomaterials and Nanomedicine (CIBER-BBN), Avenida Monforte de Lemons 3–5, 28029 Madrid, Spain
* Correspondence: rnavarro@ictp.csic.es (R.N.); amarcos@ictp.csic.es (Á.M.-F.)

Abstract: A series of non-toxic biodegradable and biocompatible polyurethanes bearing p-aminobenzoate moieties are presented. The introduction of this attractive motif was carried out by the synthesis of a novel isocyanate. These biodegradable polymers were chemically and physically characterized by several techniques and methods including bioassay and water uptake measurements. The molecular weight of the soft segment (poly-ε-caprolactone, PCL) and hard segment crystallinity dictated the mechanical behavior and water uptake. The behavior of short PCL-based polyurethanes was elastomeric, whilst increasing the molecular weight of the soft segment led to plastic polyurethanes. Water uptake was hindered for long PCL due to the crystallization of the soft segment within the polyurethane matrix. Furthermore, two different types of chain extender, hydrolyzable and non-hydrolyzable, were also evaluated: polyurethanes based on hydrolyzable chain extenders reached higher molecular weights, thus leading to a better performance than their unhydrolyzable counterparts. The good cell adhesion and cytotoxicity results demonstrated the cell viability of human osteoblasts on the surfaces of these non-toxic biodegradable polyurethanes.

Keywords: biodegradability; biocompatibility; thermoplastic polyurethane; non-toxic polyurethane; hydrolyzable chain extender

1. Introduction

The versatility of polyurethanes (PUs) becomes patent in the wide range of industrial sectors where they are commonly used, due to the possibility of tailoring their physico-chemical properties by careful selection of their components. Typically, PUs are built by the reaction of a macroglycol, a polyisocyanate and a chain extender. The macroglycol chains form the so-called soft segment (SS), while the reaction between the isocyanate and the chain extender produces the hard segment (HS). Depending on the balance and thermodynamic incompatibility between these types of segments, the polyurethane matrix can exhibit a characteristic and attractive phase separation.

In the biomedical area, the use of polyurethanes is widespread, even exceeding other polymeric materials (such as natural rubber or PVC), due to their ability to mimic the behavior of different tissues and show relatively good biocompatibility. The vast majority of polyurethanes are designed to withstand long periods of service, which require biostable materials [1]. However, there is currently great interest in the development of biodegradable polymers for certain biomedical applications, such as tissue regeneration [2,3] or controlled drug release [4,5]. Specifically, in these applications, PUs not only need to meet certain physical and/or mechanical requirements, but it is also necessary to assess whether the building blocks and their degradation products are biocompatible, non-toxic, and metabolized by the living organism. Therefore, polyurethanes obtained from the classical aromatic isocyanate 4,4′-diisocyanatodiphenyl-methane (MDI) or poly(urethane-urea)s with aromatic diamines (such as 3,3′-dichloro-4,4′-diaminodiphenylmethane, MOCA) as chain extenders, have been discarded because of the carcinogenic character of the diamines obtained as degradation products [6].

To overcome these drawbacks, several alternatives have been proposed. One alternative is the use of biodegradable chain extenders, such as amino acid derivatives, [7–9] ester-diols [10,11] or amide-diols such as N,N′-ethylene-bis(6-hydroxycaproamide) (EDA-2CL) [12]. In these cases, the introduction of labile functional groups such as esters, within the chemical structure of these alternative extenders, facilitates the degradation of the hard segment formed, without compromising the biological toxicity of the polyurethane. On the other hand, the alternative to aromatic diisocyanates is to use aliphatic diisocyanates such as 1,6-hexamethylenediisocyanate (HDI) [13] or diisocyanates of amino acids (mainly lysine) [14–16], which degrade towards non-toxic or slightly toxic diamines. However, these alternatives present new challenges that hinder the development of high-performance polyurethanes. In this regard, for example, aliphatic diisocyanates are characterized by a reduction in reactivity compared to their aromatic counterparts [17]. Another difficulty is related to the asymmetry of amino acids, which prevents the correct crystallization of the hard segment and subsequent phase separation within the polyurethane matrix. Consequently, the mechanical properties of the synthesized polyurethanes were mainly governed by the crystallization of the soft segment.

As a step forward, we intend to synthesize non-toxic biodegradable polyurethanes from an aromatic diisocyanate without adverse effects. This novel diisocyanate is based on the diamine 1,3-propanediol bis(4-aminobenzoate) (diamino-PABA), leading to a highly reactive and symmetrical aromatic diisocyanate. The precursor diamine has been rather scarcely used as a chain extender, and no diisocyanate derived from it has been described to date [17,18]. Therefore, in the present work, we describe the synthesis of a novel aromatic isocyanate derived from diamino-PABA and the synthesis and characterization of linear polyurethanes obtained from it. Polycaprolactones (PCLs) of different molecular weights were used as soft segments and, 1,4-butanediol (BD), which is not hydrolyzable, and EDA2-CL which is hydrolyzable, as chain extenders. Finally, due to the interesting features of these materials, cell adhesion and proliferation studies have also been investigated using fibroblasts and osteoblast cells.

2. Materials and Methods

2.1. Materials

1,3-Propanediol bis(4-aminobenzoate) (diamino-PABA), 1,2-ethylenediamine (EDA), stannous 2-ethylhexanoate ($Sn(Oct)_2$), and triphosgene were purchased from Sigma-Aldrich, Madrid, Spain, and were used without further purification. Polycaprolactone diols with nominal molecular weights of 530, 1250 and 2000 g·mol^{-1}, were also purchased from Sigma-Aldrich, Madrid, Spain. The molecular weight of the PCLs was accurately determined by proton NMR [8] and the values obtained were 519 (PCL519), 1100 (PCL1100) and 2054 (PCL2054) g·mol^{-1}. The PCL diols were vacuum-dried at 70 °C for at least 5 h and were stored in a desiccator under vacuum until used.

4,4′-Diphenylmethane diisocyanate (MDI) and ε-caprolactone (ε-CL) were kindly donated by Lubrizol Advanced Materials Spain SL, Montmeló, Spain. The MDI was purified by short path distillation at 110 °C and 0.4 mbar, using a vacuum sublimator connected to a cryostat at −15 °C. The sublimated MDI was stored in a desiccator under vacuum until used. This purified MDI was used within 10 days due to its tendency to dimerize.

N, N-Dimethylacetamide (DMAc), purchased from Scharlau (Barcelona, Spain), was vacuum-distilled over commercial polymeric MDI-isocyanate, to remove residual water and amines that would lead to undesired drift in the stoichiometry on the synthesis of polymers.

2.2. Characterization

Proton (^{1}H) and Carbon (^{13}C) nuclear magnetic resonance (NMR) spectra were recorded on the Bruker spectrometer Oxford 400, at room temperature in deuterated solvents ($CDCl_3$ and DMSO-d_6). NMR spectra were referenced to the residual peak of these deuterated solvents (7.26 ppm and 77.0 ppm for $CDCl_3$ and 2.50 ppm and 39.5 ppm for DMSO-d_6).

The number and weight-average molecular weights and the polydispersity index were determined by size exclusion chromatography (SEC) using a Waters apparatus (Waters Division Millipore, Madrid, Spain), equipped with a refractive index detector. A set of Styragel HR3 and HR5 Waters columns (300 × 7.8 mm, 5 µm nominal particle size) conditioned at 70 °C was used to elute the samples at a 0.7 mL min-1 flow rate. The mobile phase was N, N-dimethylformamide (DMF) with 0.1% of LiBr. Polystyrene standards (Polymer Laboratories) used for calibration.

Fourier transform infrared spectra were recorded using a Perkin-Elmer spectrometer, model Spectrum One (Perkin-Elmer, Waltham, MA, USA), coupled with an attenuated total reflection (ATR) accessory. Sixteen scans were averaged from 4000 to 450 cm^{-1} and with a resolution of 2 cm^{-1}.

Thermogravimetric analysis (TGA) was carried out in a Mettler Toledo TGA/SDTA 851 instrument (Mettler-Toledo, Schwerzenbach, Switzerland). Disc samples cut from films were heated from room temperature to 600 °C under a nitrogen atmosphere at a 10 °C min^{-1} heating rate.

Differential scanning calorimetry (DSC) was performed in a Mettler Toledo 822e calorimeter equipped with a liquid nitrogen accessory. Disc samples weighing approximately 10–15 mg were sealed in aluminum pans. The samples were initially heated from 25 to 90 °C with a heating rate of 10 °C min^{-1}, then cooled to −90 °C at the maximum rate of the instrument, maintained for 5 min at this temperature and re-heated from −90 to 200 °C at 10 °C min^{-1}. The glass transition temperature (Tg) was taken as the midpoint of the transition, whereas melting points (Mp) were taken as the maximum of the endothermic transition.

Tensile properties were measured in an MTS Synergie 200 testing machine equipped with a 100 N load cell. Test specimens were cut with the dimensions established in standard ISO37 (Type 4). A cross-head speed of 5 mm min^{-1} was used and the strain was measured from cross-head separation and referred to a 10 mm initial length. For all the synthetized polymers, a minimum of 5 specimens were analyzed.

2.3. Synthesis of 1,3-Propanediol Bis(4-Isocyanatobenzoate) (IsoPABA)

In a 250 mL three-necked round bottom flask a solution of 15.34 mmol of triphosgene with 15 mL of anhydrous toluene was cooled at 0 °C with an ice bath. A solution of 16.30 mmol of diamino-PABA in 60 mL of toluene was added dropwise to the previous solution. During the addition, the reaction temperature was kept below 5 °C. After the addition was complete, the reaction mixture was kept for one hour at 0 °C and subsequently heated at 110 °C for 10 h. Finally, the solvent was removed under reduced pressure with a rotary evaporator. The dry reaction crude was subsequently purified by sublimation, heating at 190 °C and reduced pressure (0.42 mbar). Yield 56%.

2.4. Synthesis of N, N′-Ethylene-Bis(6-Hydroxycaproamide) (EDA-2CL)

The synthesis of this chain extender has been described previously [12]. In a typical run, 0.2 mol of 1,2-ethylenediamine (EDA) and 2.0 mol of ε-caprolactone (ε-CL) were heated in an oil bath at 70 °C for 2 h. After cooling, the desired product precipitated as a white solid, which was isolated by filtration and exhaustively washed with toluene. The solid was then dried in a vacuum (yield 25%). The synthetic route is described in Scheme 1.

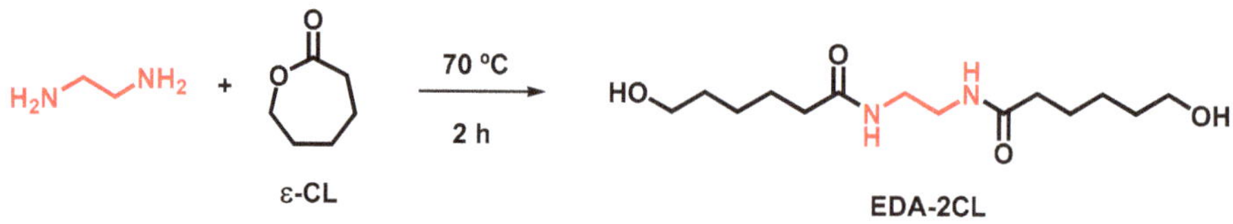

Scheme 1. Synthesis of N, N′-ethylene-bis(6-hydroxycaproamide).

2.5. Synthesis of Hard Segment Models

As an example, the reaction between isoPABA and EDA-2CL is described. In a 25 mL round-bottom flask, an equimolar mixture of isoPABA and EDA-2CL was dissolved in anhydrous DMAc (4 mL) and 2 drops (12 mg) of catalyst ($Sn(Oct)_2$) were added. This mixture was immersed in a pre-heated oil bath at 80 °C. The reaction crude was stirred for 3 h at this temperature and at room temperature overnight. The crude reaction was poured over cold water and the solid was filtered and dried under vacuum. The same procedure was followed for the hard segment model isoPABA-BD.

2.6. Synthesis of Segmented Poly(Ester-Urethane)s

The composition of these segmented polymers was set at 50% by weight of hard segment, defined as ((weight of isoPABA + weight of extender)/total weight) × 100. The method for obtaining these polymers consists of two stages: first, the formation of a prepolymer, followed by the addition of the chain extender. In a reaction flask, isoPABA and dry PCL diol in the appropriate ratio and anhydrous DMAc were charged to produce a 50:50 (w/v = weight of reactants in gram/volume of DMAc in mL) mixture. Then, two drops of the catalyst (stannous 2-ethylhexanoate) were added. The reaction mixture was blanketed with N_2 and immersed in a silicone bath preheated at 80 °C. After 3 h stirring at that temperature, the corresponding amount of chain extender (BD or EDA-2CL) was added. Heating was continued for a further three hours at 80 °C and then the reaction was warmed to room temperature and stirred overnight. The resulting viscous solution was cast onto a levelled heating plate and the solvent was slowly evaporated by heating at 60 °C. Once the solvent was removed, a homogeneous polymer film was obtained, which was stored under vacuum to remove the residual solvent.

2.7. Hydrolytic and Enzymatic In Vitro Degradation

In vitro hydrolytic degradation of the PUs was evaluated following the weight changes at certain time intervals in films submerged in phosphate buffer solution at 37 °C. The water uptake was taken from the maximum of the curve and the weight change was calculated from the equation:

$$W_{Change}(\%) = 100 * (W_m - W_0)/W_0 \quad (1)$$

where W_m is the weight of the hydrated specimen and W_0 is the initial weight of the specimen. At least three specimens were tested.

In vitro enzymatic degradation was evaluated using porcine liver esterase (PLE) (Sigma-Aldrich, Madrid, Spain). The enzymatic solution was prepared by dissolving the PLE at 30 U/mL in 0.05 M phosphate buffered saline (PBS Dulbecco's); the solutions were filtered using a 0.22 µm filter. Polyurethane discs were exposed to PLE in triplicate by immersion in the PLE solution. Enzyme activity was maintained by replacing the PLE solution every 24 h. The changes in weight of the polyurethanes were calculated using Equation (1).

2.8. Cytotoxicity Test

The materials and the discs of the negative control, Thermanox® (TMX, Labclinics SA), were immersed in 5 mL of fresh and sterile medium, in the presence and in the absence of PLE, and kept under stirring at 37 °C. After 2, 7, 14, 21 and 28 days, the medium was removed (leachates) and replaced with fresh medium.

Due to the possible use of these materials for osteochondral regeneration, human osteoblast cells (HObs, Innoprot) and articular human chondrocytes (HC-a, Innoprot) were cultured using fresh complete culture medium. HObs and HC-a were seeded in 96-well plates at a concentration of 1.1×10^5 cells/mL and maintained for 24 h at 37 °C with 5% CO_2. After 24 h, the culture medium was exchanged with the corresponding leachates (n = 8), and the plates were incubated for 24 h under the same conditions. After this time, the contents of the wells were replaced by the MTT (3-[4,5-dimethylthiazol-2-yl]-2,5-diphenyltetrazoliumbromide) solution (10% in the fresh medium). The MTT reagent was kept in contact with the cultures for 4 h at 37 °C, and then, the contents of the wells were extracted and 100 µL of dimethylsulfoxide (DMSO) was added, in order to dissolve the formazan crystals that may have formed. The optical density was read at 570 nm with a reference wavelength of 630 nm in a Biotek Synergy HT plate reader. The relative cell viability (% *CV*) was calculated with respect to the control, from Equation (2):

$$\% \, CV = \frac{OD_S - OD_B}{OD_C} \cdot 100 \quad (2)$$

where OD_S, OD_B and OD_C are the optical density measurements of the sample, the blank and the control, respectively.

2.9. Cell Adhesion and Proliferation Assay

Cell adhesion and proliferation were measured by an Alamar Blue assay at 1, 7, 14 and 19 days, using HOb and HC-a cells.

The different PUs were directly deposited on a 24-well plate. HOb and HC-a cells were seeded on the PUs at a concentration of 1.4×10^5 cells/mL in fresh medium. After the stablished time intervals, the medium was removed and a 10% solution of Alamar Blue in fresh medium, without phenol red, was added and the samples were incubated for 4 h. After this time, the medium was transferred to a 96-well plate and the fluorescence was measured at 570/630 (em/ex) in a microplate reader (Biotek Synergy HT spectrophotometer).

3. Results and Discussion

3.1. Synthesis of Aromatic Isocyanate IsoPABA

The biological interest of this aromatic isocyanate (isoPABA) lies in the structural analogy that it presents with respect to p-aminobenzoic acid. In fact, its chemical structure is very close to the local pain reliever benzocaine. In biological tests it has been shown that p-aminobenzoic esters show a very short half-life. These aromatic esters are rapidly metabolized and the corresponding metabolites are expelled through the urine [19]. Additionally, its symmetric chemical structure could favor the phase crystallization and/or phase separation of the hard segment within the polyurethane matrix, leading to an improvement in its physical performance. Furthermore, the introduction of ester groups into the isoPABA structure is expected to facilitate the hydrolysis of the hard segments by basic and acidic conditions or by the action of an esterase-enzyme. Due to the absence of adverse biological effects, the insertion of this aromatic structure would lead to high performance biodegradable materials. The synthesis route for isoPABA is depicted in Scheme 2.

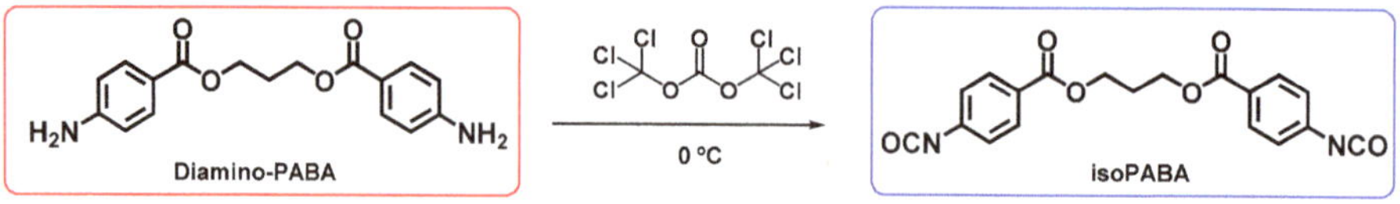

Scheme 2. Synthesis of the aromatic diisocyanate isoPABA.

The synthesis of the isocyanate isoPABA has been carried out using triphosgene as a reagent and using an anhydrous toluene to avoid side-reactions, such as the formation of aromatic ureas. Indeed, the transformation of aromatic amines into isocyanates by phosgenization with triphosgene has been shown to be very effective [20]. The chemical structure of the isoPABA monomer was corroborated by NMR and ATR–FTIR spectroscopies. As shown in Figure 1, 1H-NMR spectra from diamino-PABA and isoPABA are compared.

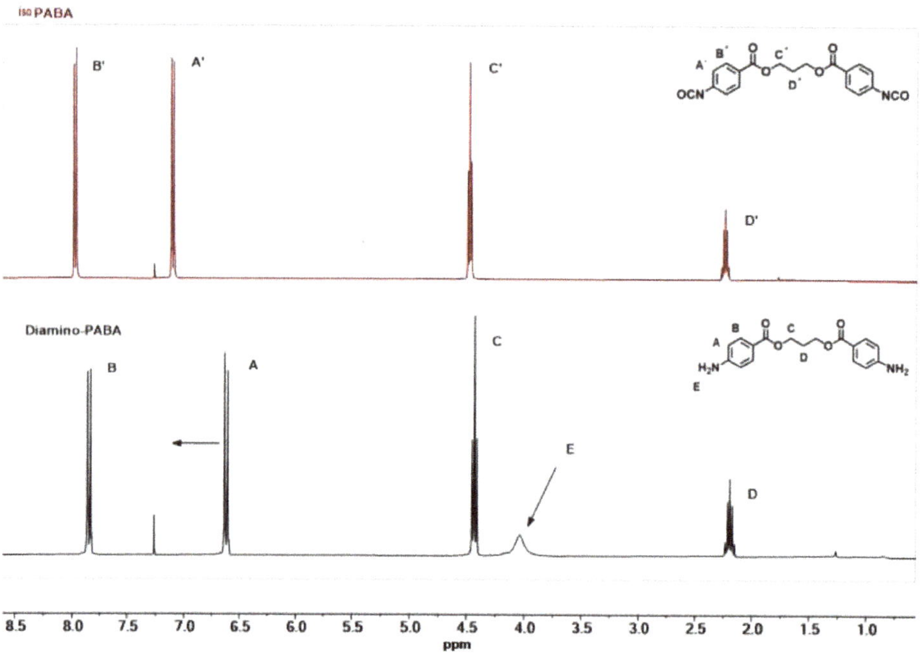

Figure 1. Comparison of 1H-NMR spectra in $CDCl_3$ of diamino-PABA (in black) and isoPABA products (in red).

The most relevant changes correspond to the signals located at 4.02 (NH_2) and 6.61 ppm (aromatic). The broad signal from the amino group (NH_2, 4.02 ppm) completely vanished after the phosgenization reaction, while the aromatic doublet (6.61 ppm) shifted to the low field (7.09 ppm) due to the electron-withdrawing nature of the isocyanate group. Additionally, the other aromatic signal slightly shifted from 7.85 to 7.97 ppm, while the signals from the propylene moiety barely changed.

Furthermore, the selective formation of the isocyanate group was confirmed by FTIR and ^{13}C-NMR spectroscopies. As shown in Figure S1, the FTIR spectrum confirmed the formation of the isocyanate group due to the intense band located at 2282 cm^{-1}, corresponding to the asymmetric stretching vibration mode of the isocyanate group. Moreover, the characteristic band associated with the ester group was detected at 1709 cm^{-1}. The presence of this ester band confirmed the stability of this functional group under the tested conditions, since this functional group could be hydrolyzed during the phosgenization reaction by the released HCl. Similarly, the formation of the isocyanate group was also confirmed by the peaks at 125.6 and 138.1 ppm in the ^{13}C-NMR spectrum, which correspond to isocyanate and aromatic carbon linked to the isocyanate group, respectively (Figure S2).

3.2. Synthesis and Characterization of Hard Segments Models

Segmented polyurethanes are characterized by two types of segments, hard and soft. The hard segment is built up by the reaction between a chain extender and a diisocyanate. To evaluate the characteristics of the hard segments (HS) derived from this novel aromatic isocyanate (isoPABA), in the following set of experiments, an equimolar ratio of isoPABA and short diol were combined, leading to a hard segment model polyurethane. Two types of diols were employed, namely BD and EDA-2CL. Both chain extenders exhibit high structural symmetry, which could lead to crystalline hard segments. Additionally, the EDA-2CL chain extender contains two hydrolyzable ester groups and its degradation by-products are non-toxic [12]. The synthesis of these HS model polyurethanes was carried out under standard conditions (3 h at 80 °C and rt overnight) and using anhydrous solvents to avoid secondary reactions. In both model PUs, FTIR spectra in the carbonyl stretching vibration region (1800–1650 cm^{-1}) presented three major spectral components (Figure S3). The strong band at higher frequencies (1729 cm^{-1}) could be assigned to "free" carbonyl groups and at lower frequencies (1710 cm^{-1}) could be due to hydrogen-bonded carbonyl groups [21]. The third intensive band at 1644 cm^{-1} could be associated to amide I. Additionally, for the isoPABA–EDA-2CL HS model, the vibration of the ester group appeared overlapped at 1710 cm^{-1}. Similarly, the amide II band located at 1533 cm^{-1} corresponded to the amide and urethane groups. The chemical structures of these HS model polyurethanes were confirmed by proton NMR. The signal assignment to its corresponding proton is shown in Figures S4 and S5.

The thermal transitions detected in these HS model polyurethanes were studied by DSC. In Figure 2, the second heating scan of the isoPABA–EDA-2CL derived polyurethane is shown. In this thermogram, first, a glass transition was detected at 72.4 °C from the amorphous phase of the polymer. Subsequently, two exothermic crystallization peaks and two endothermic melting peaks were detected alternately. This curve demonstrates that this HS can crystallize, with a complex behavior. The sum of the melting endotherms gives a substantially higher value than the sum of the crystallization exotherms, thus part of the HS crystallized during the fast-cooling step from the melt, offers indirect proof of the high capacity for crystallization of isoPABA–EDA-2CL HS. This crystallization could drive a better phase separation that would improve the performance of the segmented polyurethane, based on this aromatic isocyanate and this hydrolyzable chain extender.

In contrast, the BD-based HS model polyurethane (isoPABA–BD) (Figure S6) showed a glass transition at 58.2 °C and crystallization and melting peaks with similar normalized areas. Thus, this polyurethane was initially amorphous and, therefore, the crystallization capacity is lower than for isoPABA–EDA-2CL. The Tg value for isoPABA–EDA-2CL is significantly higher and could be explained by the restriction imposed by the rigid crystals present after cooling or by the stiffening effect of the amide groups or both. In addition, the crystals of isoPABA–EDA-2CL melt at a higher temperature than the crystals of isoPABA–BD.

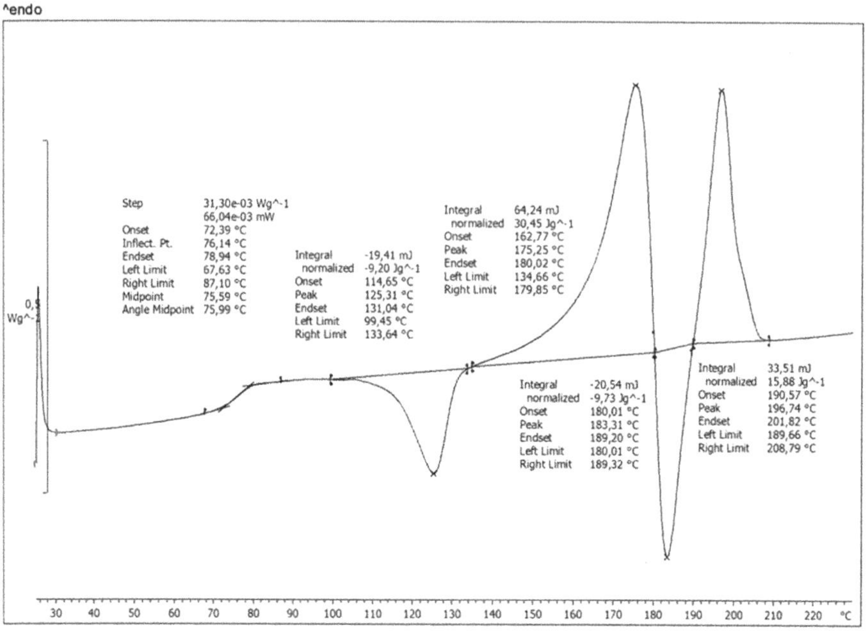

Figure 2. Thermogram of polyurethane based on isoPABA and EDA-2CL.

3.3. Synthesis of Segmented Poly(Ester-Urethane)s

Different diols based on polycaprolactone were used as the soft segment. The selection of this polyester was motivated by the fact that it is a biocompatible and hydrolytically or enzymatically degradable material. It is well-known that the use of PCL-based polymer materials is widespread in the biomedical field, such as in scaffolds [22] or amphiphilic co-networks [23–25]. In addition, its main degradation product (6-hydroxyhexanoic acid) is easily transformed into adipic acid [26]. In this manner, the three building blocks of poly(ester-urethane)s can biodegrade into non-harmful side products. For the synthesis of these materials, three PCL-diols with molecular weights ranging from 500 to 2000 g mol^{-1} were selected. Typically, linear thermoplastic polyurethanes are based on macroglycols, with a molecular weight range between 1000 and 2000 g mol^{-1}, which leads to good mechanical properties [27]. In this second set of experiments, the chain extenders described above were used, and also two polyurethanes based on common diisocyanates (MDI and HDI); PCL2054 and EDA-2CL, were prepared to compare with the corresponding polyurethane based on isoPABA. For all the poly(ester-urethane)s prepared, the hard segment content was set to 50% by weight, while maintaining an equimolar ratio of the NCO:OH groups. Synthesis schemes and images of these pol(ester-urethane)s are shown in Figure S7.

Regarding the solubility of the poly(ester-urethane)s, it was found that these materials were completely soluble in polar solvents, such as DMF, DMAc or DMSO, and in hexafluoroisopropanol (HFIP). The molecular weights of these synthesized polymers were determined by Size-Exclusion Chromatography (SEC). The number and weight average molecular weights and polydispersity index of these polymers are collected in Table 1. The poly(ester-urethane)s were named PUX-YZ-50 following this code: PU refers to poly(ester-urethane); X refers to the isocyanate (P = isoPABA, M = MDI and H = HDI); Y refers to the chain extender type (E = EDA-2CL and B = BD) and Z refers to the Mn (RMN) of the PCL diol (2 = 2054, 1 = 1100 and 5 = 519).

Table 1. SEC results for the synthetized poly(ester-urethane)s.

Entry	PU	CE	Mn (KDa)	Mw (KDa)	Đ
1	PUP-2E-50	EDA-2CL	74.4	114.4	1.54
2	PUP-1E-50	EDA-2CL	40.2	61.6	1.53
3	PUP-5E-50	EDA-2CL	94.2	142.4	1.51
4	PUM-2E-50	EDA-2CL	51.9	109.5	2.11
5	PUH-2E-50	EDA-2CL	68.6	148.3	2.16
6	PUP-2B-50	BD	32.3	47.9	1.48
7	PUP-1B-50	BD	36.0	61.9	1.72
8	PUP-5B-50	BD	56.7	85.7	1.51

The introduction of EDA-2CL as a chain extender led to an increase in the molecular weight of their polyurethanes (entries 1–3), with respect to the classical 1,4-butanediol (entries 6–8). Indeed, the influence of the chain extender on the molecular weights was much more pronounced than the length of the soft PCL segment. In this sense, comparing PUP-2E-50 (entry 1) with PUP-2B-50 (entry 6), the molecular weight was two-fold higher, while the polydispersity remained practically unchanged; something similar was observed comparing entries 3 and 8. However, an increase in the length of the soft segment did not lead to an increase in the molecular weight of the polyurethanes.

The polydispersity values were broader for the polyurethanes bearing classical isocyanates (entries 4 and 5); however, for all the PABA-polyurethanes, the polydispersity was slightly higher than 1.5, even though they reached similar molecular weights to the MDI- or HDI-based polyurethanes (entries 4–5). GPC traces of these polymers are collected in Figure S8

3.4. Characterization of Segmented Poly(ester-Urethane)s

3.4.1. Nuclear Magnetic Resonance (NMR)

The ^{1}H-NMR spectrum of segmented polyurethane PUP-2E-50 in deuterated DMSO is shown in Figure 3.

After the insertion of the PABA moiety into the polyurethane chain, the distance between the two doublets of the AB aromatic system is considerably reduced, indeed, this distance is smaller than the isoPABA or diamino-PABA products (see Figure 1), whilst the propylene chain remains nearly unchanged. Additionally, the signal positions of the methylenes, labelled **6** (4.08 ppm) and ε (3.97 ppm), demonstrate that the chain extender (EDA-2CL) and the soft segment (HO-PCL-OH) have also been inserted into the polyurethane chain. Moreover, since the isocyanate group was not detected, it confirmed that the reaction was completed and the stoichiometry between the three reactants was properly adjusted.

3.4.2. Infrared Analysis (FTIR–ATR)

Infrared spectroscopy is a very powerful tool for the characterization of polyurethanes, not only in the region of carbonyl groups, as discussed above, but valuable information can also be extracted through other bands. The FTIR spectra of isoPABA, PUP-2E-50 and diamino-PABA are shown in Figure 4.

In the case of isoPABA (red spectrum), the intense bands at 2280 and 1700 cm^{-1} were attributed to isocyanate (NCO) and ester (C=O) stretching vibrations, respectively. It should be noticed that the strong band of the isocyanate group completely vanished once the poly(ester-urethane) chain was formed. Furthermore, for PUP-2E-50, new absorption bands appeared at 3310 and 1638 cm^{-1}, which were attributed to the stretching vibration of N–H and amide I of the urethane groups, respectively. Additionally, a single band for the N–H bond was observed in the polyurethane, compared to the three characteristic bands of the aromatic amines, such as diamino-PABA (blue spectrum), demonstrating that the insertion of the PABA motif into the poly(ester-urethane) matrix was carried out through urethane bonds. Finally, the vibrations located at 1218 and 1166 cm^{-1} were attributed to the soft phase (PCL diols), which correspond to O–C–O stretching and symmetric C–O–C stretching, respectively, of this polyester [28,29].

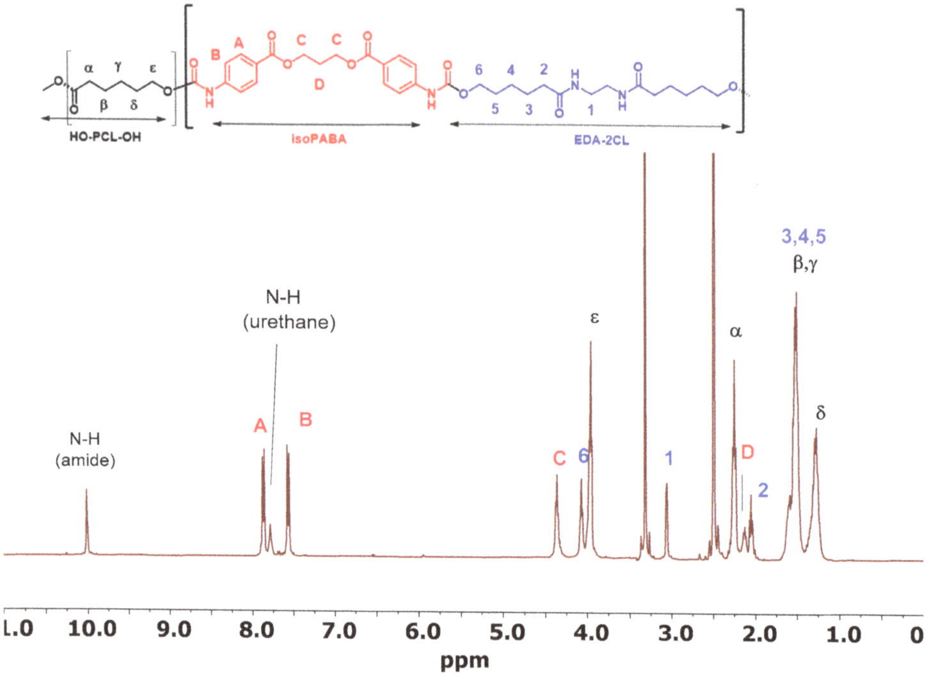

Figure 3. ^{1}H-NMR spectrum and chemical structure of poly(ester-urethane) PUP-2E-50 in deuterated DMSO.

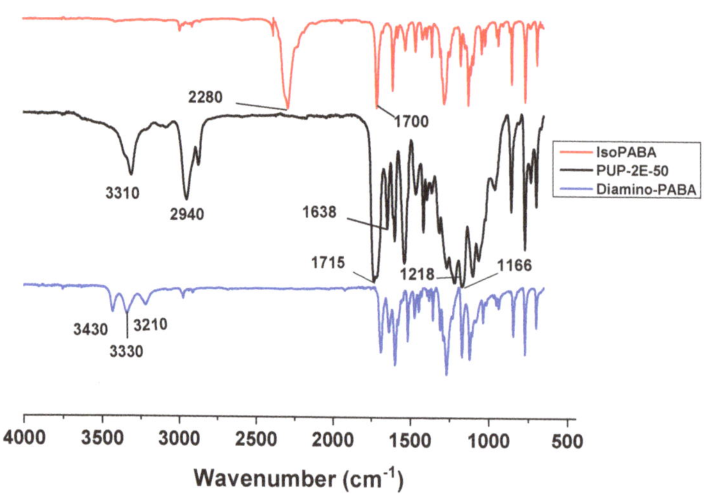

Figure 4. FTIR spectra of isoPABA, PUP-2E-50 and diamino-PABA.

3.4.3. Thermogravimetric Analysis (TGA)

Through this technique, the materials processing window can be known, and in the particular case of segmented polyurethanes, the chemical composition could be estimated. As urethane groups are more thermally labile than ester or amide counterparts, in many cases, the first drop is related to the thermal degradation of the hard segment, while the mass loss associated with the soft segment occurs at higher temperatures [30]. In isoPABA-based polyurethanes, the first weight lost at 276–288 °C, was only slightly higher than 50% of the weight (Figure 5B,C). All isoPABA-based polyurethanes behaved similarly, irrespective of the chain extender or the PCL length. However, polyurethanes based on the classical isocyanates, MDI (PUM-2E-50) or HDI (PUH-2E-50) (Figure 5A), showed a first weight lost much higher than the amount of the hard segment (50%).

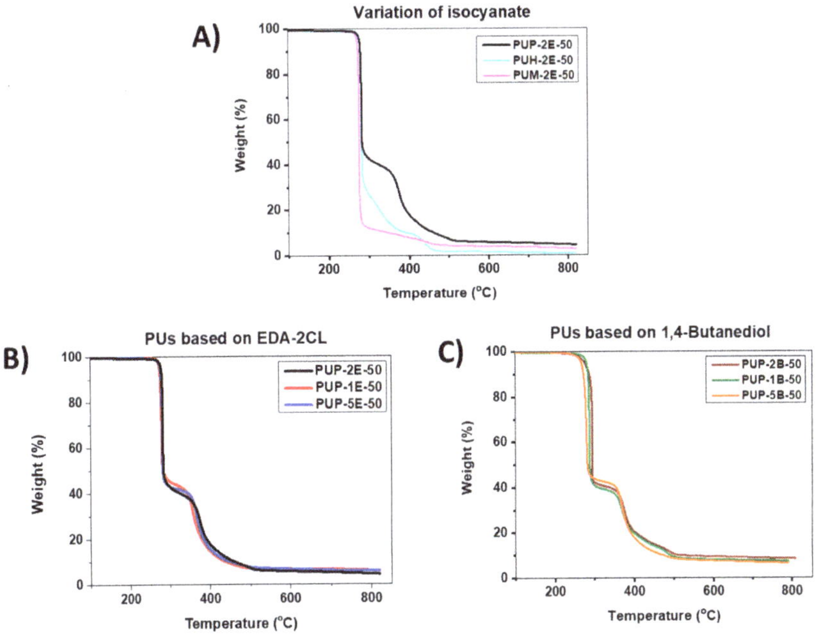

Figure 5. TGA curves of synthesized polyurethanes under inert atmosphere. (**A**) Poly(Ester-urethane)s varing the isocyanate. (**B**) PABA-Polyurethanes with hydrolysable chain-extender. (**C**) PABA-Polyurethanes using 1,4-butanediol as chain-extender.

We do not have an explanation for this difference when the initial degradation temperature is very similar for all the synthesized polyurethanes, irrespective of the diisocyanate, chain extender or PCL length (280 °C). An increase in the thermal stability of the polyurethanes when PABA motifs were incorporated has been previously reported [18], although in those polyurethanes, the PABA structure acted as a chain extender (diamino-PABA) and not as an isocyanate, and so the insertion occurred through a urea bond instead of urethane groups as in our work.

3.4.4. Differential Scanning Calorimetry (DSC)

We found in the HS models that the symmetrical structure of the chain extenders (EDA-2CL and BD) and isoPABA allowed for their crystallization. Therefore, a phase separation between the different segments within the polyurethane matrix was expected. The thermal properties of these polyurethanes obtained from the DSC curves in the second heating scan are listed in Table 2.

Table 2. Thermal properties of the synthesized poly(ester-urethane)s.

Entry	PU	CE	Tg (°C)	T_{m1} (°C)	ΔH_1 (J/g)	X_i (PCL) (%)	Tm_2 (°C)	ΔH_2 (J/g)	ΔH_2 (J/g_{HS})
1	PUP-2E-50	EDA-2CL	−46	31	3.1	4.6	167/188	5.4 + 5	20.8
2	PUP-1E-50	EDA-2CL	−20				142	6.8	13.68
3	PUP-5E-50	EDA-2CL	5				142	4.3	8.6
4	PUM-2E-50	EDA-2CL	−52	32	6.5	9.6	109	2.43	4.9
5	PUH-2E-50	EDA-2CL	−50	35	4.7	6.9	186	13.7	27.4
6	PUP-2B-50	BD	−41	35	0.2	0.3	143	1	2
7	PUP-1B-50	BD	−22				188	8.1	16.2
8	PUP-5B-50	BD	8				134/167	0.9 + 1.18	4.2

The isoPABA-derived poly(ester-urethane)s showed a clear dependence between the PCL molecular weight and the glass transition temperature of the soft segment. In this respect, as the Mn of the polyester diol decreased, the Tg value increased from −46 to 5 °C for the EDA-2CL series (entries 1 to 3) and from −41 to 8 °C for the BD series (entries 6 to 8). This increase is related, on the one hand, to the restrictions imposed on the chain ends by the hard segments and, on the other hand, to the increase in the density of urethane groups within the polymeric matrix. Likewise, for both series, the crystallization of the soft segment was only appreciated for the highest molecular weight (PCL2054), and for the rest of the PCL molecular weights (1100 and 519 g·mol^{-1}), the soft segments were amorphous.

The glass transition temperature for the polyurethanes, based on the classical isocyanates (entries 4 and 5), was slightly higher than the PCL homopolymer of 2054 g/mol [31,32], indicating that the soft segments are well separated into practically pure phases. For the polymers PUP-2E-50 and PUP-2B-50, the Tg was slightly higher, denoting that the phase separation in these polyurethanes is slightly less effective, or that the restrictions imposed by the HS crystalline phase were greater. This observation is in line with the values of the enthalpy of fusion of the crystallized soft segments. The highest enthalpies were found for the classical polyurethanes (entries 4 and 5), whilst the lowest was for PUP-2B-50 (entry 6), meaning that a higher amount of pure SS phase was present in the former and a lower amount in the latter.

All the synthesized polyurethanes showed a hard segment melting temperature that varied between 140 and 188 °C, again pointing to thermodynamic incompatibility between the segments, rendering in phase separation within the polymeric matrix. Comparing the melting temperatures of the hard segments of the polyurethanes from the aromatic isocyanates, it is observed that the PABA-derived poly(ester-urethane)s showed a higher Tm than their classical partner (PUM-2E-50). Probably, the presence of the ester groups within the isoPABA structure could increase the polar interactions within the hard segment, leading to an increase in the melting point of the hard phase. In addition, the flexibility of the propylene spacer should also be addressed, which could allow the aromatic rings to be effectively accommodated within the hard segment. However, for MDI, the methylene group between both aromatic rings could hardly provide such flexibility. This is supported by the enthalpy of fusion values (ΔH_2), which are higher for the isoPABA derivatives compared to PUM-2E-50. In the case of the polyurethane derived from the aliphatic isocyanate HDI, the enthalpy of fusion of the crystalline phase (ΔH_2) was the highest of the series, probably due to the symmetry and flexibility of the isocyanate and the absence of side groups that hinder the crystallization of this segment. Therefore, due to all these structural factors, the hard segment (HDI+EDA-2CL) of this polyurethane was the most crystalline.

3.4.5. Mechanical Properties

The mechanical properties of synthetized poly(ester-urethane)s were evaluated by tensile stress–strain tests (Figure S9). The results are summarized in Table 3. Based on these results, it could be observed that there are two types of mechanical behavior due to polymer crystallinity.

Table 3. Mechanical properties of the synthesized poly(ester-urethane)s.

Entry	PU	CE	Stress at Break (MPa)	Strain at Break (%)	Modulus
1	PUP-2E-50	EDA-2CL	24 ± 2	1060 ± 110	55 ± 4
2	PUP-1E-50	EDA-2CL	6.8 ± 0.2	220 ± 30	34 ± 2
3	PUP-5E-50	EDA-2CL	2.3 ± 1.1	2800 ± 500	4.0 ± 0.8
4	PUM-2E-50	EDA-2CL	18 ± 5	440 ± 60	65 ± 15
5	PUH-2E-50	EDA-2CL	23 ± 2	810 ± 120	64 ± 12
6	PUP-2B-50	BD	6.7 ± 0.3	120 ± 18	43 ± 5
7	PUP-1B-50	BD	5.7 ± 0.5	130 ± 5	28 ± 3
8	PUP-5B-50	BD	0.8 ± 0.2	330 ± 40	3 ± 1

On the one hand, the polyurethanes based on the shorter soft segment (PCL519) (entries 3 and 8) showed typical elastomeric behavior, with low modulus and, after rupture, the specimens practically recovered their initial shape.

On the other hand, when using longer soft segments (PCL1100 and PCL2054) (entries 1, 2, 4, 5 and 6), the polyurethanes behaved like plastics and did not recover their original shape. At low deformations, the stress grew rapidly until a yield point was reached. After this point, stress growth was lower until the break. This behavior is consistent with a co-continuous morphology, where before reaching the yield point the response was governed by the hard segment phase, while after this point the response of the material was determined by the soft segment phase until the break.

The mechanical properties depended mainly on the molecular weight of the polyurethane. As it can be seen in Table 1, when BD was used as a chain extender with isoPABA, the molecular weights were discrete, leading to poor mechanical properties. This was the case also for the polyurethane from PCL1100, isoPABA and EDA-2CL. For the other polymers, with sufficiently high molecular weights, tensile strains were high.

3.4.6. Hydrolytic and Enzymatic In Vitro Degradation

The polymer samples were immersed in a phosphate buffer solution (pH = 7.4) and incubated at 37 °C and their weight changes in the hydrated state were monitored at certain time intervals. Figure 6 shows the curves of the polymers obtained from isoPABA and EDA-2CL as hard segments (PUP-5E-50, PUP-1E-50 and PUP-2E-50). For all the studied polymers, weight change curves presented two differentiated states. At the beginning of the immersion, a maximum water absorption was reached, followed by a very slow decay of weight over time. This behavior was independent of the polymer studied. The shape of the weight loss curves in the swollen state is characteristic of surface erosion, with very low total degradation.

The maximum water absorption showed a relationship with the molecular weight of the soft segment, with the lower molecular weights (PUP-5E-50 and PUP-1E-50) reaching higher maximum absorptions than PUP-2E-50. This behavior would be due to the fact that the crystalline fraction, presented in PUP-2E-50, would hinder the inclusion of water within the polymeric structure, whereas the completely amorphous character of the SS in the polyurethanes based on PUP-5E-50 and PUP-1E-50 favors water access. This trend is in line with the DSC results described above and has been previously reported [33,34]. For all the cases studied, the maximum absorption occurred practically within the first 24 h of immersion and subsequently, the weight of the polymer decreased progressively.

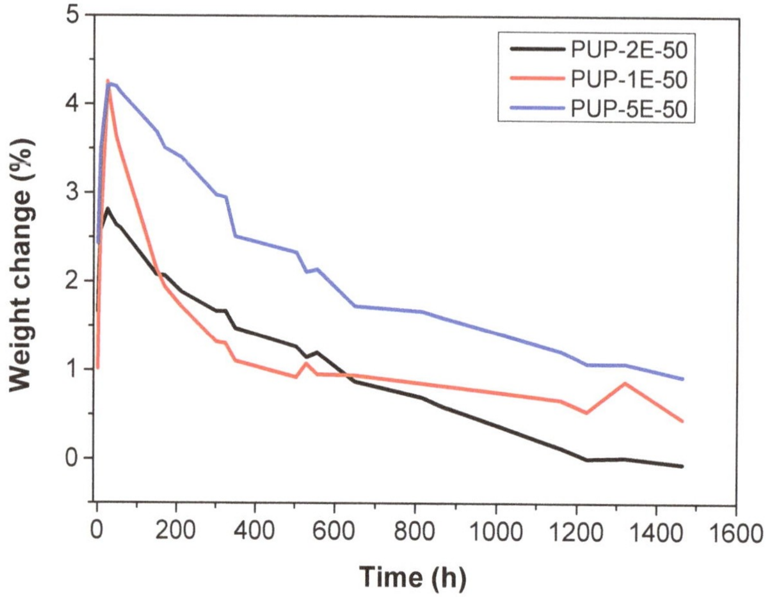

Figure 6. Variation of weight with time of immersion in PBS for polyurethanes based on isoPABA and EDA-2CL.

Regarding the nature of the isocyanate (Figure S10), it was observed that the maximum water absorption was slightly higher for the isoPABA-polyurethane (PUP-2E-50, 2.81%) compared to the classic isocyanates, HDI (PUH-2E-50, 2.18%) and MDI (PUM-2E-50, 2.16%). This increase in water absorption could be promoted by the polarity conferred by the ester groups present within the structure of the novel aromatic isocyanate. Similarly, the nature of the chain extender had a similar effect (Figure S11) because the polar (amide) groups provided by EDA-2CL would increase hydrophilicity through hydrogen bonding with water molecules, leading to increased water adsorption. Comparing both effects, it was appreciated that the nature of the chain extender was more relevant with respect to the effect provided by the isocyanate, since the maximum absorption of PUP-2E-50 amounted to 2.81%, while for PUP-2B-50, it was practically not exceeding 1.50%.

Globally, maximum water absorption was very low in these polyurethanes, ranging from 1.6 to 4.3%.

In the second stage of film degradation, the degradation rate was slightly higher for polymers with a higher swelling index, i.e., polymers based on isoPABA, EDA-2CL or with short PCL diol. This could be caused by the fact that the access of water to the hydrolyzable groups is less hindered, and the degraded products diffused more easily.

Subsequently, the enzymatic degradation of these polymers was also evaluated gravimetrically using the porcine liver esterase (PLE) enzyme.

As shown in Table 4, after 7 days of incubation in absence of an enzyme, the PABA-based polyurethanes showed a small swelling between 1.5 and 3%. However, in the presence of the PLE, the degradation of the polyurethanes is more marked. The enzyme PLE promotes the hydrolysis of ester bonds contained in the soft segment. This effect is smaller on the polyurethane with a longer PCL (PUP-2E-50) because the crystalline fraction is more resistant to enzymatic degradation, while the fully amorphous polymers (PUP-1E-50 and PUP-5E-50) present a molar ratio of PCL higher than the longer counterpart and, therefore, a higher number of hydrolysis points. Different levels of degradation are reported depending on the molecular weight, crystalline or amorphous content, or the presence of C–C strong bonds, which are resistant to attack by enzymes [35,36].

Table 4. Weight change after 7 days in presence (PLE(+)) and in absence (PLE(−)) of the esterase.

	Weight Change (7 days)	
	PLE (−)	PLE (+)
PUP-2E-50	1.67%	−1.90%
PUP-1E-50	2.13%	−3.42%
PUP-5E-50	2.94%	−4.37%

3.4.7. Cytotoxicity Test

Toxicity assays were developed in the absence and presence of PLE to study the toxicity of the degradation products. Figure 7 shows the cytotoxicity on HOb and HC-a cells.

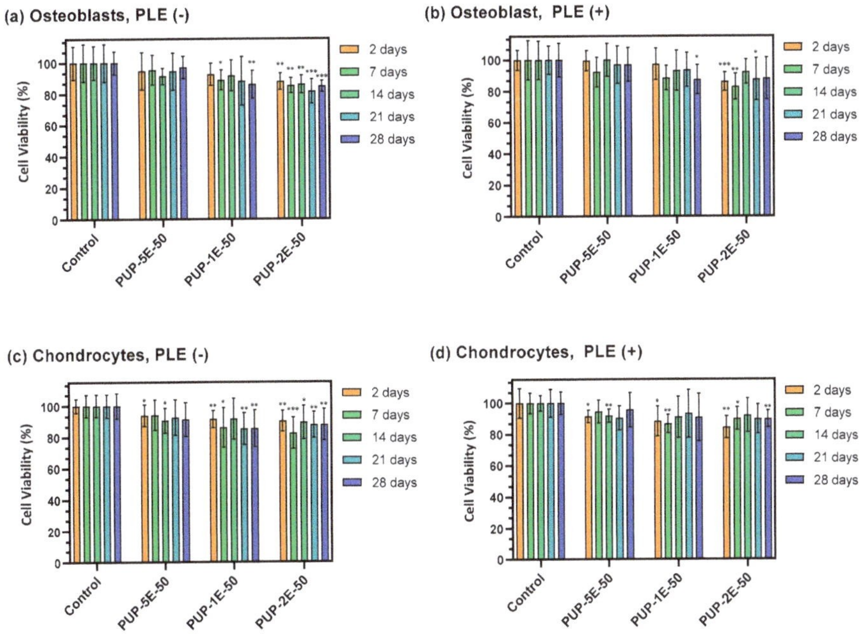

Figure 7. Cytotoxicity analysis of PABA-based polyurethanes on (**a**) HOb cells in absence of PLE, (**b**) HOb cells in presence of PLE, (**c**) HC-a cells in absence of PLE and (**d**) HC-a cells in presence of PLE. Data show the mean value and standard deviation (n = 12). Statistical differences between polyurethanes and control were analyzed by analysis of variances (ANOVA) and Tukey's Test, considering statistically significant difference when $p \leq 0.05$ (*), $p \leq 0.01$ (**) and $p \leq 0.001$ (***).

PABA-based polyurethanes showed a small but significant toxicity on both HOb and HC-a cells, depending on the molecular weight of the PCL. Low molecular weight PCL based-polyurethanes (PUP-5E-50) did not show a significant toxicity on HOb but started to show significant differences on HC-a cells. For medium and high molecular weight PCL (PUP-1E-50 and PUP-2E-50), cell viability reached close to 80%, being independent of the presence or absence of PLE. Despite the fact that aromatic diisocyanates are mentioned as cytotoxic, in this work, the contents of aromatic diisocyanates presented no cytotoxic results; other works also reported non mutagenic behavior for polyurethanes obtained with this type of diisocyanates [37].

3.4.8. Adhesion and Proliferation Assays

Adhesion and proliferation assays were made using Alamar Blue tests. Figure 8 shows the adhesion and proliferation of HOb and HC-a cells on poly(ester-urethane) films at 1, 7 and 14 days.

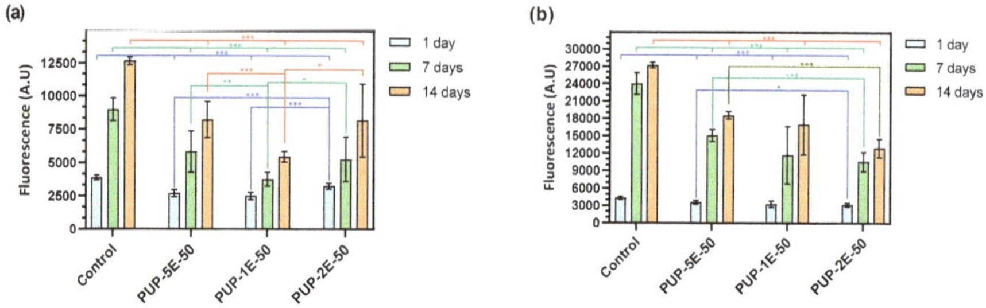

Figure 8. Adhesion and proliferation assay of (**a**) HOb cells (**b**) HC-a cells on the surface of PABA-based polyurethanes. Data show the mean value and standard deviation (n = 12). Statistical differences were analyzed by analysis of variances (ANOVA) and Tukey's Test, considering statistically significant difference when $p \leq 0.05$ (*), $p \leq 0.01$ (**) and $p \leq 0.001$ (***).

All the PUs allow the adhesion and proliferation of HOb and HC-a cells. The lower adhesion in comparison with the control is due to the different topography and hydrophilicity of the PUs; cells need time to produce extracellular matrix to adhere on the surface. The initial adhesion of the cells is significantly higher for polyurethanes with high molecular weight PCL (PUP-2E-50). This could be due to the differences in the elastic modulus of the systems; for example, HOb cells present higher adhesion efficiency for higher elastic modulus [38]. Cell proliferation increases over the time but depends on the PCL molecular weight. Proliferation for polyurethanes with high molecular weight PCL (PUP-1E-50 and PUP-2E-50) is lower than PUP-5E-50. This could be correlated with the results in Figure 7, where PUP-2E-50 showed a significant toxicity. This increase in the toxicity also has an effect on the proliferation of the cells on the polyurethanes.

Other studies on polyurethanes reported that ≥70% of cell viability is promising for biological applications [39]. These good results in cell adhesion and toxicity make this kind of novel polyurethane a good option to building extracellular matrices. In addition, the mechanical properties of these systems are tunable by the election of different soft and hard segments. There are many reports of polyurethanes with tissue engineering applications [40,41]. The literature mentioned that the rupture of ester linkages increases hydrolytic and enzymatic degradation. In addition, using molecules such as esterases increments the rate of degradation in polyurethanes. Aliphatic hard segments have been chosen in order to avoid cytotoxic degradation products, but there are studies with aromatic molecules, such as polyaniline, used to obtain polyurethanes with self-healing behaviors, that result in non-cytotoxic polyurethanes, which are shown to have excellent cellular adherence [42]. In this work, an aromatic diisocyanate (isoPABA) was prepared from aromatic diamine, that is reportedly not cytotoxic or mutagenic, according to the MTT results. On the other hand, the Alamar Blue results show good cellular adhesion and proliferation for the PUs.

4. Conclusions

A series of polyurethanes bearing p-aminobenzoate units were prepared through the synthesis of the novel symmetric aromatic isocyanate isoPABA. This diisocyanate led to polyurethanes with high molecular weights when the hydrolyzable chain extender EDA-2CL was used. The chemical structures of the biocompatible segmented polyurethanes were confirmed by several techniques (FTIR–ATR and NMR).

As shown by TGA, although the urethane groups are the thermally weakest linkages, the thermal stability of all the polymers presented was sufficiently high, with a wide processing window.

Due to the symmetrical structure of the novel isocyanate and the flexibility introduced by its propylene spacer, the hard segments of the polyurethanes crystallized easily, favoring phase separation within the poly(ester-urethane) matrix. However, the enthalpies of fusion of the crystallized soft segments were lower for the PABA-polyurethanes than those of their classical counterparts, which was probably due to a lower fraction of pure HS. Although the melting enthalpies of the isoPABA-based HS increased with respect to its aromatic MDI partner, the aliphatic isocyanate HDI was shown to be the most crystalline HS.

Soft segments derived from low molecular weight PCL (PCL519) were completely amorphous, and the corresponding polyurethanes elastomeric, recovering the initial shape after rupture. Poly(ester-urethane)s based on PCL1100 and PCL2054 showed the behavior of plastics. The polyurethanes bearing the hydrolyzable chain extender reached higher molecular weights than the BD-polyurethanes, achieving better performance and mechanical properties.

For all the poly(ester-urethane)s bearing isoPABA motifs and EDA-2CL chain extenders, cell viability after 28 days was higher than 80%, demonstrating that these polymers were non-toxic. Moreover, Alamar Blue results indicated a good cell adhesion and cytotoxicity on the surfaces of these non-toxic biodegradable polyurethanes. Ultimately, the diisocyanate isoPABA is an attractive moiety for the preparation of poly(ester-urethane)s to build extracellular matrices.

Supplementary Materials: The following supporting information can be downloaded at: https://www.mdpi.com/article/10.3390/polym14071288/s1, Figure S1: ATR-FTIR of aromatic diisocyanate isoPABA. Figure S2: 13C-NMR spectrum of isoPABA in deuterated chloroform. Figure S3: Extended FTIR-Spectrum between 2000 and 1000 cm-1 of model polyurethane isoPABA-EDA-2CL. Figure S4: 1H-NMR spectrum of model polyurethane isoPABA-BD in DMSO-d6. Figure S5: 1H-NMR spectrum of polyurethane isoPABA-EDA-2CL in deuterated DMSO. Figure S6: DSC curve of model polyurethane isoPABA-BD. Figure S7: Synthesis scheme and images of poly(ester-urethane)s. Figure S8: GPC traces of synthesised poly(ester-urethane)s. Figure S9: Stress-strain curves of synthetised poly(ester-urethane)s. Figure S10: Weight change with immersion time in phosphate buffer solution at 37 °C for hydrated poly(ester-urethane)s based on PCL2054. Figure S11: Weight change with immersion time in phosphate buffer solution at 37 °C for hydrated poly(ester-urethane)s with different chain extender.

Author Contributions: Conceptualization, Á.M.-F. and R.N.; methodology, Á.M.-F., R.N. and M.R.A.; investigation, A.R.H.-S., D.M.G.-G., L.G.-F. and R.A.R.-J.; writing—original draft preparation, A.R.H.-S. and R.N.; writing—review and editing, Á.M.-F. and R.N.; supervision, Á.M.-F., R.N. and M.R.A.; project administration, R.N. and M.R.A.; funding acquisition, Á.M.-F., R.N. and M.R.A. All authors have read and agreed to the published version of the manuscript.

Funding: This research was funded by the Spanish Ministry of Science and Innovation (MINECO) through the research projects MAT2017-87204-R, RTI2018-096636-J-100, PID2020-119047RB-I00, PID2020-114086RB-100 and PLEC2021-007793, CIBER-BBN (Instituto de Salud Carlos III, Spain) and by CSIC through the intramural project 20190E004.

Acknowledgments: The donation of MDI by Lubrizol Advanced Materials Spain SL is acknowledged.

Conflicts of Interest: The authors declare no conflict of interest.

References

1. Das, A.; Mahanwar, P. A brief discussion on advances in polyurethane applications. *Adv. Ind. Eng. Polym. Res.* **2020**, *3*, 93–101. [CrossRef]
2. Lores, N.J.; Hung, X.; Talou, M.H.; Abraham, G.A.; Caracciolo, P.C. Novel three-dimensional printing of poly(ester urethane) scaffolds for biomedical applications. *Polym. Adv. Technol.* **2021**, *32*, 3309–3321. [CrossRef]
3. Mustafov, S.D.; Sen, F.; Seydibeyoglu, M.O. Preparation and characterization of diatomite and hydroxyapatite reinforced porous polyurethane foam biocomposites. *Sci. Rep.* **2020**, *10*, 13308. [CrossRef] [PubMed]

4. Abbasnezhad, N.; Zirak, N.; Shirinbayan, M.; Kouidri, S.; Salahinejad, E.; Tcharkhtchi, A.; Bakir, F. Controlled release from polyurethane films: Drug release mechanisms. *J. Appl. Polym. Sci.* **2020**, *138*, 50083. [CrossRef]
5. Molina, G.A.; Elizalde-Mata, A.; Hernandez-Martinez, A.R.; Fonseca, G.; Cruz Soto, M.; Rodriguez-Morales, A.L.; Estevez, M. Synthesis and Characterization of Inulin-Based Responsive Polyurethanes for Breast Cancer Applications. *Polymers* **2020**, *12*, 865. [CrossRef] [PubMed]
6. Clemitson, I.R. *Castable Polyurethane Elastomers*; CRC Press: Boca Raton, FL, USA, 2008.
7. Chan-Chan, L.H.; Tkaczyk, C.; Vargas-Coronado, R.F.; Cervantes-Uc, J.M.; Tabrizian, M.; Cauich-Rodriguez, J.V. Characterization and biocompatibility studies of new degradable poly(urea)urethanes prepared with arginine, glycine or aspartic acid as chain extenders. *J. Mater. Sci. Mater. Med.* **2013**, *24*, 1733–1744. [CrossRef] [PubMed]
8. Marcos-Fernández, A.; Abraham, G.A.; Valentín, J.L.; Román, J.S. Synthesis and characterization of biodegradable non-toxic poly(ester-urethane-urea)s based on poly(ε-caprolactone) and amino acid derivatives. *Polymer* **2006**, *47*, 785–798. [CrossRef]
9. Zumbardo-Bacelis, G.A.; Meza-Villegas, L.A.; Pérez-Aranda, C.A.; Vargas-Coronado, R.; Castillo-Cruz, O.; Montaño-Machado, V.; Mantovani, D.; Cauich-Rodríguez, J.V. On arginine-based polyurethane-blends specific to vascular prostheses. *J. Appl. Polym. Sci.* **2021**, *138*, 51247. [CrossRef]
10. Tatai, L.; Moore, T.G.; Adhikari, R.; Malherbe, F.; Jayasekara, R.; Griffiths, I.; Gunatillake, P.A. Thermoplastic biodegradable polyurethanes: The effect of chain extender structure on properties and in-vitro degradation. *Biomaterials* **2007**, *28*, 5407–5417. [CrossRef]
11. Xu, C.; Hong, Y. Rational design of biodegradable thermoplastic polyurethanes for tissue repair. *Bioact. Mater.* **2021**, *15*, 250–271. [CrossRef]
12. Báez, J.E.; Ramírez, D.; Valentín, J.L.; Marcos-Fernández, Á. Biodegradable Poly(ester–urethane–amide)s Based on Poly(ε-caprolactone) and Diamide–Diol Chain Extenders with Crystalline Hard Segments. Synthesis and Characterization. *Macromolecules* **2012**, *45*, 6966–6980. [CrossRef]
13. Caracciolo, P.C.; Buffa, F.; Abraham, G.A. Effect of the hard segment chemistry and structure on the thermal and mechanical properties of novel biomedical segmented poly(esterurethanes). *J. Mater. Sci. Mater. Med.* **2009**, *20*, 145–155. [CrossRef] [PubMed]
14. Acik, G.; Karabulut, H.R.F.; Altinkok, C.; Karatavuk, A.O. Synthesis and characterization of biodegradable polyurethanes made from cholic acid and l-lysine diisocyanate ethyl ester. *Polym. Degrad. Stab.* **2019**, *165*, 43–48. [CrossRef]
15. Hassan, M.K.; Mauritz, K.A.; Storey, R.F.; Wiggins, J.S. Biodegradable aliphatic thermoplastic polyurethane based on poly(ε-caprolactone) andL-lysine diisocyanate. *J. Polym. Sci. Part A Polym. Chem.* **2006**, *44*, 2990–3000. [CrossRef]
16. Wang, Z.; Yu, L.; Ding, M.; Tan, H.; Li, J.; Fu, Q. Preparation and rapid degradation of nontoxic biodegradable polyurethanes based on poly(lactic acid)-poly(ethylene glycol)-poly(lactic acid) andl-lysine diisocyanate. *Polym. Chem.* **2011**, *2*, 601–607. [CrossRef]
17. Rubio Hernandez-Sampelayo, A.; Navarro, R.; Marcos-Fernandez, A. Preparation of High Molecular Weight Poly(urethane-urea)s Bearing Deactivated Diamines. *Polymers* **2021**, *13*, 1914. [CrossRef]
18. Oprea, S.; Potolinca, V.O.; Oprea, V. Synthesis and characterization of novel poly(urethane-urea) elastomers based on 1,3-propanediol bis(4-aminobenzoate) as chain extender. *Mater. Today Commun.* **2020**, *22*, 100860. [CrossRef]
19. Malamed, S.F. *Handbook of Local Anesthesia-E-Book*; Elsevier Health Sciences: Amsterdam, The Netherlands, 2014.
20. Navarro, R.; García, C.; Rodríguez-Hernández, J.; Elvira, C.; Marcos-Fernández, A.; Gallardo, A.; Reinecke, H. General approach to prepare polymers bearing pendant isocyanate groups. *Polym. Chem.* **2020**, *11*, 5140–5146. [CrossRef]
21. Irusta, L.; Fernandez-Berridi, M. Aromatic poly (ether-urethanes): Effect of the polyol molecular weight on the photochemical behaviour. *Polymer* **1999**, *40*, 4821–4831. [CrossRef]
22. Shor, L.; Güçeri, S.; Wen, X.; Gandhi, M.; Sun, W. Fabrication of three-dimensional polycaprolactone/hydroxyapatite tissue scaffolds and osteoblast-scaffold interactions in vitro. *Biomaterials* **2007**, *28*, 5291–5297. [CrossRef]
23. Chandel, A.K.S.; Bera, A.; Nutan, B.; Jewrajka, S.K. Reactive compatibilizer mediated precise synthesis and application of stimuli responsive polysaccharides-polycaprolactone amphiphilic co-network gels. *Polymer* **2016**, *99*, 470–479. [CrossRef]
24. Chandel, A.K.S.; Jewrajka, S.K. *Amphiphilic Polymer Co-Networks: Synthesis, Properties, Modelling and Applications*; Royal Society of Chemistry: London, UK, 2020.
25. Nutan, B.; Chandel, A.K.S.; Jewrajka, S.K. Liquid prepolymer-based in situ formation of degradable poly (ethylene glycol)-linked-poly (caprolactone)-linked-poly (2-dimethylaminoethyl) methacrylate amphiphilic conetwork gels showing polarity driven gelation and bioadhesion. *ACS Appl. Bio Mater.* **2018**, *1*, 1606–1619. [CrossRef] [PubMed]
26. Bartnikowski, M.; Dargaville, T.R.; Ivanovski, S.; Hutmacher, D.W. Degradation mechanisms of polycaprolactone in the context of chemistry, geometry and environment. *Prog. Polym. Sci.* **2019**, *96*, 1–20. [CrossRef]
27. Navarro, R.; Seoane-Rivero, R.; Cuevas, J.M.; Marcos-Fernandez, Á. A novel strategy to polyurethanes with improved mechanical properties by photoactivation of amidocoumarin moieties. *RSC Adv.* **2020**, *10*, 29935–29944. [CrossRef]
28. Elzein, T.; Nasser-Eddine, M.; Delaite, C.; Bistac, S.; Dumas, P. FTIR study of polycaprolactone chain organization at interfaces. *J. Colloid Interface Sci.* **2004**, *273*, 381–387. [CrossRef]
29. Lyu, J.S.; Lee, J.S.; Han, J. Development of a biodegradable polycaprolactone film incorporated with an antimicrobial agent via an extrusion process. *Sci. Rep.* **2019**, *9*, 20236. [CrossRef]
30. Petrović, Z.S.; Zavargo, Z.; Flyn, J.H.; Macknight, W.J. Thermal degradation of segmented polyurethanes. *J. Appl. Polym. Sci.* **1994**, *51*, 1087–1095. [CrossRef]

31. Báez, J.E.; Marcos-Fernández, Á.; Galindo-Iranzo, P. On the Effect of Alkyl End Group in Poly(ε-caprolactone) Oligomers: Preparation and Characterization. *Polym.-Plast. Technol. Eng.* **2011**, *50*, 839–850. [CrossRef]
32. Báez, J.E.; Marcos-Fernández, Á.; Navarro, R. Similarities between homopolymers and triblock copolymers derived from poly(ε-caprolactone) (PCL) macrodiols (HOPCL–E–PCLOH and HOPCL–PEG–PCLOH) and their poly(ester-ether-urethanes): Synthesis and characterization. *Chem. Pap.* **2019**, *73*, 1287–1299. [CrossRef]
33. Gharibi, R.; Agarwal, S. Polyurethanes from Hydrophobic Elastic Materials to Hydrogels with Potent Nonleaching Biocidal and Antibiofilm Activity. *ACS Appl. Polym. Mater.* **2021**, *3*, 4695–4707. [CrossRef]
34. Golmohammadian Tehrani, A.; Makki, H.; Ghaffarian Anbaran, S.R.; Vakili, H.; Ghermezcheshme, H.; Zandi, N. Superior anti-biofouling properties of mPEG-modified polyurethane networks via incorporation of a hydrophobic dangling chain. *Prog. Org. Coatings* **2021**, *158*, 106358. [CrossRef]
35. Nakajima-Kambe, T.; Shigeno-Akutsu, Y.; Nomura, N.; Onuma, F.; Nakahara, T. Microbial degradation of polyurethane, polyester polyurethanes and polyether polyurethanes. *Appl. Microbiol. Biotechnol.* **1999**, *51*, 134–140. [CrossRef] [PubMed]
36. Siracusa, V. Microbial Degradation of Synthetic Biopolymers Waste. *Polymers* **2019**, *11*, 1066. [CrossRef] [PubMed]
37. Javaid, M.A.; Zia, K.M.; Khera, R.A.; Jabeen, S.; Mumtaz, I.; Younis, M.A.; Shoaib, M.; Bhatti, I.A. Evaluation of cytotoxicity, hemocompatibility and spectral studies of chitosan assisted polyurethanes prepared with various diisocyanates. *Int. J. Biol. Macromol.* **2019**, *129*, 116–126. [CrossRef]
38. Portan, D.V.; Deligianni, D.D.; Deligianni, K.; Kroustalli, A.A.; Tyllianakis, M.; Papanicolaou, G.C. Modeling of the interaction between osteoblasts and biocompatible substrates as a function of adhesion strength. *J. Biomed. Mater. Res. Part A* **2018**, *106*, 621–628. [CrossRef]
39. Gonzalez-Garcia, D.M.; Marcos-Fernandez, A.; Rodriguez-Lorenzo, L.M.; Jimenez-Gallegos, R.; Vargas-Becerril, N.; Tellez-Jurado, L. Synthesis and in Vitro Cytocompatibility of Segmented Poly(Ester-Urethane)s and Poly(Ester-Urea-Urethane)s for Bone Tissue Engineering. *Polymers* **2018**, *10*, 991. [CrossRef]
40. Naureen, B.; Haseeb, A.; Basirun, W.J.; Muhamad, F. Recent advances in tissue engineering scaffolds based on polyurethane and modified polyurethane. *Mater. Sci. Eng. C Mater. Biol. Appl.* **2021**, *118*, 111228. [CrossRef]
41. Szczepanczyk, P.; Szlachta, M.; Zlocista-Szewczyk, N.; Chlopek, J.; Pielichowska, K. Recent Developments in Polyurethane-Based Materials for Bone Tissue Engineering. *Polymers* **2021**, *13*, 946. [CrossRef]
42. Shaabani, A.; Sedghi, R.; Motasadizadeh, H.; Dinarvand, R. Self-healable conductive polyurethane with the body temperature-responsive shape memory for bone tissue engineering. *Chem. Eng. J.* **2021**, *411*, 128449. [CrossRef]

polymers

MDPI

Article

Synthesis of Biocompatible Composite Material Based on Cryogels of Polyvinyl Alcohol and Calcium Phosphates

Rustam Sadykov, Daria Lytkina, Ksenia Stepanova and Irina Kurzina *

Faculty of Chemistry, Tomsk State University, 634050 Tomsk, Russia
* Correspondence: kurzina99@mail.ru; Tel.: +7-913-882-1028

Abstract: At the moment, the field of biomedical materials science is actively developing, which aims at creating new functional materials. A developing direction in biomedical materials science is that towards the treatment of diseases associated with bone tissue disorders, using biodegradable composite materials based on polymer and calcium phosphate materials. We developed a material based on polyvinyl alcohol cryogel, mineralized with calcium phosphate. A material based on cryogel of polyvinyl alcohol mineralized with calcium phosphate was developed. The composites were obtained by the method of cyclic freezing–thawing, and the synthesis of calcium phosphates was carried out in situ with heating, stirring, and exposure to microwave radiation. The phase composition, as well as the composition of functional groups, was determined by IR spectroscopy and X-ray phase analysis. Monocytes isolated from human blood showed higher viability compared to the controls.

Keywords: calcium phosphates; hydroxyapatite; cryogel; polyvinyl alcohol; biocompatible material

1. Introduction

Bone defects caused by various infectious diseases, trauma, and genetic factors can lead to dysfunction and problems with the musculoskeletal system. Traditional methods of treatment, such as autologous bone grafting, face such problems as high trauma and the risk of infection, which limit their clinical use [1]. However, bone tissue engineering, namely the use of scaffolds for seeding cells and bioactive factors that allow the restoration and regeneration of bone tissue, avoids these disadvantages [2]. Scaffolds used in bone engineering should have such properties as biocompatibility, bioactivity, high porosity, mechanical strength, and biomimeticity [3,4]. Based on this, we conclude that single-component scaffolds cannot fully meet the requirements for the restoration of bone defects [5].

Bone tissue is a solid and dense substance, consisting mainly of two parts: cortical bone and spongy bone [6]. According to the chemical composition, the bone comprises an organic part, which can be attributed to collagen, and an inorganic part, which includes hydroxyapatite [7]. The creation of materials for implants that combine organic and inorganic components that can provide bone tissue regeneration is one of the main problems in biomedicine. Such materials should promote cell growth, as well as have a structural and mechanical properties similar to bone tissue [8,9].

Hydroxyapatite with the general formula $Ca_{10}(PO_4)_6(OH)_2$ is the main inorganic component of bone tissue [7]. This material has high biocompatibility, bioactivity, is non-toxic [10], and is also capable of improving the mechanical properties of scaffolds [11]. The biological activity of hydroxyapatite under physiological conditions is due to the formation of a calcium phosphate layer on its surface; in such layers, calcium phosphates differ in their structure and composition due to the fact that, during precipitation, they interact with ions in the body. In addition to improving cell proliferation and adhesion [12,13], it can also promote osteogenic differentiation of mesenchymal stem cells [14].

However, despite all the advantages of hydroxyapatite, its use in bone implants is limited [15]. This is due to the low rate of bone resorption under physiological conditions, since the processes of osteogenesis and resorption are connected; when one of the processes slows down, the other one also slows down [16]. The second reason limiting the use of hydroxyapatite is its brittleness [17]; however, it retains a very high compressive strength (up to 917 MPa) [18] and has a very weak tensile property. The reason for the fragility of this material is the ionic bond between the atoms in the ceramic material: they are not capable of plastic deformation, and the fracture toughness of ceramic composites does not exceed 1.2 MPa while the strength of bones ranges from 2 to 12 Mpa [18]. For this reason, research is currently relevant, according to the results of which it will be possible to create composite materials for biomedical purposes, which will combine biological activity and mechanical properties comparable to bone tissue [19]. Composite materials of hydroxyapatite with synthetic biocompatible polymers can improve the mechanical and biocompatible properties of hydroxyapatite [20,21].

Polyvinyl alcohol (PVA) is a synthetic biocompatible and non-toxic polymer whose cryogels have a porous structure [22,23]. The main problem when using this material for bone implantation is the fixation of the material in the body, due to its biological inertness [24]. This problem can be corrected by adding bioactive and biocompatible components to the hydrogel, which include hydroxyapatite. The combination of the macroporous and elastic structure of polyvinyl alcohol cryogel, and the biologically active and biologically compatible calcium phosphate, will lead to a synergistic effect of the functional properties of biocompatible composites [21,25].

Cryogels have important advantages in bone tissue repair. Due to the fact that cryogels consist of three-dimensional hydrophilic polymer chains, which have high mechanical strength, they are able to create a nutrient medium for the growth of endogenous cells. They are also able to imitate the extracellular matrix of bone, which makes it possible to encapsulate bioactive cells and molecules. Due to their network structure, cryogels can control the release of materials as needed [26]. In addition, cryogels are able to dissolve and show high integration with body tissues, which, as a result, makes it possible to avoid difficulties during surgical removal [27]. The mechanical properties of cryogels can be adjusted, depending on the application, by changing the polymer concentration, cooling rate, and freezing time [28]. Being a porous material, the pore size of a cryogel can vary, which makes it possible to include fillers in them, including hydroxyapatite powders [29]. Based on this, the creation of new composite materials is currently becoming an important and promising approach to bone tissue regeneration. The purpose of this work is to obtain cryogel composite materials based on polyvinyl alcohol and calcium phosphates, as well as to study the physicochemical, mechanical, and functional properties.

2. Materials and Methods

To obtain materials K1 and K2, a suspension of calcium hydroxide in water (3.68 g in 20 mL) was added to a 30% polyvinyl alcohol gel in water (15 mL) and stirred for 2 h. After two hours of stirring, to obtain material K1, a solution of ammonium hydrogen phosphate (3.95 g in 10 mL) was added and stirred for 6 h, and to obtain K2, an 87% solution of phosphoric acid was added and stirred for 4 h. Further, the resulting suspensions were treated with microwave radiation (power 100 W), kept for two days to preserve the precipitate, and frozen for 12 h at −20 °C. Using the same methods, but without PVA, pure HA(K1) and HA(K2) hydroxyapatite was synthesized to compare the viability of monocytes in the presence of materials (Table 1).

Table 1. Composition of composite materials K1 and K2.

Name	Reaction Equation	PVA Content, %	HA Content, %
K1	$10Ca(OH)_2 + 6(NH_{42}HPO_4 \rightarrow Ca_{10}(PO_4)_6(OH)_2\downarrow + 6H_2O + 12NH_4OH$	10	10
K2	$10Ca(OH)_2 + 6H_3PO_4 = Ca_{10}(PO_4)_6(OH)_2\downarrow + 6H_2O + 12NH_4OH$	10	10

The phase composition of the resulting composite materials based on cryogels of polyvinyl alcohol and calcium phosphates was determined on a Rigaki diffractometer (MiniFlex 600) at 40 kV and speed 3°/min. Phase interpretation and identification was carried out using the ICDD diffraction database (PDF-2/Release 2012 RDB). The spectra of the surface layer of samples of various materials were recorded on a Nicolet 6700 IR Fourier spectrometer (Thermo Scientific). The spectra were recorded with a resolution of 4 cm^{-1} in the range 400–4000 cm^{-1} using an ATR attachment.

The surface morphology of composite materials based on cryogels of polyvinyl alcohol and calcium phosphates was studied using a HITACI II TM-3000 electron microscope at an accelerating voltage of 5 kV, without applying metals to the surface. The particle size was calculated by the linear-intercept method, in which the size measurement is made using a straight line at one selected angle for all particles, followed by plotting a diagram from the resulting data array.

To assess the solubility of HA in the composition of the composites, the total concentration of Ca^{2+} calcium ions in physiological saline at 37 °C, in which the samples were kept for 20 days to achieve saturation with respect to the solid phase, was determined by complexometric titration in the presence of Eriochrome Black T with an ammonia buffer solution, pH = 10. Titration was carried out every 7 days.

The compressive strength was measured, and the modulus of elasticity was calculated on a setup assembled on the basis of an Arduino Board Mega 2560 and a strain gauge with a sample deformation of 20%.

To determine the assessment of the effect of materials on the viability of cells of the immune system, a cytotoxicity test with an Alamar Blue indicator was used. The signal intensity was measured using a Tecan Infinite 200 microreader at a wavelength of 540 nm.

3. Results and Discussion

Analysis of the resulting composites by IR spectroscopy showed that the spectra (Figure 1) have band characteristics of both hydroxyapatite and PVA. The IR spectra of the composites contain bands corresponding to bond vibrations in the structure of PVA and HA. In the composite material, there is an absorption band in the region of 3300 cm^{-1}, which indicates the stretching vibrations of the O-H bond, which is a characteristic band for alcohols. The absorption at 2900 cm^{-1} refers to the stretching vibrations of the CH_2 bond. Absorption is observed in the region of 1600 cm^{-1}, which is characteristic of the stretching vibrations of the C=O bond: this band could appear namely as a result of the production of polyvinyl alcohol, during the saponification of polyvinyl acetate. An absorption band is also observed in the region of 1385 cm^{-1}, which is also characteristic of the bending vibrations of O-H groups, but for the K2 sample, an increase in the intensity of this band can be observed. This is due to the formation of a larger number of hydrogen bonds. In the spectrum of hydroxyapatite, an intense absorption band is observed in the region of 1000 cm^{-1}, which refers to the stretching vibrations of phosphate P-O groups, and the bands at 570 cm^{-1} refer to their deformation vibrations.

When examining samples by X-ray phase analysis (Figure 2), we can observe the presence of several phases in composites. The resulting composites are crystalline. Composite K1 is a phase of hydroxyapatite and monetite. Monetite is an acidic salt of phosphoric acid. Under physiological conditions, these types of phosphates are hydrolyzed with the formation of macroporous carbon-substituted hydroxyapatite, which is characterized by a slower resorption rate compared to hydrophosphates, and acts as a fixing agent. Composite K2 is represented only by the hydroxyapatite phase, which indicates the completion of the reaction. The average crystallite size is 38–40 nm.

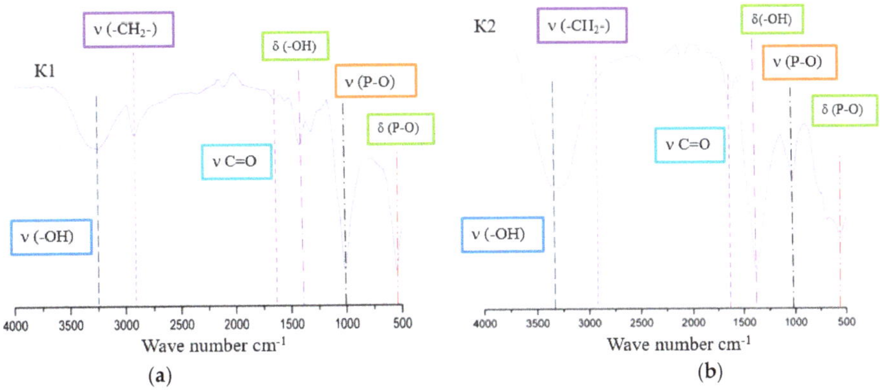

Figure 1. IR spectra of samples of composites K1 (**a**) and K2 (**b**).

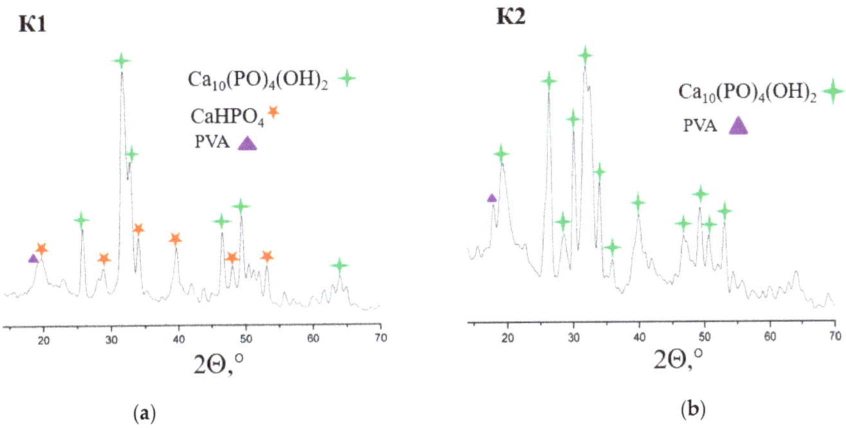

Figure 2. X-ray diffraction patterns of samples of composites K1 (**a**) and K2 (**b**).

The results of surface morphology studies by SEM showed that the K1 composite is more homogeneous compared to the K2 sample. On the histograms, we can observe that the average particle size differs slightly and is 3 μm (Figure 3). Particle size analysis was performed on the images obtained on the surface and on the cut of the sample; no significant differences in size between different particle locations were found.

Composites K1 and K2, obtained by mineralization of the polymer matrix in situ, were divided into several parts and immersed in saline, after which they were placed in a thermostat at a temperature of 37 °C and kept for four weeks. In total, four series of tests were made, and after each week, the physiological solution in which the composites were kept was titrated for the calcium content in the solution. The physiological solution was titrated with a solution of 0.01 M Trilon-B, with the indicator Eriochrome Black. From the results of the titration, we can observe that the release of calcium in the K1 composite is higher in all test series, relative to the K2 composite. For composite K1, complete diffusion occurs as early as the third week, and does not change further. The reason for the low release of calcium in the K2 composite may be the presence of only hydroxyapatite, which is a poorly soluble compound and entails low diffusion (Table 2).

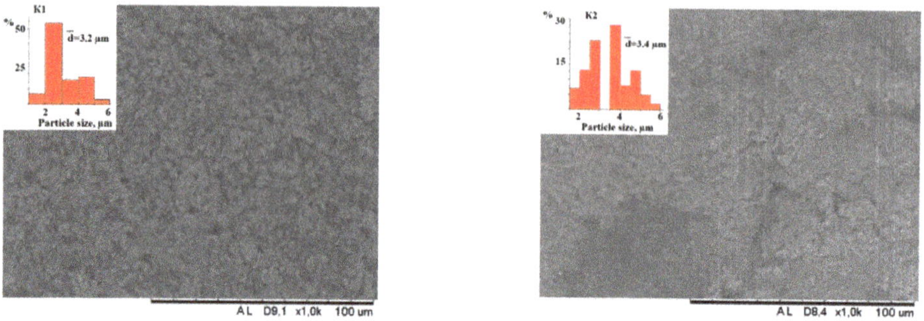

(a) (b)

Figure 3. SEM image and histograms of composites K1 (**a**) and K2 (**b**).

Table 2. Solubility of composite materials K1 and K2.

	Concentration Ca^{2+} (mmol/L)			
	1 Week	**2 Weeks**	**3 Weeks**	**4 Weeks**
K1	4.5	4.1	5.7	5.7
K2	1.9	1.3	1.2	3.5

The mechanical compressive resilience of composite materials was determined by measuring the relative stress and strain of the composites. Based on the measurement results, dependency graphs were obtained (Figure 4).

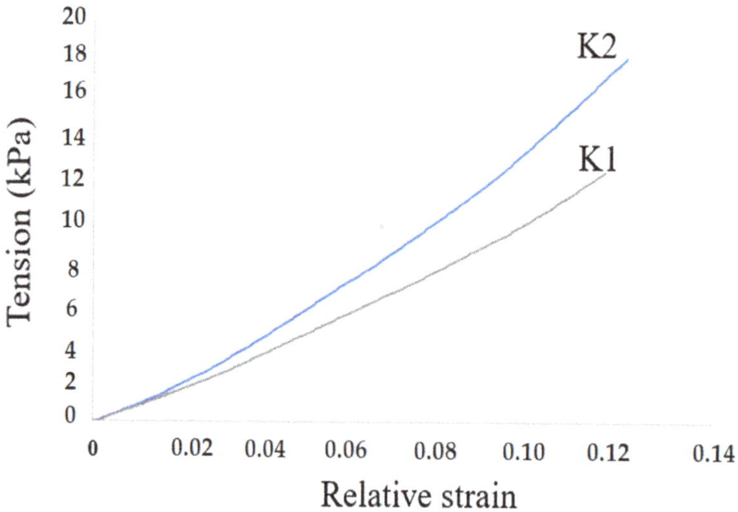

Figure 4. Dependences of the relative deformation of samples of composites K1 and K2 on stress.

0.02 Based on the results of these measurements, the graphs were processed, and the average modulus of elasticity was calculated (Table 3). Based on the results of these studies, we can observe that the average modulus of elasticity of the K2 composite is higher in comparison with the K1 composite. This is due to the fact that pure hydroxyapatite is formed in the K2 composite, which, due to its hydroxyl groups, forms a greater number of crosslinking points, which leads to an increase in the elasticity of the composite.

Table 3. Modulus of elasticity of composite materials K1 and K2.

	Elastic Modulus (kPa)
K1	101 ± 2
K2	157 ± 1

When analyzing the viability of monocytes in the presence of materials, in addition to composite materials, pure hydroxyapatite was also synthesized according to two methods (Figure 5). First, monocytes isolated from human blood were seeded on the samples. Monocytes were isolated from the blood of three donors. Next, the samples were incubated at 37 °C for 6 days. After that, the Alamar Blue indicator was added, and the samples were incubated for another three hours. At the final stage, fluorescence was measured. The study of the viability of macrophages in the presence of materials on three donors showed that, in the presence of pure HA, regardless of the method of preparation, about 60–80% of cells survive, while in the presence of pure PVA and composites, the level of viability is comparable to the control sample (cells without material) and some donors are higher than the control. This result shows that the material does not interfere with cell survival, which indicates a low cytotoxicity of the materials. The fact that, in some samples, the viability of monocytes is higher than in the control indicates that fewer cells survived in the control, and there is a more comfortable environment in the presence of the material than without it.

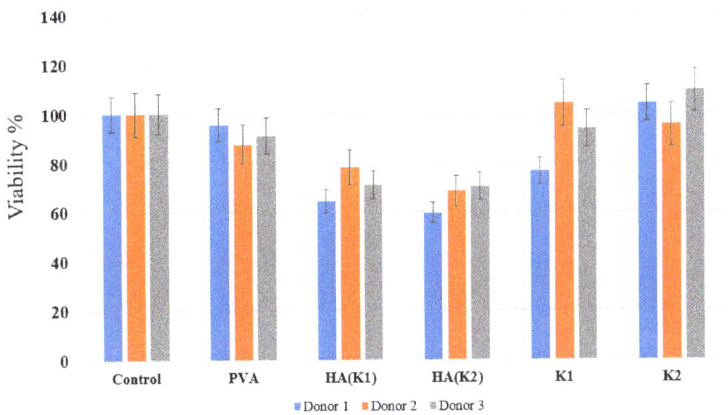

Figure 5. Study of the viability of macrophages in the presence of composite materials K1 and K2, pure hydroxyapatite HA (K1) and HA (K2), and pure PVA on three donors.

4. Conclusions

Composite materials based on a 10% solution of polyvinyl alcohol mineralized with hydroxyapatite, synthesized from a solution of calcium hydroxide, ammonium hydrogen phosphate, and phosphoric acid were obtained. It was established by X-ray phase analysis that the main phase of the mineral filler of the composite is hydroxyapatite $Ca_{10}(PO_4)_6(OH)_2$; however, in K1 there is also a phase of calcium hydroxide $Ca(OH)_2$, and in K2 there is a phase of calcium hydrophosphate $CaHPO_4$. The release of calcium in composite K2 occurs much faster due to the more soluble phase of calcium hydrogen phosphate $CaHPO_4$. For composite K2, the average modulus of elasticity is 157 kPa, which is significantly higher than that of K1, whose average modulus of elasticity is 101 kPa. The study of the surface morphology of HA/PVA showed the presence of a homogeneous structure of the K1 composite and an inhomogeneous structure of the K2 composite; the average particle size is 3 μm. The study of the viability of monocytes showed that about 60–80% of cells survive in the presence of pure HA, and in the presence of pure PVA and composites, it is comparable with the control sample. As we can see, the resulting materials do not have outstanding mechanical properties, but can be considered biocompatible since the cells of the immune response monocytes show high viability in the presence of materials. Such a material can be used to fill the voids that have formed due to pathological processes in the bone, between the bone, and the implant bearing a mechanical load.

Author Contributions: Conceptualization, I.K. and R.S.; methodology, I.K.; formal analysis, R.S.; investigation, R.S. and K.S.; writing—original draft preparation, R.S. and D.L.; writing—review and editing, R.S. and D.L. All authors have read and agreed to the published version of the manuscript.

Funding: Studies of the phase composition of materials and surface morphology were supported by grants from the President of the Russian Federation for state support of young Russian scientists-PhDs (MK-2182.2022.1.3). The study of solubility, mechanical properties and in vitro studies were supported by the Ministry of Science and Higher Education (FSWM-2020-0037).

Institutional Review Board Statement: The study was conducted according to the guidelines of the Declaration of Helsinki and approved by the Ethics Committee of Tomsk State University (Protocol of the meeting of NR TSU Bioethics Commission, Protocol No 3 from 7 March 2022). Informed consent was obtained from all subjects involved in the study. Written informed consent has been obtained from the patient to publish this paper.

Conflicts of Interest: The authors declare no conflict of interest.

References

1. Myeroff, C.; Archdeacon, M. Autogenous bone graft: Donor sites and techniques. *J. Bone Jt. Surg. Am.* **2011**, *93*, 2227–2236. [CrossRef] [PubMed]
2. Puppi, D.; Chiellini, F.; Piras, A.M.; Chiellini, E. Polymeric materials for bone and cartilage repair. *Prog. Polym. Sci.* **2010**, *35*, 403–440. [CrossRef]
3. Kim, H.D.; Amirthalingam, S.; Kim, S.L.; Lee, S.S.; Rangasamy, J.; Hwang, N.S. Biomimetic materials and fabrication approaches for bone tissue engineering. *Adv. Healthc. Mater.* **2017**, *6*, 1700612. [CrossRef] [PubMed]
4. Roseti, L.; Parisi, V.; Petretta, M.; Cavallo, C.; Desando, G.; Bartolotti, I.; Grigolo, B. Scaffolds for bone tissue engineering: State of the art and new perspectives. *Mater. Sci. Eng. C Mater. Biol. Appl.* **2017**, *78*, 1246–1262. [CrossRef]
5. Feng, P.; He, J.; Peng, S.; Gao, C.; Zhao, Z.; Xiong, S.; Shuai, C. Characterizations and interfacial reinforcement mechanisms of multicomponent biopolymer-based scaffold. *Mater. Sci. Eng. C Mater. Biol. Appl.* **2019**, *100*, 809–825. [CrossRef] [PubMed]
6. Gong, T.; Xie, J.; Liao, J.; Zhang, T.; Lin, S.; Lin, Y. Nanomaterials and bone regeneration. *Bone Res.* **2015**, *3*, 15029. [CrossRef] [PubMed]
7. Boskey, A.L. Bone composition: Relationship to bone fragility and antiosteoporotic drug effects. *BoneKEy Rep.* **2013**, *2*, 447. [CrossRef]
8. Szcze´s, A.; Hołysz, L.; Chibowski, E. Synthesis of hydroxyapatite for biomedical application. *Adv. Colloid Interface Sci.* **2017**, *249*, 321–330. [CrossRef]
9. Bueno, E.M.; Glowacki, J. Cell-free and cell-based approaches for bone regeneration. *Nat. Rev. Rheumatol.* **2009**, *5*, 685–697. [CrossRef]
10. Chocholata, P.; Kulda, V.; Babuska, V. Fabrication of scaffolds for bone-tissue regeneration. *Materials* **2019**, *12*, 568. [CrossRef]
11. Patel, P.P.; Buckley, C.; Taylor, B.L.; Sahyoun, C.C.; Patel, S.D.; Mont, A.J.; Mai, L.; Patel, S.; Freeman, J.W. Mechanical and biological evaluation of a hydroxyapatite-reinforced scaffold for bone regeneration. *J. Biomed. Mater. Res. Part A* **2019**, *107*, 732–741. [CrossRef] [PubMed]
12. Yılmaz, E.; Kabataş, F.; Gökçe, A.; Fındık, F. Production and characterization of a bone-like porous Ti/Ti-hydroxyapatite functionally graded material. *J. Mater. Eng. Perform.* **2020**, *29*, 6455–6467. [CrossRef]
13. Hu, Y.; Chen, J.; Fan, T.; Zhang, Y.; Zhao, Y.; Shi, X.; Zhang, Q. Biomimetic mineralized hierarchical hybrid scaffolds based on in situ synthesis of nano-hydroxyapatite/chitosan/chondroitin sulfate/hyaluronic acid for bone tissue engineering. *Colloids Surf. B Biointerfaces* **2017**, *157*, 93–100. [CrossRef]
14. Mucalo, M.R. Animal-bone derived hydroxyapatite in biomedical applications. In *Hydroxyapatite (HAp) for Biomedical Applications*; Woodhead Publishing: Cambridge, UK, 2015; pp. 307–342.
15. Krishnamurithy, G. A review on hydroxyapatite-based scaffolds as a potential bone graft substitute for bone tissue engineering applications. *J. Health Transl. Med.* **2013**, *16*, 22–27.
16. Clarke, B.L.; Khosla, S. Physiology of bone loss. *Radiol. Clin.* **2010**, *48*, 483–495. [CrossRef] [PubMed]
17. Khanal, S.P.; Mahfuz, H.; Rondinone, A.J.; Leventouri, T. Improvement of the fracture toughness of hydroxyapatite (HAp) by incorporation of carboxyl functionalized single walled carbon nanotubes (CfSWCNTs) and nylon. *Mater. Sci. Eng. C* **2016**, *60*, 204–210. [CrossRef]
18. Jarcho, M.; Bolen, C.H.; Thomas, M.B.; Bobick, J.; Kay, J.F.; Doremus, R.H. Hydroxylapatite synthesis and characterization in dense polycrystalline form. *J. Mater. Sci.* **1976**, *11*, 2027–2035. [CrossRef]

19. Kulinets, I. Biomaterials and their applications in medicine. In *Regulatory Affairs Biomaterials Medical Devices*; Woodhead Publishing: Cambridge, UK, 2015; pp. 1–10.
20. Martinez-Vazquez, F.J.; Perera, F.H.; Miranda, P.; Pajares, A.; Guiberteau, F. Improving the compressive strength of bioceramic robocast scaffolds by polymer infiltration. *Acta Biomater.* **2010**, *6*, 4361–4368. [CrossRef]
21. Lytkina, D.N.; Fedorishin, D.A.; Kalachikova, P.M.; Plyaskina, A.A.; Babeshin, A.R.; Kurzina, I.A. Cryo-structured materials based on polyvinyl alcohol and hydroxyapatite for osteogenesis. *J. Funct. Biomater.* **2021**, *12*, 18. [CrossRef]
22. Baker, M.I.; Walsh, S.P.; Schwartz, Z.; Boyan, B.D. A review of polyvinyl alcohol and its uses in cartilage and orthopedic applications. *J. Biomed. Mater. Res. Part B Appl. Biomater.* **2012**, *100*, 1451–1457. [CrossRef]
23. Peppas, N.A.; Merrill, E.W. Crosslinked poly (vinyl alcohol) hydrogels as swollen elastic networks. *J. Appl. Polym. Sci.* **1977**, *21*, 1763–1770. [CrossRef]
24. Kumar, A.; Han, S.S. PVA-based hydrogels for tissue engineering: A review. *Int. J. Polym. Mater. Polym. Biomater.* **2017**, *66*, 159–182. [CrossRef]
25. Chocholata, P.; Kulda, V.; Dvorakova, J.; Supova, M.; Zaloudkova, M.; Babuska, V. In Situ hydroxyapatite synthesis enhances biocompatibility of PVA/HA hydrogels. *Int. J. Mol. Sci.* **2021**, *22*, 9335. [CrossRef] [PubMed]
26. Wu, G.; Feng, C.; Quan, J.; Wang, Z.; Wei, W.; Zang, S.; Kang, S.; Hui, G.; Chen, X.; Wang, Q. In situ-controlled release of stromal cell-derived factor-1alpha and antimiR-138 for on-demand cranial bone regeneration. *Carbohydr. Polym.* **2018**, *182*, 215–224. [CrossRef] [PubMed]
27. Silva, R.; Fabry, B.; Boccaccini, A.R. Fibrous protein-based hydrogels for cell encapsulation. *Biomaterials* **2014**, *35*, 6727–6738. [CrossRef]
28. Kumar, A.; Mishra, R.; Reinwald, Y.; Bhat, S. Cryogels: Freezing unveiled by thawing. *Mater. Today* **2010**, *13*, 42–44. [CrossRef]
29. Bencherif, S.A.; Sands, R.W.; Bhatta, D.; Arany, P.; Verbeke, C.S.; Edwards, D.A.; Mooney, D.J. Injectable preformed scaffolds with shape-memory properties. *Proc. Natl. Acad. Sci. USA* **2012**, *109*, 19590–19595. [CrossRef]

polymers

Article

Antibacterial and Antifungal Properties of Modified Chitosan Nonwovens

Dominik Sikorski [1,*], Marta Bauer [2], Justyna Frączyk [3] and Zbigniew Draczyński [1]

1 Institute of Textile Materials and Polymer Composites, Lodz University of Technology, Zeromskiego 116, 90-924 Lodz, Poland; zbigniew.draczynski@p.lodz.pl
2 Department of Inorganic Chemistry, Faculty of Pharmacy, Medical University of Gdańsk, 80-416 Gdańsk, Poland; marta.bauer@gumed.edu.pl
3 Institute of Organic Chemistry, Lodz University of Technology Faculty of Chemistry, Zeromskiego 116, 90-924 Lodz, Poland; justyna.fraczyk@p.lodz.pl
* Correspondence: dominik.sikorski@dokt.p.lodz.pl

Abstract: Chitosan acquires bacteriostatic properties via protonation of its amino groups. However, much of the literature assumes that chitosan itself inhibits the growth of bacteria. This article presents a comparative study of chitosan nonwovens modified with various acids, including acetic, propionic, butyric, and valeric organic acids, as well as hydrochloric acid. The aim was to determine which acid salts influence the antibacterial and antifungal activity of chitosan-based materials. Two methods were used to modify (formation of ammonium salts) the chitosan nonwovens: First, acid vapors (gassing process) were used to find which salt of chitosan had the best antibacterial properties. Based on the results, the most effective acid was prepared in a solution in ethanol. The influence of the acid concentration in ethanol, the time of treatment of chitosan materials with acid solution, and the rinsing process of modified nonwovens on the antimicrobial activity of the modified materials was investigated. The modified materials were subjected to microbiological tests. Each of the modified materials was placed in bacterial inoculum. The cultures were tested on agar to observe their microbial activity. Toxicity to human red blood cells was also investigated. A reduction in the number of bacterial cells was observed for the *S. aureus* strain with chitosan salt modified with 10% acetic acid in ethanol. The antibacterial activity of the chitosan salts increased with the percentage of acid salts formed on the surface of the solid material (decreasing numbers of bacterial colonies or no growth). No reduction in growth was observed for the *E. coli* strain. The chitosan samples were either inactive or completely eliminated the bacterial cells. Antimicrobial activity was observed for chitosan salts with hydrochloric acid and acetic acid. Finally, ^{1}H-NMR spectroscopy and FTIR spectroscopy were used to confirm the incorporation of the acid groups to the amino groups of chitosan.

Keywords: chitosan salts; bacteriostatic; acids

1. Introduction

Chitin is one of the most common polysaccharides in the environment. It can be found in the structures of sponges and corals, as well as in the shells of marine invertebrates and the cell walls of insects and fungi [1–7]. It was isolated from fungi in 1811 by H. Braconnot, and its structure was described by A. Hofmann [8,9]. Chitosan (CS) is a biopolymer derivative of chitin. It is widely available in various forms (such as films and nonwovens), and is used in numerous industries [10]. It can be obtained by chemical or enzymatic deacetylation. The difference between chitin and CS lies in the degree of deacetylation [11]. The presence of amino and hydroxyl groups in CS makes it a biofunctional polymer that can be selectively modified. Due to its non-toxicity, antibacterial activity, bioadhesive characteristics, biodegradability, and excellent biocompatibility, CS is widely used in biomedical applications, as a drug carrier, antimicrobial, antioxidant, antitumor, and wound-dressing agent [12–22].

The mechanisms of action of CS on bacteria and fungi are well known. The main antimicrobial activity of CS is via electrostatic interactions between the cationic groups of CS and negatively charged cell walls [23–33]. Gram-positive and Gram-negative bacteria have differently structured cell walls. Gram-positive bacteria have more peptidoglycans, while Gram-negative bacteria are richer in lipopolysaccharides. These differences can affect the adherence of bacteria to the surface of the materials [34,35]. CS is a polysaccharide composed of glucosamine and acetylglucosamine units with unique physicochemical behavior. The amount of glucosamine residues depends on the degree of chitin deacetylation. CS has the ability to obtain a large number of free NH_3^+ ions under acidic conditions, which affects its ability to interact with pathogens. The macromolecules must have a high ionic charge potential to penetrate through the cell membrane. Moreover, the protonation of the amino groups in chitosan influences the possibility of obtaining lipophilic–hydrophilic properties of CS (the polar part consists of NH_3^+ and OH groups, as well as the hydrophilic part of the main biopolymer chain). Moreover, the low charge density at neutral and basic pH leads to low solubility, aggregation, and poor stability of chitosan-based formulations [36]. Chitosan itself chelates metal ions—especially those of transition metals—and also finds application as a matrix for the immobilization of enzymes [37].

The antibacterial activity of CS may also be related to the size of the CS molecules. Its potential antimicrobial effects include preventing nutrients from being taken up, and altering cell permeability. High-molecular-weight CS is unable to penetrate the cell wall and cell membrane. Low-molecular-weight CS has both extracellular and intracellular antimicrobial activity [12,38–42]. In addition, molecular weight affects the rate of dissolution of chitosan in acid as the molecular weight increases [43]. The degree of N-acetylation of chitin and CS is an important parameter determining their physicochemical properties. CS modification is related to the degree of acetylation, which is itself related to the number of N-acetylamino groups in the molecules [44].

The antimicrobial properties of CS can be improved by chemical modification of the CS structure. The two reactive groups $-NH_2$ and -OH offer vast opportunities for chemical modification. These groups allow the formation of several functional derivatives via reactions such as sulfonation, amination, and carboxymethylation [45–47]. Hydrophilic, positively charged groups (NH_3^+) cause the death of bacterial cells via interaction with and penetration through the cytoplasmic membrane in bacteria [48]. Quaternization improves the positive charge density of CS, and can also improve its antimicrobial efficacy, by enhancing its polycationic nature [32]. Previously, CS has been used in various forms, e.g., as composites, or in various solutions [49–52].

It is widely believed that chitosan has bacteriostatic properties due to the ammonium salt formation capacity of the biopolymer amino groups. However, much of the literature assumes that chitosan itself inhibits the growth of bacteria.

The aim of this study was to conduct tests on the preparation of chitosan nonwovens modified with various acids—including acetic, propionic, butyric, and valeric organic acids, as well as hydrochloric acid—and to determine which chitosan salts have antibacterial and antifungal activity. Two methods of modifying the chitosan nonwovens were planned: First, acid vapors (gassing process) were used in order to select the salt of chitosan with the best antibacterial properties. Based on the results, the most effective acid was used for research on the protonation of nonwovens reaction examples are shown in Figure 1 (formation of ammonium salts) in solution (wet method). It was assumed that an ethanolic acid solution would be used to eliminate the solubility of the chitosan salt in water. The modified materials were then subjected to microbiological tests (*E. coli* and *S. aureus*). It was also assumed that studies on the influence of modified materials on the process of hemolysis would be performed, which are particularly important from the point of view of antibacterial compounds/materials, as they quite often have adverse hemolytic activity.

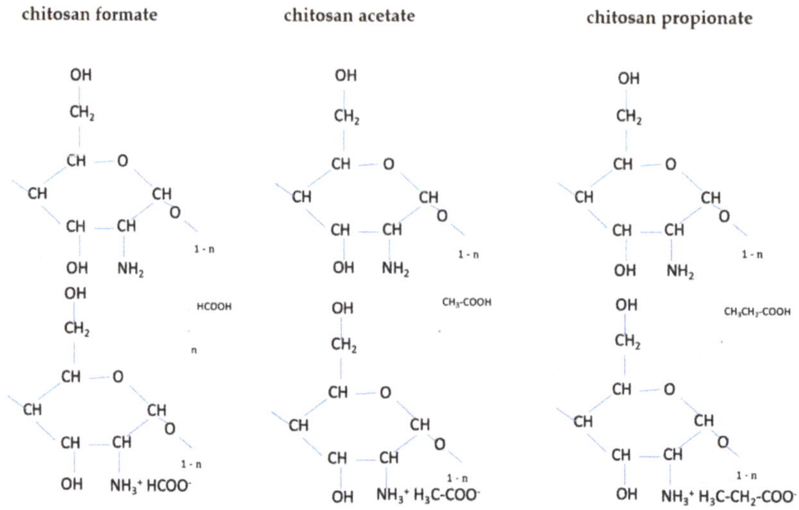

Figure 1. Example of the formation of chitosan salts in the presence of different carboxylic acids, where n is the degree of deacetylation (n = DA).

2. Materials and Methods

The medical-grade chitosan fibers used in this study were a commercial product of Hismer Biotechnology Co. Ltd., Shandong, China. The deacetylation degree of the used polymer was 91.8. Fibers of 2.02 dtex × 30 mm with a strength of 12cN/tex and relative elongation of 1.5% were used. The nonwoven structure of the biodegradable chitosan fibers was produced using a carder machine and then needled. The final product obtained from the elementary fleeces had a surface weight of 120 g/m^2 per sheet.

2.1. Modification of Chitosan Nonwovens in Different Evaporated Acid Environments

All acids were sourced from Avantor Performance Materials Poland POOCH Polish Chemicals Reagents, Gliwice, Poland. Hydrochloric (35–38% PURE P.A.), acetic (99.5–99,9% PURE P.A.), formic (98–100% PURE P.A.), propionic (99.5% PURE P.A.), butyric (PURE), and valeric acids (99%) were used to modify the chitosan nonwovens. The procedure for modification was as follows: First, a Petri dish with a diameter of 20 cm containing 20 mL of acid was placed in a desiccator of 3 dm^3 capacity. This system created an environment saturated with acid vapor after 24 h. The chitosan nonwovens were placed in the acid environments at the same time and removed at 15 min intervals for up to 120 min. The samples after treatment in the acid atmospheres were next placed in a desiccator for 24 h over solid potassium hydroxide granules (PURE P.A. POOCH Polish Chemicals Reagents, Gliwice, Poland) to remove the residual acid condensed on the nonwoven surface. They were then degassed under a vacuum pump. Finally, the nonwoven samples were dried in a normal oven at 40 °C for 2 h. The reaction process in the desiccator was based on the condensation of acid vapors on the surface of the nonwovens, whereupon a chemical reaction occurred between the acid molecules and chitosan.

2.2. Modification of Chitosan Nonwovens Using Acetic Acid Solution in Ethanol—Procedure with Rinsing

The following solutions of acetic acid dissolved in ethanol were prepared: 5%, 10%, and 15%. Then, 12 nonwoven samples with dimensions of 5 × 5 cm^2 and a weight of 1 g were added to a beaker, into which 200 mL of each solution was poured. The samples were submerged for intervals of 10 min for up to 60 min. After removal, some fleece samples were rinsed twice for 5 min with fresh ethanol, then blotted on filter paper and placed on a new filter for 24 h until they were completely dry. This procedure rinsed off any excess solution. Additionally, the nonwoven was dried in an oven at 40 °C for 2 h. The entire procedure was repeated for each of the solutions.

2.3. Modification of Chitosan Nonwovens Using Acetic Acid Solution in Ethanol—Procedure without Rinsing

The procedure without rinsing was the same but, after removal, the fleece samples were not rinsed twice for 5 min with fresh ethanol.

2.4. Titration Procedure

A standard solution of hydrochloric acid (0.1 M) and sodium hydroxide (0.1 M) was prepared. Next, 50 mL of sodium hydroxide was poured into a beaker, and 0.1 g of each nonwoven was added. The unreacted sodium hydroxide residue was titrated with an automatic pipette by adding 1 mL of hydrochloric acid every 2 min until it stabilized. The amount of acid introduced was then calculated from the conductometric diagram. The same procedure was performed for all of the samples.

Retention volume was determined using the following formulae:

Number of adsorbed moles of acid (mol) per:

$$n_x = n_{NaOH} - n_{HCL} \text{ (mol)} \tag{1}$$

where:

nNaOH—C x V;
nHCL—C x V.

Number of attached acid groups on nonwovens:

$$AAG = \frac{n_x}{m} \tag{2}$$

where:

n_x—number of adsorbed moles of acid (mol/g);
m—mass of the nonwoven fabric.

Retention volume:

$$RV = \frac{\frac{n_x}{m}}{M\ chitosan} \times 100\% \tag{3}$$

where:

n_x—number of adsorbed moles of acid (mol/g);
m—mass of nonwoven fabric;
M_{chitosan}—molar mass of chitosan.

2.5. Analysis of Acid Chitosan Salt Forms Based on FTIR Spectra

The molecular structures of the chitosan nonwovens were analyzed using a Nicolet 6700 FTIR spectrometer (Thermo Scientific, Waltham, MA, USA) in attenuated total reflectance (ATR) mode, with a diamond crystal (incidence angle 45°). The measurement conditions were as follows: the resolution was 2 cm^{-1}, the range of infrared (IR) radiation was 4000–600 cm^{-1}, and 32 scanning procedures of each sample were carried out using IR radiation to obtain the absorption spectra. The absorption spectra of the fiber preparations under investigation were plotted in the system A = f(1/l), as a basis for the interpretation of the molecular nonwoven structure.

The Equations (4) and (5) were used to determine the degree of acetylation (*DA*). Thermo Scientific[TM] OMNIC[TM] Specta Software (Thermo Scientific, Waltham, MA, USA) was used for the calculations according to the data [53].

$$DA(\%) = \frac{A_{1655}}{A_{3450}} \times 100/1.33 \tag{4}$$

$$DA(\%) = \left(\frac{A_{1320}}{A_{1410}} - 0.03822\right)/0.03133 \tag{5}$$

2.6. Analysis of Acid Chitosan Salts Based on NMR Spectroscopy

The chemical structures and DDA (degree of deacetylation) of the initial chitosan, as well as of the chitosan after acid treatment, were determined by ^{1}H-NMR spectroscopy analysis using a Bruker AM 400 (Bruker Corporation, Billerica, MA, USA) spectrometer, in a mixed solvent containing DCl (deuterium chloride) and D_2O (deuterium oxide) (1%, w/w), with DSS (sodium trimethylsilylpropanesulfonate) as a reference. The samples were prepared as follows: 50 mg of the chitosan salts was dispersed in D_2O and, after homogenization, was dissolved by adding DCl to finally obtain 1% DCl in H_2O. When the solution formed, measurements were made at 303 K using 32 scan pulse accumulations. The degree of chitosan deacetylation was calculated using the following formula, where DS is the degree of substitution (6):

$$\mathrm{DS} = \frac{\frac{1}{3}AC\,\underline{H}}{\underline{H}2} \tag{6}$$

where:

Hypothesis 1 (H1): *Signal integration of proton derived from the acidic residue of the carbon atom in the sugar ring of chitosan.*

Hypothesis 2 (H2): *Signal integration of proton at the second carbon atom in the sugar ring of chitosan.*

2.7. Antimicrobial Activity

Modified chitosan samples were cut into 1 cm^2 pieces. Two microbiological assays were conducted to evaluate the antimicrobial properties of the materials. The protocol was based on the method described previously, with slight modifications [54]. The toxicity of the chitosan was also assessed. All assays were performed in triplicate.

2.7.1. Susceptibility to Microbial Colonization

Non-sterile samples of the chitosan were placed in 24-well flat-bottomed plates containing Mueller–Hinton II broth (for bacteria) or Sabouraud broth (for fungi) to evaluate the resistance of the materials to microbial colonization. The chitosan fragments were incubated for 24 h at 37 °C (to observe the growth of bacteria) or 25 °C (to observe the growth of fungi). After incubation, 10 µL of medium from each well was seeded on the agar plates. After 24 h of incubation, the appearance of colonies was evaluated.

2.7.2. Bactericidal Activity

The second assay was performed similarly to that described above, with the difference that this time bacterial inocula were used. The aim of this study was to check the ability of the material to kill the microorganisms. Chitosan samples were soaked in 0.5 mL of bacterial suspensions with a density of 0.5 McFarland. Two reference strains were used: *Staphylococcus aureus* ATCC 25923, and *Escherichia coli* ATCC 25922. The samples were incubated for 24 h at 37 °C. After incubation, 10 μL of medium from each well was seeded on the agar plates, and the number of colonies was observed.

2.8. Hemolytic Activity of the Chitosan Samples

A hemolysis assay was conducted, using some steps from the protocols described previously [55]. Human red blood cells were obtained from fresh blood collected in test tubes, with EDTA as an anticoagulant. The blood was centrifuged and the plasma removed. After rinsing with PBS, a 4% solution of RBCs was prepared. The chitosan fragments were put into 0.5 mL of RBC solution and incubated for 1 h at 37 °C. Controls with 0% hemolysis (erythrocytes with PBS) and 100% hemolysis (erythrocytes with 1% Triton-X) were prepared. After incubation, 100 μL of the solution from each well was transferred to a 96-well plate and centrifuged (1 h, 4 °C, 7 min, 1000× *g*). The supernatants were resuspended in a new plate, and the release of hemoglobin was measured (540 nm).

3. Results

3.1. Retention Volume

To determine the amounts of attached acid groups (corresponding to the amount of ammonium groups formed in the chitosan), the reverse (acid–base) titration method was used. Two standard 0.1 M HCl and 0.1 M NaOH solutions were used as a titrant and as an auxiliary substance, respectively. Additionally, for modified chitosan materials, the retention volume was measured. Tables 1–3 show only the results for chitosan derivatives for which colony growth was reduced or absent.

Table 1. The retention volume of modified chitosan nonwovens modified by acid fumes (gass ing process).

Time of Incubation of Chitosan Material in Acid Vapor (min)	Retention Volume (%) after Treatment with Gaseous HCl	Retention Volume (%) after Treatment with CH_3COOH Vapor
15	93	92
60	86	87
120	85	88

Table 2. The retention volume of modified chitosan nonwovens after treatment in solutions of acetic acid in ethanol—procedure without rinsing.

Sample	Retention Volume (%)	Sample	Retention Volume (%)	Sample	Retention Volume (%)
5/10	88	10/10	95	15/10	92
5/20	81	10/20	92	15/20	93
5/30	82	10/30	90	15/30	94
5/40	80	10/40	93	15/40	93
5/50	87	10/50	93	15/50	94
5/60	89	10/60	95	15/60	94

5: acetic acid concentration in ethanol; 10–60: treatment time (min) of nonwovens in a bath of acetic acid in ethanol.

After 15 min in an acidic environment caused by the vaporization of different acids, the retention volume for the chitosan acetate and chitosan hydrochloride was above 90% for all of the tested chitosan nonwovens (Table 1). For chitosan hydrochloride, the retention volume was 93%, while for chitosan acetate it was 92%. After 120 min, the retention volume was lower for HCl modified chitosan (85%) than for chitosan acetate (88%).

Based on the data in Table 1, it can be concluded that in the initial period of the process the surface sorption process and the associated temporary increase in the amount of acid prevailed. After 60 min, there was an insignificant reduction in the amount of acid, which stabilized. This may be related to the establishment of an equilibrium state between the processes of sorption and desorption.

Table 3. The retention volume of modified chitosan nonwovens after treatment in solutions of acetic acid and ethanol—procedure with rinsing.

Sample	Retention Volume (%)	Sample	Retention Volume (%)	Sample	Retention Volume (%)
5/10	85	10/10	86	15/10	90
5/20	86	10/20	85	15/20	86
5/30	88	10/30	84	15/30	88
5/40	90	10/40	89	15/40	89
5/50	92	10/50	91	15/50	91
5/60	91	10/60	92	15/60	92

5: acetic acid concentration in ethanol; 10–60: time (min) of nonwovens in a bath of acetic acid and ethanol.

The data present in Table 2 show that the retention volumes for chitosan acetate were between 80% and 94% for all acid solutions in ethanol and different treatment times, in a procedure without rinsing the samples. The highest rate of retention value was in the case where a 10% solution of acetic acid in ethanol was used and the treatment time was 60 min, while the lowest value was in the case of application of 5% CH_3COOH solution in ethanol and a treatment time of 20 min.

Table 3 shows the retention volume of chitosan acetate obtained via the treatment with acetic acid when following the procedure including rising. The values of retention volume for chitosan acetate after rinsing the samples were slightly lower compared to the non-rinsing method. The observed values were above 80%. The highest rates of retention were in the cases of application 10% and 15% solutions of CH_3COOH for 60 min.

3.2. *Microbiology and the Influence of Modified Chitosan Materials on the Hemolysis Process and Toxicity to Human Red Blood Cells*

First, a screening method was used to determine the modified nonwovens with the best antimicrobial properties. The fragments were not sterilized, and were immediately placed in liquid media for fungi or bacteria. This made it possible to identify whether the materials were susceptible to colonization by microorganisms. After 24 h of incubation, the samples were seeded on solid media.

In Figures 2 and 3, "+" means that bacterial/fungal growth was observed. Mueller–Hinton agar (MHA) was the substrate used for bacteria. Sabouraud agar (SDA) was used as the fungal medium. The letter A indicates the samples with rinsing, while the letter B indicates the samples without rinsing. The samples of nonwovens 1, 2, and 3 correspond to the materials treated with 5%, 10%, and 15% acetic acid in ethanol, respectively. The numbers 10–60 refer to times of treatment, from 10 to 60 min. The most promising results were for the samples with no rinsing, as well as for the samples with rinsing but longer treatment times (50 min or 60 min). In additional observations, a sample of the chitosan hydrochloride was immediately degraded in the liquid medium (it dissolved).

The last two circles with markings from A to F in Figures 2 and 3 show antimicrobial results of chitosan-based materials treated with vapors of various acids (obtaining modified materials under gassing conditions) (see Table 4).

Table 4. Descriptions of the letters on the tested Petri dishes.

Marks on Circles	Acid Atmosphere
A	Valeric acid
B	Propionic acid
C	Formic acid
D	Butyric acid
E	Hydrochloric acid
F	Acetic acid

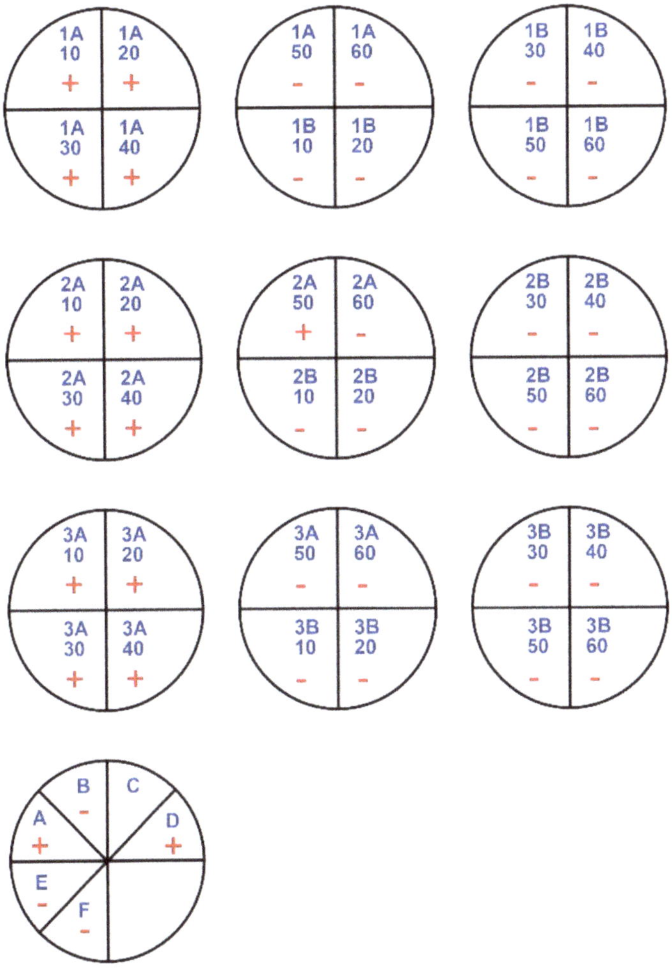

Figure 2. Growth of bacteria in medium (MHA).

Activity against bacteria was noted for the samples without rinsing. Therefore, cultures were grown from this group. Growth of organisms was observed in the unmodified blank sample.

Table 5 shows the effects on microbial growth of modifying the nonwovens using acetic acid solution in ethanol without rinsing. In 10% acetic acid solution in ethanol, the activity of the modified chitosan nonwovens increased against *S. aureus* bacterial cells (decreasing numbers of bacterial colonies or no growth). No reduction in growth was evident for the *E. coli* strain. The chitosan samples were either not active at all, or caused complete elimination of the bacterial cells.

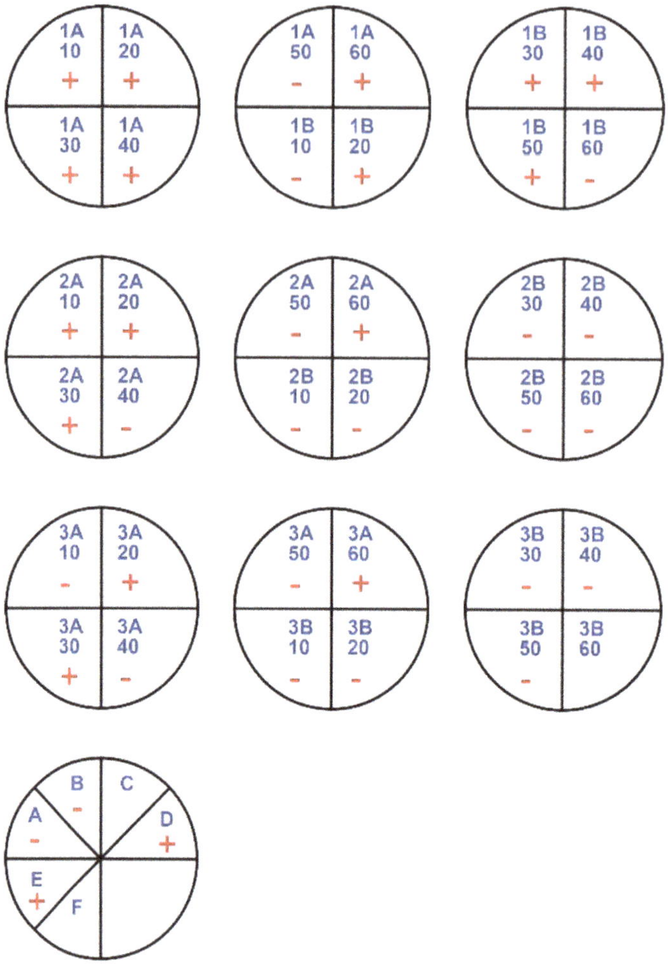

Figure 3. Growth of fungi in medium (SDA).

Table 5. Effects of modification of chitosan-based materials on microbial growth for nonwovens modified using acetic acid solution in ethanol—procedure without rinsing.

Sample	*S. aureus* ATCC 25923	*E. coli* ATCC 25922
5/10	Growth	Growth
5/20	Growth	Growth
5/30	Growth	Growth
5/40	Growth	Growth
5/50	Growth	Growth
5/60	Growth	Growth
10/10	Growth: visible growth reduction	Growth
10/20	Growth: 170 * colonies	Growth
10/30	Growth	No growth
10/40	Growth: 70 * colonies	Growth
10/50	Growth: reduction	Growth
10/60	No growth	Growth

Table 5. *Cont.*

Sample	*S. aureus* ATCC 25923	*E. coli* ATCC 25922
15/10	Growth: 14 * colonies	Growth
15/20	Growth: 12 * colonies	Growth
15/30	Growth: 30 * colonies	No growth
15/40	No growth	No growth
15/50	Growth: 9 * colonies	No growth
15/60	No growth	No growth

*—The mean values of 3 repetitions.

Antimicrobial activity of the tested materials was observed in acidic atmospheres with hydrochloric acid and acetic acid (Table 6).

Table 6. Effects of modification of chitosan-based materials on microbial growth for nonwovens modified by vapors of various acids (obtaining modified materials under gassing conditions).

Sample	*S. aureus* ATCC 25923	*E. coli* ATCC 25922
Valeric acid	Growth	Growth
Propionic acid	Growth	Growth
Formic acid	Growth: 5 * colonies	Growth
Butyric acid	Growth	Growth
Hydrochloric acid	No growth	No growth
Acetic acid	No growth	No growth

*—The mean value of 3 repetitions.

Complete absence of hemolysis was noted for materials modified with valeric acid and formic acid. Hemolysis was not observed in the unmodified blank sample. The rinsed nonwovens modified with acetic acid in ethanol solution showed significantly lower impact on hemolysis compared to the non-washed samples (Table 7). Cytotoxicity was noted for samples modified by butyric acid, hydrochloric acid, and acetic acid. The highest toxicity (21.18%) was observed in the case of treatment with acetic acid vapors (Table 8).

Table 7. Toxicity to human red blood cells measured as % of hemolysis in nonwovens modified using acetic acid solution in ethanol—procedures with and without rinsing.

Samples	% of Hemolysis in Rinsed Samples	% of Hemolysis in Non-Rinsed Samples
5/10	0.24	0.09
5/20	0.45	1.17
5/30	0.80	1.52
5/40	0.36	2.70
5/50	0.13	1.28
5/60	0.54	0.46
10/10	0.00	14.34
10/20	0.26	3.59
10/30	0.16	14.24
10/40	0.00	17.51
10/50	0.65	15.36
10/60	0.82	14.88
15/10	0.47	10.09
15/20	0.26	1.12
15/30	0.55	13.21
15/40	0.16	16.99
15/50	2.01	14.78
15/60	3.21	20.87

Table 8. Toxicity to human red blood cells measured as % of hemolysis in nonwovens modified by vapors of various acids (obtaining modified materials under gassing conditions).

Acids	Valeric Acid	Propionic Acid	Formic Acid	Butyric Acid	Hydrochloric Acid	Acetic Acid
% Hemolysis	0.00	0.26	0.00	1.04	0.88	21.18

3.3. *Analysis of Obtained Chitosan Salts Based on FTIR–ATR*

Changes in the chemical structures of the samples were analyzed using FTIR–ATR spectroscopy. Additionally, the degree of deacetylation was calculated.

Chitosan FTIR spectra showed sharp peaks at 564 cm^{-1} (out-of-plane bending NH, out-of-plane bending C–O), 711 cm^{-1} (out-of-plane bending NH), 1174 cm^{-1} (C–O–C stretching), 2871 cm^{-1} (CH_2 stretching), and 3430 cm^{-1} (−OH stretching). The vibrational mode of amide C=O stretching was observed at 1650 and 1590 cm^{-1}.

The spectra for the samples of chitosan nonwovens treated with acetic acid vapor are presented in Figure 4. Based on the spectra, the degree of deacetylation was calculated using Equations (4) and (5) proposed by Czechowska-Biskup et al. [53]. According to the first equation, the degree of deacetylation was 87%, and according to the second it was 89.7%. As can be seen in the spectra, there are differences confirming the formation of chitosan, in the form of a salt derivative of acetic acid, after 10–60 min of exposure to organic acid in the gas phase. There is a strong signal for carboxyl groups, with maxima at 1550 cm^{-1} and 1400 cm^{-1}. Adsorption of vaporized acid may occur as result of the interaction of hydrogen bonds between the hydroxyl groups of chitosan and the carboxylic group of the organic acid. After 1 h of exposure to acetic acid vapor, a peak appears with a maximum between 1600 cm^{-1} and 1200 cm^{-1}, probably due to the formation of the ammonium salt group of acetic acid and chitosan. There are strong signals for carboxyl groups, with maxima at 1549 and 1405 cm^{-1}. We assume the formation of the amine salt of acetic acid at signal 1549 cm^{-1}. Signals located in the range of 1410–1340 cm^{-1} are due to the COO^- stretching bonds in the acid residue of acetic acid.

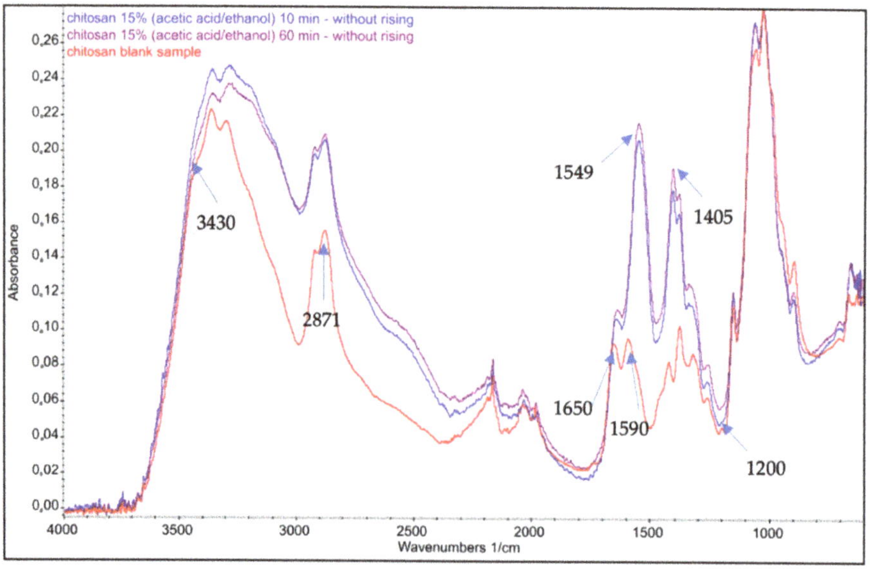

Figure 4. Example FTIR–ATR spectra for a nonwoven treated with acetic acid vapor without rinsing.

3.4. Analysis of Obtained Chitosan Salts Based on NMR Spectroscopy

Figures 5 and 6 show examples of ^{1}H-NMR spectra for chitosan-based materials modified with different acids.

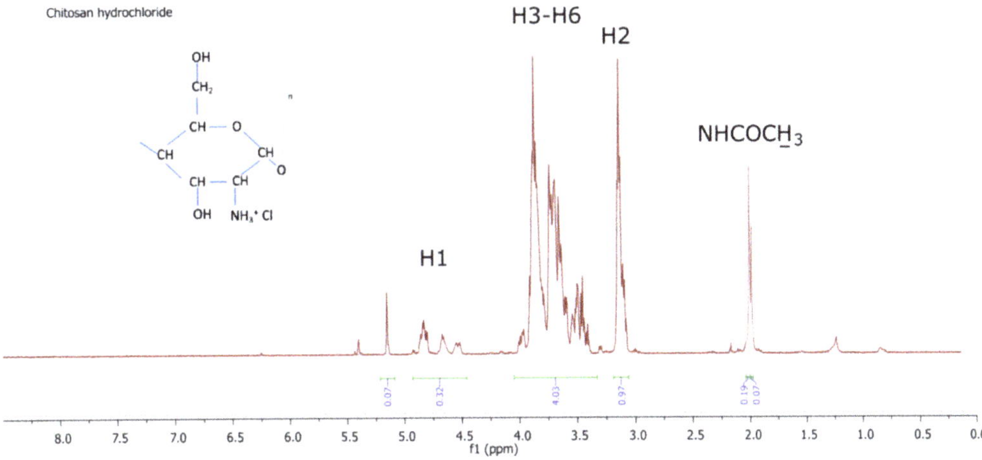

Figure 5. ^{1}H-NMR spectrum of chitosan hydrochloride.

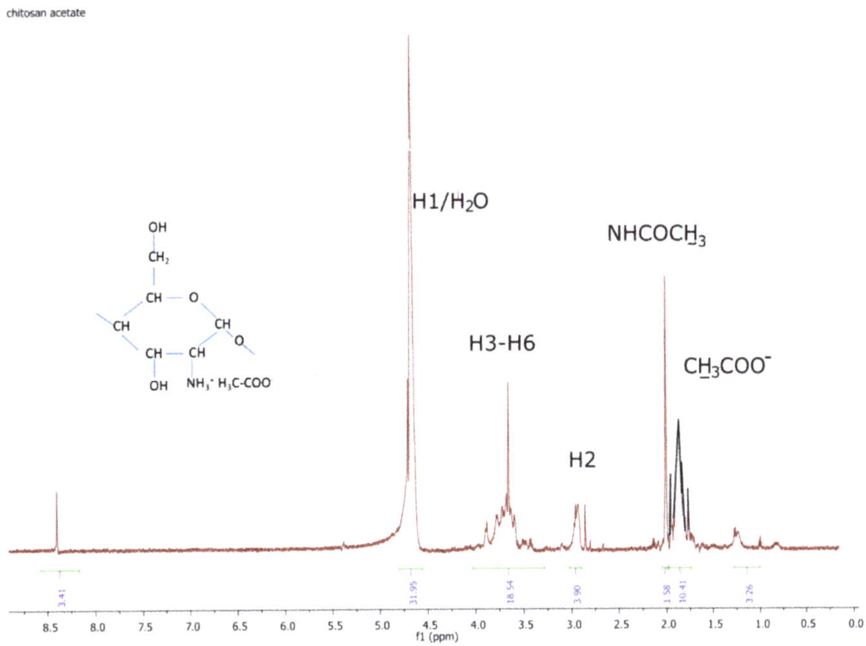

Figure 6. ^{1}H-NMR spectrum of chitosan acetate.

On the ^{1}H-NMR spectrum of chitosan hydrochloride, the following proton signals can be seen: in the range of 1.8–2.3 ppm protons of the acetamide group NHOCC$\underline{H}_3$; 3.1–3.2 ppm protons bonded to the second carbon of the sugar ring (H2); in the range of 3.3–4.1 ppm protons bonded to the carbons (C3–C6) of the sugar ring (H3–H6); in the range of 4.4–4.9 ppm protons bonded to the first carbon of the sugar ring (H1).

On the ^{1}H NMR spectrum of chitosan acetate, the following proton signals can be seen: in the range of 1.6–1.8 ppm for protons of the acetic acid salt -OOCC$\underline{H_3}$; in the range of 1.8–2.1 ppm for protons of the acetamide group NHOCC$\underline{H_3}$; 2.8–3.1 ppm for protons bonded to the second carbon of sugar ring (H2); in the range of 3.3–4.1 for protons bonded to the carbons (C3–C6) of the sugar ring (H3–H6); in the range of 4.6–4.8 for protons bonded to the first carbon of the sugar ring (H1), which is overlapped with the residues of H_2O.

As can be seen from the NMR spectra for the selected samples with good antibacterial properties (Figure 5), there were no significant changes in the chitosan spectrum, which appears normal. The proton signal from the HCl chitosan salt's hydrogen overlaps with the DCl protons. Therefore, the existence of hydrogen in the hydrochloric groups cannot be detected. In Figure 6, there is a signal from CH_3COO^-, which can be used to calculate and compare the degree of substitution. Table 9. Shows degree of substitution for chitosan salts.

Table 9. Degree of substitution for chitosan salts, calculated based on ^{1}H-NMR spectra.

Sample	Degree of Substitution (%)	Degree of Substitution (%) Based on Titration
Chitosan hydrochloride	-	85
Chitosan acetate	88.0	89

4. Discussion

The purpose of this study was to determine the bacteriostatic and bacteriological properties of nonwoven fibrous materials. In a study by Abadehie et al., chitosan/PEO nanofibers mixed with lawsone obtained by electrospinning showed improved wound-healing properties, further demonstrating biocompatibility and antibacterial properties [56]. The retention volumes of the chitosan-based materials modified with various acids show that the acids' residues may have been incorporated into the structures of the chitosan nonwovens. As can be seen in Table 1, acid vapors were deposited on the surface and within fibers. After some time, degassing occurred, and some of the residues were released from the fibers. This might be related to the uneven structure of the tested materials. At the end of the process, gas equilibrium was obtained. This may be related to the establishment of an equilibrium state between the sorption and desorption processes.

The chitosan nonwovens modified by acetic acid solution in ethanol without rinsing showed a much higher degree of substitution than the nonwovens obtained via the process with a rinsing step. The retention volume of the chitosan acetate was above 80% for both procedures (with and without the rinsing process), but the retention volume was between 10% and 15% higher in the case of the procedure without rinsing. This is probably related to the rinsing process, during which most of the acid was flushed out of the surface. Generally, the higher concentration of acetic acid and extended interaction time led to materials with a higher number of attached acid groups. This was especially noticeable in the case of the rinsed nonwovens modified with 5% and 10% acid solutions in ethanol, for which we observed an increase of 5% in retention volume. In the case of the non-rinsed fabrics, the increase was not so noticeable, because the excess acid residues remained on the surface of the modified material.

Microbiological tests were used to investigate the activity of the tested materials and the growth of the fungi and bacteria, which were related to the retention volume of the modified nonwovens. Only materials modified with acetic acid and hydrochloric acid were found to have bacteriostatic properties, as can be seen in Table 6. This is probably due to the acid strength and size of the substituents in organic acids adhered to the surface of the nonwoven fabric. Some residues may have been smaller due to the irregular structure of the nonwoven fabric, which has microcapillaries inside it that are not uniformly distributed.

Activity against bacteria was noted for the samples without rinsing (Table 5). This is because most of the acid residues were on the surface. However, it does not change the fact that only the salts of chitosan with acetic acid and hydrochloric acid have antibacterial properties. In the case of samples with rinsing, the antibacterial effects of the modified chitosan materials were also observed (Figure 2), but only for chitosan nonwovens with longer exposure times to acetic acid and hydrochloric acid.

The results for toxicity to human red blood cells measured as a percentage of hemolysis, presented in Tables 7 and 8, show much higher toxicity for the samples without rinsing. The highest toxicity was 21.18% for the sample treated with acetic acid. Complete absence of hemolysis was observed for the samples treated with valeric acid and formic acid.

The spectrogram in Figure 4 shows that the modification process did not change the degree of deacetylation in the chitosan nonwovens. FTIR–ATR spectra showed changes in the structure of the material, which confirmed that formation of the salt had taken place. Similar changes in the structure of chitosan can be seen in the ^{1}H-NMR spectra shown in Figures 5 and 6.

A major limitation of the preparation of antimicrobial (antibacterial and antifungal) bacteriostatic chitosan-based materials is obtaining them in the form of suitable acid salts. This limitation is due to the fact that using fibers treated with acids before producing the final form of the material results in complete dissolution of the chitosan salt in the medium in which bacteria are present. Therefore, the use of treatment of chitosan-based materials with acid after preparation of the final material form allows for obtaining materials where only the surface of the chitosan-based matrix is in the form of a biologically active salt. The surface conversion of chitosan amino groups to ammonium salts can be accomplished either through vapors of appropriate acids or the application of alcoholic acid solutions, leaving the core of the material in an insoluble form. The method based on the application of alcoholic acid solutions is safer from the processing point of view. However, when treated with alcoholic solutions, the morphology of the structure can be lost. With wet curing, even with an alcohol solution inert to the fibers, a reduction in fluffiness is observed, which is not observed when treated with acid vapor.

A similar study was conducted by Chung et al. [57], but using aqueous acid solutions. To investigate the antimicrobial activity, the pH was correlated with the properties of the used acids. Acids with higher pKa showed significantly better activity than those with lower pKa in control sets. As found, the application of acids in which the chitosan was dissolved was also an important factor. The best antibacterial activity was found for salts of chitosan and formic acid [57]. This observation indicates that the final antibacterial effect of chitosan derivatives is also influenced by the method of their preparation. In the conditions of chitosan gassing with formic acid vapors, we obtained chitosan formate with moderate antibacterial activity.

Chitosan has a strong affinity toward metal ions because of the presence of numerous amine and hydroxyl groups. Below pH 6.5, chitosan can interact with the bacterial cell wall to destabilize it and alter cell permeability. As was shown by Tripathi et al., chitosan–silver oxide nanocomposite films have good bacteriostatic properties against various kinds of pathogens [58]. Most studies of chitosan in the literature are related to solids and solutions [57–59].

We used a simple procedure to modify the fibers in the nonwovens, permitting us to obtain new materials with the desired antimicrobial properties and that are soluble in water. We believe that in the case of the obtained materials, polar, positively charged ammonium groups (NH_3^+) cause bacterial cell death by interacting with microorganisms and penetrating the bacterial cytoplasmic membrane. Protonation of the amino group with various acids improves the positive charge density of CS, and may also improve its antimicrobial efficacy by enhancing its polycationic nature, which is consistent with the literature data [32,45–48].

5. Conclusions

The research presented in this article enables the effective combination of acids and chitosan nonwovens to achieve antibacterial and antifungal effects, by adjusting the parameters of the solution (concentration of acid and time of treatment). No significant changes were observed in the retention volumes of the samples, showing that all of the acids were well incorporated into the nonwoven structure. Better activity against bacteria was noticed in the case of the samples obtained via procedures using an ethanolic solution of organic acid without rinsing. Therefore, cultures were grown from this group. Growth was also observed in an unmodified blank sample. The best results were obtained for materials treated with vapors of hydrochloric acid and acetic acid, and for the chitosan nonwovens modified using an acetic acid solution in ethanol without rinsing. Only these two chitosan salts showed antimicrobial properties. This was probably due to the acid strength and the size of the adhered acid residues, or the structure of the nonwoven fabric. Some residues may have been smaller due to the irregular structure of the nonwoven fabric, which has channels inside it that are not uniformly distributed. H^1-NMR and FTIR–ATR spectra confirmed the modification of the chitosan.

We intend to perform this study for other forms of chitosan, such as scaffolds, foils, and chitosan microspheres, because we expect that the method of formation of ammonium salts of chitosan in its final forms developed by us may be of a general nature, thanks to which it will be possible to obtain various chitosan materials with antibacterial properties.

Author Contributions: Conceptualization, D.S.; methodology, M.B. and D.S.; investigation, M.B. and D.S.; writing—original draft preparation, D.S.; writing—review and editing, M.B., Z.D., J.F. and D.S.; supervision, Z.D. All authors have read and agreed to the published version of the manuscript.

Funding: Not applicable.

Institutional Review Board Statement: Not applicable.

Informed Consent Statement: The protocol for the hemolysis assay was approved by the local Bioethics Committee at the Medical University of Gdańsk (NKBBN/263/2019, approval date: 10 June 2019).

Data Availability Statement: Data are stored at the Institute of Textile Materials and Polymer Composites.

Acknowledgments: This work was partially supported by the statutory research fund of the Institute of Material Science of Textiles and Polymer Composites, no. I-42/501/4-42-1-3.

Conflicts of Interest: The authors declare no conflict of interest.

References

1. Nature El-Sherbiny, I.M.; El-Baz, N.M. A Review on Bionanocomposites Based on Chitosan and Its Derivatives for Biomedical Applications. In *Eco-Friendly Polymer Nanocomposites*, 1st ed.; Thakur, V.K., Thakur, M.K., Eds.; Springer: New Delhi, India, 2015; pp. 497–531. [CrossRef]
2. Ehrlich, H. Chitin and collagen as universal and alternative templates in biomineralization. *Int. Geol. Rev.* **2010**, *52*, 661–699. [CrossRef]
3. Juárez-de La Rosa, B.A.; Quintana, P.; Ardisson, P.L.; Yáñez-Limón, J.M.; Alvarado-Gil, J.J. Effects of thermal treatments on the structure of two black coral species chitinous exoskeleton. *J. Mater. Sci.* **2012**, *47*, 990–998. [CrossRef]
4. Sikorski, D.; Gzyra-Jagieła, K.; Draczyński, Z. The kinetics of chitosan degradation in organic acid solutions. *Mar. Drugs* **2021**, *19*, 236. [CrossRef] [PubMed]
5. Grgac, S.F.; Tarbuk, A.; Dekanic, T.; Sujka, W.; Draczynski, Z. The chitosan implementation into cotton and polyester/cotton blend fabrics. *Materials* **2020**, *13*, 1616. [CrossRef]
6. João, C.F.C.; Silva, J.; Borges, J.P. Chitin-based nanocomposites: Biomedical applications. In *Eco-friendly Polymer Nanocomposites: Chemistry and Applications*; Advanced Structured Materials; Thakur, V.K., Thakur, M.K., Eds.; Springer: New Delhi, India, 2015; pp. 439–457. [CrossRef]
7. Tyliszczak, B. Animal-derived chitosans. Characteristics, comparison, application Chitozany zwierzęce. Charakterystyka, porównanie, wykorzystanie. *Przemysł Chem.* **2016**, *95*, 2059–2062. [CrossRef]
8. Wysokowski, M.; Petrenko, I.; Stelling, A.L.; Stawski, D.; Jesionowski, T.; Ehrlich, H. Poriferan chitin as a versatile template for extreme biomimetics. *Polymers* **2015**, *7*, 235–265. [CrossRef]
9. Khoushab, F.; Yamabhai, M. Chitin research revisited. *Mar. Drugs* **2010**, *8*, 1988–2012. [CrossRef]
10. Pokhrel, S.; Yadav, P.N.; Adhikari, R. Applications of Chitin and Chitosan in Industry and Medical Science: A Review. *Nepal J. Sci. Technol.* **2016**, *16*, 99–104. [CrossRef]
11. Venkatesan, J.; Kim, S.K. Chitosan composites for bone tissue engineering—An overview. *Mar. Drugs* **2010**, *8*, 2252–2266. [CrossRef]

12. Goy, R.C.; De Britto, D.; Assis, O.B.G. A review of the antimicrobial activity of chitosan. *Polimeros* **2009**, *19*, 241–247. [CrossRef]
13. Young, D.H.; Köhle, H.; Kauss, H. Effect of Chitosan on Membrane Permeability of Suspension-Cultured Glycine max and Phaseolus vulgaris Cells. *Plant Physiol.* **1982**, *70*, 1449–1454. [CrossRef]
14. Felt, O.; Buri, P.; Gurny, R. Chitosan: A unique polysaccharide for drug delivery. *Drug Dev. Ind. Pharm.* **1998**, *24*, 979–993. [CrossRef] [PubMed]
15. Mi, F.L.; Tan, Y.C.; Liang, H.F.; Sung, H.W. In vivo biocompatibility and degradability of a novel injectable-chitosan-based implant. *Biomaterials* **2002**, *23*, 181–191. [CrossRef]
16. Tamer, T.M.; Valachová, K.; Hassan, M.A.; Omer, A.M.; El-Shafeey, M.; Eldin, M.S.M.; Šoltés, L. Chitosan/hyaluronan/edaravone membranes for anti-inflammatory wound dressing: In vitro and in vivo evaluation studies. *Mater. Sci Eng. C* **2018**, *90*, 227–235. [CrossRef] [PubMed]
17. Omer, A.M.; Tamer, T.M.; Hassan, M.A.; Rychter, P.; Mohy Eldin, M.S.; Koseva, N. Development of amphoteric alginate/aminated chitosan coated microbeads for oral protein delivery. *Int. J. Biol. Macromol.* **2016**, *92*, 362–370. [CrossRef]
18. Yildirim-Aksoy, M.; Beck, B.H. Antimicrobial activity of chitosan and a chitosan oligomer against bacterial pathogens of warmwater fish. *J. Appl. Microbiol.* **2017**, *122*, 1570–1578. [CrossRef] [PubMed]
19. Valachová, K.; Tamer, T.M.; Eldin, M.M.; Šoltés, L. Radical-scavenging activity of glutathione, chitin derivatives and their combination. *Chem. Pap.* **2016**, *70*, 820–827. [CrossRef]
20. Xie, F.; Ding, R.L.; He, W.F.; Liu, Z.J.L.; Fu, S.Z.; Wu, J.B.; Yang, L.L.; Lin, S.; Wen, Q.L. In vivo antitumor effect of endostatin-loaded chitosan nanoparticles combined with paclitaxel on lewis lung carcinoma. *Drug Deliv.* **2017**, *24*, 1410–1418. [CrossRef]
21. Archana, D.; Dutta, J.; Dutta, P.K. Evaluation of chitosan nano dressing for wound healing: Characterization, in vitro and in vivo studies. *Int. J. Biol. Macromol.* **2013**, *57*, 193–203. [CrossRef]
22. Tamer, T.M.; Collins, M.N.; Valachová, K.; Hassan, M.A.; Omer, A.M.; Mohy-Eldin, M.S.; Švík, K.; Jurčík, R.; Ondruška, L'.; Biró, C.; et al. MitoQ loaded chitosan-hyaluronan composite membranes for wound healing. *Materials* **2018**, *11*, 569. [CrossRef]
23. Felt, O.; Carrel, A.; Baehni, P.; Buri, P.; Gurny, R. Chitosan as tear substitute: A wetting agent endowed with antimicrobial efficacy. *J. Ocul. Pharmacol. Ther.* **2000**, *16*, 261–270. [CrossRef] [PubMed]
24. Rabea, E.I.; Badawy, M.E.T.; Stevens, C.V.; Smagghe, G.; Steurbaut, W. Chitosan as antimicrobial agent: Applications and mode of action. *Biomacromolecules* **2003**, *4*, 1457–1465. [CrossRef] [PubMed]
25. Ke, C.L.; Deng, F.S.; Chuang, C.Y.; Lin, C.H. Antimicrobial actions and applications of Chitosan. *Polymers* **2021**, *13*, 904. [CrossRef] [PubMed]
26. Hu, Y.; Du, Y.; Wang, X.; Feng, T. Self-aggregation of water-soluble chitosan and solubilization of thymol as an antimicrobial agent. *J. Biomed. Mater. Res. Part A* **2009**, *90*, 874–881. [CrossRef]
27. Gómez-Estaca, J.; López de Lacey, A.; López-Caballero, M.E.; Gómez-Guillén, M.C.; Montero, P. Biodegradable gelatin-chitosan films incorporated with essential oils as antimicrobial agents for fish preservation. *Food Microbiol.* **2010**, *27*, 889–896. [CrossRef]
28. Kong, M.; Chen, X.G.; Xing, K.; Park, H.J. Antimicrobial properties of chitosan and mode of action: A state of the art review. *Int. J. Food Microbiol.* **2010**, *144*, 51–63. [CrossRef]
29. Hernández-Ochoa, L.; Gonzales-Gonzales, A.; Gutiérrez-Mendez, N.; Muñoz-Castellanos, L.N.; Quintero-Ramos, A. Study of the antibacterial activity of chitosan-based films prepared with different molecular weights including spices essential oils and functional extracts as antimicrobial agents. *Rev. Mex. Ing. Química* **2011**, *10*, 455–463.
30. Huang, L.; Dai, T.; Xuan, Y.; Tegos, G.P.; Hamblin, M.R. Synergistic combination of chitosan acetate with nanoparticle silver as a topical antimicrobial: Efficacy against bacterial burn infections. *Antimicrob. Agents Chemother.* **2011**, *55*, 3432–3438. [CrossRef]
31. Mohamed, N.A.; Sabaa, M.W.; El-Ghandour, A.H.; Abdel-Aziz, M.M.; Abdel-Gawad, O.F. Quaternized N-substituted carboxymethyl chitosan derivatives as antimicrobial agents. *Int. J. Biol. Macromol.* **2013**, *60*, 156–164. [CrossRef]
32. Tan, H.; Ma, R.; Lin, C.; Liu, Z.; Tang, T. Quaternized chitosan as an antimicrobial agent: Antimicrobial activity, mechanism of action and biomedical applications in orthopedics. *Int. J. Mol. Sci.* **2013**, *14*, 1854–1869. [CrossRef]
33. Avadi, M.R.; Sadeghi, A.M.M.; Tahzibi, A.; Bayati, K.H.; Pouladzadeh, M.; Zohuriaan-Mehr, M.J.; Rafiee-Tehrani, M. Diethylmethyl chitosan as an antimicrobial agent: Synthesis, characterization and antibacterial effects. *Eur. Polym. J.* **2004**, *40*, 1355–1361. [CrossRef]
34. Gan, L.; Chen, S.; Jensen, G.J. Molecular organization of Gram-negative peptidoglycan. *Proc. Natl. Acad. Sci. USA* **2008**, *105*, 18953–18957. [CrossRef] [PubMed]
35. Rohde, M. The Gram-Positive Bacterial Cell Wall. *Microbiol. Spectr.* **2019**, *7*, 1–21. [CrossRef]
36. García, O.G.Z.; Oropeza-Guzmán, M.T.; Argüelles Monal, W.M.; López-Maldonado, E.A. Design and mechanism of action of multifunctional BPE's with high performance in the separation of hazardous metal ions from polluted water Part I: Chitosan-poly(N-vinylcaprolactam) and its derivatives. *Chem. Eng. J.* **2019**, *359*, 840–851. [CrossRef]
37. Dutta, P.K.; Duta, J.; Tripathi, V.S. Chitin and Chitosan: Chemistry, properties and applications. *J. Sci. Ind. Res.* **2004**, *63*, 20–31.
38. Zubareva, A.; Shagdarova, B.; Varlamov, V.; Kashirina, E.; Svirshchevskaya, E. Penetration and toxicity of chitosan and its derivatives. *Eur. Polym. J.* **2017**, *93*, 743–749. [CrossRef]
39. Kravanja, G.; Primožič, M.; Knez, Ž.; Leitgeb, M. Chitosan-based (Nano)materials for Novel Biomedical Applications. *Molecules* **2019**, *24*, 1960. [CrossRef] [PubMed]
40. Sudarshan, N.R.; Hoover, D.G.; Knorr, D. Antibacterial Action of Chitosan. *Food Biotechnol.* **1992**, *6*, 257–272. [CrossRef]

41. Devlieghere, F.; Vermeulen, A.; Debevere, J. Chitosan: Antimicrobial activity, interactions with food components and applicability as a coating on fruit and vegetables. *Food Microbiol.* **2004**, *21*, 703–714. [CrossRef]
42. Shagdarova, B.T.; Il'ina, A.V.; Varlamov, V.P. Antibacterial Activity of Alkylated and Acylated Derivatives of Low–Molecular Weight Chitosan. *Appl. Biochem. Microbiol.* **2016**, *52*, 222–225. [CrossRef]
43. Panda, P.K.; Yang, J.M.; Chang, Y.H.; Su, W.W. Modification of different molecular weights of chitosan by p-Coumaric acid: Preparation, characterization and effect of molecular weight on its water solubility and antioxidant property. *Int. J. Biol. Macromol.* **2019**, *136*, 661–667. [CrossRef] [PubMed]
44. Aranaz, I.; Mengíbar, M.; Harris, R.; Paños, I.; Miralles, B.; Acosta, N.; Galed, G.; Heras, Á. Functional Characterization of Chitin and Chitosan. *Curr. Chem. Biol.* **2012**, *3*, 203–230. [CrossRef]
45. Rwei, S.P.; Lien, C.C. Synthesis and viscoelastic characterization of sulfonated chitosan solutions. *Colloid Polym. Sci.* **2014**, *292*, 785–795. [CrossRef]
46. Amir Afshar, H.; Ghaee, A. Preparation of aminated chitosan/alginate scaffold containing halloysite nanotubes with improved cell attachment. *Carbohydr. Polym.* **2016**, *151*, 1120–1131. [CrossRef]
47. Bukzem, A.L.; Signini, R.; dos Santos, D.M.; Lião, L.M.; Ascheri, D.P.R. Optimization of carboxymethyl chitosan synthesis using response surface methodology and desirability function. *Int. J. Biol. Macromol.* **2016**, *85*, 615–624. [CrossRef]
48. Jiang, S.; Wang, L.; Yu, H.; Chen, Y. Preparation of crosslinked polystyrenes with quaternary ammonium and their antibacterial behavior. *React. Funct. Polym.* **2005**, *62*, 209–213. [CrossRef]
49. Xu, J.; Manepalli, P.H.; Zhu, L.; Narayan-Sarathy, S.; Alavi, S. Morphological, barrier and mechanical properties of films from poly (butylene succinate) reinforced with nanocrystalline cellulose and chitin whiskers using melt extrusion. *J. Polym. Res.* **2019**, *26*, 188. [CrossRef]
50. Singh, A.; Dutta, P.K.; Kumar, H.; Kureel, A.K.; Rai, A.K. Improved antibacterial and antioxidant activities of gallic acid grafted chitin-glucan complex. *J. Polym. Res.* **2019**, *26*, 234. [CrossRef]
51. Panda, P.K.; Yang, J.M.; Chang, Y.H. Preparation and characterization of ferulic acid-modified water soluble chitosan and poly (γ-glutamic acid) polyelectrolyte films through layer-by-layer assembly towards protein adsorption. *Int. J. Biol. Macromol.* **2021**, *171*, 457–464. [CrossRef]
52. Panda, P.K.; Dash, P.; Yang, J.M.; Chang, Y.H. Development of chitosan, graphene oxide, and cerium oxide composite blended films: Structural, physical, and functional properties. *Cellulose* **2022**, *29*, 2399–2411. [CrossRef]
53. Czechowska-Biskup, R.; Jarosińska, D.; Rokita, B.; Ulański, P.; Rosiak, J.M. Determination of degree of deacetylation of chitosan-Comparision of methods. *Prog. Chem. Appl. Chitin Its Deriv.* **2012**, *2012*, 5–20.
54. Maciejewska, M.; Bauer, M.; Neubauer, D.; Kamysz, W.; Dawgul, M. Influence of amphibian antimicrobial peptides and short lipopeptides on bacterial biofilms formed on contact lenses. *Materials* **2016**, *9*, 873. [CrossRef] [PubMed]
55. Avrahami, D.; Shai, Y. A new group of antifungal and antibacterial lipopeptides derived from non-membrane active peptides conjugated to palmitic acid. *J. Biol. Chem.* **2004**, *279*, 12277–12285. [CrossRef] [PubMed]
56. Abadehie, F.S.; Dehkordi, A.H.; Zafari, M.; Bagheri, M.; Yousefiasl, S.; Pourmotabed, S.; Mahmoodnia, L.; Validi, M.; Ashrafizadeh, M.; Zare, E.N.; et al. Lawsone-encapsulated chitosan/polyethylene oxide nanofibrous mat as a potential antibacterial biobased wound dressing. *Eng Regen.* **2021**, *2*, 219–226. [CrossRef]
57. Chung, Y.C.; Wang, H.L.; Chen, Y.M.; Li, S.L. Effect of abiotic factors on the antibacterial activity of chitosan against waterborne pathogens. *Bioresour. Technol.* **2003**, *88*, 179–184. [CrossRef]
58. Tripathi, S.; Mehrotra, G.K.; Dutta, P.K. Chitosan-silver oxide nanocomposite film: Preparation and antimicrobial activity. *Bull Mater. Sci.* **2011**, *34*, 29–35. [CrossRef]
59. Yilmaz Atay, H. Antibacterial Activity of Chitosan-Based Systems. In *Functional Chitosan: Drug Delivery and Biomedical Applications*; Jana, S., Jana, S., Eds.; Springer: Singapore, 2020; pp. 457–489. [CrossRef]

polymers

MDPI

Review

Surveying the Oral Drug Delivery Avenues of Novel Chitosan Derivatives

Iyyakkannu Sivanesan [1], Shadma Tasneem [2], Nazim Hasan [2], Juhyun Shin [3], Manikandan Muthu [4], Judy Gopal [4] and Jae-Wook Oh [3,*]

1 Department of Bioresources and Food Science, Konkuk University, Seoul 143-701, Korea; isivanesan@gmail.com
2 Department of Chemistry, Faculty of Science, Jazan University, Jazan 45142, Saudi Arabia; sthaque@jazanu.edu.sa (S.T.); hhasan@jazanu.edu.sa (N.H.)
3 Department of Stem Cell and Regenerative Biotechnology, Konkuk University, Seoul 143-701, Korea; junejhs@konkuk.ac.kr
4 Department of Research and Innovation, Department of Biotechnology, Saveetha School of Engineering, Saveetha Institute of Medical and Technical Sciences (SIMATS), Thandalam, Chennai 602105, Tamil Nadu, India; bhagatmani@gmail.com (M.M.); jejudy777@gmail.com (J.G.)
* Correspondence: ohjw@konkuk.ac.kr; Tel.: +82-2-2049-6271; Fax: +82-2-455-1044

Abstract: Chitosan has come a long way in biomedical applications: drug delivery is one of its core areas of imminent application. Chitosan derivatives are the new generation variants of chitosan. These modified chitosans have overcome limitations and progressed in the area of drug delivery. This review briefly surveys the current chitosan derivatives available for biomedical applications. The biomedical applications of chitosan derivatives are revisited and their key inputs for oral drug delivery have been discussed. The limited use of the vast chitosan resources for oral drug delivery applications, speculated to be probably due to the interdisciplinary nature of this research, is pointed out in the discussion. Chitosan-derivative synthesis and practical implementation for oral drug delivery require distinct expertise from chemists and pharmacists. The lack of enthusiasm could be related to the inadequacy in the smooth transfer of the synthesized derivatives to the actual implementers. With thiolated chitosan derivatives predominating the oral delivery of drugs, the need for representation from the vast array of ready-to-use chitosan derivatives is emphasized. There is plenty to explore in this direction.

Keywords: chitosan; chitosan derivative; drug delivery; oral delivery; biomedical applications

1. Introduction

Chitosan, made up of β-(1,4)-*N*-acetyl-glucosamine [1–3], is obtained following the deacetylation of chitin. Chitin is found extensively in the exoskeletons of crustaceans and insects and in the cell walls of bacteria and fungi [4]. The quality of chitosan is influenced by the source of chitin, separation method and the degree of deacetylation [5]. The major advantages of chitosan are that it is nontoxic, mucoadhesive, hemocompatible, biodegradable and able to exhibit antioxidant, antitumor, antimicrobial properties. These properties render chitosan a highly attractive biomaterial option. The iconic characteristic of chitosan is that it does not provoke intense inflammation nor induce the body's immunity. Researchers have confirmed that chitosan with different molecular weights and degrees of deacetylation exhibit low toxicity [6–9]. The catatonic nature of chitosan gives it its bactericidal and bacteriological properties [10,11]. However, chitosan is not soluble in aqueous solutions, a major disadvantage that limits its widespread application in living systems [12].

Chitosan's surface adherence comes in handy when delivering useful molecules across mucosal pathways and adsorbs molecules that do not have any affinity for mucus [13].

Chitosan, through its permeation-related attributes, is able to open the tight epithelial junctions [14]. Chitosan also plays a role in coagulation. It accelerates the rate of wound healing by enabling interactions between amino and platelet groups [15]. These hemostatic properties are used with respect to wound healing applications. As a material for wound dressing, chitosan possesses chemoattraction, macrophage and neutrophil activation, analgesic properties, acceleration of granulation tissue/re-epithelization, limited scar formation and contraction, hemostasis and antibacterial properties [16]. The antitumor properties of chitosan and its derivatives have been well demonstrated in both in vitro and in vivo models [17]. The beneficial effects of antioxidants are well known [18], chitosan and its derivatives are able to scavenge free radicals in vitro [19,20]. The biodegradability of chitosan is yet another unique feature in biological organisms. Within the system, chitosan interacts with bioenzymes to depolymerize. The degradation breakdown products, *N*-acetyl glucose and glucosamine, are nontoxic to the human body. These degraded intermediates do not stay in the body and have no immunogenicity.

This review focusses on surveying the various novel chitosan derivatives that are available for use as drug delivery options. The milestones achieved based on the use of chitosan derivatives in the area of oral drug delivery has been comprehensively reviewed. The lack of implementing the various chitosan derivatives for oral drug delivery has been highlighted. The plausible reasons for this gap in the application of the various chitosan derivatives for oral drug delivery has been discussed. The possible accomplishments that could be achieved through utilization of the available resources has been addressed under future perspectives.

2. Comprehensive List of Novel Chitosan Derivatives

This section deals with a brief overview of the various chitosan derivatives that have been synthesized and are available for biomedical applications. The synthesis and their characterization and their applications have been elaborately reviewed by various authors [21–23]; here, we are restricted to a snapshot of their names. Figure 1 gives an overview of the various modification processes involved in the making of various chitosan derivatives.

N-(Aminoalkyl) Chitosan is a broad category of chitosan derivatives, which house many other forms. The encapsulation of calcium alginate beads with poly(L-lysine) (PLL), is the most accomplished encapsulation system for sustained delivery of bioactive agents. However, due to its high cost, large scale usage of this system for oral vaccination of animals is not possible. This is why a more economic and reliable microencapsulation chitosan and alginate system was sought after. Succinyl, Quateraminated, and Octanoyl Chitosan Porous chitosan microspheres for the delivery of antigens have been reported by Mi et al. [24]. The porous chitosan microspheres were chemically modified incorporating carboxyl, hydrophobic acyl, and quaternary ammonium groups.

Mitomycin C Conjugated *N*-succinyl Chitosan is the other class of chitosan derivatives. *N*-succinyl-chitosan, due to its carboxyl groups, has low toxicity, excellent biocompatibility and is retained in the body as a drug carrier for prolonged periods. This the reason why highly succinylated succinyl-chitosan (degree of succinylation: [25,26] can be dissolved in alkaline aqueous media, whereas chitosan cannot [27]. Succinyl-chitosan can react easily owing to the $-NH_2$ and $-COOH$ groups.

The *N*-Alkyl and Acylated Chitosan derivatives, which greatly benefit from the introduction of an alkyl or acyl chain, contribute greatly to chitosan's molecular design. This modification of chitosan with hydrophobic branches, improved its solubility properties [28,29]. The introduction of an alkyl chain to water soluble modified chitosan (*N*-methylene phosphonic chitosan) enabled the co-existence of hydrophobic and hydrophilic branches [30]. The alkyl groups in *N*-lauryl-*N*-methylene phosphonic chitosan weaken its hydrogen bonds and provide good solubility in solvents. Holding amphiphilic properties, which are typical for surfactants, this derivative has prospective demands in pharmaceutical and cosmetic fields.

Figure 1. Overview of various chemical modification processes that go into the making of chitosan derivatives.

Chitosan hydrochloride derivatives have been demonstrated for their effective in vitro release of ofloxacin from mucoadhesive erodible ocular inserts and ocular pharmacokinetics [31].Thiolated chitosans are obtained by the modification of chitosan with 2-iminothiolane [32], in order to improve the properties of chitosan as excipients in drug delivery systems. Chitosan-2-iminothiolane was obtained by grafting 2-iminothiolane onto the chitosan backbone. This exhibits excellent in situ gelling properties and improved mucoadhesive and drug releasing properties due to the thiol groups on chitosan. Phosphorylated chitosan, which is prepared by reacting chitosan with orthophosphoric acid and urea in DMF [33] or phosphorous pentoxide in methanesulphonic acid, is a water-soluble derivative of chitosan with huge potential for drug delivery.

MCC and SNOCC chitosan derivatives are a biomedically significant class. Mono-N-Carboxymethyl Chitosan (MCC), is a polyampholytic chitosan derivative, soluble at both neutral and alkaline pH [34], synthesized using glyoxolic acid in chitosan [34]. These derivatives are highly soluble and applicable for the administration of polyanionic drugs. It has also been demonstrated by the same group that MCC can improve low molecular weight heparin (LMWH) transport through Caco-2 cells.

Anionic chitosan derivatives were also attempted. N-sulfonato-N,O-carboxymethylchitosan (SNOCC) was produced [35], which retains around 50% of its nitrogen centers on the glucose subunits as free amino groups [36], which contribute to its unique biomedical characteristics.

PEGylated Chitosans are a prominent group of derivatives. Chitosan-PEG for oral peptide delivery was attempted by Prego et al. [37]. PEGylation of chitosan is apt for oral peptide/protein delivery, because generally PEGylation improves biocompatibility [38] and improves stability in GI fluid [37]. PEGylated chitosan showed enhanced solubility of hydrophobics.

3. Oral Drug Delivery by Chitosan Derivatives

Although drug delivery is a broad terminology, which is backed up by enumerable reviews when it comes to chitosan and drug delivery and good number of reviews when it comes to chitosan derivatives, this review chooses to specifically delve into oral drug delivery applications. The sections below consolidate what has been achieved in the area of oral drug delivery based on chitosan derivatives and micro/nano particulate chitosan.

3.1. Chitosan/Chitosan Derivatives

When drugs are administered orally, they must be able to survive various ranges of pH and gastrointestinal tract (GIT) secretions. The very process of oral drug absorption rests on transport (via passive diffusion, carrier-mediated transport, or pinocytosis) across the GIT membrane. This is impacted by various GIT physiological. The oral mucosa has a thin epithelium and rich vascularity, which is makes it ideally fit for buccal and sublingual administration [39]. The release of drugs from chitosan and its derivatives follows the conventional protocol that holds good for chitosan. Drug release is influenced by the hydrophilicity of chitosan and pH of the swelling solution. The chitosan-drug release mechanism involves swelling, diffusion of drugs through the polymeric matrix and polymer erosion [40] (Figure 2).

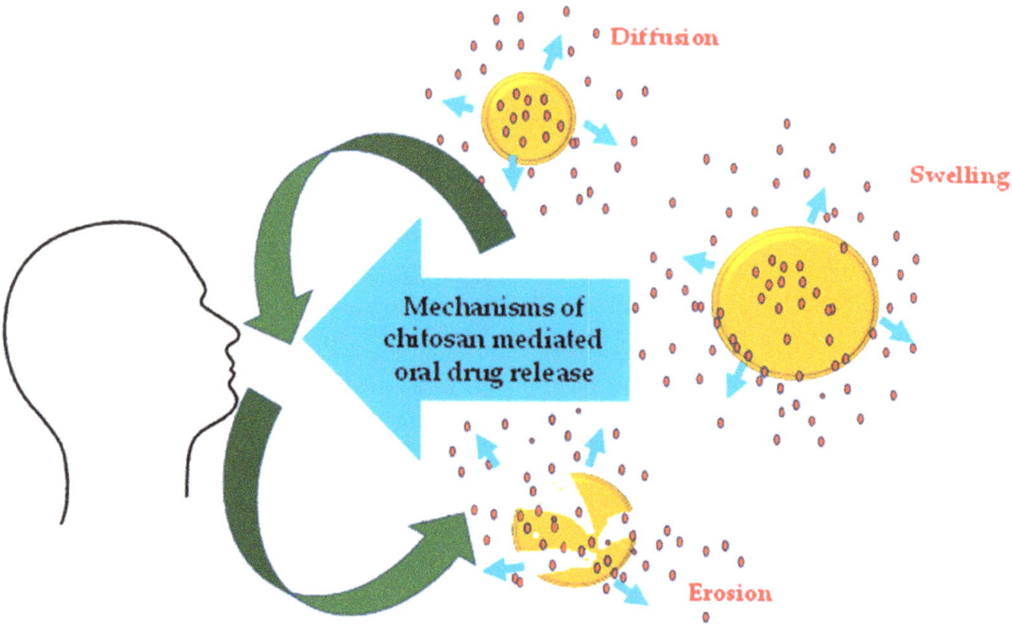

Figure 2. Mechanisms of drug release in chitosan and its derivatives.

Figure 3 lists the limitations that chitosan and its derivatives have broken, when it can to oral drug delivery. Drug delivery via the oral route is the easiest and e most convenient for patients. Chitosan because of its mucoadhesive nature, is able to protect labile drugs from GIT enzymatic degradation. Additionally, it is able to enhance absorption of administered therapeutic agent without affecting the biological system. This makes chitosan a valuable candidate as an oral delivery agent. Not only chitosan, but also chitosan micro-/nanoparticles have been demonstrated for oral drug delivery. Intestinal disinfection, suppression of *Helicobacter pylori* and dealing with ulcerative colitis, have been accomplished following treatment with antibiotic loaded chitosan particles. Amoxicillin and clarithromycin loaded into chitosan particles inhibited *H. pylori* [41,42]. The mucoadhesive properties of chitosan enabled prolonged delivery and oral bioavailability of acyclovir, an antiviral agent. This was because acyclovir chitosan microspheres could enhance drug retention in the upper GIT [43]. Protection against GIT degradation, improvement of oral bioavailability of insulin and enhancement of bioadhesion, have been reported as a result of its encapsulation into chitosan microspheres [44].

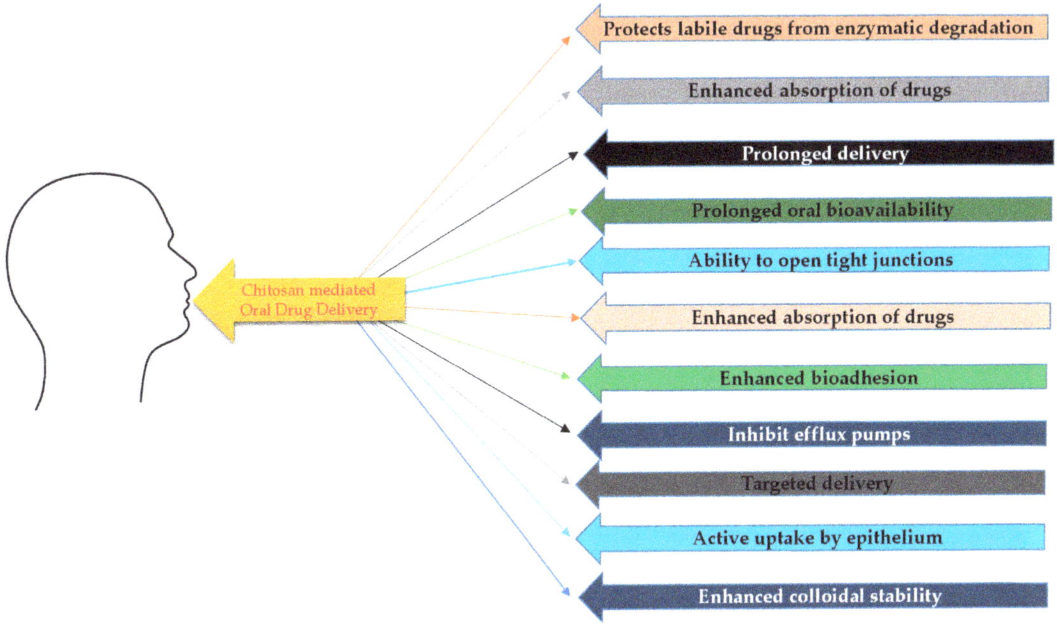

Figure 3. Listings of the limitations in oral drug delivery that chitosan and its derivatives have helped overcome.

Chitosan-based delivery systems have been applied for the protection of insulin from degradation in the upper GIT. Furthermore, it has been used to carry out the release of insulin at the colon (through degradation of the chitosan glycosidic linkage by colon microflora) [45]. Chitosan microspheres coated with cellulose acetate butyrate, loaded with 5-aminosalicylic acid (5-ASA) to treat ulcerative colitis is reported. Here, the bioadhesive nature of chitosan microspheres comes handy [46]. Another study reported localization of 5-ASA in the colon and low drug systemic bioavailability following oral administration of 5-ASA-loaded chitosan-Ca-alginate microparticles to Wistar male rats [47]. The fact that chitosan is highly soluble in the acidic medium, leading to drug burst in the stomach, has been mitigated using pH-sensitive polymer coatings [48–50].

Chitosan derivatives have also been reported for oral delivery of therapeutic peptides and proteins. Unmodified native chitosan itself has been proven for its oral peptide and protein delivery (e.g., capability to open tight junctions, mucoadhesive properties), with this being the case, how much more so with the use of chitosan derivatives. Recently, the potential of certain modified chitosans including TMC [51], thiolated chitosan [52,53] and chitosan-enzyme inhibitor conjugates [54–56] for noninvasive gene delivery has been widely reported. In addition, thiolated chitosan is able to inhibit efflux pumps, in particular P-glycoprotein (P-gp). In this way, thiolated chitosan comes handy when it comes to oral delivery of P-gp substrates [57–59]. The potential of chitosan, TMC and MCC for oral delivery of vaccine have been previously reviewed [60]. We touch on the highlights of these [61,62] reviews here.

The effect of two different trimethyl chitosans (TMC) on the oral absorption of buserelin, a peptide drug, after intraduodenal administration in rats is reported [63] Both formulations significantly enhanced buserelin plasma levels. Enhanced absorption in the presence of TMC60 (60% trimethylation) is because of the inherent ability of TMC60 to open tight junctions. The impact of TMC solutions on octreotide in vitro permeation and in vivo absorption in rats was also investigated [63]. The intrajejunally administered TMC solution led to a fivefold increase in the absorption of octreotide compared to octreotide standalone. The effect of various liquid formulations on the oral bioavailability of octreotide was studied in pigs [64]. Studies with MCC and SNOCC towards oral delivery of LMWH [34,35], confirmed that chitosan derivatives in a concentration of 3% improved the oral bioavailability of LMWH.

In vivo studies using thiolated chitosan tablets were applied using peptide drugs as well as efflux pump substrates. Enteric coated chitosan–TBA conjugated with salmon calcitonin for the oral administration to rats were tested. Besides chitosan–TBA, the tablets contained two different chitosan–enzyme inhibitor conjugates, (chitosan–BBI conjugate and chitosan–elastatinal) [65]. Oral administration of this chitosan conjugate showed decreased plasma calcium levels for several hours [66]. Another study, where stomach targeted delivery system for salmon calcitonin was investigated using tablets containing chitosan–TBA as well as chitosan–pepstatin [67]. The efficacy of chitosan–TBA/GSH for oral peptide delivery was studied using the peptide drug antide. Antide was not absorbed after oral administration; however, absorption of the drug was reported following oral administration of chitosan TBA/GSH tablets [26]. Besides peptides and proteins, oral bioavailability of efflux pump substrates was improved using thiolated chitosan tablets were used. Oral bioavailability of the P-gp substrate Rhodamine 123 (Rho-123) was reported [59]. Guggi et al., used optimized tablets comprising of chitosan-TBA with lower molecular mass (75–150 kDa instead of 400 kDa) and demonstrated a 5.5-fold increase in Rho-123 AUC in comparison to the Rho-123 buffer solution. Guggi et al. investigated the effect of various calcitonin containing tablets on the blood calcium level of rats after oral administration. Compared to tablets containing calcitonin and chitosan only, marginal reduction of the calcium level was observed after administration of chitosan–pepstatin conjugate tablets [67]. Oral insulin delivery using insulin and chitosan–aprotinin conjugate, showed reduced blood glucose level, 8 h after oral administration [68].

3.2. *Micro- and Nanoparticulate Oral Drug Delivery Systems Based on Chitosan Derivatives*

3.2.1. Microparticulate Chitosan Derivatives Oral Drug Delivery Systems

Authors reported the preparation of liposome microspheres were coated with TMC and chitosan–EDTA. In vivo studies on oral absorption of insulin, confirmed that chitosan EDTA coated liposomes decreased blood glucose [69]. Microspheres based on chitosan–succinate proved their potential for oral delivery of insulin [25]. The delivery system was tested in vivo in diabetic rats, with chitosan–succinate microspheres, the relative pharmacological efficacy showed fourfold improvement [25]. Intragastric administration of calcitonin containing liposomes coated with dodecylated chitosan was confirmed in rats. Similar results were obtained in case of chitosan–phthalate microspheres too. PEGylated chitosan was tested for oral delivery of salmon calcitonin. Alginate–chitosan microspheres with narrow size distribution were prepared by membrane emulsification technique in combination with ion (Ca^{2+}) and polymer (chitosan) solidification. The blood glucose level of diabetic rats was effectively reduced. It was made available for as long as 60 h after oral administration of the insulin-loaded alginate–chitosan microspheres. Therefore, the alginate–chitosan microspheres were found to be promising vectors showing a good efficiency in oral administration of protein or peptide drugs [70]. Chitosan microparticles prepared using the precipitation/coacervation method to obtain biodegradable chitosan microparticles. The entrapped ovalbumin was released after intracellular digestion into the Peyer's patches. The proved that the labeled chitosan microparticles could be taken up by the epithelium of the murine Peyer's patches. Since uptake by Peyer's patches is an essential step in oral vaccination, these results confirmed that the chitosan microparticles are useful when it comes to vaccine delivery system [71]. Chitosan and chondroitin sulphate microspheres were prepared and reported for controlled release of metoclopramide hydrochloride in oral administration [72]. Microparticles prepared by ionic crosslinking between tripolyphosphate (TPP) and chitosan (Cs) were applied to enable the oral bioavailability of curcumin. The developed microparticles are reported to successfully enhance the dissolution of the poorly water-soluble drug Cur, and eventually, improve its oral bioavailability effectively [73].

3.2.2. Nanoparticulate Chitosan Derivatives Oral Drug Delivery Systems

TMC-based insulin-loaded nanoparticles were investigated, it was reported that insulin-TMC polyelectrolyte complexes exhibited higher colloidal stability in simulated intestinal fluid and protected insulin from trypsinic degradation [74]. TMC nanoparticles has also been demonstrated for its oral vaccine delivery. Intragastrical (IG) administration of TMC-nanoparticles containing the model vaccine urease could result in higher IgG and IgA levels [75]. Another study reported the efficiency of TMC as vector for in vitro and in vivo gene delivery [76]. Three different TMC-based nanoparticles encapsulated pDNA encoding green fluorescent protein (GFP) were demonstrated for their successful delivery attributes. Nanoparticles based on chitosan–TGA and pDNA for oral delivery are also reported [53]. Acrylic nanoparticles with chitosan–TBA are also reported. In vivo studies with thiolated chitosan nanoparticles for oral delivery are still lacking, however, oral insulin delivery using thiolatedpoly(acrylic acid) nanoparticles [77] and intranasal gene delivery using chitosan-TGA nanoparticles have been demonstrated [52]. Fucoidan (FD) has hypoglycemic effects, TMC and FD were loaded with insulin. TMC/FD NPs are pH sensitive and defend insulin from degradation in the GIT. Moreover, they enhance the cellular transport of insulin across the intestinal barrier [78]. The delivery of insulin via glycerol monocaprylate-modified chitosan nanoparticles has also been demonstrated using TMC/FD NPs [79]. A nanoemulsion was coated with two different PEGylated chitosans. In vivo studies in rats showed, that the oral uptake of salmon calcitonin when administered in carriers coated with PEGylated chitosan was higher than the nanoemulsion alone [37]. Table 1 gives the consolidated list of chitosan derivatives that have been employed for oral drug delivery applications.

Table 1. Chitosan derivatives that have been used for oral drug delivery applications.

Chitosan Derivative	Oral Drug Delivery Application	References
TMC, thiolated chitosan	noninvasive gene delivery	[52,53]
Thiolated chitosan	oral delivery of P-glycoprotein (P-gp) substrates	[57–59]
TMC, MCC	oral vaccine delivery	[60]
Trimethyl chitosans (TMC)	oral absorption of the peptide drug buserelin after intraduodenal administration in rats	[63]
TMC	octreotide in vitro permeation and in vivo absorption in rats	[80]
TMC	oral bioavailability of octreotide in pigs	[64]
MCC, SNOCC	oral delivery of LMWH	[34,35]
Chitosan–TBA	oral administration of drug salmon calcitonin to rats	[65]
Chitosan–TBA/chitosan–enzyme inhibitor conjugate	Delivery of drug salmon calcitonin	[66]
Chitosan–TBA, chitosan–pepstatin	Stomach targeted delivery of salmon calcitonin	[67]

Table 1. *Cont.*

Chitosan Derivative	Oral Drug Delivery Application	References
Chitosan–TBA/GSH	oral peptide delivery of peptide drug antide	[26]
Chitosan–TBA	Oral bioavailability of the P-gp substrate Rhodamine 123	[59]
Chitosan–TBA	5.5-fold increase in Rho-123 AUC	[59]
Chitosan–pepstatin conjugate tablets	Reduction of blood calcium level of rats after oral administration	[67]
Chitosan–aprotinin conjugate	Oral insulin delivery	[68]
TMC–nanoparticles–Vaccine urease	Oral vaccine delivery–higher IgG and IgA levels	[75]
TMC nanoparticles	gene delivery	[76]
TMC–based nanoparticles encapsulatde pDNA encoding green fluorescentprotein (GFP)	Oral delivery	[53]
chitosan-TGA and pDNA	Oral delivery	[53]
thiolatedpoly(acrylic acid) nanoparticles	Oral insulin delivery	[77]
Chitosan–TGA nanoparticles	intranasal gene delivery	[52]
liposomes coated with dodecylated chitosan	Intragastric administration of calcitonin	[37]
TMC and chitosan–EDTA	Oral absorption of insulin	[78]
chitosan-succinate	oral delivery of insulin chitosan-succinate microspheres	[25]
Chitosan–succinate microspheres	in vivo insulin delivery in diabetic rats	[25]
TMC/FD NPs	defend insulin from degradation in the GIT, enhance transport	[79,80]

4. Future Endeavors

This review briefly surveyed the current scenario of oral drug delivery using chitosan derivatives. Drug delivery is a very appropriate subject area, which chitosan have enormously impacted. We ran a pubmed search, using keywords, chitosan and drug delivery, chitosan derivatives and oral drug delivery, chitosan derivatives and drug delivery. Backed up by a total of 10,000 odd publications as per our pubmed search from 1981–2022, chitosan has indeed generously contributed to drug delivery. Novel chitosan derivatives, which are the second-generation innovations emerging from chitosan, have a 2635 publication record when it comes to drug delivery applications.

Chitosan derivatives are well reported for their use in delivery of poorly soluble drugs, for colon-targeted drug delivery, for mucosal drug delivery, ocular drug delivery and topical delivery [81–84].

Chopra et al. [85] have extensively reviewed the advances and potential applications of chitosan derivatives as mucoadhesive biomaterials in modern drug delivery. When it comes to drug delivery, the drawbacks of chitosan have been overcome through derivatives such as carboxylated, various conjugates, thiolated, and acylated chitosan and Tan et al. have reviewed the applications of quaternized chitosan as antimicrobial agents, including their antimicrobial activity, mechanism of action and biomedical applications in orthopedics [86]. These have become an appropriate platform for sustained release at a controlled rate, prolonged residence time, improved patient compliance through reduced dosing frequency, enhanced bioavailability leading to significant improvement in therapeutic efficacy.

Currently, chitosan derivative nanoparticles are mainly used for sustained release, preparation of targeted drugs and as vectors for gene therapy. As delivery carriers, chitosan and its derivatives are usually available as microspheres, nanoparticles, micelles, and gels in delivery carriers [87,88]. Besides these options, chitosan derivative nanoparticles are also used for the delivery of polypeptides. Chitosan derivative nanoparticles interact with peptides through strong hydrogen bonds and static electricity, obtaining peptide-loaded nanoparticles. Fatty-acid-modified quaternary ammonium chitosan nanoparticles loaded with insulin have been shown to be beneficial [89]. Chitosan derivative nanoparticles have also been applied for gene delivery. Gene therapy is a promising strategy for challenging diseases. A key step in gene therapy is the successful delivery of genes [90,91]. Chitosan derivative nanoparticles, as non-viral vectors, have excellent solubility, biodegradability, biocompatibility, non-toxicity and a higher transfection rate than chitosan nanoparticles [92]. Methoxy polyethylene glycol-modified trimethyl chitosan (mPEG-TMC) has been covalently linked to doxorubicin (DOX) and cis-itaconic anhydride (CA), for better anti-tumor activities [93,94]. O-carboxymethyl chitosan inhibited tumor cell migration in vitro [95]. The poly-β-amino ester nanoparticle loading gene, after the addition of thiolated O-carboxymethyl chitosan, showed a higher cell transfection rate [96]. These are a notable few brief mention of the drug delivery potentials of chitosan derivatives, which have been dealt in detail by earlier reviews.

Chitosan derivatives upgraded to break many of the limitations that chitosan was facing, and with that reputation, it was believed that higher research curiosity and much more research interest would be evident. This expectation is well below the actual trend. As for oral drug delivery, chitosan derivatives are within the 500-article mark, which is one fifth lesser than the interest on chitosan derivatives and drug delivery. Figure 4 summarizes this trend. However, as this review points out, there is definitely a high potential contribution from chitosan derivatives in biomedical applications and drug delivery, which we stress has not been fully tapped into in terms of oral drug delivery applications. This review hopes to provoke some though and awareness towards this area of research.

Non-invasive oral drug delivery is the crown of drug delivery approaches, chitosan derivatives are the latest generation upgrades, a fusion of both these should break numerous boundaries and limitations. The fact that this is truly an interdisciplinary area, where synthetic chemists and pharmacologist need to collaborate to access the full potential of either expertise, may be the retardant. The reason for the low enthusiasm could be the interdisciplinary nature of this area of research. There is no dearth for chitosan derivatives, as pointed out by the review, diverse chitosan derivatives are in the market. Yet, as pointed out in this review, only thiolated chitosans have been predominantly applied, and few other scattered versions too. There are a whole lot of options to consider and avenues that they would open up which are yet to be looked into. This review hopes to enthuse the researchers in this direction.

Combining nanoaspects of chitosan with synthesis of chitosan derivatives is definitive progress in this area. Nanoforms have always pushed limitations of various applications, and there is surely a lot more to derive from nanostucturization of the chitosan derivatives. Oral drug delivery has benefitted greatly from the use of nanochitosan forms; combining chitosan derivatives with nano aspects could prove highly beneficial.

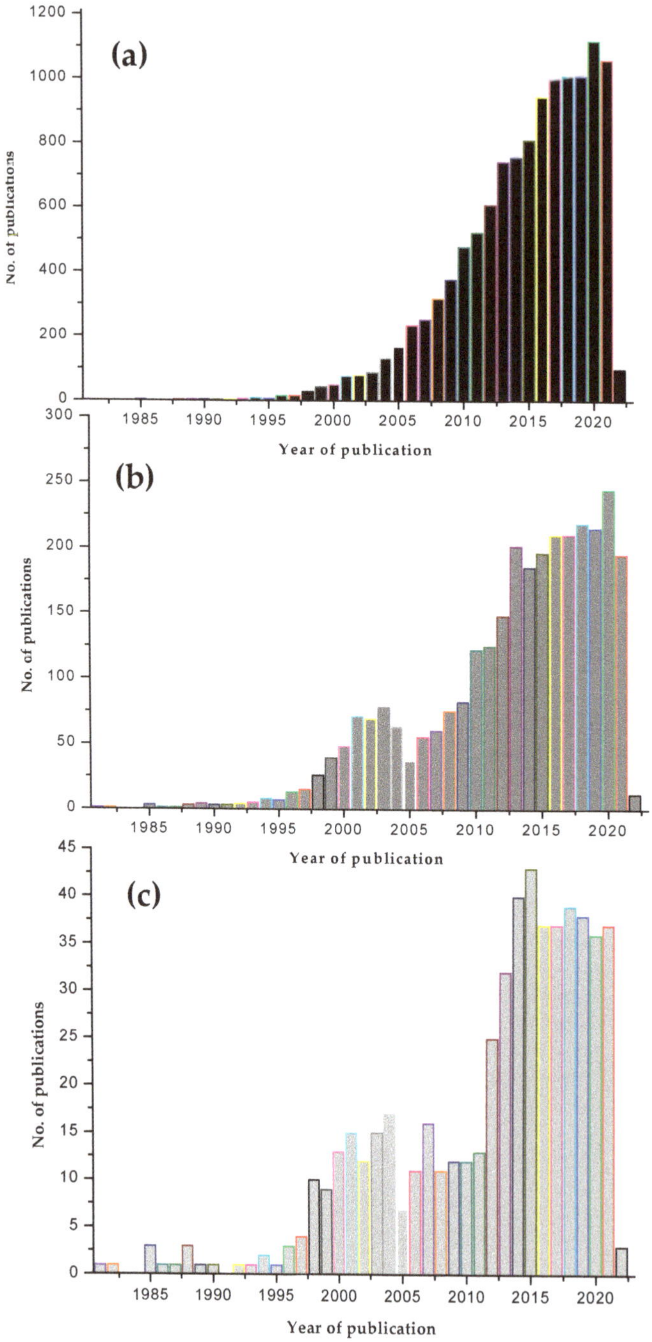

Figure 4. Comparative bar diagrams of research articles published in the area of (**a**) chitosans and drug delivery (**b**) chitosan derivatives and drug delivery (**c**) chitosan derivatives for oral drug delivery, based on our pubmed search.

5. Conclusions

The objective of this review was to showcase the wealth of available chitosan derivatives and to evaluate their achievements in the area of oral drug delivery. Numerous reviews exist in the area of chitosan and drug delivery, chitosan derivatives and drug delivery applications are also well reported. We reviewed the comparatively less-reported chitosan derivative application into oral drug delivery. During the review process, it became clear that there is no doubt as to the advantages of employing the use of chitosan derivatives for oral drug delivery purposes. However, as pointed out in the review, there is a huge gap between the available knowledge and the synthesized chitosan derivations and their oral drug delivery applications. There are so many derivatives synthesized, yet only few have been used for oral drug delivery applications. The reasons for this gap and the various reasons that could have led to this have been speculated. The need to bridge these ends have been emphasized. There is definitely much to harness and more to achieve, through proper inclusion of chitosan derivatives that have so far not been attempted for oral drug delivery applications.

Author Contributions: I.S., Revisions and funding support; M.M., S.T. and J.G., preparation of original draft; N.H. and J.S., review and revisions; J.-W.O., participated in review and revisions and funding support. All authors have read and agreed to the published version of the manuscript.

Funding: This research received no external funding.

Institutional Review Board Statement: Not applicable.

Informed Consent Statement: Not applicable.

Data Availability Statement: Not applicable.

Acknowledgments: This article was supported by the KU Research Professor Program of Konkuk University.

Conflicts of Interest: The authors declare no conflict of interest.

References

1. Kato, Y.; Onishi, H.; Machida, Y. Application of chitin and chitosan derivatives in the pharmaceutical field. *Curr. Pharm. Biotechnol.* **2003**, *4*, 303–309. [CrossRef] [PubMed]
2. Singla, A.K.; Chawla, M. Chitosan: Some pharmaceutical and biological aspects—An update. *J. Pharm. Pharmacol.* **2001**, *53*, 1047–1067. [CrossRef] [PubMed]
3. Kas, H.S. Chitosan: Properties, preparations and application to microparticulate systems. *J. Microencapsul.* **1997**, *14*, 689–711. [CrossRef] [PubMed]
4. Varshosaz, J. The promise of chitosan microspheres in drug delivery systems. *Expert Opin. Drug Deliv.* **2007**, *4*, 263–273. [CrossRef]
5. Galed, G.; Miralles, B.; Panos, I.; Santiago, A.; Heras, A. N-Deacetylation and depolymerization reactions of chitin/chitosan: Influence of the source of chitin. *Carbohydr. Polym.* **2005**, *62*, 316–320. [CrossRef]
6. Yan, C.Y.; Gu, J.W.; Hou, D.P.; Jing, H.Y.; Wang, J.; Guo, Y.Z.; Katsumi, H.; Sakane, T.; Yamamoto, A. Synthesis of Tat tagged and folate modified N-succinyl-chitosan self-assembly nanoparticles as a novel gene vector. *Int. J. Biol. Macromol.* **2015**, *72*, 751–756. [CrossRef]
7. Zhu, A.P.; Chen, T.; Yuan, L.H.; Wu, H.; Lu, P. Synthesis and characterization of N-succinyl-chitosan and its self-assembly of nanospheres. *Carbohydr. Polym.* **2006**, *66*, 274–279. [CrossRef]
8. Chien, R.C.; Yen, M.T.; Mau, J.L. Antimicrobial and antitumor activities of chitosan from shiitake stipes, compared to commercial chitosan from crab shells. *Carbohydr. Polym.* **2016**, *138*, 259–264. [CrossRef]
9. Huang, M.; Khor, E.; Lim, L.Y. Uptake and cytotoxicity of chitosan molecules and nanoparticles: Effects of molecular weight and degree of deacetylation. *Pharm. Res.* **2004**, *21*, 344–353. [CrossRef]
10. Kong, M.; Chen, X.G.; Xing, K.; Park, H.J. Antimicrobial properties of chitosan and mode of action: A state of the art review. *Int. J. Food Microbiol.* **2010**, *144*, 51–63. [CrossRef]
11. Goy, R.C.; de Britto, D.; Assis, O.B.G. A Review of the Antimicrobial Activity of Chitosan. *Polimeros* **2009**, *19*, 241–247. [CrossRef]
12. Ngo, D.H.; Vo, T.S.; Ngo, D.N.; Kang, K.H.; Je, G.Y.; Pham, H.N.; Byun, H.G.; Kim, S.K. Biological effects of chitosan and its derivatives. *Food Hydrocoll.* **2015**, *51*, 200–216. [CrossRef]
13. Bugnicourt, L.; Ladaviere, C. Interests of chitosan nanoparticles conically cross-linked with tripolyphosphate for biomedical applications. *Prog. Polym. Sci.* **2016**, *60*, 1–17. [CrossRef]

14. Yamamoto, H.; Kuno, Y.; Sugimoto, S.; Takeuchi, H.; Kawashima, Y. Surface-modified PLGA nanosphere with chitosan improved pulmonary delivery of calcitonin by mucoadhesion and opening of the intercellular tight junctions. *J. Control. Release* **2005**, *102*, 373–381. [CrossRef] [PubMed]
15. Okamoto, Y.; Yano, R.; Miyatake, K.; Tomohiro, I.; Shigemasa, Y.; Minami, S. Effects of chitin and chitosan on blood coagulation. *Carbohydr. Polym.* **2003**, *53*, 337–342. [CrossRef]
16. Busilacchi, A.; Gigante, A.; Mattioli-Belmonte, M.; Manzotti, S.; Muzzarelli, R.A.A. Chitosan stabilizes platelet growth factors and modulates stem cell differentiation toward tissue regeneration. *Carbohydr. Polym.* **2013**, *98*, 665–676. [CrossRef]
17. Tokoro, A.; Tatewaki, N.; Suzuki, K.; Mikami, T.; Suzuki, S.; Suzuki, M. Growth-Inhibitory Effect of Hexa-N-Acetylchitohexaose and Chitohexaose against Meth-a Solid Tumor. *Chem. Pharm. Bull.* **1988**, *36*, 784–790. [CrossRef]
18. Ngo, D.H.; Kim, S.K. Antioxidant effects of chitin, chitosan, and their derivatives. *Adv. Food Nutr. Res.* **2014**, *73*, 15–31. [CrossRef]
19. Younes, I.; Rinaudo, M. Chitin and chitosan preparation from marine sources. Structure, properties and applications. *Mar. Drugs* **2015**, *13*, 1133–1174. [CrossRef]
20. Park, P.J.; Je, J.Y.; Kim, S.K. Free radical scavenging activity of chitooligosaccharides by electron spin resonance spectrometry. *J. Agric. Food Chem.* **2003**, *51*, 4624–4627. [CrossRef]
21. Riaz Rajoka, M.S.; Zhao, L.; Mehwish, H.M.; Wu, Y.; Mahmood, S. Chitosan and its derivatives: Synthesis, biotechnological applications, and future challenges. *Appl. Microbiol. Biotechnol.* **2019**, *103*, 1557–1571. [CrossRef] [PubMed]
22. Confederat, L.G.; Tuchilus, C.G.; Dragan, M.; Sha'at, M.; Dragostin, O.M. Preparation and Antimicrobial Activity of Chitosan and Its Derivatives: A Concise Review. *Molecules* **2021**, *26*, 3694. [CrossRef] [PubMed]
23. Zhao, D.; Yu, S.; Sun, B.; Gao, S.; Guo, S.; Zhao, K. Biomedical Applications of Chitosan and Its Derivative Nanoparticles. *Polymers* **2018**, *10*, 462. [CrossRef]
24. Mi, F.L.; Shyu, S.S.; Chen, C.T.; Schoung, J.Y. Porous chitosan microsphere for controlling the antigen release of Newcastle disease vaccine: Preparation of antigen-adsorbed microsphere and in vitro release. *Biomaterials* **1999**, *20*, 1603–1612. [CrossRef]
25. Ubaidulla, U.; Khar, R.K.; Ahmad, F.J.; Sultana, Y.; Panda, A.K. Development and characterization of chitosan succinate microspheres for the improved oral bioavailability of insulin. *J. Pharm. Sci.* **2007**, *96*, 3010–3023. [CrossRef]
26. Bernkop-Schnurch, A.; Pinter, Y.; Guggi, D.; Kahlbacher, H.; Schoffmann, G.; Schuh, M.; Schmerold, I.; Del Curto, M.D.; D'Antonio, M.; Esposito, P.; et al. The use of thiolated polymers as carrier matrix in oral peptide delivery—Proof of concept. *J. Control. Release* **2005**, *106*, 26–33. [CrossRef]
27. Yamaguchi, R.; Arai, Y.; Itoh, T.; Hirano, S. Preparation of Partially N-Succinylated Chitosans and Their Cross-Linked Gels. *Carbohydr. Res.* **1981**, *88*, 172–175. [CrossRef]
28. Nishimura, S.I.; Miura, Y.; Ren, L.D.; Sato, M.; Yamagishi, A.; Nishi, N.; Tokura, S.; Kurita, K.; Ishii, S. An Efficient Method for the Syntheses of Novel Amphiphilic Polysaccharides by Regioselective and Thermoselective Modifications of Chitosan. *Chem. Lett.* **1993**, *22*, 1623–1626. [CrossRef]
29. Holme, K.R.; Hall, L.D. Chitosan Derivatives Bearing C-10-Alkyl Glycoside Branches—A Temperature-Induced Gelling Polysaccharide. *Macromolecules* **1991**, *24*, 3828–3833. [CrossRef]
30. Ramos, V.M.; Rodriguez, N.M.; Rodriguez, M.S.; Heras, A.; Agullo, E. Modified chitosan carrying phosphonic and alkyl groups. *Carbohydr. Polym.* **2003**, *51*, 425–429. [CrossRef]
31. Di Colo, G.; Zambito, Y.; Burgalassi, S.; Serafini, A.; Saettone, M.F. Effect of chitosan on in vitro release and ocular delivery of ofloxacin from erodible inserts based on poly(ethylene oxide). *Int. J. Pharm.* **2002**, *248*, 115–122. [CrossRef]
32. Bernkop-Schnurch, A.; Hornof, M.; Zoidl, T. Thiolated polymers-thiomers: Synthesis and in vitro evaluation of chitosan-2-iminothiolane conjugates. *Int. J. Pharm.* **2003**, *260*, 229–237. [CrossRef]
33. Sakaguchi, T.; Horikoshi, T.; Nakajima, A. Studies on the Accumulation of Heavy-Metal Elements in Biological-Systems. 20. Adsorption of Uranium by Chitin Phosphate and Chitosan Phosphate. *Agric. Biol. Chem. Tokyo* **1981**, *45*, 2191–2195. [CrossRef]
34. Thanou, M.; Nihot, M.T.; Jansen, M.; Verhoef, J.C.; Junginger, H.E. Mono-N-carboxymethyl chitosan (MCC), a polyampholytic chitosan derivative, enhances the intestinal absorption of low molecular weight heparin across intestinal epithelia in vitro and in vivo. *J. Pharm. Sci.* **2001**, *90*, 38–46. [CrossRef]
35. Thanou, M.; Henderson, S.; Kydonieus, A.; Elson, C. N-sulfonato-N,O-carboxymethylchitosan: A novel polymeric absorption enhancer for the oral delivery of macromolecules. *J. Control. Release* **2007**, *117*, 171–178. [CrossRef]
36. Hayes, E.R. N,O-carboxymethyl Chitosan and Preparative Method Therefor. U.S. Patent US4619995A, 28 October 1986.
37. Prego, C.; Torres, D.; Fernandez-Megia, E.; Novoa-Carballal, R.; Quinoa, E.; Alonso, M.J. Chitosan-PEG nanocapsules as new carriers for oral peptide delivery—Effect of chitosan pegylation degree. *J. Control. Release* **2006**, *111*, 299–308. [CrossRef]
38. Zhang, M.; Li, X.H.; Gong, Y.D.; Zhao, N.M.; Zhang, X.F. Properties and biocompatibility of chitosan films modified by blending with PEG. *Biomaterials* **2002**, *23*, 2641–2648. [CrossRef]
39. Ashford, M. Biopharmaceutical principles of drug delivery. In *Pharmaceutics: The Science of Dosage Form Design*, 2nd ed.; Aulton, M.E., Ed.; Churchill Livingstone: New York, NY, USA, 2001; pp. 211–253.
40. Younis, M.K.; Tareq, A.Z.; Kamal, I.M. Optimization Of Swelling, Drug Loading And Release From Natural Polymer Hydrogels. *IOP Conf. Ser. Mater. Sci. Eng.* **2018**, *454*, 12–17. [CrossRef]
41. Gisbert, J.P.; Torrado, G.; Torrado, S.; Olivares, D.; Pajares, J.M. Clinical trial evaluating amoxicillin and clarithromycin hydrogels (Chitosan-polyacrylic acid polyionic complex) for *H. pylori* eradication. *J. Clin. Gastroenterol.* **2006**, *40*, 618–622. [CrossRef]

42. Chang, C.H.; Lin, Y.H.; Yeh, C.L.; Chen, Y.C.; Chiou, S.F.; Hsu, Y.M.; Chen, Y.S.; Wang, C.C. Nanoparticles incorporated in pH-sensitive hydrogels as amoxicillin delivery for eradication of Helicobacter pylori. *Biomacromolecules* **2010**, *11*, 133–142. [CrossRef]
43. Dhaliwal, S.; Jain, S.; Singh, H.P.; Tiwary, A.K. Mucoadhesive microspheres for gastroretentive delivery of acyclovir: In vitro and in vivo evaluation. *AAPS J.* **2008**, *10*, 322–330. [CrossRef] [PubMed]
44. Wei, W.; Ma, G.H.; Wang, L.Y.; Wu, J.; Su, Z.G. Hollow quaternized chitosan microspheres increase the therapeutic effect of orally administered insulin. *Acta Biomater.* **2010**, *6*, 205–209. [CrossRef] [PubMed]
45. Hejazi, R.; Amiji, M. Chitosan-based gastrointestinal delivery systems. *J. Control. Release* **2003**, *89*, 151–165. [CrossRef]
46. Varshosaz, J.; Jaffarian Dehkordi, A.; Golafshan, S. Colon-specific delivery of mesalazine chitosan microspheres. *J. Microencapsul.* **2006**, *23*, 329–339. [CrossRef] [PubMed]
47. Mladenovska, K.; Raicki, R.S.; Janevik, E.I.; Ristoski, T.; Pavlova, M.J.; Kavrakovski, Z.; Dodov, M.G.; Goracinova, K. Colon-specific delivery of 5-aminosalicylic acid from chitosan-Ca-alginate microparticles. *Int. J. Pharm.* **2007**, *342*, 124–136. [CrossRef]
48. Eizatahry, A.A.; Eldin, M.S.M. Preparation and characterization of metronidazole loaded chitosan nanoparticles for drug delivery application. *Polym. Adv. Technol.* **2008**, *19*, 1787–1791. [CrossRef]
49. Jain, S.K.; Jain, A.; Gupta, Y.; Ahirwar, M. Design and development of hydrogel beads for targeted drug delivery to the colon. *AAPS PharmSciTech* **2007**, *8*, E56. [CrossRef]
50. Oosegi, T.; Onishi, H.; Machida, Y. Novel preparation of enteric-coated chitosan-prednisolone conjugate microspheres and in vitro evaluation of their potential as a colonic delivery system. *Eur. J. Pharm. Biopharm.* **2008**, *68*, 260–266. [CrossRef]
51. Li, H.Y.; Birchall, J. Chitosan-modified dry powder formulations for pulmonary gene delivery. *Pharm. Res.* **2006**, *23*, 941–950. [CrossRef]
52. Lee, D.; Zhang, W.; Shirley, S.A.; Kong, X.; Hellermann, G.R.; Lockey, R.F.; Mohapatra, S.S. Thiolated chitosan/DNA nanocomplexes exhibit enhanced and sustained gene delivery. *Pharm. Res.* **2007**, *24*, 157–167. [CrossRef]
53. Martien, R.; Loretz, B.; Thaler, M.; Majzoob, S.; Bernkop-Schnurch, A. Chitosan-thioglycolic acid conjugate: An alternative carrier for oral nonviral gene delivery? *J. Biomed. Mater. Res. A* **2007**, *82*, 1–9. [CrossRef] [PubMed]
54. Loretz, B.; Foger, F.; Werle, M.; Bernkop-Schnurch, A. Oral gene delivery: Strategies to improve stability of pDNA towards intestinal digestion. *J. Drug Target.* **2006**, *14*, 311–319. [CrossRef] [PubMed]
55. Martien, R.; Loretz, B.; Schnurch, A.B. Oral gene delivery: Design of polymeric carrier systems shielding toward intestinal enzymatic attack. *Biopolymers* **2006**, *83*, 327–336. [CrossRef] [PubMed]
56. Loretz, B.; Bernkop-Schnurch, A. In vitro evaluation of chitosan-EDTA conjugate polyplexes as a nanoparticulate gene delivery system. *AAPS J* **2006**, *8*, E756–E764. [CrossRef]
57. Foger, F.; Hoyer, H.; Kafedjiiski, K.; Thaurer, M.; Bernkop-Schurch, A. In vivo comparison of various polymeric and low molecular mass inhibitors of intestinal P-glycoprotein. *Biomaterials* **2006**, *27*, 5855–5860. [CrossRef]
58. Foger, F.; Kafedjiiski, K.; Hoyer, H.; Loretz, B.; Bernkop-Schnurch, A. Enhanced transport of P-glycoprotein substrate saquinavir in presence of thiolated chitosan. *J. Drug Target.* **2007**, *15*, 132–139. [CrossRef]
59. Foger, F.; Schmitz, T.; Bernkop-Schnurch, A. In vivo evaluation of an oral delivery system for P-gp substrates based on thiolated chitosan. *Biomaterials* **2006**, *27*, 4250–4255. [CrossRef]
60. van der Lubben, I.M.; Verhoef, J.C.; Borchard, G.; Junginger, H.E. Chitosan and its derivatives in mucosal drug and vaccine delivery. *Eur. J. Pharm. Sci.* **2001**, *14*, 201–207. [CrossRef]
61. Werle, M.; Takeuchi, H.; Bernkop-Schnurch, A. Modified Chitosans for Oral Drug Delivery. *J. Pharm Sci.* **2009**, *98*, 1643–1656. [CrossRef]
62. Werle, M.; Bernkop-Schnurch, A. Thiolated chitosans: Useful excipients for oral drug delivery. *J. Pharm. Pharmacol.* **2008**, *60*, 273–281. [CrossRef] [PubMed]
63. Thanou, M.; Florea, B.I.; Langemeyer, M.W.; Verhoef, J.C.; Junginger, H.E. N-trimethylated chitosan chloride (TMC) improves the intestinal permeation of the peptide drug buserelin in vitro (Caco-2 cells) and in vivo (rats). *Pharm. Res.* **2000**, *17*, 27–31. [CrossRef]
64. Thanou, M.; Verhoef, J.C.; Verheijden, J.H.M.; Junginger, H.E. Intestinal absorption of octreotide using trimethyl chitosan chloride: Studies in pigs. *Pharm. Res.* **2001**, *18*, 823–828. [CrossRef] [PubMed]
65. Guggi, D.; Bernkop-Schnurch, A. In vitro evaluation of polymeric excipients protecting calcitonin against degradation by intestinal serine proteases. *Int. J. Pharm.* **2003**, *252*, 187–196. [CrossRef]
66. Guggi, D.; Kast, C.E.; Bernkop-Schnurch, A. In vivo evaluation of an oral salmon calcitonin-delivery system based on a thiolated chitosan carrier matrix. *Pharm. Res.* **2003**, *20*, 1989–1994. [CrossRef]
67. Guggi, D.; Krauland, A.H.; Bernkop-Schnurch, A. Systemic peptide delivery via the stomach: In vivo evaluation of an oral dosage form for salmon calcitonin. *J. Control. Release* **2003**, *92*, 125–135. [CrossRef]
68. Werle, M.; Loretz, B.; Entstrasser, D.; Foger, F. Design and evaluation of a chitosan-aprotinin conjugate for the peroral delivery of therapeutic peptides and proteins susceptible to enzymatic degradation. *J. Drug Target.* **2007**, *15*, 327–333. [CrossRef]
69. Wu, Z.H.; Ping, Q.N.; Lei, X.M.; Li, J.Y.; Cai, P. Effects of the liposomes coated by chitosan and its derivatives on the gastrointestinal transit of insulin. *Yao Xue Xue Bao* **2005**, *40*, 618–622. [PubMed]
70. Zhang, Y.; Wei, W.; Lv, P.; Wang, L.; Guanghui, M. Preparation and evaluation of alginate—Chitosan microspheres for oral deliveryof insulin. *Eur. J. Pharm. Biopharm.* **2011**, *77*, 11–19. [CrossRef] [PubMed]

71. van der Lubben, I.M.; Verhoef, J.C.; van Aelst, A.C.; Borchard, G.; Junginger, H.E. Chitosan microparticles for oral vaccination: Preparation, characterization and preliminary in vivo uptake studies in murine Peyer's patches. *Biomaterials* **2001**, *22*, 687–694. [CrossRef]
72. Ganza-González, A.; Anguiano-Igea, S.; Otero-Espinar, F.J.; Blanco Méndez, J. Chitosan and chondroitin microspheres for oral-administration controlled release of metoclopramide. *Eur. J. Pharm. Biopharm.* **1999**, *48*, 149–155. [CrossRef]
73. Wan, S.; Sun, Y.; Sun, L.; Tan, F. Chitosan microparticles for oral bioavailability improvement of the hydrophobic drug curcumin. *Pharmazie* **2012**, *67*, 525–528. [PubMed]
74. Jintapattanakit, A.; Junyaprasert, V.B.; Mao, S.; Sitterberg, J.; Bakowsky, U.; Kissel, T. Peroral delivery of insulin using chitosan derivatives: A comparative study of polyelectrolyte nanocomplexes and nanoparticles. *Int. J. Pharm.* **2007**, *342*, 240–249. [CrossRef] [PubMed]
75. Chen, F.; Zhang, Z.R.; Yuan, F.; Qin, X.; Wang, M.T.; Huang, Y. In vitro and in vivo study of N-trimethyl chitosan nanoparticles for oral protein delivery. *Int. J. Pharm.* **2008**, *349*, 226–233. [CrossRef] [PubMed]
76. Zheng, F.; Shi, X.W.; Yang, G.F.; Gong, L.L.; Yuan, H.Y.; Cui, Y.J.; Wang, Y.; Du, Y.M.; Li, Y. Chitosan nanoparticle as gene therapy vector via gastrointestinal mucosa administration: Results of an in vitro and in vivo study. *Life Sci.* **2007**, *80*, 388–396. [CrossRef] [PubMed]
77. Deutel, B.; Greindl, M.; Thaurer, M.; Bernkop-Schnuerch, A. Novel insulin thiomer nanoparticles: In vivo evaluation of an oral drug delivery system. *Biomacromolecules* **2008**, *9*, 278–285. [CrossRef]
78. Tsai, L.C.; Chen, C.H.; Lin, C.W.; Ho, Y.C.; Mi, F.L. Development of mutlifunctional nanoparticles self-assembled from trimethyl chitosan and fucoidan for enhanced oral delivery of insulin. *Int. J. Biol. Macromol.* **2019**, *126*, 141–150. [CrossRef]
79. Gao, M.; Sun, Y.; Kou, Y.; Shen, X.; Huo, Y.; Liu, C.; Sun, Z.; Zhang, X.; Mao, S. Effect of glyceryl monocaprylate-modified chitosan on the intranasal absorption of insulin in rats. *J. Pharm. Sci.* **2019**, *108*, 3623–3629. [CrossRef]
80. Thanou, M.; Verhoef, J.C.; Marbach, P.; Junginger, H.E. Intestinal absorption of octreotide: N-trimethyl chitosan chloride (TMC) ameliorates the permeability and absorption properties of the somatostatin analogue in vitro and in vivo. *J. Pharm. Sci.* **2000**, *89*, 951–957. [CrossRef]
81. Bansal, V.; Sharma, P.K.; Sharma, N.; Pal, O.P.; Malviya, R. Applications of Chitosan and Chitosan Derivatives in Drug Delivery. *Adv. Biol. Res.* **2011**, *5*, 28–37.
82. Lai, W.F.; Lin, M.C.M. Nucleic acid delivery with chitosan and its derivatives. *J. Control. Release* **2009**, *134*, 158–168. [CrossRef]
83. Moreira, C.; Oliveira, H.; Pires, L.R.; Simoes, S.; Barbosa, M.A.; Pego, A.P. Improving chitosan-mediated gene transfer by the introduction of intracellular buffering moieties into the chitosan backbone. *Acta Biomater.* **2009**, *5*, 2995–3006. [CrossRef] [PubMed]
84. Torchilin, V.P. Nanocarriers. *Pharm. Res.* **2007**, *24*, 2333–2334. [CrossRef] [PubMed]
85. Chopra, S.; Mahdi, S.; Kaur, J.; Iqbal, Z.; Talegaonkar, S.; Ahmad, F.J. Advances and potential applications of chitosan derivatives as mucoadhesive biomaterials in modern drug delivery. *J. Pharm. Pharmacol.* **2006**, *58*, 1021–1032. [CrossRef] [PubMed]
86. Tan, H.; Ma, R.; Lin, C.; Liu, Z.; Tang, T. Quaternized Chitosan as an Antimicrobial Agent: Antimicrobial Activity, Mechanism of Action and Biomedical Applications in Orthopedics. *Int. J. Mol. Sci.* **2013**, *14*, 1854–1869. [CrossRef]
87. Rassu, G.; Gavini, E.; Jonassen, H.; Zambito, Y.; Fogli, S.; Breschi, M.C.; Giunchedi, P. New chitosan derivatives for the preparation of rokitamycin loaded microspheres designed for ocular or nasal administration. *J. Pharm. Sci.* **2009**, *98*, 4852–4865. [CrossRef]
88. Wang, F.; Zhang, Q.; Li, X.; Huang, K.; Shao, W.; Yao, D.; Huang, C. Redox-responsive blend hydrogel films based on carboxymethyl cellulose/chitosan microspheres as dual delivery carrier. *Int. J. Biol. Macromol.* **2019**, *134*, 413–421. [CrossRef]
89. Li, H.; Zhang, Z.; Bao, X.; Xu, G.; Yao, P. Fatty acid and quaternary ammonium modified chitosan nanoparticles for insulin delivery. *Colloids Surf. B* **2018**, *170*, 136–143. [CrossRef]
90. Trivedi, A.; Hoffman, J.; Arora, R. Gene therapy for atrial fibrillation—How close to clinical implementation? *Int. J. Cardiol.* **2019**, *296*, 177–183. [CrossRef]
91. Gollomp, K.L.; Doshi, B.S.; Arruda, V.R. Gene therapy for hemophilia: Progress to date and challenges moving forward. *Transfus. Apher. Sci.* **2019**, *58*, 602–612. [CrossRef]
92. Kean, T.; Roth, S.; Thanou, M. Trimethylated chitosans as non-viral gene delivery vectors: Cytotoxicity and transfection efficiency. *J. Control. Release* **2005**, *103*, 643–653. [CrossRef]
93. Mai, Q.; Shen, S.; Liu, Y.; Tang, C.; Yin, C. PEG modified trimethyl chitosan based nanoparticles for the codelivery of doxorubicin and iSur-pDNA. *Mater. Lett.* **2019**, *238*, 143–146. [CrossRef]
94. Suk, J.S.; Xu, Q.; Kim, N.; Hanes, J.; Ensign, L.M. PEGylation as a strategy for improving nanoparticle-based drug and gene delivery. *Adv. Drug Deliv. Rev.* **2016**, *99*, 28–51. [CrossRef] [PubMed]
95. Ren, L.; Xu, J.; Zhang, Y.; Zhou, J.; Chen, D.; Chang, Z. Preparation and characterization of porous chitosan microspheres and adsorption performance for hexavalent chromium. *Int. J. Biol. Macromol.* **2019**, *135*, 898–906. [CrossRef]
96. Tang, Y.; Liu, Y.; Xie, Y.; Chen, J.; Dou, Y. Apoptosis of A549 cells by small interfering RNA targeting survivin delivery using poly-β-amino ester/guanidinylated O-carboxymethyl chitosan nanoparticles. *Asian J. Pharm. Sci.* **2020**, *15*, 121–128. [CrossRef] [PubMed]

polymers

MDPI

Review

Modern Approaches in Wounds Management

Simona-Maria Tatarusanu [1,2,†], Florentina-Geanina Lupascu [1], Bianca-Stefania Profire [3,†], Andrei Szilagyi [4], Ioannis Gardikiotis [4], Andreea-Teodora Iacob [1], Iulian Caluian [1], Lorena Herciu [1], Tudor-Catalin Giscă [5], Mihaela-Cristina Baican [6], Florina Crivoi [6,*] and Lenuta Profire [1,*]

1 Department of Pharmaceutical Chemistry, Faculty of Pharmacy, University of Medicine and Pharmacy "Grigore T. Popa" of Iasi, 16 Universitatii Street, 700115 Iasi, Romania; tatarusanu.simona-maria@email.umfiasi.ro (S.-M.T.); florentina-geanina.lupascu@umfiasi.ro (F.-G.L.); andreea.panzariu@umfiasi.ro (A.-T.I.); caluian.iulian@d.umfiasi.ro (I.C.); lorena-antoanina_herciu@d.umfiasi.ro (L.H.)

2 Research & Development Department, Antibiotice Company, 1 Valea Lupului Street, 707410 Iasi, Romania

3 Department of Internal Medicine, Faculty of Medicine, University of Medicine and Pharmacy "Grigore T. Popa" of Iasi, 16 University Street, 700115 Iasi, Romania; bianca-stefania.profire@d.umfiasi.ro

4 Advanced Research and Development Center for Experimental Medicine (CEMEX), University of Medicine and Pharmacy "Grigore T. Popa" of Iasi, 16 University Street, 700115 Iasi, Romania; andrei.szilagyi@umfiasi.ro (A.S.); ioannis.gardikiotis@umfiasi.ro (I.G.)

5 Department of Obstetrics and Gynecology, Faculty of Medicine, University of Medicine and Pharmacy "Grigore T. Popa" of Iasi, 16 University Street 700115 Iasi, Romania; gisca_tudor-catalin@d.umfiasi.ro

6 Department of Pharmaceutical Physics, Faculty of Pharmacy, University of Medicine and Pharmacy "Grigore T. Popa" of Iasi, 16 University Street, 700115 Iasi, Romania; mihaela.baican@umfiasi.ro

* Correspondence: florina.crivoi@umfiasi.ro (F.C.); lenuta.profire@umfiasi.ro (L.P.)

† Authors with equal contributions.

Abstract: Wound management represents a well-known continuous challenge and concern of the global healthcare systems worldwide. The challenge is on the one hand related to the accurate diagnosis, and on the other hand to establishing an effective treatment plan and choosing appropriate wound care products in order to maximize the healing outcome and minimize the financial cost. The market of wound dressings is a dynamic field which grows and evolves continuously as a result of extensive research on developing versatile formulations with innovative properties. Hydrogels are one of the most attractive wound care products which, in many aspects, are considered ideal for wound treatment and are widely exploited for extension of their advantages in healing process. Smart hydrogels (SHs) offer the opportunities of the modulation physico-chemical properties of hydrogels in response to external stimuli (light, pressure, pH variations, magnetic/electric field, etc.) in order to achieve innovative behavior of their three-dimensional matrix (gel–sol transitions, self-healing and self-adapting abilities, controlled release of drugs). The SHs response to different triggers depends on their composition, cross-linking method, and manufacturing process approach. Both native or functionalized natural and synthetic polymers may be used to develop stimuli-responsive matrices, while the mandatory characteristics of hydrogels (biocompatibility, water permeability, bioadhesion) are preserved. In this review, we briefly present the physiopathology and healing mechanisms of chronic wounds, as well as current therapeutic approaches. The rational of using traditional hydrogels and SHs in wound healing, as well as the current research directions for developing SHs with innovative features, are addressed and discussed along with their limitations and perspectives in industrial-scale manufacturing.

Keywords: wound healing; wounds assessment tools; biopolymers; smart hydrogels; stimuli-responsive

1. Introduction

Wounds are injuries which occur when sudden, rash, and mostly unexpected accidents affect the integrity of skin. The injury may result from different causes such as cuts, crushes, thermal/radiation burns, or surgical events, and could be extended from superficial damages, which affect only the skin's layers, to deep tissues (muscles, nerves, and blood vessels) destruction [1]. The main concern in wounds management is the high risk for chronicity. Among the patients hospitalized for acute conditions, 25–50% present or develop wounds during hospitalization, with high risk of infection and chronicity [2]. The rate of infection of surgical incisions is 3–4% and causes an 5% additional increase in mortality [3,4].

It is estimated that 2% of the population in developed countries suffer from a chronic wound during their lifetime [5]. In Europe, according to the European Wound Management Association (EWMA), the prevalence of chronic wounds is 3–4/1000 (1.5–2.0 million out of 491 million inhabitants), the incidence being 4 million patients/year [6,7]. Patients suffering from chronic wounds may experience continuous pain, loss of mobility and functionality, increased stress level, social isolation, depression, anxiety, prolonged hospitalizations, impairment of work capacity, and negative financial impact. Moreover, if chronic wounds are located in the lower limbs (e.g., diabetic foot ulcer), amputation occurs in 85% of patients [8] and the mortality rate increases to 40% after 5 years [9].

Currently, a wide range of pharmaceutical products such as creams, gels, ointments, powders, pastes, and patches, with healing, antimicrobial, and/or moisturizing effects are available for wound management. The medical approach of wound care must be correlated with patients' individual reactivity and preferences in order to maximize the treatment compliance.

2. Pathophysiology of Wound Healing

Immediately after injury, a cascade of physiological reactions is triggered in order to restore the physical and functional integrity of the affected area [10]. Wound healing involves four steps: hemostasis, inflammation, proliferation, and remodeling (Figure 1).

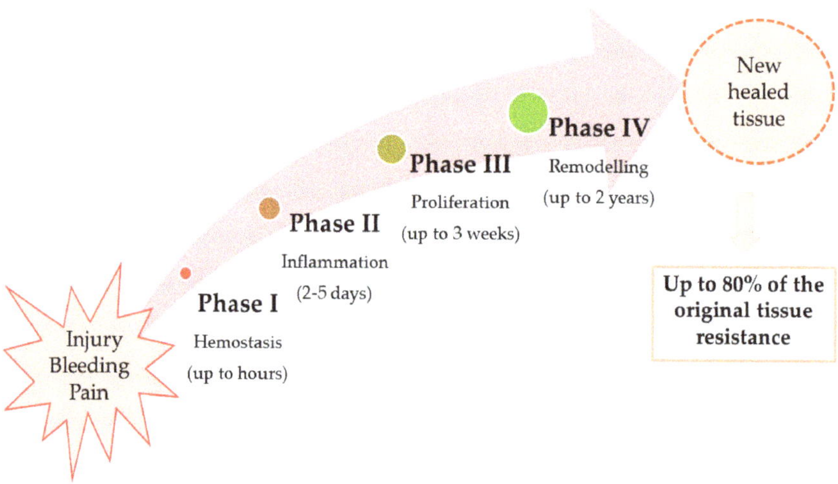

Figure 1. The phases of the wound healing process.

As a result of the inflammatory and hemostasis phases, release of cytokines, growth factors, and migration of leukocytes (in the first 2–5 days), phagocytosis and debridement (late inflammation), and vasoconstriction, platelet aggregation, and coagulation are occur [11].

The proliferative phase lasts between 2 days and 3 weeks and includes the following: *granulation* (fibroblasts fill the defect with collagen fibers; new blood vessels are formed); *contraction* (collagen fibers are reorganized, the edges of the wound come closer to each other, reducing the size of the defect), and *epithelialization* (epithelial cells proliferate, covering the defect). If adequate tissue oxygenation is not achieved, the fibroblast function and angiogenesis are disrupted and the healing process is impaired [11].

The remodeling phase continues after the proliferative phase for up to 2 years, when new, thicker collagen fibers, oriented in the direction of the force lines, are formed. The extracellular matrix is in a remodeling process and the activity of metalloproteases is increased. Epithelial cells and fibroblasts enter apoptosis and the scar that is formed at the end of this phase assures up to 80% of the strength of the original tissue [12].

The ability of the body to restore its homeostasis after a traumatic event depends on maintaining the balance between the endogenous substances and the cells that mediate the healing process. Any inconsistency between the physiological parameters can cause the disruption, prolongation, or complication of the healing phases through infection, chronicity, and expansion of the injured tissue [5,10]. The healing process is influenced by several factors such as mass index body, local anatomy, associated medication, infection, dysmetabolism, and patient compliance. In addition, the associated comorbidities (diabetes mellitus, high blood pressure, obesity, autoimmune diseases, peripheral vascular dysfunctions) interfere with healing physiological process and can impact the outcome of the treatment [13].

The common characteristic of chronic wounds is a persistent inflammatory phase, which is responsible for reducing phagocytic capacity and bactericidal effect, as well as increasing cellular and microbial residues at the site of aggression, which delays the re-epithelialization process [14]. Thus, if the healing of acute wounds is completed within 3 weeks [2], for chronic wounds the healing ranges between 1–3 months, affecting the normal recovery of the affected area [2,3].

The Wound Healing Foundation (WHF) classifies chronic wounds in several categories, depending on their etiology: pressure ulcers, venous ulcers, diabetic ulcers, and arterial insufficiency ulcers as well as radiation, surgical, and infectious wounds (Figure 2).

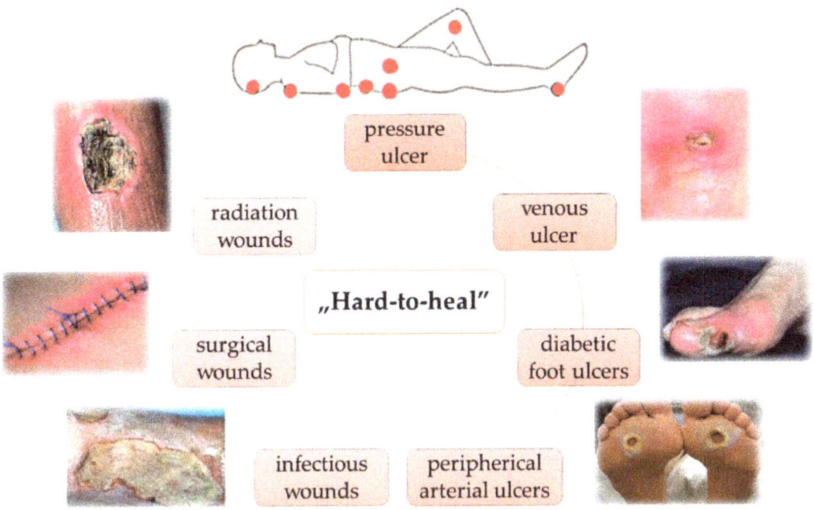

Figure 2. The most common chronic wounds, according to WHF classification.

3. Wounds Care Management Protocol

Given the diversity of causes, types of wounds, and the individual characteristics of patients, wound management is difficult to standardize. However, it is imperative that this process be carried out as objectively as possible, so that the established treatment protocol can be effective. The treatment should be closely related to type of wound and its healing phase and in correlation with the patient's comorbidity. Thus, a detailed anamnesis about the time of the wound appearance, previous treatments and therapies, the evolution until presentation in the medical unit, the associated diseases and their treatments, and other relevant information for the patient's status (e.g., nutritional screening) should be noted. It is also important to know that an impaired wound may favor the appearance of systemic pathologies, which trigger a vicious circle for the healing process [4]

The evaluation of a wound is primarily based on visual observation which emphasizes the location, size, and depth of the wound, drainage, and the type of tissue in the injured area. Secondly, the wounds must be evaluated for microbial loading. The commonly known signs of infection such as erythema, edema, pain, and fever may not always be present [15].

Two mnemonics are recommended for infected chronic wound identification NERDS, with 73% sensitivity and 80.5% specificity when three criteria are met, and STONEES, which has 90% sensitivity and 69.4% specificity when at least three signs are observed (Figure 3). When NERDS criteria are met, biofilm is present or the wound is critically colonized. If the STONEES are observed, then the chronic wound is infected [16,17].

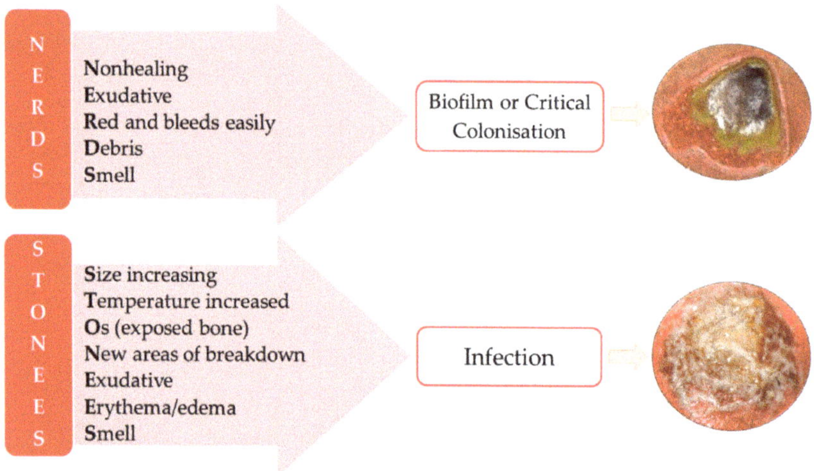

Figure 3. Mnemonics for clinical assessment and differential diagnostic between infected and biofilm-colonized chronic wounds.

The quantitative determination of the microbial load of a chronic wound is recommended to be performed by biopsy and a positive result means more than 100,000 colony forming units. Sometimes, even with the best care, some acute or chronic wounds do not respondto the treatment properly. In these cases, no sign of healing is seen even after six weeks of treatment and the physicians call them atypical wounds. These wounds could have a variety of etiologies, such as metabolic, genetic, malignant, infectious, and inflammatory [17].

To support the medical healthcare professionals, the **TIMES** acronym was implemented two decades ago, and it is still used (Figure 4). It standardizes the indicators used in wound evaluation and gives information to establish a treatment plan [18].

The first step in wound care is debridement, which consists in the removal of non-viable and dead cells and/or tissue. Studies showed that after perseverant debridement, the time healing of chronic wounds is halved [18]. In a cohort study of 154,664 patients with more than 300,000 wounds, for which the debridement was made at least once a week, an overall healing rate of 70.8% was reported [19].

The removing of biofilm is also important, even more than debridement, because it causes the blocking of the healing process in the inflammatory phase [20]. According to the National Institute of Health (NIH), biofilms are responsible for 80% of infections in chronic wounds [12]. Biofilms are composed of numerous multicellular colonies which produce an extracellular matrix to hold them together.

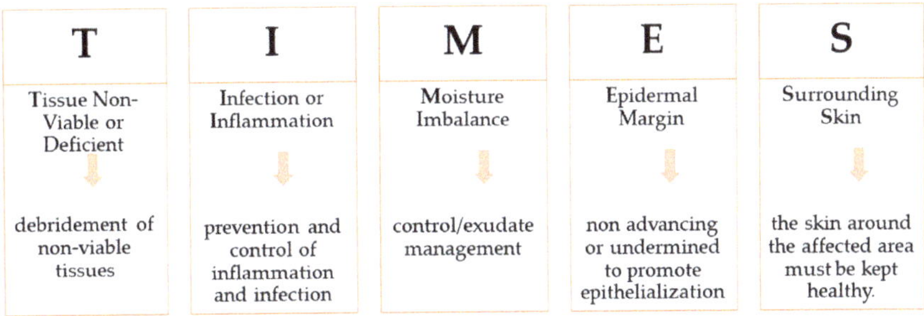

Figure 4. TIMES concept for evaluation of critical parameters during wound healing phases.

Moreover, the reestablishment of the moisture balance is mandatory. When the optimal moisture is achieved, the risk of microbial colonization is lower and the healing process is accelerated, in both acute and chronic wounds [11,12]. For a draining or a dry wound, the optimal moistness is reach by applying a proper dressing. This prevents the desiccation of a wound's bed andholds the appropriate moisture in order to accelerate the proliferation and remodeling phases [21,22].

The appropriate wound treatment is established in accordance with the principles of the TIMES protocol. Thus, the debridement, medical or surgical control of persistent inflammation, application of dressings (including negative pressure), protection of surrounding skin with dressing, and appropriate emollients will be decided upon according to the TIMES principles [12,23,24].

The Triangle of Wound Assessment recommends to extent the TIMES concept beyond wounds' edges and outlines the importance of inspecting three zones in the affected area: wound edge, wound bed, and periwound area. For the wound bed and edge, the evaluation is similar to the TIMES guideline. The examination of the periwound area gives information on skin status around the affected area such as presence of maceration, excoriation, dry skin, hyperkeratosis, callus, and/or eczema [25,26]. Keeping the skin around the wound in a healthy state is crucial for reducing the extension of the affected area and to shorten the healing time.

The best practice statement of Effective Exudate Management (EEM) was published in 2013 to help healthcare professionals in the assessment of exudate color, odor, and viscosity, in order to select the proper dressing. The guideline was focused on the understanding of the exudate role in the healing process and how its characteristic should vary based on inflammation, capillary breakage, microbial colonization, and protein content [27].

The Pressure Ulcer Scale of Healing (PUSH) is often used to assess the evolution of pressure ulcers over time in terms of dimensions of affected area, exudate amount, and tissue type. This tool has a high accuracy; it is very fast and reliable, being validated by the National Pressure Ulcer Advisory Panel [28].

4. Wound Care Treatment Approaches

A proper wound care treatment should correlate a systemic approach with topical treatment and adequate nutritional support to assure the success of healing. Systemic treatment mainly targets comorbidity or septicemia risk, while topical treatment focuses on modulation of molecular reactions which occur in pathophysiological pathways on the damaged area. The final scope of topical approach is the restoration of skin integrity with minimum scar damages (as fibrosis, hypertrophies, keloids, red fibrous lesions, contractures) [19].

Wound dressings have evolved over time, from clay tablets (2500BC) and cotton gauze (manufactured for the first time in 1891) to modern dressings, whose development began with proving (in 1948) the beneficial effect of a moist chamber for venous ulcers. Nowadays, dressings include more than 2000 pharmaceuticals products, from topical creams, gels, and ointments, to sophisticated medical devices such as bacteria-killing lasers and 3D steam cells printers [19,29]. Approximately 30% of the wound care products market is held by wounds cleansers, followed by the dressings segment [30].

An optimal wound dressing requires a balance between its benefits, safety for the patient, and cost-effectiveness. The ideal wound dressing should be biocompatible, biodegradable, and have the optimal permeability for water vapor to ensure and maintain a favorable environment during the entire healing process [31]. In addition, it should fit perfectly to the wound's shape, but not be adherent to the wound surface, ensure protection from microbial contamination, mechanical, and thermal stress, and should optimize moisture in the injured area [32]. The topical treatment for chronic wounds raises substantial costs for the healthcare system. The global annual average financial burden is estimated to rise to almost USD 100 billion per year. For example, in UK, the average cost per year is GBP 5.3 billion, which is comparable with the average 5 billion GBP/year for obesity managing, according to the National Institute for Health and Care Excellence (NICE) [24,33]. In addition, in the USA, where there are approximately 6.5 million patients diagnosed with chronic wounds, the cost on the health system is approximately 25 billion USD/year [6,7].

Wound dressings are divided intotraditional or conventionaldressings (such as cotton gauze, bandages, lint, plasters) and modern multifunctional dressings (such as foams, films, hydrocolloids, hydrogels, nanocomposites) [10,32,34]. Conventional dressings are known as passive wound dressings and are useful to cover and stabilize modern dressings. The latter dressings are defined as interactive products which modulate the entire healing cascade by controlling the damaged tissue microenvironment (moisture, temperature, pH), stimulating the granulation and reepithelization processes, and inhibiting microbial growth [35,36].

4.1. *Hydrogels as Modern Wound Dressings*

Hydrogels were documented for the first time in 1960, when Wicherke and Lim prepared a gel based on hydroxymethylacrylate for medical application [37]. Nowadays, hydrogels still fascinate biomedical researchers, having applications in tissue engineering (bone, cartilage, and muscles), cell growth, wound healing, controlled drug release, biosensors, and medical devices [34,38,39]. Hydrogels are defined as three-dimensional (3D) hydrophilic polymer networks capable of encapsulating large amounts of water or biological fluids. As a result of their hydrophilic character, oxygen permeability, ability of diffusion, cell and molecules adhesion, and ease of use, hydrogels demonstrate important beneficial effects for wound care [34,40].

In many aspects, hydrogels are considered ideal wound dressings because they mimic the skin structure, promoting the growth factor synthesis and autolysis process. Moreover, their ability to incorporate and release a wide variety of active pharmaceutical ingredients (APIs) makes hydrogels suitable for the management of draining painful wounds, radiation wounds, minor burns, or dry wounds [37].

Over time, extensive research has led to numerous types of hydrogels with varied physico-chemical properties according to several criteria. The main categories of hydrogels (Figure 5) were established according to the raw materials sources and polymeric composition, physico-chemical properties, release conditions of the APIs embedded in the 3D matrix, and other properties [34,37,41,42].

Hydrogels have several advantages, such as the following:3D structure, similar to the extracellular matrix, which plays a key role in cell proliferation, debridement, and exchange of vital substances for epidermal cells; hydrophilicity, associated with hydrophilic groups of the polymer matrix: NH_2, -COOH, -OH, -$CONH_2$, -CONH-, and -SO_3H; water absorption capacity, due the porosity degree; high degree of flexibility similar to human tissue, due the water content; biodegradability and biocompatibility [43,44]. Along with the positive aspects, some drawbacks reduce the applicability of hydrogels in medical practice. These include poor mechanical stability, high risk for microbial contamination (except chitosan-based hydrogels), toxicity potential, degradation, and variable release profiles of embedded APIs [44,45].

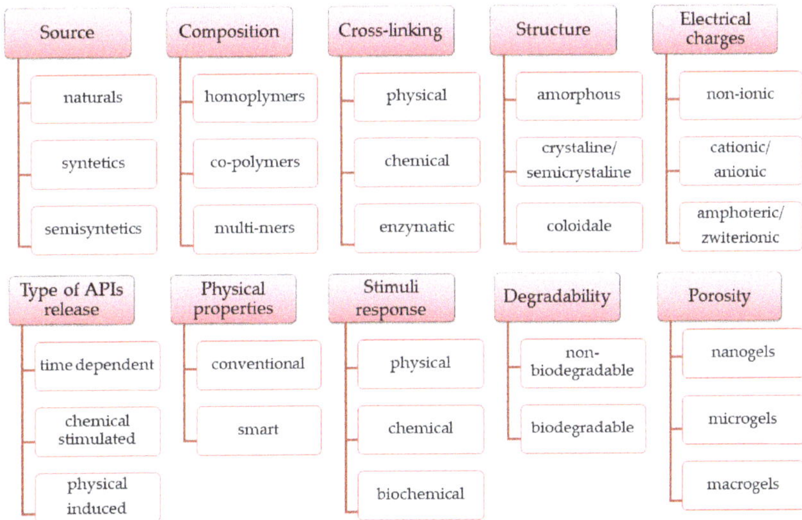

Figure 5. Hydrogels classification, based on polymers characteristics and technological manufacturing process criteria.

The mechanical properties of hydrogels are closely related to several factors, such as the cross-linking method, polymer type, ratio of polymer: cross-linking agent, structure of the embedded APIs, and storage conditions [46]. An indicator of mechanical stability of hydrogel is viscosity, which decreases over time, because its 3D structure is affected through degradation reactions (oxidation, reduction, hydrolysis, free radical interactions) [47]. Hydrogels based on covalent interactions are more stable compared to hydrogels based on hydrogen bonds or physical interactions [48]. Innovative hydrogels, based on dynamic covalent bonds (dynamic Schiff bases), combine excellent stability over time with a versatile stress behavior [49]. The dynamic covalent bonds break when shear forces (injection, application) are applied or in presence of stressful stimuli (variations in pH, temperature, electric/magnetic field), without breaking of the hydrogel's backbone. When the stress conditions are removed, the hydrogels partially or totally recover their 3D structure and mechanical performance [49–51].

The high water content of hydrogels provides a proper environment for microbial contamination (e.g., *Pseudomonas aeruginosa*, *Staphylococcus aureus*, *Escherichia coli*, *Candida albicans*), which can delay the healing process [46]. Moreover, microbial contamination causes changes in the aspects of hydrogels (color, smell) and negatively influences their mechanical stability [43]. In order to reduce the risk of contamination, different materials with antimicrobial effects (polymers, cross-linking agents, APIs, preservatives) or different cross-linking methods were used. For example, 3D matrices obtained through Schiff bases dynamic bonds have proven antifungal and antibacterial properties [50,52].

The aldehydes and epoxide compounds, used as cross-linking agents, lead to hydrogels with excellent mechanical stability, but the by-products as well as the unreacted compoundsshow high toxicity, such as genotoxicity, cytotoxicity, or carcinogenicity [53]. Moreover, these compounds may affect the healing process by extending the inflammatory phase [53–55].

In order to improve the kinetics release of APIs embedded in the 3D polymer matrix, different types of carriers (nanoparticles, nanospheres, nanofibers, liposomes, microparticles) are combined with novel cross-linking methods [56,57].

4.2. Smart Hydrogels as Innovative Wound Dressings

Extensive researchhas expanded the functional roles of hydrogels to smart materials that not only protect the wounds but have the ability to influence all phases of the healing process. The concept of smart was mentioned for the first time in 1949 by Kunh et al., in their research on polyacrylic acid 3D networks, capable of mimicking muscle contractions under acidic conditions and absorbing huge quantities of water when alkaline substances were added [58].

According to PubMed and Science Direct databases, in the last two decades, the amount of research on smart materials for biomedical fields has increased year by year, summing up to 8500 reviews and original articles in 2022, including more than 3600 on smart hydrogels (SHs) dedicated to wound care management (Figure 6).

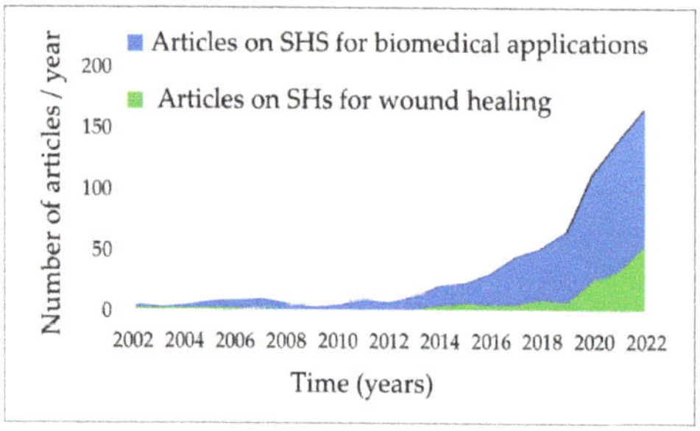

(a)

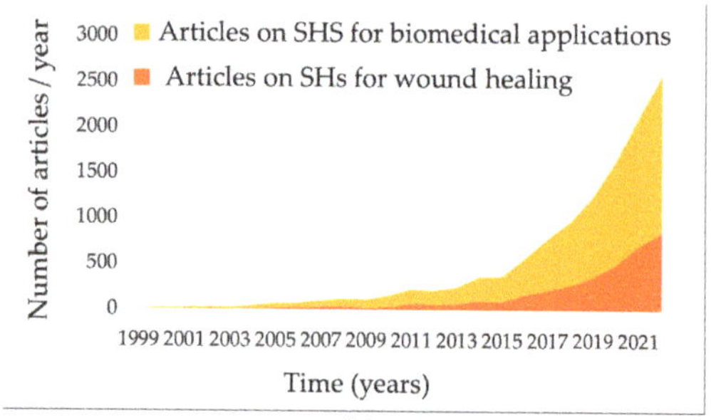

(b)

Figure 6. The articles relating to SHs from PubMed (**a**) and Science Direct (**b**) databases.

In wounds approach, SHs bring some important benefits such as shortenedhealing time and improved complication prevention, which lead to better results and increased quality of life for the patients [38,40,59].

Both natural and semi-synthetic/synthetic polymers may be successfully used for SHs formulation, using physical or chemical gelation methods [36].

Among the most widely used chemical cross-linking methods are Schiff-base reactions, Michael addition reactions, free-radical polymerization of end-functionalized macromers, high-energy radiation, and enzymatic reactions [55]. Chemical cross-linking results in highly stable structures, with process controllable features such as time gelation, degree of cross-linking, and pore size and pore density of the matrices [41]. The main drawback of these methods is toxicity risk, due the by-products resulting from cross-linking reactions [53]. In addition, the cross-linkers may interact with bioactive components (cells, pharmaceutical active proteins—enzymes, hormones, growth factors) embedded in the hydrogel matrix and affect their therapeutic target [55].

Physical cross-linking involves weak interactions such as hydrogen bonds, van der Waals forces, ionic and hydrophobic interactions, crystallizations, chain entanglements and coordination bonds, protein interactions, and freezing–thawing and heating/cooling cycles [41]. The main limitation of these methods results from gelation reversibility under a wide range of conditions which may be modulate in favor of targeted applications of the final product, or, on the contrary, may induce stability problem over time [55].

Moreover, 3D and 4D bioprinting technics (ink writing, digital light synthesis/ processing, stereolithography, fused deposition modeling) were tested for the manufacturing of SHs with excellent mechanical properties (deformability, flexibility). The low productivity due to slow speed of printing, restrictions due the biopolymers stability to heating, and expensive equipment are drawbacks that needs to be addressed before implementing these methods on a microscale and industrial scale production of SHs via 3D/4D printing [60].

The target of multifunctional SHs in wound care is to overcome the drawbacks of traditional hydrogels and to improve their therapeutic performance using the ability to reversibly modify their 3D structure in response to one or more external stimuli [61]. The comparative characteristics of traditional hydrogels versus SHs relevant for manufacturing and performance of wound care products are summarized in the Table 1

Table 1. The relevant particularities of traditional hydrogels versus SHs, designed for wound healing.

Traditional Hydrogels	SHs	References
Protection of the damaged area to external factors	Enhanced protection of the damaged area due the self-adapting ability	[36,43]
Provide moisture of the wound bed depending on hydrogel composition without regarding on wound conditions (dry, wet)	Dynamic and interactive control of moisture, based on responsive behavior to wound condition (exudate, pH) or external triggers (pressure, light)	[42,62]
Frequent changing during treatment	The need of dressing changes may be reduced and adapted to patient and wound particularities	[12,61]
Strong adhesive properties with discomfort and secondary injuries when they are removed or lack of adhesive properties with inadequate contact with the damaged tissue	Excellent adhesive properties with intimate contact to the wound bed and easiness remove without damages or discomfort for patient	[63–65]
Low or no protection against microbial contamination	Efficient protection to microorganism growth through polymers characteristics, cross-linking methods, and manufacturing design	[60,64]
No effect on physiological mechanisms of healing process	Promote, accelerate, and modulate the endogenous mediators on wound site	[35,46]
A limited number of APIs (most water-stable) could be embedded into 3D matrix	A wide variety of APIs (chemical and bioactive molecules) could be embedded using versatile carriers (vesicles, liposomes, niosomes, nanoparticles)	[45,59]
Passive release of APIs.	Controllable release of API through endogenous or exogenous stimuli, depending on wound changes	[46,66]
No possibility to evaluate the wound condition during treatment	Possibilities to real-time monitor the healing of wound	[59,67]
Low cost of production, simple design, robust manufacturing at industrial scale	Challenging manufacturing process, difficulties to reproduce the quality attributes from batch to batch	[61,68]
Safe raw materials with well-established characteristics for human use	New synthetized raw materials or modified from traditional sources which may raise stability and safety issues	[39,55]
Simple to use, easy to apply	Requires training of the patient and/or medical staff	[38,42]

4.3. Stimuli-Responsive Hydrogels

Depending on the nature of the gelling agents and the manufacturing method, the hydrogels respond to different physical (temperature, light, pressure, ultrasounds, exposure to magnetic and electric fields) or chemical/biochemical stimuli (pH variations, chemicals like reactive oxygen species, glucides, enzymes, antigens, ionic charge, solvents). Depending on their sources, the stimuli are also divided into endogenous or exogenous stimuli [39]. Endogenous stimuli result from physiological reactions or from pathological pathways: (i) presence of glucose, enzymes, and ions in the human body, (ii) pH variations in blood and exudates as result of infections, (iii) generation of reactive oxygen species in damages tissues. Exogenous stimuli (temperature, light, magnetic/electric field, ultrasounds, pressure) are used as remote tools for non-invasive temporal and spatial control on hydrogels behavior (Figure 7) [36].

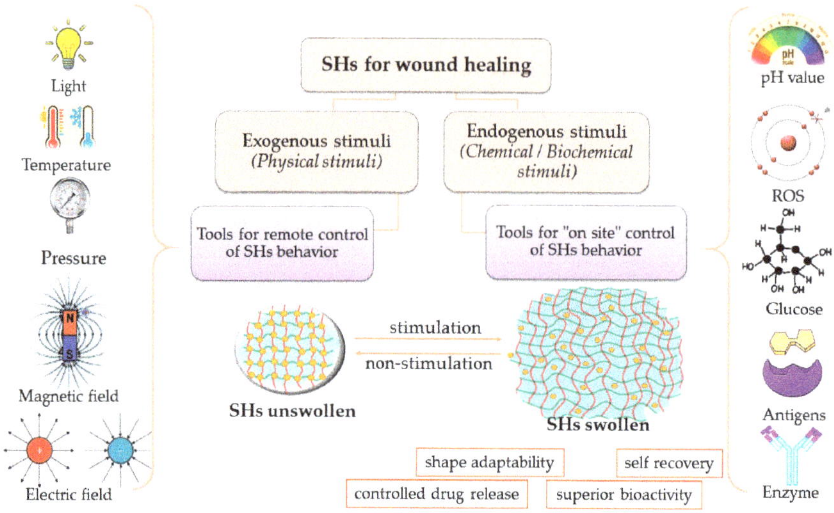

Figure 7. Stimuli that enhance the performance of SHs on the wound healing process.

Based on stimuli action, several changes in physical state and/or chemical properties such as gel–soil transition, sequential deformations, shape adapting, viscosity variations, conductivity, hydrophilicity, and degradation are produced [69]. These changes are translated into beneficial effects in medical applications such as modulation of drug release kinetics, in situ gelation, enhancing biocompatibility and over-time stability, and over-stress recovery ability. Once the stimuli are removed, the 3D networks may return to their original state, acting as smart materials with applications in tissue engineering and regenerative medicine, biosensors, drug delivery/stem cells systems, contact lenses, and implants [59,62]. In addition, SHs are able to adapt to the wound shape for a perfect fit on the affected area, which means that the damaged tissue has an intimate contact with healing care products [70,71].

The SHs may respond to one or more stimuli, which endows the product with gradually modified behavior correlated to spatial-temporal changes in each phase of wound healing.

4.3.1. Physical Stimuli-Responsive Hydrogels

Temperature-Responsive Hydrogels

Hydrogels based on natural/semi-synthetic/synthetic polymers (CS, methylcellulose, gelan gum, poloxamers, polyethylenglycol (PEG), poly(N-isopropylacrylamide) (PNIPA), poly(vinyl alcohol) (PVA), poly(N-vinylcaprolactam) and their derivatives) respond to temperature variations with sol–gel transitions due to thermal disturbance of the pre-existing equilibrium between the hydrophilic and hydrophobic units in the 3D networks [35,39]. These hydrogels, which mimic the extracellular matrix, can be easily functionalized with bioactive molecules, and may exhibit a controlled degradation rate [36].

In the wound recovery process, temperature plays a critical role in the reactions occurring in the affected area. A temperature around 37.8 °C is desired for optimal healing and deviations of ±2.2 °C indicate an ischemic or inflammatory process, which leads to tissue deterioration. The temperature-responsive hydrogels must undergo sol–gel transition at physiological temperature in order to be suitable for wound care.

A catechol modified quaternized CS, poly-lactic acid (PLA), and PEG were used as gelling agents to develop temperature responsive injectable hydrogels. The modified CS endowed the hydrogel with antibacterial properties and wet-tissue-adhesiveness. In addition, the functionalized CS lowered the gelation point to 32 °C, near the physiological value. The in vivo study showed that the thermosensitive hydrogel properly seals the wound area and promotesaccelerated healing [72]. Because thermo-responsive hydrogels behavior only lies on the amphiphilic properties of the polymers' backbones, these formulations are considered non-toxic, cost-effective, and easy obtained. However, in absence of a cross-linker agent, the stability overtime may be limited.

A thermosensitive wound dressing was obtained from combination of CS with Pluronic P123 (a triblock symmetric poloxamer) and gelatin, which was loaded with curcumin for its antioxidant and anti-inflammatory effects. At temperatures higher than 30 °C, it undergoes a sol–gel transition, depending on gelatin concentrations, and gradually releases the curcumin on the wound site. An in vivo wound model revealed that the hydrogel stimulates fibroblasts and immune cells migration and neoangiogenesis in the damaged tissues [73].

A combination of poloxamers with different molecular weights was used to develop a thermosensitive hydrogel for controlled release of *Nocardia rubra* cell wall skeleton. The formulation has proved excellent beneficial effects on in vitro and in vivo studies for diabetic wound healing, as stimulation of vascular endothelium, expression of structural proteins, activation of phosphatidylinositol, and protein kinase [74].

PNIPA and pullulan were associated to develop a 3D matrix in which silver NPs were loaded. The release profile of silver, and thus the antimicrobial activity, were depended on the wound temperature, the sustained released being maintain for 48h. In addition, in vitro tests demonstrated nontoxicity and excellent swelling properties, the hydrogels being suitable for wounds care [75].

A thermo-responsive multifunctional DNA-based hydrogel was design through cross-linking of DNA units with polyethyleneimine and black phosphorus quantum dots, and procyanidin B2 were embedded in its matrix. The finished product showed excellent uptake capacity for exudative wounds, proper adhesiveness, self-healing capacity, and antibacterial and antioxidant activity. The pro-inflammatory M1 macrophages were proliferated into repairing M2 phenotype and myeloid cells were attracted to the damaged areas. In consequence, angiogenesis was sustained and skin nerves, follicles, and hair regeneration were promoted [76].

Pressure-Responsive Hydrogels

Pressure relief is one of the main concerns of the treatment strategy in diabetic foot ulcers and pressure ulcers pathology. The pressure occurs by frictions during physical activities and environmental pressure and causes tissues necrosis and delayed healing [61]. Using pressure-sensitive dressings, secondary injures are prevented in wound care, with benefits regardinghospitalization time and the healing period.

Sodium carboxymethyl cellulose, alginate, CS, and PNIPA were successfully used in the development of hydrogels that are able modify their viscosity and swelling capacity under pressure variations [38,77]. For natural polysaccharides-based hydrogels, the reactivity to pressure was related to electrostatic repulsions that occur between the chemical groups with the same charge (e.g., NH_3^+ in CS) which control the swelling rates and the impact the osmotic pressure inside the 3D network [77]. A liquid hydrogel based on PVA and acrylamide with antibacterial and anti-inflammatory properties was designed to monitor movements and pressure status in real-time for bed-ridden patients. The hydrogel possesses very high pressure sensitivity from 9.19 kPa^{-1}, which allows to easy evaluate the pressure of the injured site and avoid exceeding the limits and secondary tissues destructions. In addition, the hydrogel shows excellent mechanical strength, enhances cell proliferation, and promotes tissue remodeling and regeneration [78].

Light-Responsive Hydrogels

An aseptic environment is crucial for wounds healing and it is a huge challenge to ensure non-contamination of the affected tissue. The exposure to different types of light (ultraviolet or visible light) was demonstrated to improve the wound healing of chronic wounds [79]. The high intensity of visible light has an efficient bactericidal effect and, followed by exposure of the wound to low intensity illumination, promotes optimal wound closure [80].

Alginate, CS, hydroxypropyl methyl cellulose, carbomers, and methacrylates were used to design near-infrared, ultra-violet, and visible-light-responsive hydrogels. Their formulation involves grafting photoactive groups (azobenzene, nitrobenzyl derivatives, diaryne, spiropyrans, photoreversible dimerization groups such as coumarin, anthracene, and pyrimidine derivatives) on polymers' backbones in order to absorb photons with different wave length energy and trigger phase transitions or chemical reactions. The absorbed light energy usual must be converted into thermal energy for the changes in hydrogels 3D networks to occur [35,38]. Light-responsive hydrogels proved to be excellent options for design matrices, with controlled release of APIs and exposure to light triggers enhancing the antimicrobial and antioxidant activity [81].

Dodecyl moieties were added to CS and tungsten disulfide nanosheets were included as a photothermal agent to develop a photo-sensitive hydrogel, which could release ciprofloxacin on wound sites when near-infrared red light is applied. The hydrogel proved efficient antimicrobial activity and antioxidant effects which prevent prolonged pro-inflammatory state in damaged tissues [82]. Poly(2-hydroxyethyl methacrylate) covalently cross-linked with polyurethane was used to develop an antimicrobial light-sensitive hydrogel. The nitric oxide (NO) photoactive donors (nitrosyls), hydrogen peroxide and methylene blue, were incorporated in hydrogels as antimicrobial agents. The data suggest the superior outcome of using the light-sensitive hydrogel over hydrogen peroxide solutions for cleaning the chronic wounds [83].

Light-sensitive polyurethane nanofibers dressing for wounds beds and bandages with tetraphenylporphyrin as photosensitizer were developed. Use of this dressing in a study in which 162 subjects with chronic diabetic foot ulcers were enrolled showed inhibition of *Staphylococcus aureus*, *Pseudomonas aeruginosa*, and *Escherichia coli* growth in the wound site proportional with visible light exposure. Furthermore, an average of 35% shrinking of wounds size and 71% wound-related pain reduction was observed on tested groups compared to the control group [84].

A nanodelivery light-responsive system with sinoporphyrin sodium (a hematoporphyrin-like photosensitizer) and fibroblast growth factor embedded into poly(lactic-co-glycolic acid) nanospheres was developed based on a carboxymethyl CS and sodium alginate hydrogel matrix. This scaffold showed promising antibacterial activity and antibiofilm effects against multidrug-resistant *Staphylococcus aureus* on severe burn wound models, which are photoirradiation-conditions-depended [85].

However, the prolonged ultra-violet light exposure of injured tissue during a potential treatment with this type of SHs may raise some concerns regarding patients' safety, due to the carcinogenicity of UV rays. In addition, some injectable light-sensitive SHs could be designed for deep wounds and the clinical applicability of these formulations is limited by lack of penetration of UV in deep tissues.

Magnetic-Responsive Hydrogels

The application of a magnetic field is a promising approach in wounds treatment strategies, being a non-invasive alternative and without significant side-effects [86]. The mechanism of healing under external applied magnetic field is still unclear. There are studies which showed a 25% faster healing of induced wounds in animal [87].

Alginates, xanthan gum, hemicellulose, and PNIPA were used to design magnetic-sensitive hydrogels with iron oxide (Fe_3O_4, Fe_2O_3), ferrites ($MnFe_2O_4$, $CoFe_2O_4$), or alloys (FePt), included in different 3D matrices as drug delivery sensing or tissue engineering and wound healing applications [38]. The main advantage of these hydrogels is the smart and precise remote response (deformation, motion, or heat generation) under magnetic exposure [86]. Controlling the magnetic pulse applied on/off, the hydrogels could enhance or inhibit drug release and protein absorption in the wound site [88].

The development of SHs gives the ability to modify the APIs behavior, as their hydrophilicity and permeability is an ongoing concern for topical formulations [59]. For example, a hydrogel based on alginate derivate and MXenes (two-dimensional inorganic compounds that consist of atomically thin layers of transition metalcarbides, nitrides, or carbonitrides)has demonstrated controllable release of antibioticsin response to photo and magnetic stimuli. This hydrogel showed the successful healing of full-thickness cutaneous and subcutaneous infected wound in a rat model [89].

Electric-Responsive Hydrogels

Alginate, agarose, HA, xanthan gum, PEG, polydimethylsiloxane, and poly(lactic-co- glycolic acid) were used to develop biocompatible electroconductive hydrogels for controlled drug delivery, wound dressing, biosensors, and tissue regeneration, which are able to modify their swelling behavior and structure flexibility upon application of an electric stimuli [35]. The electric charge could be transported by either an electrolyte in hydrogel composition, by polymers with intrinsic conductivity (gelatine, CS, polyaniline, polypyrrole), or by conductive particles (carbon, metals) embedded in a hydrogels network [90].

As is known, cell migration starts in the first days when injury occurred, and it is essential for a normal healing process. The electrical field simulates and coordinatescell migration, cell differentiation, and regeneration in chronic and non-healing wounds [91]. The mechanisms are not yet clearly understood, but the studies outlined the activation of some intercellular polyamines and cell membrane channels [92]. Using covalent or non-covalent cross-linking methods, conductive responsive hydrogels were prepared. The in vivo tests sustain the benefits of electric-responsive hydrogels in all steps of wound healing, showing a hemostatic effect, proliferation of fibroblasts, granulation stimulation, angiogenesis, and increased collagen synthesis [90].

A hydrogel based on quaternized CS-graft-polyaniline and benzaldehyde-functionalized PEGdemonstrated self-healing ability, good hemostatic properties, excellent adhesiveness, and more than a 99% bactericidal effect on *Staphylococcus aureus* and *Escherichia coli* [93]. HA functionalized with dopamine and reduced graphene oxide, as a photothermal agent, was used to prepare a conductive hydrogel that mimics the human skin and proved in vivo enhancement of antimicrobial activity of doxycycline under near-infrared radiation [94].

4.3.2. Chemical Stimuli-Responsive Hydrogels

pH-Responsive Hydrogels

The pH of wound exudates reaches values in the alkaline range (7.0–9.0) if it becomes infected, while the pH of healthy skin is in the 4.5–6.5 range. Moreover, the values of pH higher than 6.5 favor the growth of *Staphylococcus* spp., *Pseudomonas aeruginosa*, *Klebsiella* spp., and formation of biofilms, which are commonly met in impaired wound healing.

The polymers that contain acidic (carboxyl, hydroxyl) or alkaline (amino) groups (as poly(vinyl alcohol)–PVA, alginate, hyaluronic acid–HA, chitosan–CS, albumin, pectin, gelatin, polymethacrylic acid, polyacrylic acid, and phenylboronic acids) are suitable for development of pH-sensitive hydrogels, due the protons exchange between oppositely charged groups. Polymers could also be functionalized by inclusion of amino acidmoieties on the backbones to achieve pH responsiveness hydrogels [35]. For example, an alginate-based hydrogel in acidic pH has non-ionized carboxyl groups in the network and exhibits shrinkage with insignificant swelling rate. If the pH is towards alkaline (e.g., in chronic wounds), the carboxyl groups become ionized and, due the electrostatic repulsion, the hydrogel starts to uptake important amounts of fluids [95]. These types of hydrogels are most suitable for exudative wounds management such as diabetic wounds and pressure ulcers. In addition, due the swelling/deswelling behavior in response to pH variations, these formulations offer good opportunities for modulation of the drug delivery and 3D cell culture.

Wang et al., developed a pH-sensitive hydrogel to monitor the wound status using colorimetry. Based on a three-step monitoring model and using a machine learning technology, the monitoring algorithm has 94% accuracy. Moreover, the hydrogel has shape adaptability for a perfect match to the wound surface [96]. Another study reports the development of pH-sensitive hydrogels, based on carboxylated agarose and tannic acid, and zinc ions as cross-linker, which exhibited antibacterial and anti-inflammatory effects. The release of tannic acid was pH-controlled, being negligible at pH > 7 and sustained at acidic values [97].

An in situgelling injectable hydrogel, sensitive to temperature and pH, was developed based on PEG and poly (sulfamethazine ester urethane). The multiblock copolymer is a solution at room temperature with an alkaline pH and is transformed to a gel inhuman body conditions (37 °C and pH of 7.4). DNA-bearing polyplexes were included in the copolymer solution and the controlled released from the 3D matrix formed in situ was assessed. Moreover, the hydrogel showed excellent adhesive properties and was easy to remove from the affected skin, being suitable for wound healing treatment [72].

A dual-stimuli-responsive (pH and oxidative stress) phenylboronic acid-modified HA dynamically cross-linked hydrogel, loaded with tannic acid and silver NPs as APIs, was developed. The hydrogel proved cytocompatibility, self-healing, antibacterial, antioxidative, and injectable abilities, which means it is very promising for wound care applications [98]. Electron-rich furan groups were grafted on HA backbones in order to obtain a redox-responsive hydrogel based on maleimide bonds, for controlled release of APIs. Polyethylene glycol, which form disulfide bonds between HA backbones, was used as redox-responsive cross-linker.

CS grafted with carboxymethyl moiety and oxidized HA were used to develop self-healing and pH-responsive hydrogels through dynamic Schiff base interactions. Taurine was loaded into the hydrogel matrix for its anti-inflammatory effect and proved to enhance-healing by stimulating cell migration and reducing the inflammatory phase on an in vivo diabetic wound model [99].

A dual-responsive 3D network with high drug loading capacity was developed by cross-linking of alginate with glutamic acid cysteine dendrimer and polyethylene glycol. The doxorubicine hydrochloride, which was embedded in the hydrogel matrix, was released by over 76%, depending on gluthatione and pH variations [100]. Efficient antibacterial activity on *Staphylococcus aures* and *Escherichia coli* was proved by a pH-responsive physically cross-linked citric acid/alginate-based hydrogel. Vancomycin loaded in a hydrogel matrix was best released in acidic conditions, up to 86%, on zero-order kinetics [101].

A multi-stimuli-responsive PVA-based hydrogel, physically cross-linked, loaded with gentamycin and cyanine dyes (cyanine 3 and cyanine 5) as silica NPs, was developed. This hydrogel was able to detect the microbial colonization in wound beds through delivery of cyanine due to pH-responsive fluorescence resonance energy transfer, and to release the gentamycin from cleavable linkers in response to near infrared light [102]. PVA and poly 6-acrylamidohexanoic acid were dually physically cross-linked through hydrogen bonds and crystallizations in a pH-dependent self-healing 3D matrix. The hydrogel exhibits high mechanical stability, undergoes structural breakage in acidic conditions, and recovers its entire arrangement in neutral condition. In addition, it uptakes different amounts of fluids in pH variations due to the breakage and restoration of hydrogen bonds, which allows the control of moisture level on the wound healing process [103].

A double-networked multi-stimuli-responsive PNIPA and keratin-based hydrogel was designed through covalent crosslinking and ionic interactions. The hydrogel matrix was loaded with chlorhexidine, the release of which was regulated by the hydrogel's response to pH and temperature variations and the presence of reactive species of oxygen in the wound site. Moreover, the antimicrobial effect of the hydrogel loaded with chlorhexidine was superior to chlorhexidine alone, due its sustained release. In vivo evaluation of wound healing proved to be superior to commercial Tegaderm (based on polyurethane), due to neoangiogenesis and superior collagen synthesis [104,105].

Reactive-Oxygen-Species-Responsive Hydrogels

Reactive oxygen species (ROS) which act as messengers in cell migration (immunocytes and non-lymphoid cells) are involved in angiogenesis of new blood vessels on the injured area and stimulate the pathogens phagocytose. The benefit key of ROS in wound healing is maintaining equilibrium between the production and the scavenging of free radicals. In chronic and non-healing wounds, an over-expression of ROS is observed, and it contributes to prolonged inflammation [106].

ROS-responsive hydrogels could be obtained using polymers functionalized with thioethers, which due the presence of sulphur in molecules could undergo phase transitions when exposed to ROS. Thioethers are converted into sulfoxides in mild oxidative conditions or into sulfones under strong conditions, which results in modification of polymers solubility and disruption of the hydrogel network arrangement. Selenium-containing polymers act similarlyto thioethers, with even higher sensitivity to ROS, and could be oxidized to selenoxides and selenones, resulting in changing the polymer behavior from hydrophobic to hydrophilic [107].

Gallic acid was used as antioxidant in a gelatin-hydroxyphenyl propionic acid hydrogel, which significantly reduced the destruction of dermal fibroblast in H_2O_2-induced oxidative stress. L-arginine microparticles were incorporated in an ROS-responsive hydrogel to promote wound healing. Under ROS stimulation, a sustained quantity of NO is generated from L-arginine, which promotes the fibroblast and macrophage chemotaxis, collagen synthesis, and skin tissue regeneration [108].

Alginate functionalized with phenylboronic acid was used as a backbone for an injectable dual (pH and ROS)-stimuli-responsive hydrogel with self-healing ability, in which amikacin as an antibacterial, and naproxen as an anti-inflammatory drug, were embedded via preloaded micelles. The in vitro experiments showed that the hydrogel effectively killed bacteria by amikacin release, with an inhibition rate up to 90% for *Staphyloccocus aureus* and 98% for *Pseudomonas aeruginosa*. Furthermore, naproxen was also controllably released from the ROS-responsive micelles under pH 5.0 and H_2O_2 in vitro [109]. Carboxybetaine/sulfobetaine dextran were used to obtain antioxidant self-healing zwitterionic hydrogels, with the ability to scavenge free hydroxyl radicals and to inhibit *Staphylococcus aureus* and *Escherichia coli* [110]. Fluorescein isothiocyanate conjugated bovine serum albumin as API was encapsulated within HA-based hydrogel, yielded through the Diels–Alder reaction with a maleimide group containing redox-responsive PEG-based cross-linker. When 1,4-dithiothreitol was added, the disulfide bonds between HA backbones were broken, and the API loaded into the 3D matrix was rapidly released [111].

Glucose-Responsive Hydrogels

High levels of glucose, matrix metalloproteinases, and pro-inflammatory mediators are involved in impaired healing in diabetic chronic wounds. Hyperglycemia prolongs the healing processes by promoting microbial colonization, sustaining biofilms formation, and impairing tissue neovascularization. Glucose-responsive hydrogels have gained much interest in the controlled release of bioactive compounds as antioxidants and anti-inflammatory molecules, for topical wounds treatments [112].

PEG functionalized with phenylboronic acid (PBA) derivatives or lectins (concanavalin A) are promising polymers for the development of SHs that respond to elevated local glucose concentration. PBA interacts with diols in polymer backbones and results in 3D matrices. When glucose levels increase, the cross-linked network is disrupted because glucose binds to PBA at pH higher than PBA acid pKa (around 9.0). Grafting amino groups on polymers functionalized with PBA enable the scaffolds to function at physiologic pH by reducing the pKa of boronic acid derivatives, translated in modifying the swelling behavior and undergoing gel–sol phase transitions [113].

PVA and CS functionalized with PBA were used to develop a temperature-sensitive 3D matrix loaded with insulin and celecoxib, capable of quickly adapting to irregular wounds shape at 37 °C. The release of APIs is regulated by high concentrations of glucose and matrix metalloproteinase-9 (MMP-9) in exudate, resulting in their local down-regulation and inflammation decrease, cell migration, and proliferation in diabetic full-thickness-wound models. In addition, the hydrogel possesses special rheological behavior, combining a fluid-like mobility at 37 °C for the intimal cover of irregular-depth injuries with a solid-like elasticity at 25 °C, which ensures wound protection against environmental stress [114]. HA was also functionalized with PBA as a glucose-sensitive molecule and mixed with PEG diacrylates to obtain a hybrid network in which myricetin was loaded as an antioxidant active compound. Interactions between phenol groups of myricetin and functionalized HA result in dynamic bonds which endow the hydrogel with the ability to release myricetin depending on the local glucose level. In addition, the ROS-scavenging effect was higher than 80% and superoxide dismutase and glutathione levels were increased, while expression of pro-inflammatory mediators (IL-6) was reduced. Moreover, angiogenesis and tissue remodeling were accelerated, which make this hydrogel promising for diabetic wound care [113]. In a hyperglycemic wound environment, the glucose-oxidase turns glucose into glucuronic acid and hydrogen peroxidase, lowering the local pH.

A self-healing injectable hydrogel based on co-assembling of zinc ions, organic ligands, and deferoxamine mesylate was loaded with glucose-oxidase. The release of zinc ions and deferoxamine from the hydrogel improves cell proliferation, angiogenesis, reduces oxidative stress, and suppresses prolonged inflammation. In vivo evaluation on diabetic wound models showed optimal re-epithelization with collagen deposition and inhibition of microbial growth [115].

Glucose-sensitive hydrogels have shown excellent benefits for diabetic chronic wounds, but their clinical applicability is questionable because of lack of predictability of the triggered in vivo response when different molecules and/or microorganism are involved. In addition, the stage of healing in chronic wounds and patient comorbidities may also affect the responsiveness of hydrogels.

4.4. Polymers for SHs Formulation

4.4.1. Biopolymers

Polymers from Natural Sources

Natural polymers, such as CS, HA, alginates, dextran, cellulose, agarose, gelatin, collagen, cyclodextrin, fibrin, pullulan, polylysine, pectin, carrageenan, and chondroitin sulfate, may be obtained from a wide variety of natural sources and could successfully design versatile SHs. Natural-polymers-based hydrogels have superior biodegradability and biocompatibility compared to those based on synthetic polymers. In addition, natural polymersendowhydrogels with inheritance bioactivities such antimicrobial and antioxidant effects, similarities with endogenous molecules, and cytocompatility [116]. Moreover, the complex structure of natural polymers offers plenty of chemical and physical structural modulation opportunities in order to generate innovative products (Figure 8).

The use of natural polymers in hydrogels manufacturing, especially on anindustrial scale, is associated with some drawbacks such as difficulties to ensure the reproducible properties from batch to batch, high costs, lower mechanical strength over time, variable stability, and high risks of microbial contamination for some finished products. A wide variety of natural polymers were tested to design SHs, but despite favorable effects on wound healing, some critical aspects were outlined which still need to be address for successful development of innovative formulations (Table 2).

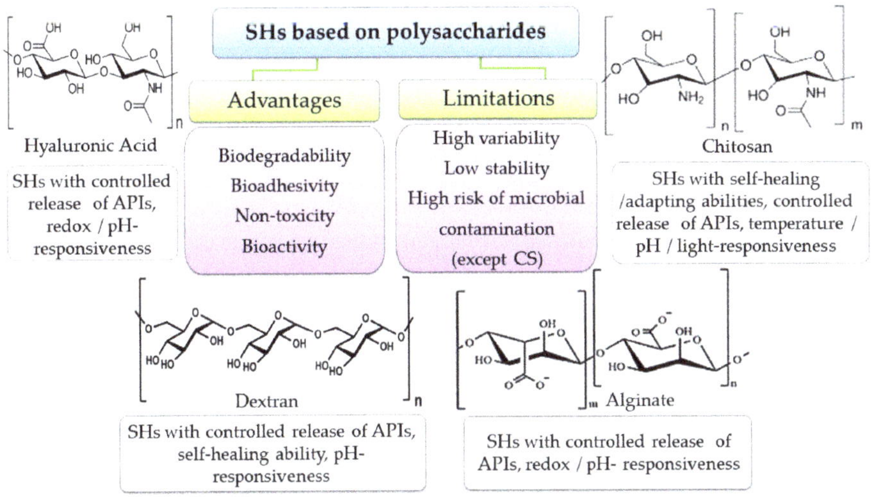

Figure 8. Natural polymers used on SHs formulation for wound managementapplications.

Table 2. Challenges innatural polymers in the development of SHs for wound healing.

Polymer	Advantages	Challenges	Polymer-Based SHs (e.g.)	References
Gelatin	Haemostatic effect, promote migration of fibroblasts on injured area	Hypersensitivity reactions; 3D structure is destroyed through bacterial contamination	Antibacterial injectable self-healing hydrogel based on gelatine and poly(ethylene glicol) bis (benzandehide) loaded with clindamycin; Antibacterial self-healing hydrogel based on gelatine methacrylate, adenine acrylate and copper(II) chloride	[117–119]
Collagen	Haemostatic effect, promote cell migration, proliferation and differentiation; stimulate formation of granulation tissue and synthesis of ECM with reduced scars; anti-inflammatory and proangiogenic effects	High cost, enzymatic degradation, poor elasticity; no gelling properties (requires additional polymers)	pH- and ROS-responsive hydrogel based on recombined collagen and HA with controlled release of APIs; Collagen/xanthan gum-based hydrogels with antibacterial effect, and controlled release of ketorolac and methylene blue	[117,120,121]
Guaran	Haemostatic effects; lack of toxicity	Risk of bacterial contamination and of depolymerization	Self-healing injectable hydrogel based on oxidized quaternized guaran/carboxymethyl CS with hemostatic and antibacterial effects	[122]
Elastin	Chemotaxis to neutrophils, monocytes, macrophages, fibroblasts and endothelial cells, protease production (upregulation of matrix metalloprotease), mimics ECM	Insoluble in water, rapid degradation, low mechanical stability; cross-linking methods affect the elastin performance	Thermoresponsive hydrogels with shape-memory ability based on elastin and elastin-like polypeptides	[123–126]

Table 2. *Cont.*

Polymer	Advantages	Challenges	Polymer-Based SHs (e.g.)	References
Silk	Stimulates cell migration, proliferation, and collagen production, wound remodeling	Insoluble in water, immunogenic and inflammatory effects (due to sericin content), high cost for sericin free silk	Self-healing hydrogels based on silk-fibroin and β-cyclodextrin; Desferrioxamine-loaded injectable silk nanofibers hydrogels for diabetic wounds	[127–130]
Cellulose	Lack of toxicity, low cost	Insoluble in water or other common solvents; no bioactivity; allergic reactions	Magnetic responsive hydrogel based on cellulose and β-cyclodextrin to control drug release by swelling behavior	[131,132]

The most used natural polymers for SHs formulation are presented below.

Dextran is an exopolysaccharide derived from the condensation of glucose (D-glucoses linked by α-(1→6) bonds, with possible branches of D-glucoses linked by α-(1→4), α-(1→3), or α-(1→2) bonds), obtained through fermentation of some species of lactic acid bacteria (*Leuconostoc mesenteroides*, *Weissella*, *Lactobacillus*, and *Streptococcus*) or by enzymatic polycondesation of glucose from sucrose via dextran saccharases. It is a neutral polysaccharide with good solubility in water, glycerol, and electrolyte solutions depending on its molecular weight and pH of media, and it is biocompatible, biodegradable, and stable in topical formulations. Moreover, dextran is nontoxic, being included in the GRAS list of substances since 1975 [133]. Food and Drug Administration (FDA) mentioned dextran on IIG as being accepted in formulation of products for topical and intravenous routes [134].

In wound healing, dextran has beneficial effects due its antioxidant, antimicrobial, and immunomodulatory ability, as well as promoting of angiogenesis in damaged tissue, properties which appear to vary with its molecular weight. For example, 40 kDa dextran from *Leuconostoc mesenteroides* showed immunomodulatory ability much better than 10 kDa and 147 kDa dextran. More exactly, with 40 kDa dextran, a 50% increasing murine macrophages proliferation and a 40% decreasing in cell release rate of nitric oxide were noted. Moreover, 40 kDa dextran showed the highest radical scavenging activity, with 70% inhibition of lipid peroxidase [135].

For dextran-based hydrogels formulation, both physical and chemical cross-linking methods are suitable. The characteristics of dextran-based hydrogels (as rheological behavior, mechanical strength, release of APIs embedded in their 3D matrix, stimuli-responsiveness) depend on the type of cross-linking method as well as on cross-linkers concentration. Using physical cross-linking methods, self-healing hydrogels with swelling behavior responsive to pH variations and promising controlled release of APIs were obtained. Chemical cross-linking based on thermal-gelation, radiation, or addition of divalent cations (Zn^{2+}, Mg^{2+}, Ca^{2+}) also leads to versatile hydrogel with superior stability over time [136].

Due their ability of modulating rheological behavior of 3D matrices, dextran and its derivatives (methacrylated, aminated, oxidized) were used to develop non-Newtonian pseudo plastic hydrogels, whose viscosity decrease when increasing the shear rate. Epichlorohydrin and basic amino acids were successfully used for cross-linking pH-responsive dextran-based hydrogels. Thus, dextran-allyl isocyanate-ethylamine hydrogel demonstrated beneficial effects for promoting neovascularization and total skin regeneration (with hair follicles and sebaceous glands) in murine burn wound model [137]. A dextran/HA-based hydrogel showed a sustained release of sanguinarine embedded in the 3D matrix, stimulated fibroblast cell proliferation, and enhanced the healing of infected burn wounds [138]. Oxidized dextran and methacrylated gelatin were also used to formulate hydrogels for embedding black phosphorus nanosheets and zinc oxide NPs, which proved immunomodulatory effects on macrophage with enhanced secretion of anti-inflammatory factors [139]. A dextran/bacterial cellulose-based hydrogel showed superior cell proliferation and skin regeneration on an in vivo wound model. Moreover, addition of dextran in the hydrogel formulation endowed the final product with non-cytotoxicity compared with unmodified cellulose [140]. Other information related to use of dextran and dextran derivatives for SHs formulation are presented in Table 3.

Table 3. SHs based on dextran and dextran derivatives for wound healing.

Polymers	APIs	SHs Features	References
Oxidized dextran/quaternized CS	-	Stimulation of myoblast proliferation under electrical field; injectability	[141]
Oxidized dextran	Sulfadiazine and tobramycin	pH-responsive injectable hydrogel with controlled release of APIs	[142]
Oxidized dextran/peptide DP7 (VQWRIRVAVIRK)	Ceftazidime	pH-responsive hydrogel with scarless healing of multidrug-resistant infected wounds	[143]
Oxidized dextran/aminated gelatin	ZnO, paeoniflorin, norfloxacin	ROS-responsive hydrogel	[144]

Chitosan (CS) is a cationic polysaccharide with semicrystalline structure resulting from repeated units of N-acetyl-D-glucose amine and D-glucose amine, linked by β-(1-4) bonds. It was discovered more than 100 years ago when it was obtained by chemical or enzymatic deacetylation of chitin harvested from different sources [145]. The sources of chitin are varied, being found both in animal and vegetable, as insects (cuticle, cocoon), crustaceae (exoskeleton of crabs, shrimps), diatomae (*Thalassiosira fluviatilis*, *Algae*), and fungi (*Mucorrouxi*, *Aspergillis nidulans*) [146]. In 1970, after a long period of extensive research and scientific doubts, the foundations were laid for the practical use of CS. Nowadays, CS is one of the most used polymers based on its characteristics and biological properties such as excellent biocompatibility, biodegradability, non-toxicity, antioxidant, antimicrobial, and hemostatic effects, and mucoadhesivity [147]. Based on primary amino groups, CS has a pKa ~ 6.5 and initiates protonation in a neutral environment, which is accentuated as the pH decreases and the degree of deacetylation increases. Thus, CS becomes soluble in aqueous solutions and has bioadhesion capacity by binding to electro-negatively charged surfaces (for example, it can interact with sialic acid, a glycoprotein presents in mucus) [146].

CS is used for a wide area of pharmaceutical applications: mucoadhesive tablets, scaffolds for controlled release of drugs, growth factors, gels, films, beads, and liposomes. The lack of toxicity (LD_{50} (mouse, oral) > 16 g/kg) and structural similarity with the proteoglycans of the extracellular matrix make CS very attractive for medical applications, including wound care [148,149]. Due to its positive charge, CS interacts with the negative ions of red blood cells and platelets, facilitating coagulation and thus homeostasis restoration, which represents a challenge in wound management [150,151].

Through the action of lysozyme from the damaged tissues, CS splits into oligomers, which stimulate the synthesis and orientation of collagen fibers to restore the components of the extracellular matrix [152]. CS also stimulates the migration of inflammatory cells, macrophages, and fibroblasts, thus accelerating the inflammatory process [153]. In this way, the inflammatory phase is reduced, and the proliferative phase starts earlier in wound healing. In addition, there is a large quantity of data which sustain the antimicrobial activity of CS over a wide variety of microorganisms involved in wound colonization, such as *Escherichia coli*, *Staphylococcus aureus*, *methicillin resistant Staphylococcus aureus*, *Salmonella typhimurium*, *Pseudomonas aeruginosa*, *Candida albicans*, and *Enterococcus faecalis* [154,155]. This effect is due to the positively charged amino groups of CS that bind to the negatively charged cell surface, disrupt the cell membrane, and cause cell death by inducing leakage of intracellular components. The inhibition of genetic material synthesis by binding to the microbial DNA is also possible due to the presence of the protonated amino groups [154]

The effectiveness of CS in the treatment of wounds depends on its physico-chemical properties: degree of deacetylation, molecular weight, viscosity, and acetyl group distribution [146,156]. These characteristics vary from batch to batch, depending on the source of chitin, the period of the year of harvesting for the same source of chitin, and technological process for converting chitin into CS. There are several types of CS for cosmetic, pharmaceutical, and food industries, with molecular weights in the range of 10^4–10^6 Da and up to 95% deacetylation degree. The degree of deacetylation modulates is directly proportional to the antimicrobial action. The greater the degree of deacetylation, the greater the number of free amino groups available for protonation and the positively charge increases [154]. In addition, the molecular weight influences the ability of microbial inhibition, with CS with low and medium molecular weight penetrating the bacterial membrane more easily than CS with high molecular weight [146].

To increase the solubility, enhance the antimicrobial effects, and modulate the hemostatic and bioadhesive properties of CS, different physical, chemical, and enzymatic methods such as quaternization, thiolation, sulfation, copolymerization, and carboxymethylation have been reported [157]. Other information related to the use of CS and CS derivatives for SHs formulation are presented in Table 4.

Table 4. SHs based on CS and CS derivatives for wound healing.

Polymers	APIs	SHs Features	References
CS/oxidized konjac glucomannan	Silver (nanoparticles)	Self-healing, self-adapting	[71]
Oxidized CS/bacterial cellulose	-	Self-healing	[158]
N,O-carboxymethyl CS/oxidized dextran	-	Self- healing, injectability	[159]
Quaternized CS/oxidized dextran	Tobramycin	Self-healing	[160]
Carboxymethyl CS/oxidized quaternized guaran	-	Self-healing	[122]
CS/oxidized CS	Fusidic acid, alantoin, coenzyme Q10	Self-healing, self-adapting	[161]
CS/polyvinylpyrrolidone/alginate/ poly(ε-caprolactone)	-	Thermo-responsive	[162]
CS/poly(aspartic acid)	Amoxicilin	pH/thermo-responsive	[163]

Alginate is an anionic natural copolymer, composed of blocks of guluronate and mannuronate, obtained by alkaline extraction from brown algaes (*Laminaria* spp., *Ascophyllumnodosum*, *Macrocysti spyrifera*) or by microbial synthesis (*Azotobacter*, *Pseudomonas*). It is characterized by several properties, such as biocompatibility, low toxicity, economical affordable, and ease of gelation, which makes alginate attractive for biomedical applications such as tissue engineering, drug delivery, biosensing, and wound healing [164,165].

Alginate salts (ammonium, potassium, calcium, sodium) and alginate derivatives (propylene-glycol alginate) are GRAS listed for 50 years, being considered safe for use in food industry in Europe. The FDA has included alginates on the IIG list for products administrated via topical and oral routes [133,134].

Several studies confirm the cytoprotective capacity of alginates due their antioxidant effects and immunoregulatory activity. Moreover, alginates oligosaccharides (guluronate and mannuronate) have proved anti-inflammatory effects through interfering with the production of nitric oxide, prostaglandins, and pro-inflammatory cytokines. The proliferation of human keratinocytes and endothelial cells was also attributed to alginates, especially due the presence of guluronic acid at the end of the backbone chain. These effects are related to molecular height, alginate sources, guluronate: mannuroanate ratio, as well as purification and extraction technologies. The alginate with the highest molecular weight (MW 38,000) and higher mannuronate: guluronate ratio (2.24) exhibited a higher effect on the stimulation of the secretion of tumor necrosis factor-α (TNF-α) in murine macrophage cell line [166,167].

Topical alginate-based products (dressings, films, foams, fibers, and hydrogels) may absorb up to 20 times their weight in moisture, promoting tissue regeneration of highly exudative wounds such as venous ulcers, pressure ulcers, and infected and large surgical wounds. The combination of alginates with different natural and synthetic polymers leads to biomimetic and biodegradable hydrogels, which could be used to achieve different targets (antibiotics and protein delivery, cell culture). Other information related to the use of alginate and alginate derivatives for SHs formulation are presented in Table 5.

Table 5. SHs based on alginate and alginate derivatives for wound healing.

Polymers	APIs	SHs Features	References
Alginate	Amikacin, naproxen	pH/ROS-responsive	[109]
Alginate/HA	Doxycycline	ROS–responsive	[60]
Oxidized alginate/gelatin	Ishophloroglucin A	Dynamic mechanical properties	[168]
Alginate/polyacrylamide	-	pH-responsive (color change)	[169]
Alginate/pluronic F127	Vascular endothelial growth factor	Thermo-responsive	[170]
Alginate/polycaprolactone	Melatonin	Controlled release of APIs	[171]
Alginate/HA	Platelet rich plasma	ROS/glucose-responsive, self-healing, injectability	[172]

Hyaluronic acid (HA) is a natural anionic glycosaminoglycan, consisting of D-glucuronic acid and D-N-acetylglucosamine units. Unlike other glycosaminoglycans (heparin, chondroitin sulfate, dermatan sulfate, keratin), HA does not contain sulfur, it is not synthesized by the action of the enzymes of the Golgi apparatus on proteins, and may have higher molecular weights up to 10^5 kDa (compared to maximum 50 kDa for other glycosaminoglycans) [173]. HA is widely distributed in nature, being found in the epithelial, connective, and nervous tissue of vertebrates, but also in the cells of microorganisms (*Streptococcus equi, Streptococcus zooepidermicus, Streptococcus equisimilis, Streptococcus pyogenes, Streptococcus uberis, Pasteurella multocida, Cryptococcus neoformans*) [85]. The human body contains approximately 215 mg HA/kg, and 50% is found in skin, dermis, and epidermis. The synthesis of HA is enzymatic controlled by hyaluronan synthetases (HASs) (HAS1, HAS2, HAS3, located on the inner surface of the cellular membrane), while hyaluronidase and oxidative stress are implicated in its degradation. The turnover is very fast (5 g/day), and the half-life is less than 24h [174].

Currently, HA is a key molecule in the medical, pharmaceutical, nutrition, and cosmetic industries and, as result, the researchers are looking for new synthesis routes and optimization of production biotechnologies or structural modulation in order to obtain new derivatives with improved features [148,175,176].

In wound healing, HA is involved in each phase, the effects being dependent to its molecular weight. For example, HA with a high molecular weight exhibits anti-inflammatory effects, while HA with a low molecular weight is responsible for mediating inflammation, with both possessing scavenging properties against oxygen free radicals [175,177,178].

In the hemostasis phase, white blood cells produce a significant amount of high molecular weight HA (4×10^2–2×10^4 kDa), which by interaction with fibrinogen forms temporary white platelets. Due to the hydrophilic character of HA, water is attracted to the area of the thrombus and edematization of the area is initiated, with the temporary blocking of the penetration of immune cells into the injured area [176]. In the inflammatory phase, HA with high molecular weight is degraded into fragments smaller than 120 kDa, with immunostimulatory properties and a proangiogenesis effect. These products stimulate the production of cytokines (IL-1 and IL-8, TNF-α) and, consequently, the infiltration, activation, and maturation of immune cells are accelerated [174,179,180]. During the proliferation phase, HA with low molecular weight is degraded to oligomers, which will interact with the CD44 and RHAMM receptors. As a result, keratinocytes, fibroblasts, and endothelial cells mature and activate, which translates into inflammation reduction, support for angiogenesis, stimulation of granulation tissue, and synthesis of type III collagen (immature collagen) [49,176,181]. Finally, the oligomers determine the synthesis of matrix-metalloproteinase and TNF-β. These mediators are necessary for remodeling the extracellular matrix through the differentiation of fibroblasts into myofibroblasts and the maturation of collagen [176,182].

Based on physical and biological characteristics, HA and its derivatives are used in the manufacturing of wound care such as hydrogels, dressings, foams, nanofibers, and sponges.

Through etherification or esterification of the hydroxyl and carboxyl groups with different chemical compounds (hydrazide, phenylboronic acid, polyphenols), increasing performance of 3D matrices was obtained, without the necessity of blending with other polymers [180]. Other information related to the use of HA and HA derivatives for SHs formulation are presented in Table 6.

Table 6. SHs based on HA and HA derivatives for wound healing.

Polymers	APIs	SHs Features	References
Phenylboronic acid-modified HA	Tannic acid–silver (nanopartices)	pH/ROS-responsive, self-healing	[98]
Adipic acid dihydrazide grafted HA	Sisomicin sulfate	pH-responsive	[183]
Tannic acid grafted HA	Silver nanoclusters and deferoxamine	pH-responsive, injectability	[184]

Biological-like Polymers

Self-assembling peptides (SAPs)

Peptides are comprised from aminoacid residues linked through peptide bonds, with a molecular weight of less than 10,000 Da. Peptides are ubiquitously present in both vegetal and animal regna and are essential for life cycles. Every physio/pathological process is regulated by various types of peptides with one or more functions. So, there are antibiotic peptides which are part of natural immune systems of prokaryotes and eukaryotes, cancer/anticancer peptides, immune/inflammatory peptides, endocrine peptides, gastrointestinal peptides, cardiovascular peptides, renal peptides, respiratory peptides, opioid peptides, and blood–brain peptides. In organisms, peptides are synthesized through ribosomal translation (antibiotics in microorganisms–microcines and bacteriocines), hormones and other signaling molecules (in the human body), or by enzymatic processes (glutathione) [185].

The use of peptides in the formulation of a 3D matrix is relatively recent and is based on the intra- and intermolecular self-assembling ability of aromatic short peptides, biomimetic peptides, amphiphilic peptides, and dendrimers. In the assembling process, super molecular structures are formed, such as spherical or worm-like structures, bilayer micelles, high ratio aspect nanofibers, twisted ribbons, nanotubes, or flat forms, through non-covalent interactions (hydrogen bonds, hydrophobic and electrostatic interactions, van der Waals forces). The resulting structures are highly biocompatible and may be used in tissues regeneration, scaffolds for drugs carriers and controlled release, and modulation of APIs to overcome adverse effects and enhance their stability [186].

The discovery of the first SAP, coded Ac-(AEAEAKAK)2-$CONH_2$ (A-alanine, E-glutamic acid, K-lysine), in a yeast protein zuotin in 1990 was regarded as a big innovation and a valuable tool to design new materials for a wide variety of applications [186]. The chemical composition of SAPs consistsof alternating hydrophobic aminoacids (alanine, valine, leucine, isoleucine, and phenylalanine) with hydrophilic aminoacids (lysine, arginine, histidine, aspartic acids, and glutamic acids), having very well-defined arrangements of negative and positive charges [187].

Hydrogels based on SAPs are sensible to multiples stimuli such as pH, temperature, ionic strength, and redox activity as a result of charging the amino and carboxyl groups on the peptide backbones [186]. Due the resembling natural extracellular matrix, these formulations are suitable for adhesion, migration, differentiation, and proliferation of cells involved in wound healing.

The weak bonds within an SAP 3D structure could be easily modulated to achieve targeted SAP performance, such as mimicking extracellular matrices, recognition of ligandreceptors, and modification of their structural architecture. For example, the SAP solution undergoes phase transition to a gel state under pH and ionic strength variations. In this way, when are applied on damaged tissues with irregular shapes and depths, an intimate contact is assured, the wound is filled with the product, and so the hemostasis is fast and the healing is accelerated.

The temperature induces different 3D organizations of the peptide aggregates which influence the water uptake and mechanical strengths of hydrogels. For instance, amyloid beta peptide (C16-KKFFVLK) exhibits different self-arrangements at 20 °C (helical ribbons and nanotubes) compared to its exposure to 55 °C (twisted tapes). In addition, peptide RADA 16-I maintains beta sheets assembling through 25 °C to 70 °C, but small-sized globular aggregates were obtained at much higher temperatures [186–188].

Targeting microorganism resistance to antibiotherapy, SAPs nanofibers with antimicrobial effects have been developed [189]. It was shown that diphenylalanine-SAP inhibits bacterial growth through cell membrane disruption, depolarization, and enhanced permeation resulting in elevated levels of ROS [190]. The addition of short SAPs to natural or synthetic polymers allows the design of SHs with improved benefits on wound healing [181]. Arg-Gly-Asp-(RGD), cytosine-guanine-adenine (CAG), Arg-Glu-Asp-Va (REDV), and Tyr-Ile-Gly-Ser-Arg (YIGSR) are short SAPs used to activate the physiological cascade of wound healing because they are recognized by integrin receptors (RGD) or endothelial cells and promote cell migration, granulation tissue formation, selective cell adhesion, and neoangiogenesis in damaged tissues [191].

PuraStat®, PuraBond®, PuraDerm®, and PuraGel® are commercial products manufactured by 3-D Matrix Europe SAS, France, based on a syringe prefilled acid solution (pH 2) of self-assembling RADA16 peptide (which is composed of alternating positively charged arginine (R), hydrophobic alanine (A), and negatively charged aspartic acid (D), that repeat periodically throughout the composition). The acid peptide solution is neutralized in contact with blood or physiological fluids and through exposure to ionic species (Na^+, K^+), resulting in clear hydrogels that mimic the extracellular matrix with a fast hemostatic effect, which serve as a mechanical barrier for exposed tissue and prevent the formation of post-operative adhesions. The viscosity of the SAP solutions allows easy application, using either a standard applicator, endoscopic catheter, or laparoscopic applicator, making the products applicable to different kinds of injuries such as surgical wounds, nasal bleedings, deep wounds (pressure ulcers, diabetic ulcers), and dental surgery [68,186,192–194].

Deoxyribonucleic Acid

Exceeding the initial focus on genetic information transmissions, synthetic deoxyribonucleic acid (DNA) has been extensively investigated for new drug discovery, modified release matrices, diagnostics tools, biomolecular therapies (including vaccines, enzymes, and hormones), and in vivo metabolism modulation up to creating an artificial living organism, as interest areas for new generation medical fields. Based on the chemical features of DNA and nucleobases (adenine, cytosine, guanine, and tyrosine), synthetic DNA is suitable for molecular and topological editing with specialized software (caDNAno, Tiamat,

Nupack, MagicDNA) in order to reproducibly and predictably design complex dynamic or static structures [195].

Tetrahedral framework nucleic acids (a static DNA tetrahedral nanostructure) were tested on multiple in vivo wounds model and proved significant anti-inflammatory and antioxidant effects, preventing tissue destruction and sustaining regeneration on skin and acute kidney injury, as well as in osteoarthritis [195,196].

Dynamic synthetic DNA-based nanostructures react with different types of biomolecules (nucleic acid, proteins) and are sensible to environment variations (pH, temperature), which make them attractive for tissue engineering. A cost-effective self-assembled DNA-based hydrogel was developed through electrostatic interactions and hydrogen bonds between DNA structure and tetrakis(hydroxymethyl) phosphoniumsulfate (THPS) molecule. The hydrogel exhibited self-healing abilities, good cytocompatibility, prevented wound microbial colonization through release of THPS, and enhanced the healing process. In another study, DNA units, polyacrylamide, and L-ascorbic acid 2-phosphate were dynamically cross-linked through hydrogen bonds in 3D hydrogel, which mimics extracellular matrices. The hydrogel was loaded with borneol to relieve pain and itching and was proved to initiate the phosphatidylinositol 3′-kinase signaling pathway, which promoted cell proliferation and regeneration in burn wounds [197].

4.4.2. Synthetic Polymers

Synthetic polymers (as PVA, PEG, ploy(acrylamides), poloxamers) offer a more reproducible physical and chemical characteristics from batch to batch of both raw materials and hydrogels, higher flexibility and mechanical strength, very good stability over time, compatibility or inertia for a wide range of APIs, and often allow higher water intake than natural-based-hydrogels. However, some drawbacks, such as poor biocompatibility, high risk of toxicity, slow biodegradability, and lack of bioactive effects, are associated with synthetic polymers (Figure 9).

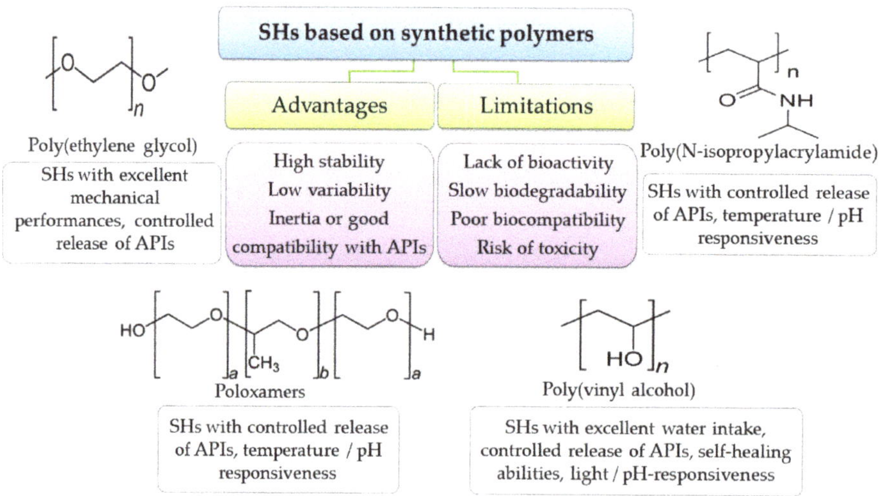

Figure 9. Synthetic polymers used in SHs formulation for wound healing management.

The combination of natural and synthetic polymers could improve the performance of hydrogels. So, the biocompatibility and bioactivity of natural structures are sustained by the robust structures of synthetic components, which result in improved therapeutic benefits, costs optimization, and support of the industrial implementation of innovative SHs manufacturing for biomedical applications. The most used synthetic polymers for SHs formulation are presented below.

Poly(ethylene glycol) (PEG) is a stable hydrophilic polymer derived from petroleum or synthesized from ethylene oxide and water in both acidic and alkaline conditions. Depending on reactants ratio and conditions, a wide variety of products with different molecular weights (3×10^2–10^7 g/mol) and different geometries of polymer chains (star, comb, branched) could be manufacturing. It is interesting to note that although the physical characteristics (density, viscosity, aggregation state) depend on structural properties, all PEGs are chemically identical. The presence of hydroxyl groups of the end of the backbones endow the PEG with the ability of attaching to multiple reactive functional groups (methoxy, amino, thiol, vinyl sulfone, azide, acetylene, and acrylate), which improve its performance. Moreover, PEG may be attached to large chemical entities such as proteins, polysaccharides, oligonucleotides, or to small molecules such as mannose or folate to improve biodegradability and biocompatibility [198].

PEGs are considered as nontoxic, non-immunogenic, and nonirritant substances, being included in the IIG database for parenteral, oral, and topical pharmaceutical products [134]. For pharmaceutical applications, PEGs may be used as ointment/suppository bases, plasticizers, solvents, emulsifiers, surfactants, tablets, and capsule lubricants. In addition, branched PEGs derivatives were investigated for their applicability in the development of biodegradable controlled-release matrices, regenerative medicine wound sealing, and wound healing. Aqueous solutions of higher-molecular-weight PEGs may form flexible, highly hydrophilic gels suitable for wound care using different cross-linking methods (radiation, free radical polymerization, condensation, click chemistry, enzymatic reactions, etc.) [199,200].

Poly(vinyl alcohol) (PVA) is a widely used highly water-soluble polymer obtained through free-radical polymerization of vinyl acetate, followed by alkaline hydrolysis of acetate groups. PVA has been shown to be chemically stable in a wide range of pH and temperatures, being suitable for various chemical and technological modulations. Because of its semicrystalline structure, it allows both the vital molecules (oxygen, glucose, and aminoacids) and metabolism products to cross the cellular membrane. In addition, PVA is biocompatible, biodegradable, and bioinert, being included in the IIG database of solutions, suspensions, creams, aerosols, foams, tablets, and capsules for topical and oral routes [134].

The physical, chemical, and biological features of PVA support its use for a wide range of biomedical applications (implantable medical devices, tissue regeneration–artificial cartilage and meniscus, contact lens, wound dressings, artificial tears). PVA-based hydrogels proved excellent mechanical strengths, bioadhesion, and high water absorption uptake.

In order to design 3D networks using PVA, both a chemical and physical method may be used. When applying repeated freezing–thawing cycles on PVA solutions, without addition of external cross-linkers, clear, pure hydrogels with enhanced biological aging resistance were obtained. The size and conformation of the hydrogel matrix are influenced by the solvents (glycerol, ethylene glycol), molecular weight, and concentrations of the PVA solutions. These characteristics are very important because they influence the performance of the hydrogel on drug delivery, moisturizing the wound bed, and the hydrogel flexibility for damaged tissue protection. It was noted that the water uptake enriched the hydrogels with higher mechanical strength, while using ethylenglycol increased the toughness of the products [201]. PVA-based SHs with versatile features such as self-healing, stimuli responsiveness, and an improved controlled drug release profile were developed.

Poloxamers is a generic name for a group of nonionic water-soluble copolymers with amphiphilic structure, due their central hydrophobic component (polyoxypropylene) and peripheral hydrophilic chains (polyoxyethylene), $HO(C_2H_4O)_x(C_3H_6O)_y\ (C_2H_4O)_xH$. Depending on the molecular weight and hydrophobic: hydrophilic ratio of the polymers, the solution of poloxamers exhibited reversible temperature-depended phase transitions with self-assembling gelation. At low temperatures and at concentrated solutions (above critical micellar concentrations), the poloxamers are liquids. Upon skin contact, the polyoxyproylene core starts the dehydration and micellization process, while the polyoxyethylene blocks remain soluble.

Poloxamers are widely used in the cosmetic and pharmaceutical industry as surfactants, emulsifiers, solubility enhancers, stabilizers, viscosifiers, cell culture media, drugs carriers, and in nanofiber synthesis [202,203]. These copolymers are included by the FDA in the IIG list, accepted for intravenous injections, inhalations, ophthalmic preparations, oral powders, solutions, suspensions, and syrups, and topical preparations [134].

Poly(N-isopropylacrylamide) (PNIPA) is a thermoresponsive water-soluble polymer synthesized through free radical polymerization of N-isopropyl acrylamide. Depending on the molecular weight and polymer dispersity, PNIPA undergoes reversible phase transition from a solution at lower temperatures to gel a state at near-body temperature (32 °C ± 5 °C), which makes it attractive for medical applications such as wound dressings, tissue engineering (artificial muscles), scaffolds for controlled release of pH-sensitive drugs, macro and microgels formulations, cell culture, and biosensors.

Functionalization of the end-chain of PNIPA with different moieties, such as pyrenil, azobenzene, thiol, and azide, endows the polymer with tunable properties, such as a modified low critical solution temperature, pH-sensitivity, and temperature responsivity [204–206].

In wound healing, the PNIPA solution with different APIs added is suitable to readily apply on damaged skin and use the body temperature to convert it into a 3D matrix with hemostatic activity. The adjustable geometries of the hydrogel matrix allow an intimate contact with the wound bed, helping to create a proper environment for proliferation and differentiation of cells involved in healing process [207].

5. Conclusions

The prevalence of acute and chronic wounds is increasing every year because of the aging population and the growing incidence of wound-related comorbidities, and remains a continuous concern of health systems worldwide. Overtime, medical specialists have developed and implemented various protocols and tools to overcome misinterpretation in wound assessment, which lead to impaired healing processes and complications. Choosing the appropriate topical approach has proved to be crucial for the entire output of the treatment.

Hydrogels have passed the test of time, being used for more than 120 years in wound healing. Because of their undeniable benefits in the healing process, hydrogels constantly have attracted the interest of researchers in the medical and pharmaceutical area and sustained efforts are made to improve and to innovate the performances of hydrogels and to widen their biomedical applications. SHs is a large category of versatile 3D matrices which has made it possible to design intelligent products with the target of overcoming the limitation of traditional hydrogels and properly responding to the increasing prevalence of wounds with impaired healing.

Althoughimpressive research work is ongoing in the SHs field and various SHs have been proved to exhibit innovative properties, some drawbacks are still to be investigated.

The controlled release of APIs is one of the top interest areas in SHs development. Despite this, in vitro release tests showed reproducible kinetics, and in vivo assessment could be affected by specific wound environment composition and irregular vascularization, which could lead to over-release with potential toxic effects or, on the contrary, to a decreased release rate which, especially in the case of antibiotics, leads to complications due to under-dosing. A future target of research could be the inclusion of biosensors in 3D matrices, where online data regarding wound bed biochemistry and the command and controlled release of the required number of APIs could be collected.

Reversible stimuli-responsive hydrogels are very attractive due their ability to overcome exposure to a wide variety of external or internal stressors (mechanical, thermal, chemical). Thus, the ability of hydrogels to undergo structure breakage when they are passed through a syringe needle and to quickly restore their matrix is an excellent feature for products designed to be injected in deep, irregular wounds. These hydrogels are more commonly based on physical interactions, which are less stable than covalently cross-linked networks. This is translated into low stability over time, which is a serious limitation of industrial-scale applicability. The dynamic covalent interactions and adequate functionalization of polymers may be successfully used to address this issue.

The biodegradability and non-cytotoxicity of SHs are widely discussed, but the behavior of the degradation products in human body is still unclear and not fully evaluated (e.g., PNIPA-based SHs may generate carcinogen moieties). Further investigations might comprehensively address this issue and the degradation pathways and resulting compounds should be carefully documented. Moreover, the SHs showed promising healing effects on various in vivo wound models, but clinical trials were not yet conducted in order to evaluate the human safety of the finished products, to establish the optimal method of application and dosage, and to assess the risk/benefits balance, in comparison with commercially available alternatives.

Author Contributions: Conceptualization, S.-M.T. and L.P.; methodology, F.-G.L., A.-T.I. and B.-S.P.; validation, L.P., and F.C.; investigation, S.-M.T., F.-G.L., B.-S.P., A.S., I.G., A.-T.I., I.C., L.H., T.-C.G. and M.-C.B.; resources, L.P.; data curation, F.-G.L., F.C. and M.-C.B.; writing—original draft preparation, S.-M.T., F.-G.L., B.-S.P., A.S., I.G., A.-T.I., I.C., L.H., T.-C.G. and M.-C.B.; writing—review and editing, L.P. and F.C.; supervision, L.P. and F.C; project administration, L.P. and F.C; funding acquisition, L.P. All authors have read and agreed to the published version of the manuscript.

Funding: This work was supported by a grant of Romanian Ministry of Education and Research, CNCS-UEFISCDI, project number PN-III-P4-ID-PCE-2020-2687 (contract nr. 244/2021) within PNCDI IIII and by European Social Fund—the Human Capital Operational Programme, Project/Grant No: POCU/993/6/13/154722.

Institutional Review Board Statement: Not applicable.

Informed Consent Statement: Not applicable.

Data Availability Statement: The data could be requested from authors.

Conflicts of Interest: The authors declare no conflict of interest.

References

1. Demidova-Rice, T.N.; Hamblin, M.R.; Herman, I.M. Acute and Impaired Wound Healing: Pathophysiology and Current Methods for Drug Delivery, Part 1: Normal and Chronic Wounds: Biology, Causes, and Approaches to Care. *Adv. Skin. Wound Care* **2012**, *25*, 304–314. [CrossRef] [PubMed]
2. Gethin, G.; McIntosh, C.; Cundell, J. The Dissemination of Wound Management Guidelines: A National Survey. *J. Wound Care* **2011**, *20*, 342–345. [CrossRef] [PubMed]
3. Zhu, X.; Olsson, M.M.; Bajpai, R.; Järbrink, K.; Tang, W.E.; Car, J. Health-Related Quality of Life and Chronic Wound Characteristics among Patients with Chronic Wounds Treated in Primary Care: A Cross-Sectional Study in Singapore. *Int. Wound J.* **2022**, *19*, 1121–1132. [CrossRef] [PubMed]
4. Yao, Z.; Niu, J.; Cheng, B. Prevalence of Chronic Skin Wounds and Their Risk Factors in an Inpatient Hospital Setting in Northern China. *Adv. Skin Wound Care* **2020**, *33*, 1–10. [CrossRef] [PubMed]
5. Järbrink, K.; Ni, G.; Sönnergren, H.; Schmidtchen, A.; Pang, C.; Bajpai, R.; Car, J. Prevalence and Incidence of Chronic Wounds and Related Complications: A Protocol for a Systematic Review. *Syst. Rev.* **2016**, *5*, 152. [CrossRef]
6. Eriksson, E.; Liu, P.Y.; Schultz, G.S.; Martins-Green, M.M.; Tanaka, R.; Weir, D.; Gould, L.J.; Armstrong, D.G.; Gibbons, G.W.; Wolcott, R.; et al. Chronic Wounds: Treatment Consensus. *Wound Repair Regen.* **2022**, *30*, 156–171. [CrossRef]
7. Sen, C.K. Human Wounds and Its Burden: An Updated Compendium of Estimates. *Adv. Wound Care* **2019**, *8*, 39–48. [CrossRef]
8. Jupiter, D.C.; Thorud, J.C.; Buckley, C.J.; Shibuya, N. The Impact of Foot Ulceration and Amputation on Mortality in Diabetic Patients. I: From Ulceration to Death, a Systematic Review. *Int. Wound J.* **2016**, *13*, 892–903. [CrossRef]
9. Cheun, T.J.; Jayakumar, L.; Sideman, M.J.; Ferrer, L.; Mitromaras, C.; Miserlis, D.; Davies, M.G. Short-Term Contemporary Outcomes for Staged versus Primary Lower Limb Amputation in Diabetic Foot Disease. *J. Vasc. Surg.* **2020**, *72*, 658–666.e2. [CrossRef]
10. Gushiken, L.F.S.; Beserra, F.P.; Bastos, J.K.; Jackson, C.J.; Pellizzon, C.H. Cutaneous Wound Healing: An Update from Physiopathology to Current Therapies. *Life* **2021**, *11*, 665. [CrossRef]
11. BuganzaTepole, A.; Kuhl, E. Systems-Based Approaches toward Wound Healing. *Pediatr. Res.* **2013**, *73*, 553–563. [CrossRef]
12. Bowers, S.; Franco, E. Chronic Wounds: Evaluation and Management. *Am. Fam. Physician* **2020**, *101*, 159–166.
13. Las Heras, K.; Igartua, M.; Santos-Vizcaino, E.; Hernandez, R.M. Chronic Wounds: Current Status, Available Strategies and Emerging Therapeutic Solutions. *J. Control. Release* **2020**, *328*, 532–550. [CrossRef] [PubMed]

14. Olsson, M.; Järbrink, K.; Divakar, U.; Bajpai, R.; Upton, Z.; Schmidtchen, A.; Car, J. The Humanistic and Economic Burden of Chronic Wounds: A Systematic Review. *Wound Repair Regen.* **2019**, *27*, 114–125. [CrossRef] [PubMed]
15. Nagle, S.M.; Stevens, K.A.; Wilbraham, S.C. Wound Assessment. In *StatPearls*; StatPearls Publishing: Treasure Island, FL, USA, 2023. Available online: https://www.ncbi.nlm.nih.gov/books/NBK482198/ (accessed on 11 August 2023).
16. Woo, K.Y.; Sibbald, R.G. A Cross-Sectional Validation Study of Using NERDS and STONEES to Assess Bacterial Burden. *Ostomy. Wound Manag.* **2009**, *55*, 40–48.
17. Calling on NERDS for Critically Colonized Wounds: Nursing. 2023. Available online: https://journals.lww.com/nursing/Citation/2007/05000/Calling_on_NERDS_for_critically_colonized_wounds.17.aspx (accessed on 2 April 2023).
18. Wound Bed Preparation: Employing the TIME Acronym. Available online: https://www.independentnurse.co.uk/content/clinical/wound-bed-preparation-employing-the-time-acronym/ (accessed on 2 April 2023).
19. Nuutila, K.; Eriksson, E. Moist Wound Healing with Commonly Available Dressings. *Adv. Wound Care* **2021**, *10*, 685–698. [CrossRef]
20. Varma, A.; Warghane, A.; Dhiman, N.K.; Paserkar, N.; Upadhye, V.; Modi, A.; Saini, R. The Role of Nanocomposites against Biofilm Infections in Humans. *Front. Cell. Infect. Microbiol.* **2023**, *13*, 1104615. [CrossRef] [PubMed]
21. Doughty, D. Dressings and More: Guidelines for Topical Wound Management. *Nurs. Clin. N. Am.* **2005**, *40*, 217–231. [CrossRef]
22. Junker, J.P.E.; Kamel, R.A.; Caterson, E.J.; Eriksson, E. Clinical Impact Upon Wound Healing and Inflammation in Moist, Wet, and Dry Environments. *Adv. Wound Care* **2013**, *2*, 348–356. [CrossRef]
23. Wound-Management-Guidelines. Available online: https://www.bcpft.nhs.uk/documents/policies/w/1444-wound-management-guidelines/file (accessed on 17 January 2023).
24. Ward, J.; Holden, J.; Grob, M.; Soldin, M. Management of Wounds in the Community: Five Principles. *Br. J. Community Nurs.* **2019**, *24*, S20–S23. [CrossRef]
25. Dowsett, C.; Swanson, T.; Karlsmark, T. A Focus on the Triangle of Wound Assessment-Addressing the Gap Challenge and Identifying Suspected Biofilm in Clinical Practice. *Clin. Pract.* **2019**, *10*, 34–39.
26. Gil, S.B. Implementing the Triangle of Wound Assessment Framework to Transform the Care Pathway for Diabetic Foot Ulcers. *J Wound Care* **2020**, *29*, 363–369. [CrossRef]
27. Dowsett, C.; Bain, K.; Bain, M. Closing the Gap between the Evidence and Clinical Practice-a Consensus Report on Exudate Management. *Wounds* **2020**, *11*, 64–68.
28. Wound Assessment Tools: An Introduction to PUSH, NPUAP and Wagner. Available online: https://www.woundsource.com/blog/wound-assessment-tools-basic-introduction-push-npuap-and-wagner (accessed on 2 April 2023).
29. Moore, Z.E.; Webster, J. Dressings and Topical Agents for Preventing Pressure Ulcers. *Cochrane Database Syst. Rev.* **2018**, *12*, CD009362. [CrossRef]
30. Advanced Wound Care Market–Table of Content, List of Tables and List of Figures. Available online: https://www.factmr.com/report/advance-wound-care-market/toc (accessed on 2 April 2023).
31. Derwin, R.; Patton, D.; Avsar, P.; Strapp, H.; Moore, Z. The Impact of Topical Agents and Dressing on PH and Temperature on Wound Healing: A Systematic, Narrative Review. *Int. Wound J.* **2022**, *19*, 1397–1408. [CrossRef]
32. Barbu, A.; Neamtu, B.; Zăhan, M.; Iancu, G.M.; Bacila, C.; Mireșan, V. Current Trends in Advanced Alginate-Based Wound Dressings for Chronic Wounds. *J. Pers. Med.* **2021**, *11*, 890. [CrossRef] [PubMed]
33. de Oliveira Ruiz, P.B.; Lima, A.F.C. Average Direct Costs of Outpatient, Hospital, and Home Care Provided to Patients with Chronic Wounds. *Rev. Esc. Enferm.* **2022**, *56*, e20220295. [CrossRef]
34. Brumberg, V.; Astrelina, T.; Malivanova, T.; Samoilov, A. Modern Wound Dressings: Hydrogel Dressings. *Biomedicines* **2021**, *9*, 1235. [CrossRef]
35. Ferreira, N.N.; Ferreira, L.M.B.; Cardoso, V.M.O.; Boni, F.I.; Souza, A.L.R.; Gremião, M.P.D. Recent Advances in Smart Hydrogels for Biomedical Applications: From Self-Assembly to Functional Approaches. *Eur. Polym. J.* **2018**, *99*, 117–133. [CrossRef]
36. El-Husseiny, H.M.; Mady, E.A.; Hamabe, L.; Abugomaa, A.; Shimada, K.; Yoshida, T.; Tanaka, T.; Yokoi, A.; Elbadawy, M.; Tanaka, R. Smart/Stimuli-Responsive Hydrogels: Cutting-Edge Platforms for Tissue Engineering and Other Biomedical Applications. *Mater. Today Bio.* **2022**, *13*, 100186. [CrossRef]
37. Thakur, S.; Thakur, V.K.; Arotiba, O.A. History, Classification, Properties and Application of Hydrogels: An Overview. In *Hydrogels*; Thakur, V.K., Thakur, M.K., Eds.; Gels Horizons: From Science to Smart Materials; Springer Singapore: Singapore, 2018; pp. 29–50, ISBN 978-981-10-6076-2.
38. Hu, C.; Yang, L.; Wang, Y. Recent Advances in Smart-Responsive Hydrogels for Tissue Repairing. *MedComm.–Biomater. Appl.* **2022**, *1*, e23. [CrossRef]
39. Li, Z.; Zhou, Y.; Li, T.; Zhang, J.; Tian, H. Stimuli-Responsive Hydrogels: Fabrication and Biomedical Applications. *VIEW* **2022**, *3*, 20200112. [CrossRef]
40. Stan, D.; Tanase, C.; Avram, M.; Apetrei, R.; Mincu, N.-B.; Mateescu, A.L.; Stan, D. Wound Healing Applications of Creams and "Smart" Hydrogels. *Exp. Dermatol.* **2021**, *30*, 1218–1232. [CrossRef]
41. Parhi, R. Cross-Linked Hydrogel for Pharmaceutical Applications: A Review. *Adv. Pharm. Bull.* **2017**, *7*, 515–530. [CrossRef]
42. Bahram, M.; Mohseni, N.; Moghtader, M. An Introduction to Hydrogels and Some Recent Applications. In *Emerging Concepts in Analysis and Applications of Hydrogels*; IntechOpen: London, UK, 2016. [CrossRef]

43. Xiang, J.; Shen, L.; Hong, Y. Status and Future Scope of Hydrogels in Wound Healing: Synthesis, Materials and Evaluation. *Eur. Polym. J.* **2020**, *130*, 109609. [CrossRef]
44. Liang, Y.; He, J.; Guo, B. Functional Hydrogels as Wound Dressing to Enhance Wound Healing. *ACS Nano* **2021**, *15*, 12687–12722. [CrossRef]
45. Ghasemiyeh, P.; Mohammadi-Samani, S. Hydrogels as Drug Delivery Systems; Pros and Cons. *Trends Pharm. Sci.* **2019**, *5*, 7–24. [CrossRef]
46. Kaith, B.S.; Singh, A.; Sharma, A.K.; Sud, D. Hydrogels: Synthesis, Classification, Properties and Potential Applications—A Brief Review. *J Polym Environ.* **2021**, *29*, 3827–3841. [CrossRef]
47. Baby, D.K. Rheology of Hydrogels. In *Rheology of Polymer Blends and Nanocomposites*; Elsevier: Amsterdam, The Netherlands, 2020; pp. 193–204, ISBN 978-0-12-816957-5.
48. Bao, Z.; Xian, C.; Yuan, Q.; Liu, G.; Wu, J. Natural Polymer-Based Hydrogels with Enhanced Mechanical Performances: Preparation, Structure, and Property. *Adv. Healthc. Mater.* **2019**, *8*, 1900670. [CrossRef]
49. Li, S.; Pei, M.; Wan, T.; Yang, H.; Gu, S.; Tao, Y.; Liu, X.; Zhou, Y.; Xu, W.; Xiao, P. Self-Healing Hyaluronic Acid Hydrogels Based on Dynamic Schiff Base Linkages as Biomaterials. *Carbohydr. Polym.* **2020**, *250*, 116922. [CrossRef]
50. da Silva, C.M.; da Silva, D.L.; Modolo, L.V.; Alves, R.B.; de Resende, M.A.; Martins, C.V.B.; de Fátima, Â. Schiff Bases: A Short Review of Their Antimicrobial Activities. *J. Adv. Res.* **2011**, *2*, 1–8. [CrossRef]
51. Gavalyan, V.B. Synthesis and Characterization of New Chitosan-Based Schiff Base Compounds. *Carbohydr. Polym.* **2016**, *145*, 37–47. [CrossRef]
52. Yin, N.; Santos, T.M.A.; Auer, G.K.; Crooks, J.A.; Oliver, P.M.; Weibel, D.B. Bacterial Cellulose as a Substrate for Microbial Cell Culture. *Appl. Environ. Microbiol.* **2014**, *80*, 1926–1932. [CrossRef]
53. Oryan, A.; Kamali, A.; Moshiri, A.; Baharvand, H.; Daemi, H. Chemical Crosslinking of Biopolymeric Scaffolds: Current Knowledge and Future Directions of Crosslinked Engineered Bone Scaffolds. *Int. J. Biol. Macromol.* **2018**, *107*, 678–688. [CrossRef]
54. Sacco, P.; Furlani, F.; De Marzo, G.; Marsich, E.; Paoletti, S.; Donati, I. Concepts for Developing Physical Gels of Chitosan and of Chitosan Derivatives. *Gels* **2018**, *4*, 67. [CrossRef]
55. Xue, X.; Hu, Y.; Wang, S.; Chen, X.; Jiang, Y.; Su, J. Fabrication of Physical and Chemical Crosslinked Hydrogels for Bone Tissue Engineering. *Bioact. Mater.* **2022**, *12*, 327–339. [CrossRef]
56. Kirschning, A.; Dibbert, N.; Dräger, G. Chemical Functionalization of Polysaccharides—Towards Biocompatible Hydrogels for Biomedical Applications. *Chemistry* **2018**, *24*, 1231–1240. [CrossRef]
57. Zarzycki, R.; Modrzejewska, Z.; Nawrotek, K. Drug Release from Hydrogel Matrices. *Ecol. Chem. Eng.* **2010**, *17*, 117–136.
58. Shahinpoor, M. Conceptual Design, Kinematics and Dynamics of Swimming Robotic Structures Using Ionic Polymeric Gel Muscles. *Smart Mater. Struct.* **1992**, *1*, 91–94. [CrossRef]
59. Bordbar-Khiabani, A.; Gasik, M. Smart Hydrogels for Advanced Drug Delivery Systems. *Int. J. Mol. Sci.* **2022**, *23*, 3665. [CrossRef]
60. AliakbarAhovan, Z.; Esmaeili, Z.; Eftekhari, B.S.; Khosravimelal, S.; Alehosseini, M.; Orive, G.; Dolatshahi-Pirouz, A.; Pal Singh Chauhan, N.; Janmey, P.A.; Hashemi, A.; et al. Antibacterial Smart Hydrogels: New Hope for Infectious Wound Management. *Mater. Today Bio.* **2022**, *17*, 100499. [CrossRef]
61. Rani Raju, N.; Silina, E.; Stupin, V.; Manturova, N.; Chidambaram, S.B.; Achar, R.R. Multifunctional and Smart Wound Dressings—A Review on Recent Research Advancements in Skin Regenerative Medicine. *Pharmaceutics* **2022**, *14*, 1574. [CrossRef] [PubMed]
62. Mantha, S.; Pillai, S.; Khayambashi, P.; Upadhyay, A.; Zhang, Y.; Tao, O.; Pham, H.M.; Tran, S.D. Smart Hydrogels in Tissue Engineering and Regenerative Medicine. *Materials* **2019**, *12*, 3323. [CrossRef]
63. Chopra, H.; Kumar, S.; Singh, I. Bioadhesive Hydrogels and Their Applications. In *Bioadhesives in Drug Delivery*; John Wiley & Sons, Ltd.: Hoboken, NJ, USA, 2020; pp. 147–170, ISBN 978-1-119-64024-0.
64. Xiong, Y.; Zhang, X.; Ma, X.; Wang, W.; Yan, F.; Zhao, X.; Chu, X.; Xu, W.; Sun, C. A Review of the Properties and Applications of Bioadhesive Hydrogels. *Polym. Chem.* **2021**, *12*, 3721–3739. [CrossRef]
65. Sundriyal, P.; Pandey, M.; Bhattacharya, S. Plasma-Assisted Surface Alteration of Industrial Polymers for Improved Adhesive Bonding. *Int. J. Adhes. Adhes.* **2020**, *101*, 102626. [CrossRef]
66. Zheng, K.; Tong, Y.; Zhang, S.; He, R.; Xiao, L.; Iqbal, Z.; Zhang, Y.; Gao, J.; Zhang, L.; Jiang, L.; et al. Flexible Bicolorimetric Polyacrylamide/Chitosan Hydrogels for Smart Real-Time Monitoring and Promotion of Wound Healing. *Adv. Funct. Mater.* **2021**, *31*, 2102599. [CrossRef]
67. Jahanmir, G.; Chau, Y. Chapter 9–Mathematical Models of Drug Release from Degradable Hydrogels. In *Biomedical Applications of Nanoparticles*; Grumezescu, A.M., Ed.; William Andrew Publishing: New York, NY, USA, 2019; pp. 233–269, ISBN 978-0-12-816506-5.
68. Vasile, C.; Pamfil, D.; Stoleru, E.; Baican, M. New Developments in Medical Applications of Hybrid Hydrogels Containing Natural Polymers. *Molecules* **2020**, *25*, 1539. [CrossRef]
69. Fan, W.; Yin, J.; Yi, C.; Xia, Y.; Nie, Z.; Sui, K. Nature-Inspired Sequential Shape Transformation of Energy-Patterned Hydrogel Sheets. *ACS Appl. Mater. Interfaces* **2020**, *12*, 4878–4886. [CrossRef]
70. Li, Y.; Wang, X.; Fu, Y.; Wei, Y.; Zhao, L.; Tao, L. Self-Adapting Hydrogel to Improve the Therapeutic Effect in Wound-Healing. *ACS Appl. Mater. Interfaces* **2018**, *10*, 26046–26055. [CrossRef]
71. Wang, Y.; Xie, R.; Li, Q.; Dai, F.; Lan, G.; Shang, S.; Lu, F. A Self-Adapting Hydrogel Based on Chitosan/Oxidized KonjacGlucomannan/AgNPs for Repairing Irregular Wounds. *Biomater. Sci.* **2020**, *8*, 1910–1922. [CrossRef]

72. Le, T.M.D.; Duong, H.T.T.; Thambi, T.; Giang Phan, V.H.; Jeong, J.H.; Lee, D.S. Bioinspired PH- and Temperature-Responsive Injectable Adhesive Hydrogels with Polyplexes Promotes Skin Wound Healing. *Biomacromolecules* **2018**, *19*, 3536–3548. [CrossRef] [PubMed]
73. Pham, L.; Dang, L.H.; Truong, M.D.; Nguyen, T.H.; Le, L.; Le, V.T.; Nam, N.D.; Bach, L.G.; Nguyen, V.T.; Tran, N.Q. A Dual Synergistic of Curcumin and Gelatin on Thermal-Responsive Hydrogel Based on Chitosan-P123 in Wound Healing Application. *Biomed. Pharmacother.* **2019**, *117*, 109183. [CrossRef]
74. Wang, J.; Zhao, B.; Sun, L.; Jiang, L.; Li, Q.; Jin, P. Smart Thermosensitive Poloxamer Hydrogels Loaded with Nr-CWs for the Treatment of Diabetic Wounds. *PLoS ONE* **2022**, *17*, e0279727. [CrossRef]
75. Paneysar, J.S.; Barton, S.; Ambre, P.; Coutinho, E. Novel Temperature Responsive Films Impregnated with Silver Nano Particles (Ag-NPs) as Potential Dressings for Wounds. *J. Pharm. Sci.* **2022**, *111*, 810–817. [CrossRef]
76. Zhou, L.; Pi, W.; Cheng, S.; Gu, Z.; Zhang, K.; Min, T.; Zhang, W.; Du, H.; Zhang, P.; Wen, Y. Multifunctional DNA Hydrogels with Hydrocolloid-Cotton Structure for Regeneration of Diabetic Infectious Wounds. *Adv. Funct. Mater.* **2021**, *31*, 2106167. [CrossRef]
77. Si, Y.; Wang, L.; Wang, X.; Tang, N.; Yu, J.; Ding, B. Ultrahigh-Water-Content, Superelastic, and Shape-Memory Nanofiber-Assembled Hydrogels Exhibiting Pressure-Responsive Conductivity. *Adv. Mater.* **2017**, *29*, 1700339. [CrossRef] [PubMed]
78. Li, D.; Fei, X.; Xu, L.; Wang, Y.; Tian, J.; Li, Y. Pressure-Sensitive Antibacterial Hydrogel Dressing for Wound Monitoring in Bed Ridden Patients. *J. Colloid Interface Sci.* **2022**, *627*, 942–955. [CrossRef]
79. de Alencar Fonseca Santos, J.; Campelo, M.B.D.; de Oliveira, R.A.; Nicolau, R.A.; Rezende, V.E.A.; Arisawa, E.Â.L. Effects of Low-Power Light Therapy on the Tissue Repair Process of Chronic Wounds in Diabetic Feet. *Photomed. Laser Surg.* **2018**, *36*, 298–304. [CrossRef]
80. Lipovsky, A.; Nitzan, Y.; Lubart, R. A Possible Mechanism for Visible Light-Induced Wound Healing. *Lasers Surg. Med.* **2008**, *40*, 509–514. [CrossRef]
81. Maiz-Fernández, S.; Pérez-Álvarez, L.; Ruiz-Rubio, L.; Vilas-Vilela, J.L.; Lanceros-Méndez, S. Chapter Twelve–Multifunctional Materials Based on Smart Hydrogels for Biomedical and 4D Applications. In *Advanced Lightweight Multifunctional Materials*; Costa, P., Costa, C.M., Lanceros-Mendez, S., Eds.; Woodhead Publishing in Materials; Woodhead Publishing: Thorston, UK, 2021; pp. 407–467, ISBN 978-0-12-818501-8.
82. Yang, N.; Zhu, M.; Xu, G.; Liu, N.; Yu, C. A Near-Infrared Light-Responsive Multifunctional Nanocomposite Hydrogel for Efficient and Synergistic Antibacterial Wound Therapy and Healing Promotion. *J. Mater. Chem. B* **2020**, *8*, 3908–3917. [CrossRef]
83. Gupta, A.; Kowalczuk, M.; Heaselgrave, W.; Britland, S.T.; Martin, C.; Radecka, I. The Production and Application of Hydrogels for Wound Management: A Review. *Eur. Polym. J.* **2019**, *111*, 134–151. [CrossRef]
84. Arenbergerova, M.; Arenberger, P.; Bednar, M.; Kubat, P.; Mosinger, J. Light-Activated Nanofibre Textiles Exert Antibacterial Effects in the Setting of Chronic Wound Healing. *Exp. Dermatol.* **2012**, *21*, 619–624. [CrossRef] [PubMed]
85. Mai, B.; Jia, M.; Liu, S.; Sheng, Z.; Li, M.; Gao, Y.; Wang, X.; Liu, Q.; Wang, P. Smart Hydrogel-Based DVDMS/BFGF Nanohybrids for Antibacterial Phototherapy with Multiple Damaging Sites and Accelerated Wound Healing. *ACS Appl. Mater. Interfaces* **2020**, *12*, 10156–10169. [CrossRef] [PubMed]
86. Lv, H.; Liu, J.; Zhen, C.; Wang, Y.; Wei, Y.; Ren, W.; Shang, P. Magnetic Fields as a Potential Therapy for Diabetic Wounds Based on Animal Experiments and Clinical Trials. *Cell Prolif.* **2021**, *54*, e12982. [CrossRef] [PubMed]
87. Ekici, Y.; Aydogan, C.; Balcik, C.; Haberal, N.; Kirnap, M.; Moray, G.; Haberal, M. Effect of Static Magnetic Field on Experimental Dermal Wound Strength. *Indian J. Plast. Surg.* **2012**, *45*, 215–219. [CrossRef]
88. Nezami, S.; Sadeghi, M.; Mohajerani, H. A Novel PH-Sensitive and Magnetic Starch-Based Nanocomposite Hydrogel as a Controlled Drug Delivery System for Wound Healing. *Polym. Degrad. Stab.* **2020**, *179*, 109255. [CrossRef]
89. Yang, X.; Zhang, C.; Deng, D.; Gu, Y.; Wang, H.; Zhong, Q. Multiple Stimuli-Responsive MXene-Based Hydrogel as Intelligent Drug Delivery Carriers for Deep Chronic Wound Healing. *Small* **2022**, *18*, 2104368. [CrossRef]
90. Xu, J.; Tsai, Y.-L.; Hsu, S. Design Strategies of Conductive Hydrogel for Biomedical Applications. *Molecules* **2020**, *25*, 5296. [CrossRef]
91. Verdes, M.; Mace, K.; Margetts, L.; Cartmell, S. Status and Challenges of Electrical Stimulation Use in Chronic Wound Healing. *Curr. Opin. Biotechnol.* **2022**, *75*, 102710. [CrossRef]
92. Tai, G.; Tai, M.; Zhao, M. Electrically Stimulated Cell Migration and Its Contribution to Wound Healing. *Burns Trauma.* **2018**, *6*, 20. [CrossRef]
93. Deng, Z.; Yu, R.; Guo, B. Stimuli-Responsive Conductive Hydrogels: Design, Properties, and Applications. *Mater. Chem. Front.* **2021**, *5*, 2092–2123. [CrossRef]
94. Liang, Y.; Zhao, X.; Hu, T.; Chen, B.; Yin, Z.; Ma, P.X.; Guo, B. Adhesive Hemostatic Conducting Injectable Composite Hydrogels with Sustained Drug Release and Photothermal Antibacterial Activity to Promote Full-Thickness Skin Regeneration during Wound Healing. *Small Weinh. Bergstr. Ger.* **2019**, *15*, e1900046. [CrossRef]
95. Zhang, S.; Ge, G.; Qin, Y.; Li, W.; Dong, J.; Mei, J.; Ma, R.; Zhang, X.; Bai, J.; Zhu, C.; et al. Recent Advances in Responsive Hydrogels for Diabetic Wound Healing. *Mater. Today Bio* **2023**, *18*, 100508. [CrossRef]
96. Wang, L.; Zhou, M.; Xu, T.; Zhang, X. Multifunctional Hydrogel as Wound Dressing for Intelligent Wound Monitoring. *Chem. Eng. J.* **2022**, *433*, 134625. [CrossRef]

97. Ninan, N.; Forget, A.; Shastri, V.P.; Voelcker, N.H.; Blencowe, A. Antibacterial and Anti-Inflammatory PH-Responsive Tannic Acid-Carboxylated Agarose Composite Hydrogels for Wound Healing. *ACS Appl. Mater. Interfaces* **2016**, *8*, 28511–28521. [CrossRef]
98. Shi, W.; Kong, Y.; Su, Y.; Kuss, M.A.; Jiang, X.; Li, X.; Xie, J.; Duan, B. Tannic Acid-Inspired, Self-Healing, and Dual Stimuli Responsive Dynamic Hydrogel with Potent Antibacterial and Anti-Oxidative Properties. *J. Mater. Chem. B* **2021**, *9*, 7182–7195. [CrossRef]
99. Zhou, Z.; Zhang, X.; Xu, L.; Lu, H.; Chen, Y.; Wu, C.; Hu, P. A Self-Healing Hydrogel Based on Crosslinked Hyaluronic Acid and Chitosan to Facilitate Diabetic Wound Healing. *Int. J. Biol. Macromol.* **2022**, *220*, 326–336. [CrossRef]
100. Li, L.; Lei, D.; Zhang, J.; Xu, L.; Li, J.; Jin, L.; Pan, L. Dual-Responsive Alginate Hydrogel Constructed by Sulfhdryl Dendrimer as an Intelligent System for Drug Delivery. *Molecules* **2022**, *27*, 281. [CrossRef]
101. Zhang, J.; Hurren, C.; Lu, Z.; Wang, D. PH-Sensitive Alginate Hydrogel for Synergistic Anti-Infection. *Int. J. Biol. Macromol.* **2022**, *222*, 1723–1733. [CrossRef]
102. Qiao, B.; Pang, Q.; Yuan, P.; Luo, Y.; Ma, L. Smart Wound Dressing for Infection Monitoring and NIR-Triggered Antibacterial Treatment. *Biomater. Sci.* **2020**, *8*, 1649–1657. [CrossRef]
103. Guo, P.; Liang, J.; Li, Y.; Lu, X.; Fu, H.; Jing, H.; Guan, S.; Han, D.; Niu, L. High-Strength and PH-Responsive Self-Healing Polyvinyl Alcohol/Poly 6-Acrylamidohexanoic Acid Hydrogel Based on Dual Physically Cross-Linked Network. *Colloids Surf. A Physicochem. Eng. Asp.* **2019**, *571*, 64–71. [CrossRef]
104. Lanzalaco, S.; Mingot, J.; Torras, J.; Alemán, C.; Armelin, E. Recent Advances in Poly(N-isopropylacrylamide) Hydrogels and Derivatives as Promising Materials for Biomedical and Engineering Emerging Applications. *Adv. Eng. Mater.* **2023**, *25*, 2201303. [CrossRef]
105. Han, X.; Yang, R.; Wan, X.; Dou, J.; Yuan, J.; Chi, B.; Shen, J. Antioxidant and Multi-Sensitive PNIPAAm/Keratin Double Network Gels for Self-Stripping Wound Dressing Application. *J. Mater. Chem. B* **2021**, *9*, 6212–6225. [CrossRef]
106. Qi, X.; Tong, X.; You, S.; Mao, R.; Cai, E.; Pan, W.; Zhang, C.; Hu, R.; Shen, J. Mild Hyperthermia-Assisted ROS Scavenging Hydrogels Achieve Diabetic Wound Healing. *ACS Macro Lett.* **2022**, *11*, 861–867. [CrossRef]
107. Gao, F.; Xiong, Z. Reactive Oxygen Species Responsive Polymers for Drug Delivery Systems. *Front. Chem.* **2021**, *9*. [CrossRef]
108. Yu, J.; Zhang, R.; Chen, B.; Liu, X.; Jia, Q.; Wang, X.; Yang, Z.; Ning, P.; Wang, Z.; Yang, Y. Injectable Reactive Oxygen Species-Responsive Hydrogel Dressing with Sustained Nitric Oxide Release for Bacterial Ablation and Wound Healing. *Adv. Funct. Mater.* **2022**, *32*, 2202857. [CrossRef]
109. Hu, C.; Zhang, F.; Long, L.; Kong, Q.; Luo, R.; Wang, Y. Dual-Responsive Injectable Hydrogels Encapsulating Drug-Loaded Micelles for on-Demand Antimicrobial Activity and Accelerated Wound Healing. *J. Control. Release* **2020**, *324*, 204–217. [CrossRef]
110. Qiu, X.; Zhang, J.; Cao, L.; Jiao, Q.; Zhou, J.; Yang, L.; Zhang, H.; Wei, Y. Antifouling Antioxidant Zwitterionic Dextran Hydrogels as Wound Dressing Materials with Excellent Healing Activities. *ACS Appl. Mater. Interfaces.* **2021**, *13*, 7060–7069. [CrossRef]
111. Bas, Y.; Sanyal, R.; Sanyal, A. Hyaluronic-Acid Based Redox-Responsive Hydrogels Using the Diels-Alder Reaction for on-Demand Release of Biomacromolecules. *J. Macromol. Sci. Part A* **2023**, *60*, 246–254. [CrossRef]
112. Webber, M.J.; Anderson, D.G. Smart Approaches to Glucose-Responsive Drug Delivery. *J. Drug Target* **2015**, *23*, 651–655. [CrossRef]
113. Xu, Z.; Liu, G.; Huang, J.; Wu, J. Novel Glucose-Responsive Antioxidant Hybrid Hydrogel for Enhanced Diabetic Wound Repair. *ACS Appl. Mater. Interfaces* **2022**, *14*, 7680–7689. [CrossRef] [PubMed]
114. Zhou, W.; Duan, Z.; Zhao, J.; Fu, R.; Zhu, C.; Fan, D. Glucose and MMP-9 Dual-Responsive Hydrogel with Temperature Sensitive Self-Adaptive Shape and Controlled Drug Release Accelerates Diabetic Wound Healing. *Bioact. Mater.* **2022**, *17*, 1–17. [CrossRef] [PubMed]
115. Yang, J.; Zeng, W.; Xu, P.; Fu, X.; Yu, X.; Chen, L.; Leng, F.; Yu, C.; Yang, Z. Glucose-Responsive Multifunctional Metal–Organic Drug-Loaded Hydrogel for Diabetic Wound Healing. *Acta Biomater.* **2022**, *140*, 206–218. [CrossRef] [PubMed]
116. Gardikiotis, I.; Cojocaru, F.-D.; Mihai, C.-T.; Balan, V.; Dodi, G. Borrowing the Features of Biopolymers for Emerging Wound Healing Dressings: A Review. *Int. J. Mol. Sci.* **2022**, *23*, 8778. [CrossRef]
117. Naomi, R.; Bahari, H.; Ridzuan, P.M.; Othman, F. Natural-Based Biomaterial for Skin Wound Healing (Gelatin vs. Collagen): Expert Review. *Polymers* **2021**, *13*, 2319. [CrossRef]
118. Chen, J.; He, J.; Yang, Y.; Qiao, L.; Hu, J.; Zhang, J.; Guo, B. Antibacterial Adhesive Self-Healing Hydrogels to Promote Diabetic Wound Healing. *Acta Biomater.* **2022**, *146*, 119–130. [CrossRef]
119. Vahedi, M.; Barzin, J.; Shokrolahi, F.; Shokrollahi, P. Self-Healing, Injectable Gelatin Hydrogels Cross-Linked by Dynamic Schiff Base Linkages Support Cell Adhesion and Sustained Release of Antibacterial Drugs. *Macromol. Mater. Eng.* **2018**, *303*, 1800200. [CrossRef]
120. Gutierrez-Reyes, J.E.; Caldera-Villalobos, M.; Claudio-Rizo, J.A.; Cabrera-Munguía, D.A.; Becerra-Rodriguez, J.J.; Soriano-Corral, F.; Herrera-Guerrero, A. Smart Collagen/Xanthan Gum-Based Hydrogels with Antibacterial Effect, Drug Release Capacity and Excellent Performance in Vitro Bioactivity for Wound Healing Application. *Biomed. Mater.* **2023**, *18*, 035011. [CrossRef]
121. Hu, C.; Liu, W.; Long, L.; Wang, Z.; Yuan, Y.; Zhang, W.; He, S.; Wang, J.; Yang, L.; Lu, L.; et al. Microenvironment-Responsive Multifunctional Hydrogels with Spatiotemporal Sequential Release of Tailored Recombinant Human Collagen Type III for the Rapid Repair of Infected Chronic Diabetic Wounds. *J. Mater. Chem. B* **2021**, *9*, 9684–9699. [CrossRef]

122. Yu, X.; Cheng, C.; Peng, X.; Zhang, K.; Yu, X. A Self-Healing and Injectable Oxidized Quaternized Guar Gum/Carboxymethyl Chitosan Hydrogel with Efficient Hemostatic and Antibacterial Properties for Wound Dressing. *Colloids Surf. B Biointerfaces* **2022**, *209*, 112207. [CrossRef]
123. Almine, J.F.; Wise, S.G.; Weiss, A.S. Elastin Signaling in Wound Repair. Birth Defects Res. *Part C Embryo Today Rev.* **2012**, *96*, 248–257. [CrossRef]
124. Kawabata, S.; Kanda, N.; Hirasawa, Y.; Noda, K.; Matsuura, Y.; Suzuki, S.; Kawai, K. The Utility of Silk-Elastin Hydrogel as a New Material for Wound Healing. *Plast. Reconstr. Surg.–Glob. Open* **2018**, *6*, e1778. [CrossRef] [PubMed]
125. Stojic, M.; Ródenas-Rochina, J.; López-Donaire, M.L.; González De Torre, I.; González Pérez, M.; Rodríguez-Cabello, J.C.; Vojtová, L.; Jorcano, J.L.; Velasco, D. Elastin-Plasma Hybrid Hydrogels for Skin Tissue Engineering. *Polymers* **2021**, *13*, 2114. [CrossRef] [PubMed]
126. Zhang, Y.; Desai, M.S.; Wang, T.; Lee, S.-W. Elastin-Based Thermoresponsive Shape-Memory Hydrogels. *Biomacromolecules* **2020**, *21*, 1149–1156. [CrossRef] [PubMed]
127. Zhang, S.; Shah, S.A.-M.; Basharat, K.; Qamar, S.A.; Raza, A.; Mohamed, A.; Bilal, M.; Iqbal, H.M.N. Silk-Based Nano-Hydrogels for Futuristic Biomedical Applications. *J. Drug Deliv. Sci. Technol.* **2022**, *72*, 103385. [CrossRef]
128. Mazurek, Ł.; Szudzik, M.; Rybka, M.; Konop, M. Silk Fibroin Biomaterials and Their Beneficial Role in Skin Wound Healing. *Biomolecules* **2022**, *12*, 1852. [CrossRef]
129. Zheng, H.; Zuo, B. Functional Silk Fibroin Hydrogels: Preparation, Properties and Applications. *J. Mater. Chem. B* **2021**, *9*, 1238–1258. [CrossRef]
130. Römer, L.; Scheibel, T. The Elaborate Structure of Spider Silk: Structure and Function of a Natural High Performance Fiber. *Prion* **2008**, *2*, 154–161. [CrossRef]
131. Chang, C.; Zhang, L. Cellulose-Based Hydrogels: Present Status and Application Prospects. *Carbohydr. Polym.* **2011**, *84*, 40–53. [CrossRef]
132. Lin, F.; Zheng, J.; Guo, W.; Zhu, Z.; Wang, Z.; Dong, B.; Lin, C.; Huang, B.; Lu, B. Smart Cellulose-Derived Magnetic Hydrogel with Rapid Swelling and Deswelling Properties for Remotely Controlled Drug Release. *Cellulose* **2019**, *26*, 6861–6877. [CrossRef]
133. SCOGS (Select Committee on GRAS Substances). Available online: https://www.cfsanappsexternal.fda.gov/scripts/fdcc/?set=SCOGS&sort=Sortsubstance&order=ASC&startrow=1&type=basic&search=ALGINATE (accessed on 3 April 2023).
134. Inactive Ingredient Search for Approved Drug Products. Available online: https://www.accessdata.fda.gov/scripts/cder/iig/index.cfm?event=BasicSearch.page (accessed on 3 April 2023).
135. Soeiro, V.C.; Melo, K.R.T.; Alves, M.G.C.F.; Medeiros, M.J.C.; Grilo, M.L.P.M.; Almeida-Lima, J.; Pontes, D.L.; Costa, L.S.; Rocha, H.A.O. Dextran: Influence of Molecular Weight in Antioxidant Properties and Immunomodulatory Potential. *Int. J. Mol. Sci.* **2016**, *17*, 1340. [CrossRef]
136. Luanda, A.; Badalamoole, V. Past, Present and Future of Biomedical Applications of Dextran-Based Hydrogels: A Review. *Int. J. Biol. Macromol.* **2023**, *228*, 794–807. [CrossRef]
137. Sun, G.; Zhang, X.; Shen, Y.-I.; Sebastian, R.; Dickinson, L.E.; Fox-Talbot, K.; Reinblatt, M.; Steenbergen, C.; Harmon, J.W.; Gerecht, S. Dextran Hydrogel Scaffolds Enhance Angiogenic Responses and Promote Complete Skin Regeneration during Burn Wound Healing. *Proc. Natl. Acad. Sci. USA* **2011**, *108*, 20976–20981. [CrossRef]
138. Zhu, Q.; Jiang, M.; Liu, Q.; Yan, S.; Feng, L.; Lan, Y.; Shan, G.; Xue, W.; Guo, R. Enhanced Healing Activity of Burn Wound Infection by a Dextran-HA Hydrogel Enriched with Sanguinarine. *Biomater. Sci.* **2018**, *6*, 2472–2486. [CrossRef] [PubMed]
139. Zhou, L.; Zhou, L.; Wei, C.; Guo, R. A Bioactive Dextran-Based Hydrogel Promote the Healing of Infected Wounds via Antibacterial and Immunomodulatory. *Carbohydr. Polym.* **2022**, *291*, 119558. [CrossRef] [PubMed]
140. Lin, S.P.; Kung, H.N.; Tsai, Y.S.; Tseng, T.N.; Hsu, K.D.; Cheng, K.C. Novel Dextran Modified Bacterial Cellulose Hydrogel Accelerating Cutaneous Wound Healing. *Cellulose* **2017**, *24*, 4927–4937. [CrossRef]
141. Zhao, X.; Li, P.; Guo, B.; Ma, P.X. Antibacterial and Conductive Injectable Hydrogels Based on Quaternized Chitosan-Graft-Polyaniline/Oxidized Dextran for Tissue Engineering. *Acta Biomater.* **2015**, *26*, 236–248. [CrossRef]
142. Zhang, M.; Chen, G.; Lei, M.; Lei, J.; Li, D.; Zheng, H. A PH-Sensitive Oxidized-Dextran Based Double Drug-Loaded Hydrogel with High Antibacterial Properties. *Int. J. Biol. Macromol.* **2021**, *182*, 385–393. [CrossRef]
143. Wu, S.; Yang, Y.; Wang, S.; Dong, C.; Zhang, X.; Zhang, R.; Yang, L. Dextran and Peptide-Based PH-Sensitive Hydrogel Boosts Healing Process in Multidrug-Resistant Bacteria-Infected Wounds. *Carbohydr. Polym.* **2022**, *278*, 118994. [CrossRef]
144. Guo, C.; Wu, Y.; Li, W.; Wang, Y.; Kong, Q. Development of a Microenvironment-Responsive Hydrogel Promoting Chronically Infected Diabetic Wound Healing through Sequential Hemostatic, Antibacterial, and Angiogenic Activities. *ACS Appl. Mater. Interfaces* **2022**, *14*, 30480–30492. [CrossRef]
145. Crini, G. Historical Review on Chitin and Chitosan Biopolymers. *Environ. Chem. Lett.* **2019**, *17*, 1623–1643. [CrossRef]
146. Kou, S.G.; Peters, L.; Mucalo, M. Chitosan: A Review of Molecular Structure, Bioactivities and Interactions with the Human Body and Micro-Organisms. *Carbohydr. Polym.* **2022**, *282*, 119132. [CrossRef]
147. Li, Y.; Wang, X.; Wei, Y.; Tao, L. Chitosan-Based Self-Healing Hydrogel for Bioapplications. *Chin. Chem. Lett.* **2017**, *28*, 2053–2057. [CrossRef]
148. Iacob, A.-T.; Drăgan, M.; Ghețu, N.; Pieptu, D.; Vasile, C.; Buron, F.; Routier, S.; Giusca, S.E.; Caruntu, I.-D.; Profire, L. Preparation, Characterization and Wound Healing Effects of New Membranes Based on Chitosan, Hyaluronic Acid and Arginine Derivatives. *Polymers* **2018**, *10*, 607. [CrossRef]

149. Azad, A.K.; Sermsintham, N.; Chandrkrachang, S.; Stevens, W.F. Chitosan Membrane as a Wound-Healing Dressing: Characterization and Clinical Application. *J. Biomed. Mater. Res. Part B Appl. Biomater.* **2004**, *69*, 216–222. [CrossRef]
150. Ong, S.-Y.; Wu, J.; Moochhala, S.M.; Tan, M.-H.; Lu, J. Development of a Chitosan-Based Wound Dressing with Improved Hemostatic and Antimicrobial Properties. *Biomaterials* **2008**, *29*, 4323–4332. [CrossRef]
151. Radwan-Pragłowska, J.; Piątkowski, M.; Deineka, V.; Janus, Ł.; Korniienko, V.; Husak, E.; Holubnycha, V.; Liubchak, I.; Zhurba, V.; Sierakowska, A.; et al. Chitosan-Based Bioactive Hemostatic Agents with Antibacterial Properties—Synthesis and Characterization. *Molecules* **2019**, *24*, 2629. [CrossRef]
152. Moeini, A.; Pedram, P.; Makvandi, P.; Malinconico, M.; Gomez d'Ayala, G. Wound Healing and Antimicrobial Effect of Active Secondary Metabolites in Chitosan-Based Wound Dressings: A Review. *Carbohydr. Polym.* **2020**, *233*, 115839. [CrossRef]
153. Guo, X.; Sun, T.; Zhong, R.; Ma, L.; You, C.; Tian, M.; Li, H.; Wang, C. Effects of Chitosan Oligosaccharides on Human Blood Components. *Front. Pharmacol.* **2018**, *9*, 1412. [CrossRef]
154. AbdEl-Hack, M.E.; El-Saadony, M.T.; Shafi, M.E.; Zabermawi, N.M.; Arif, M.; Batiha, G.E.; Khafaga, A.F.; Abd El-Hakim, Y.M.; Al-Sagheer, A.A. Antimicrobial and Antioxidant Properties of Chitosan and Its Derivatives and Their Applications: A Review. *Int. J. Biol. Macromol.* **2020**, *164*, 2726–2744. [CrossRef]
155. Dai, T.; Tanaka, M.; Huang, Y.-Y.; Hamblin, M.R. Chitosan Preparations for Wounds and Burns: Antimicrobial and Wound-Healing Effects. Expert Rev. *Anti Infect. Ther.* **2011**, *9*, 857–879. [CrossRef] [PubMed]
156. Cao, R.; Yu, H.; Long, H.; Zhang, H.; Hao, C.; Shi, L.; Du, Y.; Jiao, S.; Guo, A.; Ma, L.; et al. Low Deacetylation Degree Chitosan Oligosaccharide Protects against IL-1β Induced Inflammation and Enhances Autophagy Activity in Human Chondrocytes. *J. Biomater. Sci. Polym. Ed.* **2022**, *33*, 517–531. [CrossRef]
157. Satitsri, S.; Muanprasat, C. Chitin and Chitosan Derivatives as Biomaterial Resources for Biological and Biomedical Applications. *Molecules* **2020**, *25*, 5961. [CrossRef] [PubMed]
158. Deng, L.; Wang, B.; Li, W.; Han, Z.; Chen, S.; Wang, H. Bacterial Cellulose Reinforced Chitosan-Based Hydrogel with Highly Efficient Self-Healing and Enhanced Antibacterial Activity for Wound Healing. *Int. J. Biol. Macromol.* **2022**, *217*, 77–87. [CrossRef]
159. Li, H.; Wei, X.; Yi, X.; Tang, S.; He, J.; Huang, Y.; Cheng, F. Antibacterial, Hemostasis, Adhesive, Self-Healing Polysaccharides-Based Composite Hydrogel Wound Dressing for the Prevention and Treatment of Postoperative Adhesion. *Mater. Sci. Eng. C* **2021**, *123*, 111978. [CrossRef]
160. Huang, Y.; Mu, L.; Zhao, X.; Han, Y.; Guo, B. Bacterial Growth-Induced Tobramycin Smart Release Self-Healing Hydrogel for Pseudomonas Aeruginosa -Infected Burn Wound Healing. *ACS Nano* **2022**, *16*, 13022–13036. [CrossRef] [PubMed]
161. Tatarusanu, S.-M.; Sava, A.; Profire, B.-S.; Pinteala, T.; Jitareanu, A.; Iacob, A.-T.; Lupascu, F.; Simionescu, N.; Rosca, I.; Profire, L. New Smart Bioactive and Biomimetic Chitosan-Based Hydrogels for Wounds Care Management. *Pharmaceutics* **2023**, *15*, 975. [CrossRef] [PubMed]
162. Castillo-Henríquez, L.; Castro-Alpízar, J.; Lopretti-Correa, M.; Vega-Baudrit, J. Exploration of Bioengineered Scaffolds Composed of Thermo-Responsive Polymers for Drug Delivery in Wound Healing. *Int. J. Mol. Sci.* **2021**, *22*, 1408. [CrossRef]
163. Rusu, A.G.; Chiriac, A.P.; Nita, L.E.; Rosca, I.; Pinteala, M.; Mititelu-Tartau, L. Chitosan Derivatives in Macromolecular Co-Assembly Nanogels with Potential for Biomedical Applications. *Biomacromolecules* **2020**, *21*, 4231–4243. [CrossRef]
164. Lee, K.Y.; Mooney, D.J. Alginate: Properties and Biomedical Applications. *Prog. Polym. Sci.* **2012**, *37*, 106–126. [CrossRef]
165. Roquero, D.M.; Katz, E. "Smart" Alginate Hydrogels in Biosensing, Bioactuation and Biocomputing: State-of-the-Art and Perspectives. *Sens. Actuators Rep.* **2022**, *4*, 100095. [CrossRef]
166. Ueno, M.; Oda, T. Biological Activities of Alginate. *Adv. Food Nutr. Res.* **2014**, *72*, 95–112. [CrossRef] [PubMed]
167. Xing, M.; Cao, Q.; Wang, Y.; Xiao, H.; Zhao, J.; Zhang, Q.; Ji, A.; Song, S. Advances in Research on the Bioactivity of Alginate Oligosaccharides. *Mar. Drugs.* **2020**, *18*, 144. [CrossRef] [PubMed]
168. Kim, N.-G.; Kim, S.-C.; Kim, T.-H.; Je, J.-Y.; Lee, B.; Lee, S.G.; Kim, Y.-M.; Kang, H.W.; Qian, Z.-J.; Kim, N.; et al. Ishophloroglucin A-Based Multifunctional Oxidized Alginate/Gelatin Hydrogel for Accelerating Wound Healing. *Int. J. Biol. Macromol.* **2023**, *245*, 125484. [CrossRef] [PubMed]
169. Liu, L.; Li, X.; Nagao, M.; Elias, A.; Narain, R.; Chung, H.-J. A PH-Indicating Colorimetric Tough Hydrogel Patch towards Applications in a Substrate for Smart Wound Dressings. *Polymers* **2017**, *9*, 558. [CrossRef] [PubMed]
170. Chou, H.-Y.; Weng, C.-C.; Lai, J.-Y.; Lin, S.-Y.; Tsai, H.-C. Design of an Interpenetrating Polymeric Network Hydrogel Made of Calcium-Alginate from a Thermos-Sensitive Pluronic Template as a Thermal-Ionic Reversible Wound Dressing. *Polymers* **2020**, *12*, 2138. [CrossRef]
171. Yao, Z.; Qian, Y.; Jin, Y.; Wang, S.; Li, J.; Yuan, W.-E.; Fan, C. Biomimetic Multilayer Polycaprolactone/Sodium Alginate Hydrogel Scaffolds Loaded with Melatonin Facilitate Tendon Regeneration. *Carbohydr. Polym.* **2022**, *277*, 118865. [CrossRef]
172. Xu, K.; Deng, S.; Zhu, Y.; Yang, W.; Chen, W.; Huang, L.; Zhang, C.; Li, M.; Ao, L.; Jiang, Y.; et al. Platelet Rich Plasma Loaded Multifunctional Hydrogel Accelerates Diabetic Wound Healing via Regulating the Continuously Abnormal Microenvironments. *Adv. Healthc. Mater.* 2023, *online ahead of print*. [CrossRef]
173. Fallacara, A.; Baldini, E.; Manfredini, S.; Vertuani, S. Hyaluronic Acid in the Third Millennium. *Polymers.* **2018**, *10*, 701. [CrossRef]
174. Ke, C.; Sun, L.; Qiao, D.; Wang, D.; Zeng, X. Antioxidant Acitivity of Low Molecular Weight Hyaluronic Acid. *Food Chem. Toxicol. Int. J. Publ. Br. Ind. Biol. Res. Assoc.* **2011**, *49*, 2670–2675. [CrossRef]
175. Litwiniuk, M.; Krejner, A.; Speyrer, M.S.; Gauto, A.R.; Grzela, T. Hyaluronic Acid in Inflammation and Tissue Regeneration. *Wounds Compend. Clin. Res. Pract.* **2016**, *28*, 78–88.

176. Yang, H.; Song, L.; Zou, Y.; Sun, D.; Wang, L.; Yu, Z.; Guo, J. Role of Hyaluronic Acids and Potential as Regenerative Biomaterials in Wound Healing. *ACS Appl. Bio Mater.* **2021**, *4*, 311–324. [CrossRef] [PubMed]
177. de Paiva, W.K.V.; de Medeiros, W.R.D.B.; de Assis, C.F.; Dos Santos, E.S.; de Sousa Júnior, F.C. Physicochemical Characterization and in Vitro Antioxidant Activity of Hyaluronic Acid Produced by *Streptococcus zooepidemicus* CCT 7546. *Prep. Biochem. Biotechnol.* **2022**, *52*, 234–243. [CrossRef] [PubMed]
178. Hussain, Z.; Thu, H.E.; Katas, H.; Bukhari, S.N.A. Hyaluronic Acid-Based Biomaterials: A Versatile and Smart Approach to Tissue Regeneration and Treating Traumatic, Surgical, and Chronic Wounds. *Polym. Rev.* **2017**, *57*, 594–630. [CrossRef]
179. Ebid, R.; Lichtnekert, J.; Anders, H.-J. Hyaluronan Is Not a Ligand but a Regulator of Toll-Like Receptor Signaling in Mesangial Cells: Role of Extracellular Matrix in Innate Immunity. *ISRN Nephrol.* **2014**, *2014*, 714081. [CrossRef]
180. Ding, Y.-W.; Wang, Z.-Y.; Ren, Z.-W.; Zhang, X.-W.; Wei, D.-X. Advances in Modified Hyaluronic Acid-Based Hydrogels for Skin Wound Healing. *Biomater. Sci.* **2022**, *10*, 3393–3409. [CrossRef]
181. Makvandi, P.; Caccavale, C.; Della Sala, F.; Zeppetelli, S.; Veneziano, R.; Borzacchiello, A. Natural Formulations Provide Antioxidant Complement to Hyaluronic Acid-Based Topical Applications Used in Wound Healing. *Polymers* **2020**, *12*, 1847. [CrossRef]
182. Nyman, E.; Henricson, J.; Ghafouri, B.; Anderson, C.D.; Kratz, G. Hyaluronic Acid Accelerates Re-Epithelialization and Alters Protein Expression in a Human Wound Model. *Plast. Reconstr. Surg. Glob. Open* **2019**, *7*, e2221. [CrossRef]
183. Guan, S.; Li, Y.; Cheng, C.; Gao, X.; Gu, X.; Han, X.; Ye, H. Manufacture of PH- and HAase-Responsive Hydrogels with on-Demand and Continuous Antibacterial Activity for Full-Thickness Wound Healing. *Int. J. Biol. Macromol.* **2020**, *164*, 2418–2431. [CrossRef]
184. Deng, M.; Wu, Y.; Ren, Y.; Song, H.; Zheng, L.; Lin, G.; Wen, X.; Tao, Y.; Kong, Q.; Wang, Y. Clickable and Smart Drug Delivery Vehicles Accelerate the Healing of Infected Diabetic Wounds. *J. Control. Release* **2022**, *350*, 613–629. [CrossRef]
185. Das, S.; Das, D. Rational Design of Peptide-Based Smart Hydrogels for Therapeutic Applications. *Front. Chem.* **2021**, *9*. [CrossRef]
186. Guan, T.; Li, J.; Chen, C.; Liu, Y. Self-Assembling Peptide-Based Hydrogels for Wound Tissue Repair. *Adv. Sci.* **2022**, *9*, 2104165. [CrossRef] [PubMed]
187. Chen, J.; Zou, X. Self-Assemble Peptide Biomaterials and Their Biomedical Applications. *Bioact. Mater.* **2019**, *4*, 120–131. [CrossRef] [PubMed]
188. Ye, Z.; Zhang, H.; Luo, H.; Wang, S.; Zhou, Q.; Du, X.; Tang, C.; Chen, L.; Liu, J.; Shi, Y.-K.; et al. Temperature and PH Effects on Biophysical and Morphological Properties of Self-Assembling Peptide RADA16-I. *J. Pept. Sci.* **2008**, *14*, 152–162. [CrossRef]
189. Ng, V.W.L.; Chan, J.M.W.; Sardon, H.; Ono, R.J.; García, J.M.; Yang, Y.Y.; Hedrick, J.L. Antimicrobial Hydrogels: A New Weapon in the Arsenal against Multidrug-Resistant Infections. *Adv. Drug Deliv. Rev.* **2014**, *78*, 46–62. [CrossRef] [PubMed]
190. Schnaider, L.; Brahmachari, S.; Schmidt, N.W.; Mensa, B.; Shaham-Niv, S.; Bychenko, D.; Adler-Abramovich, L.; Shimon, L.J.W.; Kolusheva, S.; DeGrado, W.F.; et al. Self-Assembling Dipeptide Antibacterial Nanostructures with Membrane Disrupting Activity. *Nat. Commun.* **2017**, *8*, 1365. [CrossRef] [PubMed]
191. Ortega-Velázquez, R.; Díez-Marqués, M.L.; Ruiz-Torres, M.P.; González-Rubio, M.; Rodríguez-Puyol, M.; Rodríguez Puyol, D. Arg-Gly-Asp-Ser (RGDS) Peptide Stimulates Transforming Growth Factor Beta1 Transcription and Secretion through Integrin Activation. *FASEB J. Off. Publ. Fed. Am. Soc. Exp. Biol.* **2003**, *17*, 1529–1531. [CrossRef]
192. Lee, S.; Trinh, T.H.T.; Yoo, M.; Shin, J.; Lee, H.; Kim, J.; Hwang, E.; Lim, Y.B.; Ryou, C. Ryou Self-Assembling Peptides and Their Application in the Treatment of Diseases. *Int. J. Mol. Sci.* **2019**, *20*, 5850. [CrossRef]
193. Sankar, S.; O'Neill, K.; Bagot D'Arc, M.; Rebeca, F.; Buffier, M.; Aleksi, E.; Fan, M.; Matsuda, N.; Gil, E.S.; Spirio, L. Clinical Use of the Self-Assembling Peptide RADA16: A Review of Current and Future Trends in Biomedicine. *Front. Bioeng. Biotechnol.* **2021**, *9*, 679525. [CrossRef]
194. PuraStat. Available online: https://3dmatrix.com/products/purastat/ (accessed on 30 March 2023).
195. Tian, T.; Li, Y.; Lin, Y. Prospects and Challenges of Dynamic DNA Nanostructures in Biomedical Applications. *Bone Res.* **2022**, *10*, 40. [CrossRef]
196. Fu, L.; Li, P.; Zhu, J.; Liao, Z.; Gao, C.; Li, H.; Yang, Z.; Zhao, T.; Chen, W.; Peng, Y.; et al. Tetrahedral Framework Nucleic Acids Promote the Biological Functions and Related Mechanism of Synovium-Derived Mesenchymal Stem Cells and Show Improved Articular Cartilage Regeneration Activity in Situ. *Bioact. Mater.* **2022**, *9*, 411–427. [CrossRef] [PubMed]
197. Jiang, X.; Li, M.; Guo, X.; Yang, M.; Rasooly, A. Self-Assembled DNA-THPS Hydrogel as a Topical Antibacterial Agent for Wound Healing. *ACS Appl. Bio Mater.* **2019**, *2*, 1262–1269. [CrossRef] [PubMed]
198. Hutanu, D. Recent Applications of Polyethylene Glycols (PEGs) and PEG Derivatives. *Mod. Chem. Appl.* **2014**, *2*, 2. [CrossRef]
199. Handbook of Pharmaceutical Excipients 6th Edition | PDF | Magnesium | Tablet (Pharmacy). Available online: https://www.scribd.com/document/246196761/Handbook-of-Pharmaceutical-Excipients-6th-Edition (accessed on 26 March 2023).
200. Chen, S.-L.; Fu, R.-H.; Liao, S.-F.; Liu, S.-P.; Lin, S.-Z.; Wang, Y.-C. A PEG-Based Hydrogel for Effective Wound Care Management. *Cell Transplant.* **2018**, *27*, 275–284. [CrossRef]
201. Chen, Y.; Li, J.; Lu, J.; Ding, M.; Chen, Y. Synthesis and Properties of Poly(Vinyl Alcohol) Hydrogels with High Strength and Toughness. *Polym. Test.* **2022**, *108*, 107516. [CrossRef]
202. Bodratti, A.; Alexandridis, P. Formulation of Poloxamers for Drug Delivery. *J. Funct. Biomater.* **2018**, *9*, 11. [CrossRef]
203. Russo, E.; Villa, C. Poloxamer Hydrogels for Biomedical Applications. *Pharmaceutics* **2019**, *11*, 671. [CrossRef]
204. Akiyama, H.; Tamaoki, N. Synthesis and Photoinduced Phase Transitions of Poly (N-Isopropylacrylamide) Derivative Functionalized with Terminal Azobenzene Units. *Macromolecules* **2007**, *40*, 5129–5132. [CrossRef]

205. Duan, Q.; Miura, Y.; Narumi, A.; Shen, X.; Sato, S.-I.; Satoh, T.; Kakuchi, T. Synthesis and Thermoresponsive Property of End-Functionalized Poly (N-Isopropylacrylamide) with Pyrenyl Group. *J. Polym. Sci. Part Polym. Chem.* **2006**, *44*, 1117–1124. [CrossRef]
206. Yu, B.; Chan, J.W.; Hoyle, C.E.; Lowe, A.B. Sequential Thiol-Ene/Thiol-Ene and Thiol-Ene/Thiol-Yne Reactions as a Route to Well-Defined Mono and Bis End-Functionalized Poly (N-Isopropylacrylamide). *J. Polym. Sci. Part Polym. Chem.* **2009**, *47*, 3544–3557. [CrossRef]
207. Ansari, M.J.; Rajendran, R.R.; Mohanto, S.; Agarwal, U.; Panda, K.; Dhotre, K.; Manne, R.; Deepak, A.; Zafar, A.; Yasir, M.; et al. Poly(N-Isopropylacrylamide)-Based Hydrogels for Biomedical Applications: A Review of the State-of-the-Art. *Gels* **2022**, *8*, 454. [CrossRef] [PubMed]

polymers

MDPI

Article

The Analysis of Acute and Subacute Toxicity of Silver Selenide Nanoparticles Encapsulated in Arabinogalactan Polymer Matrix

Evgeniy A. Titov [1,*], Larisa M. Sosedova [1], Mikhail A. Novikov [1], Marina V. Zvereva [2], Viktor S. Rukavishnikov [1] and Oleg L. Lakhman [1]

[1] East Siberian Institute of Medical and Ecological Research, 665827 Angarsk, Russia
[2] A.E. Favorsky Irkutsk Institute of Chemistry, 664033 Irkutsk, Russia
* Correspondence: g57097@yandex.ru; Tel.: +792-470-908-89

Abstract: The acute and subacute toxicity of a newly synthesized silver selenide nanoparticles encapsulated in a natural polymeric matrix of arabinogalactan study has been studied. The nanocomposite is a promising material for the design of diagnostic and therapeutic drugs. It can also be used for the preparation of fluorescent labels and in thermal oncotherapy. The employment of binary nanocomposites enables one to unveil the potential hidden in metals which constitute these composites. The study of acute toxicity, carried out by the oral administration of nanocomposites at a dose of 2000 mg/kg, has shown that the compound belongs to low-toxic substances of the 5th hazard class. With the subacute oral administration of nanocomposites at a dose of 500 μg/kg, slight changes are observed in the brain tissue and liver of experimental animals, indicating the development of compensatory–adaptive reactions. In the kidneys, the area of the Shumlyansky–Bowman chamber decreases by 40.5% relative to the control group. It is shown that the application of the protective properties of selenium, which is contained in the composite, helps to reduce the toxicity of silver.

Keywords: nanocomposite; laboratory animals; silver selenide; toxicity; morphology; brain; liver; kidney

1. Introduction

Silver selenide is a promising material for the creation of biomedical theranostic preparations. Due to the fact that the silver selenide luminesce is in the region of electromagnetic radiation that is not absorbed by biological tissues, the application of its quantum dots (QDs) as fluorescent labels seems to be prospective [1,2]. Silver selenide can be employed in oncology to improve the efficiency of photothermal therapy, the process of destroying cancer cells using infrared radiation as a heat source [1,3,4]. Silver nanoparticles are among the strongest natural antiseptics and have pronounced bacteriostatic and bactericidal effects [5,6]. They are also capable of overcoming the blood–brain barrier and accumulating in the brain tissue [7,8]. At the same time, silver nanoparticles are known to exhibit neurotoxic [7–9] and hepatotoxic [10] action at high doses. In turn, selenium, in addition to antibacterial properties [11], has cytostatic effects [12] and can suppress tumor growth [12–14]. The use of binary nanocomposites (which contain both Ag and Se) enables the unveiling of the potential hidden in the metals which constitute these composites. All of the above effects make the silver selenide nanocomposite a promising material for theranostics.

The improvement of the bioavailability of metal nanoparticles as well as the reduction of their possible toxicity represents an urgent challenge. In this regard, the synthesis of metal-containing nanocomposites using high-molecular compounds, for example, arabinogalactan (AG) of Siberian larch [15,16], appears to be a promising direction. The natural polymer AG is a water-soluble white or cream-colored powder without taste and odor, the production technology of which was patented [17]. The AG macromolecule (Figure 1) is represented by the residues of two monosaccharides: galactose and arabinose. The structure of the polysaccharide has a main chain consisting of galactopyranosyl units connected by β-bonds (1 → 3) and a side chain representing various combinations of galactopyranosyl and arabinofuranosyl residues connected by β-bonds (1 → 6) [18].

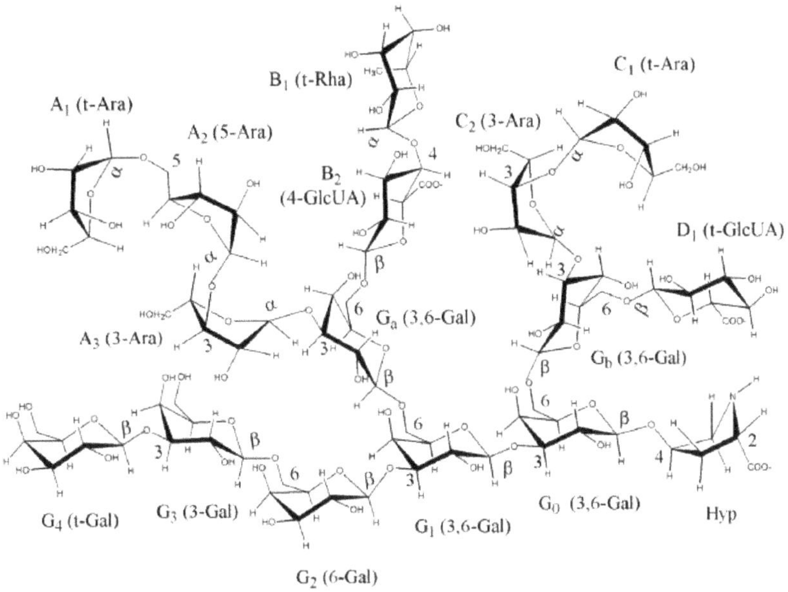

Figure 1. Fragment of AG macromolecule [18].

The polymer is isolated from larch wood, where its content reaches 15%. The specified polymer is a biologically active substance with a wide spectrum of activity. It is found that it possesses good gastroprotective, immunomodulatory, membranotropic and antioxidant properties [19–21]. It is shown that AG significantly weakens the effect of chemical toxicants on oxidation processes involving free radicals [22]. In addition, earlier AG already successfully demonstrated itself as an effective stabilizer of a number of different nanoparticles (noble metals, metal oxides, elemental chalcogenes, etc.). The obtained nanocomposites combine good water solubility, high aggregative and kinetic stability, high biocompatibility as well as a number of specific physical–chemical (optical, luminescent and magnetic) and biological (antimicrobial, antioxidant, adaptogenic, immunomodulatory and antianemic) properties due to the presence of nano-sized inorganic components. Thus, the availability of arabinogalactan, its excellent water-solubility and the expressed stabilizing properties (the z-potential of some composites is up to −70 eV) with a rather narrow molecular weight distribution Mw (42.3–45.2 kDa) allow for the water-soluble stable nanomaterials based on it to be obtained, which combine all the above features of both arabinogalactan and an inorganic nanophase [23–25].

The unique characteristics of silver selenide nanoparticles encapsulated in a natural polymer matrix of AG open up wide possibilities for its application, at the same time being possible reasons for adverse effects.

In the present work, the morphofunctional state of the tissue of the sensorimotor zone of the cerebral cortex and the hepato-renal system during the acute and subacute administration of silver selenide nanoparticles encapsulated in a natural AG polymer matrix has been evaluated.

2. Materials and Methods

2.1. Chemical Synthesis of a Water-Soluble Ag_2Se-Containing Nanocomposite Based on AG

The nanocomposite was synthesized according to the slightly modified method described in [2]. Namely, an aqueous solution containing 0.178 g $AgNO_3$ was added to 100 mL of aqueous solution containing 3.4 g of AG under vigorous stirring at room temperature. After 10 min of stirring, 225 μL of reaction medium containing Se^{2-}-anions, previously generated from powdered elemental selenium in the basic-reduction "hydrazine-hydrate-alkali" system [2], was added to the obtained transparent colorless solution (AG + $AgNO_3$). The formation of Ag_2Se nanoparticles was identified by the appearance of brown staining of the reaction medium and the absorption spectrum characteristic of Ag_2Se. The synthesis time was 20 min. The nanocomposite was isolated by precipitation from the reaction medium in a 5-fold excess of ethanol.

The optical absorption spectrum of the nanocomposite aqueous solution was recorded on a Perkin Elmer Lambda 35 spectrophotometer (Waltham, MA, USA) in a 1 cm quartz cuvette. X-ray diffractometer Bruker D8 ADVANCE (Waltham, MA, USA), (Cu radiation, Goebel mirror) was used for the X-ray diffraction study. Transmission electron microscopy (TEM) was performed on a Leo 906 E transmission electron microscope (Zeiss, Jena, Germany). The size distribution of nanoparticles was determined by statistical processing of microphotographs. The elemental composition was determined by X-ray energy dispersive microanalysis on an electron scanning microscope (SEM) Hitachi TM 3000 with X-ray detector SDD XFlash 430-4 (Tokyo, Japan)and on CHNS-analyzer Flash 2000 company Thermo Scientific (Waltham, MA, USA). AG—Found (%): C, 39.5; H, 6.39; Ag_2Se-AG—Found (%): C, 38.5; H, 5.6; Ag, 2.9%; Se, 1.1%.

2.2. Animals and Experimental Design

For the study purposes, animals were randomly selected, labeled with individual identifiers, and kept in their cages for three to four individuals each for 5 days prior to dosing to allow them to adapt in the laboratory. The absence of external signs of diseases and the homogeneity of groups by body weight (±20%) were considered as a criterion for the acceptability of randomization. Prior to dosing, the animals remained fasted for 3–4 h with free access to water. Animals of the experimental and control groups were kept under the same environmental conditions in the chambers.

Experimental animals were born by their own reproduction in the vivarium of Federal State Budgetary Scientific Institution "East Siberian Institute of Medical and Ecological Research" (FSBSI ESIMER) and kept on a standard diet. Maintenance and care of experimental animals were carried out in accordance with the interstate standard GOST 33216-2014 (Russia).

All animal experiments were approved by the ethical committee of FSBSI East-Siberian Institute of Medical and Ecological Research (identification code: E06/21; date of approval: 24 June 2021, amended/approvals every 6 months). All manipulations with experimental animals were performed in accordance with the rules of humane treatment of animals in accordance with the requirements of the International Recommendations for Biomedical Research Using Animals (WHO, Geneva, 1985).

2.3. Study of Acute Toxicity

The acute toxicity method [26] is a step-by-step procedure using a minimum number of animals of the same sex at each step. The method for determining the class of acute toxicity is based on biometric assessments with fixed doses, which are distributed over the time of administration so that it is possible to assess the substance according to the degree of danger and systematize the results in human body. Acute oral toxicity (method for determining the class of acute toxicity of the Ag_2Se nanoparticles encapsulated in a natural polymer matrix of arabinogalactan (Ag_2Se-AG)) was determined by the intragastric route of administration on outbred white male mice weighing 21–32 g (experimental $n = 6$, control $n = 6$).

The acute toxicity test of the Ag_2Se-AG nanocomposite was performed using the limiting dose (2000 mg/kg of body weight). The test substance was administered orally to white mice of the experimental group with an atraumatic probe at a dose of 2000 mg/kg in 0.9% NaCl. At the same time, the volume of the injected solution did not exceed 0.5 mL. Dosing accuracy was achieved by changing the volume of the injected solution at its constant concentration. Animals of the control group were orally administered in an equivalent volume of 0.9% NaCl (placebo). The doses were prepared immediately prior to the administration. Before the introduction of the test dose, the animals were weighed and the administered dose was calculated at the rate of 2000 mg/kg of body weight. Prior to dosing, mice remained fasted for 3–4 h with free access to water. Food was absent for another 2–3 h after introduction of substance.

Animals were observed daily for 14 days, with particular attention paid to the first 4 h after administration of the test dose. The appearance and disappearance of the external signs of poisoning were assessed: changes in the skin and coat, eyes and mucous membranes, respiratory, circulatory, autonomic and central nervous systems, as well as somatomotor activity and behavior, the appearance of tremor, convulsions, salivation, diarrhea, lethargy, sleep and coma [27]. Then the mice were sacrificed by the method of dislocation of the cervical vertebrae and macroscopic analysis of the internal organs was performed.

2.4. Study of Subacute Toxicity

The study of the subacute effect of the nanocomposite was carried out with intragastric administration of Ag_2Se-AG on white male rats weighing 200–220 g (experimental $n = 10$, control $n = 10$). In case of subacute exposure, the experimental group of animals was intragastrically injected with the studied nanocomposite at a dose of 500 mg/kg of body weight in 1 mL of distilled water for 10 days. This dose was chosen based on the results of previous investigations and was 1/4 of LD_{50}. The choice of dose is due to previous studies of the toxic properties of nanocomposites of other metals (Ag, Bi, Fe, etc.) [28–33], indicating the development of clear and persistent signs of pathology with the introduction of this dose. Control animals ($n = 10$) received 1 mL of distilled water in the same mode. For histological studies of the nervous tissue, the animals were euthanized by decapitation.

2.5. Histological Investigation

The brain, liver and kidneys were isolated from each animal under study and fixed in a neutral buffered 10% formalin solution (BioVitrum, St. Petersburg, Russia). The brain was dehydrated with isopraponol (BioVitrum, St. Petersburg, Russia) and placed in HistoMix homogenized paraffin medium for histological studies (BioVitrum, Russia). Using a HM 400 microtome (Microm, Munchen, Germany), serial frontal paraffin sections of the brain were made for subsequent staining with hematoxylin–eosin for survey microscopy. The nervous tissue of the temporo-parietal zone of the sensorimotor cortex of the brain was studied as a nerve center that provides the regulation of the basic physiological functions of the body and complex forms of behavior [9]. The liver and kidneys after fixation were examined by alcohols of increasing concentrations and embedded in paraffin. Sections 3–5 microns thick were prepared from paraffin blocks, which were stained with hematoxylin and eosin according to the generally accepted method [34]. In the brain preparations, the number of neurons per unit area, astroglial cells and dead neurons were counted, and the number of neuronophagy events were counted using the ImageScope M program (Russia). The number of Kupffer stellate macrophages and the number of polynuclear hepatocytes were counted in the liver tissue. The area of the Shumlyansky–Bowman chamber was evaluated in the kidney tissue. The obtained sections were examined using an Olympus BX 51 light-optical research microscope (Tokyo, Japan) with input of microimages into a computer using an Olympus E420 camera (Tokyo, Japan).

2.6. Statistical Analyses

Statistical analysis of the research results was carried out using the Statistica 6.1 software package. To compare the groups, the Mann–Whitney U-test was used. Changes in the studied parameters were considered statistically significant at a significance level of $p \leq 0.05$.

3. Results

3.1. Characteristics of the Synthesized Nanocomposite Ag_2Se-AG

The water-soluble Ag_2Se-containing nanocomposite based on AG with a quantitative content of the inorganic phase of 4%w was synthesized out in aqueous medium via the ion-exchange interaction of selenide anions Se^{2-} (generated from bulk powder samples of elemental selenium in a basic reduction system "hydrazine hydrate—alkali" (Equation (1)) and Ag^+ ions, according to Equation (2), in the presence of AG macromolecules.

$$2\,Se + 4\,KOH + N_2H_4 \times H_2O = 2\,K_2Se + 5\,H_2O + N_2\uparrow \quad (1)$$

$$2AgNO_3 + K_2Se \rightarrow Ag_2Se\downarrow + 2KNO_3 \quad (2)$$

The passivation of the energy-saturated surface of Ag_2Se nanoparticles and the support of their aggregative stability are probably performed by the adsorption of the polysaccharide macromolecules on their surface (steric stabilization) as well as due to the electrostatic stabilization of the Ag_2Se particle surface by the highly-polar functional AG groups (hydroxyl, terminal carbonyl). A single hybrid stable water-soluble system "nanocore—Ag_2Se/shell—polysaccharide matrix" is formed.

According to the data of transmission electron microscopy, the Ag_2Se-AG nanocomposite (4%w Ag_2Se) is formed as spherical Ag_2Se nanoparticles dispersed in the polysaccharide matrix of AG. The particle size varies between 4–16 nm with an average value of 9.6 nm. (Figure 2a). Using XRD, it was found that the obtained Ag_2Se-containing nanocomposite has a two-phase amorphous–crystalline structure. Its diffractogram is presented by a halo in the region of 10–24°, corresponding to the amorphous AG phase, and also by a set of reflexes of different intensities in the region of 33.6°, 36.1° and 45.1° (JCPDS Card No. 24-1041), characterizing the presence of silver selenide with a cubic crystal lattice (α-Ag_2Se) in the composite obtained [35]. In addition, the diffractogram shows a set of low-intensity reflexes in the region of 31–70°, corresponding to orthorhombic crystal lattice β-Ag_2Se (on the type of mineral Naumannite) [36] (Figure 2b).

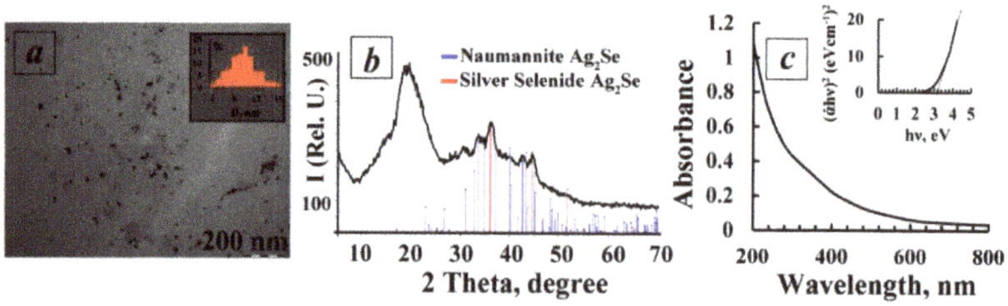

Figure 2. (**a**)—Microphotograph and particle size distribution (inset) of Ag_2Se nanoparticles in AG polysaccharide matrix; (**b**)—diffractogram of Ag_2Se-AG nanocomposite (4%w Ag_2Se); (**c**)—absorption spectrum of 0.1% of Ag_2Se-AG nanocomposite (4%w Ag_2Se) water solution and its Tauc plot of $(\alpha h\nu)^2$ vs. $(h\nu)$ (inset).

The average size of Ag_2Se nanocrystallites, calculated by the Scherrer formula, is 11.2 nm which correlates well with the TEM data (Figure 2c). The experimentally obtained value of cell parameter a (0.4962) agrees well with that of the reference sample of cubic silver selenide (a = 0.4983 nm).

The optical properties of an aqueous solution of Ag_2Se nanocomposite (4%w Ag_2Se) were studied by optical spectroscopy in the visible region of the spectrum at room temperature. It was found that the absorption spectrum is characterized by the absence of well-resolved maxima, probably due to the relatively large size of Ag_2Se nanoparticles and their wide dispersed distribution, which is confirmed by TEM data.

The value of the optical band gap energy (Eg) of the Ag_2Se nanoparticles in the AG matrix can be calculated by Tauc's plot method. The basis of this method is the suggestion of the possibility to present the absorption coefficient α in the form of the equation:

$$(\alpha \times h\nu)^{1/\gamma} = B(h\nu - Eg) \quad (3)$$

where Eg is the band gap energy, h is Planck's constant, ν is the photon frequency and B is a constant. We chose the factor γ, which depends on the nature of the electron transition, as 1/2, assuming the direct character of the transitions [37]. According to the data obtained, the optical gap energy of the synthesized Ag_2Se nanoparticles was higher (3.2 eV) than the value of 0.16 eV reported earlier for bulk Ag_2Se [38]. Presumably, the increase of the gap in Ag_2Se nanoparticles compared to the value of bulk silver selenide may be due to the decrease of the particle size and appearance of the quantum confinement effect.

3.2. Acute Toxicity Study

The measurement of body weight of laboratory animals is an integral indicator of the state of the organism. During the initial weighing of mice before the introduction of the test nanocomposite, the individuals of the studied groups almost did not differ from each other in weight, being varied within 30–35 g. When weighed one week and 2 weeks after administration, the weight of the animals either increased or remained at the same level. Thus, oral administration of the nanocomposite during observation for 14 days did not decrease the body weight in any case, $p \geq 0.05$ (Table 1).

Table 1. Increase in body weight of white mice of the experimental and control groups that survived intoxication with the Ag_2Se-AG nanocomposite.

Groups	Before Injection	One Week after Injection	Two Weeks after Injection
Experimental	32.3 ± 0.55	32.5 ± 0.56	33.5 ± 0.56
Control	31.8 ± 0.79	32.0 ± 0.77	33.0 ± 0.73

During the entire observation, there was no mortality of animals from the experimental group. As in the latter, the observation of mice from the control group also did not reveal changes in behavior, condition of wool or in the consumption of water and food.

A macroscopic investigation of the internal organs of mice euthanized after 14 days of observation showed no differences between the experimental and control groups. Mice had the correct constitution and obtained satisfactory nutrition. The coat had a neat appearance and no foci of baldness were determined. Visible mucous membranes were shiny, smooth and pale in color. Thoracic and abdominal cavities did not contain effusion. The position of the internal organs did not have any disorders. The thyroid gland was of normal size and shape and reddish in color.

The size and shape of the heart did not change. The heart muscle was moderately dense and brownish in color. The lumen of the trachea and large bronchi were uniformly wide. The lungs were easily collapsed when the chest was opened and the surface was of a uniform pale pink color. The tissue of the lungs was airy to the touch.

The stomach had the usual dimensions and its lumen was filled with food contents. The mucous membrane was folded, homogeneous and pinkish in color. Irritation and hyperemia was not observed. The shape and size of the liver did not change. The liver tissue was moderately plethoric. The pancreas was pale pink. The size and shape of the spleen corresponded to those of the control mice. The tissue of the spleen had a moderately dense consistency and dark cherry color. Kidneys were of normal color with a clearly visible cortical immedulla. The membranes of the brain were shiny, thin and smooth. No ventricular expansion was observed on the frontal sections.

It should be noted that a macroscopic investigation of the internal organs of animals from the control group also did not show any pathological changes.

Based on the results obtained, due to the absence of mortality associated with the studied nanocomposite in animals dosed at one stage, the further research seems to be unreasonable [39].

Thus, according to the data of a macroscopic study, acute oral administration of the Ag_2Se-AG nanocomposite at the maximum dose to white male mice did not cause visible changes in the internal organs, brain and tracheal and stomach mucosa. The study of acute toxicity showed that the test substance can be attributed to the 5th hazard class. Thus, this drug belongs to low-hazard substances.

3.3. Subacute Toxicity Study

The study of the subacute effect of Ag_2Se-AG on the brain tissue showed that the blood filling of the vessels of the brain substance and the state of the vascular intima were unchanged. The number of normal neurons, astroglial cells and degeneratively altered neurons (darkly stained neurons were considered degeneratively altered, without a clearly separated nucleus and cytoplasm) per unit area had no statistically significant differences from the control values. The number of neuronophagy events in animals that received the nanocomposite was statistically significantly higher than in the control group (Table 2, Figure 3).

Table 2. Morphometric parameters of the sensorimotor zone of the cerebral cortex, liver and kidney tissues during subacute administration of the Ag_2Se-AG nanocomposite to rats at a dose of 500 μg/kg of body weight for 10 days. Me (Q25–Q75).

Indicators	Experimental Group	Control Group	*p*
Sensorimotor Area of the Cerebral Cortex			
Number of normal neurons per unit area	117.0 (110.0–145.0)	152.0 (133.0–177.0)	0.2
Number of astroglial cells per unit area	140.0 (119.0–158.0)	156.0 (119.0–164.0)	0.85
Number of degeneratively altered neurons per unit area	8.0 (7.0–15.0)	5.0 (5.0–6.0)	0.08
Number of neuronophagy	4.0 (2.0–5.0) *	1.5 (1.0–2.0)	0.05
Liver			
Number of Kupffer stellate macrophages	167.5 (140.0–192.0)	145.0 (141.0–148.0)	0.55
Number of polynuclear hepatocytes	19.5 (19.0–26.0) *	16.0 (11.0–18.0)	0.04
Kidney			
Shumlyansky–Bowman capsule area	27,715.5 (24,260.9–31,714.8) *	32,556.5 (28,573.1–36,306.2)	0.02

Notes: * differences are statistically significant compared to the control according to the Mann–Whitney test, $p \leq 0.05$.

Thus, the exposure of the binary nanocomposites of silver selenide at a dose of 500 μg/kg had an insignificant effect on the number and structure of the population of nerve cells in the sensorimotor cortex of albino rats. In the liver tissue, the blood filling of the sinusoidal capillaries, central veins and veins of the portal tracts was normal. Portal tracts were not dilated and were without signs of sclerosis and inflammation. The beam-radial structure of the hepatic lobules was preserved. The number of stellate Kupffer macrophages in the sinusoidal capillaries did not differ from the control group. At the same time, the number of polynuclear hepatocytes was statistically significantly higher than in the control group (Table 2, Figure 4).

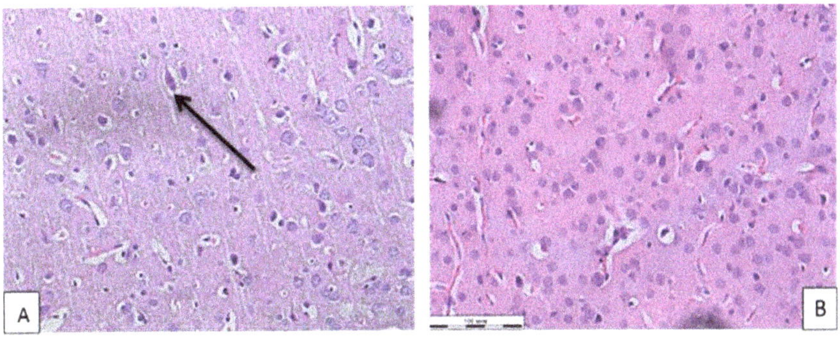

Figure 3. Microphoto of brain tissue of white rats of experimental (**A**) and control (**B**) groups. ↑—degeneratively altered neuron. Hematoxylin–eosin. Mag. ×400.

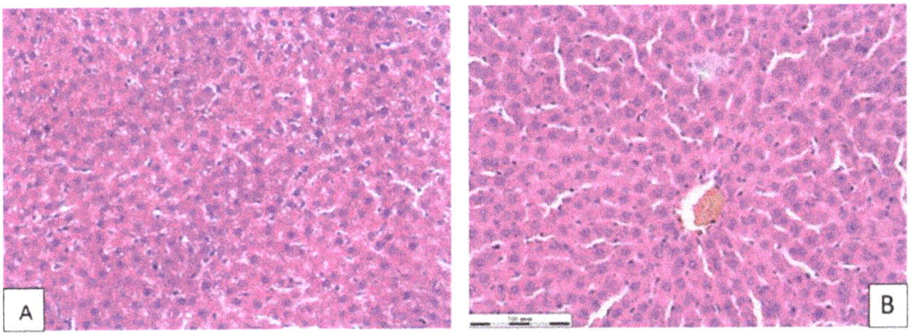

Figure 4. Microphoto of liver tissue of white rats of experimental (**A**) and control (**B**) groups. okr. Hematoxylin–eosin. Mag. ×400.

In the kidney tissue, the blood filling of the cortical and medulla of the organ was unchanged. The violations of the blood rheology in the body were not observed. The condition of the walls of the renal arteries, arterioles and interstitial space was normal. The structure of the renal glomeruli was preserved. There were no foci of inflammation or necrosis of the renal tissue. The epithelium of the distal and proximal renal tubules was intact. In the cortical substance of the kidney, a statistically significant decrease in the area of the Shumlyansky–Bowman capsule was revealed in comparison with the control group (Table 2, Figure 5).

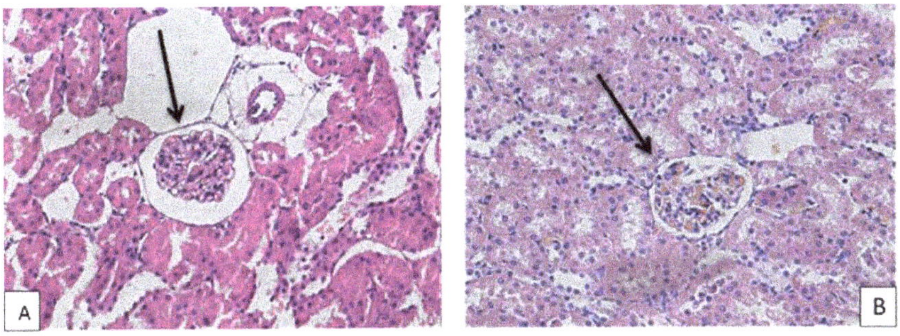

Figure 5. Microphoto of kidney tissue of white rats of experimental (**A**) and control (**B**) groups. ↑—Shumlyansky–Bowman chamber. Hematoxylin–eosin. Mag. ×400.

4. Discussion

With the acute oral administration of the nanocomposite to white mice at a maximum dose of 2000 mg/kg of body weight, no mortality was noted during the entire observation. This, together with the absence of changes in internal organs during the macroscopic investigation, allows us to classify the studied Ag_2Se-AG nanocomposite as a low-hazard substance (5th class of hazard).

Many researchers have shown that exposure to silver nanoparticles in low doses leads to the emergence and development of neurotoxic effects [7–9,11,40–42]. Skalska J. et al. [40] described the development of pathological changes in neuronal mitochondria (edema, decreased potential of the mitochondrial membrane) and, as a consequence, the induction of autophagy of the neurons themselves when exposed to silver nanoparticles at a dose of 0.2 mg/kg. The cytotoxic effect of low doses of silver nanoparticles on the main components of the blood–brain barrier (endothelial cells and astrocytes) was reported, as well as that directly on neurons, causing a disruption of the cell cytoskeleton and destruction of synaptic connections [7,41]. Our previous studies have shown that a silver nanocomposite encapsulated in an arabinogalactan polymer matrix, despite belonging to the IV low-hazard class of substances with LD_{50}, when administered intragastrically at a dose of more than 5000 mg/kg of animal weight, disorders the structure of the nervous tissue, increases the area of mitochondria, violates the structure of nerve cells and activates the apoptosis process [43]. At the same time, silver nanoparticles encapsulated in the natural biopolymer matrix of arabinogalactan are able to penetrate through the blood–brain barrier and, having remained in the nervous tissue of the brain of rats for a long time, cause structural disturbances [43,44]. Silver nanoparticles do not reduce their cytotoxicity for nervous tissue.

In turn, exposure to individual selenium nanoparticles, also encapsulated in a polymeric matrix of arabinogalactan, at a dose of 500 μg/kg of animal body weight reduced the total number of neurons and astroglial cells per unit area in the sensorimotor zone of the cerebral cortex and increased the degeneratively altered neurons and the number of neuronophagy acts, which indirectly indicated both the penetration of the nanocomposite through the blood–brain barrier and the pronounced neurotoxic effect of selenium nanoparticles [45]. Exposure to excessive amounts of selenium can lead to the disruption of the functioning of neurotransmitter systems and the development of neurodegenerative and neuropsychiatric processes [42]. Taking into account that the toxic effect of selenium is realized by the suppression of the intercellular signals transmission [46], it can be assumed that the previously established reduction in the number of normal neurons and astroglial cells in the nervous tissue disturbs the intercellular interaction.

The absence of such effects upon exposure to silver selenide encapsulated in the arabinogalactan polymer matrix is apparently due to the simultaneous presence of nanoparticles in the nanocomposite and, possibly, to their competition for binding to the cell receptors of cerebral cortex neurons.

An increase in the number of polynuclear hepatocytes upon the subacute administration of the Ag_2Se-AG nanocomposite indicates the activation and development of compensatory repair processes in the liver and stimulation of cell regeneration mechanisms, while the constant amount of Kupffer stellate macrophages evidences no inflammatory process in the liver tissue. According to the data given in [10], silver nanoparticles at a dose of 300 μg/kg disrupt the normal blood rheology in the liver, increase the number of polynuclear hepatocytes and disorder the metabolic activity of hepatocytes. A similar effect was produced by selenium nanoparticles in an arabinogalactan matrix at a dose of 500 μg/kg [46]. The hepatotoxic effect of selenium at a low dose of 10 μg mixed with lithium was shown in studies by Pinto-Vidal, F. et al. [47]. At the same time, some studies showed the hepatoprotective effect of selenium [48–50]. The inconsistency of the obtained data is apparently explained by different ways, forms and doses of selenium introduction into the body of biological objects.

In the kidney tissue, the administration of Ag_2Se-AG at a dose of 500 μg/kg showed a 40.5% decrease in the area of the Shumlyansky–Bowman capsule (a change in the area within 30% is the norm [51]), which can reduce the volume of primary urine formed, and, in turn, contribute to the difficulty of metabolic products' excretion from the body. There were no other significant structural changes in the kidney tissue exposed to the silver selenide nanocomposite. Perhaps this is due to the nephroprotective properties of selenium [52–54]. Meanwhile, it is known that silver nanoparticles have a nephrotoxic effect, causing structural changes in the renal tubules and renal glomeruli [55]. Perhaps this is the reason for the slight changes in the kidney tissue when the Ag_2Se-AG nanocomposite was administered to rats.

The administration of the nanocomposite to rats for 10 days at a dose of ¼ of LD50 had a different degree of severity depending on the place of application. Of the greatest interest was the nanocomposite effect on the tissue of the cerebral cortex. The experimental studies revealed no changes in the ratio of the cellular elements of the sensorimotor zone of the cortex. Morphological disturbances of neurons were also not observed in comparison with those in control animals. In the tissue of the cerebral cortex, attention was drawn to higher acts of neuronophagy, indicating an increase in the formation of glial nodules, through which damaged or degeneratively altered nerve cells were destroyed and removed from the body with the help of macrophages. Thus, silver selenide nanocomposites, without a pronounced neurotoxic effect, still increase the number of dead neurons. It might be possible to address this issue by studying the effects of silver selenide on brain tissue in the late post-contact period.

The conducted studies revealed that the toxicity of the silver selenide nanocomposite in the arabinogalactan polymer matrix is much less pronounced than that of silver or selenium nanoparticles alone. Apparently, this fact may be due to the competitive relationship of nanoparticles. At the same time, selenium, being a physiologically important trace element that is part of glutathione peroxidase, may be an antagonist for silver particles. The literature describes the antagonistic properties of selenium for such heavy metals as mercury, arsenic, lead and cadmium [56]. Given the great importance of selenium for the functioning of the immune, endocrine and reproductive systems, metabolism, cellular homeostasis and carcinogenesis, it can be assumed that its ability to bind and activate cell receptors is much higher than that of silver. As a result, the biological effectiveness of selenium can be more pronounced. Conversely, the biological activity of silver is suppressed. For the studied compound, it can be assumed that selenium acts as a protector, suppressing the pathological effects of silver nanoparticles and thereby protecting the cellular metabolism.

5. Conclusions

In conclusion, the study of the acute toxicity of a silver selenide nanocomposite has shown that the substance belongs to the low-hazard class. The evaluation of the subacute toxicity of a silver selenide nanoparticle encapsulated in an arabinogalactan polymer matrix to white rats does not reveal any significant changes in the tissue structure of the sensorimotor cortex and liver of animals, along with minor changes in the kidney tissue. In this connection, the silver selenide nanocomposite encapsulated in a polymer matrix is a promising preparation for further biomedical research.

Author Contributions: Conceptualization, E.A.T. and L.M.S.; methodology, E.A.T., M.A.N. and M.V.Z.; software, E.A.T. and M.A.N.; validation, E.A.T. and M.V.Z.; formal analysis, E.A.T. and M.A.N.; investigation, E.A.T., M.A.N. and M.V.Z.; resources, V.S.R. and O.L.L.; data curation, E.A.T., M.A.N. and M.V.Z.; writing—original draft preparation, E.A.T.; writing—review and editing, M.A.N. and L.M.S.; visualization, E.A.T. and M.A.N.; supervision, L.M.S. and O.L.L.; project administration, L.M.S. and O.L.L.; funding acquisition, V.S.R. and O.L.L. All authors have read and agreed to the published version of the manuscript.

Funding: This work (synthesis and characterization of nanocomposite) was partially carried out within the framework of the state task of the A. E. Favorsky Irkutsk Institute of Chemistry of SB RAS (No 121021000252-8). The acute and subacute toxicity investigations were carried out as part of the state task of the FSBSI East-Siberian Institute of Medical and Ecological Research (No AAAA-A18-118013190004-8).

Institutional Review Board Statement: The animal study protocol was approved by the Ethics Committee of FSBSI East-Siberian Institute of Medical and Ecological Research (identification code: E06/21; date of approval: 24 June 2021).

Informed Consent Statement: Not applicable.

Data Availability Statement: The data presented in this study are available from the corresponding author upon request.

Acknowledgments: The authors wish to thank the Baikal Analytical Center (Irkutsk Institute of Chemistry of the Siberian Branch of the Russian Academy of Sciences).

Conflicts of Interest: The authors declare no conflict of interest.

References

1. Shurygina, I.A.; Shurygin, M.G. Selenium nanocomposites-prospects for application in oncology. *Bull. New Med. Technol.* **2020**, *27*, 81–86. [CrossRef]
2. Lesnichaya, M.V.; Sukhov, B.G.; Shendrik, R.Y.; Sapozhnikov, A.N.; Trofimov, B.A. Synthesis of Water-Soluble Silver Selenide Quantum Dots Luminescing within the Transparency Window of Biological Tissues. *Russ. J. Gen. Chem.* **2018**, *88*, 284–287. [CrossRef]
3. Trukhan, I.S.; Dremina, N.N.; Lozovskaya, E.A.; Shurygina, I.A. Assessment of potential cytotoxicity in the framework of in vivo observation on Biostation CT. *Acta Biomed. Sci.* **2018**, *3*, 48–53. [CrossRef]
4. Shahnawaz, M.; Abou, K.; Pandey, T.; Bhaisare, M.L.; Gedda, G.; Wu, H.-F. Folic Acid navigated Silver Selenide nanoparticles for photo-thermal ablation of cancer cells. *Colloids Surf. B Biointerfaces* **2017**, *159*, 564–570. [CrossRef]
5. Kibrik, B.S.; Prokhorova, I.M.; Song, D.S.; Kreitsberg, G.N. Study of the mutagenic action of isoniazid nanocomposite and silver nanoparticles. *Tuberc. Lung Dis.* **2013**, *90*, 76–81.
6. Fastovets, I.A.; Verkhovtseva, N.V.; Pashkevich, E.B.; Netrusov, A.I. Silver nanoparticles: Toxic effect on microorganisms and interaction with higher plants. *Probl. Agrochem. Ecol.* **2017**, *1*, 51–62.
7. Zhang, B.; Liu, N.; Liu, Q.S.; Zhang, J.; Zhou, Q.; Jiang, G. Silver nanoparticles induce size-dependent and particle-specific neurotoxicity to primary cultures of rat cerebral cortical neurons. *Ecotoxicol. Environ. Saf.* **2020**, *198*, 110674. [CrossRef]
8. Boyes, W.K.; Van Thriel, C. Neurotoxicology of Nanomaterials. *Chem. Res. Toxicol.* **2020**, *33*, 1121–1144. [CrossRef] [PubMed]
9. Titov, E.A.; Sosedova, L.M.; Novikov, M.A. Alteration of the brain tissue of white rats induced by the action of a silver nanocomposite encapsulated on a polymer matrix. *Pathol. Physiol. Exp. Ther.* **2015**, *59*, 41–44.
10. Belyaeva, N.N.; Gasimova, Z.M.; Mikhailova, R.I.; Savostikova, O.N.; Alekseeva, A.V. Morphofunctional cellular evaluation of the dynamics of the impact of silver nanoparticles on the liver of rats. *Hyg. Sanit.* **2014**, *93*, 50–54.
11. Lesnichaya, M.V.; Malysheva, S.F.; Belogorlova, N.A.; Graskova, I.A.; Gazizova, A.V.; Perfilyeva, A.I.; Nozhkina, O.A.; Sukhov, B.G. Synthesis and antimicrobial activity of arabinogalactan-stabilized selenium nanoparticles from sodium bis(2-phenylethyl)diselenophosphinate. *Russ. Chem. Bull.* **2019**, *68*, 2245–2251. [CrossRef]
12. Lozovskaya, E.A.; Silkin, I.I.; Sukhov, B.G. Influence of nanopreparation "Selenium" on the functional state of Ehrlich's ascitic carcinoma cells (in vivo). *Vestn. KrasGAU* **2015**, *9*, 56–59.
13. Xuan, G.; Zhang, M.; Chen, Y.; Huang, S.; Lee, I. Design and Characterization of a Cancer-Targeted Drug Co-Delivery System Composed of Liposomes and Selenium Nanoparticles. *J. Nanosci. Nanotechnol.* **2020**, *20*, 5295–5304. [CrossRef] [PubMed]
14. Yang, J.; Pan, S.; Gao, S.; Dai, Y.; Xu, H. Anti-recurrence/metastasis and chemosensitization therapy with thioredoxin reductase-interfering drug delivery system. *Biomaterials* **2020**, *249*, 120054. [CrossRef] [PubMed]
15. Gasilova, E.R.; Toropova, A.A.; Bushin, S.V.; Khripunov, A.K.; Grischenko, L.A.; Aleksandrova, G.P. Light scattering from aqueous solutions of colloid metal nanoparticles stabilized by natural polysaccharide arabinogalactan. *Phys. Chem. B.* **2010**, *114*, 4204–4212. [CrossRef]
16. Aleksandrova, G.P.; Boymirzaev, A.S.; Lesnichaya, M.V.; Sukhov, B.G. Metal-polymer nanobiocomposites with galactose-containing stabilizing matrices: Dimensional effect in changes of molar mass parameters. *Russ. J. Gen. Chem.* **2015**, *85*, 488–496. [CrossRef]
17. Arabinogalactan Preparation Method. Available online: https://patents.google.com/patent/RU2256668C2/en (accessed on 28 February 2022).
18. Silver Nanocomposite of Sulphated Arabinogalactan Exhibiting Antimicrobial and Antithrombotic Activity and Method for Preparing It. Available online: https://patents.google.com/patent/RU2462254C2/en (accessed on 28 February 2022).
19. Zavezenova, I.V. Yoghurt fermented milk product enriched with functional additive arabinogalactan. *Basic Res.* **2014**, *6*, 29–32.

20. Medvedeva, S.A.; Hutsol, L.O.; Alexandrova, G.P. Antioxidant activity of Siberian larch arabinogalactan intoxication phenylhydrazine and ethylene. In *Advances in Chemistry and Chemical Engineering Plant Materials*; Altai State University: Barnaul, Russia, 2007; pp. 328–331.
21. Medvedeva, S.A.; Aleksandrova, G.P.; Dubrovina, V.I. Larch arabinogalactan promising polymer matrix for biogenic metals. *Butlerov Commun.* **2002**, *7*, 45–49.
22. Kolzunova, L.G.; Goldsmith, R.N.; Shaydurova-Kolzunova, E.S. Investigation of antioxidant activity of arabinogalactan electrochemical methods. In *Analytics of Siberia and the Far East*; TPU: Tomsk, Russia, 2008; pp. 132–139.
23. Trofimov, B.A.; Sukhov, B.G.; Aleksandrova, G.P.; Medvedeva, S.A.; Grichenko, L.A.; Malkina, A.G.; Feoktistova, L.P.; Sapozhnikov, A.N.; Dubrovina, V.I.; Martynovich, E.F.; et al. Nanocomposites with magnetic, optical, catalytic, and biologically active properties based on arabinogalactan. *Dokl. Chem.* **2003**, *393*, 287–288. [CrossRef]
24. Lesnichaya, M.; Perfileva, A.; Nozhkina, O.; Gazizova, A.; Graskova, I. Synthesis, toxicity evaluation and determination of possible mechanisms of antimicrobial effect of arabinogalactane-capped selenium nanoparticles. *J. Trace Elem. Med. Biol.* **2022**, *69*, 126904. [CrossRef]
25. Aleksandrova, G.P.; Grichenko, L.A.; Chetverikova, T.D.; Krasnikova, I.M.; Medvedeva, S.A. Synthesis and antianemic activity of nanosized biocomposite ferroarabinogalactan. *Russ. J. Bioorgan. Chem.* **2011**, *37*, 829–833. [CrossRef]
26. OECD. *Guidance Document on Acute Oral Toxicity/Environmental Health and Safety Monograph Series on Testing and Assessment*; OECD: Paris, France, 2000; Volume 24.
27. Khabriev, R.U. *Guidelines for the Experimental (Preclinical) Study of New Pharmacological Substances*; Medicine: Moscow, Russia, 2005; pp. 309–311.
28. Titov, E.A.; Rukavishnikov, V.S.; Sosedova, L.M.; Novikov, M.A.; Buynova, E.V. Morphofunctional changes in the tissue of the brain, liver and kidneys of white rats under the influence of selenium nanocomposite encapsulated in the polymer matrix of arabinogalactan. *Acta Biomed. Sci.* **2021**, *6*, 92–99. [CrossRef]
29. Lesnichaya, M.; Sukhov, B.; Shendrik, R.; Titov, E. Synthesis and comparative assessment of antiradical activity, toxicity, and biodistribution of k-carrageenan-capped selenium nanoparticles of different size: In vivo and in vitro study. *IET Nanobiotechnol.* **2020**, *14*, 519–526. [CrossRef] [PubMed]
30. Novikov, M.A.; Lakhman, O.L.; Titov, E.A.; Sosedova, L.M.; Rukavishnikov, V.S.; Vokina, V.A. Comparative assessment of silver nanocomposites' biological effects on the natural and synthetic matrix. *Int. J. Mol. Sci.* **2021**, *22*, 13257. [CrossRef]
31. Sosedova, L.M.; Novikov, M.A.; Titov, E.A.; Pozdnyakov, A.S.; Korzhova, S.A.; Ermakova, T.G.; Prozorova, G.F. Synthesis, antimicrobial properties, and toxicity of a nanobiocomposite base on Ag(0) particles and poly (1-vinil-1,2,4-triazole). *Pharm. Chem. J.* **2019**, *52*, 1477. [CrossRef]
32. Titov, E.A.; Sosedova, L.M.; Kapustina, E.A.; Yakimova, N.L.; Novikov, M.A.; Lisetskaya, L.G.; Lizarev, A.V. Analysis of the toxicity of a Cu_2O nanocomposite encapsulated in a polymer matrix of arabinogalactan. *Nanobiotechnol. Rep.* **2021**, *16*, 537–542. [CrossRef]
33. Rukavishnikov, V.S.; Novikov, M.A.; Titov, E.A.; Sosedova, L.M.; Vokina, V.A.; Yakimova, N.L. Estimation of toxic properties of nanocomposites containing nanoparticles of bismuth, gadolinium, and silver. *Trace Elem. Electrolytes* **2018**, *35*, 203–206. [CrossRef]
34. Korzhevsky, D.E. *Brief Summary of the Basics of Histological Technique for Physicians and Histologists*; Krof: St. Petersburg, Russia, 2005; pp. 1–48.
35. Yanling, L.; Yaoyao, Z.Y.; Lin, G.J. One-pot microwave-assisted synthesis of Ag_2Se and photothermal conversion. *Results Phys.* **2022**, *38*, 105590. [CrossRef]
36. Chougale, U.M.; Han, S.H.; Rath, M.C.; Fulari, V.J. Synthesis, characterization and surface deformation study of nanocrystalline Ag_2Se thin films. *Mater. Phys. Mech.* **2013**, *17*, 47–58.
37. Makula, P.; Pacia, M.; Macyk, W. How To Correctly Determine the Band Gap Energy of Modified Semiconductor Photo-catalysts Based on UV−Vis Spectra. *J. Phys. Chem. Lett.* **2018**, *9*, 6814–6817. [CrossRef]
38. Ferhat, M.; Nagao, J. Thermoelectric and transport properties of β-Ag_2Se compounds. *J. Appl. Phys.* **2000**, *88*, 813–816. [CrossRef]
39. *GOST 32644-2014*; Test Methods for the Effects of Chemical Products on the Human Body. Acute Oral Toxicity-Method for Determining the Class of Acute Toxicity. Standartinform: Moscow, Russia, 2014. Available online: http://www.normacs.ru/Doclist/doc/11711.html (accessed on 28 February 2022).
40. Skalska, J.; Dąbrowska-Bouta, B.; Frontczak-Baniewicz, M.; Sulkowski, G.; Strużyńska, L.A. Low Dose of Nanoparticulate Silver Induces Mitochondrial Dysfunction and Autophagy in Adult Rat Brain. *Neurotox. Res.* **2020**, *386*, 650–664. [CrossRef] [PubMed]
41. Korzeniowska, B.; Fonseca, M.P.; Gorshkov, V.; Skytte, L.; Rasmussen, K.L.; Schrøder, H.D.; Kjeldsen, F. The Cytotoxicity of Metal Nanoparticles Depends on Their Synergistic Interactions. *Part. Part. Syst. Charact.* **2020**, *37*, 2000135. [CrossRef]
42. Kaur, H.; Agarwal, S.; Agarwal, M.; Agarwal, V.; Singh, M. Therapeutic and preventive role of functional foods in process of neurodegeneration. *Int. J. Pharm. Sci. Res.* **2020**, *6*, 2882–2891.
43. Novikov, M.A.; Titov, E.A.; Vokina, V.A.; Yakimova, N.L.; Sosedova, L.M.; Korzhova, S.A.; Pozdnyakov, A.S.; Emelyanov, A.I.; Proydakova, O.A.; Ermakova, T.G.; et al. Biological effects of the new silver-containing polymer nanocomposite. *Acta Biomed. Sci.* **2012**, *2*, 121–125.
44. Titov, E.A.; Novikov, M.A.; Sosedova, L.M. Effect of silver nanoparticles encapsulated in a polymer matrix on the structure of the nervous tissue and caspase-3 expression. *Russ. Nanotechnol.* **2015**, *7–8*, 105–108. [CrossRef]

45. Powers, M.; Liu, L.; Deemer, D.; Chen, S.; Scholl, A.; Yoshinaga, M.; Liu, Z. Selenite inhibits notch signaling in cells and mice. *Int. J. Mol. Sci.* **2021**, *22*, 2518. [CrossRef]
46. Pinto-Vidal, F.; Carvalho, C.D.S.; Abdalla, F.C.; Ceschi-Bertoli, L.; Moraes Utsunomiya, H.S.; Henrique da Silva, R.; Salla, R.F.; Jones-Costa, M. Metabolic, immunologic, and histopathologic responses on premetamorphic American bullfrog (*Lithobates catesbeianus*) following exposure to lithium and selenium. *Environ. Pollut.* **2021**, *270*, 116086. [CrossRef]
47. Bai, K.; Hong, B.; He, J.; Huang, W. Antioxidant Capacity and Hepatoprotective Role of Chitosan-Stabilized Selenium Nanoparticles in Concanavalin A-Induced Liver Injury in Mice. *Nutrients* **2020**, *12*, 857. [CrossRef]
48. Hamza, R.Z.; EL-Megharbel, S.M.; Altalhi, T.; Gobouri, A.A.; Alrogi, A.A. Hypolipidemic and hepatoprotective synergistic effects of selenium nanoparticles and vitamin. E against acrylamide-induced hepatic alterations in male albino mice. *Appl. Organomet. Chem.* **2020**, *34*, e5458. [CrossRef]
49. Li, B.; Li, W.; Tian, Y.; Guo, S.; Qian, L.; Xu, D.; Cao, N. Selenium-Alleviated Hepatocyte Necrosis and DNA Damage in Cyclophosphamide-Treated Geese by Mitigating Oxidative Stress. *Biol. Trace Elem. Res.* **2020**, *193*, 508–516. [CrossRef] [PubMed]
50. Avtandilov, G.G. *Problems of Pathogenesis and Pathological Diagnosis of Diseases in the Aspect of Morphometry*; Medicine: Moscow, Russia, 1984; 288p.
51. Liu, Y.; Dong, R.; Yang, Y.; Xie, H.; Huang, Y.; Chen, X.; Zhang, Z. Protective Effect of Organic Selenium on Oxidative Damage and Inflammatory Reaction of Rabbit Kidney Induced by T-2 Toxin. *Biol. Trace Elem. Res.* **2020**, *199*, 1833–1842. [CrossRef] [PubMed]
52. Cao, L.; Zhang, L.; Zeng, H.; Wu, R.T.; Wu, T.-L.; Cheng, W.-H. Analyses of Selenotranscriptomes and Selenium Concentrations in Response to Dietary Selenium Deficiency and Age Reveal Common and Distinct Patterns by Tissue and Sex in Telomere-Dysfunctional Mice. *J. Nutr.* **2017**, *147*, 1858–1866. [CrossRef]
53. Pilarczyk, B.; Hendzel, D.; Pilarczyk, R.; Tomza-Marciniak, A.; Błaszczyk, B.; Dąbrowska-Wieczorek, M.; Bujak, T. Liver and kidney concentrations of selenium in wild boars (*Sus scrofa*) from northwestern Poland. *Eur. J. Wildl. Res.* **2010**, *56*, 797–802. [CrossRef]
54. Albrahim, T. Silver nanoparticles-induced nephrotoxicity in rats: The protective role of red beetroot (*Beta vulgaris*) juice. *Environ. Sci. Pollut. Res.* **2020**, *27*, 38871–38880. [CrossRef]
55. Belyaeva, N.N.; Nikolaeva, N.I.; Vostrikova, M.V. Effects on the liver, kidney and testes of warm-blooded animals of silver nanoparticles and silver sulfate. *Modern Sci.* **2019**, *7*, 144–146.
56. Skalny, A.V.; Rudakov, I.A. *Bioelements in Medicine*; Mir: Moscow, Russia, 2004; 272p.

polymers

MDPI

Article

The Effect of Polymers on Drug Release Kinetics in Nanoemulsion *In Situ* Gel Formulation

K. Reeta Vijaya Rani [1], Sruthi Rajan [2], Mullaicharam Bhupathyraaj [3,*], R. Krishna Priya [4], Nirmala Halligudi [3], Mohammad Abobakr Al-Ghazali [3], Sathvik B. Sridhar [5], Javedh Shareef [5], Sabin Thomas [6], Saleem M. Desai [7] and Pandurang D. Pol [8]

1 Surya School of Pharmacy, Vikravandi, Villupuram 605652, Tamilnadu, India; reeta_rani07@yahoo.co.in
2 Periyar College of Pharmaceutical Sciences, Tiruchirappalli 620021, Tamilnadu, India; sruthirajan1997@gmail.com
3 College of Pharmacy, National University of Science and Technology, Muscat 130, Oman; nirmalahalligudi@nu.edu.om (N.H.); mohammadalghazali@nu.edu.om (M.A.A.-G.)
4 College of Engineering, National University of Science and Technology, Muscat 130, Oman; krishnapriya@nu.edu.om
5 RAK College of Pharmaceutical Sciences, RAK Medical and Health Sciences University, Ras Al Khaimah 11172, United Arab Emirates; sathvik@rakmhsu.ac.ae (S.B.S.); javedh@rakmhsu.ac.ae (J.S.)
6 School of Pharmacy, College of Pharmacy & Nursing, University of Nizwa, Nizwa 616, Oman; sabin@unizwa.edu.om
7 Anjuman Arts, Science, Commerce College of PG Studies in English, Vijayapura 586101, Karnataka, India; saleem_m_desai@yahoo.co.in
8 Department of Chemistry, BHS Arts and TGP Science College, Jamkhandi 586103, Karnataka, India; polpandurang@yahoo.in
* Correspondence: mullaicharam@nu.edu.om

Abstract: Glaucoma is an ocular condition characterized by elevated intraocular pressure (IOP). Conventional treatments of glaucoma face poor corneal permeability and bioavailability. To address these issues, a nanoemulsion in situ gel of Timolol maleate was developed in this study by adding the polymer Carbopol 934p. Using Carbopol 934p, a novel ophthalmic pH-induced nanoemulsion in situ gel was formulated. The formulation was liquid at pH 4 and quickly gelled when the pH was raised to 7.4 (Lacrimal pH). The pH-triggered in situ gelling mechanism demonstrated continuous drug release over a 24 h cycle. A total of nine trial formulations were prepared (NEI_1–NEI_9) and subjected to various physicochemical and in vitro evaluations. According to the in vitro release kinetics, the drug release of Timolol maleate nanoemulsion in situ gel NEI_5 followed zero-order kinetics, with a release exponent value of 0.902, indicating that the mechanism of release was non-Fickian diffusion regulated. In vivo results showed that Timolol maleate nanoemulsion in situ gel NEI_5 provided a better-sustained release of the drug, compared with the Timolet OD eye drops. The formulation is stable in storage, with no distinguishable change in appearance, physical properties, quality, and percentage drug release. NEI_5 also reduces drug administration frequency, which improves patient compliance. Timolol maleate nanoemulsion in situ gel NEI_5 achieved the goal of controlled drug delivery with extended-release and cost-effectiveness, lowering the dosage and frequency of drug administration, and thus may improve patient compliance. In conclusion, the stable nanoemulsion in situ gel of Timolol maleate NEI_5 decreases intraocular pressure (IOP) over a prolonged period.

Keywords: Carbopol 934p; glaucoma; Timolol maleate; nanoemulsion; in situ gel

1. Introduction

Nanoemulsions are a group of dispersed particles used for pharmaceutical and biomedical aids and vehicles that show great promise for the future of drug therapies, cosmetics, diagnostics, and biotechnologies. Nanoemulsions are defined as oil-in-water (o/w) emulsions, with mean droplet diameters ranging from 50 to 1000 nm [1]. According to the second law of thermodynamics, the o/w nano-sized emulsion is subjected to various instability processes such as aggregation, flocculation, coalescence, Ostwald ripening, and hence eventual phase separation [2]. Unlike thermodynamically stable microemulsions and clear transparent liquid systems, macro (coarse)- and nano-sized emulsions are meta-stable dispersions. However, the stability of the o/w macro (coarse)- and nano-sized emulsions can substantially be improved with the help of suitable emulsifiers or emulators that are capable of forming a mono- or multilayer coating film around the dispersed oil droplets to reduce interfacial tension and to increase droplet–droplet repulsion [3].

The proper ratio of oil:water: gum, the appropriate concentration of emulsifying agents, and high-efficiency emulsification equipment that are used to make very low droplet size are the most important factors to develop the o/w nano-sized emulsion with improved stability over the desired period (in comparison with coarse emulsion) can be obtained. The advantages of the nano-sized emulsion system include natural biodegradability, sub-micrometer droplet size range, stabilizability, and substantial drug solubilization either at the innermost oil phase or the o/w interface, minimizing side effects, and improved bioavailability. Due to these advantages, the nano-sized emulsion is now recognized as a promising drug delivery vehicle or carrier for parenteral and topical (ocular and percutaneous) applications [4–6].

Initially, in situ gel drug delivery systems are in sol form. There is no gelation process occur. Once administered inside the body through any one of the many routes, such as oral, ocular, rectal, vaginal, injectable, and intraperitoneal routes, the sol form will be converted to gel form due to the gelation process. In ophthalmic products, the formation of viscoelastic gel occurs after the installation of the liquid form of in situ forming hydrogels. These hydrogels are administered through the ocular cul-de-sac route where the hydrogel goes through a phase transition [7].

Natural polymers are mostly used in the preparation of in situ gel dosage form. For example, xyloglucan, a water-soluble anionic polysaccharide of gellan gum and algin are used for ocular drug delivery system.

Many components such as ocular drugs are used to alter the function of the nervous system. Non-steroidal anti-inflammatory drugs are used to prevent growth or to kill microorganisms. The disadvantages of using conventional ocular drug delivery systems such as eye drops are poor bioavailability and poor therapeutic response.

The reason behind these disadvantages is the fast removal of the drug from the eyes due to elevated tear fluids turnover. To overcome these disadvantages, in situ gels are prepared as the ophthalmic dosage form. Sustained drug release is possible from these in situ gels. In situ gels are viscous gels and have longer pre-corneal contact times, compared with conventional eye drops.

The gellan gum undergoes changeover into the gel state due to the temperature and ionic condition (Ca^{++}) in the tear fluid. Due to this property, an aqueous solution of gellan is used in ophthalmic drug delivery.

Glaucoma is a slowly progressive pathology that can result in the loss of peripheral vision, decreased contrast sensitivity, and loss of visual acuity. Due to the asymptomatic nature of the early phases of the disease, most patients experience undiagnosed loss of vision until the advanced stages of the disease have occurred. Thus, the disease is known as the "silent thief of sight". This indolent optic neuropathy is characterized structurally by a loss of retinal ganglion cells and optic nerve axons. Glaucoma is the second leading cause of the world's blindness, with nearly 70 million cases worldwide and accounting for 12% of all cases of preventable blindness [8–10]. It is estimated that by 2020, close to 4 million Americans will have glaucoma, with 50% undiagnosed and approximately 120,000 individuals developing blindness [11,12].

This work aims to extend drug availability in glaucomatous conditions by adding a different proportion of the polymer Carbopol 934p. Carbopol is a polyacrylic acid polymer, which shows a sol-to-gel transition in an aqueous solution as the pH is raised above its PK_a

of about 5.5, and it is widely used in ophthalmology to enhance precorneal retention to the eye [13]. Moreover, Carbopol exhibits excellent mucoadhesive properties when compared with other polymers.

Carbopol 934p is also used in liquid or semisolid pharmaceutical formulations as rheology modifiers. Due to this property of Carbopol 934p, the immediate release kinetics has been modified as zero-order kinetics of the formulation, which provides the more beneficial effect of the gel formulation [14,15].

2. Materials and Methods

2.1. Materials

A gift sample of Timolol maleate (Pure drug, Madras Pharma (P) Ltd., Chennai, India. Castor oil (Lab grade), Tween 80 (Lab grade), benzalkonium chloride, and glycerol (Lab grade) were purchased from Nice Chemicals (P) Ltd., Chennai, India. Potassium dihydro orthophosphate (Lab grade) was purchased from Scientific Chemicals, Chennai, India. Sodium hydroxide (Lab grade) was purchased from Hi Pure Fine Chem Industries, Chennai, India.

2.2. Methodology

2.2.1. Formulation of In Situ Gelling System

A conventional emulsion was prepared by dissolving Timolol maleate in castor oil, and glycerol was used as a cosolvent with continuous stirring in a magnetic stirrer. The aqueous solution of Tween 80 and a sufficient amount of water was added and stirred well. The oil phase was added dropwise in continuous phase with stirring at ambient temperature and added the benzalkonium chloride as a preservative. This conventional emulsion was converted into nanoemulsions with the help of a sonication mechanism. The final step was the addition of Carbopol 943p at pH 4. A total number of nine trial batches were prepared for the optimization of process variables [16–18]. The process flowchart for nanoemulsion in situ gel is shown in Figure 1.

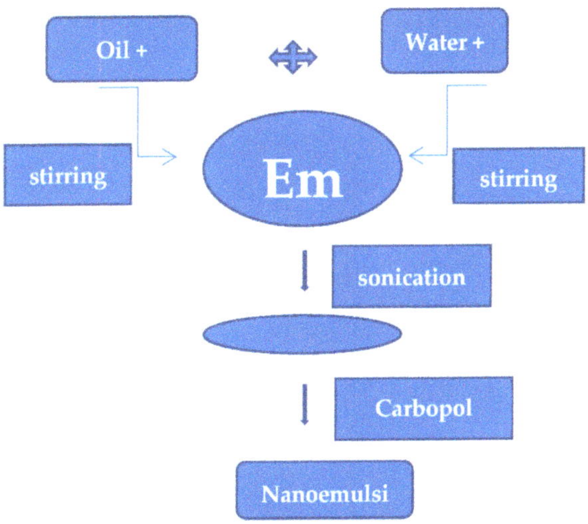

Figure 1. Preparation of nanoemulsion in situ gel by the ultra-sonication method.

Different concentrations of emulsifying agents and gelling agents were used in trial batches and studied to have a sustaining effect for 24 h. In all batches, the concentration of drug and oil were kept constant; the data are presented in Table 1.

Table 1. Compositions of the trial batch.

Ingredients	NEI_1	NEI_2	NEI_3	NEI_4	NEI_5	NEI_6	NEI_7	NEI_8	NEI_9
Timolol maleate (mg)	100	100	100	100	100	100	100	100	100
Castor oil (mL)	5	5	5	5	5	5	5	5	5
Tween 80 (mL)	3	3	3	3.5	3.5	3.5	4	4	4
Glycerol (mL)	2	2	2	2	2	2	2	2	2
Carbopol 934p	300	600	900	300	600	900	300	600	900
Benzalkonium chloride (%)	0.02	0.02	0.02	0.02	0.02	0.02	0.02	0.02	0.02
Distilled water (mL)	q.s	q.s	q.s	q.s	q.s	q.s	q.s	q.s	q.s

2.2.2. Characterization

The following parameters were evaluated for all the formulations to confirm the desired release of drug and stability of formulation: visual appearance and clarity, pH, viscosity, gelling capacity, and particle size analysis [19–23].

Visual Inspection

A visual inspection was carried out behind the dark background to observe the clarity and proper appearance of each formulation.

pH

The pH of each formulation was measured by using a digital pH meter (Elico).

Viscosity

A Brookfield viscometer (Brookfield DV-II + Pro viscometer (The Bharat Instruments & Chemicals, Ludhiana, India) with a small sample adapter, having spindle number SC4-18/13R, was used to quantify the viscosity of the prepared nanoemulsions. The gelling property was determined by mixing the 25:7 ratio of the formulation with simulated tear fluid, and the gelation was evaluated by visual examination. The time taken for the formation of gel and the time taken for dissolution was recorded.

Particle Size

Particle size distribution and the average size of particles present in the formula were determined by blue wave analytical mode by DLS method using Zetasize. Figure 2. Atomic force microscopy [24] was used to confirm the size and shape of the particles (Figure 3).

Sterilization and Sterility Testing

Moist heat sterilization is used for killing microorganisms. Autoclaving, as an efficient method to inactivate bacteria, viruses, and other biological material, is recommended for the disposal of regulated medical waste.

In this study, Timolol maleate nanoemulsion in situ gel was sterilized by moist heat sterilization. This process was carried out at 121 °C for 15 min under pressures of 15 lb/sq. inch. In this process, the moist-heat vapors at high temperatures precipitate or coagulate the cell wall proteins and destroy the microorganisms. The test for sterility is intended for detecting the presence of viable forms of bacteria, fungi, and yeast in sterilized preparations [25–28].

Content Uniformity

The vials (n = 3) containing the preparation were shaken for 2–3 min, and 100.0 μL of the preparations were transferred aseptically to sterile 25.0 mL volumetric flasks with a micropipette, and the final volume was made with phosphate buffer pH 7.4. The solution was filtered through a 0.45 μm membrane, and the concentration of Timolol maleate was determined at 295 nm, using a double beam UV spectrophotometer [29–33].

Compatibility

Drug–excipient compatibility studies were carried out by using Fourier transform infrared (FTIR) spectral analysis. The FTIR absorption spectra of the pure drug and physical admixtures of the drug with various excipients were taken in the range of 400–4000 cm^{-1} using the KBr disc method (Shimadzu IR-Prestige-21) and observed for characteristic peaks of the drug. The FTIR absorption spectra optimized formula is given in Figure 4.

In Vitro Release Study

The in vitro dissolution of the prepared in situ gel formulations was performed by diffusion method using an open embedded glass tube. A cellophane membrane pre-soaked in the dissolution media was fixed in the open end of the glass tube, considered as donor compartment that fixed inside 100 mL beaker containing 50 mL of phosphate buffer pH 7.4, which was used as receptor compartment. Then, 1 ml of the preparation was allowed to diffuse via the cellophane membrane to the receptor compartment, which was kept on a magnetic stirrer at 37 °C. Afterward, 5 mL sample was withdrawn in a specified time interval up to 24 h and analyzed by using Shimadzu Double beam UV–Visible spectrophotometer at 295 nm [34]. The cumulative % drug release in all the formulations were given in Table 5 and Figure 5.

In Vivo Studies

The intraocular pressure measurement in albino rabbits was studied in Periyar College of Pharmaceutical Sciences, Tiruchirappalli, Tamilnadu, India. (265/1/101/CPCSEA).

Intraocular Pressure Studies

The intraocular pressure study was conducted in albino rabbits (Haffkin strain) of either sex weighing between 1.8 kg and 2.5 kg. All experiments were carried out at room temperature [35].

Six rabbits were used for this study. Reduction in intraocular pressure (IOP) was measured by Schiotz tonometer. Minimum two readings of IOP were taken before administration of nanoemulsion in situ gel, which was denoted as I_o. The formulation (0.05 mL) was administered with the help of an insulin syringe in the lower cul-de-sac of one eye.

The control (0.05 mL) was administered in the right eye. Reduction in IOP at time t was denoted as I_t, and observations were recorded. The graph is plotted as I_n versus time where $I_n = I_t - I_o/I_t$. The same animal was used repeatedly, allowing a minimum of two days between two successive experiments. The results were compared with commercially available Timolet eye drops (containing 0.5% *w/v* of Timolol maleate manufactured by Sun Pharmaceuticals, Chennai, India).

Accelerated Stability

Accelerated Stability studies were carried out by exposing NEI_5 at various temperatures of 40 °C, and 2–8 °C. After a specific period of storage for stability, the in situ gel was evaluated for physical parameters, in vitro drug release, and drug content [36–39].

3. Results

The prepared in situ gel formulations were evaluated for various physicochemical evaluations such as visual appearance, clarity, gelling capacity, viscosity in pH 4 and pH 7.4, particle size, drug content, compatibility, and in vitro diffusion studies. Based on the Physicochemical and in vitro diffusion studies, formulation NEI5 has been selected and subjected to sterilization, in vivo, sterility testing, and accelerated stability studies. There was no microbial growth found for not less than 14 days at 30° to 35 °C in a fluid thioglycollate medium. The intraocular pressure effect of the Timolol maleate nanoemulsion in situ gel was compared with the effect of aqueous Timolet eye drops with 0.5% *w/v*. At 40 °C, there was a slight decrease in the consistency after three months. There was an increase in the viscosity after gelling. Additionally, the gel formed in situ maintained its integrity without dissolving or eroding for a prolonged period. Results are represented in Tables 2–8 and Figures 1 and 2.

3.1. Visual Appearance, Clarity, and Gelling Capacity

Evaluation of Visual appearance, Clarity, and Gelling Capacity carried out to find out the physicochemical properties of nine formulations having different compositions.

The status of all the nine formulations in terms of Visual appearance, Clarity, and Gelling Capacity is shown in Table 2.

Table 2. Visual appearance, Clarity, and Gelling Capacity.

Evaluation	NEI_1	NEI_2	NEI_3	NEI_4	NEI_5	NEI_6	NEI_7	NEI_8	NEI_9
Visual appearance	T	T	T	T	T	T	T	T	T
Clarity	C	C	C	C	C	C	C	C	C
Gelling capacity	+	++	+++	+	++	+++	+	++	+++

T—transparent, C—Clear, + gels slowly and dissolves; ++ gelation immediate and remains for a few hours; +++ gelation immediate and remains for an extended period.

3.2. Evaluation of Viscosity

In order to know the rheological property of the nine formulations in two different pH 4 and, pH 7.4 the viscosity was measured by using A Brookfield viscometer. The values of viscosity are given in Table 3.

Table 3. Evaluation of viscosity in pH 4 and pH 7.4.

Evaluation		Viscosity (cps)								
		NEI_1	NEI_2	NEI_3	NEI_4	NEI_5	NEI_6	NEI_7	NEI_8	NEI_9
pH	4	102	120	140	115	129	143	113	126	140
	7.4	226	260	290	230	265	299	230	260	302

3.3. Particle Size Analysis

Particle size distribution and the average size of particles was generated by using the formula were determined by blue wave analytical mode by DLS method using Zeta size. As shown in Figure 2, the report showed that the average mean diameter range was 76 nm to 1000 nm.

3.4. Atomic Force Microscopy

In order to confirm the size and shape of the particles, Atomic force microscopy was used. to confirm the size and shape of the particles (Figure 3). The surface morphology analyzed by atomic force microscopy (AFM) result showed a uniform, spherical, and discrete particle without aggregation, which was smooth in the surface and the nanosize range, at 260.4–351.8 nm.

3.5. Drug Content

The concentration of Timolol maleate of nine formulations was determined at 295 nm, using a double beam UV spectrophotometer. The percent-age of drug content in all the formulations given in Table 4.

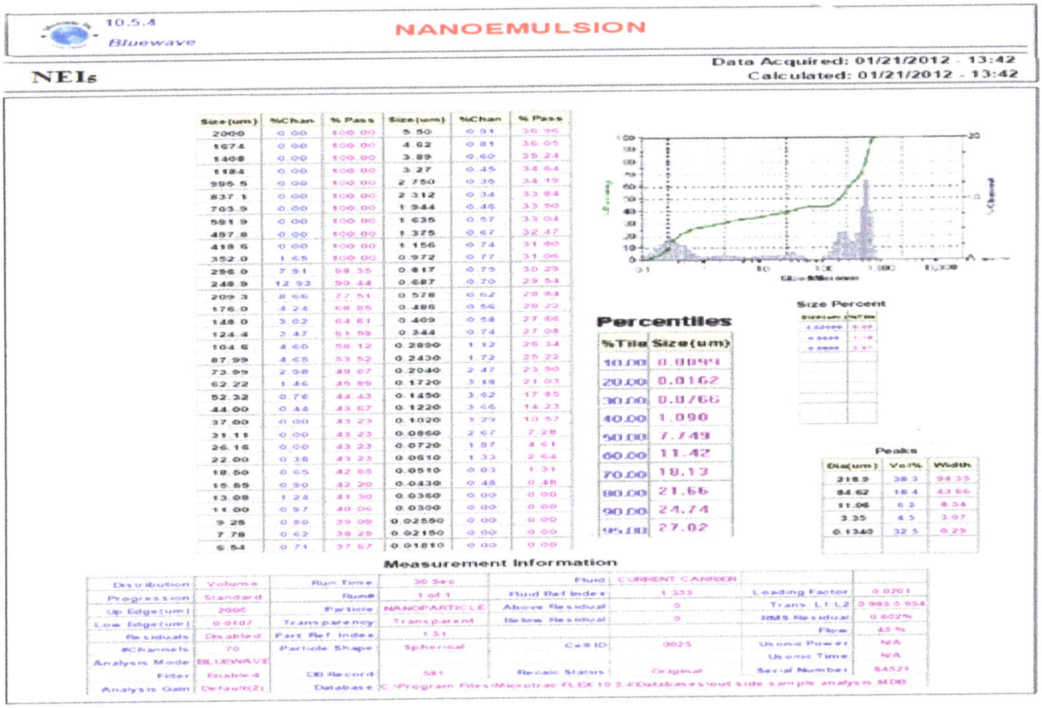

Figure 2. Particle size analysis.

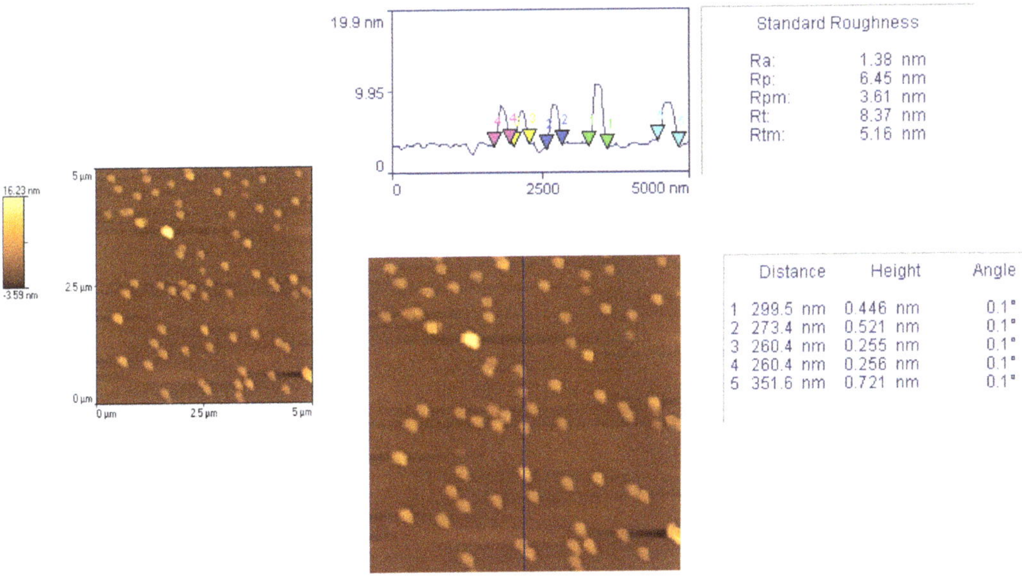

Figure 3. Atomic force microscopy.

Table 4. Drug content.

Formulation	NEI$_1$	NEI$_2$	NEI$_3$	NEI$_4$	NEI$_5$	NEI$_6$	NEI$_7$	NEI$_8$	NEI$_9$
Drug Content (%)	101.6	99.56	98.32	100.36	99.79	98.25	100.98	99.43	98.29

3.6. *Compatibility Studies*

The FTIR absorption spectra optimized formula is given in Figure 4. The FTIR spectra of formulation NEI5 showed that there were no extra peaks other than the normal peak in the spectra of the mixture of the extracts, containing active constituents, and excipients.

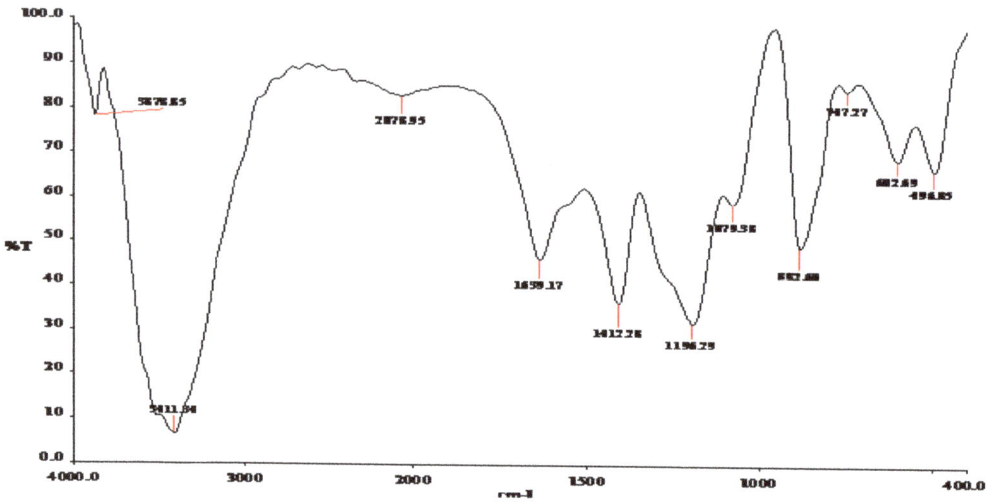

Figure 4. FTIR spectra of NEI5.

3.7. *In Vitro Diffusion Drug Release Profile*

The in vitro dissolution of the prepared in situ gel formulations was performed by diffusion method to find out the release profile of nine formulations. The cumulative % drug release in all the formulations were given in Table 5 and Figure 5.

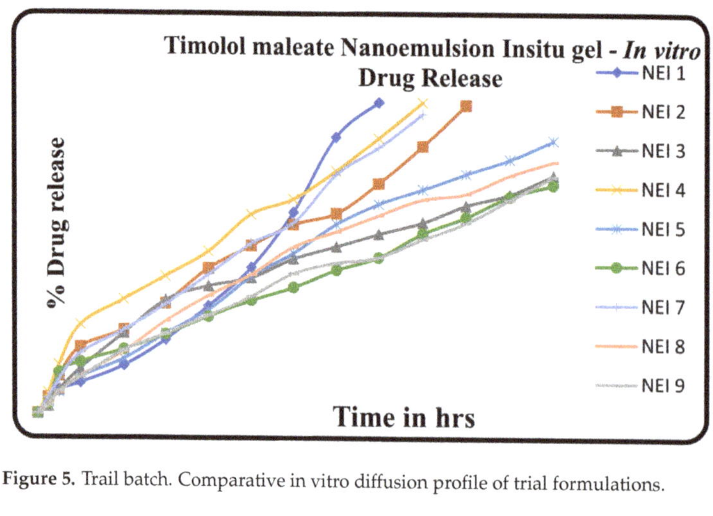

Figure 5. Trail batch. Comparative in vitro diffusion profile of trial formulations.

Table 5. Diffusion profiles of formulations trial batch: NEI_1–NEI_9.

Time (min)	Cumulative % Drug Release ± S.D. *								
	NEI_1	NEI_2	NEI_3	NEI_4	NEI_5	NEI_6	NEI_7	NEI_8	NEI_9
0	0	0	0	0	0	0	0	0	0
0.5	5.48 ± 1.1	5.11 ± 0.8	2.26 ± 0.3	6.53 ± 1.1	3.50 ± 0.4	2.97 ± 0.4	4.99 ± 0.3	3.07 ± 0.9	2.51 ± 0.14
1	7.61 ± 1.2	11.95 ± 1.5	7.35 ± 1.1	15.36 ± 1.2	6.65 ± 1.3	12.95 ± 0.2	10.78 ± 0.3	6.66 ± 1.1	6.99 ± 0.38
2	9.83 ± 0.5	21.10 ± 1.4	14.3 ± 2.2	28.31 ± 1.0	11.63 ± 1.3	16.37 ± 0.7	19.14 ± 0.5	11.9 ±0.4	11.59 ± 0.89
4	15.29 ± 0.9	26.63 ±0.7	25.54 ± 2.1	36.33 ± 0.7	17.44 ± 0.9	20.41 ± 1.1	26.52 ± 0.7	19.70 ± 1	20.02 ± 1
6	23.53 ± 0.6	35.17 ± 1.4	36.26 ± 0.09	43.80 ± 1.4	25.44 ± 0.8	25.25 ± 0.9	34.87 ± 0.9	29.42 ± 0.8	25.47 ± 0.96
8	34.37 ± 1.0	46.32 ± 1.0	40.75 ± 1.1	51.68 ± 1.0	33.16 ± 0.4	30.88 ± 0.8	44.30 ± 0.3	37.61 ± 1.1	31.25 ± 1.01
10	46.64 ± 1.1	53.56 ± 1.0	43.43 ± 1.2	63.47 ± 1.3	43.77 ± 0.5	36.16 ± 0.4	54.30 ± 0.9	44.40 ± 1.2	37.53 ± 1.08
12	64.20 ± 1.1	60.30 ± 0.09	49.43 ± 1.4	68.41 ± 1.0	51.23 ± 0.6	40.37 ± 0.4	60.84 ± 0.6	53.19 ± 1.0	44.80 ± 1.08
14	88.45 ± 0.8	64.08 ± 1.7	53.53 ± 0.9	77.54 ± 0.6	60.64 ± 1.2	46.08 ± 0.4	76.31 ± 0.5	58.31 ± 0.9	48.31 ± 0.82
16	99.53 ± 0.9	73.65 ± 1.1	57.65 ± 0.6	88.13 ± 1.8	67.15 ± 0.7	50.24 ± 0.4	85.12 ± 0.6	63.56 ± 0.8	50.22 ± 0.99
18	-	85.40 ± 1.0	61.23 ± 1.6	99.39 ± 1.2	71.95 ± 1.3	57.72 ± 0.9	95.76 ± 0.7	68.46 ± 0.9	55.91 ± 2.46
20	-	98.58 ± 1.1	66.65 ± 1.2	-	77.50 ± 0.3	62.95 ± 1.2	98.72 ± 0.5	70.37 ± 0.8	61.07 ± 0.87
22	-	-	70.28 ± 1.3	-	81.31 ± 0.9	69.50 ± 0.8	-	76.28 ± 1.0	68.54 ± 1.32
24	-	-	76.36 ± 1.32	-	87.40 ± 1.17	73.34 ± 0.7	-	80.71 ± 1.1	75.95 ± 0.48

* S.D. = standard deviation.

3.8. In Vivo Studies

Intraocular Pressure is shown is Table 6:

Table 6. Intraocular pressure (IOP mm Hg).

Time (min)		0	30	60	90	120	150	180	210	240	270	300
Formulation	R	18.7	18.7	18.7	18.7	18.7	18.7	18.7	18.7	18.7	18.7	18.7
	L	23.6	22.3	21.7	17.2	15.4	16.5	21.9	22.3	23.6	23.6	23.6
Marketed	R	18.0	18.0	18.0	18.0	18.0	18.0	18.0	18.0	18.0	18.0	18.0
	L	21.9	21.7	4.3	18.3	19.8	20.8	21.9	21.9	21.9	21.98	21.9

3.9. Accelerated Stability Study

Accelerated Stability studies were carried out by exposing NEI5 at various temperatures of 40 °C, and 2–8 °C. After a specific period of storage for stability, the in situ gel was evaluated for physi-cal parameters, in vitro drug release, and drug content.The results are showed in the Tables 7 and 8.

Table 7. Stability studies of optimized Timolol maleate nanoemulsion in situ gel.

S. No	Parameters	Initial	After 3 Months	
			40 °C	2–8 °C
2	pH	4	3.9	4
3	Viscosity	129	127	129
4			Drug Content (%)	
1	Timolol maleate	99.79	99.26	99.75

Table 8. Comparative in vitro diffusion profiles of Timolol maleate nanoemulsion in situ gel before and after storage at 3 months.

Time (h)	Cumulative % Release		
	Before Storage	After Storage	
		40 °C	2–8 °C
$\frac{1}{2}$	3.50	3.05	3.49
1	6.65	6.12	6.50
2	11.63	10.50	11.43

Table 8. *Cont.*

Time (h)	Cumulative % Release		
	Before Storage	After Storage	
		40 °C	2–8 °C
4	17.44	15.52	17.29
6	25.44	24.03	25.26
8	33.16	30.13	32.98
10	43.77	41.52	43.56
12	51.23	49.23	50.98
14	60.64	58.25	60.25
16	67.15	65.50	66.97
18	71.95	69.89	70.79
20	77.50	75.65	77.42
22	81.31	79.23	81.07
24	87.40	83.15	87.35

4. Discussion

In the present investigation, efforts were made to prepare in situ gel of Timolol maleate using surfactant and a gelling agent such as Tween 80 and Carbopol 934p, to enhance drug availability for a prolonged period and hence improve the bioavailability of ocular drugs in glaucomatous conditions. The use of the Carbopol 934p in situ gelling system is sustained by the property of its solutions to transform into stiff gels when the pH is raised. The two main prerequisites of an in situ gelling system are viscosity and gelling capacity.

To evaluate the rheological behavior, the viscosity of the formulation before and after the pH 4 to 7.4 was evaluated using a Brookfield viscometer. All selected formulations were shear thinning, exhibiting pseudoplastic behavior. All formulations were liquid at room temperature and underwent rapid gelation upon raising the pH 4 to 7.4, with Carbopol 934p formulation showing the optimum variation in viscosity. The comparative rheological properties of NEI_5 formulations at different pH conditions indicated 129 cps at pH 4 and up to 265 cps at pH 7.4.

The results of visual appearance and clarity, pH, gelling capacity, particle size analysis, and drug content are shown in Tables 2–4. The results demonstrate that all prepared formulations had a clear appearance with an acceptable pH and drug content. Moreover, the gelling capacity of NEI_1, NEI_5, and NEI_8 were found to be good, having immediate gelation, and the gel persisted for an extended period.

The particle size result showed that the average mean diameter range was 76 nm to 1000 nm. The surface morphology analyzed by atomic force microscopy (AFM) result showed a uniform, spherical, and discrete particle without aggregation, which was smooth in the surface and the nanosize range, at 260.4–351.8 nm. The FTIR spectra of formulation NEI_5 showed that there were no extra peaks other than the normal peak in the spectra of the mixture of the extracts, containing active constituents, and excipients, so no evidence was found of interaction with the drug and polymers, and therefore, they are compatible with each other.

Results of in vitro release of NEI_1–NEI_9 are illustrated in Table 5 and Figure 5, respectively. The prepared formulations such as NEI_1, NEI_2, NEI_4, and NEI_7 showed initial burst release. The regression coefficient for Timolol maleate preparation of zero-order plots was found to be 0.923, 0.982, and 0.976 from the NEI_3, NEI_5, and NEI_8. The regression values for the Timolol maleate of first-order plots were found to be 0.905, 0.835, and 0.830 from the NEI_3, NEI_5, and NEI8. When the release data were subjected to Higuchi matrix plots, it was observed that formulation for Timolol maleate with regression coefficients of 0.983, 0.990, 0.987 from the NEI_3, NEI_5, and NEI_8 suggested testing diffusion-controlled release. The "n" values obtained the from Korsemeyer–Peppas equation was found to be 0.836, 0.902, 0.839 from the NEI_3, NEI_5, and NEI_8. The diffusion exponent "n" of the Peppas model was more than 0.45, indicating the release of the drug was duc to the diffusion (non-fiction) mechanism.

The higher regression coefficient values for each formulation suggested that the formulation NEI_1–NEI_9 behaved as matrix types of drug release, with formulation NEI_5 having the maximum regression value. The result showed that formulation NEI_5 followed zero-order drug release kinetics, which is correlated with the results of the gelling capacity study, proving that NEI_5 provides immediate gelation for an extended period.

The NEI5 was sterilized by moist heat sterilization. There was no evidence of microbial growth when the formulation NEI5 was incubated for not less than 14 days at 30° to 35 °C in a fluid thioglycollate medium.

The reduction in the intraocular pressure effect of the Timolol maleate nanoemulsion in situ gel was compared with the effect of aqueous Timolet eye drops with 0.5% *w/v*. The in situ gel formulation greatly reduced the IOP, compared with the marketed conventional formulation.

Accelerated stability testing revealed that the consistency of gel was found to be the same especially at ambient temperature, but at 40 °C, there was a slight decrease inconsistency after three months. Variations were observed in pH values at all storage conditions; the pH of formulations was found to decrease slightly with time.

The maximum change was observed at 40 °C. It was revealed that fewer changes in drug content and higher drug release were observed when the formulations were stored at refrigerated temperature (2–8 °C).

5. Conclusions

The novel ophthalmic pH-triggered nanoemulsion in situ gel containing Timolol maleate was successfully formulated by using Carbopol 934p.

The formulation NEI_5 provided reasonably constant effective levels of drug within the ocular cavity for a period of 24 h, and the in vivo results clearly showed that the Timolol maleate nanoemulsion in situ gel (NEI_5) provided the best-sustained release of the drug in comparison with the marketed conventional dosage form. Timolol maleate nanoemulsion in situ gel formulation remained stable on storage conditions, with no apparent change in appearance, physical properties, drug content, and percentage drug release. This formulation (NEI_5) is an alternative to conventional eye drops for improving bioavailability through its longer precorneal residence time and ability to sustain drug release.

This formulation (NEI_5) also may reduce the frequency of drug administration, thus improving patient compliance. Timolol maleate nanoemulsion in situ gel (NEI_5) achieved the objective of controlled drug delivery with prolonged release and cost-effectiveness, which decreases dose and frequency of drug administration and hence can improve patient compliance. In conclusion, the stable nanoemulsion in situ gel of Timolol maleate (NEI_5) reduces the intraocular pressure over a prolonged period.

Author Contributions: K.R.V.R. contributed to writing original draft preparation. S.R. contributed to the methodology section of the research work. M.B. contributed to the conceptualization and editorial part of the research worked. R.K.P. contributed to the software and validation parts of the research work. N.H. contributed to in vitro evaluation studies and reviewed the writing of the article. M.A.A.-G. contributed to in vivo section of this research work. S.B.S., J.S. and S.T. contributed to the informal analysis of results. S.M.D. and P.D.P. contributed to the resources and evaluation section of this article. All authors have read and agreed to the published version of the manuscript.

Funding: This research received no external funding.

Institutional Review Board Statement: The in vivo study was conducted in in albino rabbits accordance with the Animal Eth-ical Committee of Periyar College of Pharmaceutical Sciences, Tiruchirappalli, Ta-milnadu, India, and approved by the Animal Ethical Committee." for studies involving albino rabbits. Approval Number (265/1/101/CPCSEA).

Informed Consent Statement: Not applicable.

Acknowledgments: The authors would like to acknowledge Sruthi Rajan for technical support.

Conflicts of Interest: The authors declare no conflict of interest.

References

1. Soni, H.; Sharma, S. Current Update on Nanoemulsion: A Review. *Sch. Int. J. Anat. Physiol.* **2021**, *4*, 6–13.
2. Tamilvanan, S.; Kumar, B.A.; Senthilkumar, S.R.; Baskar, R.; Sekharan, T.R. Assessment of Injectable Castor Oil-Based Nano-sized Emulsion Containing Cationic Droplets Stabilized by Poloxamer–Chitosan Emulsifier Films. *AAPS PharmSciTech* **2010**, *11*, 904–909. [CrossRef] [PubMed]
3. Grigoriev, D.O.; Miller, R. Mono- and multilayer covered drops as carriers. *Curr. Opin. Colloid Interface Sci.* **2009**, *14*, 48–59. [CrossRef]
4. Shen, J.Q.; Gan, Y.; Gan, L.; Zhu, C.L.; Zhu, J.B. Ion-sensitive nanoemulsion-in situ gel system for ophthalmic delivery of flurbiprofen axetil. *Acta Pharm. Sin.* **2010**, *45*, 120.
5. Yeung, P.; Hubbard, J.; Korchinski, E.; Midha, K. Pharmacokinetics of chlorpromazine and key metabolites. *Eur. J. Clin. Pharmacol.* **1993**, *45*, 563–569. [CrossRef] [PubMed]
6. Kassem, A.; Mohsen, M.; Ahmed, S.; Essam, M. Selfnanoemulsifying drug delivery system (SNEDDS) with enhanced solubilization of nystatin for treatment of oral candidiasis: Design, optimization, in vitro and in vivo evaluation. *J. Mol. Liq.* **2016**, *218*, 219–232. [CrossRef]
7. Lee, V.H.L.; Robinson, J.R. Topical ocular drug delivery: Recent developments and future challenges. *J. Ocul. Pharmacol.* **1986**, *2*, 67. [CrossRef]
8. Cedrone, C.; Mancino, R.; Cerulli, A.; Cesareo, M.; Nucci, C. Epidemiology of primary glaucoma: Prevalence, incidence, and blinding effects. *Prog. Brain Res.* **2008**, *173*, 3–14.
9. Quigley, H.A.; Broman, A.T. The number of people with glaucoma worldwide in 2010 and 2020. *Br. J. Ophthalmol.* **2006**, *90*, 262–267. [CrossRef]
10. Friedman, D.S.; Wolfs, R.C.; O'Colmain, B.J.; Klein, B.E.; Taylor, H.R.; West, S.; Leske, M.C.; Mitchell, P.; Congdon, N.; Kempen, J. Prevalence of open-angle glaucoma among adults in the United States. *Arch. Ophthalmol.* **2004**, *122*, 532–538.
11. National Eye Institute N.I.o.H. National Eye Institute. Available online: http://www.nei.nih.gov/ (accessed on 4 November 2011).
12. Prevent Blindness America. Prevent Blindness America. 2011. Available online: http://www.preventblindness.org/ (accessed on 4 November 2011).
13. Deshmukh, P.K.; Gattani, S.G. In vitro and in vivo consideration of novel environmentally responsive ophthalmic drug delivery system. *Pharm. Dev. Technol.* **2013**, *18*, 950–956. [CrossRef]
14. Sharma, N.; Bansal, M. Nanoemulsion: A new concept of the delivery system. *Chron. Young Sci.* **2010**, *2*, 2–6.
15. Anderson, D.R. The Optic Nerve in Glaucoma. In *Duane's Ophthalmology*, 15th ed.; Tasman, W., Jaeger, E.A., Eds.; Lippincott Williams & Wilkins: Philadelphia, PA, USA, 2009; Volume 48.
16. Nair, R.V.; Shefrin, S.; Suresh, A.; Anoop, K.R.; Nair, S.C. Sustained release timolol maleate loaded ocusert based on biopolymer composite. *Int. J. Biol. Macromol.* **2018**, *110*, 308–317. [CrossRef]
17. Hegde, R.R.; Bhattacharya, S.S.; Verma, A.; Ghosh, A. Physicochemical and pharmacological investigation of water/oil microemulsion of non-selective beta-blockers for the treatment of glaucoma. *Curr. Eye Res.* **2014**, *39*, 155–163. [CrossRef]
18. Sah, A.K.; Suresh, P.K. Medical management of glaucoma: Focus on ophthalmologic drug delivery systems of timolol maleate. *Artif. Cells Nanomed. Biotechnol.* **2017**, *45*, 448–459. [CrossRef]
19. Kulkarni, G.T.; Gowthamarajan, K.; Suresh, B. Stability testing of pharmaceutical products. *J. Appl. Pharm. Sci.* **2004**, *38*, 194–198.
20. Carlfors, J.; Edman, K.; Peterson, R.; Jornving, K. Rheological evaluation of Gelrite in situ gels for ophthalmic use. *Eur. J. Pharm. Sci.* **2006**, *6*, 113. [CrossRef]
21. Cuiné, J.F.; McEvoy, C.L.; Charman, W.N.; Pouton, C.W.; Edwards, G.A.; Benameur, H.; Porter, C.J. Evaluation of the impact of surfactant digestion on the bioavailability of danazol after oral administration of lipidic self-emulsifying formulations to dogs. *J. Pharm. Sci.* **2008**, *97*, 995–1012. [CrossRef]
22. Lopez-Montilla, J.C.; Herrera-Morales, P.E.; Pandey, S.; Shah, D. Spontaneous emulsification: Mechanisms, physicochemical aspects, modeling, and applications. *J. Dispersion. Sci. Technol.* **2002**, *23*, 219–268. [CrossRef]
23. Biradar, S.V.; Dhumal, R.S.; Paradkar, A.R. Rheological investigation of the self-emulsification process: Effect of co-surfactant. *J. Pharm. Pharm. Sci.* **2009**, *12*, 164–174. [CrossRef]
24. Akbas, E.; Soyler, U.B.; Oztop, M.H. Physicochemical and antimicrobial properties of oleoresin capsicum nanoemulsions formulated with lecithin and sucrose monopalmitate. *Appl. Biochem. Biotechnol.* **2019**, *188*, 54–71. [CrossRef] [PubMed]
25. Abraham, S.; Furtado, S.; Bharath, S.; Basavaraj, B.V.; Deveswaran, R.; Madhavan, V. Sustained Ophthalmic Delivery of Ofloxacin From An Ion Activated Insitu Galling System. *Pak. J. Pharm. Sci.* **2009**, *22*, 175. [PubMed]
26. Jain, A.; Jain, R.; Jain, S. *Basic Techniques in Biochemistry, Microbiology and Molecular Biology*; Autoclave; Springer: Berlin/Heidelberg, Germany, 2020; pp. 9–10.
27. Garibaldi, B.T.; Reimers, M.; Ernst, N.; Bova, G.; Nowakowski, E.; Bukowski, J.; Ellis, B.C.; Smith, C.; Sauer, L.; Dionne, K. Validation of autoclave protocols for successful decontamination of category a medical waste generated from care of patients with serious communicable diseases. *J. Clin. Microbiol.* **2017**, *55*, 545–555. [CrossRef]
28. Viveksarathi, K.; Kannan, K. Effect of the moist-heat sterilization on fabricated nanoscale solid lipid particles containing rasagiline mesylate. *Int. J. Pharm. Investig.* **2015**, *5*, 87.

29. Bindschedler, M.; Degen, P.; Lu, Z.L.; Jiao, X.Q.; Liu, G.Y.; Fan, F. Comparative Bioavailability of Benflumetol after Administration of Single Oral Doses of Co artemether under Fed and Fasted Conditions to Healthy Subjects. In Proceedings of the Fifth International Congress for Tropical Medicine and Malaria, Nagasaki, Japan, 17–22 November 1996; pp. 17–22.
30. Müller, R.H.; Jacobs, C.; Kayser, O. Nanosuspensions as particulate drug formulations in therapy. The rationale for the development and what we can expect for the future. *Adv. Drug Deliv. Rev.* **2001**, *23*, 3–19. [CrossRef]
31. Umapathi, P.; Ayyappan, J.; Quine, S.D. Development and validation of a dissolution test method for artemether and lumefantrine in tablets. *Trop. J. Pharm. Res.* **2011**, *10*, 643–653. [CrossRef]
32. Shanghai, B.; Aggarwal, G.; Harikumar, S. Solid self micro emulsifying drug delivery system, A review. *J. Drug Deliv. Ther.* **2013**, *3*, 168–174.
33. Gupta, E.; Barends, D.M.; Yamashita, E.; Lentz, K.A.; Harmsze, A.M.; Shah, V.P.; Dressman, J.B.; Lipper, R.A. Review of global regulations concerning biowaivers for immediate release solid oral dosage forms. *Eur. J. Pharm. Sci.* **2006**, *29*, 315–324. [CrossRef]
34. Reeshanteni, B.; Abdullah, K.; Rajermani, T. Formulation of an in-situ gelling system for ophthalmic delivery of erythromycin. *Int. J. Stud. Res. Technol. Manag.* **2017**, *5*, 1–8.
35. Campos, F.F.; Campmany, A.C.C.; Delgado, G.R.; Serrano, O.L.; Naveros, B.C. Development and Characterization of a Novel Nystatin-Loaded Nanoemulsion for the Buccal Treatment of Candidosis: Ultrastructural Effects and Release Studies. *J. Pharm. Sci.* **2012**, *101*, 3739–3752. [CrossRef]
36. Manavalan, R.; Ramasamy, C. *Physical Pharmaceutics*; Vignesh Publisher: Chennai, India, 1999; pp. 321, 322, 334–336.
37. Bott, R.F.; Oliveira, W.P. Storage conditions for stability testing of pharmaceuticals in hot and humid regions. *Drug Dev. Indus. Pharm.* **2007**, *33*, 393–401. [CrossRef] [PubMed]
38. Fathordoobady, F.; Sannikova, N.; Guo, Y. Comparing microfluidics and ultrasonication as formulation methods for developing hempseed oil nanoemulsions for oral delivery applications. *Sci. Rep.* **2021**, *11*, 72. [CrossRef] [PubMed]
39. Ng, X.W.; Liu, K.L.; Veluchamy, A.B.; Lwin, N.C.; Wong, T.T.; Venkatraman, S.S. A biodegradable ocular implant for long-term suppression of intraocular pressure. *Drug Deliv. Transl. Res.* **2015**, *5*, 469–479. [CrossRef] [PubMed]

polymers

MDPI

Article

Carbonised Human Hair Incorporated in Agar/KGM Bioscaffold for Tissue Engineering Application: Fabrication and Characterisation

Vieralynda Vitus [1,2], Fatimah Ibrahim [1,2,3], Shamsul Azlin Ahmad Shamsuddin [4], Nuguelis Razali [5], Noor Anastasha Balqis Noor Azlan [1] and Wan Safwani Wan Kamarul Zaman [1,2,*]

1 Department of Biomedical Engineering, Faculty of Engineering, Universiti Malaya, Kuala Lumpur 50603, Malaysia
2 Centre for Innovation in Medical Engineering (CIME), Department of Biomedical Engineering, Faculty of Engineering, Universiti Malaya, Kuala Lumpur 50603, Malaysia
3 Centre for Printable Electronics, Universiti Malaya, Kuala Lumpur 50603, Malaysia
4 Institute of Biological Science, Faculty of Science, Universiti Malaya, Kuala Lumpur 50603, Malaysia
5 Department of Obstetrics and Gynaecology, Faculty of Medicine, Universiti Malaya, Kuala Lumpur 50603, Malaysia
* Correspondence: wansafwani@um.edu.my; Tel.: +60-3-7967-7628

Abstract: Carbon derived from biomass waste usage is rising in various fields of application due to its availability, cost-effectiveness, and sustainability, but it remains limited in tissue engineering applications. Carbon derived from human hair waste was selected to fabricate a carbon-based bioscaffold (CHAK) due to its ease of collection and inexpensive synthesis procedure. The CHAK was fabricated via gelation, rapid freezing, and ethanol immersion and characterised based on their morphology, porosity, Fourier transforms infrared (FTIR), tensile strength, swelling ability, degradability, electrical conductivity, and biocompatibility using Wharton's jelly-derived mesenchymal stem cells (WJMSCs). The addition of carbon reduced the porosity of the bioscaffold. Via FTIR analysis, the combination of carbon, agar, and KGM was compatible. Among the CHAK, the 3HC bioscaffold displayed the highest tensile strength (62.35 ± 29.12 kPa). The CHAK also showed excellent swelling and water uptake capability. All bioscaffolds demonstrated a slow degradability rate (<50%) after 28 days of incubation, while the electrical conductivity analysis showed that the 3AHC bioscaffold had the highest conductivity compared to other CHAK bioscaffolds. Our findings also showed that the CHAK bioscaffolds were biocompatible with WJMSCs. These findings showed that the CHAK bioscaffolds have potential as bioscaffolds for tissue engineering applications.

Keywords: human hair; carbon; pyrolysis; biocompatibility; Wharton's jelly-derived mesenchymal stem cells

1. Introduction

The element carbon is widely distributed across the planet in a variety of allotropes [1]. In terms of physicochemical characteristics, carbon is superior to other materials such as metal and silicon [2]. Additionally, carbon is one of the most prevalent elements that our body uses as a building block, primarily in the form of hydrocarbon chains found in DNA components, lipids, proteins, carbohydrates, etc. [1,3]. As a result, carbon is being used more frequently as a biomaterial, especially in the creation of bioscaffolds or other technology for biomedical and tissue engineering purposes. In addition, the addition of carbon materials into the bioscaffold improves the bioscaffold's mechanical strength and biocompatibility with different cell lineages [4,5]. For example, a carbon nanofiber-incorporated collagen bioscaffold reportedly could promote neovascularisation and regeneration of cardiac tissue in a myocardial ischemia/reperfusion model in rats [6]. Another study showed that polyethylene glycol (PEG)-functionalized carbon nanotubes (CNT) incorporated into the silk fibroin (PEG-CNT/SF) bioscaffold could photoacoustically stimulate neural cells and promote neurite outgrowth [7]. Moreover, electrospun nanofibrous bioscaffolds incorporated with CNT are biocompatible with H9C2 cells [8]. Nonetheless, conventional synthesis methods and relatively expensive raw materials or non-renewable (e.g., coal, lignite, peat, SU-8 photoresist) materials used to synthesise carbon products have led to the high cost of production, which may affect their commercialisation potential [9–11]. Hence, this has encouraged researchers and manufacturers to find an alternative that is cost-effective and sustainable.

Considering these factors, the use of biomass wastes as raw materials to synthesise carbon products is rising [12]. Natural biomass wastes provide a better source of carbon precursors, as they are abundant and easily accessible. In addition, the use of biomass wastes has allowed for sound management of its accumulation, which could potentially be harmful to human well-being and the environment. Various biomass waste sources are currently being researched as potential carbon precursor materials [13–17]. Among all of these potential biomass wastes, human hair provides an inexpensive source of materials, and its usage could prevent zoonotic transmission, which can be deadly to humans. Moreover, humans shed approximately 50 to 150 strains of hair per day, and hair is one of the wastes commonly produced by human hair-based industries. This could lead to the accumulation of human hair that can negatively affect human health and well-being, and the environment. Human hair waste can potentially cause blockage in the sewage system when tufts of hair tangle in the system. Furthermore, tufts of hair could potentially be a matrix for dust to accumulate and can become a microorganism habitat in wet states. Therefore, hair waste management is important. The effort to tackle human hair waste accumulation is ongoing, and one of the strategies is synthesising carbon from human hair and using it as material to develop a functional product. To date, carbon derived from human hair has shown its potential application as electrochemical sensors [18], probes [19], and supercapacitors [20]. Despite that, the usage of carbonised human hair in bioscaffold development for tissue engineering applications is limited.

Furthermore, carbon material alone is insufficient to fabricate a bioscaffold suitable for tissue engineering applications. Therefore, combining it with other biomaterials such as natural polymers could improve the functionality of the bioscaffold, which is more suitable for tissue engineering applications. Among the existing natural polymers, konjac glucomannan (KGM) application in tissue engineering, especially in bioscaffold fabrication, is growing. KGM is a natural polysaccharide extracted from *Amorphophallus konjac* tubers [21]. KGM is largely made up of linear polymer consisting of D-glucose and D-mannose linked by β-D-1,4 bonds [22]. It is biodegradable, biocompatible, hydrophilic, and physicochemically stable, and it has film-forming properties [22,23]. Due to its beneficial properties, it has been widely applied in various fields including food industries [24,25], food packaging [26], medicine [27,28], drug carriers [29,30], and tissue engineering [31,32]. However, there is a limitation to using pure KGM film, mainly because of its hydrophilicity and weak mechanical strength [21]. These drawbacks can be overcome by blending with another polymer. Reinforcement strategies of KGM film include the blending of gum Arabic [33], zein [34], agar [35], and welan gum [36]. Agar is a polysaccharide derived from red algae, which is composed of 1,3-linked β-D-galactopyranose and 1,4-linked 3,6-anhydro-α-L-galactopyranose or agarobiose [37]. It is a renewable and easily accessible biopolymer, which has been used as gelling, thickening, water-holding, and stabilising agents in various fields including edible food and food packaging, biotechnology industries, medical fields, and pharmaceuticals [35,38–40]. Previous studies have shown positive effects of blending agar with KGM on the mechanical properties of film such as tensile strength, flexibility, and hydrophilicity [37,41].

Additionally, the size of the global scaffold technology market was estimated at USD 1.1 billion in 2020, and it is anticipated to increase at a compound annual growth rate of 8.4% from 2021 to 2028 [42]. The increasing need for three-dimensional (3D) cellular models in biological studies and translational research is primarily responsible for the market's expansion. The market for scaffold technology is predicted to expand because of the quick paradigm shift in cell culture that is needed to address the difficulties in the drug development process. The use of 3D cell cultures is expanding due to their effectiveness in simulating the physiological conditions found in vivo for the accurate representation of disease-causing microenvironmental variables. As a result, 3D cell culture techniques have become a cutting-edge tool for the early discovery of drugs as well as possible therapeutic options for the treatment of a range of disorders. By giving pertinent information about diseased cells, the 3D idea of cell cultivation enables the analysis of the phenotypic heterogeneity of cancer cells and heterotypic communication between cells, thus propelling the market [42]. Therefore, the investigation of a new bioscaffold as a cellular model with different materials or incorporating newer materials, such as carbon, to an existing developed bioscaffold is a relevant study to produce effective therapeutic tools with commercialisation potential.

The beneficial use of carbon materials in facilitating cell or tissue growth has influenced its application in tissue engineering by combining it with other biomaterials to develop a bioscaffold that has the same biological characteristics as the natural extracellular matrix (ECM) of the human body. In this regard, inexpensive and renewable carbon sources as well as facile synthesis methods to produce carbon from biomass waste (e.g., human hair) have offered non-costly carbon products. Furthermore, to the best of our knowledge, information concerning the effects of carbon derived from human hair incorporated in agar/KGM bioscaffolds and their application for cell development has not been reported. Therefore, developing this bioscaffold could help to better understand the feasibility of using carbon derived from biomass as an alternative to existing expensive carbon such as graphite. Hence, the objective of the present study was to develop a CHAK based on agar and KGM incorporated with carbon synthesised from human hair. The structural, physical, and biocompatibility of the bioscaffold were investigated. The results of this work show that human hair-derived carbon can be used to make bioscaffolds for tissue engineering applications.

2. Materials and Methods

2.1. Human Hair-Derived Carbon

Human hair waste treated with bleaching and colouring was obtained from a single female donor. The collected hair sample was rinsed with a 70% ethanol solution and distilled water to remove any adhered debris, such as dandruff and dust, and then was dried in an oven at 50 °C for 2 h. In this study, two types of carbon that are non-activated and activated carbon were synthesised.

2.1.1. Synthesis of Non-Activated Carbon

The synthesis of non-activated carbon was carried out by pyrolysing the hair sample at 900 °C in an open-ended quartz tube furnace (Nabertherm, Lilienthal, Germany) with a constant nitrogen gas flow at 2000 mL/min. Then, the pyrolysed carbon was pulverised into fine micro-powder and kept in a sealed jar for further characterisation and use. The non-activated carbon obtained was then referred to as HC.

2.1.2. Synthesis of Activated Carbon

The activated carbon synthesis was carried out by pyrolysing the hair sample at 300 °C for 1 h 30 min in the furnace with a constant nitrogen gas flow at 2000 mL/min. The cooled char was then mixed with a 1 M potassium hydroxide (KOH) solution in a 3:1 ratio. The mixture was then pyrolysed for 1 h at 800 °C in the furnace in a nitrogen environment. After that, the cooled carbon was mixed with 100 mL of 1 M HCl in a beaker and stirred for 2 h. The mixture was then filtered and rinsed with PBS (pH 7.4) until the neutral pH of the filtrate was obtained. The filtered precipitate was placed in an oven at 90 °C until it was dry. Finally, the dried carbon was pulverised into fine micro-powder and kept in a sealed jar for further characterisation and use. The activated carbon obtained was then referred to as AHC.

2.1.3. Characterisation of Carbon

Morphological characterisation of raw human hair, carbonised human hair before pulverisation, HC, and AHC were observed and analysed using tabletop scanning electron microscopy (SEM, Phenom ParticleX Steel Desktop SEM, Thermo Scientific, Waltham, MA, USA). HC and AHC were analysed using an X-ray diffractometer (XRD, Empyrean, Panalytical, Worcestershire, United Kingdom) with a Cu-Kα1 source. The X-ray generator was 40 kV/40 mA, and the detector was a two-dimensional semiconductor X-ray detector. XRD diffraction peaks were obtained in 2θ continuous scanning mode within a scanning range of 0–60° and at a scanning speed of 1° per min.

2.2. *Fabrication of Bioscaffold*

This study fabricated six types of bioscaffolds, including pure agar (A), agar/KGM (AK), 0.01 g HC/agar/KGM (1HC), 0.03 g HC/agar/KGM (3HC), 0.01 g AHC/agar/KGM (1AHC), and 0.03 g AHC/agar/KGM (3AHC).

The A bioscaffold was fabricated with agar powder (0.1 g) mixed in phosphate-buffered saline (PBS, pH 7.4, 10 mL) and heated to 98 °C on the hot plate. Moreover, the AK bioscaffold was fabricated with agar (0.1 g) and KGM (0.1 g) powder mixed in PBS (10 mL) and heated to 98 °C. Similarly, the CHAK was fabricated by mixing the composite materials consisting of agar and KGM and carbon in PBS (10 mL) and heated to 98 °C (Figure 1). Then, the resulting molten solution was slightly cooled and poured into the mould to set, which was prepared using poly (methyl methacrylate) (PMMA) glass. Three PMMA glasses were aligned at the bottom, and then two PMMA glasses with a 15 mm distance between each other were stacked on top of the three PMMA glasses. Then, the molten mixture was poured into the mould and covered with PMMA glass. The thickness of the PMMA glass was 4 mm. After cooling, the PMMA glass was removed, and the gel was cut at 1.5 cm × 1.0 cm, followed by freezing for 2 h at –20 °C. Then it was thawed at room temperature and immersed in 95% ethanol for 1 h with gentle stirring several times. Lastly, it was allowed to dry in the fume hood.

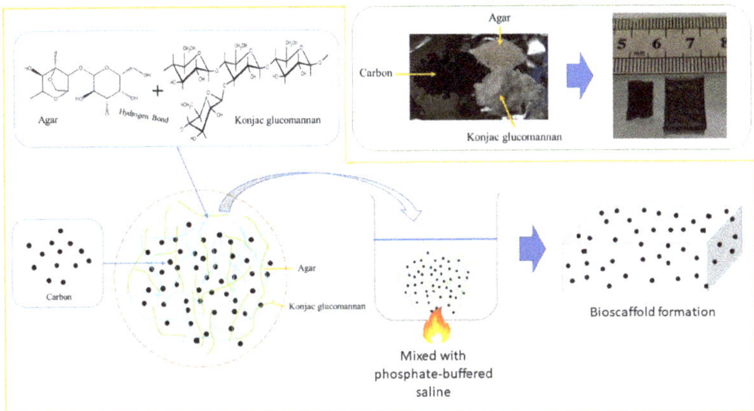

Figure 1. Schematic illustration of the bioscaffold formation.

2.3. *Characterisation of Bioscaffold*

2.3.1. Morphological Analysis

Morphological characterisation of the bioscaffold was observed using scanning electron microscopy (SEM, Phenom ParticleX Steel Desktop SEM, Thermo Scientific, Waltham,

MA, USA). The porosity of the dried bioscaffold was determined from the SEM results using ImageJ analysis software (Java 1.8.0_172 (64-bit)).

2.3.2. FTIR Analysis

FTIR analysis was performed to determine the possible interaction between carbon and the composite materials. The FTIR spectrum of agar, agar and KGM, and CHAK were analysed using a spectrometer (Nicolet™ Summit PRO, Spectrometer, Thermo Scientific, Waltham, MA, USA), whereby the bioscaffolds were analysed in the dried state. The analysis was performed at a wavelength range of 4000–500 cm^{-1} at a resolution of 2 cm^{-1} per point at room temperature and evaluated for the carbon and excipients interaction by analysing characteristic peaks.

2.3.3. Swelling and Water Uptake Analysis

In this test, the bioscaffold's dry weight (Wi) was first measured, and then the bioscaffold was immersed in PBS for 1 h at 37 °C. The excess water on the bioscaffold was removed by gentle dabbing with filter paper, and the swollen bioscaffold weight (Wf) was measured. The swelling capacity of the bioscaffold was calculated as the ratio of PBS absorbed to the initial dry weight of the scaffold, whereas the water uptake capability of the bioscaffolds was measured as the ratio of absorbed PBS to the swollen weight of the bioscaffolds. Measurements were made in triplicate.

$$\text{Degree of Swelling } (\%) = \frac{(\text{Wf} - \text{Wi})}{\text{Wi}} \times 100\% \quad (1)$$

$$\text{Water Uptake Capacity } (\%) = \frac{(\text{Wf} - \text{Wi})}{\text{Wf}} \times 100\% \quad (2)$$

2.3.4. Degradation Analysis

Initially, the weight of the bioscaffold was measured, and then the degradation analysis of the bioscaffold was carried out under static conditions for up to 28 days immersed in PBS (pH 7.4) at room temperature. The PBS was changed after 14 days of incubation for 21 days and 28 days of the group study. On analysis day, the excess PBS was removed by gently dabbing the bioscaffold on filter paper and placed it in the oven at 50 °C until completely dry. Then, the final weight of the dried bioscaffold was measured. Measurements were made in triplicate.

2.3.5. Tensile Test

The tensile test was performed using a universal testing machine (AG-X Series, SHIMADZU, Kyoto, Japan). Both ends of the bioscaffold were clamped and sprayed with phosphate buffer saline (PBS, pH 7.4) to rehydrate the bioscaffold. Then, the bioscaffold was retracted with a sample length of 10 mm and a speed of 2 mm/min. The measurement of the tensile strength was carried out on the semi-wet state of the bioscaffold.

2.3.6. Electrical Conductivity Analysis

Initially, the bioscaffold was immersed in PBS for 1 h. Then, two 70% alcohol-cleaned steel blades were placed on a non-conductive platform at a 0.35 cm distance from one another. Electrical conductivity was carried out by placing the hydrated bioscaffold between the steel blades, and the resistivity of the bioscaffold characterization was measured using a digital multimetre. The electrical conductivity was evaluated by its resistance activity. Five resistance (R) measurements were performed on each sample, and the associated conductivity (δ) of the bioscaffold was calculated using the Pouillet's law equation:

$$\delta = \frac{\text{L}}{\text{R} \times \text{A}} \quad (3)$$

where R represents the resistance of the hydrated bioscaffold, L indicates the length, and A represents the surface area of samples, $A = \pi \times (D/2)^2$, while D is the diameter of the bioscaffold [43].

2.4. Cell Culture

2.4.1. Preparation of WJMSCs

This study was approved by the Medical Research Ethics Committee of the University of Malaya Medical Centre (MREC ID No: 2021518-10145). Wharton's jelly-derived mesenchymal stem cells (WJMSCs) isolated from the human umbilical cord were used to conduct biocompatibility analysis on the fabricated bioscaffold. Initially, the collected umbilical cord was immersed in 70% ethanol for 30 s followed by rinsing twice with PBS. Then, the inner lining of the umbilical cord tissue, known as Wharton's jelly, was cut into approximately 5 mm in diameter and rinsed with PBS to remove the blood clots. Next, the tissue was transferred into a T25 flask containing complete media comprising 90% DMEM/F-12 (ATCC), 10% Gibco Foetal Bovine Serum (FBS; Thermo Fisher Scientific, Waltham, MA, USA, Europe approved, South American Origin), 1% of Gibco antibiotic–antimycotic solution (100×, Thermo Fisher Scientific), and 1% of Gibco Glutamax supplement (Thermo Fisher Scientific) and incubated at 37 °C in a humid atmosphere of 5% CO_2. The isolated cells were trypsinised and grown until passage 3.

2.4.2. Characterisation of WJMSCs

Characterisation of the WJMSCs was performed based on their morphology, surface marker, and differentiation ability. Immunophenotyping analysis to detect the surface marker of the WJMSC was conducted according to the manufacturer protocol using a BD Stemflow Human MSC Analysis Kit (BD Biosciences, San Jose, CA, USA). The differentiation ability of WJMSCs was performed using the Human Mesenchymal Stem Cell Functional Identification Kit (R&D Systems, Minneapolis, MN, USA) according to the manufacturer's protocol.

2.4.3. Biocompatibility Analysis of Bioscaffold

Before the biocompatibility study, the bioscaffolds were placed in a 24-well tissue culture plate and immersed in 70% ethanol, then left to dry in a biosafety cabinet under UV light for at least 2 h. Then the bioscaffolds were rinsed twice with PBS for 10 min each. Subsequently, the bioscaffolds were pre-soaked in 0.5 mL of complete medium containing 90% DMEM/F-12, 10% of FBS, 1% of the antibiotic–antimycotic solution, and 1% of glutamax supplement for 30 min. Then, a concentration of 3×10^4 cells per well of WJMSCs was cultured on the bioscaffold. The bioscaffold was then incubated at 37 °C in a humid atmosphere of 5% CO_2.

Biocompatibility analysis was measured using the MTT (3-(4,5-Dimethylthiazol-2-yl)-2,5-Diphenyltetrazolium Bromide) assay. The biocompatibility analysis was performed at 1 day, 3 days, and 7 days of incubation. After the period of incubation, the conditioned medium was discarded and washed with PBS, followed by the detachment of cells using trypsin-EDTA (trypsin-ethylenediamine tetraacetic acid) solution. Then, 1000 μL of complete media was added to stop the trypsin-EDTA activity. The detached cells were transferred to a centrifuge tube and centrifuged at 2000 rpm for 1 min. The cells were resuspended with 500 μL of complete media and incubated in a 24-well tissue culture plate for 45 min at 37 °C in a humid 5% CO_2 atmosphere. After incubation, the complete medium was replaced with 500 μL of a complete media containing MTT reagent (0.5 mg/mL). The cells were then incubated for 4 h at 37 °C in the dark. After 4 h, the working solution was removed, and 500 μL of dimethyl sulfoxide (DMSO; Sigma-Aldrich, Saint Louis, MO, USA) was added to dilute the formazan crystal product for 18 min. The production of the formazan crystal absorbance value was then measured by using a microplate reader at a 570 nm wavelength. The observations and measurements were performed in triplicate.

2.5. Statistical Analysis

Statistical analysis was conducted using one-way and two-way analyses of variance (ANOVA), followed by a Tukey post hoc test for multiple comparisons to confirm significant differences among samples using GraphPad Prism software (Dotmatics, San Diego, CA, USA). The data from this study were obtained in triplicate for each group, except for porosity and electrical conductivity analysis data, which were obtained in quintuplet, and the data were presented as mean value ± standard deviation (SD). The statistical significance was considered at a probability of $p < 0.05$ (* $p \leq 0.05$, ** $p \leq 0.01$, *** $p \leq 0.001$, **** $p \leq 0.0001$).

3. Results and Discussion

3.1. Carbon-Derived Human Hair

Following the synthesis of carbonised human hair, SEM micrographs of human hair and carbon samples were analysed and are presented in Figure 2. Human hair is comprised of 3 main layers, which are the outer cuticle layer, the inner cortex layer, and the innermost layer of the medulla (Figure 2A) [44]. The SEM micrograph of raw human hair showed the outer cuticle layer surrounding the hair fibre (Figure 2D), and the cross-sectional micrograph of raw hair (Figure 2B) showed the three main layers that make up human hair. After pyrolysis, the resulting carbonised human hair showed a hollow microfibre structure similar to the result obtained by Pramanick et al. (2016) (Figure 2C) [45]. The medulla layer's removal and the fusion of the cuticle and cortical layers led to the formation of a hollow carbon microfibre [45].

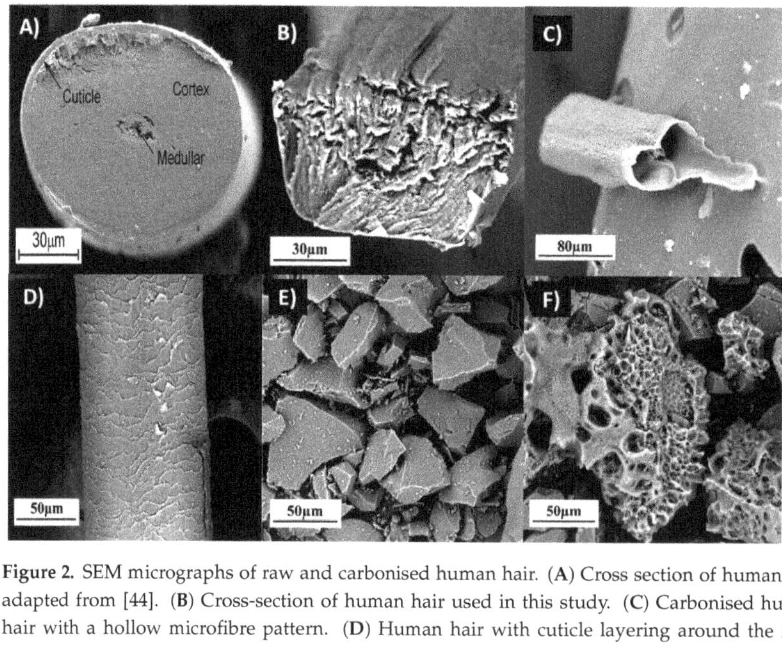

Figure 2. SEM micrographs of raw and carbonised human hair. (**A**) Cross section of human hair adapted from [44]. (**B**) Cross-section of human hair used in this study. (**C**) Carbonised human hair with a hollow microfibre pattern. (**D**) Human hair with cuticle layering around the fibre. (**E**) Non-activated carbon. (**F**) Activated carbon with porous structure.

In this study, two forms of carbon powder derived from human hair were synthesised using the pyrolysis process: (i) non-activated human hair-derived carbon (HC), and (ii) activated human hair-derived carbon (AHC). Both pulverised carbons exhibited microparticle sizes (<200 μm). The synthesised HC had a rough surface without porous features (Figure 2E), whereas the AHC showed a porous surface (Figure 2F). The resulting porous structure was due to the KOH chemical treatment. The KOH helped to remove volatile substances during activation [46]. The following reaction is a representation of the creation of porosity by KOH activation [47]:

$$6KOH + 2C \rightarrow 2K + 3H_2 + 2K_2CO_3$$

Additionally, potassium carbonate (K_2CO_3) reaction with carbons created potassium (K), potassium oxide (K_2O), carbon monoxide (CO), and carbon dioxide (CO_2), which increased the creation of pores. It was hypothesised that during the carbonisation process, KOH was converted to metallic potassium K, which entered the carbon particle and created the pores [47]. A study reported similar findings on KOH-activated pollen grains [48]. The porous carbon structure produced by the activation of carbon with an activating chemical at high temperatures resembles characteristics identified by other studies [46,49–51].

Additionally, the degree of crystallinity of the synthesised HC and AHC was evaluated through XRD analysis, and the XRD pattern is shown in Figure 3. The relative fractions of the crystalline were evaluated based on the maximum intensity of the 002 peaks. The XRD patterns had a very high baseline and a wide peak range. The diffraction angle (2θ) of the 002 peaks of crystalline carbon was around 24.03° and 22.24° for AHC and HC, respectively, whereas the human hair had 002 peaks located at 20.19°. This indicated that the micro-crystallites of hair shifted after carbonisation. Moreover, there were slight changes in the 002 peak between HC and AHC, indicating the activation process caused a shift of the micro-crystallites, thus resulting in a porous carbon structure.

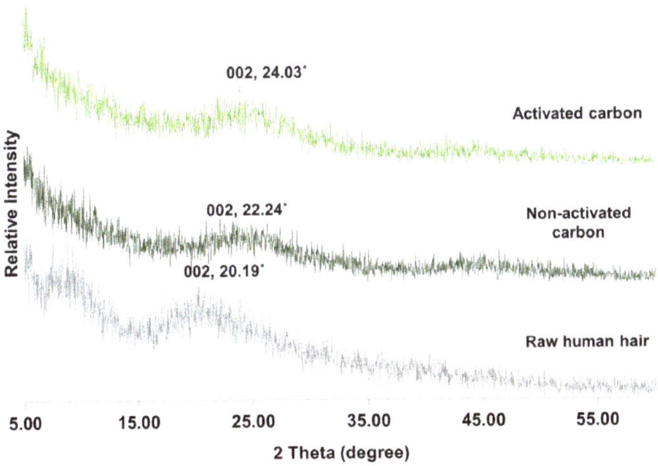

Figure 3. XRD results of raw human hair, non-activated carbon, and activated carbon.

3.2. Fabrication and Characterisation of Carbon-Based Bioscaffold

3.2.1. Morphological Analysis of Bioscaffold

The fabricated bioscaffold appeared translucent for the A bioscaffold, white-yellowish for the AK bioscaffold, and blackish for the bioscaffold incorporated with carbon. High-resolution 10 kV SEM characterization was performed to examine the surface morphology of the developed bioscaffolds, as shown in Figure 4. Based on Figure 4, the outer surface of the agar bioscaffolds showed a relatively rough surface, whereas the outer surface of the other fabricated bioscaffolds was rougher. In addition, the cross-section image of the bioscaffolds presented interconnected pores with undefined pore structure and size. The interconnected pore allowed for a better exchange of cells' metabolic waste and nutrients from the growth media [52,53]. The porosity of the bioscaffold was also important, as it influenced the infiltration and development of cells and the integration of host tissue, and it provided a structure for vascularisation [54].

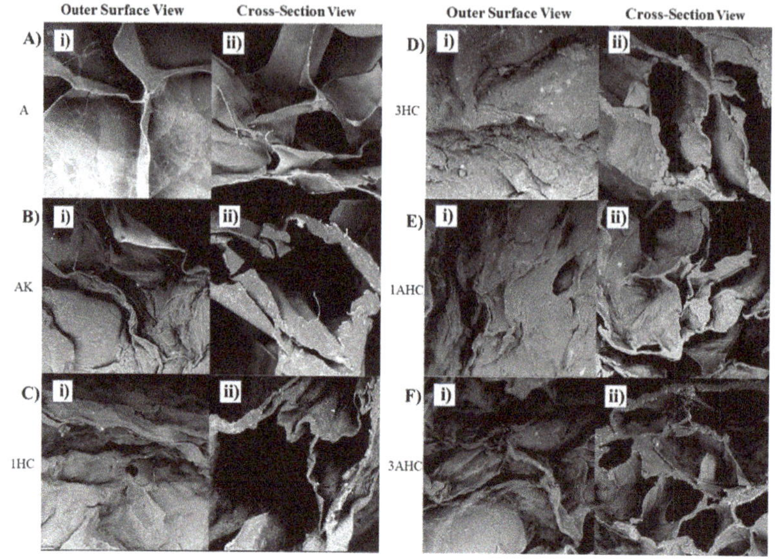

Figure 4. SEM micrographs of the fabricated bioscaffold. Each bioscaffold displayed (**i**) similar outer surface and (**ii**) cross-section features. (**A**): agar; (**B**): agar/KGM; (**C**): agar/KGM/0.01 g HC, (**D**): agar/KGM/0.03 g HC, (**E**): agar/KGM/0.01 g AHC; and (**F**): agar/KGM/0.03 g AHC.

3.2.2. Porosity Analysis of Bioscaffold

The porosity of the bioscaffold was analysed using ImageJ software based on the SEM images of each fabricated bioscaffold, and the results are depicted in Figure 5B. The percentage of the bioscaffold's porosity was evaluated from five cross-section images of the fabricated bioscaffold in dry conditions. The AK bioscaffold had the highest porosity (36.67 ± 8.1%) as compared to other fabricated bioscaffolds. Moreover, agar, 1HC, 3HC, 1AHC, and 3AHC had average percentages of porosity of 31.35 ± 7.92%, 20.77 ± 5.58%, 24.55 ± 6.33%, 19.37 ± 6.78%, and 27.1 ± 10.24%, respectively. This result indicated that the addition of carbon into the bioscaffold matrix reduced the porosity when compared to the AK bioscaffold. Overall, the series of fabrication methods used in this study successfully produced a porous bioscaffold, which endowed an alternative method when a freeze dryer was not available.

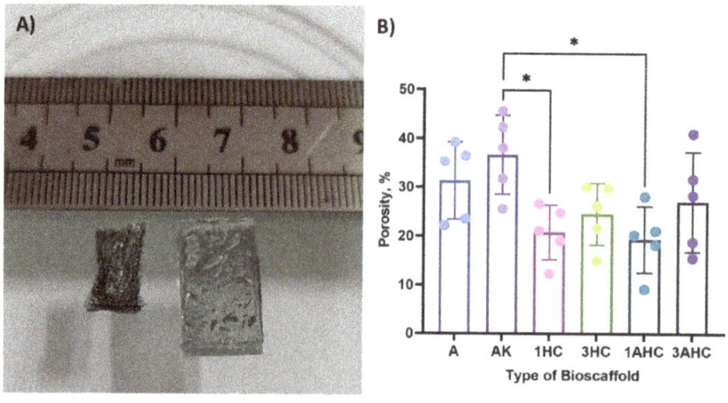

Figure 5. (**A**) Image of dried and rehydrated CHAK. (**B**) The porosity of the fabricated bioscaffolds. The dots on the graph represent individual value of each sample in the group. The statistical significance was considered at a probability of $p < 0.05$ (* $p \leq 0.05$).

3.2.3. Fourier Transform Infrared Analysis

The diffraction intensity of the bioscaffold, as analysed using FTIR, was shown with peaks in the range of 4000–500 cm^{-1}, as displayed in Figure 6. FTIR spectra of every type of bioscaffold showed characteristics absorption bands at ~3241 cm^{-1}, demonstrating the stretching of the hydroxyl (–OH) group and peaks at ~2933 cm^{-1}, indicating C–H stretching associated with the ring methine hydrogen bond. Moreover, the stretching of C–O of the hydroxyl group was found at around 1636 cm^{-1} [37,55]. In addition, at 1032 cm^{-1}, the vibration of the pyranose ring C–O–O skeleton was detected [56]. Overall, the peak properties showed no changes between A, AK, and CHAK, indicating minor chemical and structural changes in the constituent molecules of the bioscaffolds. This indicates that the combination of carbonised human hair, agar, and KGM is compatible.

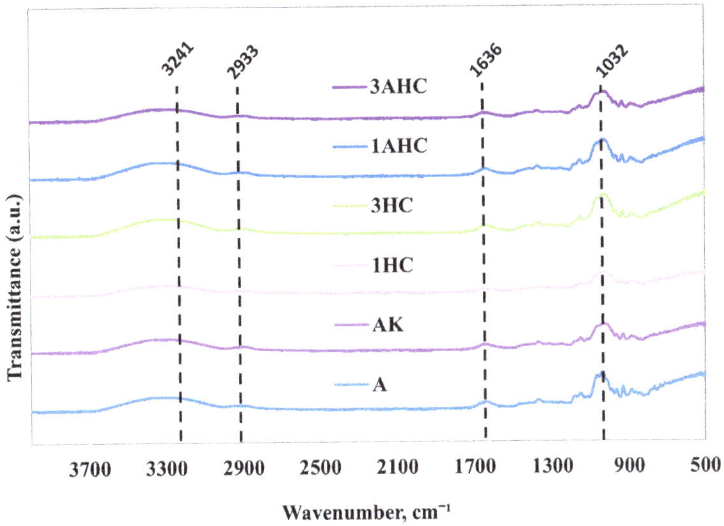

Figure 6. FTIR results of agar, agar/KGM, and CHAK.

3.2.4. Swelling Ratio Measurement

A bioscaffold's swelling ability is important for tissue engineering applications. Therefore, the fabricated bioscaffold's swelling capability was tested and is shown in Figure 7A. All of the fabricated bioscaffolds demonstrated above 1000% of swelling, similar to previous studies [56,57]. The AK bioscaffold exhibited the highest swelling capability (>2000%) among the other fabricated bioscaffolds. Based on Figure 7, the blending of KGM in the bioscaffold enhanced the swelling ability of the bioscaffold. This is because of the high hydroxyl (–OH) groups in the molecular chain of KGM, which offered robust adhesion and mechanical properties [58]. Therefore, the water absorption of the fabricated bioscaffold was enhanced. Moreover, the 3HC bioscaffold showed the lowest swelling capability (<1500%) among the other fabricated bioscaffolds.

In general, the swelling capability of the bioscaffold lessened with the incorporation of carbon, especially when compared to the AK bioscaffold. However, among CHAK, the 3AHC bioscaffold had better swelling capability in comparison to the 3HC bioscaffold, which could be because of the porous structure on the AHC surface, which could take up more liquid. In addition, some of the KGM molecules were displaced with the non-porous carbon within the 3HC bioscaffold, causing lowered HC-based bioscaffold swelling capability. In addition, the fabricated bioscaffold exhibited good water uptake capability (>90%), as shown in Figure 7B.

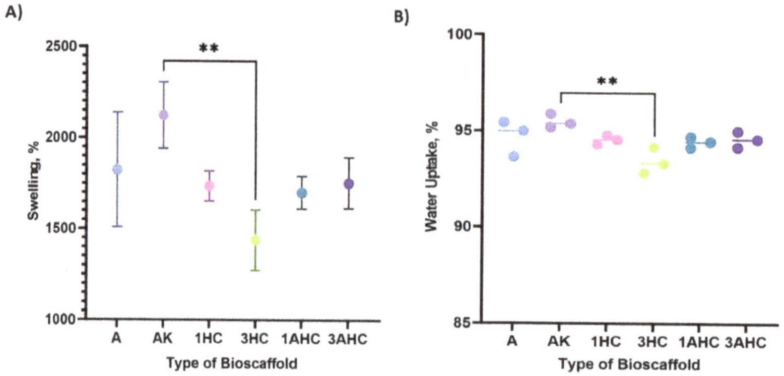

Figure 7. (**A**) Swelling capabilities and (**B**) water uptake of bioscaffold results. The dots on the graph represent individual value of each sample in the group. The statistical significance was considered at a probability of $p < 0.05$ (** $p \leq 0.01$).

3.2.5. Degradation Analysis

A bioscaffold's ability to degrade is essential for enabling the development of the cells' extracellular matrix. Before the restoration or regeneration of tissue at the implanted site is finished, a good bioscaffold should not start to break down. Therefore, the fabricated bioscaffold degradation rate was evaluated with PBS at room temperature for 28 days and is depicted in Figure 8. Based on Figure 8, the A bioscaffold showed the lowest rate of degradation among the fabricated bioscaffolds after 28 days. In contrast, the AK bioscaffold exhibited the highest degradation capability. According to the graph, the CHAK group had a slightly lower degradation rate compared to the AK bioscaffold. Apart from the A bioscaffold, the overall degradability of the fabricated bioscaffolds reached about 35% after 28 days. In comparison to the A bioscaffold, the inclusion of KGM may weaken the matrix causing the bioscaffold to be more prone to deterioration. Additionally, although it was not noticeably slower, the carbon addition slowed down the produced bioscaffold's rate of degradation.

Cells will be able to proliferate and differentiate on a bioscaffold with slow degradation before being administered to the intended treatment site. Typically, stem cells need to develop into the appropriate adult cell lineage for longer than 14 days [59]. Therefore, it is crucial to have a robust bioscaffold that gives cells enough time to mature and infiltrate in vitro as well as enough time to build tissue matrices in vivo. As a result, this work effectively manufactured a robust bioscaffold for use in tissue engineering applications based on the degradability evaluation. However, additional degradation research on the produced bioscaffold is necessary, using study parameters that mirror the pH and temperature of the human body.

3.2.6. Tensile Strength Analysis

The bioscaffold's mechanical strength is essential to ensure that the bioscaffold can withstand the compressive force produced by cells and tissue. In this study, we investigated the fabricated bioscaffold tensile strength using a universal testing machine. The results of the tensile strength test were displayed in the form of tensile strength values (kPa) and are shown in Figure 9. The bioscaffolds were sprayed with PBS to approximately 30% wet states before analysis. Based on Figure 9, the 3HC bioscaffold displayed the highest tensile strength (62.35 ± 29.12 kPa) compared to the other fabricated bioscaffolds, indicating that 3HC is the toughest scaffold. The bioscaffold tensile strength values of A, AK, 1HC, 1AHC, and 3AHC were 54.12 ± 33.53 kPa, 61.25 ± 17.60 kPa, 47.04 ± 19.42 kPa, 42.60 ± 10.69 kPa, and 53.58 ± 20.28 kPa, respectively. Among all, 1AHC exhibited the lowest tensile strength. Overall, the addition of 0.01 g of carbon weakened the tensile strength of the fabricated bioscaffold.

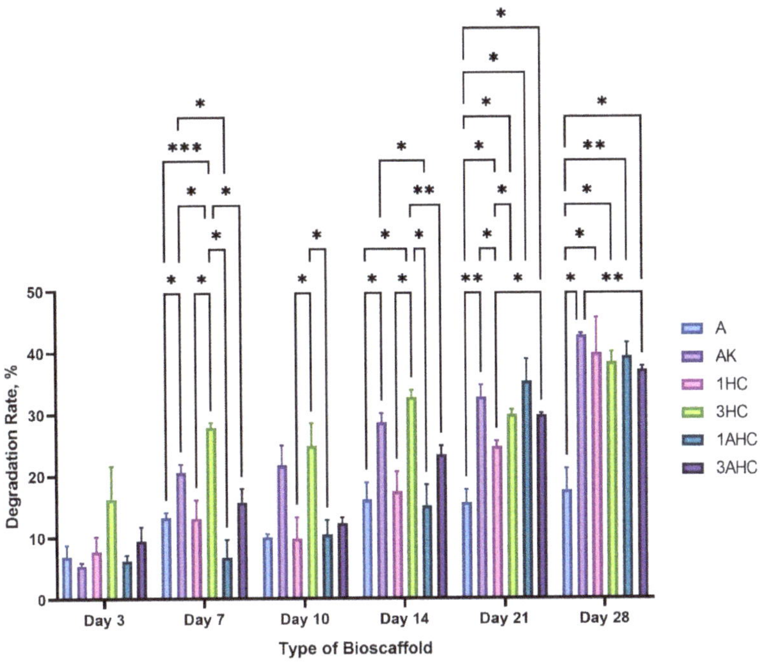

Figure 8. Degradation rate results of bioscaffolds taken on days 3, 7, 10, 14, 21, and 28. The statistical significance was considered at a probability of $p < 0.05$ (* $p \leq 0.05$, ** $p \leq 0.01$, *** $p \leq 0.001$).

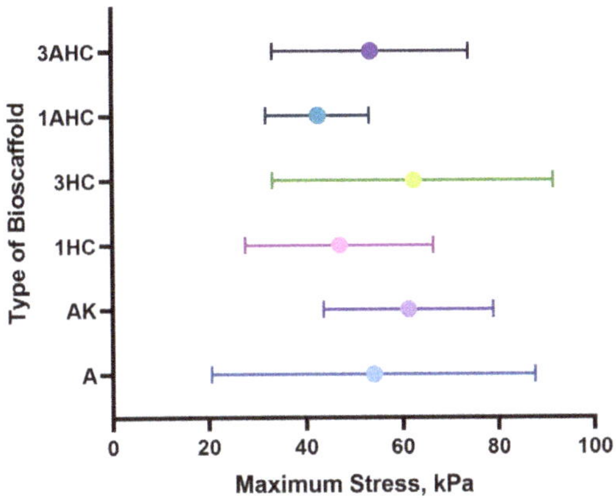

Figure 9. Tensile strength results of bioscaffolds.

Moreover, the addition of 0.03 g of carbon strengthened the tensile strength of the bioscaffold. One study reported that the concentration or proportion of materials influences the mechanical strength of the bioscaffold [60]. However, there were no significant differences in tensile strength between the fabricated bioscaffold samples analysed. In addition, a study showed that the addition of carbon improves the mechanical properties of the resulting bioscaffold [61]. Their findings are similar to those discovered in our study, particularly at higher concentrations of carbon particles.

3.2.7. Electrical Conductivity Analysis

Electrical conductive materials may influence the activity of stem cells, such as differentiation, migration, and proliferation [62]. Thus, the development of conductive bioscaffolds, which mimic the behaviour of the cell's natural microenvironment, is rising, especially for cardiac and skeletal muscle tissue engineering applications [62,63]. Hence, the electrical conductivity of the fabricated bioscaffolds was investigated and is shown in Figure 10. Without carbon and KGM, the A bioscaffold initially offered good electrical conductivity. According to a study, agar acts as a precursor to the synthesis of cobalt tungsten powders, which produce electrodes that resemble batteries, demonstrating the good electrical performance of powders made from agar [64]. The ability of pure agar hydrogel to act as a conductive substance is further supported by [65].

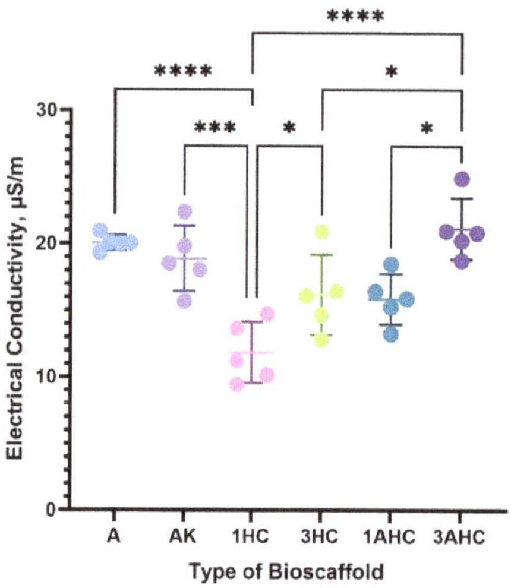

Figure 10. Electrical conductivity results of bioscaffolds. The 3AHC bioscaffold showed the highest electrical conductivity among the fabricated bioscaffolds. The statistical significance was considered at a probability of $p < 0.05$ (* $p \leq 0.05$, *** $p \leq 0.001$, **** $p \leq 0.0001$).

Additionally, the electrical conductivity of the bioscaffold was somewhat decreased by the addition of KGM; however, this effect was not particularly significant. In addition, the electrical conductivity of the bioscaffold increased with higher carbon particle concentrations when the electrical conductivity of the CHAK was compared. Significantly, the 3AHC bioscaffold exhibited the highest electrical conductivity (21.14 ± 2.29 µS/m) among the other CHAK. A, AK, 1HC, 3 HC, and 1AHC had electrical conductivities of 20.07 ± 0.57 µS/m, 18.88 ± 2.44 µS/m, 11.83 ± 2.28 µS/m, 16.18 ± 3.01 µS/m, and 15.86 ± 1.89 µS/m, respectively. The 1HC bioscaffold exhibited the lowest electrical conductivity among the fabricated bioscaffolds. According to one study, the existence of pores in activated carbon can function as an ion transfer channel, improving the bioscaffold's electroconductivity [66].

3.3. *Cell Culture*

3.3.1. Characterisation of WJMSCs

Before biocompatibility analysis, the extracted WJMSCs were characterised in accordance with the standard criteria required for distinguishing MSCs established by the International Society for Cell and Gene Therapy (ISCT) [67]. Therefore, the morphological characteristics of cells, surface markers, and differentiation ability were analysed and are depicted in Figure 11.

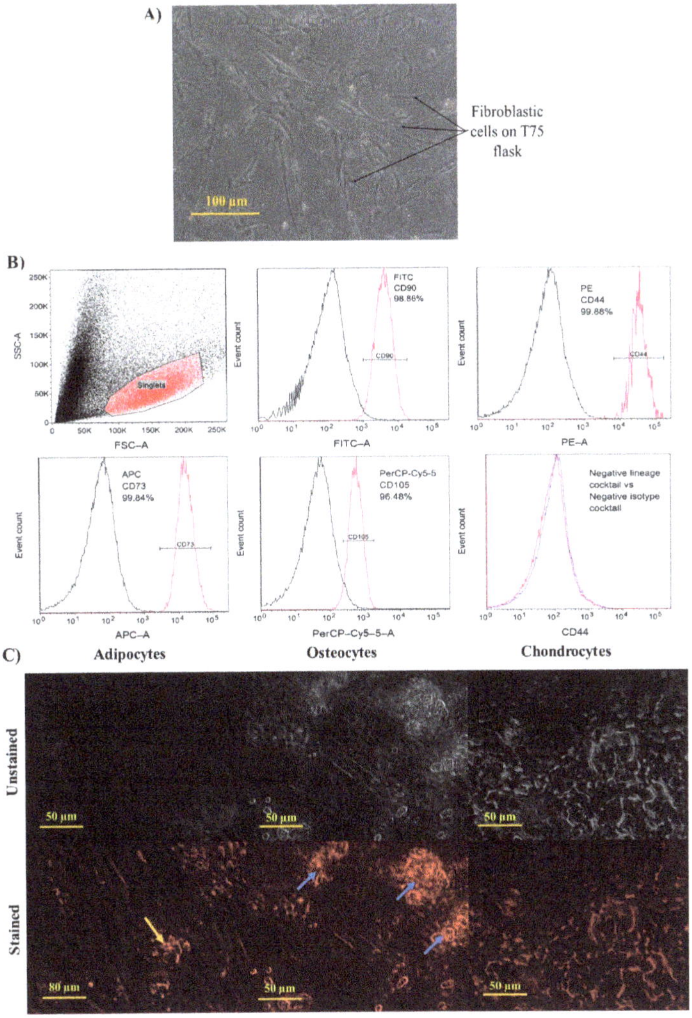

Figure 11. Characterisation of WJMSCs. (**A**) Morphology of WJMSCs showed fibroblastic characteristics. (**B**) Immnunophenotyping of WJMSC. The isolated WJMSCs displayed positive CD90, CD44, CD105, and CD73 expressions (red line). Each marker's isotype is displayed by a black line. The negative lineage cocktail (red line) against the negative isotype cocktail (black line) showed negative expression of PE fluorescent. (**C**) Differentiation analysis of WJMSCs. Adipocytes, osteocytes, and chondrocytes are depicted in the figure as both unstained and stained cells. The developed cells are indicated by red fluorescence. The lipid droplet is indicated by a orange arrow. Calcium deposits are indicated by blue arrows. The image was taken under a 40× objective lens.

Morphology of WJMSCs

The morphology of the extracted WJMSCs showed a spindle shape and fibroblastic characteristics similar to other studies, as seen in Figure 11A [68,69].

Immunophenotyping of WJMSCs

The surface marker of WJMSCs was analysed and is shown in Figure 11B. The CD90, CD44, CD105, and CD73 expressions of the WJMSCs were 96.96%, 99.16%, 97.81%, and

98.81%, respectively. In addition, the WJMSCs did not express any significant negative lineages expression. These findings were similar to those of other researchers, which showed that the expression of MSC positive markers (CD90, CD44, CD105, and CD73) was above 85% [70–72].

Differentiation Ability of WJMSCs

The differentiation ability of the extracted WJMSCs was characterised and is shown in Figure 11C. The WJMSCs were stained with anti-mFABP4, anti-hOsteocalcin, and anti-hAggrecan to stain the adipocytes, osteocytes, and chondrocytes, respectively. Under a fluorescent microscope, the WJMSCs showed fluorescent cells indicating that the WJMSCs could differentiate into adipocytes, osteocytes, and chondrocytes. Overall, the extracted WJMSCs adhere to the standard criteria set by the ISCT, indicating that it is a mesenchymal stem cell-like cell.

3.3.2. Biocompatibility Analysis

The bioscaffold's biocompatibility was investigated by WJMSCs using the MTT assay, and the viability of the cells is shown in Figure 12. After 1 day of incubation, the cell viability of A, AK, 1HC, 3HC, 1AHC, and 3AHC bioscaffold was 77.44%, 78.01%, 117.69%, 77.68%, 103.11%, and 116.52%, respectively. Overall, the cell viability outcomes of fabricated bioscaffolds showed negligible cell toxicity. After 3 days of incubation, the cell viability of A, AK, 1HC, 3HC, 1AHC, and 3AHC bioscaffold was 49.18%, 38.20%, 44.40%, 34.57%, 43.99%, and 47.78%, respectively. All the cell proliferations decreased after 3 days of incubation, which may be due to the leaching of some impurities from the bioscaffold. However, after media change and incubation for 7 days, the growth of cells improved on CHAK (1HC: 109.98%; 3HC: 103.37%; 1AHC: 101.38%, 3AHC: 111.58%) when compared to A (71.48%) and AK (62.72%) bioscaffolds. A study showed similar cell viability outcomes in the existence of hydrogel-containing carbon nanofiber, namely, that the viability of cells was reduced after 72 h, but after 96 h of incubation, the viability of the cells showed enhancement, which they suggested was due to the cells adjusting to the existence of the hydrogel materials [73].

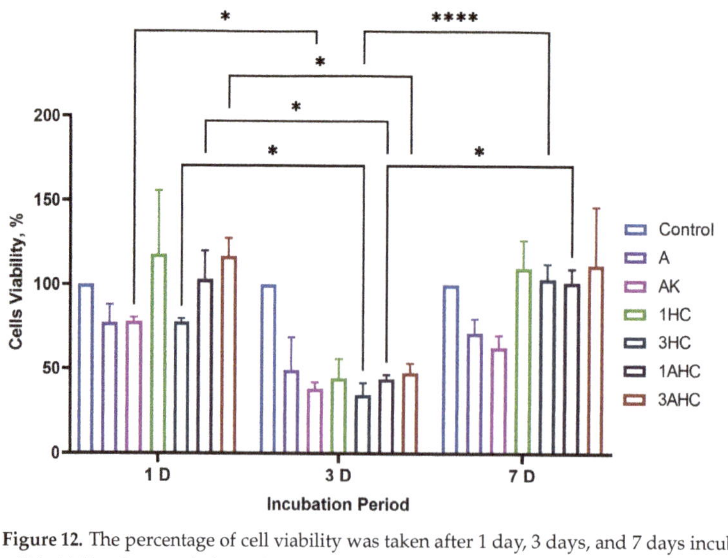

Figure 12. The percentage of cell viability was taken after 1 day, 3 days, and 7 days incubation. The cells' viability decreased after 3 days, but an improvement in cells' viability was observed after 7 days of incubation on every fabricated bioscaffold. On day 7, the CHAK bioscaffold showed better cell viability compared to the non-added carbon bioscaffold. Statistical significance was considered at a probability of $p < 0.05$ (* $p \leq 0.05$, **** $p \leq 0.0001$).

4. Conclusions

In this work, the CHAK made of carbon derived from human hair and composite materials consisting of KGM and agar via gelation, rapid freezing, and ethanol immersion processes was successfully fabricated. The fabricated bioscaffold's micrograph by SEM showed an uneven surface and asymmetrical pore shape. Our findings demonstrated that the addition of carbon reduced the porosity of the bioscaffold. The combination of carbon with agar and KGM is compatible based on the FTIR result because there were no significant changes in wavelength that could be observed between the wavelength of the A, AK, and the bioscaffold incorporated with carbon. Insignificant slight differences among each tensile strength of the bioscaffold were observed, but the 3HC bioscaffold exhibited the highest tensile strength (62.35 ± 29.12 kPa). Moreover, the fabricated CHAK showed relatively slow degradability (<50% after 28 days of incubation), good swelling (>1000%), and excellent water uptake capability (>90%). The 3AHC bioscaffold had the highest electrical conductivity (21.14 ± 2.29 μS/m) among the fabricated bioscaffold. Lastly, the biocompatibility analysis with WJMSCs showed that the CHAK is compatible (>90% cells viability of days 7). In conclusion, this study's outcomes showed the feasibility of producing carbon from human hair as a material to construct a sustainable carbon-based bioscaffold for tissue engineering applications. However, further study is needed to evaluate its efficacy as a cell culture platform.

Author Contributions: Methodology, investigation, writing—original draft preparation, V.V.; writing—review and editing, (V.V., F.I., W.S.W.K.Z., S.A.A.S., N.R. and N.A.B.N.A.); supervision and project administration, F.I. and W.S.W.K.Z.; funding acquisition, F.I. and W.S.W.K.Z. All authors have read and agreed to the published version of the manuscript.

Funding: This research was funded by the Universiti Malaya Impact-Oriented Interdisciplinary Research Grant Program, Grant Nos. IIRG007A-19HWB and IIRG007B-19HWB.

Institutional Review Board Statement: Ethics approval (MREC ID No: 2021518-10145) was obtained from the Medical Research Ethics Committee of the University of Malaya Medical Centre.

Informed Consent Statement: Not applicable.

Data Availability Statement: Not applicable.

Conflicts of Interest: The authors declare no conflict of interest. The funders had no role in the design of the study; in the collection, analyses, or interpretation of data; in the writing of the manuscript; or in the decision to publish the results.

References

1. Nasir, S.; Hussein, M.Z.; Zainal, Z.; Yusof, N.A. Carbon-Based Nanomaterials/Allotropes: A Glimpse of Their Synthesis, Properties and Some Applications. *Materials* **2018**, *11*, 295. [CrossRef] [PubMed]
2. Mishra, R.; Pramanick, B.; Maiti, T.K.; Bhattacharyya, T.K. Glassy carbon microneedles—New transdermal drug delivery device derived from a scalable C-MEMS process. *Microsyst. Nanoeng.* **2018**, *4*, 38. [CrossRef] [PubMed]
3. Hasirci, V.; Hasirci, N. Carbon as a Biomaterial. In *Fundamentals of Biomaterials*; Springer: Berlin/Heidelberg, Germany, 2018; pp. 83–94.
4. Yan, C.; Ren, Y.; Sun, X.; Jin, L.; Liu, X.; Chen, H.; Wang, K.; Yu, M.; Zhao, Y. Photoluminescent functionalized carbon quantum dots loaded electroactive Silk fibroin/PLA nanofibrous bioactive scaffolds for cardiac tissue engineering. *J. Photochem. Photobiol. B Biol.* **2020**, *202*, 111680. [CrossRef] [PubMed]
5. Gopinathan, J.; Pillai, M.M.; Shanthakumari, S.; Gnanapoongothai, S.; Dinakar Rai, B.K.; Santosh Sahanand, K.; Selvakumar, R.; Bhattacharyya, A. Carbon nanofiber amalgamated 3D poly-ε-caprolactone scaffold functionalized porous-nanoarchitectures for human meniscal tissue engineering: In vitro and in vivo biocompatibility studies. *Nanomed. Nanotechnol. Biol. Med.* **2018**, *14*, 2247–2258. [CrossRef] [PubMed]
6. Tashakori-Miyanroudi, M.; Rakhshan, K.; Ramez, M.; Asgarian, S.; Janzadeh, A.; Azizi, Y.; Seifalian, A.; Ramezani, F. Conductive carbon nanofibers incorporated into collagen bio-scaffold assists myocardial injury repair. *Int. J. Biol. Macromol.* **2020**, *163*, 1136–1146. [CrossRef]
7. Zheng, N.; Fitzpatrick, V.; Cheng, R.; Shi, L.; Kaplan, D.L.; Yang, C. Photoacoustic carbon nanotubes embedded silk scaffolds for neural stimulation and regeneration. *ACS Nano* **2022**, *16*, 2292–2305. [CrossRef] [PubMed]

8. Ahmadi, P.; Nazeri, N.; Derakhshan, M.A.; Ghanbari, H. Preparation and characterization of polyurethane/chitosan/CNT nanofibrous scaffold for cardiac tissue engineering. *Int. J. Biol. Macromol.* **2021**, *180*, 590–598. [CrossRef]
9. Yan, Y.; Nashath, F.Z.; Chen, S.; Manickam, S.; Lim, S.S.; Zhao, H.; Lester, E.; Wu, T.; Pang, C.H. Synthesis of graphene: Potential carbon precursors and approaches. *Nanotechnol. Rev.* **2020**, *9*, 1284–1314. [CrossRef]
10. Tripathi, N.; Pavelyev, V.; Islam, S.S. Synthesis of carbon nanotubes using green plant extract as catalyst: Unconventional concept and its realization. *Appl. Nanosci.* **2017**, *7*, 557–566. [CrossRef]
11. Smets, K.; De Jong, M.; Lupul, I.; Gryglewicz, G.; Schreurs, S.; Carleer, R.; Yperman, J. Rapeseed and raspberry seed cakes as inexpensive raw materials in the production of activated carbon by physical activation: Effect of activation conditions on textural and phenol adsorption characteristics. *Materials* **2016**, *9*, 565. [CrossRef]
12. Kurian, M.; Paul, A. Recent trends in the use of green sources for carbon dot synthesis–A short review. *Carbon Trends* **2021**, *3*, 100032. [CrossRef]
13. Wang, C.; Huang, J.; Qi, H.; Cao, L.; Xu, Z.; Cheng, Y.; Zhao, X.; Li, J. Controlling pseudographtic domain dimension of dandelion derived biomass carbon for excellent sodium-ion storage. *J. Power Sources* **2017**, *358*, 85–92. [CrossRef]
14. Adamson, A.; Väli, R.; Paalo, M.; Aruväli, J.; Koppel, M.; Palm, R.; Härk, E.; Nerut, J.; Romann, T.; Lust, E. Peat-derived hard carbon electrodes with superior capacity for sodium-ion batteries. *RSC Adv.* **2020**, *10*, 20145–20154. [CrossRef]
15. Li, S.C.; Hu, B.C.; Ding, Y.W.; Liang, H.W.; Li, C.; Yu, Z.Y.; Wu, Z.Y.; Chen, W.S.; Yu, S.H. Wood-Derived Ultrathin Carbon Nanofiber Aerogels. *Angew. Chem.* **2018**, *130*, 7203–7208. [CrossRef]
16. Redko, T.; Volford, A.; Marek, E.; Scott, S.; Hayhurst, A. Measurement of the times for pyrolysis and the thermal diffusivity of a pyrolysing particle of wood and also of the resulting char. *Combust. Flame* **2020**, *212*, 510–518. [CrossRef]
17. Wang, Y.; Qu, Q.; Gao, S.; Tang, G.; Liu, K.; He, S.; Huang, C. Biomass derived carbon as binder-free electrode materials for supercapacitors. *Carbon* **2019**, *155*, 706–726. [CrossRef]
18. Pramanick, B.; Martinez-Chapa, S.O.; Madou, M.; Hwang, H. Fabrication of 3D carbon microelectromechanical systems (C-MEMS). *JoVE (J. Vis. Exp.)* **2017**, *124*, e55649. [CrossRef]
19. Zhang, J.-H.; Niu, A.; Li, J.; Fu, J.-W.; Xu, Q.; Pei, D.-S. In vivo characterization of hair and skin derived carbon quantum dots with high quantum yield as long-term bioprobes in zebrafish. *Sci. Rep.* **2016**, *6*, 37860. [CrossRef]
20. Bal Altuntaş, D.; Aslan, S.; Akyol, Y.; Nevruzoğlu, V. Synthesis of new carbon material produced from human hair and its evaluation as electrochemical supercapacitor. *Energy Sources Part A Recovery Util. Environ. Eff.* **2020**, *42*, 2346–2356. [CrossRef]
21. Duan, N.; Li, Q.; Meng, X.; Wang, Z.; Wu, S. Preparation and characterization of k-carrageenan/konjac glucomannan/TiO2 nanocomposite film with efficient anti-fungal activity and its application in strawberry preservation. *Food Chem.* **2021**, *364*, 130441. [CrossRef]
22. Guo, L.; Yokoyama, W.; Chen, L.; Liu, F.; Chen, M.; Zhong, F. Characterization and physicochemical properties analysis of konjac glucomannan: Implications for structure-properties relationships. *Food Hydrocoll.* **2021**, *120*, 106818. [CrossRef]
23. Ni, Y.; Sun, J.; Wang, J. Enhanced antimicrobial activity of konjac glucomannan nanocomposite films for food packaging. *Carbohydr. Polym.* **2021**, *267*, 118215. [CrossRef] [PubMed]
24. Guo, L.; Yokoyama, W.; Chen, M.; Zhong, F. Konjac glucomannan molecular and rheological properties that delay gastric emptying and improve the regulation of appetite. *Food Hydrocoll.* **2021**, *120*, 106894. [CrossRef]
25. Du, Q.; Liu, J.; Ding, Y. Recent progress in biological activities and health benefits of konjac glucomannan and its derivatives. *Bioact. Carbohydr. Diet. Fibre* **2021**, *26*, 100270. [CrossRef]
26. Xiang, F.; Xia, Y.; Wang, Y.; Wang, Y.; Wu, K.; Ni, X. Preparation of konjac glucomannan based films reinforced with nanoparticles and its effect on cherry tomatoes preservation. *Food Packag. Shelf Life* **2021**, *29*, 100701. [CrossRef]
27. Ueno, H.; Haraguchi, N.; Azuma, M.; Shiiya, T.; Noda, T.; Ebihara, E.; Uehira, Y.; Uchida, T.; Sasaba, K.; Nakamura, M. Active Consumption of Konjac and Konjac Products Improves Blood Glucose Control in Patients with Type 2 Diabetes Mellitus. *J. Am. Nutr. Assoc.* **2021**, *29*, 1–7. [CrossRef]
28. Dai, J.; Chen, J.; Qi, J.; Ding, M.; Liu, W.; Shao, T.; Han, J.; Wang, G. Konjac Glucomannan from Amorphophallus konjac enhances immunocompetence of the cyclophosphamide-induced immunosuppressed mice. *Food Sci. Nutr.* **2021**, *9*, 728–735. [CrossRef]
29. Xu, M.; Huang, J.; Jiang, S.; He, J.; Wang, Z.; Qin, H.; Guan, Y.-Q. Glucose sensitive konjac glucomannan/concanavalin A nanoparticles as oral insulin delivery system. *Int. J. Biol. Macromol.* **2022**, *202*, 296–308. [CrossRef]
30. Wu, H.; Bu, N.; Chen, J.; Chen, Y.; Sun, R.; Wu, C.; Pang, J. Construction of konjac glucomannan/oxidized hyaluronic acid hydrogels for controlled drug release. *Polymers* **2022**, *14*, 927. [CrossRef]
31. Jiang, Y.; Li, G.; Liu, J.; Li, M.; Li, Q.; Tang, K. Gelatin/oxidized konjac glucomannan composite hydrogels with high resistance to large deformation for tissue engineering applications. *ACS Appl. Bio Mater.* **2021**, *4*, 1536–1543. [CrossRef]
32. Ghorbani, M.; Nezhad-Mokhtari, P.; Mahmoodzadeh, F. Incorporation of Oxidized Pectin to Reinforce Collagen/Konjac Glucomannan Hydrogels Designed for Tissue Engineering Applications. *Macromol. Res.* **2021**, *29*, 289–296. [CrossRef]
33. Li, Z.; Zhang, L.; Mao, C.; Song, Z.; Li, X.; Liu, C. Preparation and characterization of konjac glucomannan and gum arabic composite gel. *Int. J. Biol. Macromol.* **2021**, *183*, 2121–2130. [CrossRef] [PubMed]
34. Lai, R.; Liu, Y.; Liu, J. Properties of the konjac glucomannan and zein composite gel with or without freeze-thaw treatment. *Food Hydrocoll.* **2021**, *117*, 106700. [CrossRef]
35. Xiao, M.; Luo, L.; Tang, B.; Qin, J.; Wu, K.; Jiang, F. Physical, structural, and water barrier properties of emulsified blend film based on konjac glucomannan/agar/gum Arabic incorporating virgin coconut oil. *LWT* **2022**, *154*, 112683. [CrossRef]

36. Zhu, J.; Eid, M.; Li, J.; Geng, F.; Li, B. Synergistic interactions between konjac glucomannan and welan gum mixtures. *LWT* **2022**, *162*, 113425. [CrossRef]
37. Qiao, D.; Tu, W.; Zhong, L.; Wang, Z.; Zhang, B.; Jiang, F. Microstructure and mechanical/hydrophilic features of agar-based films incorporated with konjac glucomannan. *Polymers* **2019**, *11*, 1952. [CrossRef]
38. Chen, X.; Fu, X.; Huang, L.; Xu, J.; Gao, X. Agar oligosaccharides: A review of preparation, structures, bioactivities and application. *Carbohydr. Polym.* **2021**, *265*, 118076. [CrossRef]
39. Qiao, D.; Li, H.; Jiang, F.; Zhao, S.; Chen, S.; Zhang, B. Incorporation of κ-carrageenan improves the practical features of agar/konjac glucomannan/κ-carrageenan ternary system. *Food Sci. Hum. Wellness* **2023**, *12*, 512–519. [CrossRef]
40. Mehta, P.; Kaith, B.S. Green Synthesis of Agar/Gelatin Based Superabsorbent (BGCP) Through Gamma Radiation Cross-Linking Polymerization for Castoff as Sustained Drug Delivery Device and in Soil Treatment for Improved Water Retention. *J. Polym. Env.* **2021**, *29*, 647–661. [CrossRef]
41. Peng, S.; Zhang, J.; Zhang, T.; Hati, S.; Mo, H.; Xu, D.; Li, H.; Hu, L.; Liu, Z. Characterization of carvacrol incorporated antimicrobial film based on agar/konjac glucomannan and its application in chicken preservation. *J. Food Eng.* **2022**, *330*, 111091. [CrossRef]
42. GrandViewResearch Scaffold Technology Market Size, Share & Trends Analysis Report by Type, by Application (Stem Cell, Regenerative Medicine, Drug Discovery), by Disease Type (Cancer, Dental, Neurology), by End-Use, and Segment Forecasts, 2021–2028. Available online: https://www.grandviewresearch.com/industry-analysis/scaffold-technology-market (accessed on 6 April 2022).
43. Sahmani, S.; Shahali, M.; Ghadiri Nejad, M.; Khandan, A.; Aghdam, M.; Saber-Samandari, S. Effect of copper oxide nanoparticles on electrical conductivity and cell viability of calcium phosphate scaffolds with improved mechanical strength for bone tissue engineering. *Eur. Phys. J. Plus* **2019**, *134*, 7. [CrossRef]
44. Tohmyoh, H.; Fujita, K.; Suzuki, H.; Futada, K. Structural elasticity for tensile deformation of a single human hair and the comparison with it for the bending deformation. *J. Mech. Behav. Biomed. Mater.* **2021**, *113*, 104166. [CrossRef] [PubMed]
45. Pramanick, B.; Cadenas, L.B.; Kim, D.-M.; Lee, W.; Shim, Y.-B.; Martinez-Chapa, S.O.; Madou, M.J.; Hwang, H. Human hair-derived hollow carbon microfibers for electrochemical sensing. *Carbon* **2016**, *107*, 872–877. [CrossRef]
46. Ahmed, M.; Islam, M.A.; Asif, M.; Hameed, B. Human hair-derived high surface area porous carbon material for the adsorption isotherm and kinetics of tetracycline antibiotics. *Bioresour. Technol.* **2017**, *243*, 778–784. [CrossRef]
47. Chayid, M.A.; Ahmed, M.J. Amoxicillin adsorption on microwave prepared activated carbon from Arundo donax Linn: Isotherms, kinetics, and thermodynamics studies. *J. Environ. Chem. Eng.* **2015**, *3*, 1592–1601. [CrossRef]
48. Choi, S.W.; Tang, J.; Pol, V.G.; Lee, K.B. Pollen-derived porous carbon by KOH activation: Effect of physicochemical structure on CO_2 adsorption. *J. CO2 Util.* **2019**, *29*, 146–155. [CrossRef]
49. Rodríguez, F.; Montoya-Ruiz, C.; Estiati, I.; Saldarriaga, J.F. Removal of drugs in polluted waters with char obtained by pyrolysis of hair waste from the tannery process. *ACS Omega* **2020**, *5*, 24389–24402. [CrossRef]
50. Yagmur, E.; Gokce, Y.; Tekin, S.; Semerci, N.I.; Aktas, Z. Characteristics and comparison of activated carbons prepared from oleaster (*Elaeagnus angustifolia* L.) fruit using KOH and $ZnCl_2$. *Fuel* **2020**, *267*, 117232. [CrossRef]
51. Dubey, P.; Shrivastav, V.; Kaur, A.; Maheshwari, P.H.; Sundriyal, S. Surface and Diffusion Charge Contribution Studies of Human Hair-Derived Heteroatom-Doped Porous Carbon Electrodes for Supercapacitors. *Energy Fuels* **2021**, *36*, 626–637. [CrossRef]
52. Zopf, D.A.; Flanagan, C.L.; Mitsak, A.G.; Brennan, J.R.; Hollister, S.J. Pore architecture effects on chondrogenic potential of patient-specific 3-dimensionally printed porous tissue bioscaffolds for auricular tissue engineering. *Int. J. Pediatr. Otorhinolaryngol.* **2018**, *114*, 170–174. [CrossRef]
53. Li, J.; Huang, H.; Xu, T.; Li, J.; Guo, T.; Lu, X.; Ren, J.; Ren, X.; Mu, Y.; Weng, J. Effect of the interconnecting window diameter of hydroxyapatite scaffolds on vascularization and osteoinduction. *Ceram. Int.* **2022**, *48*, 25070–25078. [CrossRef]
54. Abbasi, N.; Hamlet, S.; Love, R.M.; Nguyen, N.-T. Porous scaffolds for bone regeneration. *J. Sci. Adv. Mater. Devices* **2020**, *5*, 1–9. [CrossRef]
55. Guo, Y.; Zhang, B.; Zhao, S.; Qiao, D.; Xie, F. Plasticized Starch/Agar Composite Films: Processing, Morphology, Structure, Mechanical Properties and Surface Hydrophilicity. *Coatings* **2021**, *11*, 311. [CrossRef]
56. Salleh, A.; Mustafa, N.; Yeit Haan, T.; Mohd Nor, F.; Fauzul Azim, K.; Ishak, A.; Mh Busra, F. Dual-Layered Approach of Ovine Collagen-Gelatin/Cellulose Hybrid Biomatrix Containing Graphene Oxide-Silver Nanoparticles for Cutaneous Wound Healing: Fabrication, Physicochemical, Cytotoxicity and Antibacterial Characterisation. *Biomedicines* **2022**, *10*, 816. [CrossRef]
57. Nokoorani, Y.D.; Shamloo, A.; Bahadoran, M.; Moravvej, H. Fabrication and characterization of scaffolds containing different amounts of allantoin for skin tissue engineering. *Sci. Rep.* **2021**, *11*, 16164. [CrossRef]
58. Guo, S.; Li, H.; Li, Y.; Han, Y.; Chen, K.; Xu, G.; Zhu, Y.; Hu, X. SiO_2-enhanced structural stability and strong adhesion with a new binder of konjac glucomannan enables stable cycling of silicon anodes for lithium-ion batteries. *Adv. Energy Mater.* **2018**, *8*, 1800434. [CrossRef]
59. Lewandowski, J.; Rozwadowska, N.; Kolanowski, T.J.; Malcher, A.; Zimna, A.; Rugowska, A.; Fiedorowicz, K.; Łabędź, W.; Kubaszewski, Ł.; Chojnacka, K. The impact of in vitro cell culture duration on the maturation of human cardiomyocytes derived from induced pluripotent stem cells of myogenic origin. *Cell Transplant.* **2018**, *27*, 1047–1067. [CrossRef] [PubMed]

60. Mohmad, M.; Abdollah, M.F.B.; Khudhair, A.Q.; Tamaldin, N.; Amiruddin, H.; Zin, M.R.B.M.; Tunggal, D. Physical-mechanical properties of palm kernel activated carbon reinforced polymeric composite: Potential as a self-lubricating material. *J. Tribol.* **2018**, *17*, 77–92.
61. Uddin, M.; Dhanasekaran, P.S.; Asmatulu, R. Mechanical properties of highly porous PEEK bionanocomposites incorporated with carbon and hydroxyapatite nanoparticles for scaffold applications. *Prog. Biomater.* **2019**, *8*, 211–221. [CrossRef]
62. Ashtari, K.; Nazari, H.; Ko, H.; Tebon, P.; Akhshik, M.; Akbari, M.; Alhosseini, S.N.; Mozafari, M.; Mehravi, B.; Soleimani, M.; et al. Electrically conductive nanomaterials for cardiac tissue engineering. *Adv. Drug Deliv. Rev.* **2019**, *144*, 162–179. [CrossRef]
63. Palmieri, V.; Sciandra, F.; Bozzi, M.; De Spirito, M.; Papi, M. 3D graphene scaffolds for skeletal muscle regeneration: Future perspectives. *Front. Bioeng. Biotechnol.* **2020**, *8*, 383. [CrossRef] [PubMed]
64. Azevêdo, H.V.S.B.; Raimundo, R.A.; Ferreira, L.S.; Silva, M.M.S.; Morales, M.A.; Macedo, D.A.; Gomes, U.U.; Cavalcante, D.G.L. Green synthesis of $CoWO_4$ powders using agar-agar from red seaweed (*Rhodophyta*): Structure, magnetic properties and battery-like behavior. *Mater. Chem. Phys.* **2020**, *242*, 122544. [CrossRef]
65. Selvalakshmi, S.; Mathavan, T.; Selvasekarapandian, S.; Premalatha, M. Effect of ethylene carbonate plasticizer on agar-agar: NH4Br-based solid polymer electrolytes. *Ionics* **2018**, *24*, 2209–2217. [CrossRef]
66. Vinayagam, M.; Babu, R.S.; Sivasamy, A.; de Barros, A.L.F. Biomass-derived porous activated carbon from Syzygium cumini fruit shells and Chrysopogon zizanioides roots for high-energy density symmetric supercapacitors. *Biomass Bioenergy* **2020**, *143*, 105838. [CrossRef]
67. Viswanathan, S.; Shi, Y.; Galipeau, J.; Krampera, M.; Leblanc, K.; Martin, I.; Nolta, J.; Phinney, D.G.; Sensebe, L. Mesenchymal stem versus stromal cells: International Society for Cell & Gene Therapy (ISCT®) Mesenchymal Stromal Cell committee position statement on nomenclature. *Cytotherapy* **2019**, *21*, 1019–1024. [PubMed]
68. Ranjbaran, H.; Abediankenari, S.; Mohammadi, M.; Jafari, N.; Khalilian, A.; Rahmani, Z.; Amiri, M.M.; Ebrahimi, P. Wharton's jelly derived-mesenchymal stem cells: Isolation and characterization. *Acta Med. Iran* **2018**, *56*, 28–33. [PubMed]
69. Alizadeh, R.; Bagher, Z.; Kamrava, S.K.; Falah, M.; Ghasemi Hamidabadi, H.; Eskandarian Boroujeni, M.; Mohammadi, F.; Khodaverdi, S.; Zare-Sadeghi, A.; Olya, A.; et al. Differentiation of human mesenchymal stem cells (MSC) to dopaminergic neurons: A comparison between Wharton's Jelly and olfactory mucosa as sources of MSCs. *J. Chem. Neuroanat.* **2019**, *96*, 126–133. [CrossRef]
70. Widowati, W.; Gunanegara, R.; Rizal, R.; Widodo, W.; Amalia, A.; Wibowo, S.; Handono, K.; Marlina, M.; Lister, I.; Chiuman, L. *Comparative Analysis of Wharton's Jelly Mesenchymal Stem Cell (WJ-MSCs) Isolated Using Explant and Enzymatic Methods*; Journal of Physics: Conference Series; IOP Publishing: Bristol, UK, 2019; Volume 1374, p. 012024.
71. Rizal; Syaidah, R.; Aqsha, Z.M.; Josephin, A.; Pakpahan, V.M. *Characterization, Differentiation, and Population Doubling Time of Wharton's Jelly Mesenchymal Stem Cells (WJ-MSCs) in Passage 5 and 8*; AIP Conference Proceedings; AIP Publishing LLC: Melville, NY, USA, 2021; Volume 2344, p. 040002.
72. Sabzevari, R.; Roushandeh, A.M.; Mehdipour, A.; Alini, M.; Roudkenar, M.H. SA/G hydrogel containing hCAP-18/LL-37-engineered WJ-MSCs-derived conditioned medium promoted wound healing in rat model of excision injury. *Life Sci.* **2020**, *261*, 118381. [CrossRef]
73. Serafin, A.; Murphy, C.; Rubio, M.C.; Collins, M.N. Printable alginate/gelatin hydrogel reinforced with carbon nanofibers as electrically conductive scaffolds for tissue engineering. *Mater. Sci. Eng. C* **2021**, *122*, 111927. [CrossRef]

polymers

MDPI

Review

Polymeric Dental Nanomaterials: Antimicrobial Action

Pavel Yudaev [1], Vladimir Chuev [2], Bogdan Klyukin [1], Andrey Kuskov [1], Yaroslav Mezhuev [1] and Evgeniy Chistyakov [1,*]

[1] Mendeleev University of Chemical Technology of Russia, Miusskaya Sq., 9, 125047 Moscow, Russia; yudaevpavel5@gmail.com (P.Y.); bourne4432@gmail.com (B.K.); a_n_kuskov@mail.ru (A.K.); valsorja@mail.ru (Y.M.)
[2] Belgorod National Research University, Pobedy Street, 85, 308015 Belgorod, Russia; mxpion@rambler.ru
* Correspondence: ewgenijj@rambler.ru

Abstract: This review aims to describe and critically analyze studies published over the past four years on the application of polymeric dental nanomaterials as antimicrobial materials in various fields of dentistry. Nanoparticles are promising antimicrobial additives to restoration materials. According to published data, composites based on silver nanoparticles, zinc(II), titanium(IV), magnesium(II), and copper(II) oxide nanoparticles, chitosan nanoparticles, calcium phosphate or fluoride nanoparticles, and nanodiamonds can be used in dental therapy and endodontics. Composites with nanoparticles of hydroxyapatite and bioactive glass proved to be of low efficiency for application in these fields. The materials applicable in orthodontics include nanodiamonds, silver nanoparticles, titanium(IV) and zinc(II) oxide nanoparticles, bioactive glass, and yttrium(III) fluoride nanoparticles. Composites of silver nanoparticles and zinc(II) oxide nanoparticles are used in periodontics, and nanodiamonds and silver, chitosan, and titanium(IV) oxide nanoparticles are employed in dental implantology and dental prosthetics. Composites based on titanium(IV) oxide can also be utilized in maxillofacial surgery to manufacture prostheses. Composites with copper(II) oxide nanoparticles and halloysite nanotubes are promising materials in the field of denture prosthetics. Composites with calcium(II) fluoride or phosphate nanoparticles can be used in therapeutic dentistry for tooth restoration.

Keywords: nanomaterials; dentistry; nanoparticles; antibacterial activity; antimicrobial action

1. Introduction

The terms nanoparticle, nanomaterial, and nanotechnology came into use in the second half of the 20th century and are identified by researchers with innovation and progress in science and industry. Currently, a high-priority area in global science is manufacturing of materials that contain particles comparable in size with molecules, which are called nanoparticles. They have found applications in medical imaging [1], biomedicine [2,3], pharmacology [4], photoelectronics [5] and optoelectronics [6,7], construction [8], photocatalysis [9], as components of heterogeneous catalysts [10,11], as a means of improving the mechanical [12] and thermal properties of polymeric materials [13], sensors [14] and biosensors [15], components of lithium-ion batteries [16], and sorbents [17].

A promising application area of composite nanomaterials is medicine, including dentistry, in which they are used for diagnosis, dental prosthetics, and prevention and treatment of diseases of the oral mucosa and hard tissues of teeth. These materials can reproduce mechanical, physicochemical, and esthetic properties of the hard tissues of teeth and often surpass them in strength characteristics. One more advantage of nanomaterials over traditional composites is that dental restoration can be made with good esthetic characteristics, identical to those of hard tooth tissues, as they possess better optical properties [18].

Decreasing the microbial action in oral cavity organs is an important issue in the prevention and treatment of caries and in restorative dentistry and dental prosthetics.

Their large surface, very small size, and high surface energy and charge density allow nanoparticles to interact with the cell membrane, easily penetrate into a pathogen cell, and induce pathogen death [19].

Nanoparticles with a positive charge interact electrostatically with the negatively charged surface of the bacterial cell wall, disrupting the permeability of the cell membrane. In the first case, the respiratory chain of electron transport is blocked (* in Figure 1). In the second, the nanoparticle destroys the membrane, entering the cytoplasm of the bacterial cell, leading to the outflow of intracellular contents and death of the bacterial cell (** in Figure 1).

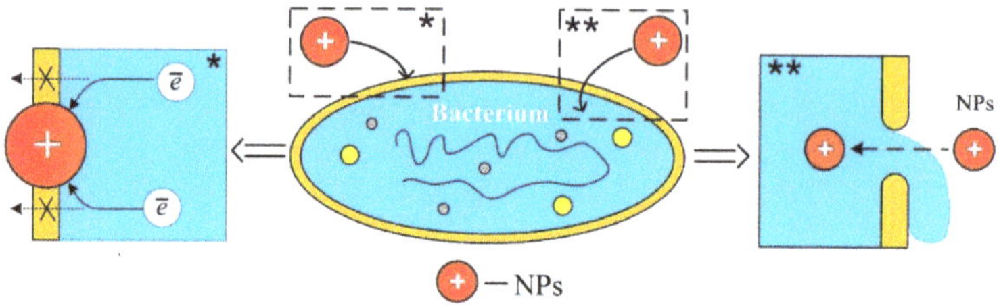

Figure 1. Mechanism of antimicrobial effect of positively charged nanoparticles (using the example of chitosan nanoparticles).

Previously, an extremely toxic amalgam was used as an antimicrobial dental material [20], which also has poor esthetic properties [21]. Since the 1970s, a less toxic glass ionomer cement (below referred to as GIC) with antimicrobial activity [22], but which has poor mechanical characteristics, has been used. The use of antibiotics leads to antibiotic resistance of microorganisms, as well as a large number of side effects on the part of the digestive system and the central nervous system [23], which limits their use in dentistry.

That is why, in the battle against pathogens, studies devoted to the development of various dental nanocomposites and devoid of most of the above disadvantages become extremely relevant and in demand. For example, studies [24,25] report that nanosized fillers of various classes impart reinforcing, antibacterial, remineralizing, self-curing, radiopaque, and esthetic properties to dental materials. Fauzi et al. [26] developed an esthetic adhesive composite material with antibacterial properties for use with esthetic orthodontic brackets.

Figure 2 shows the number of publications on polymeric dental nanomaterials with antimicrobial activity. The graph shows that most active development in this area was in 2012, 2013, 2019, and 2020. Until 2008, there were no publications on this topic.

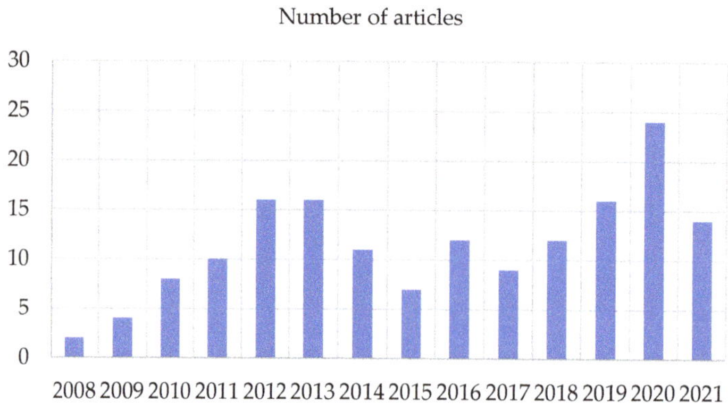

Figure 2. Number of articles on the antimicrobial activity of polymeric nanomaterials for dental applications (PubMed database).

It should be mentioned that more than 100 reviews published in the period of 2018–2021 are devoted to the use of various types of nanoparticles in dentistry. These publications consider the use of nanoparticles in implantology [27,28], dental therapy [29,30], local drug delivery to treat periodontitis [31], in the treatment of oral cavity cancer [32], and in toothpastes [33]. However, the published reviews do not give detailed descriptions of the antibacterial or antifungal action of dental polymeric nanomaterials and prospects for their application in various fields of dentistry.

The present paper gives an overview of studies published in the period of 2018–2021 that address the antimicrobial properties of nanomaterials related to their use in dentistry. The information on nanoparticles used for dental materials is summarized in Table S1 in the Supplementary Materials.

2. Nanoparticles with Antimicrobial Action Used in Dental Materials

A variety of microorganisms are known in dentistry as pathogenic (Supplementary Materials, Table S2), and nanoparticles are active suppressors of many of them (Supplementary Materials, Table S3). Nanoparticles applicable as antimicrobial components include metal, oxide, diamond, glass, and polymeric nanoparticles.

Apart from their antimicrobial action, inorganic nanoparticles are able to improve the integrity of the hybrid layer at the resin–dentin interface [34,35], inhibit enamel demineralization [36] and formation of white spot lesions [37,38] during orthodontic treatment, increase the compressive strength [39,40] and microhardness [41] of restorations based on glass ionomer cement, and reduce polymerization shrinkage of light-curing dental composites [42] and the surface roughness of dentures [43,44] and implants [45] (Supplementary Materials, Table S4).

The inorganic nanoparticles with antimicrobial action studied in recent years include silver, diamond, zinc(II) oxide, titanium(IV) oxide, zirconium(IV) oxide, magnesium(II) oxide, hydroxyapatite, glass, yttrium(III) fluoride, calcium(II) fluoride, and calcium(II) phosphate nanoparticles.

Biocompatible and biodegradable polymeric nanoparticles have shown a high therapeutic potential for controlled drug delivery techniques. The dosage forms based on polymeric nanoparticles penetrate deep into tissues, which increases the drug exposure time and efficiency and minimizes side effects.

Polymeric nanoparticles used in dentistry are subdivided into natural (chitosan, starch, sodium alginate) and synthetic ones (polylactide, poly(lactide-co-glycolide), polyethylene glycol–polylactide). Chitosan holds prospects for use in dentistry as an antimicrobial material.

Here we consider the general characteristics of nanoparticles that have antimicrobial action.

Silver nanoparticles (below referred to as Ag NPs) are spherically shaped biocompatible metal nanoparticles with a controllable size in the range from 3 to 35 nm [18].

Yaqoob et al. [46] discussed in detail chemical, physical, photochemical, and biological methods for the synthesis of silver nanoparticles. It is reported that the most simple, economical, and environmentally friendly method for the synthesis of silver nanoparticles is the biological or green method using plant extracts.

Ag NPs prepared by the reduction of silver nitrate with sodium borohydride or by green synthesis [47,48] show a long-term activity against resistant microorganisms that cause caries, periodontitis, peri-implantitis, inflammation of root canals, and oral candidiasis. However, a tendency to aggregation increases the average size of Ag NPs, which may result in the loss of antibacterial properties. The stability of Ag NPs can be enhanced by using surfactants that contain polar groups and are capable of interacting with surface atoms [49], or by synthesizing Ag NPs in a polymeric matrix [50].

Nanodiamonds (below referred to as NDs) are nanocarbon materials characterized by high strength, hardness, optical transparency, low cytotoxicity, and better chemical stability and biocompatibility than metal or metal oxide nanoparticles. The hydroxy, amino, and carboxyl groups present on the NDs' surface improve their interfacial interactions with polymethyl methacrylate (below PMMA), polyethyl and polybutyl methacrylates, and urethane dimethacrylate resin. The antimicrobial effect of NDs is caused by negatively charged acid anhydride groups present on the surface [51].

Zinc(II) oxide nanoparticles (below referred to as ZnO NPs), characterized by biocompatibility, low micro-density, and activity against *S. mutans, E. faecalis, S. mitis, Lactobacillus* spp., *P. gingivalis*, and *A. naeslundii*, have proved to be efficient in endodontic and orthodontic sealers, dental filling materials, and interim dental prostheses [34].

Titanium(IV) oxide nanoparticles (below referred to as TiO_2 NPs) have spherical shape, a smooth surface, and uniform size distribution. TiO_2 NPs possess useful properties that make them suitable for preparation of dental filling materials, such as chemical stability, biocompatibility, tooth enamel-like color, and a hydrophilic surface. TiO_2 NPs with a particle size of less than 50 nm possess photoinduced activity and can release free radicals damaging the DNA of *S. mutans* and *S. aureus* bacteria [52].

Zirconium(IV) oxide nanoparticles (below referred to as ZrO_2 NPs) are oval or spherical particles with a surface area of 9 m^2 g^{-1} and an average size of 40 nm.

ZrO_2 NPs increase biaxial flexural strength and Vickers microhardness of GIC [41] and increase the tensile and compressive strength and chemical stability of PMMA [53].

Magnesium oxide(II) nanoparticles (below referred to as MgO NPs) possessing antibacterial activity are white hygroscopic particles with an average size of 20 nm. MgO NPs are biocompatible, biodegradable, nontoxic, environmentally friendly, and cheaper than Ag NPs. However, like Ag NPs, they are prone to aggregation, which reduces the activity of nanoparticles against bacteria. Aggregation can be prevented by using cellulose, which provides a closer contact of nanoparticles with bacteria and thus bacterial growth can be inhibited by almost 100% [54].

Hydoxyapatite nanoparticles (below referred to as nHAps) have a composition and structure similar to those of dental tissue, and, hence, they can be used to coat dentin caries lesions, microcracks in teeth, and dentinal tubules, and to enhance remineralization of demineralized dentin matrix and damaged enamel; this is important for dental therapy, implantology, tissue engineering, and treatment of hypersensitivity.

nHAps also stimulate proliferation, adhesion, and differentiation of mouse odontoblast-like MDPC-23 cells [55]; they enhance the biocompatibility of silver coatings of titanium implants with human primary osteoblasts [56].

Bioactive glass nanoparticles (below referred to as BGN) consist of silicon dioxide, calcium oxide, sodium oxide, and phosphorus(V) oxide. In the oral cavity, this glass can release Ca^{2+}, $PO_4{}^{3-}$, and $CO_3{}^{2-}$ ions, thus increasing pH of the medium and inactivating bacterial enzymes. They can also form hydroxycarbonate apatite (the product of crystallization of calcium phosphate on the glass surface), which takes part in the occlusion of dentinal tubules and enamel remineralization [57]. Bioactive glass nanoparticles have an irregular morphology and are prone to aggregation.

Yttrium(III) fluoride nanoparticles (below referred to as YFN) *and calcium fluoride nanoparticles* (below referred to as $nCaF_2$) can affect the mineral layers of teeth, thus enhancing remineralization [58,59]. The average size of YFN is 60–70 nm [58], and the average size of $nCaF_2$ is 32 nm [59].

Calcium phosphate nanoparticles (below referred to as NCP) are able to continuously release calcium ions and phosphate ions into the oral cavity, providing a remineralizing effect. They are amorphous calcium phosphate with a high surface area of 17.8 m^2 g^{-1}. Such systems are synthesized by spray drying, carried out by spraying an acetic acid solution of calcium carbonate and dicalcium phosphate into a heated chamber. Average particle size is 116 nm [60].

Copper (II) oxide nanoparticles (below referred to as CuO NPs) have an average size of 18 nm. For the synthesis of copper oxide nanoparticles, ethanolic solutions of copper acetate and sodium hydroxide are used. They inhibit biofilm growth in soft denture liners in a dose-dependent manner [61].

Chitosan nanoparticles (below referred to as Cs NPs) are obtained by coagulation of the polymer from solutions of various concentrations in a treatment with acetic acid and cross-linking with sodium tripolyphosphate to form a polymer complex. The nanoparticles of chitosan—which is a biologically active, biocompatible, and biodegradable polymer—are used as carriers for targeted delivery of drugs such as doxycycline [62], amoxicillin and clavulanic acid [63], and simvastatin [64] to damaged dental tissue, which ensure prolonged drug release and thus decrease the therapeutic dose.

Halloysite nanotubes (below referred to as HNTs) are biocompatible aluminosilicate layers of a tubular structure with a diameter of several tens of nanometers and a length of approximately 200 nm [65].

3. Dental Therapy, Endodontics, and Periodontics

In dental therapy, nanomaterials are used as chemically curing, light-curing, and self-curing polymeric composites, glass-ionomer cements, insulating coatings, adhesives, and fissure sealants, while in endodontics such materials are used as root canal sealers.

3.1. Materials Containing Silver Nanoparticles

In dental therapy, composites with Ag NPs are used in dental restorative materials based on zirconia and GIC and used to treat carious lesions caused by *S. mutans*, *S. salivarius*, *L. acidophilus*, *C. albicans*, and *C. glabrata*.

A group of Polish researchers found that the addition of Ag NPs into the glass-ionomer cement Ketac Molar EasyMix, and adhesive systems Clearfil SE Bond and OptiBond Solo Plus, enhances the inhibition of growth of Gram-positive bacteria such as *S. mutans*, *S. salivarius*, and *L. acidophilus* after 48 h of observation [66].

Oh et al. [67] evaluated the activity against *S. mutans* of a filling material used for dental restoration based on zirconium dioxide coated by a glass-ceramic powder with addition of 5, 10, 15, and 20 wt. % Ag NPs or NaF and found the following:

(1) The addition of 10 wt. % and 20 wt. % Ag NPs decreases bacterial activity by 11.8% and 15.4%, respectively (Table 1);
(2) The addition of 5–15 wt. % NaF decreases the number of bacteria by 65%, but when the NaF content increases to 20 wt. %, the number of bacteria grows by 29% (Table 1);
(3) Irrespective of Ag NPs and NaF content, the viability of L929 mouse fibroblast cells exceeds 70% for all samples, which attests to the safety of the composites.

Table 1. Bacterial reduction rate for *S. mutans* on the surface of ZrO_2 disks coated by glass–ceramic powder containing Ag NPs and NaF after 24 h of observation.

Composition	Content, wt. %	Bacterial Reduction Rate, %
Ag	5	0
	10	0
	15	11.8
	20	15.4
NaF	5	4.2
	10	35.3
	15	65.4
	20	29.4

However, the authors gave no explanation for the higher rate of inhibition of the cells growth of *S. mutans* by NaF than by Ag NPs.

The mechanical endodontic treatment leaves significant areas of *E. faecalis* biofilm and necrotic tissues, which cause apical periodontitis. This problem can be solved by using Ag NP-based dual-cure adhesives, self-etch adhesive systems possessing long-term antibacterial action. Baras et al. [50] developed a dual-cure endodontic sealer based on dimethylamino hexadecyl methacrylate, a glass filler, and BTH resin (a mixture of bisphenol A glycidyl dimethacrylate, triethylene glycol dimethacrylate, 2-hydroxyethyl methacrylate, and methacryloyl oxyethyl phthalate) with addition of 0.15 wt. % Ag NPs. As a result, the concentration of *E. faecalis* bacterial cells was reduced in comparison with commercial AH-Plus sealer from $10^{7.4}$ to $10^{4.7}$ CFU mL^{-1}, thus preventing the secondary infection of the canals.

Modification of an acrylate self-etch adhesive system with an ethanol dispersion of Ag NPs resulted in increasing diameter of the inhibition zone of *S. mutans* on an agar plate from 11.6 mm to 13.8 mm. However, upon addition of the Ag NPs' dispersion to an adhesive, the curing degree decreased from 50% to 26% since ethanol diluted the adhesive system [68].

The use of orthodontic retainers that control the position of front teeth after orthodontic treatment is often accompanied by an increasing area of bacterial biofilm. This elevates the risk of periodontal inflammation. To prevent biofilm growth, the addition of Ag NPs to light-curing composite for orthodontic retainers was proposed [69]. A comparison of the activities of composites with Ag NPs (1 wt. %) and without Ag NPs against *T. denticola* showed bacterial viability two orders of magnitude lower for Ag NP- containing samples than for samples without nanoparticles ($6 \cdot 10^4$ CFU µL^{-1} and $3 \cdot 10^6$ CFU µL^{-1}, respectively).

It is known that periodontal dressing should protect the wound surface after periodontal surgery and facilitate fast healing of the wound tissue. Therefore, Ag NPs were added to increase the efficiency of the polyvinyl alcohol-based dressing, and characteristics of post-surgery periodontal inflammation in rats were estimated using the ^{99m}Tc-ciprofloxacin radiopharmaceutical. Despite the fact that the content of the ^{99m}Tc-ciprofloxacin marker at the site of dressing with Ag NPs increased after 2 days due to the body's response to the foreign material, after 4 days the proportion of inflammation decreased [70].

Thus, Ag NPs are promising additives to polymeric materials used in dentistry for periodontics. This is due, first of all, to their anti-inflammatory and wound-healing activity. In addition, silver nanoparticles stabilized by biopolymers have high biocompatibility with human gingival fibroblasts [71].

3.2. Materials Containing Zinc Oxide Nanoparticles

ZnO NPs proved to be efficient against anaerobic Gram-positive *S. mutans*, *S. mitis*, and *Lactobacillus* spp. strains under microaerophilic conditions, which mimic a carious cavity [72]. As shown in Figure 3, minimum inhibitory concentration (MIC) values for ZnO NPs were 1.2 mg mL^{-1} for *S. mitis* and 0.6 mg mL^{-1} for *S. mutans* and *Lactobacillus* spp. In addition, a slight bactericidal effect was observed at a concentration of only 0.2 mg mL^{-1}. The obtained material was meant for the use in resin filler materials.

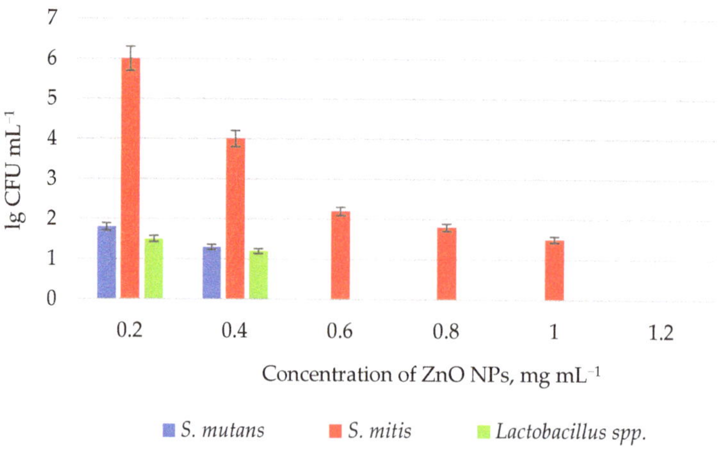

Figure 3. Effects of ZnO NPs (lg CFU mL^{-1}) on the growth of *S. mutans* (blue), *S. mitis* (red), and *Lactobacillus* spp. (green) in the thioglycolate broth after 18 h at 37 °C under microaerophilic conditions [72].

Angel Villegas et al. [72] also found that ZnO NPs in Icon methacrylate resin penetrate to a depth of up to 1020 μm from the tooth surface, which indicates a good infiltration ability of the nanomaterial.

The antimicrobial effect of ZnO NPs allowing for overcoming antibiotic resistance of *S. mutans, E. faecalis, L. fermentum*, and *C. albicans* was detected by the disk diffusion test and microdilution method [73]. In control groups, the antibiotics gentamicin and ampicillin were used for the abovementioned bacteria and fluconazole was used for the *C. albicans* fungus. The results showed an increase in the antimicrobial activity as the particle size decreased from 140 nm to 20 nm. The greatest inhibition zones against *S. mutans* were observed for 20 nm and 40 nm ZnO NPs, whereas 140 nm ZnO NPs formed the greatest inhibition zones against *S. mutans* and *E. faecalis*. The inhibition zones of *C. albicans* were the smallest for all three sizes of ZnO NPs (Table 2). In this regard, based on the data, nanoparticles were used in root canal polymer sealers. It was found that the diameters of inhibition zones against *P. gingivalis* and *A. naeslundii* were greater for ZnO NPs (18.09 mm and 12.05 mm) than for commercial light-curing AH Plus A sealer (9 mm) [74].

Table 2. Diameters of zones of growth inhibition of microorganisms, mm.

Microorganism	20 nm ZnO NPs	40 nm ZnO NPs	140 nm ZnO NPs	Control
S. mutans	16 ± 0.00	14 ± 0.00	12.03 ± 0.57	21 ± 0.00
E. faecalis	14.33 ± 0.57	13 ± 0.00	12 ± 1.00	26 ± 0.00
L. fermentum	10 ± 0.00	9.33 ± 0.57	8.33 ± 0.57	23 ± 0.00
C. albicans	7.66 ± 2.08	6 ± 0.00	6 ± 0.00	16 ± 0.00

The antimicrobial activity of nanoparticles based on ZnO NPs and ZnO NPs encapsulated in chitosan against microorganisms *B. subtilis, S. aureus, S. hemoliticus, P. aeruginosa, K. pneumoniae, E. coli* was studied (Table 3). *S. mutans* and *L. acidophilus* biofilm inhibition was tested using 3M ESPE Adper™ Single Bond Adhesive dental adhesive disks containing nanoparticles [75].

Table 3. Antibacterial activity exhibited by ZnO NPs, chitosan ZnO NPs against Gram-positive and Gram-negative bacterial strains.

Nanoparticles	Microorganism	Zone of Inhibition
ZnO NPs	*B. subtillis*	7 ± 3
	S. aureus	8 ± 3
	S. heamoliticus	9 ± 3
	P. aeroginosa	9 ± 3
	K. pneumoniae	10 ± 3
	E. coli	10 ± 3
Chitosan ZnO NPs	*B. subtillis*	9 ± 3
	S. aureus	10 ± 3
	S. heamoliticus	12 ± 3
	P. aeroginosa	10 ± 3
	K. pneumoniae	13 ± 3
	E. coli	12 ± 3

It should be mentioned that ZnO NPs with chitosan showed stronger antimicrobial activity than without the polymer. According to the authors, this is due to the synergy of the bactericidal activity of ZnO NPs and chitosan.

As for composites based on modified adhesives, the authors [75] describe their anti-caries effect from two positions: (1) preventing the development of recurrent caries on teeth previously treated for it by affecting the structural elements of dentin; (2) reduction in the number of cariogenic bacteria such as *S. mutans* and *L. acidophilus*.

According to the results obtained (Table 4), adhesives containing ZnO NPs and chitosan ZnO NPs reduced the number of cariogenic bacteria *S. mutans* and *L. acidophilus* by four orders in comparison to the control group (adhesive without nanoparticles).

Table 4. Anti-cariogenic activity of adhesive disks containing different concentrations of ZnO NPs and chitosan ZnO NPs.

Nanoparticles	Concentration of NPs, wt. %	Days	Mean CFU mL^{-1}
ZnO NPs	0 (control)	1	2.05×10^5
		3	3.26×10^5
		7	3.17×10^5
	2	1	5.44×10^1
		3	4.91×10^1
		7	3.81×10^1
	5	1	5.25×10^1
		3	3.88×10^1
		7	2.87×10^1
	10	1	3.61×10^1
		3	3.26×10^1
		7	3.28×10^1
Chitosan ZnO NPs	2	1	2.42×10^1
		3	1.85×10^1
		7	3.19×10^1
	5	1	2.54×10^1
		3	1.41×10^1
		7	2.84×10^1
	10	1	1.46×10^1
		3	1.16×10^1
		7	1.77×10^1

However, the contribution of *L. acidophilus* to the pathogenesis of carious disease is ambiguous, which is explained by their antagonistic effect on true cariogenic strains of microorganisms (*S. mutans*, *S. sanguinis*, *S. salivarius*) on the one hand, and the production of 2-hydroxypropanoic acid, which has a demineralizing effect on inorganic part of enamel and dentin, on the other hand.

3.3. Materials Containing Titanium(IV) Oxide Nanoparticles

A comparison of the antibacterial activity of GICs containing antibiotics and TiO_2 NPs [40] showed that the inhibition zone of TiO_2 NPs containing GIC against *S. mutans* was 21.2 mm, that of cetylpyridinium chloride-containing sample was 18.3 mm, and that of ampicillin-containing sample was 31.2 mm. Meanwhile, ampicillin particles deteriorate the interaction between glass particles and liquid cement and, hence, decrease the compressive strength of GIC. Conversely, TiO_2 NPs fill the voids between glass particles and thus increase the compressive strength of GIC from 140 MPa to 173 MPa.

The antibacterial activity against *S. mutans* [76] was also evaluated for an acrylic dental composite resin, Filtek Z350 XT, filled with TiO_2 NPs. The direct contact assay showed that increase in the content of nanoparticles results in decreasing bacterial growth. The incorporation of 2% TiO_2 NPs into the resin reduced the bacterial concentration in the culture broth (BHI + 1% saccharose) by 75% without deterioration of mechanical or physicochemical properties.

Thus, TiO_2 NPs can be used as antibacterial fillers for materials meant for tooth restoration.

Florez et al. [52] investigated the OptiBond Solo Plus dental acrylic adhesives containing TiO_2 NPs for activity against *S. mutans* biofilms. Determination of viable bacterial counts using bioluminescence assay showed that the antibacterial properties of the samples increase with increasing content of TiO_2 NPs, irrespective of the time of bacterial growth (3 to 24 h). The authors also showed that nitrogen-doped TiO_2 NPs are photoactive and their antibacterial properties increase upon long-term (24 h) irradiation with light. However, it is unclear how long-term irradiation of this adhesive can be carried out in an oral cavity.

3.4. Materials Containing Magnesium(II) Oxide Nanoparticles

In a study of the action of GIC modified by MgO NPs on *S. mutans* and *S. sobrinus*, the agar diffusion test demonstrated that GICs containing up to 1 wt. % MgO NPs do not suppress bacterial growth [77,78]. As the concentration of MgO NPs increases, the diameter of inhibition zones increases and reaches 9 mm for 10 wt. % MgO NPs.

The biocompatibility of MgO NPs makes them perfect candidates for clinical use in dentistry as parts of polymeric filling and restoration materials. However, as indicated above, the antibacterial activity is manifested only when the nanoparticle content in the composites is higher than 1 wt. %.

3.5. Materials Containing Hydroxyapatite Nanoparticles

The introduction of nHAps in an amount of 10 wt. % to 30 wt. % in acrylic binder based on bisphenol A glycidyl dimethacrylate and triethylene glycol dimethacrylate promotes remineralization of enamel affected by *S. mutans* biofilm [79]. For 98 days of observation, it was found that with an increase in the concentration of nanoparticles in the composite, the released amount of calcium ions and phosphate ions also increased. It is also reported that at potentially cariogenic pH = 4, more calcium and phosphate ions are released than at pH = 7 (oral pH).

However, it should be noted that with an increase in the content of nHAps in composites, their bending strength decreased. The authors explain this fact by the poor interaction between the organic matrix and mineral nHAps. A statistically significant decrease in the translucency of the composites was also observed with an increase in the content of nHAps from 20 wt. % to 30 wt. %. Unfortunately, the authors do not provide an explanation for this fact.

3.6. Materials Containing Bioactive Glass Nanoparticles

Al-Bakhash et al. [80] estimated the activity of an epoxy resin-based dental sealer, Dentsply Maillefer, modified by various nanofillers: hydoxyapatite, fluorohydroxyapatite, and BGN against Gram-positive *E. faecalis* and *S. mitis* bacteria. The most pronounced antimicrobial properties were observed for sealers doped with fluorohydroxyapatite nanoparticles, since Gram-positive bacteria with negatively charged peptidoglycans are more susceptible to hydroxyapatite than to bioactive glass (CFU mL^{-1} of *E. faecalis* and *S. mitis* decreased by 15% and 17%, while in the case of BGN, the decrease was only 2% and 4%). Thus, bioactive glass is barely useful for dental therapy.

3.7. Materials Containing Chitosan Nanoparticles

A comparison of the antibacterial properties against *S. mutans* for three composites based on Cs NPs (83 nm size), Cs NPs/ZnO NPs (186 nm size), and ZnO NPs (38 nm size) demonstrated that the introduction of ZnO NPs into the acrylic composite resin Filtek Z250 Universal Restorative provides an antibacterial effect, which is retained for up to 12 weeks, while for chitosan-containing composites it is retained only for 2 weeks. The inhibition zone was also larger for microhybrid composites containing ZnO NPs than for composites based on Cs NPs (Table 5). This may be due to the smaller size and low tendency for aggregation inherent in zinc oxide nanoparticles and, hence, to their greater surface publications.

Thus, according to published data, polymeric materials filled with Ag NPs and TiO_2 NPs are most promising for the use in dental therapy and endodontics, first of all, for the treatment of primary and secondary caries lesions. This is due to the fact that silver nanoparticles and titanium(IV) oxide nanoparticles can simultaneously improve the antimicrobial and mechanical properties of dental material. Ag NPs and TiO_2 NPs occupy voids in the GIC, acting as additional contact points between the binder and glass particles. This increases the compressive strength [39,81], flexural strength, and Vickers microhardness of the composites and the micro-shear bond strength to dentin [82].

Table 5. Width of the inhibition zone against *S. mutans*, mm.

Sample	24 h	2 Weeks	6 Weeks	12 Weeks
Filtek Z250 microhybrid composite resin	6.0 ± 0.0	6.0 ± 0.0	6.0 ± 0.0	6.0 ± 0.0
Resin + ZnO NPs	18.0 ± 0.71	16.0 ± 1.58	14.0 ± 0.71	8.0 ± 0.71
Resin + Cs NPs	18.0 ± 1.87	10.0 ± 1.0	6.0 ± 0.0	6.0 ± 0.0
Resin + Cs NPs/ZnO NPs	15.80 ± 1.48	10.0 ± 0.71	6.0 ± 0.0	6.0 ± 0.0
p	<0.001 *	<0.001 *	<0.001 *	<0.001 *

* Statistically significant at $p \leq 0.05$.

3.8. Materials Containing Calcium Phosphate Orcalcium Fluoride Nanoparticles

Most modern composite dental materials are formed in the process of photopolymerization of low molecular weight binders. In this regard, there is a problem of shrinkage, which often leads to damage to fillings and restorations in the form of microcracks, which contributes to their colonization by microorganisms. A study was conducted aimed at reducing the shrinkage of composites [83]. The composite included urethane dimethacrylate, thriethylene glycol divinylbenzyl ether, and dimethylaminohexadecyl methacrylate. NCP (20 wt. %) and silanized barium boroaluminosilicate glass particles (43 wt. %) were used as fillers.

The resulting material had a 40% lower polymerization stress compared to the commercial Heliomolar composite (Ivoclar, Ontario, Canada). In addition, the developed composite material protected tooth enamel from demineralization caused by *S. mutans* biofilm by 5.17 ± 0.48 mmol L^{-1}. The cytotoxicity of the composite material in relation to the human gingival fibroblast cell was similar to that of the composite based on bisphenol A glycidyl dimethacrylate, which indicated that the developed material is suitable for clinical use.

Also of interest are composite materials containing fluoride anions that release them into the environment surrounding the material. The authors of [59] fabricated a composite based on acrylic resin containing 15 wt. % $nCaF_2$ as filler and 3 wt. % dimethylaminohexadecyl methacrylate. The CFU values of biofilm grown on composite disks decreased by four orders of magnitude compared to the commercial composite Heliomolar (Ivoclar Vivadent, Mississauga, ON, Canada). The release of fluoride ions after 70 days of observation was 0.20 ± 0.03 mmol L^{-1}, calcium ions 0.18 ± 0.005 mmol L^{-1}, while for the commercial composite the release of calcium and fluoride ions was close to zero. The developed material also significantly (by 60%) reduced the production of lactic acid by bacteria compared to the commercial composite.

However, the authors of works [59,83] did not take into account the remineralizing effect of saliva and used only one type of bacteria, *S. mutans*.

Summarizing, we can conclude that materials filled with zinc oxide and chitosan nanoparticles, as well as nanoparticles of calcium phosphate and fluoride, are highly active against the main caries bacteria (*S. mutans*). Zinc oxide nanoparticles are also active against bacteria causing periodontitis and apical periodontitis (*P. gingivalis, A. naeslundii*) and can be used as additives to fillers and sealers for filling root canals and for treatment of periodontitis.

Restorative materials filled with magnesium oxide, bioactive glass, and hydroxyapatite nanoparticles have low efficacy against pathogens and are fluoride inapplicable as antimicrobial additives.

4. Orthodontics

In orthodontics, nanoparticles are added to orthodontic adhesives and acrylic resins and are used as coatings for orthodontic appliances, in particular, orthodontic brackets.

Orthodontic appliances such as metallic and esthetic braces, rings, arcs, and bands complicate oral hygiene and create favorable conditions for the growth of *S. mutans*, *S. aureus*, *S. sobrinus*, *S. sanguis*, *P. gingivalis*, *E. coli*, and *L. acidophilus* biofilms, which increases the risk of tooth surface lesions.

In the field of orthodontics, most prospective are materials containing silver, titanium(IV) oxide, zinc(II) oxide, bioactive glass, and yttrium(III) fluoride nanoparticles and nanodiamonds.

4.1. Materials Containing Silver Nanoparticles

Orthodontic adhesives containing Ag NPs decrease bacterial adhesion at the brace/enamel interface.

The adhesive composite materials meant for attachment of orthodontic brackets should also possess antimicrobial action in order to prevent the formation of white spots. Lee et al. [84] tested an antimicrobial resin based on the *Transbond XT* primer with the addition of Ag NPs. The antibacterial activity of this resin was evaluated in vitro against two oral pathogens, *S. mutans* and *S. sobrinus*. In both cases, the percentages of viable bacterial cells considerably decreased.

A comparison of the antibacterial activities of this resin against *S. mutans* and *L. acidophilus* indicated that the diameter of the inhibition zones against *S. mutans* growth was greater than that against *L. acidophilus* growth [85].

The authors of [86] studied the antimicrobial effect of composite resin disks based on acrylic composite resin Flow Tain (Reliance, Houston, TX, USA), containing 1 wt. %, 2 wt. %, and 5 wt. % Ag NPs, against the bacteria *S. mutans*, *S. sanguis*, and *L. acidophilus*. The results obtained are presented in Table 6.

Table 6. Results of biofilm inhibition tests for *S. mutans*, *S. sanguis*, and *L. acidophilus* for composite disks containing Ag NPs and control group.

Sample	Microorganism	CFU	CFU mL^{-1} Decrease (%) Compared to the Control Group
Control	*S. mutans*	56,666 ± 30,550	-
	S. sanguis	446,666 ± 117,189	-
	L. acidophilus	146,666 ± 32,145	-
Composite disks containing 1 wt. % Ag NPs	*S. mutans*	7000 ± 1000	87.64
	S. sanguis	8333 ± 1527	98.13
	L. acidophilus	27,000 ± 7549	81.59
Composite disks containing 2 wt. % Ag NPs	*S. mutans*	2000 ± 1000	96.47
	S. sanguis	2333 ± 1527	99.47
	L. acidophilus	13,333 ± 2516	90.9
Composite disks containing 5 wt. % Ag NPs	*S. mutans*	133 ± 57	99.76
	S. sanguis	300 ± 100	99.93
	L. acidophilus	566 ± 251	99.61

According to the data obtained, composites containing AgNPs reduce the number of colonies of microorganisms in a dose-dependent manner. The largest decrease in CFU was observed for the biofilm of oral bacteria *S. sanguis* compared to *S. mutans* and *L. acidophilus*.

However, the zone of inhibition was observed only for composites with the highest content of nanoparticles, equal to 5 wt. %. The authors do not provide an explanation for this fact.

The resulting composites, according to the authors, after clinical trials can be used in orthodontics as orthodontic appliances.

4.2. Materials Containing Titanium(IV) Oxide Nanoparticles

In a comparison of antibacterial activities against *S. mutans* in three groups of Transbond XT acrylate composites modified with Ag NPs, ZnO NPs, and TiO_2 NPs, the viable bacterial count decreased in all cases. Nevertheless, the antibacterial activity was higher for the TiO_2 NP-containing group than for the Ag NPs and ZnO NP-containing groups, which is due to smaller size of TiO_2 NPs (25 nm) compared with Ag NPs (80 nm) and ZnO NPs (50 nm) [87].

The antibacterial properties of composite disks based on a thriethylene glycol dimethacrylate, diurethane dimethacrylate resin mixture (50:50 wt./wt.) and nitrogen-doped TiO_2 NPs were studied in relation to cariogenic bacteria *S. mutans* [26]. The nanoparticles were doped with nitrogen to prevent discoloration of the resin composite containing the nanoparticles after exposure to visible light. The antibacterial effect was assessed by the metabolic activity of bacterial cells under illumination conditions (indicator—tetrazolium salt). A decrease in absorption was observed with the addition of TiO_2 NPs, and with increasing concentration of nanoparticles, the decrease in absorption increased (Table 7). As can be seen from Table 7, the surface treatment of the composite with both polishing and plasma contributed to the improvement in antibacterial action compared to untreated composites.

Table 7. The mean absorbance observed for resin disk samples with different wt. % powder concentrations (0–9 wt. %), exposed to different surface treatments (unpolished, plasma treated, polished, and polished with plasma treatment) under light conditions.

Surface Treatment	Concentration of Nitrogen-Doped TiO_2 NPs, wt. %	Absorbance (a.u.)
Untreated	0	0.072 ± 0.002
	1	0.069 ± 0.005
	3	0.066 ± 0.002
	5	0.060 ± 0.004
	7	0.058 ± 0.004
	9	0.056 ± 0.004
Plasma treated	0	0.072 ± 0.002
	1	0.067 ± 0.004
	3	0.066 ± 0.005
	5	0.060 ± 0.004
	7	0.056 ± 0.004
	9	0.055 ± 0.004
Polished treated	0	0.072 ± 0.001
	1	0.067 ± 0.005
	3	0.056 ± 0.004
	5	0.055 ± 0.004
	7	0.052 ± 0.004
	9	0.049 ± 0.004
Polished with plasma treated	0	0.072 ± 0.001
	1	0.066 ± 0.004
	3	0.052 ± 0.004
	5	0.047 ± 0.004
	7	0.044 ± 0.004
	9	0.041 ± 0.004

In the future, the authors plan to investigate the effect of other microorganisms on the inhibitory effect of TiO_2 NPs doped with nitrogen.

4.3. Materials Containing Nanodiamond

It was shown by Mangal et al. [88] that composites based on Ortho-Jet orthodontic acrylic resin meant for the manufacture and repair of orthodontic appliances, containing 0.1, 0.3, and 0.5 wt. % of ND powder, are active against *C. albicans* fungi. The CFU count decreased almost to zero upon the addition of only 0.1 wt. % NDs to the resin. The biofilm thickness and weight decreased for all concentrations of NDs. However, further studies are required to establish the mechanism of interaction of composites with the microbes. It is also noteworthy that the addition of NDs deteriorated the optical properties of the composite.

In work [89], the antibacterial and tribological properties of a composite based on PMMA containing 0.1 wt. % ND powder were studied by Mangal et al. For this, samples were obtained by 3D printing and studied in vitro. A polymer without nanoparticles served as control sample.

The severity of the antibacterial effect of the composite was assessed by the resistance to the formation of biofilms based on *S. mutans* on these materials for 48 h. It was found that the biofilm thickness decreased from 200 μm (control group) to 150 μm (nanocomposite), while biofilm biomass decreased from 120 μm^3 μm^{-2} (control group) to 30 μm^3 μm^{-2} (composite).

In addition, composite increased Vickers microhardness and wear resistance, as well as reduced the coefficient of friction, compared with the control group. The pronounced antibacterial effect and good performance characteristics of diamond composites allow them to be recommended for use in orthodontics for the manufacture of orthodontic appliances.

4.4. Materials Containing Zinc(II) Oxide Nanocomposites

Pourhajibagher et al. [90] evaluated the antimicrobial properties against *S. mutans*, *S. sobrinus*, and *L. acidophilus* for the Transbond XT acrylic orthodontic adhesive containing cationic curcumin-doped ZnO NPs. The results, summarized in Table 8, indicate than the samples retained antimicrobial properties after 180 days of observation.

Table 8. Sizes of the inhibition zones of bacterial growth by orthodontic adhesives containing photoactivated 7.5 wt. % ZnO NPs/cCur, mm.

Microorganisms	Days							
	1	15	30	60	90	120	150	180
S. mutans	13	13	13	11	7	7	6	6
S. sobrinus	13	12	12	11	7	7	6	6
L. acidophilus	10	10	10	5	5	5	5	5

4.5. Materials Containing Bioactive Glass Nanoparticles

Nam et al. [36] modified the low-viscosity acrylic orthodontic bonding resin Transbond Supreme LV with BGN particles. Samples with 3 wt. % and 5 wt. % BGN were evaluated for the antibacterial activity against Gram-positive *S. mutans* bacteria.

For both concentrations of nanoparticles, the antibacterial activity against *S. mutans* after 48 h of culturing of BGN samples in the BHI broth was higher than that for unfilled bonding resin.

4.6. Materials Containing Yttrium(III) Fluoride Nanoparticles

Yttrium(III) fluoride nanoparticles can also be used in orthodontics as a component of orthodontic polymeric adhesive. Asiry et al. [58] evaluated the adhesive strength and antibacterial effect of Transbond XT acrylic orthodontic composite resins mixed with yttrium fluoride nanoparticles, with an average particle size of 60–70 nm. The presence of YFN in the resin in concentration of 1 wt. % induced a considerable antibacterial effect against *S. mutans*, as indicated by CFU decrease from 75.85 to 2.24. There was no statistically significant decrease in the bracket–enamel adhesive strength (11.61 MPa in the control group and 11.44 MPa in the test group). The antibacterial action of YFN was attributed [57] to the formation of metal fluoride complexes with bacterial peroxidases. It is noteworthy that an increase in the nanoparticle concentration to 2% decreased the antibacterial activity; however, the authors did not give an explanation for this fact.

An advantage of YFN over all other nanoparticles considered above is the presence of remineralization activity apart from the antibacterial activity.

However, compared to composites based on YFN, the introduction of zinc(II) oxide composites into orthodontic acrylic resins improves both antimicrobial and mechanical properties of orthodontic appliances, such as flexural strength and Vickers hardness [88]. This makes them most appropriate for orthodontic dentistry applications.

It is worth noting that most of the publications considered in this section address the antimicrobial activity against only one or two types of microorganisms; meanwhile, the oral microflora is quite diverse.

5. Dental Implantology and Dental Prosthetics

Modification of the surface of dental implants with nanoparticles reduces the probability of post-implantation infection owing to their antimicrobial action.

5.1. Materials Containing Silver Nanoparticles

After dental implantation or prosthetics, an important task is to prevent *C. albicans* c and *C. glabrata* fungal infections in order to avoid repeated surgery because of infection. Silver NPs exhibit higher antimicrobial activity than antifungal drugs such as fluconazole, griseofulvin, itraconazole, and miconazole; therefore, silver nanoparticles are used to modify heat-curing and self-curing acrylic dental composites.

Since dental prostheses made of PMMA are actively colonized by fungi, De Matteis et al. [91] attempted to control fungal infection by adding Ag NPs/sodium citrate of 20 nm diameter to the PMMA-based Paladon 65 material. The negatively charged Ag NPs/sodium citrate particles decreased roughness, porosity, and hydrophobicity of the PMMA surface, thus decreasing adhesion and colonization of *C. albicans* on the polymer.

The metabolic activity of *C. glabrata* was studied in the presence of composites based on Lucitone 550 acrylic resin containing Ag NPs/sodium citrate [92]. It was found that 0.5 to 5 vol. % nanoparticle concentrations in the composite resulted in the formation of aggregates. However, the authors noted that a decrease in the Ag NPs concentration to 0.05 vol. % decreases the metabolic activity of *C. glabrata* biofilms by 43%.

5.2. Materials Containing Chitosan Nanoparticles

In order to prevent complications of dental prosthetics such as stomatitis and chronic atrophic candidiasis, it was proposed to deposit a tissue acrylic conditioner containing 40 nm to 100 nm Cs NPs on the complete denture surface [93]. Samples containing 2.5 wt. % to 10 wt. % Cs NPs completely inhibited the growth of *C. albicans* fungi and *S. mutans*, *P. aeruginosa*, and *E. faecalis* bacteria after incubation for 24 h. The inhibitory action of Cs NPs was attributed to the interaction of chitosan, which is a polycation, with anionic components of the cell walls of microorganisms [93].

5.3. Materials Containing Nanodiamonds

Fouda et al. [51] estimated the adhesion of *C. albicans*, the main causative agent of the denture stomatitis, to the cured Major Base 20 acrylic resin depending on the content of NDs (0.5–1.5 wt. %). The lowest CFU count (~290 per μL of the resin) was observed for 1% NDs. For the control group, this count was ~1300 per μL of the resin. In turn, the addition of 0.5 wt. % NDs to the resin enhanced the mechanical properties of the composite and decreased the surface roughness [43]. The decrease in roughness deteriorated the microbial adhesion to the denture and improved the esthetic quality of the denture and patients' comfort.

Acrylic resins with quaternized nanodiamonds (below QNDs) can markedly decrease the number of viable *S. mutans* bacterial cells. After testing, a non-modified polymer disk was mainly covered with viable bacteria, while in case of QND-containing resins, both viable and dead cells were detected on the disk. Furthermore, the area of viable cells decreased with increasing QND concentration. When 1.0 wt. % and 1.5 wt. % QNDs was introduced into the resin, the number of viable bacteria on the disks decreased to 37.9% and 25.3%, respectively. These QND-containing resins are intended for the manufacture of tooth dentures [94].

5.4. Materials Containing Zinc(II) Oxide Nanoparticles

The antifungal properties of tissue conditioners such as GC Soft-liner, based on plasticized polyether methacrylate and containing various amounts of ZnO NPs, were evaluated [95]. It is reported that air conditioners containing 15 wt. % ZnO NPs reduced by an order of magnitude the number of *C. albicans* fungi cells compared with the control group after 7 and 14 days of observation. However, air conditioners containing 5 wt. % and 10 wt. % ZnO NPs did not show a statistically significant reduction in the number of *C. albicans* compared to the control group. The authors explain this fact by the possible leaching of ZnO NPs due to the absence of bonds between nanoparticles and the polymer matrix of the air conditioner. In the future, the authors plan to investigate the cytotoxicity of the resulting materials.

5.5. Materials Containing Titanium(IV) Oxide Nanoparticles

In the work of Cascione et al. [96], to solve the problem of adhesion of microorganisms to the surface of dentures based on PMMA and the low strength of such dentures, two composites were developed and studied based on PMMA and TiO_2 NPs, and the same polymer with HNTs.

Material properties were assessed in terms of physical parameters such as Young's modulus, roughness, and wettability, and the reduction in the concentration of the diploid fungus *C. albicans* was evaluated. Both types of material showed an improvement in these characteristics in comparison with unmodified composites. The improvement in properties correlated with the concentration of added nanoparticles. At the same time, the authors of [96] found that materials containing TiO_2 NPs had higher Young's modulus and, consequently, stiffness, in comparison with materials containing HNTs (Figure 4).

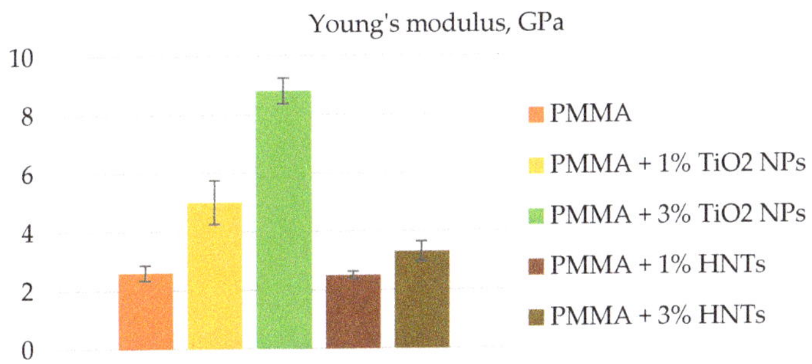

Figure 4. The mean value and its standard deviation of Young's modulus, evaluated for control (PMMA), PMMA + 1% TiO_2 NPs, PMMA + 3% TiO_2 NPs, PMMA + 1% HNTs, and PMMA + 3% HNTs, are represented. Reported results were considered statistically significant respect to control (PMMA) for a p-value < 0.005.

For *C. albicans*, it was found that composites containing HNTs reduced the rate of colonization by diploid fungi significantly more than materials containing TiO_2 NPs (Figure 5). However, the authors did not explain this fact.

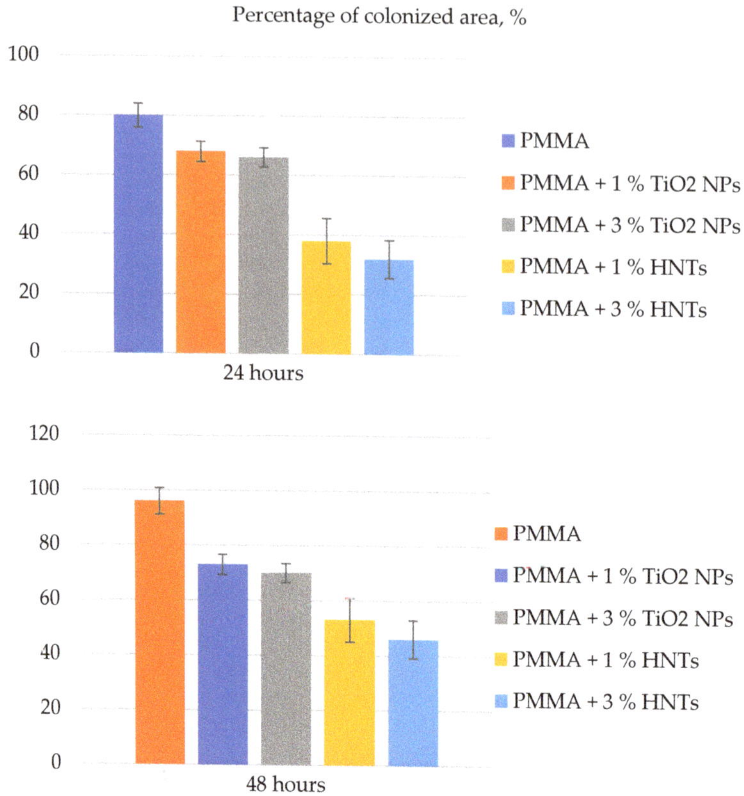

Figure 5. Histograms reporting the colonization assay measurements of *C. albicans* on (PMMA) and different PMMA-based substrates. The colonized area was expressed as a percentage rate representing the area covered by *C. albicans* in respect to the entire surface at two time points (24 h and 48 h). Reported results were calculated as average ± SD for three different areas of each sample, and the values were considered statistically significant in respect to control (PMMA at corresponding time point) for a p-value < 0.005.

5.6. Materials Containing Copper(II) Oxide Nanoparticles

In a study of the antibiofilm activity of the self-cured acrylic-based GC Soft-liner containing CuO NPs in relation to the *C. albicans* biofilm, it was found that the material inhibits the growth of the *C. albicans* biofilm in a dose-dependent manner (in particular, up to 75% at a concentration of CuO NPs 500 μg mL^{-1}) [61].

A comparison was also made of the antimicrobial activity of PMMA-based denture composites containing CuO NPs (2.5 wt. % and 7.5 wt. %) and TiO_2 NPs (2.5 wt. % and 7.5 wt. %) against fungi of the genus *Candida*—*C. albicans*, *C. dubliniensis*— and bacteria of the genus *Streptococcus (S. mutans*, *S. sobrinus*, *S. salivarius*, *S. sanguis)* [97].

All samples had antimicrobial activity against *S. salivarius*, *S. sanguis*, and *C. dubliniensis*. However, only samples containing 2.5 wt. % and 7.5 wt. % CuO NPs were active against *S. mutans* bacteria. In relation to bacteria *S. sobrinus*, samples with 2.5 wt. % and 7.5 wt. % of TiO_2 NPs and with 7.5 wt. % of CuO NPs had activity, and in relation to fungi *C. albicans*, only samples with 7.5 wt. % of TiO_2 NPs showed activity.

Unfortunately, the authors of this work studied composites with only two concentrations of CuO NPs and TiO_2 NPs.

Summarizing the above, we can conclude that materials containing silver, chitosan, titanium(IV) oxide and copper(II) oxide nanoparticles, halloysite nanotubes, and nanodiamonds have activity against fungi of the genus *Candida*, the main pathogens in the field of dental implantology and dental prosthetics. Materials based on zinc(II) oxide nanoparticles are effective against *C. albicans* only if their content in the composite is high (more than 15 wt. %).

6. Maxillofacial Surgery

The introduction of nanofillers into a polymer matrix for the manufacture of maxillofacial prostheses is aimed at improving the physicomechanical properties of prostheses and increasing the compressive strength, flexural strength, and microhardness of materials. However, some of them also possess a high antimicrobial activity, which is important as well.

6.1. Materials Containing Titanium(IV) Oxide Nanoparticles

A fundamental goal of maxillofacial prostheses is the restoration of the lost esthetic appearance of the upper and lower jaws. Silicone elastomers filled with TiO_2 NPs act as an ideal impression material for the elimination of maxillofacial defects owing to the antibacterial properties of TiO_2 NPs against *S. aureus* and *E. coli*. For example, Salih et al. [98] reported a 2.5-fold increase in the antibacterial activity of a material based on Versiltal RTV silicone elastomer and PMMA with the addition of 0.1 wt. % TiO_2 NPs.

6.2. Materials Containing Bioactive Glass Nanoparticles

Composites based on zein and silver-doped BGN were proposed for jaw bone tissue engineering. The antibacterial properties of the composites against *E. coli* and *S. aureus* were studied in vitro. The experiments were carried out [99] using porous zein/BGN and silver-doped BGN 3D scaffolds. It was noted that the scaffolds containing BGN without silver exhibited no antibacterial activity, unlike the scaffolds containing silver-doped BGN, as indicated by the absence of inhibition zones of bacterial growth. However, the authors did not report the diameters of inhibition zones, which is a considerable drawback of the study.

Over the past two years, there were almost no publications devoted to antimicrobial properties of nanomaterials for maxillofacial prostheses. The researchers' attention was focused on the cytotoxicity [100], physicomechanical properties [101,102], and osseointegration [103] because these materials should have a mechanical strength similar to bone tissue strength and should be nontoxic on contact with biological fluids and living tissues of the human body. Nevertheless, microorganisms can form biofilms on the surface of maxillofacial prostheses, which gives rise to purulent inflammatory processes; therefore, it would be appropriate to study the effect of nanoparticles on antimicrobial characteristics of materials used in this field of dentistry.

7. Prospects for the Application of Dental Materials Containing Nanoparticles of Various Classes

The application of polymeric nanomaterials in dentistry may increase the efficiency of treatment of diseases of hard dental, gingival, and periodontal tissues.

Materials containing silver nanoparticles show promise to be used in filling and restorative dental materials in dental therapy, adhesives and sealers in endodontics, acrylic resins for the manufacture of dentures, and in periodontal materials. The broad scope of applications of materials containing silver nanoparticles in dentistry is due to their lack of toxicity, anti-inflammatory action, and ability not only to endow dental materials with antibacterial and antifungal properties but also to improve their mechanical properties.

Zinc oxide nanocomposites can be used in dental therapy, periodontics, and orthodontics.

However, the antibacterial activity of zinc oxide nanoparticles decreases with aging of dental materials [90]. In addition, zinc oxide nanoparticles of any size and concentration exhibit low antifungal activity.

Titanium dioxide nanocomposites are promising restorative materials and composites for the manufacture of orthodontic appliances and maxillofacial prostheses.

Magnesium oxide nanoparticles deteriorate the mechanical properties of filling materials [76]. They have low efficiency against bacteria, while their antifungal properties are unexplored.

Hydroxyapatite nanoparticles show antimicrobial activity only at high concentrations in dental material. Therefore, hydroxyapatite nanoparticles can be used as additives to filler materials in dental therapy only in combination with silver ions [104].

Zirconium dioxide nanoparticles can adversely affect the physical and esthetic properties of dentures, for example, by increasing surface roughness and decreasing the transparency of PMMA [105]. Therefore, zirconium dioxide nanoparticles are of no interest as dental prosthetics.

Bioactive glass nanoparticles can be used as fillers for polymeric orthodontic materials since, apart from their antibacterial effect, they increase the materials' hardness [36].

Materials filled with chitosan nanoparticles are less promising for dental therapy and endodontics than materials containing inorganic nanoparticles, e.g., zinc oxide nanoparticles, since they are less efficient against microorganisms. Furthermore, chitosan nanoparticles are cytotoxic to normal human dental pulp cells, substantially reducing their proliferation and viability [106].

Materials containing halloysite nanotubes are promising materials for the manufacture of partial and complete removable dentures in the field of dental prosthetics.

However, most studies of dental materials containing nanoparticles have been based on in vitro experiments, while in vivo studies allow working with multi-species biofilms and are considered more accurate than in vitro ones. Therefore, there is growing need of further experiments in vivo for implementation of these materials into clinical practice.

One of the problems that may limit the use and implementation of dental nanomaterials in clinical practice is their potential toxicity to patient tissues during prolonged exposure to the oral cavity. Toxicity is due to the leaching of nanosized fillers from the material and their effect on the cells of surrounding biological tissues. Toxicity and biocompatibility studies are of great importance for the clinical application of dental nanomaterials, especially in the field of dental implantology and dental prosthetics, since implants and prostheses come into contact with living tissues. However, it should be noted that the toxicity and biocompatibility of dental nanomaterials have not been studied in most of the considered works.

8. Conclusions

The presented literature review shows the prospects for the use of polymeric dental materials containing nanoparticles of various classes in such areas of dentistry as dental therapy, orthodontics, dental implantology, dental prosthetics, and maxillofacial surgery in terms of their antimicrobial activity. The innovations described in the review set a new vector of development in relation to restorative dental materials and composites, aimed at improving the quality of life of patients.

Researchers are continuously working on nanoparticle-filled dental materials capable of preserving the natural color of hard dental tissue without change over time, antimicrobial action, low cytotoxicity, high (compressive, tensile, and shear) strength, microhardness, impact strength, radiopacity, and color stability, and also high chemical stability. For example, composites filled with hybrid nanoparticles based on phosphazenes containing carboxyl groups [107,108] and methacrylic groups [109] are promising filler materials.

Supplementary Materials: The following supporting information can be downloaded at: https://www.mdpi.com/article/10.3390/polym14050864/s1, Table S1: Antimicrobial action of nanoparticles and their fields of application in dentistry; Table S2: Major dental pathogens; Table S3: Minimum inhibitory concentration (MIC); Table S4: Surface roughness values (R_a).

Author Contributions: Conceptualization, E.C. and P.Y.; methodology, E.C.; validation, E.C., V.C., B.K., A.K., P.Y., Y.M.; writing—original draft preparation, P.Y.; writing—review and editing, E.C.; visualization, P.Y.; supervision, E.C.; funding acquisition, Y.M.; project administration, E.C. All authors have read and agreed to the published version of the manuscript.

Funding: The researchers received financial support from the Ministry of Science and Higher Education of the Russian Federation as part of the state task for the FSSM-2020-0004 project.

Institutional Review Board Statement: Not applicable.

Informed Consent Statement: Not applicable.

Data Availability Statement: Not applicable.

Conflicts of Interest: The authors declare no conflict of interest.

Abbreviations

Ag NPs	silver nanoparticles
BGN	bioactive glass nanoparticles
BHI	brain heart infusion broth
CFU	colony-forming unit
Cs NPs	chitosan nanoparticles
CuO NPs	copper oxide nanoparticles
DNA	deoxyribonucleic acid
GIC	glass ionomer cement
HNTs	halloysite nanotubes
MDPC-23	odontoblast-like cell line
MIC	minimal inhibitory concentration
MgO NPs	magnesium oxide nanoparticles
$nCaF_2$	calcium fluoride nanoparticles
L929	mouse fibroblast cell line
NCP	calcium phosphate nanoparticles
NDs	nanodiamonds
nHAps	hydroxyapatite nanoparticles
PMMA	poly(methyl methacrylate)
TiO_2 NPs	titanium(IV) oxide nanoparticles
QND	quaternized nanodiamond
YFN	yttrium fluoride nanoparticles
ZnO NPs	zinc oxide nanoparticles
ZrO_2 NPs	zirconium(IV) oxide nanoparticles

References

1. Zhidkov, I.S.; Kurmaev, E.Z.; Cholakh, S.O.; Fazio, E.; D'Urso, L. XPS study of interactions between linear carbon chains and colloidal Au nanoparticles. *Mendeleev Commun.* **2020**, *30*, 285–287. [CrossRef]
2. Strutynska, N.; Livitska, O.; Prylutska, S.; Yumyna, Y.; Zelena, P.; Skivka, L.; Malyshenko, A.; Vovchenko, L.; Strelchuk, V.; Prylutskyy, Y.; et al. New nanostructured apatite-type (Na^+, Zn^{2+}, CO^{32-})-doped calcium phosphates: Preparation, mechanical properties and antibacterial activity. *J. Mol. Struct.* **2020**, *1222*, 128932. [CrossRef]
3. Owonubi, S.J.; Ateba, C.N.; Revaprasadu, N. Co-assembled ZnO-Fe_2O_3x-CuOx nano-oxide materials for antibacterial protection. *Phosphorus Sulfur Silicon Relat. Elem.* **2020**, *195*, 981–987. [CrossRef]
4. Aleksandrova, G.P.; Sapozhnikov, A.N.; Boymirzaev, A.S.; Sukhov, B.G.; Trofimov, B.A. Nanobiocomposites of Pharmacophoric Iron and Bismuth Oxides with Arabinogalactan Matrix. Russ. *J. Gen. Chem.* **2020**, *90*, 672–679. [CrossRef]
5. Mikhlin, Y.; Nasluzov, V.; Ivaneeva, A.; Vorobyev, S.; Likhatski, M.; Romanchenko, A.; Krylov, A.; Zharkov, S.; Meira, D.M. Formation, evolution and characteristics of copper sulfide nanoparticles in the reactions of aqueous cupric and sulfide ions. *Mater. Chem. Phys.* **2020**, *255*, 123600. [CrossRef]

6. Kazaryan, S.A.; Starodubtsev, N.F. Study of the Optical and Luminescent Properties of Carbon Nanoparticles Using the Microphotoluminescence Method. *Inorg. Mater. Appl. Res.* **2020**, *11*, 243–256. [CrossRef]
7. Kuzub, L.I.; Guriev, L.L.; Khodos, I.I.; Badamshina, E.R. Influence of Precursor Concentration on the Formation of Silver Nanoparticles with Oligostyrylmonocarboxylate Ligands in ED-20 Epoxy Oligomer. *Polym. Sci. Ser. B* **2020**, *62*, 299–305. [CrossRef]
8. Dahlan, A.S. Impact of Nanotechnology on High Performance Cement and Concrete. *J. Mol. Struct.* **2020**, *1223*, 128896. [CrossRef]
9. Hosny, N.M.; Sherif, Y.E. Novel synthesis and optical properties of $CaSb_2O_6$ nanoparticles via thermal decomposition of thioacetate precursor. *Phosphorus Sulfur Silicon Relat. Elem.* **2019**, *195*, 60–64. [CrossRef]
10. Vibhute, S.P.; Mhaldar, P.M.; Shejwal, R.V.; Pore, D.M. Magnetic Nanoparticles-Supported Palladium catalyzed Suzuki-Miyaura Cross Coupling. *Tetrahedron Lett.* **2020**, *61*, 151594. [CrossRef]
11. Harrison, A.; Zeevi, M.P.; Vasey, C.L.; Nguyen, M.D.; Tang, C. Accelerated Reaction Rates within Self-Assembled Polymer Nanoreactors with Tunable Hydrophobic Microenvironments. *Polymers* **2020**, *12*, 1774. [CrossRef]
12. Kim, H.-S.; Joo, S.-R.; Shin, U.S.; Kim, S.-H. Recyclable CNT-Chitosan nanohybrid film utilized in copper-catalyzed aerobic ipso-hydroxylation of arylboronic acids in aqueous media. *Tetrahedron Lett.* **2018**, *59*, 4597–4601. [CrossRef]
13. Korobov, I.I.; Kovalev, D.Y.; Vinokurov, A.A.; Nadkhina, S.E.; Kalinnikov, G.V.; Konovalikhin, S.V.; Khomenkob, N.Y.; Shilkin, S.P. Synthesis of Titanium Diboride Nanoparticles via the Reaction of TiCl4 with NaBH4 in NaCl-KCl Ionic Melt. *Russ. J. Gen. Chem.* **2020**, *90*, 924–926. [CrossRef]
14. Kazemi, H.; Adelkhani, H.; Didehban, K.; Kazazi, M.; Ajami, N. Electrophoretic as New Method for Deposition of Polyaniline Derivatives Nanostructure Coatings. *Polym. Sci. Ser. B* **2019**, *61*, 835–845. [CrossRef]
15. Shumyantseva, V.V.; Bulko, T.V.; Kuzikov, A.V.; Masamrekh, R.A.; Pergushov, D.V.; Schacher, F.H.; Sigolaeva, L.V. Electrochemical fingerprint of cytochrome c on a polymer/MWCNT nanocomposite electrode. *Mendeleev Commun.* **2020**, *30*, 299–301. [CrossRef]
16. Bhardwaj, R.; Jha, R. Trisodium citrate assisted morphology-controlled synthesis of nickel sulphide nanoparticles with enhanced cyclic stability as carbonaceous free electrode material. *Mater. Chem. Phys.* **2020**, *255*, 123581. [CrossRef]
17. Neskoromnaya, E.A.; Burakov, A.E.; Melezhik, A.V.; Babkin, A.V.; Burakova, I.V.; Kurnosov, D.A.; Tkachev, A.G. Synthesis and Evaluation of Adsorption Properties of Reduced Graphene Oxide Hydro- and Aerogels Modified by Iron Oxide Nanoparticles. *Inorg. Mater. Appl. Res.* **2020**, *11*, 467–475. [CrossRef]
18. Bapat, R.A.; Joshi, C.P.; Bapat, P.; Chaubal, T.V.; Pandurangappa, R.; Jnanendrappa, N.; Gorains, B.; Khurana, S.; Kesharwani, P. The use of nanoparticles as biomaterials in dentistry. *Drug Discov. Today* **2019**, *24*, 85–98. [CrossRef]
19. Kishen, A.S. *Nanotechnology in Endodontics. Current and Potential Clinical Applications;* Springer Science & Business Media: Cham, Switzerland, 2015. [CrossRef]
20. Lorscheider, F.L.; Vimy, M.J.; Summers, A.O. Mercury exposure from "silver" tooth fillings: Emerging evidence questions a traditional dental paradigm. *FASEB J.* **1995**, *9*, 504–508. [CrossRef]
21. Chan, K.H.; Mai, Y.; Kim, H.; Tong, K.C.; Ng, D.; Hsiao, J. Resin composite filling. *Materials* **2010**, *3*, 1228–1243. [CrossRef]
22. Lohbauer, U. Dental glass ionomer cements as permanent filling materials?—Properties, limitations and future trends. *Materials* **2010**, *3*, 76–96. [CrossRef]
23. Qiu, W.; Zhou, Y.; Li, Z.; Huang, T.; Xiao, Y.; Cheng, L.; Peng, X.; Zhang, L.; Ren, B. Application of antibiotics/antimicrobial agents on dental caries. *BioMed Res. Int.* **2020**, *2020*, 5658212. [CrossRef] [PubMed]
24. Wang, Y.; Zhu, M.; Zhu, X.X. Functional fillers for dental resin composites. *Acta Biomater.* **2021**, *122*, 50–65. [CrossRef] [PubMed]
25. Barot, T.; Rawtani, D.; Kulkarni, P. Nanotechnology-based materials as emerging trends for dental applications. *RAMS* **2021**, *60*, 173–189. [CrossRef]
26. Fauzi, N.A.; Ireland, A.J.; Sherriff, M.; Bandara, H.M.H.N.; Su, B. Nitrogen doped titanium dioxide as an aesthetic antimicrobial filler in dental polymers. *Dent. Mater.* **2022**, *38*, 147–157. [CrossRef] [PubMed]
27. Yazdani, J.; Ahmadian, E.; Sharifi, S.; Shahi, S.; Dizaj, S.M. A short view on nanohydroxyapatite as coating of dental implants. *Biomed. Pharmacother.* **2018**, *105*, 553–557. [CrossRef]
28. Wang, Z.; Wang, X.; Wang, Y.; Zhu, Y.; Liu, X.; Zhou, Q. NanoZnO-modified titanium implants for enhanced antibacterial activity, osteogenesis and corrosion resistance. *J. Nanobiotechnol.* **2021**, *19*, 1–23. [CrossRef]
29. Imazato, S.; Kohno, T.; Tsuboi, R.; Thongthai, P.; Xu, H.H.; Kitagawa, H. Cutting-edge filler technologies to release bio-active components for restorative and preventive dentistry. *Dent. Mater. J.* **2020**, *39*, 69–79. [CrossRef]
30. Ferrando-Magraner, E.; Bellot-Arcís, C.; Paredes-Gallardo, V.; Almerich-Silla, M.; García-Sanz, V.; Fernández-Alonso, M.; Montiel-Company, J.M. Antibacterial Properties of Nanoparticles in Dental Restorative Materials. A Systematic Review and Meta-Analysis. *Medicina* **2020**, *56*, 55. [CrossRef]
31. Rajeshwari, H.R.; Dhamecha, D.; Jagwani, S.; Rao, M.; Jadhav, K.; Shaikh, S.; Puzhankara, L.; Jalalpure, S. Local drug delivery systems in the management of periodontitis: A scientific review. *J. Control Release* **2019**, *307*, 393–409.
32. Ketabat, F.; Pundir, M.; Mohabatpour, F.; Lobanova, L.; Koutsopoulos, S.; Hadjiiski, L.; Chen, X.; Papagerakism, P. Controlled Drug Delivery Systems for Oral Cancer Treatment-Current Status and Future Perspectives. *Pharmaceutics* **2019**, *11*, 302. [CrossRef] [PubMed]

33. Carrouel, F.; Viennot, S.; Ottolenghi, L.; Gaillard, C.; Bourgeois, D. Nanoparticles as Anti-Microbial, Anti-Inflammatory, and Remineralizing Agents in Oral Care Cosmetics: A Review of the Current Situation. *Nanomaterials* **2020**, *10*, 140. [CrossRef] [PubMed]
34. Kati, F.A. Effect of the incorporation of zinc oxide nanoparticles on the flexural strength of auto-polymerized acrylic resins. *J. Oral. Res.* **2019**, *8*, 37–41. [CrossRef]
35. Leitune, V.C.B.; Schiroky, P.R.; Genari, B.; Camassola, M.; Samuel, S.M.W.; Collares, F.M. Nanoneedle-like zinc oxide as a filler particle for an experimental adhesive resin. *Indian J. Dent. Res.* **2019**, *30*, 777. [CrossRef] [PubMed]
36. Nam, H.J.; Kim, Y.M.; Kwon, Y.H.; Yoo, K.H.; Yoon, S.Y.; Kim, I.R.; Park, B.S.; Kim, Y.I. Fluorinated Bioactive Glass Nanoparticles: Enamel Demineralization Prevention and Antibacterial Effect of Orthodontic Bonding Resin. *Materials* **2019**, *12*, 1813. [CrossRef]
37. Jasso-Ruiz, I.; Velazquez-Enriquez, U.; Scougall-Vilchis, R.J.; Lara-Carrillo, E.; Toral-Rizo, V.H.; López-Castañares, R.; Morales-Luckie, R.A. Synthesis and Characterization of Silver Nanoparticles on Orthodontic Brackets: A New Alternative in the Prevention of White Spots. *Coatings* **2019**, *9*, 480. [CrossRef]
38. Behnaz, M.; Dalaie, K.; Mirmohammadsadeghi, H.; Salehi, H.; Rakhshan, V.; Aslani, F. Shear bond strength and adhesive remnant index of orthodontic brackets bonded to enamel using adhesive systems mixed with TiO_2 nanoparticles. *Dental Press J. Orthod.* **2018**, *23*, 43. [CrossRef] [PubMed]
39. Barandehfard, F.; Rad, M.K.; Hosseinnia, A.; Rashidi, A.; Tahriri, M.; Tayebi, L. The evaluation of the mechanical characteristics of the synthesized glass-ionomer cements (GICs): The effect of hydroxyapatite and fluorapatite nanoparticles and glass powders. *J. Aust. Ceram. Soc.* **2019**, *55*, 507–517. [CrossRef]
40. Hamid, N.; Telgi, R.L.; Tirth, A.; Tandon, V.; Chandra, S.; Chaturvedi, R.K. Titanium Dioxide Nanoparticles and Cetylpyridinium Chloride Enriched Glass-Ionomer Restorative Cement: A Comparative Study Assessing Compressive Strength and Antibacterial Activity. *Int. J. Clin. Pediatr. Dent.* **2019**, *43*, 42–45. [CrossRef]
41. Alobiedy, A.N.; Alhille, A.H.; Al-Hamaoy, A.R. Mechanical properties enhancement of conventional glass ionomer cement by adding zirconium oxide micro and nanoparticles. *J. Eng.* **2019**, *25*, 72–81. [CrossRef]
42. Raorane, D.V.; Chaughule, R.S.; Pednekar, S.R.; Lokur, A. Experimental synthesis of size-controlled TiO_2 nanofillers and their possible use as composites in restorative dentistry. *Saudi Dent. J.* **2019**, *31*, 194–203. [CrossRef]
43. Al-Harbi, F.A.; Abdel-Halim, M.S.; Gad, M.M.; Fouda, S.M.; Baba, N.Z.; Al Rumaih, H.S.; Akhtar, S. Effect of nanodiamond addition on flexural strength, impact strength, and surface roughness of PMMA denture base. *Int. J. Prosthodont.* **2019**, *28*, 417–425. [CrossRef] [PubMed]
44. Ergun, G.; Sahin, Z.; Ataol, A.S. The effects of adding various ratios of zirconium oxide nanoparticles to poly(methylmethacrylate) on physical and mechanical properties. *J. Oral Sci.* **2018**, *60*, 304–315. [CrossRef]
45. Azzawi, Z.G.; Hamad, T.I.; Kadhim, S.A.; Naji, G.A.H. Osseointegration evaluation of laser-deposited titanium dioxide nanoparticles on commercially pure titanium dental implants. *J. Mater. Sci. Mater. Med.* **2018**, *29*, 96. [CrossRef] [PubMed]
46. Yaqoob, A.A.; Umar, K.; Ibrahim, M.N.M. Silver nanoparticles: Various methods of synthesis, size affecting factors and their potential applications—A review. *Appl. Nanosci.* **2020**, *10*, 1369–1378. [CrossRef]
47. Jadhav, K.; Dhamecha, D.; Dalvi, B.; Patil, M. Green Synthesis of Silver Nanoparticles Using Salacia chinensis: Characterization and its Antibacterial Activity. *Part. Sci. Technol.* **2015**, *33*, 445–455. [CrossRef]
48. Jadhav, K.; Dhamecha, D.; Bhattacharya, D.; Patil, M. Green and ecofriendly synthesis of silver nanoparticles: Characterization, biocompatibility studies and gel formulation for treatment of infections in burns. *J. Photochem. Photobiol. B* **2016**, *155*, 109–115. [CrossRef] [PubMed]
49. Lee, S.H.; Jun, B.H. Silver Nanoparticles: Synthesis and application for nanomedicine. *Int. J. Mol. Sci.* **2019**, *20*, 865. [CrossRef]
50. Baras, B.H.; Melo, M.A.S.; Sun, J.; Oates, T.W.; Weir, M.D.; Xie, X.; Bai, Y.; Xu, H.H.K. Novel endodontic sealer with dual strategies of dimethylaminohexadecyl methacrylate and nanoparticles of silver to inhibit root canal biofilms. *Dent. Mater.* **2019**, *35*, 1117–1129. [CrossRef]
51. Fouda, S.M.; Gad, M.M.; Ellakany, P.; Al-Thobity, A.M.; Al-Harbi, F.A.; Virtanen, J.I.; Raustia, A. The effect of nanodiamonds on Candida albicans adhesion and surface characteristics of PMMA denture base material-an in vitro study. *J. Appl. Oral Sci.* **2019**, *27*, e20180779. [CrossRef] [PubMed]
52. Florez, F.L.E.; Hiers, R.D.; Larson, P.; Johnson, M.; O'Rear, E.; Rondinone, A.J.; Khajotia, S.S. Antibacterial dental adhesive resins containing nitrogen-doped titanium dioxide nanoparticles. *Mater. Sci. Eng. C* **2018**, *93*, 931–943. [CrossRef] [PubMed]
53. Al-Karam, L.Q.; Majeed, S.M. Evaluation the mechanical properties of PMMA/ZrO_2 nanoparticles for dental application. *Int. J. Pharm. Sci.* **2019**, *10*, 2002–2007. [CrossRef]
54. Safaei, M.; Taran, M.; Imani, M.M.; Moradpoor, H.; Rezaei, F.; Jamshidy, L.; Rezaei, R. Application of Taguchi method in the optimization of synthesis of cellulose-MgO bionanocomposite as antibacterial agent. *Polish J. Chem. Technol.* **2019**, *21*, 116–122. [CrossRef]
55. Li, N.; Wu, G.; Yao, H.; Tang, R.; Gu, X.; Tu, C. Size effect of nano-hydroxyapatite on proliferation of odontoblast-like MDPC-23 cells. *Dent. Mater. J.* **2019**, *38*, 534–539. [CrossRef] [PubMed]
56. Salaie, R.N.; Besinis, A.; Le, H.; Tredwin, C.; Handy, R.D. The biocompatibility of silver and nanohydroxyapatite coatings on titanium dental implants with human primary osteoblast cells. *Mater Sci. Eng. C* **2020**, *107*, 110210. [CrossRef]
57. Bae, J.; Son, W.S.; Yoo, K.H.; Yoon, S.Y.; Bae, M.K.; Lee, D.J.; Ko, C.C.; Choi, Y.K.; Kim, Y.I. Effects of Poly (Amidoamine) Dendrimer-Coated Mesoporous Bioactive Glass Nanoparticles on Dentin Remineralization. *J. Nanomater.* **2019**, *9*, 591. [CrossRef]

58. Asiry, M.A.; Alshahrani, I.; Alqahtani, N.D.; Durgesh, B.H. Efficacy of yttrium (III) fluoride nanoparticles in orthodontic bonding. *J. Nanosci. Nanotechnol.* **2019**, *19*, 1105–1110. [CrossRef]
59. Mitwalli, H.; Balhaddad, A.A.; AlSahafi, R.; Oates, T.W.; Melo, M.A.S.; Xu, H.H.; Weir, M.D. Novel CaF_2 nanocomposites with antibacterial function and fluoride and calcium ion release to inhibit oral biofilm and protect teeth. *J. Funct. Biomater.* **2020**, *11*, 56. [CrossRef]
60. Al-Qarni, F.D.; Tay, F.; Weir, M.D.; Melo, M.A.; Sun, J.; Oates, T.W.; Xie, X.; Xu, H.H. Protein-repelling adhesive resin containing calcium phosphate nanoparticles with repeated ion-recharge and re-releases. *J. Dent.* **2018**, *78*, 91–99. [CrossRef]
61. Ansarifard, E.; Zareshahrabadi, Z.; Sarafraz, N.; Zomorodian, K. Evaluation of Antimicrobial and Antibiofilm Activities of Copper Oxide Nanoparticles within Soft Denture Liners against Oral Pathogens. *Bioinorg. Chem. Appl.* **2021**, *2021*, 9939275. [CrossRef]
62. Zegan, G.; Toma, V.; Cernei, E.R.; Anistoroaei, D.; Carausu, E.M.; Moscu, M. Study on Antibiotic Loaded Nanoparticles for Oral Infection Treatment. *Rev. Chim.* **2019**, *70*, 1712–1714. [CrossRef]
63. Zegan, G.; Anistoroaei, D.; Carausu, E.M.; Cernei, E.R.; Golovcencu, L. Amoxicilin and Clavulanic Acid Intercaled Nanostructures for Dentistry Uses. *Mater. Plast.* **2019**, *56*, 396. [CrossRef]
64. Delan, W.K.; Zakaria, M.; Elsaadany, B.; ElMeshad, A.N.; Mamdouh, W.; Fares, A.R. Formulation of simvastatin chitosan nanoparticles for controlled delivery in bone regeneration: Optimization using Box-Behnken design, stability and in vivo study. *Int. J. Pharm.* **2020**, *577*, 119038. [CrossRef] [PubMed]
65. Nair, B.S.; Surendra Kumar, G.P.; Mithra, A.; Rao, S.; Murthy, T. To evaluate and compare the mechanical properties and wetting ability of heat polymerized polymethyl methacrylate resin after reinforcement with halloysite nanotubes and Titania nanoparticles. *Int. J. Appl. Dent. Sci.* **2021**, *7*, 374–379. [CrossRef]
66. Porenczuk, A.; Grzeczkowicz, A.; Maciejewska, I.; Gołaś, M.; Piskorska, K.; Kolenda, A.; Gozdowski, D.; Kopeć-Swoboda, E.; Granicka, L.; Olczak-Kowalczyk, D. An initial evaluation of cytotoxicity, genotoxicity and antibacterial effectiveness of a disinfection liquid containing silver nanoparticles alone and combined with a glass-ionomer cement and dentin bonding systems. *Adv. Clin. Exp. Med.* **2019**, *28*, 75–83. [CrossRef]
67. Oh, G.J.; Kim, J.W.; Ji, M.K.; Yim, E.K.; Vu, V.T.; Kang, B.M.; Park, S.W.; Yang, H.S.; Moon, B.K.; Lee, K.K.; et al. Antibacterial Activity and Fibroblast Cell Viability of Zirconia Coated with Glass Ceramic Containing Ag and NaF Nanoparticles. *J. Nanosci. Nanotechnol.* **2019**, *19*, 1035–1037. [CrossRef]
68. Mohammed, H.F.; Riad, M.I. The effect of silver nanoparticles incorporation in the self-etch adhesive system on its antibacterial activity and degree of conversion: An in-vitro study. *F1000Research* **2019**, *8*, 244. [CrossRef]
69. Santoso, J.; Purbiati, M. Antibacterial activity of silver nanoparticles on fixed retainer adhesive toward Treponema denticola. *Int. J. Pharm.* **2019**, *11*, 198–200. [CrossRef]
70. Prasetyo, B.C.; Sugiharti, R.J.; Mahendra, I.; Halimah, I.; Widyasar, E.M.; Rusminah, N.; Mustika, I. Evaluation of Silver Nanoparticles Addition in Periodontal Dressing for Wound Tissue Healing by 99mTc-ciprofloxacin. *J. Young Pharm.* **2019**, *11*, 17–20. [CrossRef]
71. Craciunescu, O.; Seciu, A.M.; Manoiu, V.S.; Trif, M.; Moisei, M.; Nicu, A.I.; Zarnescu, O. Biosynthesis of silver nanoparticles in collagen gel improves their medical use in periodontitis treatment. *Particul. Sci. Technol.* **2019**, *37*, 757–763. [CrossRef]
72. Angel Villegas, N.; Silvero Compagnucci, M.J.; Sainz Ajá, M.; Rocca, D.M.; Becerra, M.C.; Fabián Molina, G.; Palma, S.D. Novel Antibacterial Resin-Based Filling Material Containing Nanoparticles for the Potential One-Step Treatment of Caries. *J. Healthc. Eng.* **2019**, *2019*, 6367919. [CrossRef] [PubMed]
73. Mirhosseini, F.; Amiri, M.; Daneshkazemi, A.; Zandi, H.; Javadi, Z.S. Antimicrobial Effect of Different Sizes of Nano Zinc Oxide on Oral Microorganisms. *Front. Dent.* **2019**, *16*, 105. [CrossRef] [PubMed]
74. Wang, J.; Du, L.; Fu, Y.; Jiang, P.; Wang, X. ZnO nanoparticles inhibit the activity of Porphyromonas gingivalis and Actinomyces naeslundii and promote the mineralization of the cementum. *BMC Oral Health* **2019**, *19*, 84. [CrossRef] [PubMed]
75. Javed, R.; Rais, F.; Fatima, H.; ul Haq, I.; Kaleem, M.; Naz, S.S.; Ao, Q. Chitosan encapsulated ZnO nanocomposites: Fabrication, characterization, and functionalization of bio-dental approaches. *Mater. Sci. Eng. C* **2020**, *116*, 111184. [CrossRef] [PubMed]
76. Dias, H.B.; Bernardi, M.I.B.; Bauab, T.M.; Hernandes, A.C.; de Souza Rastelli, A.N. Titanium dioxide and modified titanium dioxide by silver nanoparticles as an anti- biofilm filler content for composite resins. *Dent. Mater. J.* **2019**, *35*, 36–46. [CrossRef]
77. Noori, A.J.; Kareem, F.A. Setting time, mechanical and adhesive properties of magnesium oxide nanoparticles modified glass-ionomer cement. *J. Mater. Res. Technol.* **2020**, *9*, 1809–1818. [CrossRef]
78. Noori, A.J.; Kareem, F.A. The effect of magnesium oxide nanoparticles on the antibacterial and antibiofilm properties of glass-ionomer cement. *Heliyon* **2019**, *5*, e02568. [CrossRef]
79. Jardim, R.N.; Rocha, A.A.; Rossi, A.M.; de Almeida Neves, A.; Portela, M.B.; Lopes, R.T.; dos Santos, T.M.P.; Xing, Y.; da Silva, E.M. Fabrication and characterization of remineralizing dental composites containing hydroxyapatite nanoparticles. *J. Mech. Behav. Biomed. Mater.* **2020**, *109*, 103817. [CrossRef]
80. Al-Bakhash, J.A.B.; Shafiei, F.; Pourhajibagher, M.; Shekofteh, K.; Hashemian, A.; Behroozibakhsh, M. The antibacterial activity of an epoxy resin-based dental sealer containing bioactive glass, hydroxyapatite, and fluorohydroxyapatite nanoparticles against Enterococcus faecalis and Streptococcus mitis. *Nanomed. J.* **2020**, *7*, 13–20. [CrossRef]
81. Abdelrehim, M.M.; Mohy El Din, M.H.; El-Shabrawy, S.M.; Fahmy, A.E.; Abdelhamid, S.M.; Ramadan, H.S. Synthesis and characterization of metallic and polymeric nanoparticles and their effect on the antibacterial properties of microhybrid composite resin. *Alex. Dent. J.* **2019**, *44*, 39–45. [CrossRef]

82. Jowkar, Z.; Jowkar, M.; Shafiei, F. Mechanical and dentin bond strength properties of the nanosilver enriched glass ionomer cement. *J. Clin. Exp. Dent.* **2019**, *11*, 275–281. [CrossRef] [PubMed]
83. Bhadila, G.; Wang, X.; Zhou, W.; Menon, D.; Melo, M.A.S.; Montaner, S.; Sun, J.; Xu, H.H. Novel low-shrinkage-stress nanocomposite with remineralization and antibacterial abilities to protect marginal enamel under biofilm. *J. Dent.* **2020**, *99*, 103406. [CrossRef] [PubMed]
84. Lee, S.J.; Heo, M.; Lee, D.; Han, S.; Moon, J.-H.; Lim, H.-N.; Kwon, I.K. Preparation and characterization of antibacterial orthodontic resin containing silver nanoparticles. *Appl. Surf. Sci.* **2018**, *432*, 317–323. [CrossRef]
85. Soans, C.; Jenitta, E.P.; Murali, P.S.; Shetty, A.V.; Ravi, M.S.; Nayak, U.S. Evaluation of Antimicrobial Properties of Orthodontic Adhesive Mixed with Silver Nanoparticles: An in Vitro Study. *Indian J. Public Health Res. Dev.* **2019**, *10*, 32–38. [CrossRef]
86. Mirhashemi, A.; Bahador, A.; Sodagar, A.; Pourhajibagher, M.; Amiri, A.; Gholamrezayi, E. Evaluation of antimicrobial properties of nano-silver particles used in orthodontics fixed retainer composites: An experimental in-vitro study. *J. Dent. Res. Dent. Clin. Dent. Proscpects.* **2021**, *15*, 87–93. [CrossRef]
87. Hailan, S.Y.; Al-Khatieeb, M.M. Antimicrobial efficacy of silver, zinc oxide, and titanium dioxide nanoparticles incorporated in orthodontic bonding agent. *J. Baghdad Coll. Dent.* **2019**, *31*, 10–16. [CrossRef]
88. Mangal, U.; Kim, J.Y.; Seo, J.Y.; Kwon, J.S.; Choi, S.H. Novel Poly (Methyl Methacrylate) Containing Nanodiamond to Improve the Mechanical Properties and Fungal Resistance. *Materials* **2019**, *12*, 3438. [CrossRef]
89. Mangal, U.; Min, Y.J.; Seo, J.Y.; Kim, D.E.; Cha, J.Y.; Lee, K.J.; Kwon, J.-S.; Choi, S.H. Changes in tribological and antibacterial properties of poly (methyl methacrylate)-based 3D-printed intra-oral appliances by incorporating nanodiamonds. *J. Mech. Behav. Biomed. Mater.* **2020**, *110*, 103992. [CrossRef]
90. Pourhajibagher, M.; Vaziri, A.S.; Takzaree, N.; Ghorbanzadeh, R. Physico-mechanical and antimicrobial properties of an orthodontic adhesive containing cationic curcumin doped zinc oxide nanoparticles subjected to photodynamic therapy. *Photodiagnosis Photodyn. Ther.* **2019**, *25*, 239–246. [CrossRef]
91. De Matteis, V.; Cascione, M.; Toma, C.C.; Albanese, G.; De Giorgi, M.L.; Corsalini, M.; Rinaldi, R. Silver Nanoparticles Addition in Poly (Methyl Methacrylate) Dental Matrix: Topographic and Antimycotic Studies. *Int. J. Mol.* **2019**, *20*, 4691. [CrossRef]
92. De Souza Neto, F.N.; Sala, R.L.; Fernandes, R.A.; Xavier, T.P.O.; Cruz, S.A.; Paranhos, C.M.; Monteiro, D.R.; Barboss, D.B.; Delbem, A.C.B.; de Camargo, E.R. Effect of synthetic colloidal nanoparticles in acrylic resin of dental use. *Eur. Polym. J.* **2019**, *112*, 531–538. [CrossRef]
93. Mousavi, S.A.; Ghotaslou, R.; Kordi, S.; Khoramdel, A.; Aeenfar, A.; Kahjough, S.T.; Akbarzadeh, A. Antibacterial and antifungal effects of chitosan nanoparticles on tissue conditioners of complete dentures. *Int. J. Biol. Macromol.* **2018**, *118*, 881–885. [CrossRef] [PubMed]
94. Zhang, Y.; Wang, X.; Li, Q.; Xiao, Y.; Li, P.; Wang, L.; Ye, Z.; Xing, X. Novel resin-based dental material with anti-biofilm activity and improved mechanical property by incorporating hydrophilic cationic copolymer functionalized nanodiamond. *J. Mater. Sci. Mater. Med.* **2018**, *29*, 162. [CrossRef]
95. Homsiang, W.; Kamonkhantikul, K.; Arksornnukit, M.; Takahashi, H. Effect of zinc oxide nanoparticles incorporated into tissue conditioner on antifungal, physical, and mechanical properties. *Dent, Mater. J.* **2021**, *40*, 481–486. [CrossRef] [PubMed]
96. Cascione, M.; De Matteis, V.; Pellegrino, P.; Albanese, G.; De Giorgi, M.L.; Paladini, F.; Corsalini, M.; Rinaldi, R. Improvement of PMMA Dental Matrix Performance by Addition of Titanium Dioxide Nanoparticles and Clay Nanotubes. *Nanomaterials* **2021**, *11*, 2027. [CrossRef]
97. Giti, R.; Zomorodian, K.; Firouzmandi, M.; Zareshahrabadi, Z.; Rahmannasab, S. Antimicrobial Activity of Thermocycled Polymethyl Methacrylate Resin Reinforced with Titanium Dioxide and Copper Oxide Nanoparticles. *Int. J. Dent.* **2021**, *2021*, 6690806. [CrossRef]
98. Salih, S.I.; Oleiwi, J.K.; Ali, H.M. Development the Physical Properties of Polymeric Blend (SR/PMMA) by Adding various Types of Nanoparticles, Used for Maxillofacial Prosthesis Applications. *J. Eng. Technol.* **2019**, *37*, 120–127. [CrossRef]
99. El-Rashidy, A.A.; Waly, G.; Gad, A.; Roether, J.A.; Hum, J.; Yang, Y.; Detsch, R.; Hashem, A.A.; Sami, I.; Goldmann, W.H.; et al. Antibacterial activity and biocompatibility of zein scaffolds containing silver-doped bioactive glass. *Biomed. Mater.* **2018**, *13*, 065006. [CrossRef]
100. Tukmachi, M.S.; Safi, I.N.; Ali, M.M.M. Evaluation of mechanical properties and cytotoxicity of maxillofacial silicone material after incorporation of zirconia nanopowder. *Mater. Today Proc.* **2021**, *42*, 2209–2217. [CrossRef]
101. Shakir, D.A.; Abdul-Ameer, F.M. Effect of nano-titanium oxide addition on some mechanical properties of silicone elastomers for maxillofacial prostheses. *J. Taibah Univ. Medical. Sci.* **2018**, *13*, 281–290. [CrossRef]
102. Abdul-Ameer, F.M. Impact of a mixture of nanofiller and intrinsic pigment on tear strength and hardness of two types of maxillofacial silicone elastomers. *J. Dent. Res. J.* **2020**, *17*, 251. [CrossRef]
103. Ibrahim, M.; Xue, Y.; Ostermann, M.; Sauter, A.; Steinmueller-Nethl, D.; Schweeberg, S.; Krueger, A.; Cimpan, M.R.; Mustafa, K. In vitro cytotoxicity assessment of nanodiamond particles and their osteogenic potential. *J. Biomed. Mater. Res. A* **2018**, *106*, 1697–1707. [CrossRef] [PubMed]
104. Erdem, U.; Dogan, M.; Metin, A.U.; Baglar, S.; Turkoz, M.B.; Turk, M.; Nezir, S. Hydroxyapatite-based nanoparticles as a coating material for the dentine surface: An antibacterial and toxicological effect. *Ceram. Int.* **2020**, *46*, 270–280. [CrossRef]

105. Gad, M.M.; Rahoma, A.; Abualsaud, R.; Al-Thobity, A.M.; Akhtar, S.; Helal, M.A.; Al-Harbi, F.A. Impact of different surface treatments and repair material reinforcement on the flexural strength of repaired PMMA denture base material. *Dent. Mater. J.* **2020**, *39*, 471–482. [CrossRef] [PubMed]
106. Alhomrany, R.; Zhang, C.; Chou, L. Cytotoxic effect of chitosan nanoparticles on normal human dental pulp cells. *Nanosci. Nanotechnol.* **2019**, *3*, 1. [CrossRef]
107. Chistyakov, E.; Kolpinskaya, N.; Posokhova, V.; Chuev, V. Dental Composition Modified with Aryloxyphosphazene Containing Carboxyl Groups. *Polymers* **2020**, *12*, 1176. [CrossRef]
108. Chistyakov, E.M.; Panfilova, D.V.; Kireev, V.V. Carboxyl derivatives of phosphazenes. *Russ. J. Gen. Chem.* **2017**, *87*, 997–1006. [CrossRef]
109. Kireev, V.V.; Chistyakov, E.M.; Filatov, S.N.; Tupikov, A.; Panfilova, D.V.; Chetverikova, A.I. Polymeric dental composites modified with carboxy phosphazene methacrylates. *Russ. J. Appl. Chem.* **2015**, *88*, 866–870. [CrossRef]

MDPI AG
Grosspeteranlage 5
4052 Basel
Switzerland
Tel.: +41 61 683 77 34

Polymers Editorial Office
E-mail: polymers@mdpi.com
www.mdpi.com/journal/polymers

www.ingramcontent.com/pod-product-compliance
Lightning Source LLC
Chambersburg PA
CBHW061545220926
51850CB00005B/256
* 9 7 8 3 7 2 5 8 5 6 7 1 8 *